Biological Control in Latin America and the Caribbean: Its Rich History and Bright Future

CABI INVASIVES SERIES

Invasive species are plants, animals or microorganisms not native to an ecosystem, whose introduction has threatened biodiversity, food security, health or economic development. Many ecosystems are affected by invasive species and they pose one of the biggest threats to biodiversity worldwide. Globalization through increased trade, transport, travel and tourism will inevitably increase the intentional or accidental introduction of organisms to new environments, and it is widely predicted that climate change will further increase the threat posed by invasive species. To help control and mitigate the effects of invasive species, scientists need access to information that not only provides an overview of and background to the field, but also keeps them up to date with the latest research findings.

This series addresses all topics relating to invasive species, including biosecurity surveillance, mapping and modelling, economics of invasive species and species interactions in plant invasions. Aimed at researchers, upper-level students and policy makers, titles in the series provide international coverage of topics related to invasive species, including both a synthesis of facts and discussions of future research perspectives and possible solutions.

Titles Available

1. *Invasive Alien Plants: An Ecological Appraisal for the Indian Subcontinent*
Edited by J.R. Bhatt, J.S. Singh, R.S. Tripathi, S.P. Singh and R.K. Kohli

2. *Invasive Plant Ecology and Management: Linking Processes to Practice*
Edited by T.A. Monaco and R.L. Sheley

3. *Potential Invasive Pests of Agricultural Crops*
Edited by J.E. Peña

4. *Invasive Species and Global Climate Change*
Edited by L.H. Ziska and J.S. Dukes

5. *Bioenergy and Biological Invasions: Ecological, Agronomic and Policy Perspectives on Minimizing Risk*
Edited by L.D. Quinn, D.P. Matlaga and J.N. Barney

6. *Biosecurity Surveillance: Quantitative Approaches*
Edited by F. Jarrad, S.L. Choy and K. Mengersen

7. *Pest Risk Modeling and Mapping for Invasive Alien Species*
Edited by R.C. Venette

8. *Invasive Alien Plants: Impacts on Development and Options for Management*
Edited by C. Ellison, K.V. Sankaran and S. Murphy

9. *Invasion Biology: Hypotheses and Evidence*
Edited by J.M. Jeschke and T. Heger

10. *Invasive Species and Human Health*
Edited by G. Mazza and E. Tricarico

11. *Parthenium Weed: Biology, Ecology and Management*
Edited by S. Adkins, A. Shabbir and K. Dhileepan

12. *Biological Control In Latin America and the Caribbean: Its Rich History and Bright Future*
Edited by J.C. van Lenteren, V.H.P. Bueno, M. Gabriela Luna and Y.C. Colmenarez

Biological Control in Latin America and the Caribbean: Its Rich History and Bright Future

Edited by

Joop C. van Lenteren

Vanda H.P. Bueno

M. Gabriela Luna

and

Yelitza C. Colmenarez

CABI

CABI is a trading name of CAB International

CABI
Nosworthy Way
Wallingford
Oxfordshire OX10 8DE
UK

Tel: +44 (0)1491 832111
Fax: +44 (0)1491 833508
E-mail: info@cabi.org
Website: www.cabi.org

CABI
745 Atlantic Avenue
8th Floor
Boston, MA 02111
USA

Tel: +1 (617)682-9015
E-mail: cabi-nao@cabi.org

A catalogue record for this book is available from the British Library, London, UK.

Library of Congress Cataloging-in-Publication Data

Names: Lenteren, J. C. van, editor. | Bueno, Vanda H. P., editor. | Colmenarez, Yelitza C., editor. | Luna, M. Gabriela (Maria Gabriela), editor.
Title: Biological control in Latin America and the Caribbean : its rich history and bright future / editors: Joop C. van Lenteren, Vanda H.P. Bueno, Yelitza C. Colmenarez and M. Gabriela Luna.
Other titles: CABI invasive species series ; 12.
Description: Boston, MA : CAB International, [2020] | Series: CABI invasives series ; 12 | Includes bibliographical references and index. | Summary: "The book summarises the history of biological control in Latin America and the Caribbean"-- Provided by publisher.
Identifiers: LCCN 2019047071 (print) | LCCN 2019047072 (ebook) | ISBN 9781789242430 (hardback) | ISBN 9781789242447 (ebk) | ISBN 9781789242454 (epub)
Subjects: LCSH: Pests--Biological control--Latin America. | Pests--Biological control--Caribbean Area.
Classification: LCC SB975 .B562 2020 (print) | LCC SB975 (ebook) | DDC 632/.96--dc23
LC record available at https://lccn.loc.gov/2019047071
LC ebook record available at https://lccn.loc.gov/2019047072

ISBN-13: 9781789242430 (hardback)
9781789242447 (ePDF)
9781789242454 (ePub)

Commissioning Editor: David Hemming
Editorial Assistant: Emma McCann
Production Editor: Tim Kapp
Cartographer: Kathryn Reynolds

Typeset by SPi, Pondicherry, India
Printed and bound in the UK by Bell and Bain Ltd, Glasgow

Contents

Supplementary material:

(1) PDFs of reports and publications in Spanish
(2) A list of areas under biocontrol per crop per country
(3) A list of all organisms mentioned in the book with author names, order, family, common name and chapter in which the organism is mentioned
(4) List of countries with classical biocontrol introductions
(5) List of introductions for classical biocontrol per decade
(6) List of addenda and corrections

These resources can be found at: https://www.cabi.org/openresources/42430

About the Editors

The editors of this book all have many years of experience with biological control in the Latin American region.

Vanda H.P. Bueno is Full Professor of Biological Control of Pests at the Federal University of Lavras, Minas Gerais, Brazil. She was initially working on natural enemies of forest pests and pests in coffee plantations, but subsequently switched to augmentative biocontrol in protected cultures. She was a postdoctoral fellow at UC Berkeley, USA. Currently, she studies Neotropical predatory mirids for pests such as *Tuta absoluta* and whitefly in tomato. She has supervised many Brazilian, Colombian, Cuban and Dutch BSc, MSc and PhD students, published 150 peer-reviewed papers and 19 book chapters and produced two editions of the book *Biological Control of Pests: Mass Production and Quality Control*. She is an active member of the Neotropical Regional Section of IOBC (IOBC-NTRS) where her previous roles were president and treasurer; she is now vice-president of IOBC-Global.

Yelitza C. Colmenarez is a specialist in sustainable production – she is heavily involved in biological control and plant protection techniques, with more than 15 years of experience working with multidisciplinary teams, developing international cooperation projects and establishing sustainable production programmes in Latin America and the Caribbean. She has published 15 peer-reviewed papers and ten book chapters, edited a book on *Biological Control of Plant Diseases in Latin America and the Caribbean*, developed a *Field Guide for Identification of Natural Enemies* and conducted 35 international cooperation projects in Latin America and the Caribbean. Currently, she is the CABI Latin America Centre Director and also works as past president and advisor of IOBC-NTRS.

María Gabriela Luna is a researcher at CEPAVE (CONICET-UNLP-asociado CIC-BA), La Plata, Argentina, and professor at UNLP and UNSAdA, Argentina. Her interest in agroecology, entomology and biological control led to studies of parasitoid insects and to evaluating their potential as natural enemies of pests in field and protected crops. She was a postdoctoral fellow at UC Irvine, USA. Currently, she is involved in projects to develop augmentative biological control for *Tuta absoluta* by means of parasitoid species. She has supervised ten MSc and PhD students and published over 30 peer-reviewed papers and two book chapters. She acts in Academic and Counseling committees at UNLP, UNSAdA, CONICET and IOBC-NTRS. She was the secretary of the Executive Committee of IOBC/NTRS from 2010 to 2018.

Joop C. van Lenteren is Emeritus Professor of Entomology at Wageningen University, The Netherlands. He started working on biocontrol of whiteflies and other pests in greenhouses in the 1970s, which later resulted in widely applied IPM programmes comprising many species of natural enemies and microbial control agents for all major pests and several diseases in protected cultivation. He became involved in Latin American biocontrol when he represented Wageningen University in international collaboration programmes starting in the 1990s. He now works on the development of evaluation criteria for selection of promising natural enemies and on Neotropical mirids. He has supervised 260 MSc and 84 PhD students from Europe, Asia, Africa, North and South America. He has published more than 200 peer-reviewed papers and edited several books on IPM and biocontrol. From 2000 to 2016, he fulfilled several different functions in the Executive Committee of IOBC-Global, as well as in the European section of IOBC (IOBC-WPRS).

List of Contributors

Hugo César Arredondo-Bernal, Centro Nacional de Referencia de Control Biológico, Dirección General de Sanidad Vegetal. Km 1.5 Carretera Tecomán-Estación, C.P. 28110, Tecomán, Colima. México. E-mail: hugo.arredondo@senasica.gob.mx

Jorge L. Ayala Sifontes, Dirección Provincial de Sanidad Vegetal, Sancti Spiritus, Cuba. E-mail: jlayala@svssp.co.cu

Anobel Barba, Plant Protection Laboratory, Instituto de Investigación Agropecuaria de Panamá (IDIAP). E-mail: anovelbarba@gmail.com

Lorena Barra-Bucarei, Instituto de Investigaciones Agropecuarias, INIA-Quilamapu, Chile. E-mail: lbarra@inia.cl

César Basso, Unidad de Entomología, Facultad de Agronomía, Universidad de la República. Av. Garzón 780, 12900 Montevideo, Uruguay. E-mail: cbasso.bis@gmail.com

Wagner Bettiol, Embrapa Environment, PO Box 69, 13820-000 Jaguariúna, São Paulo, Brazil. E-mail: wagner.bettiol@embrapa.br

Fermin Blanco, Organismo Internacional Regional de Sanidad Agropecuaria (OIRSA); Show-grounds, Belmopan, Belize. E-mail: fblanco@oirsa.org

Helga Blanco-Metzler, Crop Protection Research Centre (CIPROC), University of Costa Rica, San José, Costa Rica. E-mail: helga.blanco@ucr.ac.cr

Vanda H.P. Bueno, Laboratory of Biological Control, Department of Entomology, Federal University of Lavras, PO Box 3037, 37200-000 Lavras, Minas Gerais, Brazil. E-mail: vhpbueno@ufla.br

Claudia Carolina Cabral Antúnez, Laboratory of Entomology, Facultad de Ciencias Agrarias, Universidad Nacional de Asunción, Paraguay. E-mail: claudia.cabral@agr.una.py; claudia_c_cabral@hotmail.com

Guillermo Cabrera Walsh, FUEDEI (Fundación para el Estudio de Especies Invasivas), Hurlingham, Argentina. E-mail: gcabrera@fuedei.org

Carmen C. Castillo, Plant Protection Department, Santa Catalina Research Station, INIAP (National Institute for Agriculture Research), Panamerica Sur km 1, Quito, Ecuador. E-mail: carmen.castillo@iniap.gob.ec

Patricia Castillo, Centro de Investigación y Reproducción en Control Biológico (CIRCB), Facultad de Ciencias y Tecnología, Universidad Nacional Autónoma de Nicaragua (UNAN)-León, Nicaragua. E-mail: castillopatri@yahoo.es

Charlotte E. Causton, Charles Darwin Foundation, Puerto Ayora, Santa Cruz Island, Galapagos, Ecuador. E-mail: charlotte.causton@fcdarwin.org.ec

Ronald D. Cave, University of Florida, Fort Pierce, Florida, USA. E-mail: rdcave@ufl.edu

Willy Chiaravalle, Entoagro. Roberto Koch 4194, 11700 Montevideo, Uruguay. E-mail: wchiaravalle52@gmail.com

Ximena Cibils, Entomología, Protección Vegetal, Instituto Nacional de Investigación Agropecuaria, Ruta 50 km 11, 70000 Colonia, Uruguay. E-mail: xcibils@inia.org.uy

Matthew J.W. Cock, CABI, Bakeham Lane, Egham TW20 9TY, UK. E-mail: m.cock@cabi.org

Yelitza C. Colmenarez, CABI-UNESP- FEPAF, Rua José Barbosa de Barros, 1780, Botucatu, São Paulo, CEP: 18610-307, Brazil. E-mail: y.colmenarez@cabi.org

Alba Marina Cotes, Corporación Colombiana de Investigación Agropecuaria (Agrosavia), Centro de Investigación Tibaitata, Km 14 vía Mosquera, Cundinamarca, Colombia. E-mail: amcotes@agrosavia.co

Luis V. Crespo, Fundación PROINPA, Oficina principal, Regional Centro. Av. Meneces s/n. Km 4, Zona El Paso. Cochabamba, Bolivia. E-mail: l.crespo@proinpa.org

Luis Devotto Moreno, Instituto de Investigaciones Agropecuarias, INIA-Quilamapu, Chile. E-mail: ldevotto@inia.cl

Francisco Ferrer, Independent Entomologist Advisor, Barquisimeto, Lara State, Venezuela.

Andrés France Iglesias, Instituto de Investigaciones Agropecuarias, INIA-Quilamapu, Chile. E-mail: afrance@inia.cl

Javier P. Franco, CABI Plantwise-Perú. Los Cerezos. 338. Apartamento 103, Surco, Lima, Perú. E-mail: j.franco@cabi.org

Fermín Fuentes, Laboratorio Provincial de Sanidad Vegetal, Havana, Cuba. E-mail: nereida.garcia@infomed.sld.cu;

Fernando Gallardo-Covas, Entomology, Department of Agro-environmental Sciences, University of Puerto Rico at Mayaguez, Puerto Rico. E-mail: fernando.gallardocovas@upr.edu

Patricio G. Gallegos, Entomologist, independent consultant. Francisco Farfan N11-120. Barrio Los Arupos, Conocoto, Quito, Ecuador. E-mail: patricio23.gallegos@gmail.com

Victor Adolfo Gómez López, Laboratory of Entomology, Facultad de Ciencias Agrarias, Universidad Nacional de Asunción, Paraguay. E-mail: victor.gomez@agr.una.py

Nancy M. Greco, CEPAVE, CONICET-UNLP (Centro de Estudios Parasitológicos y de Vectores), La Plata, Argentina. E-mail: ngreco@cepave.edu.ar; lunam@cepave.edu.ar

Lidcay Herrera, Facultad de Ciencias Agropecuarias, Universidad Central Martha Abreu, Villa Clara, Cuba. E-mail: lidcayhi@uclv.edu.cu

Leopoldo Hidalgo, Centro Nacional de Sanidad Agropecuaria-CENSA, apartado 10, Mayabeque, Cuba. E-mail: leopoldoj.hidalgo@gmail.com;

Wendy-Ann P. Isaac, Department of Food Production, University of the West Indies, St Augustine Trinidad. E-mail: wendy-ann.isaac@sta.uwi.edu

Maitrie Jagroep, Agricultural Experiment Station, Ministry of Agriculture, Animal Husbandry and Fisheries, Letitia Vriesdelaan 8, Paramaribo, Suriname. E-mail: maitjag@mail.com

Ayub Khan, Department of Life Sciences, University of the West Indies, St Augustine, Trinidad. E-mail: ayub.khan@sta.uwi.edu

Takumasa Kondo, Corporación Colombiana de Investigación Agropecuaria (Agrosavia), Centro de Investigación Palmira, Valle del Cauca, Colombia. E-mail: tkondo@agrosavia.co

Maria Gabriela Luna, CEPAVE, CONICET-UNLP (Centro de Estudios Parasitológicos y de Vectores), La Plata, Argentina. E-mail: ngreco@cepave.edu.ar; lunam@cepave.edu.ar

Maria R. Manzano, Universidad Nacional de Colombia, sede Palmira, Departamento de Ciencias Agrícolas, Palmira, Valle del Cauca, Colombia. E-mail: mrmanzanom@unal.edu.co

María Elena Márquez, Universidad de La Habana, Havana, Cuba. E-mail: marquezmariaelena18@gmail.com

José Morales Sanchez, Universidad Centroccidental Lisandro Alvarado, Department of Biological Sciences, Barquisimeto, Lara State, Venezuela.

Rossy Morera-Montoya, Graduate Programme in Agricultural Sciences and Natural Resources, University of Costa Rica, San José, Costa Rica. E-mail: rossymorera@gmail.com

Norma Mujica, Departamento de Entomología, Universidad Nacional Agraria – La Molina, POB 12-056, Av. La Molina s/n, Lima 12, Peru. E-mail: nmujica@lamolina.edu.pe

José Roberto Postali Parra, Laboratory of Biology of Insects, Department of Entomology and Acarology, ESALQ/USP, Piracicaba, São Paulo, Brazil. E-mail: jrpparra@usp.br

Abelino Pitty, Escuela Agrícola Panamericana, Zamorano, Honduras. E-mail: apitty@zamorano.edu

Karina Punschke, Registro de Agentes de Control Biológico, División Control de Insumos, Dirección General de Servicios Agrícolas, Ministerio de Ganadería, Agricultura y Pesca, Millán 4703, 12900 Montevideo, Uruguay. E-mail: kpunschke@mgap.gub.uy

Mariangie Ramos, Sustainable Agriculture Program, Department of Agricultural Technology, University of Puerto Rico at Utuado, Puerto Rico. E-mail: mariangie.ramos@upr.edu

Mayra Ramos, Instituto Superior de Tecnologías y Ciencias Aplicadas-InSTEC, Havana, Cuba. E-mail: mramos@instec.cu

Olgaly Ramos-Rodriguez, Entomology and Pest Management, Department of Agricultural Technology, University of Puerto Rico at Utuado, Puerto Rico. E-mail: olgaly.ramos@upr.edu

Gloria Resquín Romero, Laboratory of Phytopatology, Facultad de Ciencias Agrarias, Universidad Nacional de Asunción, Paraguay. E-mail: gloresqx@agr.una.py

Adela Ribeiro, Unidad de Entomología, Estación Experimental 'Dr. M.A. Cassinoni', Facultad de Agronomía, Universidad de la República. Ruta 3 km 363, 60000 Paysandú, Uruguay. E-mail: adelaribeiro57@gmail.com

Mayra G. Rodríguez, Centro Nacional de Sanidad Agropecuaria-CENSA, apartado 10, Mayabeque, Cuba. E-mail: mrguez@censa.edu.cu

Beatriz Rodríguez-Vélez, Centro Nacional de Referencia de Control Biológico, Dirección General de Sanidad Vegetal. Km 1.5 Carretera Tecomán-Estación, C.P. 28110, Tecomán, Colima, México.

Philippe Ryckewaert, CIRAD, UR Hortsys, Campus Agro-environnemental Caraïbes, Petit Morne 97232 - Le Lamentin, Martinique (French West Indies). E-mail: philippe.ryckewaert@cirad.fr

Colmar Serra, Instituto Dominicano de Investigaciones Agropecuarias y Forestales (IDIAF), Centro de Tecnologías Agrícolas (CENTA), C/ Rafael Augusto Sánchez 89, 10147 Santo Domingo, D.N., Dominican Republic. E-mail: colmar.serra@gmx.net

Michelle A. Sherwood, Crop and Plant Protection Unit, Research and Development Division, Ministry of Industry of Agriculture and Fisheries, Old Harbour, St Catherine, Jamaica. E-mail: masherwood@micaf.gov.jm

Edwin E. Sosa, Organismo Internacional Regional de Sanidad Agropecuaria (OIRSA), Showgrounds, Belmopan, Belize. E-mail: esosa@oirsa.org

Rogelio Trabanino, Escuela Agrícola Panamericana, Zamorano, Honduras. E-mail: rtrabanino@zamorano.edu; apitty@zamorano.edu

Joop C. van Lenteren, Laboratory of Entomology, Wageningen University, PO Box 16, 6700 AA, The Netherlands. E-mail: joop.vanLenteren@wur.nl

Alies van Sauers-Muller, Agricultural Experiment Station, Ministry of Agriculture, Animal Husbandry and Fisheries, Letitia Vriesdelaan 8, Paramaribo, Suriname. E-mail: aliesmuller@yahoo.com

Carlos Vásquez, Technical University of Ambato. Faculty of Agricultural Sciences. Campus Querochaca. Province of Tungurahua, Ecuador. E-mail: ca.vasquez@uta.edu.ec; cvasqz@yahoo.com

Jean-François Vayssières, CIRAD, UR Hortsys, Campus International de Baillarguet 34398 - Montpellier Cedex 5, France

Luis L. Vázquez, Instituto de Investigaciones de Sanidad Vegetal-INISAV, Havana, Cuba. E-mail: llvazquezmoreno@yahoo.es

Mary Whu, Asociación de Control Biológico del Perú. E-mail: marywhup@gmail.com

Bruno Zachrisson, Entomology and Biological Control of Insect-Pest Laboratory, Instituto de Investigación Agropecuaria de Panamá (IDIAP), Chepo, Panamá. E-mail: bruno.zachrissons@idiap.gob.pa

Foreword

Although overviews of the history and the current state of affairs in biological control have been published for several world regions, such as Europe, North America and Australia, this information has been lacking for Latin America and the Caribbean. Two of the editors of the current book – Bueno and Van Lenteren – wrote a review in 2003 with the title 'Augmentative Biological Control of Arthropods in Latin America' (*BioControl* 48, 123–139), and the last lines in this paper were: 'It has taken us a lot of time to obtain data on the use of biological control in Latin America, and we are convinced that this survey is still not complete. We encourage readers to send us up to date information, so we can provide a more reliable overview in the near future.' As a result of this request, we received a good amount of reactions, and we began to understand that there was a large quantity of information in Spanish and Portuguese publications and reports. However, this information was either hard to access, or could not be understood easily by an international readership. We also realized that the amount of material was too large and too interesting for a simple update of the above-mentioned review. Clearly, producing an entirely new book would be a better solution. Ideas for an English language publication on biocontrol in Latin America and the Caribbean were discussed with members of the Executive Committee of NeoTropical Regional Section of the International Organisation for Biological Control (IOBC-NTRS), and potential authors in the region were contacted. By the end of 2016, we had succeeded in obtaining authors for most countries with significant biocontrol activities and we could start with writing and editing the country-specific chapters.

Why do we think it important to have this information available in English? First of all, the large amount of historical knowledge about biocontrol in this region deserves the attention of those working in biocontrol elsewhere in the world. We document both successes and failures. Facts about successes in certain regions may help researchers elsewhere in finding candidate natural enemies for the same or similar pests in their area. Furthermore, the lists of biocontrol agents found in a certain country or region of Latin America and the Caribbean which we provide in this book might help to facilitate Access and Benefit Sharing procedures, compulsory since the ratification of the Nagoya Protocol in 2014 (see Chapter 1 for details). We are also convinced that knowledge about projects that failed is essential to prevent making the same mistakes in the future.

Secondly, by providing an extensive overview of the current situation in biocontrol for many Latin American and Caribbean countries, we aim not only to show what is happening in this region, but also to help in networking and collaboration between regions with similar pest problems.

Thirdly, during the data collection phase for the book, we were astonished by the amount of practical biocontrol applied in this region, which was often undocumented or not easily available.

The new data, presented in the final chapter of this book, show that Latin America and the Caribbean might currently have the largest area under biocontrol worldwide.

With 30 country-specific chapters and more than 60 authors involved, this was a somewhat ambitious enterprise, but with the invaluable assistance of the authors, we think we have succeeded in the production of an interesting and important book. We have also decided to produce a Spanish version of the book for those students, researchers, extension workers and farmers who do not read English. We aim to have this Spanish version published within a year of the English version.

We hope that the book will play a positive role in future developments of biocontrol, sustainable food production and protection of biodiversity, both in the Latin American region and worldwide.

Preface

I am absolutely delighted that this volume focusing on biological control in Latin America and the Caribbean has been put together. It is a very important piece of scholarship that highlights an inspiring amount and breadth of work that has been going on in this region over the past century. The editors include luminaries in the area of biological control in Latin America and internationally, and they have gone to great lengths to provide a comprehensive treatment of biological control in countries ranging from Mexico and the Caribbean Islands to the tip of Tierra del Fuego – more than 30 countries are represented in all! What is so important about this compendium is that it brings to light information that was previously either inaccessible or very difficult to find. In our current age of instant gratification I worry that many scientists would not do the work to dig up information even in the 'difficult to find' category and so this volume is all the more appreciated.

Biological control remains the best way to combat many invasive pests, diseases and weeds, and for ensuring a healthy and sustainable food supply. With the recent improvements in safety protocols and commercial production of effective agents, there is really no reason for biological control not to expand to meet growing problems in invasive species and food security worldwide. Biological control is particularly important in regions like Latin America, where some farmers cannot afford pesticides in the first place, and where some of the compounds that are available pose very serious risks to human and environmental health. Latin America also contains some of the world's most celebrated and pristine natural areas and is the region with the highest biodiversity in the world, some of which is under threat from invasive species. The capacity to engage these problems on a regional basis is there but needs to be nurtured and improved. This book is an important part of this process, as it not only documents a rich history of biological control in the region but also points a way forward to a Latin America under diminished pressure of invasive species and pesticides.

George E. Heimpel

President, International Organisation for Biological Control (IOBC) Global,
and Professor of Entomology, University of Minnesota, USA.

Dedication and Acknowledgements

The editors would like to dedicate this book first of all to the pioneers of biological control in this region. The first pioneers, the indigenous peoples of the Americas, considered the majority of insects as non-harmful, and reduced the impact of harmful species by inter- and multi-cropping, periods of fallow, use of optimal production sites, and even a form of field selection of resistant cultivars. Farmers recognized the contribution to pest reduction by spiders, aquatic nymphs of dragon flies, lizards, snakes, squirrels, weasels and skunks. In Nicaragua, for example, ancient people recognized the value and abundance of insects as natural enemies of pests, and coined them *náhuatl* (Dávila, 1992: *Glosario de nombres náhuatl de plantas, pájaros y algunas otras especies, con descripción de su etimología y comentarios del autor* (*Glossary of náhuatl names for plants, birds and other species, with etymological descriptions and commentaries from the author*). Fondo Editorial Centro de Investigación de la Realidad de América Latina (CIRA). Managua, Nicaragua, 47pp.)

Many of the second group of pioneers, i.e. those who started using biocontrol at the beginning of the 20th century, are mentioned in the country-specific chapters. Although native biocontrol agents were often used, in many other cases natural enemies were imported from other regions in the world, either because the crops with their pests originated from these regions, or because information was available about the success rate of certain natural enemies.

We would also like to dedicate the book to those who were able to persevere successfully with earlier developed biocontrol projects, or even to develop and implement new applications after 1945, when the use of chemical control increased dramatically. This was a difficult period for biocontrol in most Latin American countries, and many effective projects were either terminated due to market pressure by the pesticide industry, or because they could no longer function due to negative side effects of synthetic chemical pesticides.

The book is also dedicated to everyone who is currently studying, teaching and practising biological control in Latin America and the Caribbean. Although biocontrol is now applied widely in this region, researchers and producers of biocontrol agents continue to be confronted with the overwhelming dominance of the chemical pesticide industry. We know it requires an enormous amount of stamina to keep convincing farmers, policy makers and governments that biocontrol is the most sustainable, economic and safest way to reduce pests, diseases and weeds!

Finally, we dedicate the book to the next generation with an interest in the production of sufficient and healthy food, and in restoring biodiversity in this region. We, the editors, can assure you that the study of biocontrol, which actually is an extremely interesting combination of applied microbiology, behaviour, ecology and environmental sciences, is always fascinating and very rewarding.

Authors for the book were initially identified through the IOBC-NTRS network (www.iobcntrs.org) with help of the Executive Committees of IOBC-NTRS. Later, additional authors were found via correspondence and internet searches. We would like to thank all the authors – many of them members of IOBC-NTRS – for collecting and translating Spanish/Portuguese information into English, and for the important chapters they wrote for the book.

Also, we thank David Hemming, Emma McCann and Tim Kapp for the editorial support from CABI, Val Porter for copy editing all chapters and Alice Sparks for proofreading. M.J.W. Cock and Y.C. Colmenarez gratefully acknowledge the core financial support from the member countries (and lead agencies) of CABI (an international intergovernmental organization) including the United Kingdom (Department for International Development), China (Chinese Ministry of Agriculture), Australia (Australian Centre for International Agricultural Research), Canada (Agriculture and Agri-Food Canada), the Netherlands (Directorate-General for International Cooperation), and Switzerland (Swiss Agency for Development and Cooperation). See https://www.cabi.org/what-we-do/how-we-work/cabi-donors-and-partners/ for full details.

Joop C. van Lenteren, Vanda H.P. Bueno,
M. Gabriela Luna and Yelitza C. Colmenarez

Glossary

Acronym	Details
ABC	augmentative biological control
ABCBio	Brazilian Association of Biocontrol Companies
ABRAF	Associação Brasileira de Produtores de Florestas Plantadas
ABS	Access and Benefit Sharing [Nagoya Protocol]
ACMNPV	*Autographa californica* multiple nucleopolyhedrovirus
ACP	Asian citrus psyllid
ADEXVO	Asociación Dominica de Exportadores de Vegetales Orientales
AgMNPV	*Anticarsia gemmatilis* multiple nucleopolyhedrovirus
AgNPV	*Anticarsia gemmatilis* nucleopolyhedrovirus
AGROCALIDAD	Ecuadorian Agency to assure quality in Agriculture
Agrosavia	Corporation for Agricultural Research [Colombia] (formerly Corpoica)
ALAM	Asociación Latinoamericano de Microbiología
ALF	Asociación Latinoamericana de Fitopalogía
AMID	Agricultural Marketing Information Division
ANACAFÉ	Asociación Nacional del Café [Guatemala]
ANAO	Asociación Nacional de Agricultura Orgánica [National Association of Organic Agriculture]
ANAPO	Asociación de Productores de Oleaginosas y Trigo
ANCA	National Cotton Association [Venezuela]
ANCUPA	Association of Oil Palm Growers
ANECAFE	Association of Coffee Exporters of Ecuador
ANII	National Agency for Research and Innovation
APCB	Asociación Peruana de Control Biológica
APHIS	Animal and Plant Health Inspection Service [USDA]
APM	agroecological pest management
Ascolflores	Asociación Colombiana de Exportadores de Flores
asl	above sea level
BCL	Biological Control Laboratory
BIOESA	Biológicos Ecuatorianos S.A.
BRGM	Banco de Recursos Genéticos Microbianos [Microbial Genetic Resources Bank]
Bt	*Bacillus thuringiensis*
BYDV	barley yellow dwarf virus

Continued

Acronym	Details
CABI	Centre for Agriculture and Biosciences International
CARDI	Caribbean Agricultural Research and Development Institute
CARICOM	Caribbean Community
CATIE	Centro Agronómico Tropical de Investigación y Enseñanza [Tropical Agriculture Research and Higher Education Center]
CBB	coffee berry borer
CBBC	Commonwealth Bureau of Biological Control
CBC	classical biological control
CBD	Convention on Biological Diversity [Rio]
CCBBM	Center for Biological Control and Molecular Biology
CCBCA	Centro para el Control Biológico en Centroamérica
CChRGM	Collección Chilena de Recursos Genéticos Microbianos [Chilean Collection of Microbial Genetic Resources]
CCIPA	Comité de Control Integrado de Plagas del Algodonero
CDF	Charles Darwin Foundation
CEA	Centro de Entomoligía Aplicada [Applied Entomology Center]
CEA	Consejo Estatal del Azucar [State Sugar Council]
CEAZA	Centro de Estudios Avanzados en Zonas Arida [Center for Advanced Studies in Arid Zones]
CEBIOF	Center for Forest Bioservices
CECTEC	Centro de Educación, Capacitación y Tecnología Campesina [Center of Education, Training and Peasant Technology]
CEDAF	Centro de Desarrollo Agropecuario y Forestal [Center for Agricultural and Forestry Development]
CEMUBIO	Center for Reproduction of Biocontrol Agents
CENA	Centro de Energia Nuclear na Agricultura
CENAPROVE	Centro Nacional de Protección Vegetal [National Center for Plant Protection]
CENETROP	Centro Nacional de Enfermedades Tropicales
CENIAP	National Center for Agricultural Research
Cenicafé	National Coffee Research Center
CENSA	Centro Nacional de Sanidad Agropecuaria
CENTA	Centro de Tecnologías Agrícolas / Centro Nacional de Tecnologia Agropecuaria
CEPAVE	Centro de Estudios Parasitológicos y de Vectores
CEQIS	Center for Excellence in Quarantine and Invasive Species
CESDA	Centro Sur de Desarrollo Agropecuarios
CFF	carambola fruit fly
CFS	Colombian fluted scale
CGA	Citrus Growers Association
CGIAR	Consortium of International Agricultural Research Centers
CIA	Central Intelligence Agency
CIAC	Central Research Center of IDIAP
CIAT	International Center for Tropical Agriculture
CIB	Coffee Industry Board
CIBC	Commonwealth Institute for Biological Control
CICIU	Centro de Introducción y Cría des Insectos Utiles [Center for Introduction and Rearing of Useful Insects]
CIMCA	Centre for Research and Improvement of Sugarcane
CIMIC	Center for Microbiological Research
CIMMYT	International Maize and Wheat Improvement Center
CINCAE	Sugar Cane Research Center
CIP	International Potato Center
CIPROC	Crop Protection Research Center
CIRA	Centro de Investigación de la Realidad de América Latina
CIRAD	Centre de coopération internationale en recherche agronomique pour le développement

Continued

Acronym	Details
CIRCB	Centro de Investigación Reproducción Control Biológico
CIRPON	Centro de Investigaciónes sobre Regulación de Poblaciones de Organismes Nacivos
CITTCA	Centro de Investigación y Transferencion de Tecnológia para la Caña de Azúcar [Center for Research and Technology Transfer for Sugar Cane]
CNRCB	Centro Nacional de Referencia de Control Biologico
CNSV	Centro Nacional de Sanidad Vegetal
CODOCAFÉ	Consejo Dominicano del Café
CONAF	Corporación Nacional Forestal [National Forerst Corporation]
CONIAF	Consejo Nacional de Investigaciones Agropecuaria y Forestales [National Council of Agriculture and Forestry Research]
CONICET	National Council of Scientific and Technical Research
CONQUITO	Economy Promotion Agency, Quito Municipality
ConsBC	conservation biological control
CORBANA	National Banana Corporation
CoSAVE	Comité de Sanidad Vegetal [Plant Health Committee]
CPF	Forestry Pest Control Company
CpGV	*Cydia pomonella* granulovirus
CPHST	Center for Plant Health Science and Technology
CRE	Centros Reproductores de Entomófagos
CREE	Entomophagous and Entomopathogens Mass Rearing Centers
CREI	Citrus Research and Education Institute
CSIRO	Commonwealth Scientific and Industrial Research Organization
CTCB	Biological Control Technological Center
CTV	citrus tristeza (clostero)virus
CURLA	Regional University Center of the Atlantic Coast
CVA	Venezuelan Agrarian Corporation
DAJ	Department of Agriculture, Jamaica
DANE	Departamento Administrativo Nacional de Estadística
DFP	Departmento de Fitopatologia
DGSA	General Directorate of Agricultural Services
DGSV	Dirrección General de Sanidad Vegetal [General Directorate of Plant Health]
DIECA	División de Investigación en Caña de Azúcar [Sugar Cane Research and Extension Division]
DOF	Diario Oficial de la Federación [Official Journal of the Federation]
DOMEX	Dominica Export
DPNG	Dirección del Parque Nacional Galápagos
DPV	Departamento de Protección Vegetal
DSV	Department of Plant Health
ECI	Exportando Calidad e Inocuidad
EF1α1	elongation factor 1α1
EMATER-PR	Instituto Paranaense de Assistência Técnica e Extensão Rural
EMBRAPA / Embrapa	Empresa Brasileira de Pesquisa Agropecuária
ENA	Encuesta Nacional Agropecuaria [National Agricultural Survey]
ENEE	National Electrical Company
EPA	Environmental Protection Agency (US)
EPF	entomopathogenic fungi
EPN	entomopathogenic nematodes
ERPE	Escuelas Radiofónicas Populares del Ecuador
ESALQ	Escola Superior de Agricultura ‘Luiz de Queiroz’
ESPAC	Encuesta de Superficcie y Producción Agropecuaria Continua
ESPOCH	Polytechnic University of Chimborazo Province
ETPP	Territorial Stations of Plant Health

Continued

Acronym	Details
FAO	Food and Agriculture Organization of the United Nations
FAPESP	São Paulo Research Foundation
FBC	fortuitous biocontrol
FCA-UNA	Facultad de Ciencias Agrarias, Universidad Nacional de Asunción [Faculty of Agricultural Sciences]
FCQ-UNDA	Facultad de Ciencias Químicas, Universidad Nacional de Asunción [Faculty of Chemical Sciences]
FDA	Fundacion de Desarrollo Agropecuario [Center for Agricultural Development]
FEPAF	Foundation for Agricultural and Forestry Studies and Research
FFS	Farmer Field Schools
FHTET	Forest Health Technology Enterprise Team
FIA	Fundación para la Innovación Agraria
FINTRAC / Fintrac	(private US-based non-governmental organization)
FONAIAP	National Agricultural Research Fund (currently INIA)
FONDAFA	Development Fund for Agriculture, Fishing, Forestry and Related Activities
FONDEF	Fondo de Desarrollo Cientifico y Tecnológico [Scientific and Technological Development Fund]
FREDON	Fédérations Régionale de Défense contre les Organismes Nursibles [Regional Federation of Protection Against Damaging Organisms]
FUEDEI	Fundación para el Estudio de Especies Invasivas
FUNCEMA	National Resources Unit for Control of Sirex Wood Wasp
FUNDACAÑA	Sugar Foundation for Development, Productivity and Research
FUNDAR	(an Ecuador NGO)
FUNICA	Fundación para el Desarrollo Tecnológico Agropecuaria, Forestal de Nicaragua
FUSAGRI	Fundación Servicio para el Agricultor
GAP	Good Agricultural Practices
GAR	Government of the Argentine Republic
GBIF	Global Biodiversity Information Facility
GRDB	Guyana Rice Development Board
GTZ	Deutsche Gesellschaft für Technische Zusammenarbeit [German Cooperation Agency] [Sociedad Alamana de Cooperación Técnica]
HLB	huanglongbing [lit. yellow dragon disease] = citrus greening
HzSNPV	*Helicoverpa zea* nucleopolyhedrovirus
IAEA	International Atomic Energy Agency
IANAS	Inter-American Network of Academies of Sciences
IASA	Andean Institute of Agriculture
IBA	Indústria Brasileira de árvores
IBGE	Institute Brasileiro de Geografia e Estatistica
IBSP	Instituto Biológica São Paulo
ICA	Instituto Colombiano Agropecuario [Colombian Agricultural Institute]
ICAFE	Instituto del Café de Costa Rica
ICDF	International Cooperation Development Fund
ICIDCA	Institute de Investigaciones de Derivados de la Caña de Azúcar
IDA	International Depository Authority
IDIAF	Instituto Dominicano de Investigaciones Agropecuarias y Forestales
IDIAP	Instituto de Investigación Agropecuaria de Panamá
IE	Institute of Ecology
IFAD	International Fund for Agricultural Development
IFPA	Institute for the Promotion of Sugar Production
IIBC	International Institute of Biological Control
IICA	Instituto Interamericano de Cooperación para la Agricultura [Inter-American Institute for Cooperation on Agriculture]
IITA	International Institute of Tropical Agriculture
IMO	Institute of Marketology
IMTA	Mexican Institute of Water Technology

Continued

Acronym	Details
IMyZA	Instituto de Microbiología y Zoología Agricola [Microbiology and Agricultural Zoology Institute]
INA	Instituto Nacional de Aprendizaje [National Learning Institute]
INBIO	Instituto Nacional de Biotecnología [National Institute of Biotechnology]
INCE	National Education Training Institute
INE	Instituto Nacional de Estadistica
INEGI	Instituto Nacional de Estadística y Geografía
INEI	Instituto Nacional de Estadistica e Informática
INFOR	Instituto de Investigaciones Forestal
INIA/INIAP	Instituto Nacional de Investigaciones Agrarias/Instituto Nacional de Investigaciones Agropecuarias [National Institute of Agricultural Research]
INISAV	Instituto Nacional de Investigaciones de Sanidad Vegetal [Plant Health Research Institute]
INRA	Institut National de la Recherche Agronomique
INSAI	National Institute of Integral Agricultural Health
INTA	Instituto Nacional de Innovación Transferencia en Tecnología Agropecuaria [National Institute for Agricultural Technology]
IOBC	International Organisation for Biological Control
IOBC/NTRS	IOBC/NeoTropical Regional Section
IPADE	Instituto Panamericano de Alta Dirección de Empresa
IPBES	Intergovernmental Science Policy Platform on Biodiversity and Ecosystem Services
IPDM	integrated pest and disease management
IPE	Peruvian Asparagus Institute
IPL-GTZ	(Dominican/German project) German Society for Technical Cooperation
IPM	integrated pest management
IPM-CRSP	Integrated Pest Management, Collaborative Research Support Program
IPPC	International Plant Protection Convention
IPTA	Institut Paraguayo de Tecnología Agraria [Paraguayan Institute of Agricultural Technology]
IRD	Institut de Recherche pour le Développement
ISA	Instituto Superior de Agricultura
ISCAMEN	Instituto de Sanidad y Calidad Agropecuaria de Mendoza
ISPM	International Standards for Phytosanitary Measures
JAD	Junta Empresarial Dominicana
JICA	Japan International Cooperation Agency [Agencia Internacional de Colaboración del Japán]
LABOCOBI	Laboratorio de Control Biológico
LAICA	Liga Agrícola Industrial de la Caña de Azúcar [Sugarcane Industry Association]
LAPROSAV	Laboratorios Provinciales de Sanidad Vegetal
LSF	liquid state fermentation
MA	Ministry of Agriculture
MAD	Medio Abiente y Desarrollo
MAG	Ministerio de Agricultura y Ganaderia [Ministry of Agriculture and Livestock]
MAGAP	Ministerio de Agricultura, Ganadería, Acuacultura y Pesca
MALMR	Ministry of Agriculture, Land and Marine Resources
MAPA	Ministério da Agricultura, Pecuária e Abastecimento
MARNDR	Ministère del'Agriculture, des Ressources Naturelles et de Développement Rural
MAT	male annihilation technique
MEJMO	Museum of Entomology José M. Osorio
MESCyT	Ministry of Higher Education, Science and Technology
MGAP	Ministry of Livestock, Agriculture and Fisheries
MICAF	Ministry of Industry, Commerce, Agriculture and Fisheries
MIDA	Ministerio de Desarrollo Agropecuaria

Continued

Acronym	Details
MINAG	Ministry of Agriculture
MINAGRI	Ministerio de Agricultura y Riego
Minagri	Ministry of Agribusiness [Argentina]
MIPH	Manejo Integrado de Plagas en Honduras
MoAF	Ministry of Agriculture and Fisheries
MOSCAMED / Moscamed	mosca de la fruta del Mediterráneo [Mediterranean fruit fly] [= medfly, *C. capitata*]
MRL	maximum residue level
NAREI	National Agricultural Research and Extension Institute
NC	natural control
NGO	non-governmental organization
NPV	nuclear polyhedrosis virus
NRI	Natural Resources Institute
OCM	Organización Mundial de Comercío [= World Trade Organization, WTO]
ODA	Overseas Development Agency
ODEPA	Oficina de Estudios y Politicas Agrarias
OEA	Organización de los Estados Americanos [Organization of American States]
OECD	Organization for Economic Co-operation and Development
OIRSA	Organismo Internacional Regional de Sanidad Agropecuaria [International Regional Organization for Plant Protection and Animal Health]
PIOJ	Planning Institute of Jamaica
PLANUSA	Plan International USA
PNAO	Programa Nacional de Agricultura Organica [National Program of Organic Agriculture]
PNCB	National Program of Biological Control
PNCVM PIVCVM	National Program for Control of Wood Wasp
PNMIP	Programa Nacional de Manejo Integrado de Plagas
PNV	polyhedrosis nuclear virus – *see* NPV
PRECODEPA	Programa Regional Cooperative de Papa
PROBIOMA	Productivity Biosphere and Environment
PROBIOTEC	Center for Research, Diagnosis and Production of Biocontrol Agents for Control of Pests and Diseases
PROCOBI	National Program for Biological Control
PRODESA	Development Program of Agricultural Health
PROIMI	Pilot Plant of Industrial Microbiological Processes
PROINPA	Fundación para la Promoción e Investigación de Productos Andinos [Potato Research Project]
PROMECAFE	Regional Cooperative Program for Technological Development and Modernization of the Coffee Industry
PROMIB	Programa de Manejo Integrado de la Broca del Cafe
PROMSA	Program for Modernization of Agricultural Services
PROTEF / IPEF	Programa de Proteção Florestal / Instituto de Pesquisas e Estudos Florestais
RCC	Regional Research Center
REDCAHOR	Red Colaborative de Investigación, Desarrollo de las Hortalizos para América Central [Collaborative Vegetable Research and Development Network]
RKN	root-knot nematode
SAC	Sociedad de Agricoltores de Colombia
SAG	Servicio Agricola y Ganadero
SAGARPA	Secretaría de Agricultura y Desarrollo Rural [Ministry of Agriculture, Livestock, Rural Development, Fishers and Food]
SASA	Autonomous Service of Agricultural Health
SCB	Subdirección de Control Biológica
SCBD	Secretariat of the Convention on Biological Diversity
SEA	Secretariía de Estado de Agricultura

Continued

Acronym	Details
SEB	Sociedad Entomológica do Brasil [Entomological Society of Brazil]
SENASA	Servicio Nacional de Sanidad y Calidad Agroalimentaria
SENASAG	National Service of Agricultural Health and Food Safety
SENASICA	Servicio Nacional de Sanidad, Inocuidad y Calidad Agroalimentaria [National Service of Health, Safety and Agrifood Quality]
SENAVE	Servicio Nacional de Calidad y Sanidad Vegetal y de Semillas
SeNPV	*Spodoptera exigua* nucleopolyhedrovirus
SERVBIO	Biological Control Service Company
SESA	Ecuadorian Service for Agricultural Health
SFE	Servicio Fitosanitario del Estado
SfMNPV	*Spodoptera frugiperda* nucleopolyhedrovirus
SIAP	Servicio de Información Agroalimentario y Pesquera
SICGAL	Galapagos Inspection and Quarantine System
Siconbiol	Symposium of Biological Control
SINAGAP	Sistema de Información Nacional de Agricultura, Ganadería, Acuacultura y Pesca
SIRDI	Sugar Industry Research and Development Institute
SIT	sterile insect technology / sterile insect technique
SMBC	Sociedad Mexicana de Control Biológico
SNI – SENACYT	Sistema Nacional de Investigación – Secretaría Nacional de Ciencia, Technlogía e Innovación
SPF	Forestry Producers Association
SRL	Sociedad de responsabilidad limitada [equiv. of 'limited liability company']
SSF	solid-state fermentation
TL50	lethal time
TnNPV	*Trichoplusia ni* nucleopolyhedrovirus
UAGRM	Autonomous University Gabriel Rene Moreno
UASD	Universidad Autónoma de Santo Domingo
UAV	unmanned aerial vehicle [= drone]
UCLA	Universidad Centro Occidental 'Lisandro Alvarado'
UCV	Central University of Venezuela
UdelaR	Universidad de la Republíca [Republic University of Uruguay]
UENF	Universidade Estadual do Norte Fluminense Darcy Ribeiro
UFV	Universidade Federal de Viçosa
UNA	Universidad Nacional Agraria
UNAN	Universidad Nacional Autónoma de Nicaragua
UNDP	United Nations Development Programme
UNESP-FCA	Universidad Estadual Paulista – Faculdade de Ciéncias Agronómicos
UN-IAEA	see IAEA
UNPHU	Universidad Nacional Pedro Henríquez Ureña
UPAVE	Venezuelan Sugarcane Producers Association
USAID	United States Agency for International Development
USDA	United States Department of Agriculture
USDA ARS	USDA Agriculture Research Service
USDA-NRCS	USDA National Resource Conservation Service
USP	Universidad de São Paulo
VG	granulosis virus
VIFINEX	Vigilancia Fitosanitaria en Cultivos de Exportación no Tradicionales
VPNAg	see AgNPV
WDF	water-diluted granules
WHO	World Health Organization
WIPO	World Intellectual Property Organization
WREP	Whitefly Research and Extension Project

1 Biological Control in Latin America and the Caribbean: Information Sources, Organizations, Types and Approaches in Biological Control

Joop C. van Lenteren[1]*, Vanda H.P. Bueno[2], M. Gabriela Luna[3] and Yelitza C. Colmenarez[4]

[1]Laboratory of Entomology, Wageningen University, The Netherlands; [2]Laboratory of Biological Control, Department of Entomology, Federal University of Lavras, Minas Gerais, Brazil; [3]CEPAVE, CONICET-UNLP (Centro de Estudios Parasitológicos y de Vectores), La Plata, Argentina; [4]CABI-UNESP-FEPAF, Botucatu, São Paulo, Brazil

Abstract

Biological control with arthropod natural enemies and microbial control agents has been applied since the year 1895 in Latin America and the Caribbean and is currently used on a very large scale. Sources about the history and current situation of biocontrol in this region were not easy to trace and are, therefore, presented in this chapter. Next, organizations working on biocontrol in this region are listed. This is followed by a description of natural, conservation, classical and augmentative biocontrol with some regional examples. Then, an approach to find, evaluate and use biocontrol agents is sketched, as guidance for research projects. Often, tens to a hundred biocontrol candidates are found in association with a pest. A well organized research approach using evaluation criteria allows for quick exclusion of unsuitable or problematic candidate species. Biocontrol research has limited funding and early elimination of poor candidates results in spending more money on promising candidates. Regulations concerning import and release of agents that have been implemented during the past 30 years are summarized. Effects of these regulations are that prospecting for exotic natural enemies is now very difficult and that fewer new biocontrol agents have become available. Finally, the structure of the book is explained.

1.1 Introduction

Biological control (biocontrol) is, simply said, the use of an organism to reduce the population density of another organism. It is the most successful, most cost effective and environmentally safest system for pest, disease and weed management (Bale *et al.*, 2008). It is nature's own way to keep numbers of pest organisms at low levels. Biocontrol is present in all ecosystems, both natural and man-made. The result of natural biocontrol is that the earth is green and that plants can produce sufficient biomass to sustain other forms of life. Without biocontrol, the production of energy by plants would be a tiny fraction of what is generated currently.

* E-mail: joop.vanlenteren@wur.nl

Developments in biocontrol have been summarized for several world regions and in various handbooks. However, few publications provide historical detail about the development and the current situation of biocontrol in Latin America and the Caribbean. Nonetheless, biocontrol with arthropod natural enemies and microbial control agents has been applied since 1895 on large areas in Latin America and the need was felt for a well documented overview. The main aims of this book are threefold: first to summarize the history; next to describe the current situation; and, finally, to speculate about the future of biocontrol. The history of biocontrol is presented in 30 chapters; most chapters are about individual countries, but some deal with groups of smaller countries and islands. The history was fragmented until now and if information was available it was often in local reports and publications in Spanish or Portuguese that were difficult to obtain. We asked authors of the book chapters to translate and summarize information about developments in biocontrol in their country, and by presenting this knowledge in the current book, we aspire to offer a reasonably complete picture of important historical developments in this region. When authors referred to abstracts, unpublished reports, information leaflets or symposium papers, we asked them to provide pdf versions of this material, so that readers of this book can consult the original material upon which parts of the country-specific chapters were based. These pdf files can be obtained from a website made available by CABI. An ample amount of text, tables and references about this history is provided, because in order to make progress in biocontrol it is essential to know what was done, which projects succeeded, but also which projects failed and the main limitations faced by those projects. Documentation of the pest species, crops, biocontrol agents, locations and type of biocontrol programmes tested will hopefully help to prevent making the same mistakes and stimulate initiation of new projects with biocontrol agents that have been successfully used elsewhere. After summarizing the history, the current situation is described and then the authors present their ideas about the future of biocontrol in their country.

During the data-collection phase for the book, we were astonished by the amount of often unknown practical biocontrol applied in Latin America and the Caribbean. Van Lenteren and Bueno (2003) estimated that the area under augmentative biocontrol in this region was about 4,350,000 ha in the year 2000, but realized that this might be a serious underestimate (Table 1.1). Reliable data for areas and crops protected by classical biocontrol were even more difficult to obtain than those for augmentative biocontrol. In the final chapter of this book, new estimates are given based on data presented in each chapter. The 2018 estimates for areas under augmentative biocontrol alone, amount to more than 31,300,000 ha (see Table 32.1) and are, not surprisingly, much larger than known earlier. Still, these areas are underestimates, as for quite a number of projects up-to-date data were not available. The newly

Table 1.1. Major augmentative biological control programmes in Latin America in the year 2000 (after van Lenteren and Bueno (2003), with additions; areas of < 10,000 ha not included).

Natural enemy	Pest and crop	Area under control (ha)
Trichogramma spp.	Lepidopteran pests in maize, cotton, sugarcane, tobacco	1.5 million, Mexico
Trichogramma spp.	Lepidopteran pests in cereals, cotton, sugarcane, pastures	1.2 million, Latin America
AgMNPV virus	Soybean caterpillar in soybean	1 million, Brazil
Entomopathogenic fungi	Lepidopteran pests in pastures, cassava and vegetables	0.695 million, Latin America
Entomopathogenic fungi	Coffee berry borer in coffee	0.55 million, Colombia
Entomopathogenic fungi	Lepidopteran pests in palm oil plantations	0.13 million, Colombia
Cotesia sp.	Sugarcane borers in sugarcane	0.4 million, Latin America
Egg parasitoids	Soybean stink bugs in soybean	0.03 million, Latin America
Orgilus sp.	Pine shoot moth in pine plantations	0.05 million, Chile

collected data presented in this book indicate that Latin America and the Caribbean may currently have the largest area under biocontrol worldwide.

In this introductory chapter, we first present information on the literature about biocontrol in the region. Then we provide an overview of organizations working on biocontrol in Latin America and the Caribbean. Next, different types of biocontrol are described; an approach to find, evaluate and use biocontrol agents is sketched; and regulations concerning import and release of agents are summarized. Finally, the structure of the book is explained.

1.2 Literature on Biological Control in Latin America and the Caribbean

The Inter-American Network of Academies of Sciences (IANAS Regional Report, 2017) recently mentioned that the Latin America region is a biodiversity superpower that includes five of the world's ten most biodiverse countries – Brazil, Colombia, Ecuador, Mexico and Peru – as well as the most biologically diverse area in the world: the Amazon rainforest. South America alone is home to more than 40% of the earth's biodiversity and over a quarter of its forests, 30% of its freshwater and nearly 30% of its arable land, which makes the region a genetic reserve and supplier for the planet.

It is, therefore, not surprising that Latin America has provided many biocontrol agents for other parts of the world, as well as having a rich history in biocontrol in the area itself. However, the history and current situation of biocontrol in Latin America and the Caribbean are, in many cases, difficult to trace and hidden in local reports that are not written in English. Still, for some countries the information has been well summarized in books (e.g. Brazil, Caribbean, Chile, Colombia, Cuba, Mexico and Venezuela) (Table 1.2), though often either in Spanish and Portuguese, which makes this information less accessible to an international readership. Also, several review papers have appeared over the years, partly in English (Table 1.2). Although it is clear that many biocontrol projects have been and are executed in this region, it appeared very difficult to estimate the area under biocontrol based on the material presented in these books and review papers. Most information is qualitative, providing insight into research projects and pests, diseases and weeds for which biocontrol programmes have been developed. From now on, we will often use the word 'pest' as defined by FAO/IPPC (1997), which includes animal pests, weeds and diseases.

1.3 International and Regional Organizations working on Biological Control in Latin America and the Caribbean

Several organizations have been active in this region to initiate and coordinate activities on biocontrol. Some of these organizations worked only on biocontrol, while others dealt with biocontrol as part of integrated pest management (IPM) or sustainable agriculture. Table 1.3 gives the names and websites of these organizations, details of which are described in the following sections.

1.3.1 The Centre for Agriculture and Biosciences International (CABI)

The oldest organization coordinating research and application of biocontrol in this region, particularly in the Caribbean, is probably CABI. Here follows a short description of the history of the Trinidad station of CABI which we quote from Cock (1985, p. x):

> ... in November 1928 when J.G. Myers and his wife (I.H. Myers) were sent to the West Indies by the Farnham House Laboratory (subsequently to become the Imperial Parasite Service, the Commonwealth Institute of Biological Control (CIBC), International Institute of Biological Control (IIBC) and now known as CABI). They were 'to study the possibility of biological control for the main pests of agriculture in the British Colonies of tropical America'. Myers visited and worked in all the countries covered in this review except Bermuda and Belize and his work and recommendations, followed by those of the sugar cane entomologist H.E. Box, led to a sharp increase in biological control activity which was only brought to a halt by World War II. In September 1946, F.J. Simmonds of the

Table 1.2. Books and review papers concerning biological control in Latin America and the Caribbean.

(**a**) In chronological order

Year	Regional Latin American and Caribbean reviews and books in chronological order
1973	Hagen, K.S., and Franz, J.M. A history of biological control. In: Smith, R.F., Mittler, T. and Smith, C.N. (eds) *History of Entomology*. Annual Reviews Inc., Palo Alto, pp. 433–476 [ENGLISH]
1984	Bennett, D. and Street, G. The Commonwealth Institute of Biological Control in integrated pest management programs in Latin America. In: Allen, G. and Rada, A. (eds) *Proceedings of the International Symposium: The Role of Biological Control in Pest Management*. Ottawa University Press, Santiago, Chile, pp. 41–53 [ENGLISH]
1985	Cock, M.J.W. (ed.) *A review of biological control of pests in the Commonwealth Caribbean and Bermuda up to 1982*. Technical Communication No. 9, Commonwealth Institute of Biological Control. Commonwealth Agricultural Bureaux, Farnham Royal, UK [ENGLISH]
1989	DeLoach, C., Cordo, H.A. and Crouzel, I.S. *Control biológico de malezas*. Editorial El Ateneo, Buenos Aires, Argentina [SPANISH]
1989	Altieri, M.A., Klein-Koch, C., Trujillo, J., Gold, C.S., Campos, L.S. and Quezada, J.R. El control biologico clasico en America Latina en su context historico. [Classical biological control in Latin America in its historical context]. *Manejo Integrado de Plagas Costa Rica*, No. 12, pp. 87–107 [SPANISH]
1996	Zapater, M.C. (ed.) *El Control Biológico en América Latina*. IOBC/NTRS, Buenos Aires, Argentina [SPANISH]
1999	Altieri, M.A. and Nichols, C.I. Classical biological control in Latin America. In: Bellows, T.S. and Fisher, T.W. (eds) *Handbook of Biological Control*. Academic Press, San Diego, California, pp. 975–991 [ENGLISH]
2003	van Lenteren, J.C. and Bueno, V.H.P. Augmentative biological control of arthropods in Latin America. *BioControl* 48, 123–139 [ENGLISH]
2008	Alves, S.B. and Lopes, R.B. (eds) *Controle microbiano de pragas na América Latina*. Fapesp/Fealq. São Paulo, Brazil [PORTUGUESE]
2008	Barreto, R.W. Latin American weed biological control science at the crossroads. In: Julien, M.H., Sforza, R., Bon, M.C., Evans, H.C., Hatcher, P.E., Hinz., H.L. and Rector, B.G. (eds) *Proceedings of the XII International Symposium on Biological Control of Weeds*. CAB International, Wallingford, UK, pp. 109–121 [ENGLISH]
2008	Hilje, L. and Saunders, J.I. *Manejo integrado de plagas en Mesoamérica: aportes conceptuales* [*Integrated pest management in Mesoamerica: conceptual contributions*]. Editorial Tecnológica de Costa Rica, Cartago, Costa Rica [SPANISH]
2009	Bettiol, W. and Morandi, M.A.B. Biocontrole de doenças de plantas: uso e perspectivas [*Biological control of plant disease: use and perspectives*]. Embrapa Meio Ambiente, Jaguariúna, São Paulo, Brazil. [PORTUGUESE]
2012	Fuentes, F., Ferrer, F.R. and Salas, J.L. *Reseña Histórica del Control Biológico en Centroamérica y el Caribe* [*History of biological control in Central America and the Caribbean*]. Ed. Académica Española, LAP LAMBERT Academic Publishing GmbH& Co, Saarbrucken, Germany [SPANISH]
2014	Bettiol, W., Rivera, M.C., Mondino, P., Montealegre, J.R. and Colmenárez, Y.C. *Control biológico de enfermedades de plantas en América Latina y el Caribe* [*Biological control of plant diseases in Latin America and the Caribbean*]. Facultad de Agronomia, Universidad de la Republica, Montevideo, Uruguay [SPANISH]
2018	Cotes, A.M. (ed.) *Control biológico, de fitopatógenos, insectos y ácaros* [*Biological control of phytopathogens, insects and mites*]. *Vol 1. Applicaciones y perspectivas* [*Applications and perspectives*]. *Vol 2. Agentes de control* biológico [*Biological control agents*]. AgroSavia Editores, Mosquera, Colombia [SPANISH]

Continued

Table 1.2. Continued.

(b) By country

Country	Country reviews and books
Argentina	Cabrera Walsh G., Briano, J., Enrique de Briano, A., and Anderson, F.E. (2014) Malezas e invasoras de la Argentina [Invasive weeds in Argentina]. In: Fernández O.A., Leguizamón E.S. and Acciaresi H.A. (eds) *Control Biológico de Malezas* [*Biological Control of Weeds*]. Tomo I, Ecología y manejo. Editorial de la Universidad Nacional del Sur. Ediciones. Bahía Blanca, Argentina, pp. 801–821 [SPANISH] See country-specific chapters in Zapater (1996)[a] [SPANISH] and Bettiol *et al.* (2014)[b] [SPANISH]
Barbados	See sections in Cock (1985) [ENGLISH][c]
Belize	See sections in Cock (1985) [ENGLISH][c]
Bolivia	Rogg, H.W. (2000a) Manual de Entomología Agrícola de Bolivia [*Handbook of Agricultural Entomology of Bolivia*]. Abya-Yala, Quito, Ecuador. 926 pp. Rogg, H.W. (2000b) Manual Manejo Integrado de Plagas en Cultivos Tropicales [*Handbook of Integrated Pest Management in Tropical Crops*]. Abya-Yala, Quito, Ecuador. 117 pp. [SPANISH] See country-specific chapters in Zapater (1996) [ENGLISH] and Bettiol *et al.* (2014) [SPANISH]
Brazil	Bueno, V.H.P. (ed.) (2000/2009) *Controle Biológico de Pragas: Produção Massal e Controle de Qualidade* [*Biological pest control: mass production and quality control*]. Editora UFLA, Lavras, Brazil. 1st edition, 207 pp.; 2nd edition, 429 pp. [PORTUGUESE] Parra, J.R.P., Botelho, P.S.M., Corrêa-Ferreira, B.S. and Bento, J.M.S. (eds.) (2002) *Controle Biológico no Brasil. Parasitóides e Predadores (Biological Control in Brazil. Parasitoids and Predators).* Ed.Manole, São Paulo, 635pp. [PORTUGUESE] See country-specific chapters in Zapater (1996) [SPANISH] and Bettiol *et al.* (2014) [SPANISH]
Chile	Rojas, S. (2005) *Control biológico de plagas en Chile. Historia y avances* [*Biological control of pests in Chile. History and advances*]. Instituto de Investigaciones Agropecuarias. Centro Regional de Investigación La Platina. Edit. Ograma, La Cruz, Chile [SPANISH] See country-specific chapters in Zapater (1996) [SPANISH] and Bettiol *et al.* (2014) [SPANISH]
Colombia	Palacios, F. (ed.) (1993) *Control biológico en Colombia: historia, avances y proyecciones* [*Biological control in Colombia: history, progress and projections*]. Universidad Nacional de Colombia, Palmira, Colombia. 282 pp. [SPANISH] Carreño, A.M. (2001) *Fundamentos de control biológico de plagas* [*Fundamentals of biological pest control*]. Universidad Nacional de Colombia sede Medellín, Colombia [SPANISH] See country-specific chapters in Zapater (1996) [SPANISH] and Bettiol *et al.* (2014) [SPANISH]
Costa Rica	See country-specific chapters in Zapater (1996) [SPANISH] and Bettiol *et al.* (2014) [SPANISH]
Cuba	Vázquez, L.L. and Pérez, N. (2016) Control biológico [Biological control]. In: Funes, F. and Vázquez, L.L. (eds). *Avances de la agroecología en Cuba*. Estación experimental de Pastos y Forrajes Indio Hatuey. Matanzas. pp. 169–182 [SPANISH] See country-specific chapter in Bettiol *et al.* (2014) [SPANISH]
Dominica	See sections in Cock (1985) [ENGLISH]

Continued

Table 1.2. Continued.

(b) By country

Country	Country reviews and books
Dominican Republic	Pérez-Gelabert, D. (2008) Arthropods of Hispaniola (Dominican Republic and Haiti): a checklist and bibliography. *Zootaxa* 1831, 1–530 [ENGLISH] See country-specific chapter in Bettiol *et al.* (2014) [SPANISH]
Ecuador	Klein-Koch, C. (1989) El control biológico de plagas en Ecuador. [Biological control of pests in Ecuador]. Ministerio de Agricultura y GTZ, Quito, Ecuador. *Sanidad Vegetal*, 4 (4), 5–20. [SPANISH] See country-specific chapters in Zapater (1996) [SPANISH] and Bettiol *et al.* (2014) [SPANISH]
El Salvador	No review papers/books available
Guatemala	See country-specific chapter in Zapater (1996) [SPANISH]
French Guiana, Guadeloupe, Martinique	Ryckewaert P. and Rhino B. (2017) *Insectes et acariens des cultures maraîchères en milieu tropical humide: reconnaissance, bio-écologie et gestion agro-écologique* [*Insects and mites of vegetable crops in humid tropical environment: recognition, bio-ecology and agro-ecological management*]. Ed. Quae, Versailles, France, 152 pp. [FRENCH]
Guyana	See sections in Cock (1985) [ENGLISH]
Haiti	Pérez-Gelabert, D. (2008) Arthropods of Hispaniola (Dominican Republic and Haiti): a checklist and bibliography. *Zootaxa* 1831, 1–530. [ENGLISH] No review papers/books available
Honduras	See country-specific chapters in Zapater (1996) [SPANISH] and Bettiol *et al.* (2014) [SPANISH]
Jamaica	See sections in Cock (1985) [ENGLISH]
Mexico	Arredondo-Bernal, H.C. and Rodríguez-del-Bosque, L.A. (eds) (2008) *Casos de Control Biológico en México* [*Cases of biological control in Mexico*]. Ed. Mundi-Prensa, Mexico. 423 pp. [SPANISH] Arredondo-Bernal, H.C. and Rodríguez-del-Bosque, L.A. (2015) *Casos de Control Biológico en México* [*Cases of biological control in Mexico*]. Vol. 2. Biblioteca Básica de Agricultura, Colegio de Postgraduados, México [SPANISH] See country-specific chapter in Bettiol *et al.* (2014) [SPANISH]
Nicaragua	See country-specific chapter in Zapater (1996) [SPANISH]
Panama	See country-specific chapter in Bettiol *et al.* (2014) [SPANISH]
Paraguay	See country-specific chapter in Bettiol *et al.* (2014) [SPANISH]
Peru	Aguilar, P. (1980) Apuntes sobre el control biológico y el control integrado de las plagas agrícolas en el Perú [Notes on biological control and integrated control of agricultural pests in Peru]. *Revista Peruana de Entomología,* 23(1), 83–110. [SPANISH] Beingolea, O. (1990) Sinopsis sobre el control biológico de plagas insectiles en el Perú, 1909–1990 [Synopsis on the biological control of insect pests in Peru, 1909–1990]. *Revista Peruana de Entomología,* 33, 105–112. [SPANISH] See country-specific chapters in Zapater (1996) [SPANISH] and Bettiol *et al.* (2014) [SPANISH]
Puerto Rico	Gallardo-Covas, F. (2017) Biological control of insect pests in Puerto Rico. *Journal of Agriculture of the University of P.R.* 101, 153–163 [ENGLISH]
Remaining Caribbean islands	See sections in Cock (1985) [ENGLISH] and Bettiol *et al.* (2014) [SPANISH]
Suriname	van Dinther, J.B.M. (1960) Insect pests of cultivated plants in Suriname. *Bulletin Agricultural Experiment Station*, Suriname 76, 1–159 [ENGLISH]
Trinidad and Tobago	See sections in Cock (1985) [ENGLISH]

Continued

Table 1.2. Continued.

(b) By country

Country	Country reviews and books
Uruguay	Bentancourt, C.M. and Scatoni, I.B. (2001) *Enemigos naturales: Manual ilustrado para la agricultura y la forestación* [*Biological control agents: Illustrated manual for agriculture and forestry*]. Universidad de la República, Facultad de Agronomía, Montevideo, Uruguay [SPANISH] See country-specific chapters in Zapater (1996) [SPANISH] and Bettiol *et al.* (2014) [SPANISH]
Venezuela	Ferrer, F. (2001) Biological control of agricultural insect pests in Venezuela; advances, achievements, and future perspectives. *Biocontrol News and Information* 22.3, 67–74. (ENGLISH) See country-specific chapter in Bettiol *et al.* (2014) (SPANISH)

[a]Zapater (1996) mainly summarized arthropod biocontrol
[b]Bettiol *et al.* (2014) summarized biocontrol of diseases
[c]Cock (1985) summarized arthropod and weed biocontrol

Table 1.3. International organizations and websites.

Acronym	Full name	Website
CABI	Centre for Agriculture and Biosciences International	https://www.cabi.org
CARDI	Caribbean Agricultural Research and Development Institute	http://www.cardi.org
CATIE	Centro Agronómico Tropical de Investigación y Enseñanza (Tropical Agriculture Research and Higher Education Center)	http://catie.ac.cr/en/
CGIAR	Consortium of International Agricultural Research Centers	www.cgiar.org
CIAT, Colombia	International Center for Tropical Agriculture	http://ciat.cgiar.org
CIMMYT, Mexico	International Maize and Wheat Improvement Center	www.cimmyt.org
CIP, Peru	International Potato Center	http://cipotato.org/
FAO	Food and Agriculture Organization of the United Nations	http://www.fao.org/americas/en/
IICA	Inter-American Institute for Cooperation on Agriculture	http://www.iica.int/
IOBC Global	International Organization for Biological Control	www.iobc-global.org
IOBC/NTRS	Neotropical Regional Section of the International Organization for Biological Control	http://www.iobcntrs.org/
OIRSA	International Regional Organization for Plant Protection and Animal Health	https://www.oirsa.org

CIBC was stationed at the Imperial College of Tropical Agriculture (which became the St Augustine Campus of the University of the West Indies) to study the natural enemies for control of the weed *Cordia curassavica* (Jacq.) R. & S. on the island of Mauritius. This led to the establishment of the West Indian Station CIBC. He was joined by F.D. Bennett in 1952 and in 1958 when Simmonds became director of CIBC, Bennett became entomologist in charge of the West Indian Station. Ever since Simmonds arrived in Trinidad the CIBC has played a dominant role in biological control in the region. During the intensive programme against sugar cane stem borers the CIBC ran a substation in Barbados. However, as is to be expected, the region is developing its own expertise and in recent years the Caribbean Agricultural Research and Development Institute (CARDI) has become involved in the rearing, shipment

> and release of parasites in the Lesser Antilles from its unit in Barbados which developed from the CIBC substation.

The CABI station in Trinidad and Tobago was for a long time the centre arranging the introduction of most biocontrol agents into the Caribbean region. In 1985, CABI published an extensive summary of biocontrol projects in the Caribbean and Bermuda (Cock, 1985). Contributors to this book have used information from Cock (1985) to summarize the history of biocontrol for a number of Caribbean islands.

CABI is an international not-for-profit organization and has partners in 49 member countries across the world. Currently, CABI has two units in the region: one in Trinidad and Tobago and another in Brazil. The Trinidad and Tobago centre (CABI, 2019a):

> ... researches and identifies agricultural pests and diseases, and works to mitigate the threats of invasive species. Farmers are supported in their integrated pest management (IPM) choices, and encouraged to implement sustainable crop management and production strategies ... The centre also collaborates with Ministries of Agriculture in the region, and provides information to guide policy.

The Brazil centre (CABI, 2019b) 'operates across the whole of Latin America, providing ... scientific knowledge, information and expertise to the Latin American nations'. It also implements projects related to IPM, biocontrol of invasive weeds, and agricultural and forest pests. Examples of current involvement of CABI in the region are biocontrol of pink hibiscus mealybug in the Caribbean, fall armyworm in Latin America, rubber vine in Brazil and blackberry on the Galapagos islands.

1.3.2 The Caribbean Agricultural Research and Development Institute (CARDI)

Next to CABI activities in the region, the English-speaking Caribbean islands and mainland areas have collaborated in a regional research system, starting in 1946 and coordinated by the Imperial College of Tropical Agriculture (ICTA), then in 1955 by the Regional Research Centre (RCC) and since 1975 by CARDI. CARDI's objectives (CARDI, 2019) are 'providing for the research and development needs of the agriculture of the region as identified in national plans and policies, as well as providing an appropriate research and development service to the agricultural sector of member countries'. Member countries of CARDI are Antigua and Barbuda, Barbados, Belize, Dominica, Grenada, Guyana, Jamaica, Montserrat, St Kitts–Nevis, St Lucia, St Vincent and the Grenadines and Trinidad and Tobago. They provide the funding for the Institute's core budget, while funding for projects comes also from donor agencies. Parts of CARDI's projects concern biocontrol and these are mentioned in the country-specific chapters later in the book.

1.3.3 The Consortium of International Agricultural Research Centers (CGIAR)

The CGIAR was founded in 1971 and its 15 international centres constitute the core of the organization (www.cgiar.org). CGIAR's vision is to: (i) reduce poverty and hunger; (ii) improve human health and nutrition; and (iii) enhance ecosystem resilience through high-quality international agricultural research, partnership and leadership. Three of its centres are located in Latin America: the International Maize and Wheat Improvement Center (CIMMYT, Mexico, established in 1966), the International Center for Tropical Agriculture (CIAT, Colombia, established in 1967) and the International Potato Center (CIP, Peru, established in 1971). These three centres, as well as sub-units of other CGIAR centres, have activities in the field of biocontrol within the framework of IPM and sustainable agriculture.

1.3.4 The Inter-American Institute for Cooperation on Agriculture (IICA)

Since 1942, the Inter-American Institute for Cooperation on Agriculture (IICA), with its headquarters in Costa Rica, has supported the efforts of its Member States (34, all over the American hemisphere) to achieve agricultural development and rural well-being. Some of the projects executed under the umbrella of IICA concern aspects of biocontrol related to sustainable agricultural production.

1.3.5 The Tropical Agriculture Research and Higher Education Center (CATIE)

The Centro Agronómico Tropical de Investigación y Enseñanza (The Tropical Agriculture Research and Higher Education Center, CATIE) was also founded in Costa Rica in 1942. Its mandate is research and education in agriculture and natural resources in the American tropics. Today, CATIE is an international, non-for-profit institution dedicated to research, higher education and outreach in agricultural sciences, natural resources and related topics in the American tropics. CATIE has 14 member countries: Belize, Bolivia, Brazil, Colombia, Costa Rica, the Dominican Republic, El Salvador, Guatemala, Honduras, Mexico, Nicaragua, Panama, Paraguay and Venezuela. CATIE's mission is to 'Increase sustainable and inclusive human well-being in Latin America and the Caribbean, promoting education, research and innovation for development, sustainable management of agriculture and conservation of natural resources'. Projects of CATIE may involve research and teaching in biocontrol.

1.3.6 The International Regional Organization for Plant Protection and Animal Health (OIRSA)

Founded in 1953 in El Salvador, the objective of the International Regional Organization for Plant Protection and Animal Health (OIRSA) is to support the efforts of the member states, to achieve development of their animal and plant health plans and to strengthen their quarantine systems. OIRSA's member countries are Belize, Costa Rica, the Dominican Republic, El Salvador, Honduras, Guatemala, Mexico, Nicaragua and Panama. An example of involvement of OIRSA in biocontrol is given in Chapter 4 (Belize) under classical biocontrol of the pink hibiscus mealybug.

1.3.7 The United Nations Food and Agriculture Organization Regional Office for Latin America and the Caribbean (FAO)

FAO's Regional Office has been located since 1955 in Chile, with sub-regional offices in Panama and Barbados. It 'works on a series of priority areas in order to move towards the total eradication of hunger in Latin America and the Caribbean' (FAO, 2019). The Latin American and Caribbean Office has been and is still involved in activities related to biocontrol, like the programme on biocontrol of the pink hibiscus mealybug.

1.3.8 The Neotropical Regional Section of the International Organisation for Biological Control (IOBC/NTRS)

The Neotropical Regional Section of IOBC (IOBC/NTRS) is a regional branch of the International Organisation for Biological Control (IOBC/Global) and was founded in 1989 in Argentina. It is a not-for-profit organization that aims to promote the development and utilization of biocontrol in Latin America as a way to reduce or avoid losses inflicted by noxious animals and plants. IOBC/NTRS has a working group on parasitoids of the Neotropical Region. In addition, IOBC/NTRS assists in the organization of courses in biocontrol, as well as in networking among the biocontrol researchers in the neotropics.

1.3.9 National universities and research institutes

Information on many universities and national research institutes working on biocontrol can be found in the country chapters and will, therefore, not be listed here.

1.3.10 National biological control, entomological, microbiological and phytopathological societies

Many countries in the region have entomological societies that organize regular national meetings, which include sections or symposia on biocontrol (Table 1.4). Mexico and Peru have specific societies for biocontrol, and the Brazilian Entomological Society organizes biannual meetings on biocontrol with participation by scientists of other Latin American and Caribbean countries. Biocontrol issues may also be addressed

Table 1.4. Scientific societies involved in biocontrol in Latin America.

Name of society	Website
Sociedad Entomológica de Argentina	http://www.seargentina.com.ar/
Sociedad Boliviana de Entomologia	https://sociedadbe.webs.com/
Sociedade Brasileira de Entomologia (SBE)	http://www.sbe.ufpr.br/
Sociedade Entomológica do Brasil (SEB)	https://seb.org.br/
Sociedad Chilena de Entomologia	http://www.insectachile.cl/
Sociedad Colombiana de Entomologia	http://www.socolen.org.co/
Sociedad Entomológica Ecuatoriana	http://entomologia.ec/
Sociedad Entomológica del Perú	http://sepperu.com.pe/
Asociación Peruana de Control Biológico (APCB)	https://www.facebook.com/Asociaci %C3%B3n-Peruana-de-Control-Biol%C3%B3gico-APCB-660619824001577/
Sociedad Mexicana de Control Biológico (SMCB)	https://www.smcb-mx.org
Sociedad Mexicana de Entomologia	http://www.socmexent.org/
Sociedad Venezolana de Entomologia	https://ojs3.entomotropica.org/

during the regional meetings of phytopathological societies (united in the Asociación Latinoamericana de Fitopatología (ALF)) and microbiological societies (united in Asociación Latinoamericana de Microbiología (ALAM)).

1.4 Types of Biological Control

Biological control is one of the most environmentally safe and economically profitable pest management methods (Cock *et al.*, 2010, and references therein). In biocontrol, parasitoids, predators, pathogens, herbivores and antagonists are used to reduce populations of pests, diseases and weeds. Several types of biocontrol can be distinguished (Table 1.5). In this book we will mainly use the terms that are most often applied in the literature: natural, conservation, classical and augmentative biocontrol.

1.4.1 Natural control

Natural control (NC) or natural biological control is an ecosystem service (Millennium Ecosystem Assessment, 2005) whereby pest organisms are reduced by naturally occurring beneficial organisms. This type of control occurs in all of the world's ecosystems, whether natural or agroecosystems, and on land as well as in water. In economic value, it is the greatest contribution to agriculture (Waage and Greathead, 1988). However, due to pesticide applications, the full benefit of NC is often curtailed. Many chapters in this book report on prospecting for native natural enemies and the role they play in NC. An example is NC of the diamondback moth, *Plutella xylostella* (L.), in Jamaica. Sampling of diamondback moth populations in several locations during a 5-year period resulted in finding 34 species of natural enemies: five parasitoids, 11 insect predators, 15 species of spiders and three species of entomophagous fungi. These natural enemies together caused high pest mortality (see Chapter 20: Jamaica).

1.4.2 Conservation biological control

In conservation biological control (ConsBC) farmers try to protect and stimulate the performance of naturally occurring natural enemies (DeBach, 1974). ConsBC currently receives a lot of attention in Latin America and the Caribbean (Wyckhuys *et al.*, 2013). An example is ConsBC of the spiralling whitefly *Aleurodicus dispersus* Russell in banana in Costa Rica. This pest was mainly a problem in plantations treated with nematicides, which produce vapours that eliminate natural enemies. Natural enemies of the spiralling whitefly were sampled, resulting in the choice of, among others, four species of parasitoids and seven species of predators. Use of a selective nematicide increased natural enemy survival and reduced pest problems (see Chapter 9: Costa Rica).

Table 1.5. Types of biological control.

Type of biocontrol (reference)	Synonym (reference)	Description
Natural control (NC) (DeBach, 1964)	Natural biological control	Form of pest control whereby pests are reduced by naturally occurring beneficial organisms
Conservation biological control (ConsBC) (DeBach, 1974)		Human actions that protect and stimulate the performance of naturally occurring beneficial organisms
Classical biological control (CBC) (Greathead, 1994)	Inoculative control (van Lenteren, this chapter); Importation control (Nordlund, 1996)	Introduction of relatively low numbers of beneficial organisms from the area of origin of the pest with the aim to obtain permanent control
Augmentative biological control (ABC) (DeBach, 1974)		Mass production and periodic release of beneficial organisms without aim to obtain permanent control
Inundative control (van Lenteren, 1986)		Periodic release of large numbers of organisms to obtain immediate control of the pest in crops with a short production cycle
Seasonal inoculative control (van Lenteren, 1986)		Periodic release of relatively low numbers of organisms to obtain control during several generations of pests in crops with a long production cycle
Special cases of classical biological control:		
Fortuitous control (FBC) (DeBach, 1974)		Control of a pest by an accidentally introduced beneficial organism
Neoclassical biological control (Lockwood, 1993)	New association control (Hokkanen and Pimentel, 1989)	Use of exotic beneficial organisms to control a native pest

1.4.3 Classical biological control

Classical biological control (CBC) is the introduction of relatively low numbers (generally fewer than 1000) of beneficial organisms, usually from a pest's area of origin, to control a pest in an area where it has invaded. Once introduced, the aim is that the biocontrol agent will become established, reproduce, spread and have a self-sustaining effect on the target pest. CBC has the highest benefit–cost ratios, because financial involvement in research costs is usually not very large, while the profits accumulate each year after release of a successful natural enemy (Cock *et al.*, 2010). CBC is most effective in perennial crops where the pest and natural enemy can co-exist indeterminately. Many historical and current CBC projects, for control of both insect pests and weeds, are mentioned in this book. Latin America and the Caribbean have also provided many species of beneficial insects for CBC projects in other world regions and these are mentioned in the country-specific chapters. As CBC was the first type of biocontrol to be widely practised, it is often called 'classical' biocontrol (Greathead, 1994). However, the word classical does not explain what method and aim are involved, so several synonyms have been proposed. One synonym is 'importation' biocontrol, because CBC often refers to importation and release of an exotic natural enemy to control an accidentally introduced pest (Nordlund, 1996; Heimpel and Mills, 2017). We prefer not to use the term importation, because many exotic natural enemies have also been imported for augmentative forms of biocontrol. In our opinion, the term 'inoculative' biocontrol would be more suitable, as the aim is to obtain permanent control of a pest (whether exotic or native) by releasing a relatively limited number of beneficial organisms (whether exotic or native). In this book we will for pragmatic reasons use the term classical biocontrol. Two early 20th century examples of CBC in Latin American are the introduction in many countries of the coccinellid predator *Rodolia cardinalis* (Mulsant) for control of cottony cushion scale (*Icerya purchasi* Maskell) and the release of the hymenopteran parasitoid *Encarsia berlesei* Howard for control of the white peach scale (*Pseudaulacaspis pentagona* (Targioni Tozzetti). Offspring of the biocontrol agents of the then-released predators and parasitoids are still around in many countries in this region and keep reducing pest populations. Two recent successes in Latin America and the Caribbean are control of the pink hibiscus mealybug with the parasitoid *Anagyrus kamali* Moursi and control of the Asian citrus psyllid *Diaphorina citri* Kuwayama with the parasitoid *Tamarixia radiata* (Waterston).

1.4.4 Augmentative biological control

Augmentative biological control (ABC) is the production and release of native or exotic biocontrol agents to obtain direct pest control, but the agents are not expected to persist from one cropping cycle to the next. Usually two types of ABC are distinguished: (i) 'inundative' control in short-cycle crops (e.g. vegetables) of up to a few months, whereby biocontrol agents are introduced in very large numbers (hundreds of thousands) per hectare to obtain immediate control; and (ii) 'seasonal inoculative' control in crops with a production cycle of up to many months (e.g. ornamentals), whereby biocontrol agents are introduced in relatively low numbers (thousands to tens of thousands) per hectare to obtain control during several generations of the pest (van Lenteren, 1986). ABC is applied over large areas in various cropping systems in Latin America and the Caribbean. Well known regional examples are the use of species of the egg parasitoid genus *Trichogramma* for management of Lepidoptera in various crops and the use of *Cotesia* parasitoids against sugarcane borer (see Chapter 6: Brazil). In addition to the use of parasitoids and predators, Latin America is applying microbial control agents in ABC projects on a large scale, such as viruses for control of caterpillars in soybean, fungi for control of pests in coffee, cotton and sugarcane, and nematodes for control of soil-borne pests.

1.4.5 Earliest activities in biological control in Latin America and the Caribbean

All types of biocontrol that have been described above are used in Latin America and the Caribbean. Some countries were very early in the development and application of biocontrol, whereas others started after the Second World War (Table 1.6). Although classical biocontrol was most often used in most of the countries during the early period, it is remarkable that demonstration of natural control, application of augmentative biocontrol and treatments with entomopathogenic bacteria and fungi were all implemented before 1920. Table 1.6 presents an overview of the earliest biocontrol activities for each country in the region, and more details about these activities can be found in the country-specific chapters.

1.5 Finding, Evaluation and Utilization of Biological Control Agents

Approaches for finding and evaluation for invertebrate biocontrol agents can be found in Cock *et al.* (2010) and for microbial biocontrol agents in Ravensberg (2011). Many ideas about biocontrol agent selection have been published, varying from easy approaches such as the collection and release of all natural enemies that might attack the pest (the 'hit-or-miss approach') (DeBach, 1964) to time-consuming research programmes that include behavioural and ecological studies, as well as environmental risk assessments (van Lenteren, 1980; van Driesche and Bellows, 1996; Stiling and Cornelissen, 2005; Heimpel and Mills, 2017; McEvoy, 2018). The hit-or-miss approach was often used in Latin America and the Caribbean, and in particular the many inter-island movements of natural enemies by CABI illustrate this tactic. Due to guidelines and regulations applied since the 1990s, this approach is no longer used.

It generally takes about 10 years to find, evaluate, select and eventually release/market a biocontrol agent, which is similar to the time needed to find a new synthetic pesticide. Recently, a list of criteria for evaluation of biocontrol

Table 1.6. Earliest activities in biological control using arthropod natural enemies or microbial control agents for control of arthropod pests and weeds in each country in Latin America and the Caribbean.

Year	Country	Biological control activity
1884	Venezuela	Tests using a hymenopteran parasitoid for control of migratory locusts
1895	Puerto Rico	Documentation of NC of sugarcane borers by hymenopteran parasitoids
1899	Argentina	Export of a phytophagous coleopteran for CBC of snake weed in the USA
1902	Mexico	ABC of Mexican boll weevil with native predatory mite
1903	Chile	Import of a coleopteran predator for CBC of olive black scale
1904	Peru	Import of hymenopteran parasitoids and coleopteran predator for CBC of cotton white scale
1911	Uruguay	Import of an entomopathogenic bacterium for ABC of locusts
1913	Colombia	ABC of locusts with an entomopathogenic bacterium
1913	Suriname	Documentation of native neuropteran and hemipteran predators of cocoa thrips
1915	Costa Rica	ABC of locusts with an entomopathogenic bacterium
1915	Trinidad and Tobago	ABC of froghoppers with hymenopteran parasitoids, neuropteran predators and entomopathogenic fungi
1917	Remaining Caribbean islands	Demonstration of NC of froghoppers by an entomopathogenic fungus
1918	Jamaica	Import of coleopteran predators for CBC of banana weevil
1919	Barbados	ABC of sugarcane borer with hymenopteran parasitoids
1921	Brazil	Import of a hymenopteran parasitoid for CBC of white peach scale
1927	Dominican Republic	Finding of a fungal pathogen killing important weeds
1928	Cuba	Import of coleopteran predator for CBC of cottony cushion scale
1930	Haiti	Import of hymenopteran parasitoid for CBC of citrus blackfly
1931	Panama	Import of hymenopteran parasitoid for CBC of citrus blackfly
1933	Guyana	Import of tachinid parasitoid for CBC of sugarcane borer
1937	Ecuador	Import of hymenopteran parasitoid for CBC of woolly apple aphid
1938	French Guiana etc.	Import of tachinid parasitoid for CBC of sugarcane borer
1947	Nicaragua	Documentation of hymenopteran parasitoids of fall armyworm
1950	Bolivia	Import of hymenopteran parasitoids and coleopteran predators for CBC of cottony cushion scale, woolly apple aphid and olive scale
1950	Dominica	Import of coleopteran predators for CBC of banana weevil
1967	El Salvador	Documentation of NC of saturniid butterflies on fruit trees by dipteran and hymenopteran parasitoids
1969	Belize	Import of hymenopteran parasitoids for CBC of fruit flies
1969	Honduras	Import of hymenopteran parasitoids for CBC of fruit flies
1985	Paraguay	ABC of soybean caterpillar with an entomopathogenic virus, and sugarcane borer with hymenopteran parasitoids
1993	Guatemala	ABC of diamondback moth in cruciferous crops with hymenopteran parasitoid

agents has been published (van Lenteren *et al.*, 2020) and an adapted version of these criteria is presented in Table 1.7. When searching for biocontrol agents, it is not unusual to find up to tens or even hundreds of species attacking a certain pest. This large number of potential candidate species stresses the need for evaluation criteria. The main aim of using these criteria is not to find the best natural enemy, but mostly to exclude unsuitable or problematic species quickly.

Most criteria concern biological characteristics of natural enemies (1–10), others relate to experience obtained in earlier biocontrol projects (11 and 13), economics of biocontrol (12 and 14), or to difficulties in importation and registration of exotic natural enemies (15). The relevance of these criteria is determined by the type of biocontrol programme one aims to develop. For example, in classical biocontrol, one will start with criterion 15 (how realistic it is to

Table 1.7. Criteria for pre-introduction evaluation of biocontrol agents (adapted from van Lenteren *et al.*, 2020).

No.	Criterion
1	Seasonal synchronization with pest (mainly for classical biocontrol)
2	Developmental synchronization with pest (mainly for classical biocontrol)
3	Climatic adaptation to location where agent needs to be applied
4	Capable of searching for pest on target crop and establishing a population
5	No negative side effects
6	Preference for pest species
7	High pest-kill ability
8	Good pest-finding capacity
9	Responsiveness to changes in pest density
10	Able to survive on alternative food in absence of pest (mainly for augmentative biocontrol)
11	Same or similar species effective in same or similar crop and climate elsewhere
12	Cost-effective mass production (mainly for augmentative biocontrol)
13	Reliable performance
14	Market potential (mainly for augmentative biocontrol)
15	Potential for importation and/or registration of biocontrol agent

obtain a permit for importation and release of an exotic natural enemy), then criterion 11 (same or related species effective elsewhere), followed by collecting literature information about biological criteria 1–9, while little attention is paid to issues 10 (ability to survive on alternative food), 12 (costs of mass rearing) and 14 (market potential). In the case of an augmentative biocontrol programme, evaluation starts with criteria 5 (not causing side effects), 12, 14 and 15 (costs of mass rearing, market potential, complexity of regulations).

Information obtained with computer searches and simple experiments related to criteria 1–6 is often sufficient to reduce the number of potential candidates for biocontrol to fewer than ten species. After having excluded problematic species that may cause negative effects, or species that are clearly ineffective, research money can then be spent on the most promising candidates. Eventually, if testing under realistic production conditions reveals one or more effective biocontrol agents, these can be released in the case of classical biocontrol projects. However, if the agents will be used in augmentative biocontrol, it will be necessary to develop mass production, storage, shipment and release methods (van Lenteren and Tommasini, 2003). An important point for attention in mass production of biocontrol agents is quality control, and quality control protocols have been developed for a number of agents.

When a pest is of exotic origin, a foreign prospecting expedition may have to be organized. This was relatively easy in the past and many examples of using exotic natural enemies are presented in the country-specific chapters. However, for several decades many governments have demanded an environmental risk assessment before registering a new exotic biocontrol agent. More recently, since the adoption of the Nagoya protocol in 2014 (see below), very complicated and time-consuming procedures have to be followed before exotic species can be collected, imported and applied. As a result, prospecting for biocontrol agents now often starts with a search 'at home'.

The country-specific chapters in this book provide examples about prospecting for and evaluation of new biocontrol agents, as well as mass production and release methods.

1.6 Regulations Concerning the Use of Biological Control Agents

During the past 40 years a number of regulations have been developed that are related to import and release of biocontrol agents. Most countries, including those in the Latin American region, will have to comply with these regulations, in particular regulations developed and implemented by the Convention on Biological Diversity (SCBD, 2011). Many biocontrol researchers do not yet realize the serious impact of these regulations and so in this section we will discuss how new regulations delay or even seriously frustrate development and application of biocontrol.

Accidental introduction of exotic organisms has been occurring at an ever-growing rate during the past 150 years (Seebens *et al.*, 2017) and increasing travel, trade and tourism will continue to result in the introduction of new pests. In contrast, deliberate introductions of many exotic biocontrol agents have caused remarkably few problems, while they have often resulted in permanent control of the unintentionally introduced pests (Cock *et al.*, 2010). Many examples of this approach in biocontrol will be presented in this book, and especially during the period 1880–1970 many classical biocontrol projects were executed in Latin America and the Caribbean. Until a few years ago, prospecting for new, exotic biocontrol agents after unintended introductions of exotic pests was possible and usually occurred with the consent of the country where prospecting took place. However, due to recent developments concerning regulation and registration of biocontrol agents, and particularly as a result of the Nagoya protocol pertaining to exchange of biological resources, prospecting for biocontrol agents has practically come to a standstill (Cock *et al.*, 2016).

During the first period of 'modern' biocontrol from 1880 to 1970, hardly any regulations existed with regard to import and use of exotic agents, though many researchers were well aware of the risk of importing certain types of organisms, in particular generalist predators that might prey on non-target organisms. Two often cited and early problematic cases of biocontrol in the Latin American and Caribbean area – release of a mammal (mongoose) for control of rats and snakes and introduction of an amphibian (the giant toad) for control of insects – were caused by amateurs, not by biocontrol experts. Most of the early introductions into the region concerned natural enemies that were earlier introduced for classical biocontrol in other countries, in particular the USA and countries of the former British Empire, and had not shown negative side effects. The Trinidad and Tobago station of the predecessors of CABI played an important role in importing and releasing a number of exotic natural enemies in the Caribbean region and also organized many inter-island exchanges. All these releases, even if they consisted of somewhat polyphagous predators such as coccinellids, have apparently not led to negative effects, though it should be stressed that post-release environmental assessments have seldom been made. Still, we would classify them now as the hit-or-miss approach and we suppose that most countries would no longer allow such releases without environmental risk assessments.

Non-target effects of biocontrol agents were first considered in weed biocontrol, because of the risk that imported exotic herbivores might eat not only the weed, but also related plant species, including crops. It has since long been common practice to apply risk analyses in weed biocontrol and many countries demand such an analysis before a weed biocontrol agent can be imported and released (Wapshere, 1974; Sheppard *et al.*, 2003). Risk analyses for candidate agents to control arthropod pests were developed much later and have only been applied since the 1980s. The reason is probably that very few problems had been reported about non-target effects caused by exotic invertebrates for arthropod biocontrol (Lynch *et al.*, 2001). However, since the 1980s, when commercial augmentative biocontrol became popular and the number of exotic species applied in biocontrol strongly increased (see Fig. 2 in van Lenteren, 2012), many non-experts in the field of biocontrol started to release exotic agents. Thus, the need was felt for pre-release environmental risk assessments for new biocontrol agents to prevent non-target effects.

The first step towards risk assessments was the design a code of conduct for import and release of biocontrol agents by the Food and Agriculture Organization of the United Nations (FAO), together with CABI and IOBC (FAO, 1996). Since its development, the code of conduct has been used in CABI projects in the Caribbean and by a number of Latin American countries. Examples can be found in Kairo *et al.* (2003). Next, IOBC developed a set of standard methods to perform risk assessments, including practical guidance on how to measure and evaluate effects needed to draw conclusions about risks and benefits of biocontrol agents (Bigler *et al.*, 2006). Environmental risk assessments are being demanded by a growing number of countries and are characterized by questions about: (i) the identity of the biocontrol agent; (ii) potential human health risks; (iii) potential environmental risks; and (iv) efficacy in controlling the pest (van Lenteren *et al.*, 2006). Implementation of environmental risk assessment has resulted in a slowdown of newly marketed exotic biocontrol agents for augmentative biocontrol (see Figs 1 and 2 in van Lenteren, 2020) and introductions for classical biocontrol (see Fig. 1 in Cock *et al.*, 2016). Preparation of extensive application dossiers also caused higher developmental costs, but did not bring prospecting for new non-native species to an end. This first phase of development of regulations was aimed at improving biocontrol, preventing potential negative effects and increasing confidence in biocontrol. Regulations for import and release, demands for environmental risk assessments and procedures for registration vary widely in Latin America and the Caribbean (see country-specific chapters for details). Regional harmonization of these regulations might speed up registration and application of biocontrol agents. Still, harmonization is not expected to be an easy and quick process. Colmenarez *et al.* (Y.C. Colmenarez, Botucatu, Brazil, 2019, personal communication) propose the formation of a regional platform for harmonization of procedures related to biocontrol. In South America, the Plant Health Committee (Comité de Sanidad Vegetal (COSAVE)), a Regional Plant Protection Organization, might host such a platform.

A new and more recent phase of regulations was not developed to improve the science of biocontrol, but dealt with the question of who owns biocontrol agents. This question is related to one of the objectives of the Rio Convention on Biological Diversity (CBD, 1993): the fair and equitable sharing of the benefits arising from the utilization of genetic resources. Genetic resources are defined by CBD as genetic material, i.e. material containing functional units of heredity that is of actual or potential value, so this includes all biocontrol agents taken from one country (provider) to another (recipient) (Cock *et al.*, 2010). As a result of this CBD objective, the Nagoya Protocol on Access and Benefit Sharing (ABS) was developed, which provides a framework for implementation of the fair and equitable sharing of benefits arising from the utilization of genetic resources (SCBD, 2011). The Nagoya Protocol came into force in October 2014 and is a potentially serious threat to the use of biocontrol, because there are no clear guidelines on how to develop agreements between the providing and recipient countries. The result is that prospecting for exotic natural enemies has currently been suspended due to lack of clear CBD and ABS procedures in many countries. The IOBC Global Commission on Access and Benefit Sharing (IOBC, 2019) made an appeal to develop ABS regulations that support the biocontrol sector by facilitating the exchange of biocontrol agents, including clear guidelines. These guidelines should also include fast-track procedures for finding and applying biocontrol agents in case of humanitarian or emergency situations, such as after unintentional export of an invasive pest to a new area. The IOBC also strongly recommended that biocontrol agents should be considered as a special case with respect to an ABS regime under the CBD (Cock *et al.*, 2010). The IOBC recently prepared a best-practice guide for exchange of biocontrol genetic resources to assist the biocontrol community to demonstrate due diligence in complying with ABS requirements (Mason *et al.*, 2018). This IOBC best-practice guide contains a section concerning gaining access to biocontrol agents and a draft ABS agreement for collection and study of biocontrol agents that can be used for scientific research and non-commercial release into nature by countries having signed the Nagoya Protocol (Mason *et al.*, 2018). The draft agreement is designed to promote non-commercial activities, such as research in taxonomy, ecology and genetics, to foster biological conservation and the environmentally sound use of biocontrol agents. The objective is to provide a sound basis for cooperation, transparency, communication and trust between the parties and it will hopefully result in renewed prospecting for biocontrol agents to control invasive exotics. Similar to the issues of registration, risk assessment and regulations for import and release, the way in which the Latin American and Caribbean countries deal with the Nagoya protocol also shows great differences.

1.7 Structure of the Book

After this introductory chapter, the history and current situation of biocontrol are presented for each Latin American or Caribbean country in the following 30 chapters. We introduce each country by mentioning the number of inhabitants and the major agricultural activities. This information often originated from three sources: (i) the IANAS 2017 Regional Report (IANAS, 2017), which contains a lot of information about the agricultural situation in the Americas; (ii) the website of the Central Intelligence Agency (CIA) of the USA (CIA, 2019), which has the most recent information about inhabitants and agricultural products for each country; and (iii) Wikipedia if data for inhabitants and agriculture could not be found in the first two sources.

Although the editors have tried to harmonize the structure of the chapters in which the countries present their history, current situation and future of biocontrol, different approaches can be observed. For example, authors of some countries discuss developments of biocontrol per crop (or group of crops), whereas other countries do this either per pest (or group of pests), or per type of biocontrol agent (e.g. predators, parasitoids, pathogens, microbial control agents and weed biocontrol agents). The editors asked the authors to estimate the area under biocontrol in their country and we have received estimates for many countries, but these were often said to be incomplete. For a few countries, data about areas under biocontrol could not be obtained. Thus, the figures mentioned in the country chapters, as well as in the summarizing chapter, will almost always be underestimates.

In the final chapter, we summarize achievements, discuss current factors stimulating and limiting the development of biocontrol and speculate about the future of biocontrol in Latin America and the Caribbean.

In order to help users of the book in finding projects related to certain species of biocontrol agents, pests and crops, there is a supplementary index that lists all scientific names of species with author names, order and family, as well as their common names and the countries in which the species were mentioned.

References

Aguilar, P. (1980) Apuntes sobre el control biológico y el control integrado de las plagas agrícolas en el Perú [Notes on biological control and integrated control of agricultural pests in Peru]. *Revista Peruana de Entomología* 23(1), 83–110.

Altieri, M.A. and Nichols, C.I. (1999) Classical biological control in Latin America. In: Bellows, T.S. and Fisher, T.W. (eds) *Handbook of Biological Control.* Academic Press, San Diego, California, pp. 975–991.

Altieri, M.A., Klein-Koch, C., Trujillo, J., Gold, C.S., Campos, L.S. and Quezada, J.R. (1989) El control biologico clasico en America Latina en su context historico [Classical biological control in Latin America in its historical context]. Manejo Integrado de Plagas, Costa Rica, No. 12, pp. 87–107.

Alves, S.B. and Lopes, R.B. (eds) (2008) *Controle microbiano de pragas na América Latina* [*Microbial control of pests in Latin America*]. Fapesp/Fealq, Piracicaba, SP, Brazil.

Arredondo-Bernal, H.C. and Rodríguez-del-Bosque, L.A. (eds.) (2008) *Casos de Control Biológico en México* [*Cases of Biological Control in Mexico*]. Ed. Mundi-Prensa México, Rio Pánuco, México DF.

Arredondo-Bernal, H.C. and Rodríguez-del-Bosque, L.A. (2015) *Casos de Control Biológico en México* [*Cases of Biological Control in Mexico*], Vol. 2. Biblioteca Básica de Agricultura, Colegio de Postgraduados, México DF.

Bale, J., van Lenteren, J.C. and Bigler, F. (2008) Biological control and sustainable food production. *Philosophical Transactions of the Royal Society, Series B* 363 (1492), 761–776. DOI: 10.1098/rstb.2007.2182

Barreto, R.W. (2008) Latin American weed biological control science at the crossroads In: Julien, M.H., Sforza, R., Bon, M.C., Evans, H.C., Hatcher, P.E., Hinz., H.L. and Rector, B.G. (eds) *Proceedings of the XII International Symposium on Biological Control of Weeds*. CAB International, Wallingford, UK, pp. 109–121.

Beingolea, O. (1990) Sinopsis sobre el control biológico de plagas insectiles en el Perú, 1909–1990 [Synopsis on the biological control of insect pests in Peru, 1909–1990]. *Revista Peruana de Entomología* 33, 105–112.

Bennett, D. and Street, G. (1984) The Commonwealth Institute of Biological Control in integrated pest management programs in Latin America. In: Allen, G. and Rada, A. (ed.) *Proceedings of the International Symposium: The role of biological control in pest management*. IOBC/WPRS, Ottawa University Press, Ottawa, Canada, pp. 41–53.

Bentancourt, C.M. and Scatoni, I.B. (2001) *Enemigos naturales: Manual ilustrado para la agricultura y la forestación* [*Biological control agents: Illustrated manual for agriculture and forestry*]. Universidad de la República, Facultad de Agronomía, Montevideo, Uruguay.

Bettiol, W. and Morandi, M.A.B. (2009) *Biocontrole de doenças de plantas: uso e perspectivas* [*Biological control of plant disease: use and perspectives*]. Embrapa Meio Ambiente, Jaguariúna, São Paulo, Brazil.

Bettiol, W., Rivera, M.C., Mondino, P., Montealegre, J.R. and Colmenárez, Y.C. (2014) *Control biológico de enfermedades de plantas en América Latina y el Caribe* [*Biological control of plant diseases in Latin America and the Caribbean*]. Universidad de la Republica, Montevideo, Uruguay. Available at: http://www.fagro.edu.uy/index.php/ documentos/category/46-libros-fagro.html (accessed 2 April 2019).

Bigler, F., Babendreier, D. and Kuhlmann, U. (2006) Environmental Impact of Invertebrates for Biological Control of Arthropods: Methods and Risk Assessment. CAB International, Wallingford, UK.

Bueno, V.H.P. (ed.) (2000) *Controle Biológico de Pragas: Produção Massal e Controle de Qualidade* [*Biological pest control: mass production and quality control*], 1st edn. Editora UFLA, Lavras, Brazil.

Bueno, V.H.P. (ed.) (2009) *Controle Biológico de Pragas: Produção Massal e Controle de Qualidade* [*Biological pest control: mass production and quality control*], 2nd edn. Editora UFLA, Lavras, Brazil.

CABI (2019a) CABI centres: Trinidad and Tobago. Available at: https://www.cabi.org/about-cabi/cabi-centres/trinidad-and-tobago (accessed 7 July 2019).

CABI (2019b) CABI centres: Brazil. Available at: https://www.cabi.org/about-cabi/cabi-centres/brazil (accessed 7 July 2019).

Cabrera Walsh, G., Briano, J., Enrique de Briano, A. and Anderson, F.E. (2014) Malezas e invasoras de la Argentina (Invasive weeds in Argentina). In: Fernández, O.A., Leguizamón, E.S. and Acciaresi, H.A. (eds). *Control Biológico de Malezas* [*Biological Control of Weeds*]. Tomo I, Ecología y manejo. Editorial de la Universidad Nacional del Sur. Ediciones. Bahía Blanca, Argentina, pp. 801–821.

CARDI (2019) Welcome to CARDI. Available at: http://www.cardi.org/ (accessed 7 July 2019).

Carreño, A.M. (2001) *Fundamentos de control biológico de plagas plagas [Fundamentals of biological pest control*]. Universidad Nacional de Colombia sede Medellín, Colombia.

CBD (Convention on Biological Diversity) (1993) Convention on Biological Diversity (with annexes). Concluded at Rio de Janeiro on 5 June 1992. United Nations – Treaty Series 1760(30619), pp. 142–382.

CIA (2019) The World Factbook. Available at: https://www.cia.gov/library/publications/the-world-factbook (accessed 7 July 2019).

Cock, M.J.W. (ed) (1985) *A review of biological control of pests in the Commonwealth Caribbean and Bermuda up to 1982*. Technical Communication No. 9, Commonwealth Institute of Biological Control. Commonwealth Agricultural Bureaux, Farnham Royal, UK.

Cock, M.J.W., van Lenteren, J.C., Brodeur, J., Barratt, B.I.P., Bigler, F., Bolckmans, K., Consoli, F.L., Haas, F., Mason, P.G. and Parra, J.R.P. (2010) Do new access and benefit sharing procedures under the convention on biological diversity threaten the future of biological control? *BioControl* 55, 199–218.

Cock, M.J.W., Murphy, S.T., Kairo, M.T.K., Thompson, E., Murphy, R.J. and Francis, A.W. (2016) Trends in the classical biological control of insect pests by insects: an update of the BIOCAT database. *BioControl* 61, 349–363. DOI: 10.1007/s10526-016-9726-3.

Cotes, A.M. (ed.) (2018) *Control biológico, de fitopatógenos, insectos y ácaros* [*Biological control of phytopathogens, insects and mites*]. *Vol 1. Applicaciones y perspectivas* [*Applications and perspectives*]. *Vol 2. Agentes de control biológico* [*Biological control agents*]. AgroSavia Editores, Mosquera, Colombia, 566. Available at: http://hdl.handle.net/20.500.12324/33829 and http://hdl.handle.net/20.500.12324/33519 (accessed 2 April 2019)

DeBach, P. (1964) *Biological Control of Insect Pests and Weeds*. Chapman and Hall, London, UK.

DeBach, P. (1974) *Biological Control by Natural Enemies*. Cambridge University Press, Cambridge, UK.

DeLoach, C., Cordo, H.A. and Crouzel, I.S. (1989) *Control biológico de malezas* [*Biological control of weeds*]. Editorial El Ateneo, Buenos Aires, Argentina.

FAO (1996) Code of conduct on the import and release of biological control agents. FAO, Rome. Available at: www.fao.org (accessed 2 April 2019).

FAO (2019) Priorities. Available at: http://www.fao.org/americas/en/ (accessed 2 April 2019).

FAO/IPPC (1997) International Plant Protection Convention. FAO, Rome. Available at: www.fao.org (accessed 2 April 2019).

Ferrer, F. (2001) Biological control of agricultural insect pests in Venezuela: advances, achievements, and future perspectives. *Biocontrol News and Information* 22(3), 67–74.

Fuentes, F., Ferrer, F.R. and Salas, J.L. (2012) *Reseña Histórica del Control Biológico en Centroamérica y el Caribe* [*History of biological control in Central America and the Caribbean*]. Ed. Académica Española, LAP LAMBERT Academic Publishing GmbH & Co., Saarbrucken, Germany.

Gallardo-Covas, F. (2017) Biological control of insect pests in Puerto Rico. *Journal of Agriculture of the University of P.R.* 101, 153–163.

Greathead, D.J. (1994) History of biological control. *Antenna* 18, 187–199.

Hagen, K.S. and Franz, J.M. (1973) A history of biological control. In: Smith, R.F., Mittler, T.E. and Smith, C.N. (eds) *History of Entomology*. Annual Reviews Inc., Palo Alto, California, pp. 433–476.

Heimpel, G.E. and Mills, N.J. (2017) *Biological Control: Ecology and Applications*. Cambridge University Press, Cambridge, UK.

Hilje, L. and Saunders, J.I. (2008) *Manejo integrado de plagas en Mesoamérica: aportes conceptuales* [*Integrated pest management in Mesoamerica: conceptual contributions*]. Editorial Tecnológica de Costa Rica, Cartago, Costa Rica.

Hokkanen, M.T. and Pimentel, D. (1989) New associations in biological control: theory and practice. *Canadian Entomologist* 121, 829–840.

IANAS Regional Report (2017) *Challenges and Opportunities for Food and Nutrition Security in the Americas. The View of the Academies of Sciences*. The Inter-American Network of Academies of Sciences (IANAS-IAP), Ciudad de México, Mexico. Available in English and Spanish at: www.ianas.org (accessed 2 April 2019).

IOBC (2019) IOBC Global Commission on Biological Control and Access and Benefit Sharing. Available at: http://www.iobc-global.org/global_comm_bc_access_benefit_sharing.html (accessed 7 July 2019).

Kairo, M.T.K., Cock, M.J.W. and Quinlan, M.M. (2003) An assessment of the use of the Code of Conduct for the Import and Release of Exotic Biological Control Agents (ISPM No. 3) since its endorsement as an international standard. *Biocontrol News and Information 2003* 24(1), 15N–27N.

Klein-Koch, C. (1989) El control biológico de plagas en Ecuador [Biological control of pests in Ecuador]. Ministerio de Agricultura y GTZ, Quito, Ecuador. *Sanidad Vegetal* 4(4), 5–20.

Lockwood, J.A. (1993) Environmental issues involved in biological control of rangeland grasshoppers (Orthoptera: Acrididae) with exotic agents. *Environmental Entomology* 22, 503–518.

Lynch, L.D., Hokkanen, H.M.T., Babendreier, D., Bigler, F., Burgio, G., Gao, Z.H., Kuske, S., Loomans, A.J.H.M., Menzler-Hokkanen, I., Thomas, M.B., Tommasini, M.G., Waage, J., van Lenteren, J.C. and Zeng, Q.Q. (2001) Indirect effects in the biological control of arthropods with arthropods. In: Wajnberg, E., Scott, J.C. and Quimby, P.C. (eds.) *Evaluating Indirect Ecological Effects of Biological Control*. CAB International, Wallingford, UK, pp. 99–125.

Mason, P.G., Cock, M.J.W., Barratt, B.I.P, Klapwijk, J., van Lenteren, J.C., Brodeur, J., Hoelmer, K.A. and Heimpel, G.E. (2018) Best practices for the use and exchange of invertebrate biological control genetic resources relevant for food and agriculture. *BioControl* 63, 149–154. DOI: 10.1007/s10526-017-9810-3

McEvoy, P.B. (2018) Theoretical contributions to biological control success. *BioControl* 63, 87–103. DOI: 10.1007/s10526-017-9852-6

Millennium Ecosystem Assessment (2005) *Ecosystems and Human Well-being: Synthesis*. Island Press, Washington, DC.

Nordlund, D.A. (1996) Biological control, integrated pest management and conceptual models. *Biocontrol News and Information* 17, 35N–44N.

Palacios, F. (ed.) (1993) *Control biológico en Colombia: historia, avances y proyecciones* [*Biological control in Colombia: history, progress and projections*]. Universidad Nacional de Colombia, Palmira, Colombia.

Parra, J.R.P., Botelho, P.S.M., Corrêa-Ferreira, B.S. and Bento, J.M.S. (2002) *Controle Biológico no Brasil. Parasitóides e Predadores* [*Biological Control in Brazil. Parasitoids and Predators*]. Ed.Manole, São Paulo, Brazil.

Pérez-Gelabert, D. (2008) Arthropods of Hispaniola (Dominican Republic and Haiti): A checklist and bibliography. *Zootaxa* 1831, 1–530.

Ravensberg, W.J. (2011) *A Roadmap to the Successful Development and Commercialization of Microbial Pest Control Products for Control of Arthropods*. Springer, Dordrecht, Netherlands.

Rogg, H.W. (2000a) *Manual de Entomología Agrícola de Bolivia* [*Handbook of Agricultural Entomology of Bolivia*]. Abya-Yala, Quito, Ecuador.

Rogg, H.W. (2000b) *Manual Manejo Integrado de Plagas en Cultivos Tropicales* [*Handbook of Integrated Pest Management in Tropical Crops*]. Abya-Yala, Quito, Ecuador.

Rojas, S. (2005) *Control biológico de plagas en Chile. Historia y avances* [*Biological control of pests in Chile. History and advances*]. Instituto de Investigaciones Agropecuarias. Centro Regional de Investigación La Platina. Edit. Ograma, La Cruz, Chile.

Ryckewaert, P. and Rhino, B. (2017) *Insectes et acariens des cultures maraîchères en milieu tropical humide: reconnaissance, bio-écologie et gestion agro-écologique* [*Insects and mites of vegetable crops in humid tropical environment: recognition, bio-ecology and agro-ecological management*]. Ed. Quae, Versailles, France.

SCBD (Secretariat of the Convention on Biological Diversity) (2011) *Nagoya protocol on access to genetic resources and the fair and equitable sharing of benefits arising from their utilization to the Convention on Biological Diversity: text and annex*. Convention on Biological Diversity, United Nations, Montreal, Canada.

Seebens, H., Blackburn, T.M., Dyer, E.E., Genovesi, P., Hulme, P.E. *et al.* (2017) No saturation in the accumulation of alien species worldwide. *Nature Communications* 8, 14435.

Sheppard, A.W., Hill, R.L., DeClerck-Floate, R.A., McClay, A., Olckers, T., Quimby, P.C. and Zimmermann, H.G. (2003) A global review of risk-benefit-cost analysis for the introduction of classical weed biological control agents against weeds: a crisis in the making? *Biocontrol News and Information* 24(4), 91N–108N.

Stiling, P. and Cornelissen, T. (2005) What makes a successful biocontrol agent? A meta-analysis of biological control agent performance. *Biological Control* 34, 236–246. DOI: 10.1016/j.biocontrol.2005.02.017.

van Dinther, J.B.M. (1960) Insect pests of cultivated plants in Surinam. *Bulletin Agricultural Experiment Station Suriname* 76, 1–159.

van Driesche, R.D. and Bellows, T.S. (1996) *Biological Control*. Chapman & Hall, New York.

van Lenteren, J.C. (1980) Evaluation of control capabilities of natural enemies: does art have to become science? *Netherlands Journal of Zoology* 30, 369–381.

van Lenteren, J.C. (1986) Parasitoids in the greenhouse: successes with seasonal inoculative release systems. In: Waage, J.K. and Greathead, D.J. (eds) *Insect Parasitoids*. Academic Press, London, pp. 341–374.

van Lenteren, J.C. (2012) The state of commercial augmentative biological control: plenty of natural enemies, but a frustrating lack of uptake. *BioControl* 57, 71–84 + supplemental material. DOI: 10.1007/s10526-011-9395-1

van Lenteren, J.C. (2020) Will the 'Nagoya Protocol on Access and Benefit Sharing' put an end to biological control? In: Hendrichs, J., Pereira, R. and Vreysen, M.J.B. (eds) *Area-wide Integrated Pest Management: Development and Field Application.* FAO/IAEA, Vienna, Austria.

van Lenteren, J.C. and Bueno, V.H.P. (2003) Augmentative biological control of arthropods in Latin America. *BioControl* 48, 123–139.

van Lenteren, J.C. and Tommasini, M.G. (2003) Mass production, storage, shipment and release of natural enemies. In: van Lenteren, J.C. (ed.) *Quality Control and Production of Biological Control Agents: Theory and Testing Procedures*. CAB International, Wallingford, UK, pp. 181–189.

van Lenteren, J.C., Bale, J., Bigler, F., Hokkanen, H.M.T. and Loomans, A.J.M.(2006) Assessing risks of releasing exotic biological control agents of arthropod pests. *Annual Review of Entomology* 51, 609–634 + supplemental material. DOI: 10.1146/annurev.ento.51.110104.151129

van Lenteren, J.C., Bueno, V.H.P., Burgio, G., Lanzoni, A., Montes, F.C., Silva, D.B., de Jong, P.W. and Hemerik, L. (2019) Pest kill rate as aggregate evaluation criterion to rank biological control agents: a case study with Neotropical predators of *Tuta absoluta* on tomato. *Bulletin of Entomological Research* 109. DOI: 10.1017/S0007485319000130

Vázquez, L.L. and Pérez, N. (2016) Control biológico [Biological control]. In: Funes, F. and Vázquez, L.L. (eds) *Avances de la agroecología en Cuba*. Estación experimental de Pastos y Forrajes Indio Hatuey, Matanzas, pp. 169–182.

Waage, J.K. and Greathead, D.J. (1988) Biological control: challenges and opportunities. *Philosophical Transactions of the Royal Society, London, Series* B 318, 111–128.

Wapshere, A.J. (1974) A strategy for evaluating the safety of organisms for biological weed control. *Annals of Applied Biology* 77, 201–211.

Wyckhuys, K.A., Lu, Y., Morales, H., Vazquez, L.L., Legaspi, J.C., Eliopoulos, P.A. and Hernandez, L.M. (2013) Current status and potential of conservation biological control for agriculture in the developing world. *Biological Control*, 65, 152–167.

Zapater, M.C. (ed.) (1996) *El Control Biológico en América Latina* [*Biological Control in Latin America*]. IOBC/NTRS, Buenos Aires, Argentina.

2 Biological Control in Argentina

Nancy Mabel Greco[1]*, Guillermo Cabrera Walsh[2] and María Gabriela Luna[1]

[1]CEPAVE, CONICET-UNLP (Centro de Estudios Parasitológicos y de Vectores), La Plata, Argentina; [2]FUEDEI (Fundación para el Estudio de Especies Invasivas), Hurlingham, Argentina

* E-mail: ngreco@cepave.edu.ar

Abstract

Biological control in Argentina has a longstanding tradition with records of natural enemy introductions since the beginning of the 20th century, mainly between 1900 and 1940. Eight predators, 70 parasitoids and seven pathogens have been introduced for arthropod control, plus eight weed biocontrol agents. Argentina has also provided 22 arthropod species to Africa, Australia, Canada and the USA for arthropod pest biocontrol. At least 26 agents from Argentina have been released against 24 weeds of South American origin around the world, notably for freshwater invaders. Fruit production and pine plantations still have the largest areas under some degree of classical biocontrol, yet the current impact of the agents is not well known. Citrus fruit flies are under experimental augmentative biocontrol with one parasitoid species and 1 million hectares under an IPM regime that includes cultural control, trapping and SIT technology. As for private initiatives, a small fraction of greenhouse tomatoes and peppers are under augmentative biocontrol. Augmentative releases are also made in sugarcane plantations and citrus groves. Finally, an extensive part of the Argentine territory is affected by thistles and skeleton weed, and several artificial and natural water bodies invaded by native aquatics are subject to classical biocontrol, although they are still important weeds in many areas. Despite the government's explicit endorsement, resources are scarce and applied biocontrol in all its forms is still sorely undeveloped in Argentina.

2.1 Introduction

Argentina has an estimated population of more than 44 million (July 2017) (CIA, 2017) and is, according to Bianchi *et al.* (2017, pp. 33 and 39):

> ... a major producer of cereals, such as wheat, maize, sorghum, rice, barley; oilseeds such as soybeans and sunflowers; industrial crops such as cotton, sugar, mate, tobacco and tea; and fruits and vegetables. The country also plays a key role in livestock production, mainly beef and dairy products. In several of these products, Argentina is a major global producer and consequently also a top exporter ... The size of its territory and its diversity of climates mean that Argentina possesses significant forest wealth. This in turn favors climate regulation, biodiversity, water basin protection, soil conservation, water supply and ecosystem maintenance. The country boasts 1.2 million ha of forest plantations and 50 million ha of native forests. Implanted forests are dominated by rapidly growing species such as pine and eucalyptus.

Argentina has a continental area of 280 million hectares, of which around 34 million hectares are cultivated mainly with soybean, wheat, maize, sunflower, sorghum and rice; approximately 500,000 ha with vegetables (leafy vegetables, tomato, pepper, onion, potato, eggplant, artichoke and others) and legumes (peas dry and green); and 2,800 ha with flower crops (mainly rose, carnation and chrysanthemum) and ornamental plants. An area of approximately 526,000 ha is used for the production of fruit (lemon, orange, tangerine, apple, pear, peach, plum, grape, olives) and 42,000 ha for forestry production (conifers, eucalyptus, Salicaceae). The agricultural sector recently exceeded 100 million tonnes of grain (53% being oilseeds, the rest cereals and other grains), 8–10 million t of vegetables, 3.3 million tonnes of citrus, 13.4 million hectolitres of wine, nearly 3 million tonnes of beef, 2 million tonnes of poultry meat, 441,000 tonnes of pork and 11 billion litres of bovine milk. Also, the production of honey is important, with 23,000 producers officially registered. Nowadays, Argentina has 1157 certified organic production establishments, with a production area in 2017 of 3.2 million hectares; about 2.9 million hectares were used for livestock production and 203,000 ha for vegetable production (49% cereals and oilseeds, 32% industrial crops, 14% fruit, 6% vegetables and legumes). Semi-industrial commercial aquaculture in Argentina produces around 3700 t. In total, 23 species are produced for human consumption, including fish, bivalve molluscs, reptiles and amphibians. The rainbow trout *Oncorhynchus mykiss* (Walbaum) in the Patagonian region and the pacú *Piaractus mesopotamicus* (Holmberg) in the north-eastern region of Argentina are the most important species, which together represent almost 90% of aquaculture production. The information presented in this paragraph was obtained from Villanova *et al.* (2013), Andrade (2017), Leguizamón (2018), GAR (2019a, b) and SENASA (2019a, b).

2.2 History of Biological Control in Argentina

2.2.1 Period 1900–1969

There are records of biocontrol projects in Argentina since the beginning of the 20th century showing that introductions of natural enemies were very frequent between 1900 and 1940, before the popularity of organochlorine pesticides caused the interruption of biocontrol programmes. The main crops protected through biocontrol were fruit orchards (peaches, apples, citrus). Due to the absence of government agencies devoted specifically to the promotion, application and regulation of these activities, the information available about the biocontrol programmes during this period is sketchy and records come mostly from communications between researchers (Crouzel, 1983; Coulson and Zapater, 1992). According to these records, nine species of parasitoids, one predator and one entomopathogenic fungus were released before 1940 (Table 2.1).

Argentina also provided several species of natural enemies to Africa, Australia, Canada and the USA for arthropod pest biocontrol during this period (Table 2.2). As with arthropod biocontrol, but to a greater extent, Argentina has been an important donor of weed biocontrol agents that were released in many countries around the world. At least 26 agents from Argentina have been released against 24 weeds of South American origin (Julien and Griffiths, 1998; Cabrera Walsh *et al.*, 2014). Between 1920 and 1930, one of the most resounding successes in biocontrol was achieved in Australia with the release of the Argentine moth *Cactoblastis cactorum* (Berg) against invasive *Opuntia* spp. Since 1962, several Argentine insects have been released around the world against water hyacinth *Eichornia crassipes* (Mart.) Solms and alligator weed *Alternanthera philoxeroides* (Mart.) Griseb (Table 2.3).

2.2.2 Period 1970–2000

Biological control of agricultural pests

During this period biocontrol regained the attention of researchers and agricultural institutions, such as the National Institute for Agricultural Technology (INTA), which had been created in 1956. Biocontrol was aimed at pests of alfalfa, olives, sugarcane, wheat and sorghum, as well as of fruit trees (peaches, apples, citrus). Classical biocontrol was the main strategy adopted and a total of 35 species were introduced (33 parasitoids, one predator and one pathogen) (Table 2.1). However, very few species could be mass reared and therefore many of the imported species were never released, such as, for example, the parasitoid *Genea* (= *Jaynesleskia*) *jaynesi* (Aldrich) imported from Venezuela to control *Diatraea saccharalis* (Fabricius) (Crouzel, 1983).

Biological control of weeds

Biocontrol of weeds can be considered to have begun in Argentina in 1974, with the use of the weevil *Neochetina bruchi* Hustache, native to the Paraná-Paraguay basin, to control *Eichhornia crassipes* (Mart.) Solms in the Los Sauces reservoir, province of La Rioja, also in Argentina (Deloach and Cordo, 1983). Regardless of the impending discussion on whether agent and weed should be considered native or not to La Rioja, it could be considered a case of classical biocontrol, as neither the agent nor the weed occurred naturally in the region.

From the late 1970s to the early 1990s other weed biocontrol initiatives were undertaken by different Argentine state institutions. INTA worked on thistles (Asteraceae: Cynareae), skeleton weed *Chondrilla juncea* L., *Geoffroea decorticans* (Gillies ex Hook. and Arn.) Burkart and some aquatic weeds such as water fern (*Azolla* sp.). The native weeds *Prosopis ruscifolia* Griseb. and *Flaveria bidentis* (L.) Kuntze were studied at the Center for Research on the Regulation of Populations of Harmful Organisms (Centro de Investigaciones sobre Regulación de Poblaciones de Organismos Nocivos) (CIRPON), province of Tucumán (Table 2.4).

Seven species of thistles in the genera *Carduus* spp., *Silybum* spp., *Cirsium* spp., *Onopordum spp.* and *Cynara* spp. have been introduced into Argentina from Eurasia since early colonial days. The species *Carduus acanthoides* L. and *C. thoermeri* L. were particularly noxious in pastures, crops and roadsides. Between 1981 and 1982, the INTA at Castelar, province of Buenos Aires,

Table 2.1. Exotic species introduced into Argentina from other countries for arthropod biocontrol.

Agent species	Year	Origin	Pest	Crop	Established	Reference
Aceratoneuromyia (= *Synthomosphyrum*) *indicum* (P)	1961	Mexico	*Anastrepha fraterculus*	Citrus	No	Turica, 1968
Agathis diversa (P)	1936	USA	*Cydia molesta*	Peaches	Yes	Crouzel, 1983
Ageniaspis citricola (P)	1996	Spain-Peru	*Phylocnistis citrella*	Citrus	Yes	SENASA, 2017a,b
	1997	USA			Yes	
Aphelinus mali (P)	1920	Uruguay	*Eriosoma lanigerum*	Apples	Yes	Crouzel, 1983
	1921					
Aphidius colemani (P)	1950	Uruguay	*Schizaphis graminum*	Cereals	Yes	Coulson and Zapater, 1992
Aphidius ervi (P)	1972	USA	*Acyrtosiphon pisum*	Alfalfa	Yes	Crouzel, 1983
	1980	USA	*Metopolophium dirhodum*		Yes	Coulson and Zapater, 1992
Aphidius plagiator (P)	1978	USA	*Acyrtosiphon kondoi*	Alfalfa	Yes	Crouzel, 1983
Aphidius rhopalosiphi (P)	1980	USA	*Metopolophium dirhodum*	Alfalfa	Yes	Coulson and Zapater, 1992
Aphidius smithi (P)	1972	USA	*Acyrtosiphon pisum*	Alfalfa	Yes	Crouzel, 1983
Aphidius uzbekistanicus (P)	1980	Brazil	*Sitobium avenae*	Cereals	Yes	Coulson and Zapater, 1992
Aphytis holoxanthus (P)	1970	USA	*Chrysomphalus ficus*	Citrus	Yes	Crouzel, 1983
Aphytis lepidosaphes (P)	1973	–	*Lepidosaphes beckii*	Citrus	Yes	Crouzel, 1983
Aphytis lingnanensis (P)	1961	USA	*Aonidiella aurantii*	Citrus	Yes	Crouzel, 1983
	1966					
	1967					
	1971					
	1976	China	*Unaspis citri*	Citrus	Yes	Crouzel, 1983
Aphytis maculicornis (P)	–	USA	*Parlatoria oleae*	Olives	No	Crouzel, 1983
Aphytis melinus (P)[a]	1966	USA	*Aonidiella aurantii*	Citrus	Yes	Crouzel, 1983
		India				Coulson and Zapater, 1992
	1967	USA				
	1971					
Aphytis mytilaspidis (P)	1958	USA	*Quadraspidiotus perniciosus*	Apples	Yes	Coulson and Zapater, 1992
Aphytis yanonensis (P)	1961	Chile	*Aonidiella aurantii*	Citrus	Yes	Coulson and Zapater, 1992
Ascogaster quadridentata (P)	1936	USA	*Cydia molesta*	Peaches	Yes	Crouzel, 1983
	2003	Chile	*Cydia pomonella*	Apples, pears	Yes	SENASA, 2017a,b
Beauveria bassiana (F)	2012	Mexico	*Diaphorina citri*	Lemons	Yes	SENASA, 2017a,b

Blaesoxipha aculeata (P)	1937–1939	Canada	Locusts	Pastures	–	Clausen, 1978b
Blaesoxipha atlantis (P)						Clausen, 1978b
Blaesoxipha hunteri (P)	1940		*Dichroplus pratensis* *D. maculipennis* *D. vittatus*		–	Crouzel, 1983
Blaesoxipha opifera (P)	1937–1939		Locusts		–	Clausen, 1978b
Blaesoxipha reversa (P)	1940		*Dichroplus pratensis* *D. maculipennis* *D. vittatus*		–	Crouzel, 1983
Carpovirus (V)	2000	France	*Cydia pomonella*	Apples, pears	Yes	SENASA, 2017a,b
Cleruchoides noackae (P)	2010	Australia	*Thaumastocoris peregrinus*	Eucalyptus	–	SENASA, 2017a,b
	2013	Uruguay			–	Andorno *et al*., 2016
Coccinella septenpunctata (D)	1989	USA	*Schizaphis graminum*, *Colias lesbia*	Sorghum, alfalfa	Yes	Salto *et al*., 1990
Coccophagus caridei (P)	–	Chile	*Saissetia oleae*	Olives	Yes	Crouzel, 1983
Coccophagoides utilis (P)	–	USA	*Parlatoria oleae*		No	Crouzel, 1983
Comperiella bifasciata (P)	1971[a]	USA	*Aonidiella aurantii*	Citrus	Yes	Crouzel, 1983
Cotesia (=*Apanteles*) *flavipes* (P)	1980	Brazil	*Diatraea saccharalis*	Sugarcane	Yes	Crouzel, 1983
	1982	Bolivia				Coulson and Zapater, 1992
	2000 2005 2006 2007	Brazil				SENASA, 2017a,b
Deladenus siricidicola (N)	1998	Brazil	*Sirex noctilio*	Pine	Yes	SENASA, 2017a,b
Diachasmimorpha longicaudata (P)[b]	1961	Mexico	*Anastrepha fraterculus* *Ceratitis capitata*	Citrus	Yes	Schliserman *et al*., 2003; Turica, 1968
	1963					Coulson and Zapater, 1992
	1999	Mexico			Yes	Schliserman *et al*., 2003
Diachasmimorpha tryoni (P)	1999	Mexico	*Ceratitis capitata*	Fruit crops	Yes	Ovruski *et al*., 2003
Diadegma molesta (P)	1932	USA	*Cydia molesta*	Peaches	Yes	Crouzel, 1983
Encarsia berlesei (P)[c]	1908 1914 1915	Italy	*Pseudaulacaspis pentagona*	Peaches	Yes	Crouzel, 1983
Encarsia formosa (P)	1981	–	*Trialeurodes vaporariorum*	Vegetables	Yes	De Santis, 1981
Encarsia perniciosi (P)[d]	1971	USA	*Aonidiella aurantii*	Citrus	Yes	Crouzel, 1983

Continued

Table 2.1. Continued.

Agent species	Year	Origin	Pest	Crop	Established	Reference
Ephedrus plagiator (P)	1980 1981 1982	Brazil	Complex of aphids	Wheat	No	Crouzel, 1983
Eretmocerus mundus (P)	2007	Netherlands	*Bemisia tabaci* and *Trialeurodes vaporariorum*	Peppers	Yes	SENASA, 2017a,b
Fusarium episphaerea f. *coccophila* (F)	1900	USA	*Aonidiella aurantii*	Citrus	Yes	Crouzel, 1983
Glypta rufiscutellaris (P)	1936	USA	*Cydia molesta*	Peaches	Yes	Crouzel, 1983
Harmonia axyridis (D)	1990s	France	*Myzus persicae*	Peaches	Yes	Saini, 2004
Iphiseius (=Amblyseius) degenerans (D)	2009	Belgium	*Frankliniella occidentalis*	–	–	SENASA, 2017a,b
Isaria (=Paecilomyces) fumorosorea (F)	2012	Mexico	*Diaphorina citri*	Lemon	–	SENASA, 2017a,b
Lindorus lophantae (D)	1909	USA	*Pseudaulacaspis pentagona*	Peaches	Yes	Clausen, 1978[b]
Lixophaga diatraeae (P)	1976	Brazil	*Diatraea saccharalis*	Sugarcane	No	Crouzel, 1983
Lysiphlebus testaceipes (P)	1984	USA	*Schizaphis graminum*	Sorghum	Yes	Botto *et al.*, 1991
Macrocentrus ancylivorus (P)	1936	USA	*Cydia molesta*	Peaches	No	Crouzel, 1983
	1946				Yes	SENASA, 2017a,b
	2009					
Mastrus ridibundus (P)[e]	2005	China	*Cydia pomonella*	Apples, pears	Yes	SENASA, 2017a,b
Megarhyssa nortoni (P)	1999	Tasmania	*Sirex noctilio*	Pine	Yes	SENASA, 2017a,b
	2004	New Zealand				
Metagonistylum minense (P)	1975	Brazil	*Diatraea saccharalis*	Sugarcane	No	Crouzel, 1983
	1977	Venezuela				
Metaphycus helvolus (P)	–	Chile	*Saissetia oleae*	Olive	Yes	Crouzel, 1983
Metaphycus lounsburyi (P)	–	Chile	*Saissetia oleae*	Olive	Yes	Crouzel, 1983
Metaphycus lounsburyi (P)	1942	Peru			–	Clausen, 1978b
Metarhizium anisopliae (F)	2012	Mexico	*Diaphorina citri*	Lemon	–	SENASA, 2017a,b
Neoseiulus californicus (P)[f]	–	–	*Panonychus ulmi*	Apples	Yes	Crouzel, 1983
	2007	Netherlands	*Tetranychus urticae*	–	Yes	SENASA, 2017a,b
Opius oophilus (P)	1963	Mexico	*Anastrepha fraterculus* *Ceratitis capitata*	Citrus	No	Crouzel, 1983
Opius crawfordi (P)	1963	Mexico	*Anastrepha fraterculus* *Ceratitis capitata*	Citrus	No	Crouzel, 1983

Orius insidiosus (D)	2007	Netherlands	*Frankliniella occidentalis*	Pepper	Yes	SENASA, 2017a,b
			Thrips tabaci			
	2009	Belgium				
	2011					
Pachycrepoideus vindemiae (P)	1963	Mexico	*Anastrepha fraterculus*	Citrus	Yes	Crouzel, 1983
			Ceratitis capitata			
Palpozenillia palpalis (P)	1982	Bolivia	*Diatraea andina*	–	–	Coulson and Zapater, 1992
			D. rufescens			
Paranosema locustae (F, formerly classified as Microsporidia)	1978	USA	Locusts	Pastures	Yes	Lange and Cigliano, 2005
	1982					
	2014		*Dichroplus maculipennis*	Pastures and field crops	Yes	SENASA, 2017a,b
			D. elongatus			
			Scotussa lemniscata			
Paratheresia claripalpis (P)	1969	Peru	*Diatraea saccharalis*	Sugarcane	Yes	Crouzel, 1983
	1972					
	1973					
	1982	Bolivia				Coulson and Zapater, 1992
Patasson nitens (P)[g]	1927[g]	South Africa	*Gonipterus gibberus*	Eucalyptus	Yes	Crouzel, 1983
			Gonipterus platensis			
Praon gallicum (P)	1980	Brazil	Complex of aphids	Wheat	No	Crouzel, 1983
	1981					
	1982					
Rhyssa persuasoria (P)	1999	Tasmania	*Sirex noctilio*	Pine	Yes	SENASA, 2017a,b
Rodolia cardinalis (D)	1922	Uruguay	*Icerya purchasi*	Citrus	Yes	Clausen, 1978a
	1929					Crouzel, 1983
	1932					
Scutellista cyanea (P)	1939	Peru	*Saissetia oleae*	Olive	–	Clausen, 1978b
Spalangia cameroni (P)	1999	USA	*Haematobia irritans*		Yes	SENASA, 2017a,b
Spalangia endius (P)	1999	USA	*H. irritans*		Yes	SENASA, 2017a,b
Syneura cocciphila (D)	–	–	*Icerya purchasi*	Citrus	–	Coulson and Zapater, 1992
Systoechus vulgaris (D)	1937–39	Canada	Locusts	Pastures	–	Clausen, 1978b
Telenomus basalis (P)	1981[g]	Australia	*Nezara viridula*	–	Yes	Coulson and Zapater, 1992
Telenomus sp. nr. *alecto* (P)	1981	Bolivia	*Diatraea rufescens*	–	–	Coulson and Zapater, 1992
Tetrastichus gallerucae (P)	–	–	*Xanthogaleruca luteola*	–	–	Botto, 1996
Trichogramma australicum (P)	1981	Bolivia	*Diatraea saccharalis*	–	–	Coulson and Zapater, 1992

Continued

Table 2.1. Continued.

Agent species	Year	Origin	Pest	Crop	Established	Reference
Trichogramma cacoeciae (P)	2003 2004 2005	Chile	*Cydia pomonella*	Apples and pears	–	Botto *et al.*, 2005
Trichogramma euproctidis (P)	1932	USA	*Cydia molesta*	Peaches	Yes	Crouzel, 1983
Trichogramma nerudai (P)	–	Chile	–	–	Yes	Querino and Zucchi, 2003
Trichogramma perkinsi (P)	1981	Bolivia	*Diatraea saccharalis*	–	–	Coulson and Zapater, 1992
Trichogrammatoidea armigera (P)	1981	Bolivia	Lepidoptera	–	–	Coulson and Zapater, 1992
Trichogrammatoidea bactrae (P)	1999	Chile	*Pectinophora gossypiella*	Cotton	–	Riquelme Virgala and Botto, 2010
Trissolcus basalis (P)	1981	Australia	*Nezara viridula*	Soybean	Yes	Crouzel, 1983

Notes: –, unknown; D, predator; F, fungus; N, nematode; P, parasitoid; V, virus.
[a]Periodical releases were made in Corrientes between 1975 and 1978 (Zanelli, unpublished)
[b]Entered as *Opius* (= *Biosteres*) *longicaudata* (Crouzel, 1983)
[c]Entered as *Prospaltella berlesei*
[d]Entered as *Prospaltella perniciosi*
[e]Re-described as *Mastrus ridens* in 2009 (D'Hervé and Aquino, 2015)
[f]Entered as *Amblyseius chilensis*;
[g]Several introductions during ensuing years.

Table 2.2. Species introduced from Argentina to other countries for arthropod biocontrol.

Biocontrol agent	Year	Destination	Pest	Crop	Established	Reference
Blaesoxipha caridei (P)	1938–39	Canada	*Camnula pellucida* *Melanoplus* spp.	Pastures, cereals	Yes	Clausen, 1978b
Blaesoxipha australis (P)					No	Clausen, 1978b
Blaesoxipha neuquenensis (P)					Yes	Clausen, 1978b
Coccophagus caridei (P)	1934–35	USA	*Saissetia oleae*	Citrus, olives	–	Clausen, 1978b
Coccophagus lycimnia (P)	1935, 1954	USA	*Saissetia oleae*	Citrus, olives	–	Clausen, 1978b
Azya orbigera (P)	1935				–	Clausen, 1978b
Agathis stigmatera (P)	1928	USA	*Diatraea saccaralis*	Sugarcane	Yes	Clausen, 1978b
Trichopoda giacomelli (P)	1996	Australia	*Nezara viridula*	Pecan	Yes	Sands and Coombs, 1999
Epidinocarsis lopezi (P)	1985	Africa	*Phenacoccus manihoti*	Cassava	Yes	De Santis and Ras, 1998
Hister bruchi (D)	1992?	USA	Cattle flies		–	Fincher, 1996
Lydinolydella melallica (P)	1940–43	USA	*Epilachna varivestis*	Legumes	No	Clausen, 1978b
Ontherus sulcator (C)	1987–89	USA	Cattle flies		–	Fincher, 1996
Philonthus quadraticeps (D)	1992?	USA	Cattle flies		–	Fincher, 1996
Sulcophanaeus menelas (C)	1987–89	USA	Cattle flies		–	Fincher, 1996
Thersilochus argentinensis (P)	1942–45	USA	*Listroderes costirostris*	Vegetable crops	–	Clausen, 1978b
Thersilochus parkeri (P)						Clausen, 1978b
Epiplagiops littoralis (P)						Clausen, 1978b
Triaspis sp.						Clausen, 1978b
Thersilochus argentinensis (P)	1957–58	Australia	*Listroderes costirostris*	Vegetable crops	–	Clausen, 1978b
Thersilochus parkeri (P)						Clausen, 1978b
Epiplagiops littoralis (P)						Clausen, 1978b

Notes: –, establishment not reported; C, competitor; D, predator; P, parasitoid

Table 2.3. Argentine weed biocontrol agents released around the world (retrieved from Cabrera Walsh *et al.*, 2014).

Weed	Biocontrol agent	Year / release country
Amaranthaceae		
Alternanthera philoxeroides	*Agasicles hygrophila* (Coleoptera: Chrysomelidae)	1964 / USA
		1977 / Australia
		1981 / Thailand
		1982 / New Zealand
		1986 / China
	Amynothrips andersoni (Thysanoptera: Phlaeothripidae)	1967 / USA
	Arcola malloi (Lepidoptera: Pyralidae)	1971 / USA
		1977 / Australia
		1984 / New Zealand
Araceae		
Pistia stratiotes	*Neohydronomus affinis* (Coleoptera: Curculionidae)	1987 / USA
Asteraceae		
Gutierrezia spp.	*Heilipodus ventralis* (Coleoptera: Curculionidae)	1899 / USA
Parthenium hysterophorus	*Conotrachelus albocinereus* (Coleoptera: Curculionidae)	1995 / Australia
Cactaceae		
Acanthocereus pentagonus	*Hypogeococcus festerianus* (Hemiptera: Pseudococcidae)	1979 / Australia
Cereus jamacaru	*Alcidion cereicola* (Coleoptera: Cerambycidae)	1990 / South Africa
	Hypogeococcus festerianus	1983 / South Africa
Harrisia bonplandii	*Alcidion cereicola*	1974 / Australia
	Hypogeococcus festerianus	1982 / Australia
Harrisia martinii	*Alcidion cereicola*	1974 / Australia
		1990 / South Africa
	Hypogeococcus festerianus	1975 / Australia
		1983 / South Africa
Harrisia tortuosus	*Alcidion cereicola*	1976 / Australia
	Hypogeococcus festerianus	1976 / Australia
Opuntia aurantiaca	*Cactoblastis cactorum* (Lepidoptera: Pyralidae)	1926 / Australia
		1933 / South Africa
		1933 / Australia
	Dactylopius austrinus (Hemiptera: Dactylopiidae)	1935 / South Africa
		1935 / Australia
	Tucumania tapiacola (Lepidoptera: Pyralidae)	1935 / Australia
Opuntia ficus-indica	*Cactoblastis cactorum*	1933 / South Africa
		1950 / USA (Hawaii)
Opuntia lindheimeri	*Cactoblastis cactorum*	1960 / Antigua
Opuntia streptacantha	*Cactoblastis cactorum*	1926 / Australia
Opuntia stricta	*Cactoblastis cactorum*	1926 / Australia
		1970 / Cayman Islands
		1957 / Nevis
		1933 / New Caledonia
Opuntia triacantha	*Cactoblastis cactorum*	1957 / Nevis
		1960 / Antigua
		1960 / Montserrat
Opuntia tuna	*Cactoblastis cactorum*	1950 / Mauritius
Opuntia vulgaris	*Cactoblastis cactorum*	1950 / Mauritius
Opuntia spp.	*Cactoblastis cactorum*	1971 / St Helena

Continued

Table 2.3. Continued.

Weed	Biocontrol agent	Year / release country
		1973 / Ascencion
	Dactylopius sp.	? / New Caledonia
Fabaceae		
Parkinsonia aculeata	*Penthobruchus germaini* (Coleoptera: Bruchidae)	1995 / Australia
	Eupithecia cisplatensis (Lepidoptera: Geometridae)	2015 / Australia
Prosopis spp.	*Evippe* sp. (Lepidoptera: Gelechiidae)	1998 / Australia
	Prosopidopsilla flava (Homoptera: Psillidae)	1998 / Australia
Sesbania punicea	*Neodiplogrammus quadrivittatus* (Coleoptera: Curculionidae)	1984 / South Africa
	Rhyssomatus marginatus (Coleoptera: Curculionidae)	1984 / South Africa
Pontederiaceae		
Eichhornia crassipes	*Neochetina bruchi* (Coleoptera: Curculionidae)	1974 / USA
		1979 / Sudan
		1984 / India
		1989 / South Africa
		1993 / Uganda
	Neochetina eichhorniae (Coleoptera: Curculionidae)	1970 / Mexico
		1972 / USA
		1974 / South Africa
		1975 / Australia
		1977 / Fiji
		1978 / Sudan
		1979 / Indonesia
		1979 / Thailand
		1980 / Myanmar
		1983 / India
		1983 / Malaysia
		1986 / Papua New Guinea
		1988 / Sri Lanka
		1990 / Honduras
		1991 / Benin
		1993 / Kenya
		1993 / Nigeria
		1993 / Uganda
		1994 / Ghana
		1996 / China
	Niphograpta albiguttalis (Lepidoptera: Pyralidae)	1977 / Australia
		1980 / Sudan
		1990 / South Africa
		1995 / Thailand
		1996 / Malaysia
		1997 / USA
	Xubida infusella (Lepidoptera: Pyralidae)	1996 / Australia
	Megamelus scutellaris (Hemiptera: Delphacidae)	2010 / USA
Solanaceae		
Solanum sisymbriifolium	*Gratiana spadicea* (Coleoptera: Chrysomelidae)	1994 / South Africa
Solanum viarum	*Gratiana boliviana* (Coleoptera: Chrysomelidae)	2003 / USA

Table 2.4. Releases of weed biocontrol agents in Argentina (retrieved from Cabrera Walsh *et al.*, 2014).

Weed	Agent	Release year	Results
Eichhornia crassipes	*Neochetina bruchi* (Coleoptera)	1974	Established, successful
	N. bruchi and *N. eichhorniae*	2014	Established
Carduus spp.	*Rhinocyllus conicus* (Coleoptera)	1981–83	Established
	Trichocirocalus horridus (Coleoptera)	1983	Not established
Chondrilla juncea	*Cystiphora schmidti* (Diptera)	1983–87	Not established
	Bradyrrhoa gilveolella (Lepidoptera)	1992	Not established
	Aceria chondrillae (Eriopyidae)	1983–1991	Established
	Puccinia chondrillina (Uredinales)	1983–84	Established
Pistia stratiotes	*Neohydronomus affinis* (Coleoptera)	2012	Established, successful
	Lepidelphax pistiae (Hemiptera)	2012	Established, successful
Hypericum perforatum	*Chrysolina quadrigemina* (Coleoptera)	?	Established
Rubus fruticosus	*Phragmidium violaceum* (Uredinales)	?	Established
Aquatic weeds	*Ctenopharyngodon idella* (Cypriniformes)	?	Established

imported the weevil *Rhinocyllus conicus* Frölich from the USA and New Zealand (Table 2.4). This weevil had already been used successfully to control thistles in the USA, Canada, New Zealand, Australia and South Africa. The weevil is native to eastern Asia, Europe and northern Africa. The larvae feed upon the flower heads, preventing seed development, and the adults are defoliators. There were several releases in 26 sites in the provinces of Buenos Aires, Chubut, Córdoba and La Pampa. It has established and dispersed in many different environments of temperate Argentina and Uruguay, with varied control levels. However, it shows a clear latitudinal limit (roughly 32° S), since it has not established in subtropical areas (Enrique de Briano *et al.*, 2013).

Skeleton weed *C. juncea* is a perennial Eurasian weed that was first discovered in Argentina in 1977. It currently covers more than 4 million hectares of pastures and wheat/sunflower croplands. Between 1982 and 1992 the INTA released the gall midge *Cystiphora schmidti* (Rübsaamen) from Australia, the eriophyid mite *Aceria chondrillae* (Canestrini) from Australia and the USA, the moth *Bradyrrhoa gilveolella* (Treitschke) from Greece and the rust *Puccinia chondrillina* (Bubak and Syd.) from the USA (Table 2.4). Impact studies have not been reported to date.

This brief period of little over 10 years was not adequate to develop effective weed biocontrol projects and the imposition in the 1990s of technological packages reliant exclusively on the use of pesticides deactivated the few weed biocontrol initiatives and took management of environmental weeds off the local scientific agenda.

2.3 Current Situation of Biological Control in Argentina

2.3.1 Introduction

Traditionally, pest control in Argentina has been based on the employment of chemical pesticides and their use has dramatically increased since the 1980s, along with other Latin American countries. In the 1990s, at the height of the use of high chemical pesticides, this situation seemed to have reached a breakpoint when the INTA prepared an extensive report that critically revised local crop protection programmes and stated the need for a shift to more sustainable agriculture (INTA, 1991). One of the main objectives pursued was to increase crop production by means of providing alternatives to conventional synthetic pesticides through integrated pest management (IPM) (Zapater, 1996). However, nearly 30 years after this document, chemical control is still virtually the exclusive tool for agricultural pest management. Below we describe the status of biocontrol in Argentina and supply clues to its progress, based on information from published sources, unpublished academic papers and reports, proceedings and personal communications.

2.3.2 Classical biological control

Biological control of agricultural pests

In the late 1990s, a new age began for biocontrol in Argentina, after the country adhered to international plant protection regulations in collaboration with neighbouring countries. Evidence of this is the creation of the Plant Protection Committee (Comité de Sanidad Vegetal) (COSAVE) formed by Brazil, Chile, Peru, Bolivia, Uruguay and Paraguay. Since then, classical biocontrol programmes have been evaluated by state organizations, such as the National Plant and Animal Protection Service (Servicio Nacional de Sanidad Animal y Vegetal) (SENASA) and the national and provincial ministries of agriculture and environment. It is important to emphasize that Argentine regulations reproduce international counterparts regarding the environmental risk of the introduction of exotic organisms (Bigler *et al.*, 2006). The Servicio Nacional de Sanidad y Calidad Agroalimentaria (SENASA, 2017a) authorizes and keeps records of the organisms approved for release. These records show that during the past 26 years, 15 species of parasitoids, three predators, one nematode, one microsporidia (currently considered fungi), one virus and three entomopathogenic fungi have been imported for use against pest arthropods (Table 2.1). These organisms entered the following quarantines: (i) Centro de Investigaciónes sobre Regulación de Poblaciones de Organismes Nacivos (CIRPON), in the province of Tucumán, which existed until 1999; and (ii) Microbiology and Agricultural Zoology Institute (Instituto de Microbiología y Zoología Agrícola) (IMyZA) at the INTA in Castelar, province of Buenos Aires, which exists to this day. There are also records of the rejection of several petitions based on specificity concerns, often derived from inadequate information in the petitions.

Most of the imported biocontrol agents came from the USA, though the native range of both the agents and the target pests could be different. Some agents have been introduced several times (Table 2.1), such as the parasitoids *Aphytis lingnanensis* Compere, used against citrus scales, *Cotesia flavipes* Cameron for control of the sugarcane borer, and *Diachasmimorpha longicaudata* (Ashmead) for control of fruit flies in citrus and other fruit trees. Also, some accidental introductions, possibly due to their similarity to deliberate introductions, have been detected, such as *Macrocentrus delicatus* Cresson, a parasitoid of *Cydia molesta* (Busck), which might have entered as *Macrocentrus ancylivorus* (Rohwer).

Some species were imported even though they were already present in Argentina, such as the predators *Orius insidiosus* (Say) and *Neoseiulus californicus* (MacGregor), which were obtained from Belgium. These cases should not be considered as classical or neoclassical biocontrol (the release of exotic natural enemies to control native pests, also known as biocontrol by 'new association'), because the species were not exotics. However, when liberating individuals belonging to lineages or local populations from another region, some researchers argue that their introduction could generate the same environmental or economic risks as the importation of exotic species. A similar case occurred with the introduction of the pupal parasitoids of Diptera, *Spalangia cameroni* Perkins and *S. endius* Walker, both considered native in Argentina, imported from the USA together with *S. gemina* Boucek to control houseflies and the horn fly *Haematobia irritans* (L.), between 1992 and 1999 (SENASA Res 758/97; D. Crespo, Buenos Aires, 2004, personal communication).

The establishment of the agents released has been successful in general. However, there have been few post-release evaluations and additional management strategies are still necessary for several pests. Likewise, there are no evaluations of non-target effects of the agents released in Argentina. For instance, the predator *Harmonia axyridis* (Pallas), deemed of high risk to the environment in several countries, is assumed to pose a threat to Argentine coccinelid species assemblages, similar to the reduction in population size of native predators observed in Chile (Grez *et al.*, 2016), but this has not been confirmed.

On the other hand, 17 parasitoid species, two predators and two dung-fly competitors native to Argentina have been released in other countries for biocontrol of arthropods, in both classical and neoclassical endeavours, yet establishment of only five species has been confirmed (Table 2.2).

Biological control of weeds

Two exotic specific agents have been found in Argentina on Eurasian weeds: rust for blackberries (*Rubus* spp.) (Cabrera Walsh *et al.*, 2014) and the leaf beetle *Chrysolina quadrigemina* (Suffrian) against St John's wort *Hypericum perforatum* L. (Turienzo, 2006), but no release or research reports exist for either of them, suggesting that they may have entered the country accidentally from Chile. The generalist grass carp *Ctenopharyngodon idella* (Valenciennes), native to Asia, has been released to control several different aquatic weeds in irrigation canals, but its exact release date and site were not reported either (Table 2.4).

In Argentina, there are also a few instances of classical biocontrol using native insects against native plants invading water bodies isolated from their usual distribution areas. These have been local initiatives aimed at solving local problems in closed water bodies, as in the case of the Los Sauces reservoir described above, and two other projects in the province of Buenos Aires. One project was to control water lettuce *Pistia stratiotes* L. in a small lake of the Vicente Lopez natural reserve; the other was to control water hyacinth in the El Ojo lake, San Vicente county (Cabrera Walsh *et al.*, 2017).

Classical biocontrol of weeds with pathogens is a growing discipline. Pathogens can be very specific; however, biocontrol of grasses (Poaceae) has always been met with distrust, possibly because of the history of pathogen-related famines. Two South American grasses, *Nassella* (= *Stipa*) *trichotoma* Nees and *N. neesiana* Trin. and Rupr., which are important weeds in Australia and New Zealand, were among the first grasses to be considered for biocontrol. The National Council of Scientific and Technical Research (CONICET) and the Commonwealth Scientific and Industrial Research Organization (CSIRO), from Australia, began collaborative research in 1999, and the rust *Uromyces pencanus* (Diet. and Neg.) was considered safe for introduction into New Zealand in 2015 (Anderson *et al.*, 2011). However, the rust has not been exported yet because of difficulties in obtaining export permits from local authorities (F. Anderson, Buenos Aires, 2017, personal communication).

2.3.3 Augmentative biological control

Augmentative biological control with invertebrates

Commercial augmentative biocontrol has not yet developed to a large extent in Argentina. One company, Brometán SRL, sells the predator *O. insidiosus* and the parasitoid *Aphidius colemani* Viereck. In recent years, this company has signed collaboration agreements with the INTA to carry out experiments on management of vegetable crop pests (S. Cáceres, Corrientes, 2014, personal communication).

Since 2008, the parasitoid *C. flavipes* has been mass reared in the sugarcane grower and production plant Seaboard: Energías Renovables y Alimentos SRL (formerly Tabacal Agroindustria) for releasing in their own sugarcane fields against *Diatraea* spp. (S. Barrera and A. Fonollat, Salta, 2019, personal communication). Previous to this there was a mass-rearing facility for commercial production of this parasitoid at CIRPON, which functioned between 2001 and 2009 and aimed to provide the parasitoid for all of Northwest Argentina. The citrus company Citrícola Ayuí SA (province of Entre Ríos) developed a mass-rearing facility for rearing parasitoids in the genus *Aphytis* until 2014, to control *Aonidiella aurantii* (Maskell). This project resulted in a dramatic improvement in orange production (G. Toller, Entre Ríos, 2016, personal communication). The company discontinued the parasitoid mass rearing because of changes in business administration issues.

In the area of livestock facilities, the biofactory Insectarios SRL sells *S. endius* to control flies in pig and poultry farms.

In vegetable crops, the increase in the interest for augmentative biocontrol on behalf of producers and specialists has led to experimental releases of natural enemies during the past decade. Joint field assessments were carried out by INTA and Brometán SRL to evaluate the efficacy of *O. insidiosus* to control *Frankliniella occidentalis* (Pergande) in strawberries (Lefebvre *et al.*, 2013) and sweet peppers in greenhouses (L. Viglianchino, Santa Fe, 2013, personal communication).

Neoseiulus californicus imported from the Netherlands was released on sweet peppers (S. Cáceres, Corrientes, 2012, personal communication) and strawberries in 2007 and 2008 to control *Tetranychus urticae* Koch, but these releases were not considered successful (D. Kirschbaum, Tucumán, 2012, personal communication). The parasitoid *Encarsia formosa* Gahan, reared at the IMyZA, was released to control *Bemisia tabaci* (Gennadius) on sweet pepper, cucurbits, tomato and ground cherries *Physalis* sp. in greenhouses in the province of Corrientes. Results indicate that it did not sufficiently reduce the *B. tabaci* complex (S. Cáceres, Corrientes, 2004, personal communication). The INTA also made experimental releases of *Trichogramma nerudai* Pintureau and Gerding and *Trichogrammatoidea bactrae* Nagaraja to control *Tuta absoluta* (Meyrick) on tomatoes with variable parasitism rates (11.4–7.8%). *T. nerudai* was the most collected species in a post-release evaluation, while *T. bactrae* was collected in very low numbers (M. Riquelme Virgala, Buenos Aires, 2007, personal communication).

In fruit trees, a medium-scale rearing protocol for the parasitoids *D. longicaudata* and *D. tryoni* (Cameron), introduced into Argentina in 1999 (Table 2.1), was developed by Ovruski *et al.* (2003) to control the fruit fly *Ceratitis capitata* Wied. Subsequently, large numbers of *D. longicaudata* were released in fig trees in the province of San Juan in 2012, as a result of an agreement between the Pilot Plant of Industrial Microbiological Processes (PROIMI, CONICET) and the provincial authorities. Results are promising and inoculative releases continue to date (S. Ovruski, Tucumán, 2017, personal communication).

INTA has made experimental releases of the natural enemies *Ascogaster quadridentata* Wesmael, *Chrysoperla externa* (Hagen), the native species *Goniozus legneri* Gordh, *Mastrus ridens* Horstmann and two species of *Trichogramma* to control *Cydia pomonella* (L.) in apple and pear groves of the province of Río Negro. Currently, inoculative releases continue, mainly of *G. legneri* (Garrido *et al.*, 2007). This parasitoid species is mass-reared at the CEMUBIO (Center of Biocontrol Agents Reproduction), INTA Río Negro, General Roca. *M. ridens* has also been reared at the Instituto de Sanidad y Calidad Agropecuaria de Mendoza (ISCAMEN), in the province of Mendoza, and released against the same pest in pear, quince, walnut and apple orchards. Establishment was successful, but *C. pomonella* parasitism only reached 10% (Tortosa *et al.*, 2014).

Augmentative biological control with microbial agents

Several entomopathogenic agents have been developed for biocontrol in Argentina, mainly bacteria, fungi and viruses, and to a lesser extent protozoa and nematodes. The bacterium *Bacillus thuringiensis* Berliner is utilized in Argentina to control *Anticarsia gemmatalis* Hübner, *Epinotia aporema* (Walsingham), *T. absoluta*, *C. pomonella* and *Anthonomus grandis* Boheman. Thirteen companies sell pathogen-based products in Argentina; seven of these are based on *B. thuringiensis*, six on *B. thuringiensis* var. *kurstaki* and one on *B. thuringiensis* var. *aizawai* (SENASA, 2017b).

Argentina has a large bank of native strains of entomopathogenic fungus, mostly *Beauveria bassiana* (Balsamo) Vuillemin, *Lecanicillium lecanii* (Zimmerman), *Metarhizium anisopliae* (Metschnikoff) and *Nomurea rileyi* (Farlow) Samson (Sosa-Gómez *et al.*, 2010; Nussenbaum and Lecuona, 2012; Gutierrez *et al.*, 2017; Lopéz Lastra and Lecuona, 2019). However, no commercial products have been registered to date.

As for viruses, commercial virus products based on the *C. pomonella* granulovirus (CpGV) are used in Argentina to control this pest on pome fruits and walnuts, as Carpovirus Plus (INTA – NPP/Arysta Life Science - AgroRoca SA) and Madex (Andermatt Biocontrol) (R. Lecuona, Buenos Aires, 2011, personal communication; Haase *et al.*, 2015).

2.3.4 Conservation biological control of agricultural pests

In Argentina, three main areas of interest can be identified related to conservation biocontrol: (i) a deepening of understanding of plant diversity useful for the development, survival and fecundity of parasitoids and predators in different agricultural systems; (ii) developmental of protocols for the reduction of the frequency of applications with broad-spectrum pesticides and searching for selectivity; and (iii) exploring alternatives or supplementary foods for parasitoids and predators.

Regarding the first area of interest, habitat management through plant diversity has been determined to be the key element for the conservation and increase of natural enemies in some crops. There is also evidence that certain plant combinations favour natural enemy establishment. Indeed, the design of floriferous and aromatic field borders can improve the survival conditions of the coccinellid *Eriopis connexa* (Germar) and the control of cereal aphids and *Helicoverpa zea* (Boddie) in maize (M. Tulli, Misiones, 2015, personal communication). Zumoffen *et al.* (2015) studied the possibility of implementing an open breeding system of *Lysiphlebus testaceipes* (Cresson) for the management of *Aphis craccivora* Koch in alfalfa and horticultural crops in the province of Santa Fe. This system proved to be easy to maintain based on the use of *Araujia* spp. as alternative host plant where *L. testaceipes* attacks another, non-pest aphid. Fernández *et al.* (2008) found in apple orchards that the number of natural enemies and their diversity increased with the presence of cover crops and decreased with the utilization of agrochemicals. In olive groves, the presence of parasitoids and predators of mealybugs and whiteflies is favoured through techniques of adjacent vegetation management (M. Holgado, Misiones, 2015, personal communication).

Regarding protected crops, the value of different plant species of spontaneous growth, inside and outside the greenhouse, has been evaluated to favour the presence of natural enemies of common pests. In pepper, *Portulaca oleracea* L., *Cynodon dactilon* (L.) Pers. and *Anoda cristata* L. are potential host species of *O. insidiosus*, a biocontrol agent used in augmentative releases (L. Viglianchino, Misiones, 2015, personal communication). Gugole Ottaviano *et al.* (2015) found that, in autumn and winter, *Urtica urens* L., *Lamium amplexicaule* L. and *Sonchus oleraceous* L. would promote *N. californicus* persistence, when *T. urticae* density is low in strawberry crops, offering them thrips and pollen as alternative food; whereas in summer, the presence of *Convolvulus arvensis* L. and *Galega officinalis* L. allows the predator to survive when the crop cycle is ending and prey are scarce. Salas Gervassio *et al.* (2016) identified one cultivated (*Solanum melongena* L.) and four wild Solanaceae (*Nicotiana glauca* Graham, *Salpichroa origanifolia* (Lam.) Baill., *Solanum americanum* Mill. and *S. sysimbriifolium* Lam.) to be attacked by *T. absoluta*, the main pest in tomato crop. Yet they also support parasitoids, resulting in natural parasitism rates of up to 65% in the Horticultural Belt of La Plata, located north-east of the province of Buenos Aires. In addition, a strong seasonal synchronization between *T. absoluta* and one of its main enemies, the larval endoparasitoid *Pseudapanteles dignus* Muesebeck, occurs in these plants.

Experiences with banker plants (e.g. *Avena sativa* L., *Calendula officinalis* L., *Matricaria chamomilla* L.), demonstrated the potential of these species for the management of entomophages in horticultural systems (A. Andorno, Buenos Aires, 2014, personal communication). Also, implanting alfalfa field borders close to soybean crops provides alternative source of hosts, nectar and pollen for the parasitoids (J. Zubiaurre, Misiones, 2015, personal communication).

With regard to the second area of interest, changes in the frequency of the use of pesticides or use of selective pesticides for the conservation of natural enemies, specific knowledge of each system is being generated. For instance, there is a programme for the control of *T. urticae* in strawberry crops in the province of Buenos Aires with the predatory mite *N. californicus* that comprises monitoring both species from the beginning of the crop cycle and the use of a decision table before applying acaricides. The components of the programme for the monitoring protocol and the decision table are: (i) a model that relates the density of mites per leaflet with the presence–absence of both species based on the prey–predator spatial coincidence; (ii) the rate of increase of the pest; and (iii) the time that the population takes to reach the level of economic damage, starting from different initial densities of both populations. This management plan promotes a significant reduction in chemical applications, favouring the survival of predators and enhancing biocontrol (Greco *et al.*, 2011).

Also related to these objectives, the selectivity of some insecticides has been evaluated through lethal and sub-lethal effects on several natural enemies. For example, all stages of the coccinellid *E. connexa* (Fogel *et al.*, 2013, 2016) and the pupal and adult stages of *Eretmocerus mundus* Mercet were found to be susceptible to several insecticide compounds (Francesena *et al.*, 2012, 2017). On the other hand, the lacewing *C. externa* showed natural tolerance to several pyrethroids (Haramboure *et al.*, 2010, 2013).

Concerning the third area of interest, the provision of nutritional supplements has been tested for Neuroptera associated with pear groves in Río Negro, Argentina. This methodology is currently in use in organic production for the control of the mealybugs *Pseudococcus viburni* (Signoret) and *Diaspidiotus perniciosus* (Comstock) through the supply of sugary solutions enriched with beer yeast to feed lacewings (S. Garrido, Misiones, 2015, personal communication).

Finally, it should be pointed out that the pest control exercised by spider communities present in soybean agroecosystems has also gained attention in Argentina. Numerous studies describe the specific richness, the guilds that make up the community, the availability of prey, the presence of shelters and the susceptibility to different agrochemicals. Indeed, in several of them, guidelines are given for spider protection (Benamú *et al.*, 2010).

2.4 Conclusions and New Developments of Biological Control in Argentina

In Argentina, eight predators, 70 parasitoids and seven pathogens (including five entomopathogenic fungi) have been introduced for arthropod biocontrol since 1900, which adds up to 0.72 species per year. These biocontrol agents have been released in many agricultural systems found throughout the diverse agroclimatic conditions in Argentina. Yet the implementation of biocontrol is not as widespread as expected. Based on the data presented above and in Table 2.5, we estimate that at least 4.3 million hectares are under classical biocontrol and 26,600 ha are under augmentative biocontrol.

Fruit production and pine plantations have the largest areas under some degree of classical biocontrol (Table 2.5). Pine plantations could eventually benefit from biocontrol agents that have been released in many other countries. An example of current use in Argentina is the application of a nematode and three parasitoid species for control of the *Sirex* wood wasp on almost 650,000 ha (Table 2.5). Another example is the use of the parasitoid *T. radiata* for control of the Asian citrus psyllid, *Diaphorina citri* Kuwayama (INTA, 2019a, b).

The fruit flies *C. capitata* and *Anastrepha fraterculus* (Wiedemann) are under experimental inundative biocontrol with the parasitoid *D. longicaudata* in citrus production in the province of San Juan (Suárez *et al.*, 2018). Another 1 million hectares are under an IPM regime that includes cultural control, trapping and sterile insect technology (SIT) (SENASA, 2019c).

Table 2.5. Estimated areas under biological control in Argentina.

Biocontrol agent	Pest / crop or weed	Type of biocontrol	Area under biocontrol (ha)
Tamarixia radiata	*Diaphorina citri* / citrus	Classical	150,000
Tamarixia radiata	*Diaphorina citri* / citrus	Augmentative	1,500
Beddingia (= *Deladenus*) *siricidicola* *Ibalia leucospoides* *Megarhyssa nortoni* *Rhyssa persuasoria*	*Sirex noctilio* / pine trees	Classical	648,000
Orius insidiosus *Bacilus thuringiensis*	*Frankliniella occidentalis, Tuta absoluta,* other lepidopteran pests / horticultural crops	Augmentative	128
Cotesia flavipes	*Diatraea saccharalis* / sugarcane	Augmentative	25,000
Rhynocillus conicus *Cystiphora schmidti* *Puccinia chondrilina* *Chrysolina quadrigemina*	*Carduus* spp., *Chondrilla juncea* *Hypericum perforatum* / weeds in pastures	Classical	3,500,000
Neochetina bruchi *Neohydronomus affinis* *Lepidelphax pistiae*	*Eichhornia crassipes* *Pistia stratiotes* / aquatic weeds	Classical	150

Horticultural production is very important in Argentina; however, only a small fraction of greenhouse tomatoes and peppers is under biocontrol (C. Silvestre, Buenos Aires, 2018, personal communication, and Table 2.5). Some areas in the Seaboard (formerly Tabacal) sugarcane plantations and citrus groves belonging to the company Ayuí in the province of Entre Ríos are treated with augmentative releases of biocontrol agents (Table 2.5).

As for biocontrol of environmental and pasture weeds, an extensive part of Argentina affected by thistles and skeleton weed is under some degree of biocontrol, though they are still important weeds in many areas (Cabrera Walsh *et al.*, 2014). Also, several artificial and natural water bodies are under aquatic weed biocontrol (Cabrera Walsh *et al.*, 2017, Table 2.5).

Currently, new biocontrol products are pending for approval, like commercial formulations of *Beauveria*, *Metharizhium* and *Trichoderma* (M.F. Achinelly, La Plata, 2018, personal communication). In addition, several research groups are evaluating native strains of these microorganisms for use in Argentine agriculture. Another current development concerns the production of the fungus *Paranosema locustae* Camming, which started at the end of 2018 after the creation of the Plant Health Laboratory by the Ministry of Agribusiness of the province of Buenos Aires (Minagri, Buenos Aires Province). The fungus will be used to control grasshoppers and locusts (Acridoidea), important pests of pastures and crops in this province.

Biocontrol in Argentina has a longstanding tradition and comprises a community of highly trained scientists who deal with every aspect of the discipline. However, the applied aspects of biocontrol need to be developed further in Argentina. Essentially, we need more basic and applied research and funds related to the evaluation of native natural enemies, and more knowledge about how the official plant protection strategy should be implemented by policy makers and agricultural technicians. We hope that as public demand increases for sustainable food production and pest control techniques, and awareness for the threats of invasive species increases, biocontrol will gain importance in the pest manager's toolbox.

2.5 Acknowledgements

The following persons are thanked for providing unpublished information: personnel of the Plant Protection Dept. of the SENASA (National Agricultural Sanitary and Quality Service), M. Benamú (CUR, Centro Universitario de Rivera, Univ. de la República de Uruguay), C. Silvestre and C. Cáceres (Brometán), S. Ovruski (PROIMI-CONICET), F. Anderson (CERZOS-CONICET), L. Viglianchino (UNL), M.C. Tulli (UNMdP), M. Holgado (UNCuyo), J. Zubiaurre, S. Garrido, D. Crespo, D. Kirschbaum, S. Cáceres, A. Andorno, M. Riquelme Virgala, R. Lecuona and A. Saluso (INTA), M.F. Achinelly (CEPAVE-CONICET), and N. Toller (Citrícola Ayuí SA Concordia).

References

Anderson, F., Gallego L., Barton, J, and McLaren, D.A. (2011) Biological control of Chilean needle grass (*Nassella neesiana*, Poaceae) in Australasia. Completion of host range testing. In: *13th International Symposium on Biological Control of Weeds*. Waikoloa, Hawaii, pp. 26–32.

Andorno, A., Hernández, C., Cuello, E. and Botto, E. (2016) Control biológico de plagas en eucaliptos: avances en el empleo de enemigos naturales exóticos [Biological control of eucalyptus pests: advances in the use of exotic natural enemies]. In: López, S.N., Viscarret, M., Sauka, D. and Segura, D. (eds) *Resúmenes de la Jornada de actualización e intercambio en control biológico, comportamental y genético de plagas agropecuarias*. Ediciones INTA, Buenos Aires, Argentina, pp. 1–2.

Andrade, F. (2017) Los desafíos de la agricultura Argentina [Challenges for the agriculture of Argentina]. INTA Ediciones.

Benamú, M.A, Schneider, M. and Sánchez N. (2010) Effects of the herbicide glyphosate on biological attributes of *Alpaida veniliae* (Araneae, Araneidae), in laboratory. *Chemosphere* 78, 871–876.

Bianchi, E.D., Anon, M.C., Pagano, E., Pineiro, M., Szpak, G., Trigo, E., Vaudagn, S. and Bianchi, F. (2017) Food and nutrition security in Argentina. In: *Challenges and Opportunities for Food and Nutrition Security in the Americas. The View of the Academies of Sciences*. IANAS, IAP and BMBF, México DF, pp. 30–51. [Free public access of this publication in English and Spanish at: www.ianas.org]

Bigler, F., Babendreier, D. and Kuhlmann, U. (2006) *Environmental Impact of Invertebrates for Biological Control of Arthropods. Methods and risk assessment*. CAB International, Wallingford, UK.

Botto, E.N. (1996) Control biológico de plagas en la Argentina: informe de la Situación actual [Biological control in Argentina: current situation]. In: Zapater, M.C. (ed.) *El control biológico en América Latina*. OICB SRNT Buenos Aires, Argentina, pp. 1–8.

Botto, E.N., Monetti, N.C. and de Saluso, A.R. (1991) Introduction, colonization and establishment of *Lysiphlebus testaceipes* (Cresson) (Hymenoptera: Aphidiidae), in Argentina. *Entomophaga* 36, 323–324.

Botto, E., Garrido, S., Saez, T., Hernández, C., Lafalce, C., Cichón, L. and Fernández, D. (2005) Evaluación de *Trichogramma* spp. (Hymenoptera: Trichogrammatidae) para el control biológico de *Cydia pomonella* (L.) (Lepidoptera: Tortricidae) en Alto Valle de Río Negro, Argentina. [Evaluation of *Trichogramma* spp. for biological control of *Cydia pomonella* in Alto Valle de Río Negro, Argentina]. In: *Libro de resumenes, XXVII Congreso Nacional Entomología de Chile* [*Proceedings of XXVII National Congress of Entomology of Chile*], p. 44.

Cabrera Walsh, G., Briano, J., Enrique de Briano, A. and Anderson, F.E. (2014) Control biológico de malezas [Biological control of weeds]. In: Fernández, O.A., Leguizamón, E.S. and Acciaresi, H.A. (eds) *Malezas e invasoras de la Argentina: Ecología y manejo*. Editorial de la Universidad Nacional del Sur. Ediuns. Bahía Blanca, Argentina, pp. 801–821.

Cabrera Walsh, G., Hernández, M.C., McKay, F., Oleiro, M., Guala, M. and Sosa, A. (2017) Lessons from three cases of biological control of native freshwater macrophytes isolated from their natural enemies. *Aquatic Ecosystem Health and Management* 20, 353–360.

CIA (2017) The World Factbook: Argentina. Available at: https://www.cia.gov/library/publications/the-world-factbook/geos/ar.html (accessed 7 July 2019).

Clausen, C.P. (1978a) Biological control of citrus insects. In: *The Citrus Industry*. University of California, Division of Agricultural Science, Berkeley, California.

Clausen, C.P. (1978b) Introduced parasites and predators of arthropod pests and weeds: a world review. Agricultural Research Service, United States Department of Agriculture, Agriculture Handbook N° 480. Washington, DC.

Coulson, J.R. and Zapater, M.C. (1992) Opportunities for implementation of Biocontrol in Latin America. In: *Proceedings of the International Organization for Biological Control (IOBC), Workshop on Opportunities for implementation of BioControl in Latin America*, Rio de Janeiro, Brazil. IOBC-NTRS.

Crouzel, I.S. (1983) El control biológico en Argentina (Biological control in Argentina). In: *Actas del IX Congreso Latinoamericano de Zoología (CLAZ)*, Arequipa, Perú, pp. 169–174.

Deloach, C.J. and Cordo, H.A. (1983) Control of water hyacinth by *Neochetina bruchi* (Coleoptera: Curculionidae: Bagoini) in Argentina. *Environmental Entomology* 12, 19–23.

De Santis, L. (1981) Catálogo de los Himenópteros Calcidoideos de América al Sur de los Estados Unidos. Primer Suplemento. [Catalogue of the chalcidoid hymenopterans of America South of the United States]. *Revista Peruana de Entomología* 24, 1–38.

De Santis, L. and Ras, N.P. (1998) Control biológico de la cochinilla *Phenacoccus manihoti* en África (Insecta) [Biological control of *Phenacoccus manihoti* in Africa]. *Academia Nacional de Agronomía y Veterinaria* XLII (7), 5–11.

D'Hervé, F.S. and Aquino, D.A. (2015) Detección de hiperparasitoides de *Mastrus ridens* (Hymenoptera: Ichneumonidae) y *Ascogaster quadridentata* (Hymenoptera: Braconidae) en el Alto Valle del Río Negro, Argentina [Detection of hyperparasitoids of *Mastrus ridens* and *Ascogaster quadridentata* in Alto Valle del Río Negro]. *Revista del Museo Argentino de Ciencias Naturales* 17, 153–158.

Enrique de Briano, A.E., Acciaresi, H.A. and Briano, J.A. (2013) Establishment, dispersal, and prevalence of *Rhinocyllus conicus* (Coleoptera: Curculionidae), a biological control agent of thistles, *Carduus* species (Asteraceae), in Argentina, with experimental information on its damage. *Biological Control*, 67, 186–193.

Fernández, D.E., Cichon, L.I., Sanchez, E.E., Garrido, S.A. and Gittins, C. (2008) Effect of different cover crops on the presence of arthropods in an organic apple (*Malus domestica* Borkh) orchard. *Journal of Sustainable Agriculture* 2, 197–211.

Fincher, G.T. (1996) Biological control of dung-breeding flies affecting pastured cattle. In: *Great Plains Beef Cattle Handbook BCH-3810*, pp. 13. Available at: http://www.iowabeefcenter.org/bch/DungBreedingFlyControl.pdf (accessed 27 July 2019).

Fogel, M., Schneider, M., Desneux, N., González, B. and Ronco, A. (2013) Impact of the neonicotinoid acetamiprid on immature stages of the predator *Eriopis connexa* (Coleoptera: Coccinellidae). *Ecotoxicology* 22, 1063–1071.

Fogel, M., Schneider, Rimoldi, F., Ladux, L., Desneux, N. and Ronco, A. (2016) Toxicity assessment of four insecticides with different modes of action on pupae and adults of *Eriopis connexa* (Coleoptera:

Coccinellidae), a relevant predator of the Neotropical Region. *Environmental Science and Pollution Research* 23, 14918–14926.

Francesena, N., Haramboure, M., Smagghe, G., Stadler, T. and Schneider, M.I. (2012) Preliminary studies of effectiveness and selectivity of Movento on *Bemisia tabaci* and its parasitoid *Eretmocerus mundus*. *Communications in Agricultural and Applied Biological Sciences* 77, 727–733.

Francesena, N., Desneux, N., Campos, M.R. and Schneider, M.I. (2017) Side effects of spirotetramat on pupae and adults of a Neotropical strain of *Eretmocerus mundus* (Hymenoptera: Aphelinidae): Effects on the life parameters and demography. *Environmental Science and Pollution Research* 24, 17719–17730.

GAR (2019a) Data estimations. Government of the Argentine Republic. Available at: http://datosestimaciones.magyp.gob.ar/reportes.php?reporte=Estimaciones (accessed 7 July 2019).

GAR (2019b) Agro-industries. Government of the Argentine Republic. Available at: https://www.agroindustria.gob.ar/sitio/areas/ss_mercados_agropecuarios/areas/frutas/_archivos/000030_Informes/000042_Anuario%20de%20Frutas%20-%202018.pdf (accessed 7 July 2019).

Garrido, S., Cichón, L. and Fernández, D. (2007) Control biológico de carpocapsa [Biological control of carocapsa]. *Fruticultura & Diversificación* 54, 27–33.

Greco, N.M., Liljesthröm, G.G., Gugole Ottaviano, M.F., Cluigt, N., Cingolani, M.F., Zembo, J.C. and Sánchez, N.E. (2011) Pest management plan for *Tetranychus urticae* based on the natural occurrence of *Neoseiulus californicus* (Acari: Tetranychidae, Phytoseiidae) in strawberry. *International Journal of Pest Management* 57, 299–308.

Grez, A.A., Zaviezo, T., Roy, H.E., Brown, P.M.J. and Bizama, G. (2016) Rapid spread of *Harmonia axyridis* in Chile and its effects on local coccinellid biodiversity. *Diversity and Distributions* 22, 982–994.

Gugole Ottaviano, M.F., Cédola, C.V., Sánchez, N.E. and Greco, N.M. (2015) Conservation biological control in strawberry: effect of different pollen on development, survival, and reproduction of *Neoseiulus californicus* (Acari: Phytoseiidae). *Experimental and Applied Acarology* 67, 507–521.

Gutierrez, A.C., Tornesello-Galván, J., Manfrino, R.G., Hipperdinger, M., Falvo, M., D'Alessandro, C. and López Lastra, C.C. (2017) Organización y conservación de la colección de hongos patógenos y simbiontes de insectos y otros artrópodos del CEPAVE (CONICET-UNLP), La Plata, Argentina [Organization and conservation of pathogenic and symbionts fungi]. *Revista Argentina de Microbiología* 49, 183–188.

Haase, S., Sciocco-Cap, A. and Romanowski, V. (2015) Baculovirus insecticides in Latin America: historical overview, current status and future perspectives. *Viruses* 7, 2230–2267.

Haramboure, M., Mirande, L., Smagghe, G., Pineda, S. and Schneider, M.I. (2010) Compatibility of a *Melia azedarach* extract with *Eriopis connexa* (Coleoptera: Coccinellidae). *Communications in Agricultural and Applied Biological Sciences* 75, 373–378.

Haramboure, M., Francesena, N., Reboredo, G.R., Smagghe, G., Alzogaray, R.A. and Schneider, M.I. (2013) Toxicity of Cypermethrin on the Neotropical lacewing *Chrysoperla externa* (Neuroptera: Chrysopidae): behavior disruption and recovery capacity. *Communications in Agricultural and Applied Biological Sciences* 72, 339–344.

INTA (1991) Juicio a Nuestra Agricultura: hacia un desarrollo sostenible [Our Agriculture in Question: towards sustainable development]. Editorial Hemisferio Sur, Buenos Aires, Argentina.

INTA (2019a) Serie tecnica manejo integrado de plagas forestales [Integrated management of forest pests] Available at: https://inta.gob.ar/documentos/serie-tecnica-manejo-integrado-de-plagas-forestales (accessed 27 July 2019).

INTA (2019b) Control biológica de Carpocapsa [Biological control of codling moth]. Available at: https://inta.gob.ar/sites/default/files/script-tmp-fyd54_cbiol.pdf (accessed 27 July 2019).

Julien, M.H. and Griffiths, M.W. (1998) *Biological Control of Weeds. A World Catalogue of Agents and their Target Weeds*, 4th edn. CAB International, Wallingford, UK.

Lange, C.E. and Cigliano, M.M. (2005) Overview and perspectives on the introduction and establishment of the grasshopper biocontrol agent *Paranosema locustae* (Microsporidia) in the western Pampas of Argentina. *Vedalia* 12, 61–84.

Lefebvre, M., Reguilón, C. and Kirschbaum, D. (2013) Evaluación del efecto de la liberación de *Orius insidiosus* (Hemiptera:Anthocoridae), como agente de control biológico de trips en el cultivo de frutilla [Evaluation of *Orius insidiosus* release, biological control agent of thrips in strawberry crop]. *RIA* 39, 273–280.

Leguizamón, E.S. (2018) *Historia de la Horticultura* (*History of Horticulture*), 1st edn. Ediciones INTA, Ciudad Autónoma de Buenos Aires.

López Lastra, C.C. and Lecuona, R.E. (2019) *Micopatología de artrópodos: hongos entomopatógenos para ser usados como bioinsumos en el control microbiano de plagas* [*Mycopathology of Arthropods: Entomopathogenic fungi as a tactic for use in microbial control of pests*]. Ediciones INTA, Ciudad Autónoma de Buenos Aires, Argentina.

Nussenbaum, A.L. and Lecuona, R.E. (2012) Selection of *Beauveria bassiana* sensu lato and *Metarhizium anisopliae* sensu lato isolates as microbial control agents against the boll weevil (*Anthonomus grandis*) in Argentina. *Journal of Invertebrate Pathology* 110, 1–7.

Ovruski, S.M., Colin, C., Soria, A., Oroño, L.E. and Schliserman, P. (2003) Introducción y producción en laboratorio de *Diachasmimorpha tryoni* y *Diachasmimorpha longicaudata* (Hymenoptera: Braconidae) para el control biológico de *Ceratitis capitata* (Diptera: Tephritidae) en la Argentina [Importation and experimental mass rearing of *Diachasmimorpha tryoni* and *Diachasmimorpha longicaudata* for biocontrol of *Ceratitis capitata* in Argentina]. *Revista de la Sociedad Entomológica Argentina* 62, 49–59.

Querino, R.B. and Zucchi, R.A. (2003) Caracterização morfológica de dez espécies de *Trichogramma* (Hymenoptera: Trichogrammatidae) registradas na América do Sul. [Morphological characterization of ten species of *Trichogramma* registered in South America]. *Neotropical Entomology* 32, 597–613.

Riquelme Virgala, M.B. and Botto, E.N. (2010) Estudios Biológicos de *Trichogrammatoidea bactrae* Nagaraja (Hymenoptera: Trichogrammatidae), parasitoide de huevos de *Tuta absoluta* Meyrick (Lepidoptera: Gelechiidae) [Biological studies of *Trichogrammatoidea bactrae*, egg parasitoid of *Tuta absoluta*]. *Neotropical Entomology* 39, 612–617.

Saini, E.D. (2004) Presencia de *Harmonia axyridis* (pallas) (Coleoptera: Coccinelidae) en la provincia de buenos aires. Aspectos biológicos y morfológicos. [Presence of *Harmonia axyridis* in Buenos Aires. Biological and morphological aspects]. *RIA* 33, 151–160.

Salas Gervassio, N., Luna, M.G., Lee, S., Salvo, A. and Sánchez, N. (2016) Trophic web associated with the South American tomato moth *Tuta absoluta*: implications for its conservation biological control in Argentina. *Agricultural and Forest Entomology* 18, 137–144.

Salto, C.E., Imwinkelried, J.M., Bertolaccini, I. and Botto, E.N. (1990) Introducción y liberación en Argentina de *Coccinella septenpunctata* L. (Coleoptera: Coccinellidae). [Introduction and release of *Coccinella septenpunctata* in Argentina]. Información para extensión N° 114. INTA Estación Experimental Agropecuaria Rafaela.

Sands, D.P.A. and Coombs, M. (1999) Evaluation of the Argentinian parasitoid, *Trichopoda giacomellii* (Diptera: Tachinidae), for biological control of *Nezara viridula* (Hemiptera: Pentatomidae) in Australia. *Biological Control* 15, 19–24.

Schliserman, P., Ovruski, S. and De Coll, O.R. (2003) The establishment of *Diachasmimorpha longicaudata* (Hymenoptera: Braconidae) in Misiones, northeastern Argentina. *Florida Entomologist* 86, 491–492.

SENASA (2017a) Listado de agentes de control biológico y organismos benéficos evaluados. [List of biological control agents and beneficial organisms]. Available at: http://www.senasa.gob.ar/cadena-vegetal/forrajes/produccion-primaria/control-biologico/listado-de-agentes-evaluados (accessed 27 July 2017).

SENASA (2017b) Registro Nacional de Terapéutica Vegetal [National Registry of Plant Therapeutics]. Available at: https://www.argentina.gob.ar/senasa/programas-sanitarios/productos-veterinarios-fitosanitarios-y-fertilizantes/registro-nacional-de-terapeutica-vegetal (accessed 28 October 2019).

SENASA (2019a) Bosques argentinos, actividad forestal y economías regionales [Argentine forests, forestry activity and regional economies]. Available at: http://www.senasa.gob.ar/senasa-comunica/noticias/bosques-argentinos-actividad-forestal-y-economias-regionales (accessed 27 July 2019).

SENASA (2019b) Hortalizas [Vegetables]. Available at: https://www.argentina.gob.ar/senasa/programas-sanitarios/cadenavegetal/hortalizas (accessed 28 October 2019).

SENASA (2019c) Mosca de los frutos [Fruit fly]. Available at: http://www.senasa.gob.ar/cadena-vegetal/frutales/produccion-primaria/programas-fitosanitarios/mosca-de-los-frutos-0) (accessed 27 July 2019).

Sosa-Gómez, D.R., López Lastra, C.C. and Humber, R.A. (2010) An overview of arthropod-associated fungi from Argentina and Brazil. *Mycopathologia* 170, 61–76.

Suárez, L., Bilbao, M., Murúa, F., Carta Gaena, S., Sánchez, G., Pantano, V., Molina, D., Laria, O. and Ovruski, S. (2018) Producción masiva del parasitoide exótico *Diachasmimorpha longicaudata* (Hymenoptera: Braconidae) para el control biológico aumentativo de *Ceratitis capitata* (Diptera: Tephritidae) en la Provincia de San Juan. [Mass production of the exotic parasitoid *Diachasmimorpha longicaudata* for the biological control of *Ceratitis capitata* in San Juan]. *Acta zoológica lilloana* 62, 139–141.

Tortosa, O.E., Carmona, A., Martínez, E., Manzano, P. and Giardina, M. (2014) Release and establishment of *Mastrus ridens* (Hymenoptera: Ichneumonidae) for the control of *Cydia pomonella* (Lepidoptera: Tortricidae) in Mendoza, Argentina [in Spanish]. *Revista de la Sociedad Entomológica Argentina* 73, 109–118.

Turica, A. (1968) Lucha biológica como medio de control de las moscas de los frutos. [Biological control of fruit fly]. *IDIA* 241, 29–38.

Turienzo, P. (2006) First record in Argentina of a plant-insect association, with interest in biological control. *Ecología Austral* 16, 95–98.

Villanova, I., Brieva, S. and Ceveiro, R. (2013) Producción y comercialización de flores de corte en el AMBA [Production and commercialization of cut flowers in the AMBA]. Estudios Socioeconómicos de los Sistemas Agroalimentarios y Agroindustriales N° 13. Available at: https://inta.gob.ar/sites/default/files/script-tmp-inta_produccin_y_comercializacin_de_flores_de_corte_e.pdf (accessed 27 July 2019).

Zapater, M.C. (1996) El control biológico en América Latina [Biological control in Latin America]. OICB SRNT, Buenos Aires, Argentina.

Zumoffen, L., Tavella, J., Signorini, M. and Salvo, A. (2015) Laboratory and field studies to evaluate the potential of an open rearing system of *Lysiphlebus testaceipes* for the control of *Aphis craccivora* in Argentina. *BioControl* 61, 23–33.

Please see the supplementary file "Addenda and Corrections" for several species added to Tables 2.2 and 2.3.

3 Biological Control in Barbados

Joop C. van Lenteren[1]* and Yelitza C. Colmenarez[2]

[1]Laboratory of Entomology, Wageningen University, Wageningen, The Netherlands; [2]CABI-UNESP-FEPAF, Botucatu, São Paulo, Brazil

* E-mail: joop.vanLenteren@wur.nl

Abstract

Early classical biocontrol successes in Barbados, some in combination with natural control, were the control of: sugarcane borers, sugarcane mealybugs and West Indian cane fly in sugarcane; cottony cushion scale and citrus blackfly in citrus; coconut whitefly in palm; fall armyworm in vegetables and field crops; diamondback moth in cruciferous crops; and green scale and whitefly on fruit and ornamental trees. Recent successes concern classical biocontrol, often in combination with natural control, of: the pink hibiscus mealybug in various crops and ornamentals; sago palm scale on cycads and ornamental palm; and the citrus leaf miner and the Asian citrus psyllid in citrus. Natural control included that of: papaya mealybug in papaya; chilli thrips in various crops; and red palm mite in coconut palm, ornamentals and bananas. Parasitoids were most often used, followed by predators, while microbial agents were rarely used. Barbados has regularly served as provider of natural enemies for other islands in the Caribbean. The island has faced at least 25 arthropod invasions of pests since 2000, stressing the need for biocontrol solutions.

3.1 Introduction

Barbados has an estimated population of slightly more than 290,000 (July 2017) and its main agricultural products are sugarcane, vegetables and cotton (CIA, 2017).

3.2 History of Biological Control in Barbados

The text of this section is a summary of information presented in Cock (1985).

3.2.1 Period 1830–1969

Many classical and a few augmentative and conservation biocontrol activities took place in Barbados during this period. They are summarized for the main crops below and in Table 3.1.

Biological control of pests in sugarcane

Sugarcane was the most important crop in this period and biocontrol was attempted against major pests of this crop with a number of major successes.

HARD-BACK BEETLES. The larvae of hard-back beetles *Phyllophaga* spp., *Clemora smithi* Arr., commonly known as white grubs, feed on roots of sugarcane and other crops. *Bufo marinus* L., the giant toad, was introduced from Guyana in about 1830. It was said to have reduced pest populations, but due to lack of breeding sites for the toad, its population went down. During the 1910s releases of the major parasitoid of *C. smithi*, *Tiphia parallela* Smith, were made but without clear success. New attempts to control the pest were proposed in the 1920s consisting of releases of *T. parallela* as well as planting of boraginaceous shrubs (*Cordia curassavica* (Jacq.) Roem. and Schult.) that provide food for the parasitoids, provision of breeding sites for *B. marinus* and introduction of other parasitoids. In the 1930s, new natural enemies were imported (the parasitoid tiphiids *Myzinum ephippium* (F.) (= *M. xanthonotus* (Rohw.)), *M. haemorrhoidalis* F., and the scoliids *Campsomeris tricincta* F. and *C. trifasciata* (F.)) from Puerto Rico, but none became established. The predator *Ignelater luminosus* (Illiger), also introduced in this period from Puerto Rico, established well but did not control *C. smithi*.

SUGARCANE MOTH BORERS. Sugarcane moth borers, *Diatraea* spp., form the most important pests of sugarcane in the Caribbean region. In Barbados, only *Diatraea saccharalis* (Fabricius) is important; on other islands three other *Diatraea* species may cause problems as well. During the intensive programme against sugarcane stem borers the Centre for Agriculture and Biosciences International (CABI, at that time Commonwealth Institute of Biological Control (CIBC)) ran a substation in Barbados. From 1919 to 1959, inundative releases of *Trichogramma* spp. (supposedly *T. fuentesi* Torres and *T. exiguum* Pinto and Platner) were used for control of the borer. Early observations in the 1910s and 1920s showed that natural parasitism by *T. exiguum* and *Telenomus alecto* Crawford was too low for borer control. In 1929 mass rearing and inundative releases of *Trichogramma* commenced with strains obtained from the USA, Mexico, Antigua, St Lucia and Montserrat. Rearing took

Table 3.1. Overview of major biocontrol activities in Barbados.

Biocontrol agent / exotic (ex), native (na)	Pest / crop	Type of biocontrol[a] / since	Effect / established or not / area (ha) under biocontrol[b]
PERIOD 1830–1999[c]			
Bufo marinus / ex	Hard-back beetles (white grubs) sugarcane	CBC / 1830	Some control, not established
Tiphia parallela / na		CBC / 1910–20	Not established
Myzinum ephippium / ex		CBC / 1930	Not established
M. haemorrhoidalis / ex		CBC / 1930	Not established
Campsomeris tricincta / ex		CBC / 1930	Not established
C. trifasciata, / ex		CBC / 1930	Not established
Ignelater luminosus / ex		CBC / 1930	No control, established
Trichogramma spp. / na	Sugarcane moth borers, sugarcane	NC / 1919–59	Poor control
Telenomus alecto / na		NC / 1919–59	Poor control
Trichogramma spp. / ex		ABC / 1930–59	Poor control
Ipobracon grenadensis / ex		CBC / 1920s	Poor control, temp established
I. puberuloides / ex		CBC / 1920s	Not established
Agathis stigmatera / ex		CBC / 1920s	Poor control, established
Agathis sp. / ex		CBC / 1920s	Not established
Lixophaga diatraeae / ex		CBC / 1930–60	No control, established
Paratheresia claripalpis / ex		CBC/ 1934–1960s	No control, not established
Cotesia flavipes / ex		CBC / 1966	Good control, established / 1,733 ha
Lixophaga diatraeae / ex		CBC / since1930	Good control, established / 1,733 ha
Tetrastichus haitiensis / ex	Sugarcane root borer, sugarcane	CBC / 1931 + 1973	No control, not established
Hololepta quadridentate / ex		CBC / 1950–51	No control, not established
Fidiobia citri / ex		CBC / 1974 + 1976	?, ?
Brachyufens osborni / ex		CBC/ 1976	No control, not established
Plagioprospherysa trinitatis / na	Jumping borer, sugarcane	NC / 1960s	Poor control
Macrocentrus sp. / ex		CBC / 1973	No control, not established
Cryptolaemus montrouzieri / ex	Mealybugs, sugarcane, various other crops	CBC / 1968–69	Partial control, established
Hyperaspis sp. / ex		CBC / 1968–69	No control, ?
Nephus sp. / ex		CBC / 1968–69	No control, ?
Anagyrus saccharicola / ex		CBC / 1970	Control, established / 1,733 ha
Herpestes auropunctatus / ex	Rats, sugarcane	CBC / 1872	Bad side effects, established
Rodolia cardinalis / ex	Cottony cushion scale, citrus	CBC / 1943	Good control, established / part of 243 ha
Anicetus sp. / ex		CBC / 1961	?, ?
Brethesiella abnormicornis / ex		CBC / 1961	?, ?

Continued

Table 3.1. Continued.

Biocontrol agent / exotic (ex), native (na)	Pest / crop	Type of biocontrol[a] / since	Effect / established or not / area (ha) under biocontrol[b]
Eretmocerus serius / ex	Citrus blackfly, citrus	CBC / 1965–65	Good control, established / part of 243 ha
Encarsia opulenta / ex		CBC / 1965–65	Good control, established / part of 243 ha
Pseudoazya trinitatis / ex	Coconut scale, coconut palm	CBC / 1940s	? / established
Cryptognatha nodiceps / ex		CBC / 1940s	? / ?
C. simillima / ex		CBC / 1940s	? / ?
Pentilia insidiosa / ex		CBC / 1940s	? / ?
Zagloba aenipennis / ex		CBC / 1940s	? / ?
Encarsiella noyesi / ex	Coconut whitefly, coconut palm	CBC / 1950–51	Good control, established / 550 ha
Nephaspis amnicola / ex		CBC / 1951	? / established
N. nigra / ex		CBC / 1951	? / established
Trichogramma fasciatum / na	Pink bollworm, cotton	ABC / 1930	Control, area unknown
Bracon greeni / ex		CBC / 1970s	?, ?
Bracon gelechiae / ex		CBC / 1970s	?, ?
Apanteles angeleti / ex		CBC / 1970s	?, ?
Rogas aligarhensis / ex		CBC / 1970s	?, ?
Compsilura concinnata / ex	Armyworms, vegetables and field crops	CBC / 1931–32	No control, not established
Archytas marmoratus / ex		CBC / 1952	?, established
Telenomus remus / ex		CBC / 1968	Control, established / part of 600 ha
Blaesoxipha filipjevi / ex	Locusts, vegetables and field crops	CBC / 1970s	No control, not established
Scelio aegyptiacus / ex		CBC / 1970s	No control, not established
Scelio sp. nr. *serdangensis* / ex		CBC / 1970s	No control, not established
Entomophthora parvispora / ex	Thrips, vegetables and field crops	ABC / 1973–76	No control, not established
Native natural enemies	Agromyzid leaf miners, vegetables	NC	Partial control
Diglyphis minoeus / ex		CBC / 1972–75	No control, not established
D. sp. ?isaea / ex		CBC / 1972–75	No control, not established
Chrysocharis sp. / ex		CBC / 1972–75	No control, not established
Opius sp. / ex		CBC / 1972–75	No control, not established
Synopeas sp. / ex	Tomato flower midge, vegetables	CBC / 1974–75	No control, not established
Cotesia glomeratus / ex	Cabbage butterfly, cruciferous crops	CBC / 1970 & 1981	No control, not established
Diadegma pierisae / ex		CBC / 1970 & 1981	No control, not established
Pteromalus puparum / ex		CBC / 1970 & 1981	No control, established
Compsilura concinnata / ex		CBC / 1981	No control, not established
Trichogramma sp. / na	Loopers, cotton and cruciferous crops etc	NC	Control on cotton, not on cabbage

Continued

Table 3.1. Continued.

Biocontrol agent / exotic (ex), native (na)	Pest / crop	Type of biocontrol[a] / since	Effect / established or not / area (ha) under biocontrol[b]
Cotesia sp. / na		NC	Control on cotton, not on cabbage
Apanteles sp. / na		NC	Control on cotton, not on cabbage
Euplectrus platyhypenae / na		NC	Control on cotton, not on cabbage
Litomastix sp. / na		NC	Control on cotton, not on cabbage
Brachymeria sp. / na		NC	Control on cotton, not on cabbage
Winthemia sp. / na		NC	Control on cotton, not on cabbage
W. sp. ?*pyrrhopyga* / na		NC	Control on cotton, not on cabbage
Litomastix sp. *truncatella* / ex	Loopers, vegetables	CBC / 1975	Partial control, established, part of 600 ha
Trichospilus pupivora / ex		CBC / 1982	?, established
Apanteles sp. / na	Diamondback moth	NC	Insufficient control
Trichogramma sp. / na		NC	Insufficient control
Spilochalcis hirtifemora / na		NC	Insufficient control
Cotestia plutellae / ex		CBC / 1968–1976	Partial control, established / part of 600 ha
Tetrastichus sokolowskii / ex		CBC / 1968–1976	Good control, established / part of 600 ha
Diadromus collaris / ex		CBC / 1968–1976	?, established
Macromalon orientale / ex		CBC / 1968–1976	?, not established
P. xylostella polyhedrosis virus / ex		CBC / 1968–1976	?, ?
Goniozus sp. *punctulaticeps* / ex	Pigeon pea pod borers, field crops	CBC / 1952–1974	No control, not established
Apanteles etiellae / ex		CBC / 1952–1974	No control, not established
Bracon cajani / ex		CBC / 1952–1974	No control, not established
B. thurberiphagae / ex		CBC / 1952–1974	No control, not established
Phanerotoma bennetti / ex		CBC / 1952–1974	No control, not established
Eiphosoma dentator / ex		CBC / 1952–1974	No control, not established
Agathis sp / na	Sweet potato leaf roller, field crops	NC	Insufficient control
Nemorilla sp. / na		NC	Insufficient control
Eiphosoma dentator / ex		CBC / 1974	No control, established
Phanerotoma sp. / ex		CBC / 1974	No control, not established
Trichogrammatoidea bactrae / ex		CBC / 1975	No control, not established
Ardalus scutellatus / ex	Arrow root leaf roller, field crops	CBC / 1982	No control, not established
Dinarmus vagabundus / ex	Legume seed weevils, field crops	CBC / 1974–75	?, ?
D. basalis / ex			
Cephalosporium lecanii / na	Scales, fruit and ornamental trees	NC / 1910	Control / part of 243 ha
Pachylister chinensis / ex	House and stable flies	CBC / 1950	No control, not established
Muscidifurax spp. / ex		ABC / 1969–70	?, ?

Continued

Table 3.1. Continued.

Biocontrol agent / exotic (ex), native (na)	Pest / crop	Type of biocontrol[a] / since	Effect / established or not / area (ha) under biocontrol[b]
Spalangia nigra / ex		ABC / 1969–70	?, ?
Sphegigaster sp. / ex		ABC / 1969–70	?, ?
Pachycrepoideus vindemiae / ex		ABC / 1969–70	?, ?
Melanagromyza cuscutae / ex	Love vine weeds	CBC / 1967–68	No control, not established
Smicronyx roridus / ex		CBC / 1967–68	No control, not established
Athesapeuta cyperi / ex	Nutgrass weed, vegetables and field crops	CBC / 1974–76	No control, not established
Bactra spp. / ex		CBC / 1974–76	No control, not established
PERIOD 2000–NOW[d]			
Predators			
Allograpta exotica	Aphids, various crops	NC	Control, ?
Amblyseius sp.	Red palm mite, various palms	CBC	Control, established
Chrysoperla sp.	Lepidopterans, aphids, mites, thrips etc., various	NC	Control, ?
Chrysoperla externa	Chilli thrips, other thrips, red palm mite, various	NC	Control, ?
Cryptolaemus montrouzieri	Mealybugs, lepidopterans, thrips, aphids, various	NC	Control, ?
Cryptolaemus montrouzieri	Pink hibiscus mealybug, various plants	NC	Control, large areas
Cycloneda sanguinea	Lepidopterans, aphids, mites, thrips, various	NC	Control, ?
Cryptolaemus montrouzieri	Papaya mealybug, papaya	NC	Control, ?
Cycloneda sanguinea		NC	Control, ?
Lestodiplosis sp.		NC	Control, ?
Cybocephalus nipponicus	Sago palm scale, ornamental palms	CBC	Control, established, ?
Franklinothrips vespiformis	Chilli thrips and other thrips, various crops	NC	Control, ?
Haplothrips gowdeyi	Thrips, vars crops	NC	?, ?
Orius insidiosus	Chilli thrips and other thrips, mites, various	NC	Control,?
Telsimia sp.	Red palm mite, various palms	NC	Control, ?
Parasitoids			
Acerophagus papaya	Papaya mealybug, papaya	NC	Control, ?

Continued

Table 3.1. Continued.

Biocontrol agent / exotic (ex), native (na)	Pest / crop	Type of biocontrol[a] / since	Effect / established or not / area (ha) under biocontrol[b]
Ageniaspis citricola	Citrus leaf miner, citrus	CBC	Control, established / part of 243 ha
Anagyrus loecki		NC	Control / part of 243 ha
Anagyrus kamali	Pink hibiscus mealybug, various plants	CBC	Control, established / large areas
Coccobius fulvus	Sago palm scale, ornamental palms	CBC	Control, established, ?
Euplectrus sp.	*Alabama argillacea*, cotton	NC	Control, ?
Tamarixia radiata	Asian citrus psyllid, citrus	CBC	Control, established / part of 243 ha
Trichogramma chilonis	*Helicoverpa/ Heliothis* spp., various crops	NC	?, ?

[a]Type of biocontrol: ABC = augmentative, CBC = classical, ConsBC = conservation biological control, NC = natural control
[b]Area of crop harvested in 2016 according to FAO (2018)
[c]All information based on Cock (1985)
[d]All information based on Colmenarez *et al*. (2014, 2016); and Ian Gibbs and Yelitza Colmenarez, St. Thomas, Barbados, October 2018, personal communication

place on *Sitotroga cerealella* Oliver and cards with parasitized eggs were put into sugarcane fields. Millions of *Trichogramma* were released, but no evidence was found that they gave any control.

Next, during the 1920s, parasitoids (the braconids *Ipobracon grenadensis* Ashmead, *I. puberuloides* Myers, *Agathis stigmatera* (Cresson) and *Agathis* sp.) were imported from Guyana, Argentina and Venezuela, and *C. curassavica* was planted to provide food for the parasitoids. *I. grenadensis* temporarily established and *A. stigmatera* was recovered in 1935.

Also the dipteran parasitoid *Lixophaga diatraeae* (Townsend) was released in large numbers on several occasions during the period 1930–1960, with populations obtained from Antigua, Cuba, Dominican Republic and Jamaica. Recoveries were made, but *L. diatraeae* did not control *D. saccharalis* on that occasion. Attempts to establish *Paratheresia claripalpis* Van der Wulp from 1934 to the mid-1960s failed.

Due to the poor success obtained with the above-mentioned natural enemy introductions, the Asian parasitoid *Cotesia flavipes* (Cameron) was imported from India in 1966. The species was recovered in 1967 and a mass rearing was initiated for releases in 1968 and 1969, resulting in island-wide establishment. During this period, *L. diatraeae* became more abundant and the two parasitoids have produced continuous sugarcane borer control.

SUGARCANE ROOT BORER. Sugarcane root borer *Diaprepes abbreviatus* (L.) had no native parasitoids in Barbados. Therefore, the egg parasitoid *Tetrastichus haitiensis* Gahan was imported from Haiti and Puerto Rico in 1931 and was mass reared and released for several years, but without success. In 1950–1951, the beetle *Hololepta quadridentata* (Olivier) was shipped several times from Trinidad and Tobago and released, but did not recover. In 1973, *T. haitiensis* was introduced again, this time from Montserrat, but did not establish. In 1974 and 1976, the egg parasitoid *Fidiobia citri* (Nixon) was introduced from Jamaica. In 1976, the egg parasitoid *Brachyufens osborni* (Dozier) was imported from Florida, USA, but did not establish.

SUGARCANE MEALYBUGS. Of the two sugarcane mealybugs *Saccharicoccus sacchari* (Ckll.) and *Dysmicoccus boninsis* (Kuwana), generally found wherever sugarcane is grown, only *S. sacchari* is causing temporary problems. Coccinellid predators (*Cryptolaemus montrouzieri* Mulsant, *Hyperaspis* sp. and *Nephus* sp.) were imported from India in 1968–1969 and released, but information about establishment is not available. In 1970 the parasitoid *Anagyrus saccharicola* Timb. was imported from East Africa, reared and released. It established, became widespread and is believed to have reduced mealybug populations.

JUMPING BORER. Jumping borer *Elasmopalpus lignosellus* Zell. became a problem with the introduction of pre-harvest burning at the end of the 1960s. The parasitoid *Plagioprospherysa trinitatis* Thomps. was imported from Trinidad and Tobago in 1973, but appeared to be already present in Barbados. Its effect on jumping borer populations is not known. In the same period, several shipments of *Macrocentrus* sp. were received from Trinidad and Tobago but it has not been recovered.

WEST INDIAN CANE FLY. The West Indian cane fly *Saccharosydne saccharivora* (Westw.) is under natural control in Barbados.

RATS. The small Indian mongoose *Herpestes auropunctatus* (Hodgson) was introduced to Barbados from Jamaica somewhere after 1872 for control of rats. With the exception of Jamaica, import and releases of the mongoose on many of the Caribbean island is now considered a serious mistake. Insectivorous lizards (*Ameiva* spp.) have become rare or extinct; also, mongooses are pests of poultry.

Biological control of pests in citrus

COTTONY CUSHION SCALE. Cottony cushion scale *Icerya purchasi* Maskell was found on the island in 1938 and developed to pest status by 1941. Not only citrus, but also pigeon pea, casuarina and other garden plants were attacked. Through CIBC (now CABI), *Rodolia cardinalis* (Mulsant) was introduced in 1943, which established and successfully controlled the scale. Surveys in the 1960s showed that the pest was still under good biological control by *R. cardinalis*. In 1961 the encyrtid parasitoids *Anicetus* sp. and *Brethesiella abnormicornis* (Gir.) were sent from Trinidad and Tobago to Barbados, but results of the introduction are unknown.

CITRUS BLACKFLY. Citrus blackfly *Aleurocanthus woglumi* Ashby was first found on the island in 1964. Apparently the citrus blackfly had already been present a few years, because dead and weakened trees were found where blackfly was present. *Eretmocerus serius* Silvestri was obtained in 1964 from Jamaica and *Encarsia opulenta* (Silvestri) in 1964–1965 from Mexico. The parasitoids were released, established, spread rapidly and successfully controlled the pest within a year. Later studies showed that *E. opulenta* had replaced *E. serius*. Complete biocontrol of blackfly with the parasitoids prevented expensive chemical control.

Biological control of pests in coconut palm

COCONUT SCALE. After outbreaks of coconut scale *Aspidiotus destructor* Sign. at the end of the 1940s, coccinellid predators, including *Pseudoazya trinitatis* (Marshall), *Cryptognatha nodiceps* Marshall, *C. simillima* Sic, *Pentilia insidiosa* Mulsant and *Zagloba aenipennis* (Sicard), were introduced from Trinidad and Tobago. In 1954 *P. trinitatis*, *Prodilis* sp. and *Scymnus* sp. were found in Barbados.

COCONUT WHITEFLY. Coconut whitefly *Aleurodicus cocois* (Curt.) is a pest of coconut and ornamental plants in Barbados. Large numbers of the parasitoid *Encarsiella noyesi* Hayat were introduced from Trinidad in 1950 and 1951; the parasitoid rapidly established and provided excellent control of the whitefly. Also several scymnine coccinellids, including *Nephaspis amnicola* Wingo and *N. nigra* Gordon, were introduced in 1950 from Trinidad and these coccinellids were reported to have established in 1951.

Biological control of pink bollworm in cotton

Pink bollworm *Pectinophora gossypiella* (Saund.), native to India, was found in Barbados in 1920 and developed into a major pest. In the 1930s, the pest became less of a problem, which was thought to be the result of mass releases of *Trichogramma fasciatum* (Perkins) (= *minutum* Auct.) made against *D. saccharalis* (F.) in sugarcane, because this egg parasitoid also attacks *P. gossypiella*.

Biological control of armyworms on vegetables and field crops

The following species of armyworms occur in the region: *Spodoptera frugiperda* (J.E. Smith), *S. latifascia* (Wlk.), *S. dolichos* (F.), *S. eridania* (Cram.), *S. exigua* (Hb.), *S. sunia* (Gn.), *Helicoverpa zea* (Boddie) and *H. virescens* (F.). *Compsilura concinnata* (Meigen), a European tachinid which has a wide host range, was introduced from Massachusetts into Barbados in 1931–1932 to control lepidopterous pests of crops including sweet potato, maize, cotton and cover crops, but did not establish. In 1952, tachinids, including *Archytas marmoratus* Tns. have been imported from Trinidad and Tobago; *A. marmoratus* was found in Barbados in surveys from 1969 onwards. Since 1968, a large number of natural enemies have been imported into Barbados from Pakistan and Trinidad. Of these, *Telenomus remus* Nixon became established and rates of parasitism of more than 80% were observed on several crops; the parasitoid substantially reduced *Spodoptera* populations.

Biological control of green scale and whitefly on fruit and ornamental trees

Green scale *Coccus viridis* (Green) is a pest of several fruit and ornamental trees. In 1910, the fungus *Cephalosporium* (= *Lecanicillium*) *lecanii* Zimm. was found attacking *Saissetia nigra* (Nietn.) on hibiscus. Branches with fungus-infested scales were attached to mango, cherry and ornamental trees with *C. viridis*, *C. mangiferae* (Green), *Pulvinaria pyriformis* (Ckll.), *Vinsonia stellifera* (Westw.) and other scales. *Coccus* spp. and *P. pyriformis* were being killed by the fungus and became rare. Another fungus that resembled the 'cinnamon fungus in Florida' was found on *C. viridis* on coffee. Spores of this fungus were then sprayed on *C. viridis* and a whitefly on *Ipomea* sp. and seemed to control both effectively.

Biological control of house and stable flies

Attempts to control houseflies (*Musca* spp.) and stable flies (*Stomoxys* spp.) started in 1950 by importing the predator *Pachylister chinensis* Quensel from Trinidad and Tobago, but it was not recovered after release. In 1969–1970 many individuals of *Muscidifurax* spp., *Spalangia nigra* Latreille and *Sphegigaster* sp., originating from California and reared in Trinidad, were released in Barbados, as well as the Trinidad species *Pachycrepoideus vindemiae* Rond. Results of the releases are not available.

Biological control of love vine weeds

The semi-parasitic love vines *Cuscuta americana* L. and *C. indecora* Choisy were kept under good control by legally enforced measures for many years, but once vigilance was relaxed the vines spread rapidly. In 1967–1968 an agromyzid *Melanagromyza cuscutae* Hering and a seed-feeding weevil *Smicronyx roridus* Mshl. were imported from Pakistan, reared and released, but did not establish. In 1971, *Smicronyx rufovittatus* Anderson was imported from Pakistan, but there is no information whether it established.

Barbados as provider of natural enemies

During the period up to 1969, Barbados provided several natural enemies to other islands in the region. The giant toad *B. marinus* was moved to most of the Caribbean islands, as well as to Bermuda, after 1830. *Rodolia cardinalis* was sent on several occasions to other islands after its establishment in Barbados in the 1940s. *Cotesia flavipes* and *Lixophaga diatraeae*, which had been shown to reduce sugarcane borers effectively in the 1960s in Barbados, have been exported to other islands in the region and established. Barbados provided the parasitoid *A. saccharicola*, a natural enemy of sugarcane mealybugs, to islands in the region in the 1970s, where it also became established. *Telenomus remus*, an effective parasitoid of *Spodoptera* species in Barbados since 1968, has been sent to other islands.

3.2.2 Period 1970–2000

Biological control of pests in sugarcane

Cotesia flavipes, imported and released at the end of the 1960s (see above), and the native *L. diatraeae* have resulted in continuous control of sugarcane borer *D. saccharalis*. In 1970 the parasitoid *A. saccharicola* was imported from East Africa; it was reared and released, established and became widespread and is believed to have reduced *S. sacchari* mealybug populations. The West Indian cane fly *S. saccharivora* is under natural control in Barbados.

Biological control of pests in citrus

Since the introduction of parasitoids in 1964 (see above), citrus blackfly has been under effective control by *E. opulenta* in Barbados. Cottony cushion scale *I. purckasi* has been under effective classical biocontrol in Barbados since the introduction of *R. cardinalis* in 1943 (see above) (Cock, 1985).

Biological control of coconut whitefly

The coconut whitefly *A. cocois* has been under effective biocontrol in Barbados since the import and release of *E. noyesi* in 1951 (see above) (Cock, 1985).

Biological control of pink bollworm in cotton

In 1974, 1975 and 1976 four species of parasitoids (*Bracon greeni* Ashmead, *Bracon gelechiae* Ashmead, *Apanteles angeleti* Mues., *Rogas aligarhensis* (Quadri)) and the predator *Coranus spiniscutus* Reuter were introduced from Asia to Barbados. However, it seems they did not become established as no recoveries have been made since their release in the field. Ingram (1980) reported the parasitoid *Perisierola nigrifemur* Ashmead attacking *P. gossypiella* in Barbados. This parasitoid is very common in pink bollworm larvae towards the end of the season, but this timing is considered to be too late to be of economic value. Another biocontrol agent reported attacking diapausing larvae of the pink bollworm was the predatory mite *Pyemotes ventricosus* (Newport). This is very common at the end of the cotton crop and during the close season. Ingram (1980) supposed that this predatory mite exerts considerable control of long-cycle larvae at the field level.

Biological control of pests in vegetables and other field crops

ARMYWORMS. Since 1968, large numbers of natural enemies have been imported into Barbados from Pakistan and from Trinidad and Tobago. Of these, *T. remus* became established and substantially reduced *Spodoptera* populations on several crops

LOCUSTS AND GRASSHOPPERS. *Schistocerca pallens* (Thnb.) attacks crops as well as grasslands. In the early 1970s attempts to establish *Blaesoxipha filipjevi* Rhod. from East Africa and the scelionids *Scelio aegyptiacus* Priesne and *Scelio* sp. nr. *serdangensis* (Timb.) from Pakistan failed.

THRIPS. *Thrips tabaci* Lind. is a serious pest in Barbados. The fungus *Entomophthora parvispora* MacLeod & Carl was imported from Switzerland on several occasions during 1973–1976 and infested thrips were released in the field, but not recovered.

TOMATO FLOWER MIDGE. Larvae of the tomato flower midge *Contarinia lycopersici* Felt cause wilting, flower shed and distorted fruit in tomato. The parasitoid *Synopeas* sp. was imported from Trinidad and Tobago in 1974–1975 and released but did not establish.

AGROMYZID LEAF MINERS. Agromyzid leaf miners *Liriomyza sativae* (Blanch.) and *L. trifolii* (Burgess) are pests of various vegetables. Native natural enemies do reduce leaf-miner populations, but for better control the parasitoids *Diglyphus minoeus* (Wlk.) and a *Diglyphus* sp. similar to *D. isaea* (Wlk.), along with *Chrysocharis* sp. and *Opius* sp., were imported from Pakistan from 1972 to 1975 and released, but no recoveries were reported.

PIGEON PEAS POD BORERS. The pigeon peas pod borers *Fundella pellucens* Zell. (= *cistipennis* Dyar) and to a lesser extent *Ancylostoma stercorea* (Zell.) cause problems in Barbados. Parasitoids of *A. stercorea* (*Goniozus* sp. *punctulaticeps* group, *Apanteles etiellae* Vier., *Bracon cajani* Mues., *B. thurberiphagae* (Mues.), *Phanerotoma bennetti* Mues. and *Eiphosoma dentator* (F.)) were introduced from Trinidad between 1952 and 1974, but no recoveries were reported from pigeon peas.

SWEET POTATO LEAF ROLLER. Sweet potato leaf roller *Syllepte helcitalis* (Wlk.) is a minor pest in Barbados. The native parasitoids *Agathis* sp. and *Nemorilla* sp. cause only 5% parasitism. The parasitoids *E. dentator* (from Trinidad and Tobago) and *Phanerotoma* sp. (1974) and *Trichogrammatoidea bactrae* Nagaraja (1975), both from India, were introduced. Only *E. dentator* was recovered, but parasitism of the leaf roller was very low.

ARROWROOT LEAF ROLLER. Arrowroot leaf roller *Calpodes ethlius* (Stoll) is the principal pest of arrowroot (*Maranta arundinacea* L.). The parasitoid *Ardalus scutellatus* (How.) was introduced from St Vincent in 1982, but no results of the introduction were reported.

LEGUME SEED WEEVILS. Legume seed weevils *Callosobruchus chinensis* (L.) and *C. maculatus* (F.) attack black-eye peas and the parasitoids *Dinarmus vagabundus* (Timb.) and *D. basalis* (Rond.) were introduced from India in 1974 and 1975 to control these pests. No results of these releases have been published.

Biological control of pests on fruit and ornamental trees

GREEN SCALES. Green scale and several other scale species that are pests of fruit and ornamental trees have been controlled by a native fungus, *C. lecanii*, since the 1910s (see above). In 1972, the parasitoid *Adelencyrtus moderatus* (How.), or *A. odonaspidis* Fullaway, of the yam scale *Aspidiella hartii* (Ckll.) was imported from Trinidad for control trials on various scale species, though the parasitoid was known to be present already. Results of the trials are not known.

ORTHEZIA SCALES. *Orthezia insignis* Browne and/or *O. praelonga* (Dgl.) cause problems in citrus, coffee, croton and other ornamentals. In 1976 and 1977, coccinellid predators (*Hyperaspis distinguenda* (Muls.), *H. donzeli* (Muls.) and *H. jucunda* (Muls.) were imported from Trinidad, but no recoveries have been reported.

Biological control of pests of cruciferous crops

CABBAGE BUTTERFLY. The cabbage butterfly *Ascia monuste* (L.) causes problems on crucifers in Barbados. In 1970–1971, the parasitoids *Apanteles glomeratus* (L.), *Diadegma pierisae* (Rao) and *Pteromalus puparum* (L.) were imported from Pakistan, but did not establish. In 1981, the same three species together with the tachinid *C. concinnata* were imported from Pakistan and some pupae parasitized by *P. puparum* were recovered.

CABBAGE LOOPERS. The cabbage loopers *Trichoplusia ni* (Hb.) and *Pseudoplusia* (= *Chrysodeixis*) *includens* (Wlk.) feed on a wide range of host plants, including legumes, crucifers, cotton, okra and solanaceous crops. They are parasitized by a number of native natural enemies, including: *Trichogramma* sp.; *Apanteles (Cotesia)* sp. poss. *marginiventris* (Cress.); *Apanteles (Glyptapanteles)* sp.; *Euplectrus platyhypenae* How.; *Litomastix* sp. nr. *truncatella* (Dalm.); *Brachymeria* sp.; *Winthemia* sp. nr. *pinguis* F.; and a *Winthemia* sp. similar to *W. pyrrhopyga* (Wied.). On cotton they are normally kept in check by their natural enemies, but on cabbage they are serious pests. *Litomastix* sp. *truncatella* group was imported from India in 1975, reared and released. The species established in many crops and levels of parasitism rose from 5% before the introduction to 25% (on cabbage) and 79% (on tomato) in 1982. In 1981, *Trichospilus pupivora* (= *pupivorus*) Ferriere, introduced to control *Spodoptera* spp., was found to attack *P. includens* in Barbados.

DIAMONDBACK MOTH. The diamondback moth *Plutella xylostella* (L.) became a problem in the Caribbean during the 1950s and1960s. Native parasitoids (*Apanteles* sp., *Trichogramma* sp. and the hyperparasitic chalcidid *Spilochalcis hirtifemora* (Ashmead.)) attack the moth, but insufficiently to be able to control it. Between 1968 and 1976, the parasitoids *Cotesia* (*Apanteles*) *plutellae* (Kurdjumov), *Tetrastichus sokolowskii* Kurd., *Diadromus collaris* (Grav.) and *Macromalon orientale* Kerrich and a sample of a *P. xylostella* polyhedrosis virus were shipped to Barbados from India. Recoveries of *C. plutellae* and *D. collaris* were reported. Further releases of *C. plutellae*, reared in Trinidad and Tobago and in Barbados, were made between 1968 and 1973. Surveys showed that *C. plutellae* obtained up to 52% parasitism. Additional releases of *T. sokolowskii* reared from stocks obtained from Montserrat were made in 1973. *T. sokolowskii* established and resulted in levels of parasitism of 68–100% in 1976.

Biological control of nutgrass weed

Nutgrass *Cyperus rotundus* L., native to Pakistan and India, is now a worldwide pest and occurs in crops including sugarcane, cotton and vegetables in the Caribbean. The weevil *Athesapeuta cyperi* Mshl. and two species of the tortricid genus *Bactra* were imported in 1974–1976 from Pakistan, but did not establish.

Barbados as provider of natural enemies

Barbados provided *C. plutellae* during this period to several other islands in the region

3.3 Current Situation of Biological Control in Barbados

3.3.1 Classical biological control of pink hibiscus mealybug

The pink hibiscus mealybug *Maconellicoccus hirsutus* (Green) was introduced into Barbados in 2000 and attacks different ornamentals and agricultural crops. It is a very prolific pest that causes severe distortion of leaves, new shoots and fruits. Initial use of chemical and cultural control was ineffective. Two natural enemies of the mealybug were tested for classical biocontrol: *C. montrouzieri*, which was sourced and released locally, and *Anagyrus kamali* Moursi, which was brought from Grenada and Trinidad and Tobago through CABI. Use of these natural enemies resulted in effective biocontrol of the pest (Ian Gibbs, St Thomas, Barbados, October 2018, personal communication).

3.3.2 Natural biological control of the papaya mealybug

The papaya mealybug *Paracoccus marginatus* Williams and Granara de Willink was introduced into Barbados in 2000. High population densities of the mealybug cause deformation of new growth, leaf yellowing, leaf curl and early fall of fruits. With the objective of establishing a biocontrol programme of this pest, the Ministry of Agriculture of Barbados and the State University Paulista (UNESP-FCA, Jaboticabal, Brazil) determined the complex of natural enemies of the pest in Barbados. The most efficient parasitoids found were *Acerophagus papaya* Noyes & Schauff and *Anagyrus loecki* Noyes & Menezes, and the most important predators were *Lestodiplosis* sp., *Cycloneda sanguinea* Linnaeus and *C. montrouzieri*. The pest was successfully controlled using these indigenous natural enemies. Currently the above-mentioned parasitoids and predators are commonly found attacking the pest in the field (Ian Gibbs and Yelitza Colmenarez, Botocatu, Brazil, October 2018, personal communication).

3.3.3 Classical biological control of the sago palm scale

The sago palm scale *Aulacaspis yasumatsui* Takagi was introduced into Barbados in 2003, infesting cycads and other ornamental palms. Initial damage appears as chlorotic spots. Highly infested cycads are heavily coated with a white crust that includes scales of live and dead insects. Biocontrol had earlier been used successfully to manage the sago palm scale (Cave, 2006). The Ministry of Agriculture of Barbados collaborated with R. Cave from the University of Florida and imported the parasitoid *Coccobius fulvus* (Compere and Annecke) and the predatory beetle *Cybocephalus nipponicus* Endrödy-Younga. Both species were reared at the field level at different locations on the island, resulting in successful biocontrol. Currently, these natural enemies are collected from fields where biocontrol works well and are then released in new areas affected by the pest (Ian Gibbs, St Thomas, Barbados, October 2018, personal communication).

3.3.4 Classical biological control of the citrus leaf miner

The citrus leaf miner *Phyllocnistis citrella* Stainton was introduced into Barbados in 2000. In collaboration with the University of Florida, *Ageniaspis citricola* Logvinovskaya was introduced, but the initial introduction did not result in sufficient control of the pest, as the citrus plantations were not pruned and the parasitoids had difficulty in finding the preferred early larval stages. When parasitoids were released after pruning, excellent biocontrol was obtained.

An indigenous parasitoid, a species of the genus *Cirrospilus*, was reported to attack the citrus leaf miner (Ian Gibbs and Yelitza Colmenarez, Botocatu, Brazil, October 2018, personal communication).

3.3.5 Classical biological control of Asian citrus psyllid

The Asian citrus psyllid *Diaphorina citri* Kuwayama, native to southern Asia, is a vector of the most serious citrus disease worldwide, the bacterium *Candidatus liberibacter*, commonly referred to as citrus greening or huanglongbing (HLB) (literally, yellow dragon disease). According to a report by FAO (2013), the psyllid is present in Barbados, as well as the disease causing *Candidatus liberibacter asiaticus*, but not *Candidatus liberibacter americanus*. The report mentioned a number of management methods, including biocontrol of the psyllid with the parasitoid *Tamarixia radiata* (Waterston). Currently *T. radiata* is mass reared at field level in collaboration with the University of Florida and CABI. This methodology allowed the establishment of the Asian citrus psyllid/HLB biocontrol programme, despite the limited laboratory infrastructure on the island. Parasitoid mass rearing and releases strongly reduced psyllid populations and avoided the presence of the HLB disease on the island for many years (Ian Gibbs, St Thomas, Barbados, October 2018, personal communication)

3.3.6 Natural biological control of the chilli thrips

The chilli thrips *Scirtothrips dorsalis* Hood was introduced into Barbados in 2005 and attacks different crops. It has a great reproductive potential together with the ability to adapt easily to new areas. In Barbados, different predators attack the chilli thrips, e.g. *Franklinothrips vespiformis* Crawford, *Orius insidiosus* (Say) and *Chrysoperla externa* (Hagen) (Ian Gibbs, St Thomas, Barbados, and Yelitza Colmenarez, Botocatu, Brazil, October 2018, personal communications).

3.3.7 Natural biological control of the red palm mite

The red palm mite *Raoiella indica* Hirst was introduced into Barbados in 2010. This prolific invasive species attacks different hosts, including *Cocos nucifera* (L.), ornamentals and *Musa* sp. Colmenarez *et al.* (2014) studied the population trends of this species on different Caribbean islands and found entomopathogenic fungi, of which three isolates of the genus *Simplicillium* are most interesting as they might have potential for biocontrol of the red palm mite in the Caribbean. In Barbados the pest was also found to be attacked by a predatory mite belonging to the genus *Amblyseius* and by other predators, e.g. the coccinellid *Telsimia* sp. and neuropteran *C. externa* (Ian Gibbs, St. Thomas, Barbados, October 2018, personal communication).

3.3.8 Natural enemies of cotton pests

Colmenarez *et al.* (2016) studied natural enemies of pests of 'West Indian Sea Island Cotton' (*Gossypium barbadense* L.) by weekly monitoring of the crop during two production seasons (2009–2011). Seven species of predators and two parasitoids were found. However, cotton farmers frequently apply chemical control and often confuse natural enemies with pests, treating them with pesticides. Training of farmers and development of IPM programmes using biocontrol were suggested by Colmenarez *et al.* (2016) to prevent indiscriminate use of pesticides and increase the use of biocontrol agents in the island.

3.3.9 Areas under biological control in Barbados

Based on data about areas of agricultural products harvested in 2016 (FAO, 2018) and the successful cases of biocontrol listed in Table 3.1, the area under biocontrol in Barbados is estimated to be at least 3,000 ha, with about 300 ha under natural control, 2,700 ha under classical biocontrol and less than 10 ha under augmentative biocontrol.

Table 3.2. Overview of key pests introduced and established in Barbados (source: M. James, Barbados National Plant Protection Officer, October 2018, personal communication).

Common name	Scientific name	Year of introduction
Pink hibiscus mealybug	*Maconellicoccus hirsutus*	2000
Papaya mealybug	*Paracoccus marginatus*	2000
Citrus leaf miner	*Phyllocnistis citrella*	2000
Giant African snail	*Achatina fulica*	2000
Tomato russet mite	*Aculops lycopersici*	2000
Broad mite	*Polyphagotarsonemus latus*	2000
Gliricidia moth	*Azeta melanea*	2000
West Indian fruit fly	*Anastrepha obliqua*	2001
Pickle worm	*Diaphania nitidalis*	2002
Asian citrus psyllid	*Diaphorina citri*	2003
Varroa mite	*Varroa destructor*	2003
Sago palm scale	*Aulacaspis yasumatsui*	2003
Chilli thrips	*Scirtothrips dorsalis*	2005
Soybean scale	*Crypticerya genistae*	2006
Cotton stainer	*Dysdercus discolor*	2006
Dendrobium/hibiscus midge	*Contarinia maculipennis*	2006
Ficus thrips	*Gynaikothrips uzeli*	2006
Fig whitefly	*Singhiella simplex*	2007
Red palm mite	*Raoiella indica*	2010
Cardin's whitefly	*Metaleurodicus cardini*	2011
Croton scale	*Phalacrococcus howertoni*	2011
Erythrina gall wasp	*Quadrastichus erythrinae*	2012
Duges wax scale	*Ceroplastes dugesii*	2014
Crepe myrtle aphid	*Sarucallis kahawaluokalani*	2015
Avocado lace bug	*Pseudacysta perseae*	2015

3.4 New Developments of Biological Control in Barbados

The constant movement of people and intense international trade that Barbados experiences result in introduction and establishment of new pests and invasive species in the country. Recent invasions are summarized in Table 3.2. Several biocontrol programmes have recently been successfully developed as a sustainable approach to face those challenges and these have been summarized above. Other programmes are still in development, like biocontrol of chilli thrips, red palm mite and pests in cotton.

3.5 Acknowledgements

We would like to thank the Barbados Ministry of Agriculture for providing annual reports, and I. Gibbs, M. James and B. Taylor (all of the Barbados Ministry of Agriculture) for providing important information. CABI is an international intergovernmental organization, and Y.C. Colmenarez gratefully acknowledges the core financial support from CABI's member countries (see https://www.cabi.org/what-we-do/how-we-work/cabi-donors-and-partners/ for full details).

References

Cave, R.D. (2006) Biological control agents of the cycad aulacaspis scale, *Aulacaspis yasumatsui*. *Proceedings of the Florida State Horticultural Society* 119, 422–424.

CIA (2017) The World Factbook: Barbados. Available at: https://www.cia.gov/library/publications/the-world-factbook/geos/bb.html (accessed 7 July 2019).

Cock, M.J.W. (ed.) (1985) *A Review of Biological Control of Pests in the Commonwealth Caribbean and Bermuda up to 1982*. Technical Communication No. 9, Commonwealth Institute of Biological Control. Commonwealth Agricultural Bureaux, Farnham Royal, UK.

Colmenarez, Y., Moore, D., Polar, P. and Vasquez, C. (2014) Population trends of the Red Palm Mite, *Raoiella indica* Hirst (Acari: Tenuipalpidae) and associated entomopathogenic fungi in Trinidad, Antigua, St Kitts and Nevis and Dominica. *Acarologia* 54(4), 433–442.

Colmenarez, Y., Gibbs, I.H., Ciomperlik, M. and Vasquez, C. (2016) Biological control of agents of cotton pests in Barbados. *Entomoptropica* 31 (18), 146–154.

FAO (2013) National action plan for HLB management in Barbados. Report TCP/RLA/3401, 20 pp. Available at: http://www.ipcinfo.org/fileadmin/user_upload/rlc/docs/HLB_BRB.pdf. (accessed 7 July 2019).

FAO (2018) Area harvested, Barbados, 2016. Available at: http://www.fao.org/faostat/en/#data/QC (accessed 27 November 2018).

Ingram, W.R. (1980) Studies of the pink bollworm, *Pectinophora gossypiella*, on sea island cotton in Barbados. *Tropical Pest Management* 26 (2), 118–137.

4 Biological Control in Belize

Edwin E. Sosa[1]*, Fermin Blanco[1] and Joop C. van Lenteren[2]

[1]Organismo Internacional Regional de Sanidad Agropecuaria (OIRSA), Showgrounds, Belmopan, Belize; [2]Laboratory of Entomology, Wageningen University, Wageningen, The Netherlands

* E-mail: esosa@oirsa.org

Abstract

Parasitoids were first introduced into Belize in 1969 for control of *Anastrepha* spp. fruit flies, but although this and other classical biocontrol attempts sometimes resulted in establishment, control was insufficient. During the same period, natural control of the West Indian cane fly was documented. In 2003, the International Regional Organization for Health in Agriculture (OIRSA) built a laboratory for the mass production of the parasitoid *Anagyrus kamali* for biocontrol of the pink hibiscus mealybug. The pest was successfully brought under classical biocontrol in the entire country. The laboratory also provides the parasitoids to OIRSA member countries. Recently, classical biocontrol of the Asian citrus psyllid has been initiated and a mass rearing of *Tamarixia radiata* was started at the OIRSA laboratory for releases in Belize, as well as in other countries. An entomopathogenic fungus is currently being tested for control of the sugarcane froghopper.

4.1 Introduction

Belize has an estimated 360,000 inhabitants (July 2017) and its main agricultural products are bananas, cacao, citrus, sugar, fish, cultured shrimp and lumber (CIA, 2017).

4.2 History of Biological Control in Belize

4.2.1 Period 1880–1969

Classical biological control of fruit flies

Fruit flies of the genus *Anastrepha* spp., mainly *Anastrepha ludens* (Lw.), are serious pests of grapefruit, and to a lesser extent orange, in the Stan Creek Valley. Native parasitoids were not found during a survey for natural enemies of this pest on citrus. Consequently, at the end of 1969, the West Indian Station of the Commonwealth Institute of Biological Control (CIBC, now CABI) arranged for the introduction and release of three parasitoids: *Biosteres longicaudatus* Ashm. and *Aceratoneuromyia (= Syntomosphyrum) indica* (Silv.), both obtained from Mexico, and *Pachycrepoideus vindemiae* (Rond.) from Trinidad and Tobago. A few specimens of an '*Opius* sp.', presumably *Doryctobracon cereus* (Gah.), were included in one shipment from Trinidad and Tobago. Although no recoveries were made from citrus, *A. indica* and *P. vindemiae* were recovered from *Anastrepha obliqua* (Macq.) in other fruits (Cock, 1985).

Natural biological control of West Indian cane fly

Sugarcane is an important crop in Belize. The West Indian cane fly *Saccharosydne saccharivora* (Westw.) is an occasional pest and has a well developed natural enemy complex. During the 1960s, intensive studies were carried out for natural enemies in the region and several egg parasitoids were found, including *Anagrus flaveolus* Waterh., *Tetrastichus* sp. and *Tetrastichus* sp. nr. *vaquitarum* Wolc. in Belize. These parasitoids were introduced into Jamaica in 1966, but no indication of establishment was found (Cock, 1985).

4.2.2 Period 1970–2000

Classical biological control of diamondback moth

The diamondback moth *Plutella xylostella* (L.) became problematic during the 1950s and 1960s in the Caribbean region. CIBC (now CABI) Trinidad and Tobago succeeded in rearing two strains of *Apanteles* (*Cotesia*) *plutellae* Kurd (one obtained from India in 1970 and another from the Netherlands in 1971) and *Diadromus collaris* (Grav.) (obtained from India in 1972). Both *Apanteles* strains were introduced and released in Belize in 1971–1972, but no recoveries have been reported of the parasitoids (Cock, 1985).

Classical biological control of the mahogany shoot borer

Mahogany (*Swietenia macrophylla* G. King and *S. mahagoni* Jacq.), the American cedar (*Cedrela odorata* L. (= *C. mexicana* Roem.)) and crappo (*Carapa guianensis* Aubl.) are forest trees of commercial value in the Caribbean region. Attempts in Belize, among other countries, to establish plantations of these trees have failed because of damage inflicted by the mahogany shoot borer *Hypsipyla grandella* (ZeM.), which is the most serious pest of these timber trees in the Neotropics. CIBC along with the Forest Division of Trinidad and Tobago surveyed for parasitoids in the region and also obtained parasitoids from abroad during 1968–1977 and 1980–1982. Funds provided by the UK Overseas Development Administration allowed for supplying parasitoids to Belize in 1974–75 and further shipments were sent to Belize in 1974–75 under funding from the Netherlands Bureau of Technical Assistance. The following species were introduced into Belize: *Antrocephalus renalis* Wtstn (1969), *Phanerotoma* sp. (1970–1975), *Tetrastichus spirabilis* Waterston (1969–1972) and *Trichogrammatoidea robusta* Nagaraja (as *T. nana* Zhnt.) (1970–1975); but no recoveries have been reported from Belize (Cock, 1985).

4.3 Current Situation of Biological Control in Belize

4.3.1 Classical biological control of the pink hibiscus mealybug

The pink hibiscus mealybug *Maconellicoccus hirsutus* (Green) is a polyphagous pest of Asian origin and was detected in 1999 in Baja California, Mexico. Mexico then imported and released the parasitoids *Anagyrus kamali* Moursi and *Gyranusoidea indica* Shafee, Alam and Agarwal, which were imported from Puerto Rico, and the parasitoids became established (Santiago-Islas *et al.*, 2008). In the same year, the mealybug was also found in Belize and the same parasitoids were liberated at the border of Quintana Roo (Mexico) to establish an ecological barrier and slow down introduction of this pest from Belize into Mexico and also to prevent spread to other Central American countries. Due to its rapid dispersion, the pest disseminated in urban and rural areas, affecting mostly ornamental plants mainly in backyards such as hibiscus (*Hibiscus rosa-sinensis* L.), sorrel (*Hibiscus sabdariffa* L.), soursop (*Annona muricata* L.) and mango (*Mangifera indica* L.). At present, the pink hibiscus mealybug is found in the following Central American and Caribbean countries: Anguilla (since 1996), Antigua and Barbuda, Aruba, Bahamas (2000), Barbados (2000), Belize (1999), British Virgin Islands (1997), Dominica (2001), Dominican Republic (2002), Grenada (1994), Guadeloupe (1998), Guatemala (2013), Haiti (2002), Jamaica, Martinique (1999), Montserrat (1998), Netherlands Antilles (1996), Puerto Rico (1997), St Kitts and Nevis (1995), St Lucia (1996), St Vincent and Grenadines (1997), Trinidad and Tobago (1995) and US Virgin Islands (1997) (OIRSA, 2018).

In order to lower the levels of the pest in Belize, the predator *Cryptolaemus montrouzieri* Mulsant was imported from CABI Trinidad and Tobago and released in areas of high infestations. The populations of the pest lowered and then *A. kamali* was introduced, also obtained from CABI, to maintain biocontrol of the mealybug. In 2000, the International Regional Organization for Health in Agriculture (Organismo Internacional Regional de Sanidad Agropecuaria) (OIRSA) agreed to provide financial support for the construction of a laboratory in Belize that would serve as a centre for rearing *A. kamali*. In 2003, the regional laboratory was built at the Ministry of Agriculture property in Belmopan, located at the National Agriculture Showgrounds. The programme received technical assistance from the United States Department of Agriculture (USDA) to develop the rearing technology.

The biocontrol programme is considered a success, since the pest is under control in the entire country. As a result, strict integrated pest management (IPM) measures were enforced to control the levels of infestation on its primary host, hibiscus. Hibiscus plants serve as natural green fences in backyards and are appreciated for their beautiful flowers. Inundative releases of *A. kamali*, along with an effective surveillance programme led to low levels of mealybug infestation.

Thousands of parasitoids were released in Benque Viejo del Carmen, the western region of the country, creating a natural biological barrier to slow down the rapid dissemination of the pest into neighbouring Guatemala. At a regional level, Mexico received 397,045 *A. kamali* individuals in 2004 and 2005 for releases to control *M. hirsutus*, as well as to initiate a mass rearing of the parasitoid. At present, the laboratory in Nayarit, Mexico, has one of the largest mass-production units of *A. kamali*. The OIRSA laboratory provided 259,000 wasps to Costa

Rica in 2012 and 2013 for control of the pink hibiscus mealybug in coffee and hibiscus, 180,000 parasitoids to El Salvador and Honduras in 2014 and 488,000 parasitoids to Guatemala (OIRSA, 2018). Since its existence, the Regional OIRSA pink hibiscus mealybug laboratory has produced over 10 million *A. kamali*, 13% of which were exported to the region and the remaining 87% were used in Belize for releases and rearing (Belize Ag Report, 2011, 2015).

4.3.2 Classical biological control of the Asian citrus psyllid

OIRSA, along with the support of the International Cooperation Development Fund ICDF-Taiwan and the Ministry of Agriculture in Belize, conducted a 5-year project for control of huanglongbing (HLB) disease and its vector, the Asian citrus psyllid *Diaphorina citri* Kuwayama, in Belize. HLB was causing severe citrus losses in the region. For its control the parasitoid *Tamarixia radiata* (Waterston), well known for its efficiency in reducing psyllid populations, was introduced to Belize from Mexico in 2014. Parasitoids were released in abandoned groves and were provided to the IPM programme of the Citrus Growers Association (CGA) and Citrus Research and Education Institute (CREI, 2009). Parasitism of the psyllid was determined in different plots around Stann Creek district and 13.9% parasitism was found on average. At the same time, parasitoids were reared in small cages in the fields.

4.4 New Developments of Biological Control in Belize

4.4.1 Classical biological control of the pink hibiscus mealybug

The OIRSA laboratory will continue to mass produce *A. kamali* for releases in Belize and Guatemala, and, if needed in other regional countries. As a result of this project, pink hibiscus mealybug is under control in Belize, thereby avoiding massive losses to the agricultural economy of the country. During surveillance programmes, no major pink hibiscus mealybug pest invasions in crops of economic importance have been reported.

4.4.2 Classical biological control of the Asian citrus psyllid

Recently OIRSA received a mandate for a regional feasibility study to determine which organisms are most needed to be produced and commercialized by its laboratory.

4.4.3 Augmentative biological control of the sugarcane froghopper

Froghoppers *Aeneolamia varia saccharina* (Dist.) are currently considered a major pest in sugarcane in Belize, demanding increasing frequencies and amounts of chemical control, which interfere with profits and do not comply with the standards set by the Fairtrade market. Therefore, an integrated pest and disease management (IPDM) project was initiated with the assistance of the European Union, the Government of Belize and the Sugar Industry Research and Development Institute (SIRDI). One of the elements of this IPDM project is the construction of a laboratory for the production of biocontrol agents, including a natural enemy of the sugarcane froghopper, the entomopathogenic fungus *Metarhizium anisopliae* (Metchnikoff) Sorokin, which started in 2017. The entomopathogen will be produced by a culture method based on whole-grain rice. It will be available not only to sugarcane farmers, but also to producers of vegetables, fruit, citrus, poultry and livestock (Cob *et al.*, 2017).

In conclusion, Belize had and has a number of interesting biocontrol programmes. For example, the West Indian cane fly is under natural biocontrol. Fruit fly parasitoids introduced in 1969 are still supposed to be present. New programmes for control of the Asian citrus psyllid and the sugarcane froghopper have recently started, and the pink hibiscus mealybug has been brought under classical biocontrol in the entire country, resulting in the protection of various plants and crops from this pest. Based on data on biocontrol activities provided in Table 4.1 and hectares of the crops in which natural enemies were released, we estimate that at least

Table 4.1. Overview of major biocontrol activities in Belize.

Biocontrol agent / exotic (ex), native (na)	Pest / crop	Type of biocontrol[a] / since	Effect / area under biocontrol[b]	Reference
Biosteres longicaudatus / ex	Fruit flies / fruit incl. citrus	CBC / 1969	No control, established	Cock, 1985
Aceratoneuromyia indica / ex		CBC / 1969	No control, established	
Anagrus flaveolus / na	West Indian cane fly / sugarcane	NC	Some	Cock, 1985
Tetrastichus spp. / na		NC	Some	
Cotesia plutellae / ex	Diamondback moth / vegetables	CBC / 1970–1972	No control, not established	Cock, 1985
Diadromus collaris / ex		CBC / 1970–1972	No control, not established	
Antrocephalus renalis / ex	Mahogany shoot borer / mahogany, cedar	CBC / 1969	No control, not established	Cock, 1985
Phanerotoma sp. / ex		CBC / 1970–1975	No control, not established	
Tetrastichus spirabilis / ex		CBC / 1969–1972	No control, not established	
Trichogrammatoidea robusta / ex		CBC / 1970–1975	No control, not established	
Cryptolaemus montrouzieri / ex	Pink hibiscus mealybug / ornamental shrubs	CBC / 2000	Control, established / ?ha	OIRSA, 2018
Anagyrus kamalil / ex		CBC / 2000	Control, established	
Tamarixia radiata / ex	Asian citrus psyllid / citrus	CBC / 2014	Control, established / 16,000 ha	OIRS, 2018
Metarhizium anisopliae / ex	Sugarcane froghopper / sugarcane	ABC / 2017	Testing phase	Cob *et al.*, 2017

[a]Type of biocontrol: ABC = augmentative, CBC = classical, ConsBC = conservation biocontrol; NC = natural control
[b]Area of crop harvested in 2016 according to FAO (http://www.fao.org/faostat/en/#data/qc)

16,000 ha are under classical biocontrol in Belize.

4.5 Acknowledgements

The Ministry of Agriculture of Belize is thanked for allowing the implementation of IPM and biocontrol programmes in Belize, the Belize Agricultural Health Authority for working together with OIRSA to implement biocontrol programmes and the OIRSA member countries for funding of OIRSA programmes since 2000. L. Chi is thanked for information about the SIRDI EU IPDM project.

References

Belize Ag Report (2011) Regional pink hibiscus mealybug project. Available at: http://agreport.bz/2011/09/regional-pink-hibiscus-mealybug-project-phmb/ (accessed 7 August 2018).

Belize Ag Report (2015) Pink hibiscus mealybug. Available at: http://agreport.bz/pinkhibiscusmealybug/ (accessed 27 October 2019).

CIA (2017) The World Factbook: Belize. Available at: https://www.cia.gov/library/publications/the-world-factbook/geos/bh.html (accessed 7 July 2019).

Cob, J., Gongora, J. and Chi, L. (2017) Strengthening of integrated pest and disease management in the sugar industry. SIRDI EU IPDM project-*Metarhizium anisopliae* production: an integrated approach to control froghopper in sugarcane. Newsletter SIRDI EU IPDM *M. anisopliae*. 3 pp. Available at: http://agreport.bz/from-cane-to-cattle-pastures/ (accessed 27 October 2019).

Cock, M.J.W. (ed.) (1985) *A Review of Biological Control of Pests in the Commonwealth Caribbean and Bermuda up to 1982*. Technical Communication No. 9, Commonwealth Institute of Biological Control. Commonwealth Agricultural Bureaux, Farnham Royal, UK.

CREI (2009) How CREI found HLB. Available at: http://www.belizecitrus.org/index.php?option=com_content&task=view&id=63&Itemid=84 (accessed 13 September 2018).

OIRSA (2018) Informe técnico sobre la situación actual de la CRH *Maconellicoccus hirsutus* (Green) OIRSA-Belize [Technical information about the actual situation of *Maconellicoccus hirsutus*]. Available at: https://www.oirsa.org/busqueda.aspx?q=cochinella%20rosada (accessed 27 October 2019).

Santiago-Islas, T., Zamora-Cruz, A., Fuentes-Temblador, E.A., Valencia-Luna, L. and Arredondo-Bernal, H.C. (2008) Cochinilla Rosada del Hibiscus, *Maconellicoccus hirsutus* (Hemiptera: Pseudococcidae) [Pink hibiscus mealybug, *Maconellicoccus hirsutus* (Hemiptera: Pseudococcidae)]. In: Arredondo-Bernal, H.C. and Rodríguez-del-Bosque, L.A. (eds.) *Casos de Control Biológico en México* [*Cases of biological control in Mexico*]. Ed. Mundi-Prensa, Mexico, pp.177–191.

5 Biological Control in Bolivia

Javier P. Franco[1]*, Luis V. Crespo[2], Yelitza C. Colmenarez[3] and Joop C. van Lenteren[4]

[1]*CABI Plantwise-Perú, Surco, Lima, Perú;* [2]*Fundación PROINPA, Oficina principal, Regional Centro, Cochabamba, Bolivia;* [3]*CABI-UNESP-FEPAF, Botucatu, São Paulo, Brazil;* [4]*Laboratory of Entomology, Wageningen University, Wageningen, The Netherlands*

* E-mail: j.franco@cabi.org

Abstract

A number of introductions of parasitoids and predators were carried out in the 1950s for classical biological control of olive scale, woolly apple aphid, white peach scale, Mediterranean fruit fly and *Anastrepha* fruit fly, with control of cottony cushion scale by the coccinellid *Rodolia* being a particular success. In 1963, dipteran natural enemies were introduced for the control of sugarcane borers, *Diatraea* spp. Since 1963 native hymenopteran and dipteran parasitoids have been field collected and re-released for control of the borers. In 1969 an IPM programme of sugarcane borers was started, and biocontrol in sugarcane in the period 1970–2000 mainly consisted of augmentative releases of hymenopteran and tachinid parasitoids. Another successful IPM programme dealt with control of potato moth species with a product – now commercially available – that contains a native strain of the granulosis virus *Baculovirus phthorimaea* and a native strain of *Bacillus thuringiensis*. Coffee berry borer was brought under biocontrol in the 1990s by releasing a hymenopteran parasitoid and application of an entomopathogenic fungus. An increased demand for organic products since 2000 has stimulated work on isolation, characterization, mass production, formulation and certification of a number of microbial control agents. These are used in many crops and examples are microbial control of pest in potato and quinoa. Many of the quinoa pests are kept under natural control by predators and parasitoids, which has been well documented during the past 10 years. Currently most pests in sugarcane and soybean are under a combination of natural, augmentative and classical biocontrol.

5.1 Introduction

Bolivia has an estimated population of slightly more than 11 million (July 2017) (CIA, 2017). According to Vélez *et al.* (2017, pp. 56 and 76):

> ... the main crop groups planted during the 2013 summer season, 43.4% of the area was cultivated with oilseeds and industrial crops (999,369 ha with soybean and the rest with sunflower, sugarcane and peanuts). A total of 31.9% of the area was used for cereal cultivation (390,668 ha with maize, and the remainder with grain sorghum, paddy rice, quinoa and wheat). Tubers and roots were planted on 7.5% of the area (170,447 ha with potato), vegetables on 3.9%, fruit trees on 5.8%, fodder on 6.1% and stimulants on 1.4% (http://www.paginasiete.bo/economia). Arable land comprises 7,837,864 ha, and 2,349,062 ha of cultivated pastures; 1,635,898 ha of resting land and 1,089,665 ha of fallow land; 27,132,304 ha are non-agricultural land. Of this total, forests or mountains account for 13,775,113 ha, natural pastures 11,053,246 ha, other lands 2,153,726 ha and forest plantations 150,219 ha. Livestock farms in Bolivia constitute an essential resource for the food security of peasant families ...

According to the Instituto Nacional de Estadistica (INE, 2017), Bolivia's main food exports from 2010 to 2015 were: soybean, soybean by-products, quinoa and Brazil nuts.

Traditional agriculture is practised in the Altiplano and Valleys (departments of Cochabamba, La Paz and Oruro) and modern agriculture is practised in the East, and in some areas in the North and South (departments of Santa Cruz, Beni and Tarija). In traditional agriculture, land is tilled with animal traction and harvesting is by shovels, hoes and sickles. Hardly any artificial irrigation is used, fertilizers are natural, planting and harvesting dates (one annual harvest) are fixed and there is a rigorous system of land rotation. Modern agriculture by medium and large entrepreneurs depends largely on the external market and incorporates industrial inputs, use of machinery, certified seeds, fertilizers, artificial irrigation and pest control.

5.2 History of Biological Control in Bolivia

5.2.1 Period 1880–1969

Biological control of pests in sugarcane

Sugarcane *Saccharum officinarum* (L.) covers the largest crop acreage in Bolivia due the international demand for its by-products. The first attempts in biocontrol date from the 1930s in order to control the sugarcane spittlebug *Mahanarva spectabilis* (Distant), when a farmer from the department of Santa Cruz, Mr A. Arredondo, organized the purchase of toads (*Bufo* spp.) and introduced them into Bolivia (Pruett, 1996; Rogg, 2000a). According to Pruett (1996), several natural enemies were imported between

1949 and 1952: *Scutellista cyanea* Motsch. for control of olive scale *Saissetia oleae* (Olivier); *Aphelinus mali* Hald. for control of the woolly apple aphid *Eriosoma lanigerum* Haus. in apples; and *Prospaltella berlesi* How. against the white peach scale *Pseudaulacaspis pentagona* (Targioni-Tozzetti). The three natural enemies established, but results about their effectiveness have not been published.

Dr F. Simmonds of the Centre for Agriculture and Biosciences International (CABI) Trinidad and Tobago proposed in 1963 to introduce natural enemies for control of sugarcane borers *Diatraea* spp. and several species were imported and released, of which only *Lydella* (= *Metagonistylum*) *minense* Townsend established (Pruett, 1996). In 1963, the sugar mill La Bélgica (department of Santa Cruz) employed the Peruvian entomologist S. Risco to start the first laboratory to mass rear natural enemies for the control of sugarcane borers *Diatraea* spp. (Pruett, 1996). From 1963 to 1965 about 35,000 egg masses of *Diatraea rufescens* Box parasitized by *Telenomus* sp. were collected from dead buds; the emerged parasitoids were released, which was equivalent to the release of 600,000 wasps. Also around 23,000 individuals of the native tachinid *Palpozenillia diatraeae* Townsend were released, collected previously from parasitized borer larvae in dead buds. In addition, the tachinid *Paratheresia claripalpis* Wulp was imported from Peru in 1963, but it is unknown whether this parasitoid was mass reared and released, although Pruett (1996) mentioned that it established in Bolivia. Further, *Trichogramma fasciatum* Perkins was reared in the laboratory, but information about releases is lacking. Unfortunately, this laboratory closed in 1966 (Rogg, 2000a).

In 1969, Mr P. Terán, a pioneer in the control of sugarcane pests in Santa Cruz, started an integrated pest management (IPM) programme of *Diatraea* spp. for the mill La Bélgica. From September 1969 to March 1970, larvae and pupae of dead buds with *Diatraea* spp. were collected and a total of 948,000 *Telenomus* sp. were released to control *D. rufescens*. At the same time, the tachinid *P. diatraeae* was released, obtained from larvae in dead buds. Borer infestation, expressed as drilled cane internodes, dropped from 48% to 27% in 1970. The programme ended in 1970 because the property was sold to cotton growers (Pruett, 1996).

Various other early biological control projects

A number of introductions and releases of parasitoids and predators were carried out in the 1950s (Pruett, 1996; Rogg, 2000a, b). Between 1949 and 1952, the parasitoid *Scutellista cyanea* Mötsch was imported for control of olive scale *S. oleae*; *Aphelinus mali* Hald was imported to control *Eriosoma lanigerum* (Haus.) in apples; also *P. berlesi* was imported against the scale *Pseudaulacapsis pentagona* (Targ.-Tozz.). All three parasitoids established, but results of their effect on the pests have not been published (Pruett, 1996). Several parasitoid species of the Mediterranean fruit fly *Ceratitis capitata* (Wiedemann) and *Anastrepha* spp. fruit flies have been introduced from Hawaii into citrus orchards. In 1952, the coccinellid predator *Rodolia cardinalis* (Mulsant) was introduced for the control of the cottony cushion scale *Icerya purchasi* Maskell (Pruett, 1996). The predator established and controlled this important citrus pest; both the pest and the predator can still be found today at very low densities. In Bolivia in 2015–2016, an area of 54,413 ha was planted with citrus, with tangerines and oranges on 26,796 ha and 22,864 ha, respectively (INE, 2017). No information could be obtained about the total area currently treated with biocontrol in citrus.

Natural enemies were also introduced to control pests in olives, apples, citrus and sugarcane. In 1963, Simmonds made proposals for the introduction of natural enemies to control three important pests: *Heliothis* spp. in maize, sugarcane borers in sugarcane and fruit flies in various fruits. In 1967, the UK's Overseas Development Administration (ODA) provided funds for the importation of several natural enemies, but it is unknown which species have been introduced and if they established (Rogg, 2000a, b).

5.2.2 Period 1970–2000

Bennett and Squire (1972) concluded that, if Bolivia wanted to continue trying to achieve biocontrol, it was necessary to make larger releases for a longer time, with laboratory facilities (insectaries) and people trained to multiply beneficial insects, to avoid the high cost of continuous importation from other countries. In the same publication, the authors presented the results on biocontrol of some insect pests in Bolivia in key crops such as potato and sugarcane, among others. Teaching of IPM and biocontrol started in 1992 at the Institute of Agricultural Research 'El Vallecito', Faculty of Agricultural Sciences, Universidad Autónoma 'Gabriel René Moreno' in Santa Cruz, when its programme of Agricultural Entomology was initiated. During this period, the following organizations also worked on biocontrol programmes: CIAT (Centre for Research in Tropical Agriculture) in Santa Cruz, the non-governmental organization (NGO) PROBIOMA (Productivity Biosphere Environment) in Santa Cruz, PROINPA (Potato Research Project) in Torralapa/Cochabamba, the Universidad Mayor de San Simón in Cochabamba, the Institute of Ecology (IE) of the Universidad Mayor de San Andrés in La Paz, CABI (Centre for Agriculture and Biosciences International), the FAO (Food and Agriculture Organization) in La Paz and some other small NGOs. Their projects concerned biocontrol of the most important pests on local regional crops like potato, quinoa, corn and sugarcane.

Biological control of pests in sugarcane

The tachinid *L. minense*, which was imported and released in the 1960s, was recovered from the sugarcane borers *Diatraea saccharalis* (Fabricius) and *Eoreuma morbidella* (Dyar). Laboratory work showed that *D. rufescens* is neither an appropriate host for the tachinids *L. minense* and *Lixophaga diatraeae* (Towns), nor suitable for the hymenopteran *Cotesia flavipes* (Cameron) (Rogg, 2000a).

From 1979 to 1983, different species of *Trichogramma* spp. were imported from the Colombian company Centro Biológico to be released for sugarcane borer control, but with limited success. The Centre for Research and Improvement of Sugarcane (CIMCA), with the assistance of international entomologists, worked from the 1970s on the identification of pests and on mass production of their natural enemies, and released natural enemies until the laboratory was closed in 1992. In 1992, the sugarcane mill UNAGRO started a mass rearing of *Trichogramma* spp. The Autonomous University Gabriel René Moreno (UAGRM), together with the Institute of Agricultural Research 'El Vallecito', began researching biocontrol and rearing of beneficial insects. In conclusion, during the period 1970–2000 biocontrol in sugarcane mainly consisted of augmentative releases of *Trichogramma* sp., *Telenomus* sp., *P. claripalpis* and, most importantly, *P. diatraeae* against the borer *D. rufescens* (Rogg, 2000a). CIMCA released the parasitoid *C. flavipes* against *Diatraea* spp. from 1980 to 1992 without achieving establishment. CIMCA also imported many exotic natural enemies for classical biocontrol of Crambidae borers, but without success (Rogg, 2000a).

Biological control of potato moths

In Bolivia, two species of moths, *Phthorimaea operculella* (Zeller) and *Symmetrischema tangolias* (Gyen), cause severe damage to the stored potato tubers of both *Solanum tuberosum* ssp. *tuberosum* and ssp. *andigena* (Calderón *et al.*, 2002). PROINPA extensively studied the biology and occurrence of *P. operculella* and its natural enemies in different production systems in the 1990s. To control the pest, sex pheromones, a baculovirus (*Baculovirus phthorimaea*) and the parasitoid *Copidosoma* sp. were tested. Pheromones and releases of *Copidosoma* sp. did not significantly reduce moth populations. The use of an IPM programme comprised recommendations of cultural practices (e.g. hilling up, harvesting on time) and application of a baculovirus (see below) to the tubers before storing them, as the most important tuber damage occurs during storage. This resulted in a 90% reduction of *P. operculella* and it proved to be more effective than chemical control (Calderón *et al.*, 2002).

After the discovery in 1991 of *P. operculella* larvae infested with the granulosis virus (VG) *Baculovirus phthorimaea* in a farmer's warehouse in the Mizque area (Barea *et al.*, 2002), studies were conducted to determine the potential of the virus as a biocontrol agent, using methodologies developed by the International Potato Center (CIP). Artisan production of the virus was started in the 1993–1994 season in the entomology laboratory in Toralpa (Calderón *et al.*, 2002). In 1994, to meet farmers' demands, the virus was produced in a pilot plant at the Toralapa Service and Production Center with financing from Swiss Development Cooperation's project FIS-BALPAG. The project resulted in the formulation and use of Matapol, which was registered by the National Service of Agricultural Health and Food Safety (SENASAG).

After the success obtained with Matapol for *P. operculella* control, the same virus was tested on *S. tangolias*, but without success. Next, a specific virus of *S. tangolias* and other biocontrol agents were identified, but neither the virus nor the biocontrol agents were effectively controlling the pest (PROINPA, 2004a). However, a local strain of *Bacillus thuringiensis* was found that sufficiently controlled *S. tangolias*. This local strain was multiplied in the laboratory of PROINPA and incorporated into the formulation of Matapol, resulting in Matapol Plus, and registered by SENASAG (Calderón *et al.*, 2002). The bioproduct is used within an IPM strategy that includes components such as cleaning and disinfection of the warehouse and selection and treatment of the tubers with Matapol Plus. During the treatment, the tubers must be completely covered with the product and then stored, resulting in about 90% control of the pest.

Biological control of fruit flies in citrus

In 1992, the IE in La Paz and the Institute of Agricultural Research 'El Vallecito' in Santa Cruz independently developed an IPM programme for fruit flies, which included destruction of fallen fruits, trapping fruit flies with a hydrolysed protein and release of parasitoids of the fruit flies like *Opius oophilus* Fullaway, *O. incise* Silvestri, *Biosteres formosanus* Fullaway, *B. vandenboschi* (Fullaway), *B. compensans* Silvestri and *B. longicaudatus* Ashmead (Rogg, 2000b). This IPM programme was presented to the Ministry of Agriculture to be included in the national fruit fly control programme called 'Promosca', which was used from 2007 to 2012 and coordinated by SENASAG.

Biological control of coffee berry borer

The coffee berry borer *Hypothenemus hampei* (Ferrari), native to Africa, was accidentally introduced from Ecuador and Peru to Bolivia in the 1980s in the Yungas and soon endangered about 95% of national coffee production. Yungas coffee growers are small producers who usually do not use pest control. The only native biocontrol agent is the entomopathogenic fungus *Beauveria bassiana* (Vuillemin), but it causes only around 5% mortality of *H. hampei*. In 1992, the IE initiated a biocontrol project with massive applications of the fungus *B. bassiana* and, in collaboration with the NGO Quana (Irupana, Sudyungas), mass produced and released the parasitoid *Cephalonomia stephanoderis* Betrem in combination with the fungus. The result was a 90% reduction in borer populations and the combined use of the fungus and the parasitoid is still practised today (Rogg, 2000a).

Biological control in cotton

Trichogramma (semifumatum) pretiosum Riley was marketed in Bolivia by a Colombian producer for control of lepidopteran pests in the 1980s. However, *Trichogramma* spp. never emerged from eggs of *Heliothis* spp. nor of *Alabama argillacea* (Hübner) collected in cotton fields, though a native species of *Trichogrammatoidea* was often found and resulted in high percentages of parasitism (Rogg, 2000a).

Biological control of the large kissing bug, the vector of Chagas disease

The large kissing bug *Triatoma rubrofasciata* (DeG.), vector of *Trypanosoma cruzi* Chagas, which causes Chagas disease, is widespread in Bolivia. The University of San Simon (Cochabamba), with funds from the World Health Organization (WHO), imported the parasitoid *Gyron triatomae* Msn. from Singapore in 1969. However, it appeared that this egg parasitoid was not able to penetrate the egg shell of the kissing bug. *Telenomus fariai* Costa Lima, a native egg parasitoid of *Triatoma* spp., caused about 2% parasitism in the field, but much higher percentages of parasitism were observed in the laboratory of the Centro Nacional de Enfermedades Tropicales (CENETROP) (Pruett, 1996). A check of the website of CENETROP (2018) showed that the Center is still working on Chagas disease, but whether biocontrol of the insect vector is still studied could not be found.

5.3 Current Situation of Biological Control in Bolivia

Overuse of chemical products has often disturbed the balance in insect communities and, therefore, some insects, initially without economic importance, have become key pests. In addition, the overuse of pesticides has resulted in insect resistance, environmental pollution and disruption of links in trophic chains. Pesticide residues and risks to human health have resulted in an increasing demand for semi-organic and organic products, with a better price for the farmer. Within this new type of agricultural production, biocontrol can play an essential role. The next sections will provide detailed information about how biocontrol has been developed in Bolivia. First, we summarize an important initiative for development of microbial control agents. Next, we describe the situation in the Altiplano and Valles Interandinos, with potato (*Solanum tuberosum* ssp. *andigena* and other native species) and quinoa (*Chenopodium quinoa* Willd.) as the most important crops. Finally, we discuss research conducted in the Valles Mesotérmicos and Cálidos del Oriente, where biocontrol research mainly relates to sugarcane, soybean, citrus and coffee crops.

5.3.1 Development of microbiological control agents and bioinsecticides

To generate a more sustainable agriculture, PROINPA formed an alliance with BIOTOP SRL, a company that sells and promotes organic products. The objective was to work on the development of pest management methods that are harmless to human health, do not affect the environment and are accessible to low-income farmers. Work was concentrated on isolation, characterization, mass production, formulation and certification of microbial control agents, so that they could be used within the international standards of organic production. Examples of microbial control agents formulated by BIOTOP (Ortuño *et al.*, 2011) are:

- MATAPOL PLUS, a powdery formulation of a combination of native strains of *Baculovirus phthorimaea* granulovirus and *B. thuringiensis* d-endotoxin for control of potato tuber moths.
- BAUVETOP, based on native entomopathogenic fungal strains of *Beauveria* spp., for control of a wide array of pests, including weevils in maize, potato, etc., moths, bugs, cicadas, and coffee and banana borers.
- BIOBAT, a powdery or liquid bioinsecticide based on the endotoxin of a native strain of *B. thuringiensis* and mainly used for control of lepidopterans in various crops.

PROBIOMA in Santa Cruz is another institution producing the following microbial agents:

- PROBIONE, based on the entomopathogenic nematode *Heterorhabditis bacteriophora* (Poinar), for control of fall armyworm, coccids, aphids and whiteflies in sugarcane, melon, potato, vegetables, soybeans, watermelon, rice, coffee, citrus, bananas and ornamentals.
- PROBIOBASS, based on *B. bassiana* for control of the banana weevil and other pests in banana, sugarcane, chilli, soybean, citrus, corn, sweet potato, potato, rice, tomato, watermelon and fruit trees.
- PROBIOMET, based on *Metarhizium anisopliae* (Metschnikoff), for control of white cabbage butterfly and other pests in vegetables, fruit trees, wheat, peas, rice and ornamentals.
- PROBIOVERT, based on *Verticillium lecanii* (Zimmerman) for control of scale insects, coccids, aphids, whitefly and stinkbugs in, among others, vegetables, fruit trees, wheat, peas, rice, ornamentals, barley and watermelon.

5.3.2 Control of pests in the Altiplano and Valles Interandinos

Although there is a great diversity of native and introduced crops in these regions, potato (*S. tuberosum* ssp. *andigena* and other native species) and quinoa (*C. quinoa*) are native crops considered as the most important staple foods.

Microbial control of potato weevils and potato tuber moth

Earlier, biocontrol of two moth species causing problems in potato was described. Here, the control of several species of the Andean potato weevils is summarized. The Andean potato weevils comprise several genera and species, including *Premnotrypes latithorax* (Pierce), *Rhigopsidius tucumanus* Heller and *Phyrdenus* sp., commonly known as white potato larvae, potato larvae, potato weevil or, simply, white larvae. These pests were extensively studied in Bolivia (e.g. Calderón *et al.*, 2004) and mainly cause damage to the tubers.

To control the potato weevils, firstly 181 native isolates of *B. thuringiensis* were tested, but the toxicity of the Cry 3A protein was found to be low for the *R. tucumanus* larvae and null for *P. latithorax* (Calderon *et al.*, 2004). On the other hand, *Beauveria* spp. appeared to attack *Premnotrypes* spp. and *R. tucumanus* in warehouses and in the field. Initially, eight native strains of *Beauveria brongniartii* (Saccardo) were tested in the laboratory. Next, experiments were done in greenhouses, warehouses and in the field with the Ayopaya isolate of *B. brongniartii* with promising results to control the pests, particularly during rainy periods and at relative humidity above 50%. During dry periods, infection of *B. brongniartii* did not progress satisfactorily against any of the weevil species.

The major biocontrol programme in Bolivian potatoes is currently targeting *P. operculella* (Zeller), the potato tuber moth, with around 1,800 ha treated with Matapol (BioTop, October 2018, personal communication).

Natural biological control of lepidopteran pests in quinoa

Quinoa (*C. quinoa*) is a native species of the Andean zone, mainly from Bolivia and Peru, where it was domesticated by ancestral cultures such as Aymara and Quechua and has been cultivated for thousands of years. Quinoa has recently acquired worldwide popularity due to its high nutritive quality, its rich composition of essential amino acids, its gluten-free condition and easy digestion. Quinoa is attacked by many pest species, but the most important are those that feed on leaves and grains, such as noctuid larvae and the quinoa moth complex. In Bolivia, the noctuid complex consists of *Helicoverpa quinoae* Pogue and Harp, *Helicoverpa titicacae* Hardwick and *Copitarsia incommoda* (Walker). In the Bolivian Altiplano, *H. quinoa* is the most important pest, while *C. incommoda* is particularly causing problems in the area near Lake Titicaca. Use of biopesticides based on *B. thuringiensis* to control the noctuids is common in organic production. Further, PROINPA found dead noctuid larvae with clear symptoms of infection caused by a nuclear polyhedrosis virus (NPV) in the region of La Paz, which was confirmed by specialists from the UK's Natural Resources Institute (NRI) (Saravia and Quispe, 2006).

The quinoa moth complex, *Eurysacca melanocampta* Meyrick, *Eurysacca quinoae* Povolný and *Eurysacca media* Povolný, cause major problems. Larvae of *E. quinoae* and *E. melanocampta* are initially found between the apical leaves of plants, causing damage to the panicle in formation. Greater damage occurs during grain formation and maturity, when larvae feed mainly on the tender leaves and the immature and mature grains, causing yield decreases between 15% and 60% (Ortíz, 1993; PROINPA, 2008). Both *E. melanocampta* and *E. quinoae* have a complex of natural enemies. Natural parasitism of *E. melanocampta* and *E. quinoae* by several species of parasitoids ranges between 5% and 60% in Bolivia (PROINPA, 2004b; Saravia and Quispe, 2006; Saravia *et al.*, 2008). A study conducted in 2016 reported that the average natural parasitism in the North and Central Altiplano was 37.6% in 2013 and 12.9% in 2014, with *Meteorus* sp. and *Copidosoma* sp. reported in the first year and *Deleboeae* sp. and *Copidosoma* sp. in the second year (Barrantes, 2016). According to research conducted in Peru, an undescribed species of the genus *Phytomyptera* is the most dominant parasitoid of the complex of the quinoa moth (Rasmussen *et al.*, 2001) and less important are *Copidosoma gelechiae* Howard and *Diadegma* sp. Nine species of parasitoids have been found in the agroecological zones of the Salares and the Altiplano, two dipteran tachinids (*Phytomyptera* sp. and *Dolichostoma* sp.) and seven Hymenoptera (*Meteorus* sp., *Apanteles* sp., *Microplitis* sp., *Deleboeae* sp., *Venturia* sp., *Diadegma* sp. and *Copidosoma* sp.) (Saravia and Quispe, 2006). The species *Venturia*, *Deleboeae* and *Meteorus* are most often found in the organic quinoa production areas of the Altiplano (Saravia and Quispe, 2006; PROINPA, 2013), whereas in the Inter-Andean Valleys the dominant parasitoid is

Phytomyptera sp. (Figueroa *et al.*, 2013). Sampling of larvae in 12 communities of the Bolivian Altiplano in 2012–2013 showed an average natural parasitism of 28% (PROINPA, 2013).

Predators also constitute important natural enemies of the quinoa moth complex and include species of the orders Coleoptera [Carabidae (*Notiobia schnusei* Van Emden and *N. laevis bolivianus* (Van Emden), Cicindelidae and Coccinellidae (*Eriopis* sp., *Hippodamia* sp. and *Harmonía* sp.)], Hymenoptera (Sphecidae), Diptera (Asilidae) and Neuroptera (*Hemerobius tolimensis* Banks (Hemerobiidae) (Saravia and Quispe, 2006; PROINPA, 2013).

To date there are no reports of entomopathogens attacking species of the quinoa moth complex.

5.3.3 Control of pests in the Valles Meso térmicos and Cálidos del Oriente

Land areas used for agricultural production have increased in recent decades in the Santa Cruz department, which is the main supplier of several types of food for other departments of the country, e.g. soybean and its derivatives for national and export sales. The most important crops grown in this area are cereals (maize, rice, wheat), tubers (cassava), vegetables (tomato, carrot), stimulating products (cocoa, coffee), industrial crops (soybean, cotton, sunflower, sugarcane) and fruit (banana, pineapple, orange, tangerine, apple, mango, papaya, peach). Disappointingly, the expansion of production areas did not result in a concurrent investment in research on biocontrol. Still, biocontrol projects were executed for various type of fruit, (including citrus), sugarcane, coffee, wheat and soybean (Rogg, 2000a, b) and these projects are described below.

Natural, augmentative and classical biological control of sugarcane pests

An interesting biocontrol programme for several sugarcane pests and diseases has now been developed and is being applied. The sugar mill UNAGRO produces *T. pretiosum* for releases in 4,500 ha to control corn borers. Mole crickets *Scapteriscus* spp. and *Neocurtilla hexadactyla* (Perty) have been kept at a low level by releasing a complex of natural enemies: the carabid *Pheropsophus aequinoctialis* (L.), the reduviid *Sirentha carinata* (Fabricius), the tachinid *Ormia depleta* (Wiedemann) and the wasps *Larra bicolor* Fabricius, *L. transandina* Williams and an unidentified *Larra* species. The sugarcane coccid *Saccharicoccus sacchari* (Cockerell) is controlled by several coccinellid predators (the most important are *Brachiacantha* sp. and *Hyperaspis* spp.); the entomopathogen *Aspergillus flavus* Link is killing up to 90% of the pest in the rainy season; and the imported parasitoid *Anagyrus saccharicola* Timberlake (originally from Africa, but probably introduced from Peru) causes up to 40% parasitism. Further, a serious disease, the sugarcane smut *Ustilago scitaminea* Sydow, is kept at a low density by two coleopteran species (*Phalacrus obscurus* Sharp and *Brachytarsus* sp.) that eat fungal spores. As well as the above biocontrol agents, *Telenomus remus* Nixon is used against eggs of *D. rufescens* (Box), and *P. diatraeae* is used against larvae of *D. rufescens*. Also entomopathogens are used, like a baculovirus against *Diatraea* spp., and *B. bassiana* and *M. anisopliae* against borers, Crambinae, weevils, spittlebug and defoliators. Finally, bioinsecticides based on *B. thuringiensis* are applied against lepidopteran defoliators (Rogg, 2000a, b).

According to the Center for Research and Technology Transfer for Sugar Cane (Centro de Investigación y Transferencia de Tecnología para la Caña de Azúcar) (CITTCA), the following species of stem borers occurred in sugarcane in the northern zone in 1998: *D. rufescens*, *Myelobia bimaculata* (Box) (which is currently more damaging than *D. rufescens*), *E. morbidella* and *D. saccharalis* (which is also causing damage to corn and rice). For control of *D. saccharalis*, *Telenomus alecto* Crawford and *Trichogramma galloi* Zucchi are released. Larvae of all four stem borer species are attacked by the naturally occurring dipteran parasitoid *P. claripalpis* and the highest percentages of parasitism was in *M. bimaculata* (36%), followed by *D. rufescens* (16%), followed by *E. morbidella* (3%) and *D. saccharalis* (1%) (Rogg, 2000a).

Natural and augmentative biological control of soybean pests

Soybean *Glycine max* (L.) is attacked by many pests in Bolivia. The larvae of the soybean stalk weevil

Sternechus subsignatus Boheman attack early plantings and are controlled by applying *B. bassiana* (ANAPO, 2011). Larvae of *Hypsonotus* sp., *Agrotis ipsilon* and *Spodoptera* sp. are attacked by naturally occurring biocontrol agents, such as the carabids *Calosoma* sp. and *Pterostichus* sp., the tiger beetles *Megacephala carolina* (L.), *Megacephala chilensis* (Laporte de Castelnau), the dermapteran *Labidura riparia* (Pallas) and the tachinid parasitoids *Archytas marmoratus* (Townsend) and *Gonia peruviana* (Townsend). A baculovirus is used for the control of *Anticarsia gemmatalis* Hübner. A bioinsecticide based on *B. thuringiensis* is applied for control of *Omiodes (Hedylepta) indicata* (Fabricius), but this pest is also attacked by a number of naturally occurring beneficials, such as predators (*Cycloneda sanguinea* L., *Hippodamia convergens* Guérin-Meneville, *Polystes* sp.) and parasitoids (*Carinodes* sp., *Microcharops* sp.) (Rogg, 2000a; ANAPO, 2011).

In the Santa Cruz department several species of parasitoids of the genera *Trissolcus* and *Telenomus* (*Trissolcus urichi* (Crawford), *Trissolcus basalis* (Wollaston), *Trissolcus brochymenae* (Ashmead), *Telenomus podisi* Ashmead and *Ooencyrtus submetallicus* (Howard)) have been found attacking stink bug eggs; the predominant species found was *T. basalis*. In order to develop biocontrol of stink bugs, a mass rearing of *Trissolcus* spp. was developed on host eggs of the stink bug *Euschistus heros* (Fabricius). Releases of the mass-reared egg parasitoids are made around the soybean flowering period, when adult stink bugs attack the crop (Rogg, 2000a).

5.3.4 Areas under biological control in Bolivia

Based on information provided above and in Table 5.1, it is estimated that at least 760,000 ha are under natural control, 54,000 ha are under classical and 1,130,000 ha are under augmentative biocontrol.

5.4 New Developments of Biological Control in Bolivia

Several factors limit the introduction of augmentative biocontrol in Bolivia and are similar to those in many other Latin American countries. First of all, the pesticide industry is actively advertising its conventional products. They are just now starting to market biological products, but this is still very limited. Next, there is a lack of long-term financial support for research and development of biocontrol, and also there are no governmental efforts to implement biocontrol programmes. Further, there is insufficient collaboration among biocontrol researchers and lack of information on biocontrol in educational programmes offered by extension services. In addition, there is inadequate transfer of knowledge from universities to farmers and, finally, there is still little consumer interest in pesticide-free food within the country. Also, there are currently only a few public and private institutions having extension services dedicated to the development of IPM. A public-funded one is the Center for Research, Diagnosis and Production of Biocontrol agents for control of Pests and Diseases (PROBIOTEC, Santa Cruz). A private one is BioTop (Ortuño, *et al.*, 2011) in Cochabamba. They provide natural enemies and growth regulators to organic agriculture.

However, the high biodiversity in Bolivia and the opportunities offered by small biocontrol companies to market natural enemies are helping farmers, including those in remote rural areas, to have access to biocontrol agents. Improved cooperation between researchers and these companies is a critical factor to ensure a higher level of adoption of biocontrol. Recent practical results obtained in biocontrol (e.g. the fall armyworm biocontrol studies implemented by CIAT and CABI in Santa Cruz, the IPM-biocontrol programme for potato moths and biocontrol of the coffee berry borer) may stimulate increased use in the future. Further, training courses and better-informed advisory services are essential to introduce and help farmers to understand and apply the principles of biocontrol and sustainable food production.

5.5 Acknowledgements

We thank the staff of the PROINPA Foundation (Cochabamba), G. Rivadeneira (CIAT, Santa Cruz) and J. J. Lagrava (Autonomous University Gabriel René) for providing valuable information. CABI is an international intergovernmental organization, and Y.C. Colmenarez gratefully acknowledges the core financial support from CABI's member countries (see https://www.cabi.org/what-we-do/how-we-work/cabi-donors-and-partners/ for full details).

Table 5.1. Overview of major biocontrol activities in Bolivia.

Biocontrol agent / exotic (ex), native (na)	Pest / crop	Type of biocontrol[a] / since	Effect /area (ha) under biocontrol[b]	Reference
Bufo sp. / ex	Sugarcane spittlebug / sugarcane	CBC / 1930	? / ?	Rogg, 2000a
Beauveria bassiana / na		ABC / 2000	Control / 150,000	
Metarhizium anisopliae / na		ABC / 2000	Control / 150,000	
Microseramisia sphenophori / ex	Sugarcane borers / sugarcane	CBC / 1960s	? / ?	Rogg, 2000a
Telenomus sp. / na		ABC / 1963–65	? / ?	
Palpozenillia diatraeae / na		ABC / 1960s–2000	Partial control / ?	
Paratheresia claripalpis / ex		ABC / 1963–2000	Partial control / ?	
Telenomus remus. / na		ABC / 1969–2000	Partial control / ?	
Lydella minense / ex		CBC / 1969–1980/81	? / ?	Pruett, 1996
Lixophaga diatraeae / ex		CBC / 1968–1981	? / ?	
Cotesia flavipes / ex		CBC / 1969–1992	No control, not established	
Bracon chinensis (Szeph.) / ex		CBC / 1971	? / ?	
Itoplectus narangae Ashm. / ex		CBC / 1971	? / ?	
Pediobius furvus (Gahn.)		CBC / 1971–1981	? / ?	
Palpozenillia palpalis Ald. / ex		CBC / 1981	? / ?	
Campletis chlorideae / ex		CBC / 1981	? / ?	
Allorhogas pyralophagous / ex		CBC / 1982	? / ?	
Rhaconotus rosiliensis Lal. / ex		CBC / 1982–1985	? / ?	
Goniozus natalensis Gord / ex		CBC / 1984	? / ?	
Euplectrus platyhypenae / ex		CBC / 1984	? / ?	
Trichosphilus diatraea / ex		CBC / 1984	? / ?	
Descampsina sesamiae / ex		CBC / 1984–1986	? / ?	
Iphiaulax kimballi Kirk / ex		CBC / 1985	? / ?	
Macrocentrus prolificus / ex		CBC / 1986	? / ?	
Trichogramma spp. / ex		ABC / 1979–2000	Partial control / ?	Rogg, 2000a
Trichogramma pretiosum / ex		ABC / 1992–now	Good control / 4,500	
Baculovirus diatraea / na		ABC / 1985–2000	Control / ?	
Beauveria bassiana / na		ABC / 2000	Control / ?	
Metarhizium anisopliae / na		ABC / 2000	Control / ?	
Telenomus alecto/ ?		ABC / 2000	Control / ?	
Trichogramma galloi / ?		ABC / 2000	Control / ?	
Pheropsophus aequinoctalis / na	Mole crickets / sugarcane	CBC / 2000	Good control / ?	Rogg, 2000a,b
Sirentha carinata / na		CBC / 2000	Good control / ?	
Ormia depleta / na		CBC / 2000	Good control / ?	
Larra bicolor / na		CBC / 2000	Good control / ?	
Larra sp. / na		CBC / 2000	Good control / ?	
Brachicantha sp. / na	Sugarcane coccid / sugarcane	NC	Good control / ?	Rogg, 2000a,b
Hyperaspis spp. / na		NC	Good control / ?	
Aspergillus flavus / na		NC	Good control / ?	
Anagyrus saccharicola / ex		CBC / 2000	partial control, ?	
Phalacrus obscurus / na	Sugarcane smut / sugarcane	NC	Good control / ?	Rogg, 2000a,b

Continued

Table 5.1. Continued.

Biocontrol agent / exotic (ex), native (na)	Pest / crop	Type of biocontrol[a] / since	Effect /area (ha) under biocontrol[b]	Reference
Brachytarsus sp. / na		NC	Good control / ?	
Drino inhertis (Wied.)	Corn earworm / maize	ABC/CBC / 1970	? / ?	Pruett, 1996
Eucarcelia illota (Curran)		ABC/CBC / 1970	? / ?	
Ecphoropsis perdistinctus Vier.		ABC/CBC / 1970	? / ?	
Eriborus sp.		ABC/CBC / 1970	? / ?	
Bracon hebetor Say		ABC/CBC / 1970	? / ?	
Goniophthalmus halli Mesnil		ABC/CBC / 1970	? / ?	
Eucelotoria sp.		ABC/CBC/ 1970	? / ?	
Cotesia marginiventris (Cresson)		ABC/CBC/ 1970	? / ?	
Opius concolor siculus Sppl. / ex	Fruit flies / citrus	CBC / 1960s	? / ?	Pruett, 1996
Opius longicaudatus (Ash.) / ex		CBC / 1969	? / ?	
Pachycrepoideus vindemiae (Rond.) / ex		CBC / 1969	? / ?	
Syntomosphyrum indicum Silv./ex		CBC / 1969	? / ?	
Dirhinus giffardii (Silv.) / ex		CBC / 1971	? / ?	
Opius oophilus / ?		ABC / 1992 – 2012	Control / 54,000	Rogg 2000a,b
O. incisi / ?		ABC / 1992 – 2012	Control / 54,000	
Biosteres formosanus / ?		ABC / 1992 – 2012	Control / 54,000	
B. vandenboschi / ?		ABC / 1992 – 2012	Control / 54,000	
B. compensans / ?		ABC / 1992 – 2012	Control / 54,000	
B. longicaudatus / ?		ABC / 1992 – 2012	Control / 54,000	
Rodolia cardinalis / ex	Cottony cushion scale / citrus	CBC / 1950s	Good control / 54,000	Rogg, 2000a
Trichogramma evanescens / ex	Potato tuber moths / potato	ABC / 1984	? / ?	Pruett, 1996
Baculovirus phthorimaea / na		ABC / 1990s	Good control / 1,800	Calderon *et al.*, 2002
Copidosoma sp. / na		ABC / 1990s	? / ?	
Bacillus thuringiensis / na		ABC / 2000	Good control / 1,800	
B. thuringiensis / na	Potato weevils / potato	ABC / 2000	No control	Calderon *et al.*, 2004
Beauveria brongniartii / na		ABC / 2000	Partial control / 180,000	
Beauveria bassiana / na	Coffee berry borer / coffee	NC	Insufficient	Rogg, 2000a
B. bassiana / na		ABC / 1992	Control / 23,000	
Cephalonomia stephanoderis / ?		ABC / 1992	Control / 23,000	
Trichogramma pretiosum / ex	Lepidopterans / cotton	ABC / 1980s	No control, not established	Rogg, 2000a
Trichogrammatoidea sp. / na		NC	partial control / 140,000	

Continued

Table 5.1. Continued.

Biocontrol agent / exotic (ex), native (na)	Pest / crop	Type of biocontrol[a] / since	Effect /area (ha) under biocontrol[b]	Reference
B. thuringiensis / na	Noctuid lepidopterans / quinoa	ABC / 2000	Control / organic quinoa ?	PROINPA, 2013
Parasitoid and predator complex / na	Quinoa moths	NC	Partial control / 118,000	PROINPA, 2013
B. bassiana / na	Soybean stalk borer / soybean	ABC / 2010	Control / part of 1,300,000	Anapo, 2011
Parasitoid and predator complex / na	Lepidopterans / soybean	NC	Control / part of 1,300,000	Anapo, 2011
Baculovirus *Anticarsia gemmatalis* / na	*Anticarsia gemmatalis* / soybean	ABC / 2010	Control / part of 1,300,000	
B. thuringiensis / na	*Omiodes indicate* / soybean	ABC / 2010	Control / part of 1,300,000	
Parasitoid and predator complex / na		ABC / 2010	Control / part of 1,300,000 ha	
Parasitoid complex / na	Stink bugs / soybean	NC	Partial control / 1,300,000	Rogg, 2000a
Trissolcus spp. / na		ABC / 2000	Control / part of 1,300,000	
Scutellista cyanea / ex	Olive scale / olive	CBC / 1950s	? / established	Pruett, 1996
Aphelinus mali / ex	Woolly apple aphid / apple	CBC / 1950s	? / established	Pruett, 1996
Prospaltella berlesi / ex	White peach scale / fruit	CBC / 1950s	? / established	Pruett, 1996
Telenomus nawaii / ex	*Spodoptera* spp. / various	ABC / 1985	? / ?	Pruett, 1996
Telenomus remus / ex		ABC / 1981–1989	? / ?	
Cordyceps sobolifera Berk / ex	Cicadellids / asparagus	? / 1987	? / ?	Pruett, 1996
Spalangia endius / ex	Housefly / stable flies	ABC / 1986	? / ?	Pruett, 1996
S. cameroni / ex		ABC / 1986	? / ?	

[a]Type of biocontrol: ABC = augmentative, CBC = classical, ConsBC = conservation biocontrol; NC = natural control
[b]Area of crop harvested in 2016 according to FAO (http://www.fao.org/faostat/en/#data/qc)

References

(References with grey shading are available as supplementary electronic material)

ANAPO (2011) Technical dissemination cards: Soybean Pest – Integrated Pest Management. Project: Production of Responsible Soy in Bolivia. Asociación de Productores de Oleaginosas y Tigo, Santa Cruz de la Sierra, Bolivia.

Barea, O., Herbas, J., Calderón, R., Crespo, L., Bejarano, C. and Lino, V. (2002) Desarrollo de alternativas para el manejo integrado de la polilla [Development of alternatives for integrated moth management].

In: Calderón, R., Bares, O., Ramos, J., Crespo, L., Bejarano, C., Herbas, J. and Lino, V. (eds) Desarrollo de Componentes de Manejo Integrado de las Polillas de la Papa (*Phthorimaea operculella y Symmetrischema tangolias*) en Bolivia y el Bioinsecticida Baculovirus (MATAPOL) [Development of Integrated Management of Potato Tuber Moth (*Phthorimaea operculella* and *Symmetrischema tangolias*) in Bolivia and the Baculovirus Bioinsecticide (MATAPOL)]. PROINPA (Fundación para la Promoción e Investigación de Productos Andinos [Potato Research Project]), Cochabamba, Bolivia, pp. 45–71.

Barrantes, C.M.A. (2016) Cría en condiciones controladas de la polilla de la quinua (*Eurysacca melanocampta*) y sus niveles de parasitismo natural en comunidades del Altiplano Centro y Norte [Rearing in controlled conditions of the quinoa moth (*Eurysacca melanocampta*) and its levels of natural parasitism in communities of the Central and Northern Highlands]. BSc thesis. Universidad Mayor de San Andrés, La Paz, Bolivia.

Bennett, F.D. and Squire, F.A. (1972) Investigations on the biological control of some insect pests in Bolivia. *International Journal of Pest Management* 18, 459–467.

Calderón, R., Bares, O., Ramos, J., Crespo, L., Bejarano, C., Herbas, J. and Lino, V. (2002) Desarrollo de Componentes de Manejo Integrado de las Polillas de la Papa (*Phthorimaea operculella y Symmetrischema tangolias*) en Bolivia y el Bioinsecticida Baculovirus (MATAPOL) [Development of Integrated Management of Potato Tuber Moths (*Phthorimaea operculella and Symmetrischema tangolias*) in Bolivia and the Baculovirus Bioinsecticide (MATAPOL)]. PROINPA, Cochabamba, Bolivia.

Calderón, R., Franco, J., Barea, O., Crespo, L., Esprella, R., Bejarano, C., Ramos, J., Iporre, G. and Casso, R. (2004) Desarrollo de Componentes del Manejo Integrado del Gorgojo de los Andes en el Cultivo de la Papa en Bolivia [Development of Integrated Pest Management Strategies for the Andean Potato Weevil in Bolivia]. Documento de trabajo 25. Fundación PROINPA – Proyecto PAPA ANDINA, Cochabamba, Bolivia.

CENETROP (2018) Centro Nacional de Enfermedades Tropicales: Projects. Available at: http://www.cenetrop.org.bo (accessed 27 October 2019).

CIA (2017) The World Factbook: Bolivia. Available at: https://www.cia.gov/library/publications/the-world-factbook/geos/bl.html (accessed 7 July 2019).

Figueroa, I., Ríos, B., Crespo, L., Saravia, R. and Quispe, R. (2013) Parasitoides de larvas de polilla de la quinua (*Eurysacca quinoae* R). Perspectiva de control biológico en quinua orgánica (Parasitoids of quinoa moth larvae (*Eurysacca quinoae* R) perspective of biological control in organic quinoa). In: Vargas, M. (ed) *Proceedings of the Quinoa Scientific Congress*, La Paz, Bolivia, pp. 359-369.

INE (Instituto Nacional de Estadistica) (2017) Estadísticas sobre la evolución de la superficie, producción y rendimiento de los principales cultivos agrícolas. Available at: http://www.ine.gob.bo/ (accessed 26 August 2018).

Ortíz, R. (1993) Insectos plaga en Quinua [Insect pests in Quinoa]. Cultivos Andinos. FAO, Oficina Regional para América Latina y el Caribe, Santiago, Chile.

Ortuño, N., Navia, O., Meneces, E., Barja, D., Villca, S., Plata, G., Claros, M., Arandia, W. and Crespo, L. (2011) Catálogo de Bioinsumos [Catalogue of bio-inputs]. Fundación PROINPA-BioTop, La Paz, Bolivia.

PROINPA (2004a) Final Technical Report. Integrated Management of Major Insect Pests of Potatoes in Hillside Systems in the Cochabamba Region of Bolivia R 8044 (ZA0485). Fundación para la Promoción e Investigación de Productos Andinos [Potato Research Project], La Paz, Bolivia.

PROINPA (2004b) Estudio de los Impactos Sociales, Ambientales y Económicos de la Promoción de la Quinua en Bolivia [Study of the Social, Environmental and Economic Impacts of the Promotion of Quinoa in Bolivia]. Fundación para la Promoción e Investigación de Productos Andinos [Potato Research Project], La Paz, Bolivia.

PROINPA (2008) Herramientas para el Desarrollo del Manejo Integrado de Plagas en la Producción de Quinua Orgánica [Tools for the Development of Integrated Pest Management in the Production of Organic Quinoa] (Nov. 2007, June 2008). Informe Proyecto. Fundación AUTAPO. La Paz, Bolivia.

PROINPA (2013) Informe Anual 2012–2013 del Proyecto: Desarrollo y validación participativa de las innovaciones tecnológicas que mejoren las estrategias para manejo sostenible del sistema centrado en quinua en el Altiplano boliviano [Annual Report 2012–2013 of the Project: Development and participatory validation of technological innovations that improve strategies for sustainable management of the system centered on quinoa in the Bolivian Altiplano]. Fundación McKnight, La Paz, Bolivia.

Pruett, C.J.H. (1996) Biological control in Bolivia: history and development. In: Zapater, M.C. (ed.) *El Control Biológico en América Latina*. IOBC/NTRS, Buenos Aires, Argentina, pp. 17–24.

Rasmussen, C., Lagnaoui, A. and Delgado, R. (2001) *Phytomyptera* sp. (Diptera: Tachinidae): an important natural control agent of the quinoa moth, *Eurysacca quinoae* (Lepidoptera: Gelechiidae) in central Peru. *Tachinid Times* 14, 5–6.

Rogg, H.W. (2000a) *Manual de Entomología Agrícola de Bolivia* [*Handbook of Agricultural Entomology of Bolivia*]. Abya-Yala, Quito, Ecuador.

Rogg, H.W. (2000b) *Manual Manejo Integrado de Plagas en Cultivos Tropicales* [*Handbook of Integrated Pest Management in Tropical Crops*]. Abya-Yala, Quito, Ecuador.

Saravia, R. and Quispe, R. (2006) Manejo Integrado de las Plagas Insectiles del Cultivo de la Quinua Fascicule 4 (Integrated Management of Insect Pests of Quinoa Cultivation). In: PROINPA and FAUTAPO (eds) Series of Modules published in Sustainable Production Systems in the Cultivation of Quinoa (2006, La Paz, BO). Module 2. Agronomic Management of Organic Quinoa. Fundación PROINPA, Fundación AUTAPO, Embajada Real de los Países Bajos. La Paz, Bolivia.

Saravia, R., Mamani, A., Bonifacio, A. and Alcon, M. (2008) Diagnóstico de los enemigos naturales de las plagas del cultivo de quinua [Diagnosis of the natural enemies of the insect pests of the quinoa crop]. Annual Report 2008–2009, Fundación PROINPA, Rubro Granos Altoandinos. Cochabamba, Bolivia.

Vélez, E.T., García, M.E.A., Moraes, R.M., Gálvez, F.B., Ocampo, D.R.L.V., del Carpio, A.T., Arnéz, L.L., Lara, G.A., Kraljevic, J.B., Padilla, C.A.M. *et al.* (2017) Food and nutrition security in Bolivia. A country of incalculable wealth. In: *Challenges and Opportunities for Food and Nutrition Security in the Americas. The View of the Academies of Sciences*. IANAS, IAP and BMBF, México DF, pp. 52–75. [Free public access of this publication in English and Spanish at www.ianas.org]

6 Biological Control in Brazil

Vanda Helena Paes Bueno[1]*, José Roberto Postali Parra[2], Wagner Bettiol[3] and Joop C. van Lenteren[4]

[1]*Laboratory of Biological Control, Department of Entomology, Federal University of Lavras, Lavras, Minas Gerais, Brazil;* [2]*Laboratory of Biology of Insects, Department of Entomology and Acarology, ESALQ/USP, Piracicaba, São Paulo, Brazil;* [3]*Embrapa Environment, Jaguariúna, São Paulo, Brazil;* [4]*Laboratory of Entomology, Wageningen University, Wageningen, The Netherlands*

* E-mail: vhpbueno@ufla.br

Abstract

Classical biological control attempts from 1921 to 1944 were not effective. During the 1960s, an important success was obtained by controlling the rhodesgrass scale in thousands of hectares of pastures with an introduced parasitoid. Also biocontrol of wheat aphids by introduction of parasitoids and predators appeared effective. Further, classical biocontrol of sirex wood wasps in pine plantations was achieved with parasitoids and entomopathogenic nematodes. Augmentative biocontrol of the sugarcane borer by native dipteran parasitoids started in the 1960s, later followed by importation and release of *Cotesia flavipes* parasitoids. In the 1980s, biocontrol of soybean caterpillars was realized on more than 2 million hectares by application of on-farm produced AgMNPV virus. Predatory mites are used for augmentative biocontrol of spider mites in apple orchards and greenhouse crops; predators and parasitoids are used for control of lepidopterans in eucalyptus plantations and field crops such as sugarcane. Parasitoids are released in soybean for control of stink bugs. *Trichoderma* spp. are applied on 5.5 million hectares for control of soil-borne diseases in many crops. Recent successful classical biocontrol programmes deal with control of cassava mealybug, citrus leaf miner, and Asian citrus psyllid. Brazil has become one of the pioneer countries worldwide in the production and use of microbial control agents and natural enemies to control pests and diseases on millions of hectares. Brazil currently has 26 facilities producing microbial agents and 21 for mass rearing natural enemies. More than 80 microbial products are registered for control of arthropods, while fewer than ten natural enemies have been registered to date.

6.1 Introduction

Brazil has an estimated population of almost 207,360,000 (July 2017) (CIA, 2017). The areas of main crops in Brazil as at June 2018 are: soybean 34,765,926 ha; maize 16,531,306 ha; sugarcane 9,411,207 ha; cereals (including wheat, oat, barley and triticale) 2,515,714 ha; rice 1,941,458 ha; coffee 1,916,382 ha; cotton 1,150,745 ha; citrus (orange) 626,611 ha; cocoa 597,556 ha; and tobacco 387,902 ha (IBGE, 2018). Brazil has 6.66 million hectares of commercial forest, which is expected to increase to 9 million hectares in 2020 (SNIF, 2018). The Brazilian Agricultural Research Corporation, Empresa Brasileira de Pesquisa Agropecuária (Embrapa), reported that Brazil has 167 million hectares of cattle pastures. According to Vilela *et al.* (2017, pp. 77, 78, 86, 87 and 97):

> Brazil has become an example of a food secure country and one of the world's most important agricultural export countries ... Brazil's geographic area is one of the largest in the world, totaling 8,515,767 km^2 ... Brazil is among the most biodiverse countries in the world. Brazilian flora is the most diverse with approximately 55,000 species accounting for a quarter of the world's total number of species ... The main crops where Brazil has wild relatives present ... are *Arachis*, *Manihot*, *Anacardium*, *Hevea*, *Oryza*, *Ipomoea*, *Solanum* and several tropical fruits such as passion fruit ... Brazil is a major producer of animal protein and the world's largest beef exporter ... Brazilian beef and milk production systems are almost exclusively based on pastures, resulting in a comparative advantage through relatively low production costs as well as a competitive advantage from farming 'green cattle', which is a safe product, with quality features highly valued by the market ... Over the past four decades, Brazil eventually became self-sufficient in food production ...

6.2 History of Biological Control in Brazil

6.2.1 Period 1880–1969

Classical and augmentative biological control of white peach scale, rhodesgrass scale and sugarcane borer

The first attempted case of classical biocontrol took place in 1921, with the importation and release of the parasitoid *Encarsia berlesei* Howard from the USA to control white peach scale *Pseudaulacaspis pentagona* (Targioni), but the project was not successful. Importation of natural enemies into Brazil continued until 1944 (Table 6.1), but none of these introductions resulted in sufficient control, supposedly because these few isolated biocontrol programmes were conducted by individual researchers and did not have any scientific back-up (Parra, 2014). Following the last introduction of natural enemies in 1944,

Table 6.1. Overview of major biological control projects in Brazil.

Natural enemy	Origin	Started	Target pest and crop	Type of biocontrol[a]	Success	Area under biocontrol[b]
Period 1880–1969						
Encarsia berlesei	USA	1921	*Pseudaulacaspis pentagona* white peach scale in orchards	CBC	No	N.A.
Prorops nasuta	Africa	1923	*Hypothenemus hampei* coffee berry borer in coffee plantations	CBC	No	N.A.
Aphelinus mali	Uruguay	1928	*Eriosoma lanigerum* woolly apple aphid in apple orchards	CBC	No	N.A.
Tetrastichus giffardianus	USA	1937	*Ceratitis capitata* Mediterranean fruit fly in fruit orchards	CBC	No	N.A.
Macrocentrus ancylivorus	USA	1944	*Grapholita molesta* Oriental fruit moth in fruit orchards	CBC	No	N.A.
Neodusmetia sangwani	USA	1967	*Antonina graminis* rhodesgrass scale in pastures	CBC	Yes	100,000s of ha
Period 1970–now						
Cotesia flavipes	Trinidad + Pakistan	1970s	*Diatraea saccharalis* sugarcane borer in sugarcane	ABC	Yes	3,500,000 ha
Parasitoids and predadors (Table 6.2)	Many countries	1970s	Wheat aphids in wheat	CBC	Yes	1,000,000 ha
AgMNPV virus	Native	1980–2010	Soybean caterpillars in soybean	ABC	Yes	> 2,000,000 ha
Deladenus siricidicola	Australia	1989	*Sirex noctilio* sirex woodwasp in pine plantations	CBC	Yes	450,000 ha
Beauveria bassiana	Native	1990	*Hedypathes betulinus, Bemisia tabaci* and other pests	ABC	Yes	> 4,000,000 ha
Ibalia leucospoides	Accidental introd	1990	*Sirex noctilio* sirex woodwasp in pine plantations	CBC/FBC	Partial	450,000 ha
Neoseiulus californicus	Native	1990s	Spider mites in apple orchards	ABC	Yes	1,800 ha
Podisus nigrispinus	Native	1980s	Lepidopterans in eucalyptus	ABC	Yes	> 20,000 ha
Trichogramma pretiosum	Native + Colombia	1990s	Lepidopterans in various crops	ABC	Yes	200,000 ha
Trissolcus basalis	Native	1990–2005	*Nezara viridula* stink bugs in soybean	ABC	Yes	20,000 ha
Trichoderma spp.	Native	1991	Soilborne plant pathogens in soybean, beans, cotton etc.	ABC	Yes	5,500.000 ha
Diachasmimorpha longicaudata	USA	1994	*Ceratitis capitata* and *Anastrepha* spp in fruit orchards	CBC	No data	No data
Acerophagus coccois, Anagyrus diversicornis and Aenasius vexans	Colombia	1994/95	*Phenacoccus herreni* cassava mealybug in cassava	CBC	Yes	1,000,000 ha
Megarhyssa nortoni, Rhyssa persuasoria,	Australia	1996	*Sirex noctilio* sirex woodwasp in pine plantation	CBC	No	N.A.
Ageniaspis citricola	USA	1998	*Phyllocnistis citrella* in citrus	CBC	Yes	450,000 ha
Xenostigmus bifasciatus	USA	2001	*Cinara atlantica, C. pinivora* in pine plantations	CBC	Yes	No data
Psyllaephagus pilosus	Accidental introd	2005	*Ctenarytaina spatulata*	CBC/FBC	Yes	No data

Tamarixia radiata	Accidental introd	2005	*Diaphorina citri* vector of huanglongbing (HLB) in citrus	CBC/FBC	Yes	12,000 ha[c]
Telenomus podisi	Native	2005	*Euschistus heros* stink bugs in soybean	ABC	Yes	20,000 ha
Neoseiulus californicus	Native	2006	Pest in orchards, greenhouse vegetables and ornamentals	ABC	Yes	
Phytoseiulus macropilis, P. longipes	Native	2006	Greenhouse vegetables and ornamentals	ABC	Yes	15,000 ha
Stratiolaelaps scimitus	Native	2006	Greenhouse vegetables and ornamentals, mushrooms	ABC	Yes	
AgMNPV virus	Native	2010	Soybean caterpillars in soybean	ABC	Yes	100,000 ha
Helicoverpa zea SNPV virus	Native + exotic	2010	*Helicoverpa armigera* and other *H.* species	ABC	Yes	1,300,000 ha
Metarhizium anisopliae	Native	2010	*Mahanarva fimbriolata* spittle bug in sugarcane	ABC	Yes	4,000,000 ha
Trichogramma galloi	Native	2010	*Diatraea saccharalis* in sugarcane	ABC	Yes	1,700,000 ha
Cleruchoides noackae	Australia	2012	*Thaumastocoris peregrinus* bronze bug in eucalyptus	CBC	Yes	No data
Selitrichodes neseri	South Africa	2014	*Leptocybe invasa* gallwasp in eucalyptus	CBC	Yes	No data
Purpureocillium lilacinum	Native	2017	*Meloidogyne spp.* in soybean, corn	ABC	Yes	500,000 ha
Biopesticide *Bacillus thuringiensis*	Native + exotic	1990	Many lepidpterans, coleopterans etc. in many crops	ABC	Yes	5,000,000

[a]Type of biocontrol: CBC = classical biological control, ABC = augmentative biological control, FBC = fortuitous biological control
[b]N.A. = Not applicable
[c]See text, special type of application

Brazil passed through a period where large amounts of synthetic pesticides were used against pests.

A new, large-scale classical biocontrol project was started in 1967, when the parasitoid *Neodusmetia sangwani* (Subba Rao) was introduced from Texas (Schuster and Boling, 1971) for control of the rhodesgrass scale *Antonina graminis* Maskell, a serious pest in pastures in Brazil. In 1969 the parasitoid was recovered in the areas where it was first released (the State of Bahia), demonstrating its establishment (Costa *et al.*, 1970). In 1981, 14 years after the introduction, surveys indicated the establishment of the parasitoid in São Paulo State and in several other areas with pastures where it was released (Batista Filho *et al.*, 2018). Gabriel (2017) reported the predominance of *N. sangwani* in the complex of parasitoids of *A. graminis* during a survey more than 30 years after its release in São Paulo State. This was the first Brazilian classical biocontrol project demonstrating successful establishment and effective permanent control of a pest by an introduced parasitoid (Batista Filho *et al.*, 2018). It is difficult to provide an estimate of the area where the parasitoid controls the rhodesgrass scale, but everywhere where the scale occurs in Brazil its natural enemy *N. sangwani* also occurs, indicating that the surface under classical biocontrol must be hundreds of thousands of hectares. However, the damage caused by this pest became less important due to replacement of rhodesgrass by other grass species.

An augmentative biocontrol programme to control the sugarcane borer *Diatraea saccharalis* (F.) was started in the 1960s by Dr D. Gallo (ESALQ/University of Sao Paulo) and concerned the use of two native tachinid parasitoids *Lydella minense* (Town.) (amazon fly) and *Billaea claripalpis* Wulp (Parra, 2014). Mass rearing of *D. saccharalis* as host for the parasitoids was initially done on a natural diet and later continued by using an artificial diet developed by Hensley and Hammond (1968).

Augmentative biological control of pests and diseases with microbial control agents

The first reports of an entomopathogenic organism date back to the 1920s, when *Metarhizium anisopliae* (Metchnikoff) Sorokin was observed to have attacked spittlebugs, and in 1925 an isolate of the same fungus was introduced from Trinidad and Tobago and tested against the cercopid *Mahanarva fimbriolata* (Stal), but without success (Li *et al.*, 2010). In the 1930s, several observations indicated that the fungus *Hirsutella verticillioides* Charles was responsible for decimating populations of lace bugs (*Leptopharsa heveae* Drake & Poor) over wide areas of rubber tree plantations (Charles, 1937). Li *et al.* (2010) mentioned a number of other cases where natural outbreaks of fungi like *Lecanicillium* (= *Verticillium*) *lecanii* Zare & Gams, *M. anisopliae* and *Beauveria bassiana* (Bals.-Criv) Vuill resulted in significant reduction of several important agricultural pests in the period from 1930 to 1969. Entomopathogenic nematodes were studied in the 1920s and 1930s, but they were not yet applied in Brazilian agriculture (Dolinski *et al.*, 2017).

Activities in the area of biocontrol of plant diseases in Brazil started in 1950 with research on *Trichoderma* spp. (see section 'Current situation', below, for more information). Microbial agents for disease control were first developed to control the 'Tristeza' virus by pre-immunization or cross-protection in 1959 by G.W. Muller and A.S. Costa at the Agronomic Institute of São Paulo State, where a mild strain from severely injured orchards was selected and used to infect healthy citrus plants. Plants that had been pre-immunized with these mild strains grew well and showed practically no symptoms of 'Tristeza' or stem pitting. These plants produced good yields compared with plants from the same clones that either had not been pre-immunized or had been inoculated with severe isolates (Costa and Muller, 1980; Bettiol, 1996). This technique is still in use today.

6.2.2 Period 1970–2000

Classical biological control of arthropods in agriculture and forestry

Several new classical biocontrol programmes have been developed since 1970. The first one concerned biocontrol of wheat aphids and started in 1978 by targeting the aphid pest species *Metopolophium dirhodum* (Walker), *Schizaphis graminum* (Rondani) and *Sitobion avenae*

(Fabricius). This project is of particular interest because at least 16 different natural enemies were imported from eight countries (Table 6.2). Initially, from 1978 to 1982, large numbers of 14 species of hymenopteran parasitoids, among which six species were from the genus *Aphidius* and including *Aphidius colemani* Viereck, were imported from different countries and released for the control of wheat aphid complex in southern Brazil (Gassen and Tambasco, 1983). These releases resulted in complete termination of chemical pesticide sprays for control of *S. graminum*, the main aphid pest in wheat in Paraná State. *A. colemani* appeared to be the most effective of the released exotic parasitoid species (Gassen and Tambasco, 1983). Until 1992, 12 aphidiine species were introduced in five States (Rio Grande do Sul, Santa Catarina, Paraná, Mato Grosso do Sul and São Paulo). The programme was considered very successful, because the high aphid populations observed on wheat crops in the 1970s were observed to have been drastically reduced in the 1980s (Gassen and Tambasco, 1983; Salvadori and Salles, 2002). A total of 20,396,000 parasitoids were released on approximately 1 million hectares of wheat. The success of this programme can be explained by several observations: parasitoid establishment has been observed; a larger complex of parasitoids was present in wheat fields than before the releases and this complex was active and survived during critical climatic periods; percentages of parasitism increased and populations of *M. dirhodum* and *S. avenae* declined, as well as damage caused by these aphids (Salvadori and Salles, 2002).

Table 6.2. Species of natural enemies introduced for biological control of wheat aphids by CNPT/Embrapa, Passo Fundo, Brazil, 1983 (retrieved from Salvadori and Salles, 2002).

Natural enemy	Aphid host/prey	Country providing natural enemy
PARASITOIDS		
Hymenoptera — Apheliniidae		
Aphelinus abdominalis Dalman	*Metopolophium dirhodum*	Chile
Aphelinus asychis Walker	*Metopolophium dirhodum*, *Sitobion avenae*	France
Aphelinus flavipes Forster	*Schizaphis graminum*	France
Aphelinus varipes Forster	*Metopolophium dirhodum, Schizaphis graminum*	Hungary, France
Hymenoptera — Aphidiidae		
Aphidius colemani Viereck	*Metopolophium dirhodum*, *Sitobion avenae*	France, Israel
Aphidius ervi Haliday	*Acyrthosiphon pisum, A. kondoi, Macrosiphum carnosum*	France, Czech Republic
Aphidius pascuorum Marshall	*Schizaphis graminum*	France
Aphidius picipes Ness	*Schizaphis graminum*	Czech Republic, Italy, Hungary
Aphidius rhopalosiphi De Stefani	*Sitobion avenae*; *Schizaphis graminum*, *Metopolophium dirhodum*	Chile, Czech Republic, France
Aphidius uzbekistanicus Luzhetzki	*Sitobion avenae*; *Metopolophium dirhodum*	Italy
Ephedrus plagiator Nees	*Sitobion avenae*; *Metopolophium dirhodum*	France, Czech Republic
Lysiphlebus testaceipes Cresson	*Schizaphis graminum*	Chile
Praon gallicum Stary	*Metopolophium dirhodum*	France
Praon volucre Haliday	*Metopolophium dirhodum*	France, Czech Republic, Spain
PREDATORS		
Coleoptera — Coccinellidae		
Hippodamia quinquesignata Kirby	Several aphid species	USA
Coccinella septempunctata Linnaeus	Several aphid species	Israel

The next classical biological programme dealt with biocontrol of cassava pests and started in the 1990s. Cassava (*Manihot esculenta* Crantz) is among the top five most important sources of carbohydrates for human consumption in the tropics, and Brazil produced cassava on about 1,700,000 hectares in 2002 (Souza and Fialho, 2003). The mealybug *Phenacoccus herreni* Cox & Williams was first observed damaging cassava plants in North-eastern Brazil in the early 1980s. By the mid-1990s this pest was found in 57 municipalities belonging to six states in north-eastern Brazil, making cassava production uneconomic in some areas, with yield losses of up to 80% (Bento *et al.*,1999). Because of the social and economic importance of cassava in this area, a project was initiated in mid-1994 for biocontrol of *P. herreni*, with the introduction of three Hymenoptera of the family Encyrtidae: *Apoanagyrus diversicornis* (Howard), *Aenasius vexans* (Kerrich) and *Acerophagus coccois* Smith, from Colombia and Venezuela. By the end of 1996, *A. diversicornis* was recovered at 130, 234, 304 and 550 km from its release site after 6, 14, 21 and 33 months, respectively (Bento *et al.*, 2000). *A. coccois* was recovered at 180 km from its release site 9 months after release. *A. vexans*, however, did not disperse at all, despite being consistently recovered at its release site. The overall mean population of *P. herreni* densities reduced progressively from the beginning to the end of the period of study. After 1997, it was clear that the concerted action of the three introduced parasitoids and native natural enemies was sufficient to reduce *P. herreni* to low population levels. Again, it is difficult to provide an estimate of the area where this project was successful.

Another classical biocontrol programme involved the citrus miner *Phyllocnistis citrella* Stainton, which was first reported in the State of São Paulo in 1996. One of its parasitoids, *Ageniaspis citricola* (Logvinovskaya), originating from Vietnam, was brought to Brazil in 1998 from Dr Hoy's laboratory (University of Florida, Gainesville, USA). After a quarantine period in the Costa Lima Quarantine Laboratory at Embrapa Environment Jaguariúna/SP, the insects were mass reared and releases began in October 1998, using a technique developed by Chagas and Parra (2000). About 2 million parasitoids were released at different sites in 75 municipalities that were representative of the citrus industry in São Paulo State. Parasitoids were recovered in some localities 3 months after their release. Subsequent releases were conducted in the States of Paraná (PR), Minas Gerais (MG), Rio Grande do Sul (RS), Santa Catarina (SC), Goiás (GO), Bahia (BA), Rio de Janeiro (RJ) and Piauí (PI). A significant reduction of the pest was observed on 450,000 ha of citrus in São Paulo; and today, even with the application of insecticides to control the Asian citrus psyllid *Diaphorina citri* Kuwayama, the natural parasitism rate is on average 50%, indicating that the parasitoid has become fully established in Brazil (Chagas *et al.*, 2002).

During the period 1970–1999 a classical biocontrol project for a forest pest, the Sirex wood wasp *Sirex noctilio* Fabricius, was developed. *S. noctilio* was recorded for the first time in *Pinus* plantations in Brazil in 1988 in the State of Rio Grande do Sul. This pest is now present on about 450,000 ha, in the States of Rio Grande do Sul, Santa Catarina, Paraná, São Paulo (SP) and Minas Gerais. One of the parasitoids of *S. noctilio*, *Ibalia leucospoides* (Hochenwarth), was introduced accidentally together with its host in 1990 in Rio Grande do Sul and it is now found in all *Pinus* plantations attacked by the wood wasp and provides partial control (Embrapa, 2018). Since 1996, the National Program for Control of the Wood Wasp (PNCVM) has reared and released the parasitoids *I. leucospoides*, *Rhyssa persuasoria* (L.) and *Megarhyssa nortoni* (Cresson) to increase wood wasp mortality. The latter two parasitoid species were introduced in 1996–1997 and in 2003 from Tasmania, in a cooperative project between the Brazilian Agricultural Research Corporation (Embrapa Foresty, Brazil), the Commonwealth Scientific and Industrial Research Organization (CSIRO, Australia), the International Institute of Biological Control (IIBC, now CABI, UK) and the Forestry Service of the United States Department of Agriculture (USDA). Establishment of the two parasitoids has not yet been confirmed (Iede *et al.*, 2012).

Augmentative biological control of arthropods by macrobial control agents in agriculture and forestry

Importation of natural enemies for augmentative biological was initiated after the 'Costa Lima' quarantine system was established in 1991 by Embrapa Environment, Jaguariúna/SP (Parra, 2014). Since the establishment of this quarantine unit, around 773 species of organisms for biocontrol (parasitoids, predators and antagonists) and other purposes (inoculants, biofertilizers and microorganisms in general) were imported until 2016 (Sá *et al.*, 2016). Examples of the imported species can be found on a CD Rom available at the Laboratório de Quarentena Costa Lima (2005) at Embrapa Environment. An additional factor contributing to expansion of biocontrol in Brazil was the release of the 'Fourth Catalogue of Insects that Live on Plants from Brazil: Their Parasites and Predators' by Silva *et al.* (1968). Progress in biocontrol was further stimulated by: (i) teaching of entomology and biocontrol at universities from the end of the 1960s; (ii) changes in the entomologists' mindset from just killing insects to managing pest populations in IPM programmes; (iii) initiation of research programmes in biocontrol, insect mass rearing and insect nutrition; and (iv) establishment of the Entomological Society of Brazil (SEB) in 1972, which regularly organizes the Symposium of Biological Control (Siconbiol) (SEB, 2018).

All these developments resulted in the creation of insect mass-rearing facilities initially financed by the federal government and national programmes and later by private companies to supply the expanding market for biocontrol agents. Brazil became very active during this period in the field of augmentative biocontrol and had 47 mass production facilities for arthropod biocontrol agents; 26 of these produced microbial control agents and 21 produced macrobial control agents.

A survey of parasitoids associated with *Musca domestica* L. and *Stomaxis calcitrans* L. in Brazil revealed the presence of seven native parasitoids: *Muscidifurax raptorellus* Kogan & Legner, *Muscidifurax uniraptor* Kogan & Legner, *Nasonia vitripennis* (Walker), *Pachycrepoideus vindemmiae* (Rondani), *Spalangia cameroni* Perkins, *Spalangia endius* Walker, *Spalangia gemina* Bouček and *Tachinaephagus zealandicus* Ashmead (Berti-Filho *et al.*, 1989; Costa *et al.*, 2004). Mass rearings of the parasitoids *M. uniraptor*, *P. vindemmiae*, *S. cameroni*, *S. endius* and *S. gemina* were developed from 1984 to 1990, and mixed releases of these parasitoids resulted in good control of flies in poultry production with a total flock of more than 5 million laying hens.

The coccinellid predator *Cryptolaemus montrouzieri* Mulsant, imported from Chile, was used for biocontrol of *Planococcus citri* Risso and *Pseudococcus longispinus* (Targioni Tozzetti) in commercial citrus orchards in Brazil. The predator was mass reared and released from 1990 to 2005 on 400 ha in São Paulo State and effectively controlled the pests. However, due to the costs of mass producing the predator, the programme was terminated (Gravena, 2005).

Another augmentative biocontrol project was initiated to control the fruit flies *Ceratitis capitata* (Wied.) and *Anastrepha* spp. Many species of native parasitoids of fruit flies were found in Brazil, but their mass rearing was not successful (Garcia and Ricalde, 2013). Therefore, the parasitoid *Diachasmimorpha longicaudata* (Ashmead) was introduced in 1994 from Florida and first releases of 42,000 individuals in total were made in 1995. Ten weeks after the first releases, the parasitoid was recovered in orchards with guava, star fruit (carambola), pitanga and mango. From 1995 to 2004 a total of 35,118,915 adult parasitoids and 1,321,870 pupae of *D. longicaudata* were released in several regions of Brazil (Carvalho and Nascimento, 2002; Walder *et al.*, 2009; Garcia and Ricalde, 2013).

Some work was done during the period 1970–1990 on insect-killing nematodes. An imported formulation of *Steinernema carpocapsae* (Weiser) was tested by Arrigoni *et al.* (1986) for control of the sugarcane pest *Migdolus fryanus* (Westwood). Later, Schmitt *et al.* (1992) evaluated another imported formulation of the same species, to control the banana root borer *Cosmopolites sordidus* (Germar).

Since the 1980s studies have been done on the possible use of native species of generalist predators such as *Montina confusa* (Stal) and *Podisus* spp. to control lepidopteran pests in eucalyptus forests (Zanuncio *et al.*, 2002; Pires *et al.*, 2015). They occur in a wide variety of (agro-) ecosystems (e.g. Freitas *et al.*, 1990) and have been mass reared and released to control developing populations of various lepidopteran defoliaters of eucalyptus trees. Several successful releases, including areas of up to 20,000 ha, were reported in Torres *et al.* (2006).

In 1989 the National Resources Unit for Control of the Sirex Wood Wasp (FUNCEMA) in *Pinus* plantations was created, integrating the activities of private companies and public institutes to give support to the PNCVM. The programme involves the use of the nematode *Deladenus* (= *Beddingia*) *siricidicola* Bedding against the wood wasp. This nematode sterilizes the female wasps and may reach levels of infection near to 100%. The nematode was introduced in Brazil in 1989–1990 from CSIRO (Australia) and Embrapa Forest in Paraná State started the large-scale mass rearing of the nematode. Application is by doses of 20 ml (= 1 million nematodes) for treatment of ten *Pinus* trees. However, the initially imported strain of nematode, which was released after mass production in 1990, appeared ineffective, showing levels of infection of only 2–6%. Two strategies were followed to improve the situation: (i) importation and release in 1995 of a strain Kamona (re-isolated by CSIRO from Tasmania Island); and (ii) release of new strains, re-isolated in Brazil from field collections in areas where the nematode showed good efficiency. The best Brazilian strain was isolated from the municipality of Encruzilhada do Sul, RS State; this strain was probably best adapted to the local climate. Since 1995, more new strains were re-isolated and released every year, which reached infection levels up to 91% in 1999–2000 (Iede *et al.*, 2012).

Augmentative biological control of arthropods by microbial control agents in agriculture and forestry

With the occurrence of the spittlebug *Mahanarva posticata* (Stal) in pastures and sugarcane in the northern states of Brazil, studies of *M. anisopliae* were initiated by institutes involved in sugar production and in pest control in sugarcane, and grower cooperatives in this region started production of this fungus. Later, microbial agents to control pests, including *M. anisopliae*, *B. bassiana* and *Bacillus* spp., were initiated in the 1980s by S.B. Alves (ESALQ/USP) and were summarized (as well as studies in other Latin American countries) in Alves and Lopes (2008).

Of specific interest is the use of the *Anticarsia gemmatalis* multiple nucleopolyhedrovirus (AgMNPV) against soybean caterpillar *Anticarsia gemmatalis* Hubner on more than 2 million hectares during the 1980s and 1990s (Moscardi, 1999), because the virus was produced *in vivo* on farms. Unfortunately, AgMNPV use strongly decreased as a result of changes in soybean production and use of genetically modified soybean.

Microbial control also started to be used in forestry, for example for the control of the mate tree borer *Hedypathes betulinus* (Klug), the main pest of mate tree (*Ilex paraguariensis* A. St.-Hil) plantations. The total area with mate trees in Brazil is about 54 million hectares. Embrapa Forestry screened different strains of the entomopathogenic fungus *B. bassiana* to identify an efficient strain to control this pest and 'Strain CG 716' selected in 1990 appeared to be the most efficient. In partnership with the company Turfal (Indústria e Comércio de Produtos Biológicos e Agronômicos Ltda. (Novozymes)), a biological product based on *B. bassiana* named Bovemax EC was developed.

Biological control of plant diseases

The first Brazilian meeting on biocontrol of diseases took place in 1986. The first product that became available through Embrapa Clima Temperado was *Trichoderma viride* Pers. (Valdebenito-Sanhueza, 1991) for the control of *Phytophthora cactorum* (Lebert & Cohn) in apple trees in 1991. This project stimulated the selection of strains of this genus for the control of other diseases. Associated with the potential of *Trichoderma* to control soil-borne pathogens and in combination with the limited availability of chemical fungicides to control these pathogens, a rapid expansion of *Trichoderma*-based biopesticides took place to control soilborne plant pathogens in crops such as soybeans, beans, cotton and tomato.

Biological control of weeds

There are no cases of classical biocontrol of weeds in Brazil, although Latin America, including Brazil, has played a major role in weed biocontrol since the start of the 20th century as a provider of natural enemies of invasive weeds throughout the world (see e.g. Barreto, 2008).

6.3 Current Situation of Biological Control in Brazil

6.3.1 Classical and augmentative biological control of forest pests

Today forestry, mainly consisting of eucalyptus and pine plantations, is an economically important activity in Brazil and contributes 17% of all wood produced worldwide (Schühli *et al.*, 2016). Strongly increased trade in wood has led to movement of pests all over the globe. Once introduced to a new region, establishment and spread of these pests is made easy because similar tree species and cultivars are planted everywhere in vast monocultures for wood production.

The first eucalyptus plantations in Brazil date back to the 1920s, but the area with eucalyptus expanded only in the 1960s and now covers 7 million hectares (Indústria Brasileira de Árvores, 2015). The main pests of eucalyptus are currently longicorn beetles of the genus *Phoracantha* Newman, weevils belonging to the genus *Gonipterus* Schoenherr, several Psylloidea species, two species of gall-making wasps and the bronze bug *Thaumastocoris peregrinus* Carpintero & Dellapé (Schühli *et al.*, 2016). On eucalyptus, the psyllid *Ctenarytaina spatulata* Taylor is controlled by the accidentally introduced parasitoid *Psyllaephagus pilosus* Noyes (Kurylo *et al.*, 2010). The psyllid *Ctenarytaina eucalypti* (Maskell) occurs on several species of eucalyptus. The parasitoid *P. pilosus* is used in forestry biocontrol in Uruguay; it was accidentally introduced in Brazil and is now present in the State of Rio Grande do Sul.

The bronze bug *T. peregrinus* was detected in eucalyptus plantations in 2008 in Brazil and the egg parasitoid *Cleruchoides noackae* Lin & Hubr (Lin *et al.*, 2007) was imported from Australia in 2012 and released. Releases of the parasitoid have been realized by enterprises associated with the Programa de Proteção Florestal/Instituto de Pesquisas e Estudos Florestais (PROTEF/IPEF), located in the States of Bahia, Espírito Santo, Goiás, Maranhão, Mato Grosso do Sul, Minas Gerais, Paraná, Rio Grande do Sul, São Paulo and Tocantins, and resulted in the establishment of the parasitoid; parasitism has recently reached rates of 25–50% (Wilcken and Oliveira, 2015; Wilcken *et al.*, 2015; Barbosa *et al.*, 2017a). According to Barbosa *et al.* (2017a), the recovery of *C. noackae* from *T. peregrinus* eggs collected in the field shows that this parasitoid reproduced and dispersed over a distance of more than 10 km from its initial release point after 1 year. Another biocontrol programme in eucalyptus is in progress and aims at releasing the parasitoid *Selitrichodes neseri* Kelly & La Salle from South Africa for control of the gall wasp *Leptocybe invasa* Fisher & La Salle (Garlet *et al.*, 2013).

The exotic wood wasp *S. noctilio* is still the most important pest on *Pinus* spp. in Brazil. Its populations, occurring on 450,000 ha of the total 1,840,000 ha of pine in Brazil, are strongly reduced (> 70%) by inoculative injections in the trees of the nematode *D. siricidicola* and by the parasitoid *I. leucospoides* (Iede *et al.*, 2012; Fischbein and Corley, 2015; Embrapa, 2018). Interestingly *I. leucospoides* was first accidentally imported, probably resulting from its transportation as a by-product within wood commodities attacked by *S. noctilio* (Fischbein and Corley, 2015).

Since 1996, some species belonging to the giant conifer aphid genus *Cinara* (*Cinara atlantica* (Wilson) and *Cinara pinivora* (Wilson)) have been found attacking pine trees. A biocontrol programme was started in Brazil in 2001, with the parasitoid *Xenostigmus bifasciatus* Ashmed imported from the USA, mass reared by Embrapa Forest and released in the years 2002–2004. *X. bifasciatus* releases were carried out in pine plantations of up to 2 years of age that were attacked by the giant pine-aphid in Paraná, Santa Catarina and in Itapeva-SP. The parasitoid established and reached levels of parasitism of up to 100% (Vilela and Zucchi, 2015). *X. bifasciatus* is able to spread up to 80 km from a release site in one year. The parasitoid is now present in all the states of the South, South-east and Midwest with *Pinus* plantations. Other important pine pests are the Monterey pine aphid *Essigella californica* (Essig), the aphid *Eulachnus rileyi* (Williams), the pine woolly aphid *Pineus boerneri* Annand and the banded pine weevil *Pissodes castaneus* (De Geer).

6.3.2 Classical and augmentative biological control of arthropods by macrobial control agents in agriculture

One of the largest and very successful augmentative biocontrol programmes in Brazil concerns the control of the main pests of sugarcane: the sugarcane borer *D. saccharalis* and the spittlebug *M. fimbriolata*. To control *D. saccharalis*, 3.5 million hectares (39% of the sugarcane area in Brazil) are currently being treated with *C. flavipes*. In 2017, *Trichogramma galloi* Zucchi was used on 550,000 ha of sugarcane to control the eggs of the sugarcane borer and in 2018 the treated area increased to 1.7 million hectares. *M. fimbriolata* is currently controlled on an area of 2.5 million hectares with the fungus *M. anisopliae*. *Beauveria bassiana* is used in 100,000 ha of sugarcane.

Another *Trichogramma* species, *T. pretiosum* Riley, was used on about 200,000 ha in 2017 for control of a number of lepidopteran pests (*A. gemmatalis*, *Chrysodeixis includes* (Walker), *Helicoverpa zea* (Boddie), *Spodoptera frugiperda* (J.E. Smith), *Neoleucinodes elegantalis* (Guenée) and *T. absoluta*) was used in various field crops, such as soybean, cotton, maize, beans, millet and tomatoes. Multi- and interdisciplinary research on *Trichogramma* species for control of different agricultural pests has been an important issue at the University of São Paulo's Luiz de Queiroz College of Agriculture (ESALQ/USP) since 1984 (Parra *et al.*, 2015).

The egg parasitoid *Trissolcus basalis* (Wollaston) was used for control of the green stink bug *Nezara viridula* (Linnaeus) on 20,000 ha of soybean (Corrêa-Ferreira, 2002) and is registered as a biocontrol product in Brazil (Table 6.3). However, due to a change in composition of the stink bug pest species in soybean (Panizzi, 2015), the egg parasitoid *Telenomus podisi* Ashmead was released on 20,000 ha in 2017 to control the most abundant species currently, the Neotropical brown stink bug *Euschistus heros* (Fabricius).

Since 2006, the predatory mite *N. californicus* has been commercially available for control of spotted mite *Tetranychus urticae*. Today, this predatory mite species and *Phytoseiulus macropilis* Banks, *Phytoseiulus longipes* Evans and *Stratiolaelaps scimitus* Womersley are also used for control of fungus gnats (*Bradysia* spp.) and thrips in cotton and soybean, fruit orchards (plum, apple, nectarine, grape and peach), ornamentals (e.g. anthurium, chrysanthemum, gerbera and rose) and vegetables (e.g. lettuce, eggplant, strawberries, cucumber, tomato and mushrooms) (Watanabe *et al.*, 1994; Monteiro, 2002; Sato *et al.*, 2007; Poletti and Omoto, 2012; Souza-Pimentel *et al.*, 2014). Bueno and Poletti (2009) reported that two releases of five *N. californicus* per square metre were enough to reduce the population of *T. urticae* from six mites per leaflet to 1 mite per leaflet. Barbosa *et al.* (2017b) demonstrated that the release of *P. macropilis* in rose crops reduced the population of *T. urticae* in this crop from 80 mites per leaflet to 16 mites per leaflet, and the combined release of *N. californicus* + *P. macropilis* in strawberry crops resulted after 7 weeks in total control of the *T. urticae* population. The only non-chemical option for control of fungus gnats *Bradysia matogrossensis* in Brazil is by releasing *S. scimitus*, which is done for several ornamentals in greenhouses and also in mushroom cultivation. Bueno and Poletti (2009) showed that releases of this predatory mite on azalea seedlings resulted in the control of fungus gnats. Barbosa *et al.* (2017b) found that use of this predator resulted in a significant reduction of fungus gnats 6 weeks after its release in mushroom cultivation. About 300 growers of mushrooms, producing some 12,000 t per annum, use biocontrol against fungus gnats (Barbosa *et al.*, 2017b). The Brazilian market for mushrooms is growing by around 20% a year. Also, studies are being conducted with the native mite species *Amblyseius tamatavensis* Blommers for control of whitefly in Brazil (Cavalcante *et al.*, 2017). Predatory mite species are being used on an area of 15,000 ha in several crops.

The parasitoid *Habrobracon hebetor* Say is released inside 1,500 tobacco (*Nicotiana tabacum* L.) warehouses by small growers to control *Ephestia elutella* (Hübner).

The recent introduction of *Helicoverpa armigera* (Hübner) into Brazil may be another landmark in the history of biocontrol, as there is no chemical available for its control. Growers have begun to implement IPM to control *H. armigera* by adopting crop-free periods, establishing crop refuges for transgenic plants and using selective chemicals, as well as biological products, viruses and *T. pretiosum*. Nevertheless, the

Table 6.3. Registered biological control agents in Brazil in 2019 (retrieved from Agrofit, 2019)

Biological control agent	Target pest/disease	No. products registered
	Target pest	
Autographa californica multiple nucleopolyhedrovirus (AcMNPV)	*Helicoverpa armigera*	2
Bacillus thuringiensis (strictly speaking, this is not a biocontrol agent but a biopesticide)	*Alabama argillacea, Anticarisa gemmatalis, Argyrotaenia sphaleropa, Ascia monuste orseis, Brassolis astyra astyra, Brossolis sophorae, Colias lesbia pyrrhothea, Condylorrhiza vestigialis, Chrysodeixis includens, Cryptoblades gnidiella, Diaphania hyalinata, Diaphania nitidalis, Diatraea saccharalis, Dione juno juno, Eacles imperialis magnifica, Erinnys ello, Grapholita molesta, Gymnandrosoma aurantianum, Helicoverpa armigera, Helicoverpa zea, Helicoverpa sp., Heliothis virescens, Manduca sexta paphus, Mocis latipes, Neoleucinodes elegantalis, Opsiphanes invirae, Plutella xylostella, Pseudaletia sequax, Rachiplusia nu, Spodoptera frugiperda, Strymon basalides, Thyrinteina arnobia, Trichoplusia ni, Tuta absoluta*	22
Baculovirus *Anticarsia*	*Anticarsia gemmatalis, Helicoverpa armigera*	5
Baculovirus *Chrysodeixis includens*	*Chrysodeixis includens, Helicoverpa armigera*	4
Baculovirus *Helicoverpa armigera*	*Helicoverpa*	1
Baculovirus *Spodoptera frugiperda*	*Spodoptera frugiperda*	3
Beauveria bassiana	*Bemisia tabaci* strain B, *Cosmopolites sordidus, Dalbulus maidis, Diaphorina citri, Gonipterus scutellatus, Hedypathes betulinus, Hypothenemus hampei, Tetranychus urticae, Diabrotica speciosa, Gonipterus scutellatus, Coccus viridis, Deois flavopicta, Euschistus heros*	23
Ceratitis capitata strain SIT (strictly speaking, this is not biocontrol)	*Ceratitis capitata*	1
Condylorrhiza vestigialis nucleopolyhedrovirus	*Condylorrhiza vestigialis*	2
Cotesia flavipes	*Diatraea saccharalis*	28
Cryptolaemus montrouzieri	*Maconellicoccus hirsutus*	1
Diachasmimorpha longicaudata	Tephitidae fruit flies	1
Deladenus siricidicola	*Sirex noctilio*	1
Helicoverpa zea nucleopolyhedrovirus (HzSNPV)	*Helicoverpa armigera, Helicoverpa* sp.	1
Isaria fumosorosea	*Diaphorina citri, Helicoverpa armigera*	2
Metarhizium anisopliae	*Deois flavopicta, Mahanarva fimbriolata, Notozullia entreriana, Zulia entreriana*	34
Neoseiulus californicus	*Tetranychus urticae*	4
Orius insidiosus	*Frankliniella occidentalis*	1
Paecilomyces fumosoroseus	*Bemisia tabaci* strain B	1
Phytoseiulus macropilis	*Tetranychus urticae*	1
Stratiolaelaps scimitus	*Bradysia matogrossensis*	1
Spodoptera frugiperda multiple nucleopolyhedrovirus (SfMNPV)	*Spodoptera frugiperda*	1
Steinernema puertoricense	*Sphenophorus levis*	1
Trichogramma galloi	*Diatraea saccharalis*	6

Continued

Table 6.3. Continued.

Biological control agent	Target pest/disease	No. products registered
Trichogramma pretiosum	*Anticarsia gemmatalis, Crysodeixis includens, Helicoverpa zea, Spodoptera frugiperda, Tuta absoluta*	6
Trissolcus basalis	*Nezara viridula*	1
	Target disease	
Aspergillus flavus NRRL21882	*Aspergillus flavus*	1
Bacillus amyloliquefaciens + Trichoderma harzianum	*Rhizoctonia solani, Sclerotinia sclerotiorum*	1
Bacillus amyloliquefaciens	*Botrytis cinerea, Phyllosticta citricarpa, Sclerotinia sclerotiorum, Sphaerotheca fuliginea, Pratylenchus brachyurus*	7
Bacillus firmus	*Meloidogyne incognita, Meloidogyne javanica, Pratylenchus brachyurus*	3
Bacillus licheniformis + Bacillus subtilis	*Meloidogyne incognita, Meloidogyne javanica, Pratylenchus brachyurus, Pratylenchus zeae*	2
Bacillus methilotrophicus	*Meloidogyne javanica, Pratylenchus brachyurus*	2
Bacillus pumilus	*Alternaria porri, Botrytis cinerea, Cryptosporiopsis perennans, Sphaerotheca macularis*	1
Bacillus subtilis	*Alternaria dauci, Alternaria porri, Botrytis cinerea, Colletotrichum acutatum, Colletotrichum gloeosporioides, Cryptosporiopsis perennans, Hemileia vastatrix, Mycosphaerella fijensis, Pythium ultimum, Sclerotinia sclerotiorum, Sphaerotheca macularis*	3
Bacillus subtilis	*Meloidogyne javanica, Pratylenchus brachyurus, Rhizoctonia solani, Xanthomonas vesicatoria*	1
Paecilomyces lilacinus = Purpureocillium lilacinum	*Meloidogyne incognita, Meloidogyne javanica, Pratylenchus brachyurus*	3
Pasteuria nishizawae	*Heterodera glycines*	2
Pochonia chlamydosporia	*Meloidogyne javanica*	1
Trichoderma asperellum	*Fusarium solani* f. sp. *glycines, Fusarium solani* f. sp. *phaseoli, Rhizoctonia solani, Sclerotinia sclerotiorum*	4
Trichoderma harzianum	*Fusarium solani* f. sp. *phaseoli, Sclerotinia sclerotiorum, Rhizoctonia solani*	12
Trichoderma koningiopsis	*Meloidogyne incognita, Pratylenchus brachyurus, Heterodera glycines*	1
Trichoderma stromaticum	*Moniliophthora perniciosa*	1
Trichoderma harzianum + Bacillus amyloliquefacies	*Rhizoctonia solani* and *Sclerotinia sclerotiorum*	1

amount of biocontrol agents available for use on *H. armigera*-infested areas is currently insufficient, because this pest attacks more than 180 hosts. Recently, companies have started producing high-quality natural enemies for combating this pest, i.e. *T. pretiosum* and a virus.

The Asian citrus psyllid (ACP) *D. citri* is native to southern Asia and is a vector of the most serious citrus disease worldwide, known as huanglongbing (HLB). In Brazil, almost 50 million citrus plants have been eliminated since 2004 due to HLB. ACP was first reported in Brazil in 1938 and the disease HLB was found for the first time in 2004. As there are no curative control measures for the disease, its management is focused on vector control, mainly with pesticide sprays. Since the 1970s, several countries have been applying biocontrol programmes for ACP based on the use of the specific ectoparasitoid, *Tamarixia radiata* (Waterston), native to north-western India (Grafton-Cardwell *et al.*, 2013). The parasitoid was found parasitizing *D. citri* in commercial groves in Brazil in 2005 (Gomez-Torres *et al.*, 2006) and was supposedly introduced accidentally. Releases of the parasitoid have reduced the pest population by up to 90% in the State of São Paulo, which produces 75% of the citrus in Brazil. A mass production method for the parasitoid has been developed (Parra *et al.*, 2016) and there are now six *Tamarixia* rearing facilities on large citrus farms producing parasitoids for their own use and one institute producing parasitoids for research. Parasitoids are released in areas near commercial orchards, because these orchards use large amounts of insecticides, which would kill the parasitoids. This is a new approach, as the primary inocula of the disease (HLB) are located outside the groves in areas abandoned due to HLB attack, areas with *Murraya paniculata* (L.) Jack (a host of the pest), in organic groves, and in backyards. This approach has proved to be highly efficient as a component of IPM in citrus and has been employed on more than 12,000 ha outside commercial citrus groves where the primary inocula of the disease are located; actual data about the area of citrus protected in this way are not yet available. Since the first releases in 2014 almost 9.5 million parasitoids have been released in the State of São Paulo. The number of infected psyllids reaching citrus groves (measured by the number of psyllids collected in adhesive traps) is reduced by 80% and the dispersal radius of the parasitoid has been estimated at 40 km per year (Parra *et al.*, 2016).

Studies on biocontrol of pests in greenhouses started in 1999 (Bueno, 1999) and resulted in experimental applications of heteropteran predators (Anhocoridae and Miridae), aphid parasitoids (Braconidae) and predatory mites (Phytoseiidae) (Bueno, 1999; Bueno and van Lenteren, 2011). *Orius insidiosus* (Say) was evaluated and released against the western flower thrips *Frankliniella occidentalis* (Pergande) in greenhouses with chrysanthemum, rose, gerbera and strawberry, and resulted in effective control (Silveira *et al.*, 2004; Bueno *et al.*, 2009); since 2017 the bug has been registered as a biocontrol agent in Brazil for control of thrips. Currently, three mirid predators (*Campyloneuropsis infumatus* (Carvalho), *Engytatus varians* (Distant) and *Macrolophus basicornis* (Stal)) are being evaluated for control of *T. absoluta* and other tomato pests (Bueno *et al.*, 2013; van Lenteren *et al.*, 2017, 2018a). The biology and capacity of parasitism of the aphid parasitoids *Lysiphlebus testaceipes* (Cresson), *Aphidius colemani* Viereck, *Aphidius ervi* Haliday and *Praon volucre* (Haliday) were evaluated when exposed to various aphid pests of greenhouse crops (Bueno and Sampaio, 2009). The parasitoid *L. testaceipes* Glover was released in a commercial greenhouse of 600 m^2 with two chrysanthemum cultivars and successfully controlled *Aphis gossypii* on both cultivars (Rodrigues *et al.*, 2005). Also, experimental releases by using banker plants in greenhouses were made with *L. testaceipes* against *A. gossypii* in sweet pepper (Rodrigues *et al.*, 2001) and with *O. insidiosus* (Say) against *F. occidentalis* in rose crops (Bueno *et al.*, 2009). The egg parasitoid *T. pretiosum* was released on an area of about 2,600 ha of open-field tomatoes against *Tuta absoluta* (Meyrick) (Haji *et al.*, 2002) and the predatory mite *Neoseiulus californicus* (McGregor) against the European red spider mite *Panonychus ulmi* (Koch) in apple orchards on about 7,800 ha (Monteiro, 2002).

The predator *C. montrouzieri* is being evaluated for control of the rose scale *Maconellicoccus hirsutus* Green and the prickly pear cochineal *Dactylopius opuntiae* (Cockerell). Since 2015, *C. montrouzieri* has been registered as a biocontrol agent to control scales (Peronti *et al.*, 2016).

Biocontrol of the fruit flies *Ceratitis capitata* (Wied.) and *Anastrepha* spp., which started in the period 1970–2000, is ongoing. The parasitoid *D. longicaudata* is reared by the biofactory Moscamed Brasil (Moscamed Brasil, 2018) and has recently been registered as a biocontrol agent (Table 6.3). Currently, Moscamed also studies the parasitoid *Fopius arisanus* (Sonan) for control of the carambola fruit fly *Bactrocera carambolae* Drew and Hancock and *C. capitata* with promising results.

6.3.3 Augmentative biological control of invertebrates by microbial control agents in agriculture and forestry

Microbial control of pests occurs on a large scale in Brazil (Table 6.3), with currently more than

80 products registered for control of arthropods, of which 60% are fungal products, 29% bacteria, 10% baculoviruses and 1% nematodes (Mascarin *et al.*, 2018). The most often used fungal products for arthropod control are based on *M. anisopliae* and *B. bassiana*. *M. anisopliae* is mainly used to control spittlebugs in sugarcane fields and *B. bassiana* for control of whiteflies in row crops. Other important microbial control programmes are the use of *Anticarsia gemmatalis* nucleopolyhedrosis virus (AgMNPV) for management of soybean caterpillars. An example of nematode use is application of *D. siricidicola*, for control of *S. noctilio* in *Pinus*. Twenty-two registered commercial microbial control products are based on *Bacillus* spp. (Table 6.3) and form a special category, as in most cases the living organism is not used but the endotoxins that they produce. Thus, according to the definition of biocontrol (use of an organism to reduce the population density of another organism) the endotoxin of *Bacillus* spp. is not considered to be a biocontrol agent. However, these products are often discussed under microbial control agents.

Fungal-based products

The entomopathogenic fungus *M. anisopliae* has been used since the 1970s against spittlebugs (Cercopidae) in sugarcane in Brazil and represents the largest microbial control programme using a mycoinsecticide worldwide, with its application on about 2 million hectares annually. This implies that over 20% of the 9 million hectares of Brazilian sugarcane are treated with this fungus, and, according to Mascarin *et al.* (2018), the demand is increasing. An important incentive to use this fungus is the much lower costs: it took only about US\$19 million to control the spittlebug on 1 million hectares in 2014, while the cost of chemical control for control of this pest on 1 million hectares was US\$68 million (IBSP, 2014).

Fungal entomopathogens have been tested against whiteflies worldwide for many years. In Brazil, the use of *B. bassiana* to manage *Bemisia tabaci* (Gennadius) has recently increased strongly, with currently an estimated 1.5 million hectares of soybean treated annually (Mascarin *et al.*, 2018); the strong use of this fungus is reported to be related to the whitefly's development of resistance to pesticides. In addition, three commercial products based on *B. bassiana* are used for control of the coffee berry borer (CBB) *Hypothenemus hampei* (Ferrari), the most important insect pest of coffee in Brazil and worldwide. In another project, an attract-and-kill strategy is being tested that uses a trap with the aggregation pheromone of the banana weevil *C. sordidus*; weevils are then killed in a paste of dry aerial conidia of *B. bassiana* mixed with vegetable oil. Small and organic banana producers use another attract-and-kill approach, in which traps are composed of pieces of pseudostem treated with *B. bassiana*: the decaying plant volatiles attract adult weevils, which are subsequently infected. The banana weevil is a problem in all banana-producing areas of Brazil. It is estimated that the attract-and-kill strategy is applied on 6,000 ha (J.E.M. Almeida, Campinas, 2018, personal communication).

B. bassiana is also used in eucalyptus plantations to manage the snout beetle *Gonipterus scutellatus* Gyllenhal in combination with releases of the egg parasitoid *Anaphes nitens* Girault. In 2014, 11,000 ha of eucalyptus were infected by snout beetles (Wilcken and Oliveira, 2015).

Bovemax EC, a product based on *B. bassiana* and developed for control of the mate tree borer *H. betulinus*, was registered in 2011. This bioinsecticide has an oil-based formulation and is effective in reducing the economic losses caused by the attack of the mate tree borer (Embrapa, 2018); data about the area treated with Bovemax could not be retrieved.

Recently a new commercial product based on the fungus *Isaria fumosorosea* Wize for control of *D. citri*, the vector of HLB, came on the market, resulting from research by I. Delalibera (Ausique *et al.*, 2017). This entomopathogen can safely be used together with the parasitoid *T. radiata*, as the fungus does not kill the parasitoid. Currently, two products based on *I. fumosorosea* are registered in Brazil (Table 6.3)

Another recent development is the use of the fungus *Purpureocillium lilacinum* (Thon) Luangsa-ard, Howbraken, Hywel-Jones & Samson for control of nematodes (*Meloidogyne* spp.) on an estimated 500,000 ha of soybean and maize, as well as on about 10,000 ha of coffee plantations, fruit orchards and vegetable crops (J.E.M. Almeida, Campinas, 2018, personal communication).

Mascarin *et al.* (2018) mentioned another Brazilian biocontrol programme in which, interestingly, an unidentified entomopathogenic fungus has been used since 2010 on about 15,000 ha annually to control *L. heveae*, a pest of rubber trees (see above for information about attacks on the same pest by *H. verticillioides*). The same authors also mentioned several fungal-based control programmes that are in development, such as: (i) control of stink bugs (Pentatomidae), important pests in many Brazilian crops, with *M. anisopliae*; (ii) use of *B. bassiana* to control *D. citri*; and (iii) control of insect vectors of human pathogens, such as *Aedes aegypti* L., with *Metarhizium* species, testing of *B. bassiana* and *Metarhizium acridum* (Driver & Milner) J.F. Bisch, Rehner & Humber against the sand fly *Lutzomyia longipalpis* Lutz and Neiva, and use of *B. bassiana and M. anisopliae* against kissing bugs, *Triatoma* spp. Entomopathogenic fungi of the genus *Metarhizium* are also tested for control of dog and cattle ticks (*Rhipicephalus* spp.).

Baculovirus-based products

The largest and worldwide known success of the use of entomopathogenic viruses is probably the application of the *A. gemmatalis* multiple nucleopolyhedrovirus (AgMNPV) to control soybean caterpillars, with the virus produced *in vivo* on farms (Moscardi, 1999). This virus is now also available commercially.

After the arrival of the invasive pest *H. armigera* in Brazil in 2013, emergency registration was achieved for, among others, the *H. zea* nucleopolyhedrovirus (HzSNPV). Heliothinae-active baculoviruses were used in 2013–2014 on 1.3 million hectares of various crops, including soybean, cotton and corn, due to the recent outbreaks of *H. armigera*, and HzSNPV and AgMNPV are now the most used baculoviruses in soybean (Sosa-Gómez, 2017). In fact, the commercially available material was completely sold out during the 2013–2014 season.

Another virus, the *S. frugiperda* nucleopolyhedrovirus (SfMNPV), has been tested for use for control of the fall armyworm *S. frugiperda* in maize and commercial products have been developed (Valicente *et al.*, 2013; Sosa-Gómez, 2017), of which one product is now registered (Table 6.3) (Agrofit, 2019).

Nematode-based products

As described earlier in this chapter, the nematode *D. siricidicola* is used for control of *S. noctilio* in *Pinus*. This nematode was registered as biological product named 'Nematec' and its application results in 70–100% parasitism of sirex wood wasp. Dolinski *et al.* (2017) summarized other projects on entomopathogenic nematodes and mentioned that most of these nematode studies were started after 2000. A number of prospecting projects were executed, resulting in the isolation and identification of many nematodes, including two new species. Further, the biology and ecology of several species were studied in detail. Laboratory and field tests with nematodes for control of agricultural pests and insect and snail vectors of human and cattle diseases were described by Dolinski *et al.* (2017), as well as their practical application. An example of practical application is the control of the Guava weevil *Conotrachelus psidii* Marshall based on the infected cadaver technique. Basic and applied research coordinated by Dr C. Dolinski (Universidade Estadual do Norte Fluminense Darcy Ribeiro (UENF), Campos dos Goytacazes, RJ) showed effectiveness of a strain of *Heterorhabditis baujardi*. A partnership was established between the UENF laboratory with 20 farmers organized in the Association of Guava Producers of Cachoeiras de Macacu in Rio de Janeiro State. The Association paid to register the nematode *H. baujardi* and applied the nematodes in their guava orchards. Dolinski *et al.* (2017) concluded that the scarcity of companies producing these nematodes might be the main factor limiting wider use of entomopathogenic nematodes in the field. The lack of a nematode production industry resulted in a group coordinated by Dr L. Leite (Instituto Biológico, Agência Paulista de Tecnologia em Agronegócios, Campinas, SP), in partnership with the company BioControle and supported by collaboration from the USDA, which has developed a product based on the nematodes *Heterorhabditis indica* Poinar, Karunakar & David and *Steinernema brazilense* for control of the sugarcane billbug *Sphenophorus levis* Vaurie. Application of *S. brazilense* resulted in an increase of up to 17 t of sugarcane per hectare.

Bacterial-based products

Although they are not considered to be biocontrol agents, as the 'dead' toxin is used for control, we shortly summarize the use of Bt *Bacillus thuringiensis* (Berliner) for arthropod control in Brazil. Bt bioinsecticides are of particular interest as their use can often be combined with application of predators and parasitoids, and pollinators, because there are hardly any or no negative side effects on these beneficial organisms. Bt bacteria produce proteins with insecticidal properties that are toxic to different insect orders such as Lepidoptera, Coleoptera, Hymenoptera, Homoptera, Orthoptera and Mallophaga, as well as to nematodes. In addition to control of agricultural pests, Bt is used to combat insect vectors of human and animal diseases. Recent overviews on research and application of Bt in Brazil can be found in Fiuza *et al.* (2017). In 1988 Embrapa began prospecting for Bt strains and they started the first large Bt research project in 1993, for control of the fall armyworm *S. frugiperda* in maize (Polanczyk *et al.*, 2017). Until 2000, Bt-based products were recommended for the control of 26 pests in forests, vegetables and row crops, but they were only used on a limited scale. Around 2000, the citrus fruit borer *Gymnandrosoma aurantianum* Todd M. Gilligan and Marc E. Epstein was controlled with Bt on 50,000 ha of citrus. Brazil became the largest world market for Bt use during the 2013–2014 crop season, due to the *H. armigera* invasion and the sudden increase of *C. includens* in soybean. During this season, Bt was used on 9 million hectares of, among others, soybean, cotton, maize and citrus (Polanczyk *et al.*, 2017). Due to a strong decrease in lepidopteran pest populations in soybean during the 2014–2015 and later seasons, considerably less Bt was used (Sosa-Gómez, 2017), but even with this recent decline in sales of Bt, Polanczyk *et al.* (2017) considered the future Brazilian market for Bt to be very promising.

6.3.4 Augmentative biological control of plant diseases

The history of biocontrol of plant diseases in Brazil is relatively recent and, in several aspects, differs from that of biocontrol of insects. For biocontrol of insects, introduction of exotic natural enemies was for a long time prioritized, while for biocontrol of plant pathogens introduction of antagonists was not. Instead, researchers aimed to isolate and select antagonists from Brazilian agroecosystems. Registration of the first commercial biofungicide based on *Trichoderma harzianum* Rifai to control *Rhizoctonia solani* Kuhn and *Fusarium solani* f. sp. *phaseoli* Kendricht & Snyder in bean occurred only in 2008 (Bettiol *et al.*, 2014). Despite studies with a number of other species of antagonists (*Clonostachys rosea* (Link) Schroers, *Pseudomonas fluorescens* Flügge, *Acremonium persicinum* (Nicot.) W. Gams, *Dicyma pulvinata* (Berk. & Curt.) von Arx, *Lecanicillium lecanii*, *Pasteuria penetrans* (ex Thorne) Sayre & Starr, *Pochonia chlamydosporia* (Goddard) Zare & Gams, *Cladobotryum amazonense* Bastos, Evans & Samson, *Purpureocillium* (= *Paecilomyces lilacinus* (Thom) Samson and other antagonists), the two most important biocontrol agents for plant pathogens in Brazil are still *Trichoderma* spp. and *Bacillus* spp. (Bettiol, 2011). In February 2019, 26 species of microorganisms were registered as biofungicides, bionematicides or entomopathogens and more than 120 products were registered (Table 6.3) (Agrofit, 2019).

Although commercial production of *Trichoderma* started in 1991, the large breakthrough in production and use of *Trichoderma* took place after the year 2000. This beneficial fungus completely changed the picture of biocontrol of plant diseases in Brazil. The first registration of a commercial *Trichoderma* product took place in 2008 with an application on 600,000 ha of soybean for control of *Sclerotinia sclerotiorum* (Lib.) de Bary. In 2010, more than 1.2 million hectares of soybean were treated with *Trichoderma* and this rapidly increased to more than 5 million hectares in 2015 (Table 6.4). *Trichoderma* exhibits a suite of working mechanisms (parasitism, antibiosis and competition with other fungi) and is able to induce resistance to diseases in crop plants.

Another interesting disease control agent is *Bacillus subtilis* Cohn, which was registered in 2014. Its working mechanism is by antibiosis and competition of the disease as well as by inducing resistance to disease in the crop. Some 180,000 litres were sold in 2014. Sixteen products based on *Bacillus amyloliquefaciens* Priest *et al.*, *Bacillus firmus* Bredemann & Werner, *B. subtilis* plus *Bacillus licheniformis* (Weigmann) Chester, *Bacillus methylotrophicus* Madhaiyan *et al.* and *B. subtilis* were

Table 6.4. Crop surfaces treated with biological control agents in Brazil.

Crop	Pest	Natural enemy	Origin	Type of biocontrol[a]	Used since	Area (ha) under biocontrol
Apple orchards	Spider mites	*Neoseiulus californicus*	Native	ABC	1990s	1,800
Banana	Banana weevil	*Beauveria bassiana*	Native	ABC	2015	6,000
Cassava	*Phenacoccus herreni* cassava mealybug	Parasitoids 3 spp	Colombia	CBC	1994/95	1,000,000
Citrus	*Diaphorina citri*	*Beauveria bassiana*	Native	ABC	1994	Part of 4,000,000
	Phyllocnistis citrella	*Ageniaspis citricola*	USA	CBC	1998	450,000
	Diaphorina citri vector HLB	*Tamarixia radiata*	Exotic acc	CBC	2005	12,000[b]
	Gymnandrosoma aurantianum	*Bacillus thuringiensis* biopesticide	Native + exotic	ABC	2000	50,000
Coffee	Coffee berry borer	*Beauveria bassiana*	Native	ABC	2000	Part of 4,000,000
Corn	*Helicoverpa armigera,* other *H.*	*Helicoverpa zea* SNPV virus	Native + exotic	ABC	2010	Part of 1,300,000
	Meloidogyne spp.	*Purpureocillium lilacinum*	Native	ABC	2017	500,000
	Spodoptera frugiperda	*S. frugiperda* SfMNPV	Native	ABC	2017	No data
	Spodoptera frugiperda	*Bacillus thuringiensis* biopesticide	Native + exotic	ABC	1990	Part of 5,000,000
Cotton	*Helicoverpa armigera,* other *H.* spp.	*Helicoverpa zea* SNPV virus	Native + exotic	ABC	2010	Part of 1,300,000
	Lepidopterans	*Bacillus thuringiensis* biopesticide	Native + exotic	ABC	2010	Part of 5,000,000
Eucalyptus	Lepidopterans	*Podisus nigrispinus*	Native	ABC	1980s	> 20,000
	Gonipterus scutellatus	*Beauveria bassiana*	Native	ABC	2014	11,000
	Ctenarytaina spatulata	*Psyllaephagus pilosus*	Exotic acc	CBC	2005	No data
	Thaumastocoris peregrinus	*Cleruchoides noackae*	Australia	CBC	2012	No data
	Leptocybe invasa gall wasp	*Selitrichodes neseri*	S Africa	CBC	2014	No data
Fruit orchards	*Ceratitis capitata, Anastrepha* spp	*Diachasmimorpha longicaudata*	USA	ABC	1995	No data
Greenhouse crops, mushrooms	Spider mites	*Neoseiulus californicus, Phytoseiulus* sp.	Native	ABC	2006	15,000 (both rows)
	Bradysia matogrossensis gall midge	*Stratiolaelaps scimitus*	Native	ABC	2006	
Mate trees	*Hedypathes betulinus*	*Beauveria bassiana*	Native	ABC	1990	Part of 4,000,000
Pastures	*Antonina graminis* rhodesgrass scale	*Neodusmetia sangwani*	USA	CBC	1967–2000	100,000s
Pine	*Sirex noctilio* sirex woodwasp	*Ibalia leucospoides*	Exotic acc	CBC	1990-96	450,000
	Sirex noctilio sirex woodwasp	*Deladenus siricidicola* nematode	Australia	CBC	1989	450,000
	Cinara atlantica, C. pinivora	*Xenostigmus bifasciatus*	USA	CBC	2001	No data
Soybean	*Anticarsia gemmatalis* soybean caterpillars	AgMNPVvirus	Native	ABC	1980–2010	> 2,000,000

Continued

Table 6.4. Continued.

Crop	Pest	Natural enemy	Origin	Type of biocontrol[a]	Used since	Area (ha) under biocontrol
	Anticarsia gemmatalis soybean caterpillars	AgMNPVvirus	Native	ABC	2017	700,000
	Nezara viridula	*Trissolcus basalis*	Native	ABC	1990–2005	20,000
	Euschistus heros	*Telenomus podisi*	Native	ABC	2005	20,000
	Bemisia tabaci + *Anticarsia gemmatalis*	*Beauveria bassiana*	Native	ABC	1990	Part of 4,000.000
	Helicoverpa armigera, other *H.*	Helicoverpa zea SNPV virus	Native + exotic	ABC	2010	Part of 1,300,000
	Meloidogyne spp.	*Purpureocillium lilacinum*	Native	ABC	2017	Part of 500,000
	Fusarium and other soilborne plant diseases	*Trichoderma* spp.	Native	ABC	1991	5,500,000
	Chrysodeixis includens + *H. armigera*	*Bacillus thuringiensis* biopesticide	Native + exotic	ABC	2012	Part of 5,000,000
Sugarcane	*Diatraea saccharalis* sugarcane borer	*Cotesia flavipes*	Trinidad	ABC	1970s	3,500,000
	Diatraea saccharalis sugarcane borer	*Trichogramma galloi*	Native	ABC	2010	1,700,000
	Mahanarva spp. leafhoppers	*Metarhizium anisopliae*	Native	ABC	1990	4,000,000
	Sphenophorus levis billbug	*Steinernema* spp.	Native + exotic	ABC	2015	No data
Various crops, soybean, cotton, maize +	Lepidopterans in soybean + others	*Trichogramma pretiosum, T. atopivirilia*	Native/Colombia	ABC	1990s	200,000
	Many lepidpterans, coleopterans etc in many crops	*Bacillus thuringiensis*-based biopesticide	Native + exotic	ABC	1990	5,000,000
	Pratylenchus brachyurus and others	*Bacillus amyloliquefaciens*	Native	ABC	2015	No data
	Meloidogyne spp.	*Bacillus firmus*	Native	ABC	2015	No data
	Meloidogyne spp. and others	*B. subtilis* + *Bacillus licheniformis*	Native	ABC	2015	No data
	Meloidogyne and *Pratylenchus*	*Bacillus methilotrophicus*	Native	ABC	2015	No data
	Botrytis and other diseases	*Bacillus pumilis*	Native	ABC	2015	No data
	Botrytis and other diseases	*Bacillus subtilis*	Native	ABC	2015	No data
Wheat	Wheat aphid spp.	Predators and parasitoid spp.	Many countries	CBC	1970s	1,000,000

[a]Type of biocontrol: CBC = classical biological control, ABC = augmentative biological control
[b]See text, special biocontrol programme

registered in 2019 for the control of several nematodes (*Meloidogyne incognita* (Kofoid & White) Chitwood, *Meloidogyne javanica* (Treub) Chitwood, *Pratylenchus brachyurus* (Godfrey) Filipjev & S. Stekhoven and *Pratylenchus zeae* Graham). The area treated with these products for the control of nematodes is expected to grow significantly, since they are used in large-scale crops such as sugarcane and soybean.

6.3.5 Biological control of weeds

Biocontrol of native Brazilian weeds has remained restricted to the study of inundative releases of fungi. The fungus *Bipolaris euphorbiae* (Han) Muc Sfordhovej & Carvalho was used as a mycoherbicide against wild poinsettia *Euphorbia heterophylla* L. (Yorinori and Gazziero, 1989), but this project was terminated at the discovery of biotypes of the weed that were resistant to the fungus (Barreto, 2008). Most work on weed biocontrol in Brazil takes place at the Departamento de Fitopatologia, Universidade Federal de Viçosa (DFP/UFV). Here, surveys to discover fungal pathogens attacking more than ten different species of weeds in Brazil have been conducted and the mycobiota of these weeds have been described (e.g. Pereira *et al.*, 2007). These mycobiota studies provide information about potential biocontrol agents for use in Brazil or abroad. For example, two of these fungi have been used for weed control in Hawaii and Australia (Barreto, 2008). Also nematodes and bacteria are studied in UFV (Viçosa) as potential weed biocontrol agents.

6.3.6 Mass production and registration of natural enemies and microbial control agents

Arthropods

Before 2000, insect mass rearing took place at facilities initially financed by the federal government and national programmes, and later by private companies to supply the expanding market for biocontrol agents. The number of available officially registered natural enemies is limited to fewer than ten species of insects and mites. Because few companies are at present marketing these natural enemies, adequate quantities of natural enemies are available for sugarcane only; also because mills that produce sugar and ethanol have their own natural-enemy production laboratories.

Biocontrol in Sertãozinho/SP, the largest commercial producer of *C. flavipes*, reared more than 1 trillion individuals to treat 400,000 ha of sugarcane in 2017. The largest non-commercial producer, Coprodia in Campo Novo do Parecis/MT, a growers' cooperative with 46 cooperatives, released 195,414,000 individuals of *C. flavipes* in an area of 30,927 ha of sugarcane in 2017. The way in which *C. flavipes* is mass produced is quite labour intensive, as the host caterpillars have to be offered individually by hand to the parasitoid to obtain optimal parasitism. This mass production method has been published in Portuguese by Bueno (2000, 2009) and recently also in English (van Lenteren and Bueno, 2019). The majority of *Trichogramma* in Brazil was produced by Bug Agentes Biológicos Company, now part of Koppert Biological Systems Brazil. Mass production of predatory mites was initiated in 2006, when the company Promip started to rear the predatory mite *N. californicus* for control of spotted mite *T. urticae*.

Most arthropod natural enemies are reared on one of their field hosts or on factitious hosts/prey. A well known factitious host for *Trichogramma* egg parasitoids is *Ephestia kuehniella* Zeller. Host/prey are in a number of cases reared on artificial media for efficiency reasons; an example is the artificial medium used for the culture of *D. saccharalis*, the host of *C. flavipes*. In order to further economize mass production, *in vitro* rearing was studied for the egg parasitoids *T. pretiosum*, *T. galloi* and *Trichogramma atopovirilia* Oatman and Platner, as well as for the ectoparasitoid *H. hebetor* (Parra, 2014 and references therein).

Entomopathogenic nematodes

Initially, nematodes were multiplied *in vivo* mainly on larvae of greater wax moth *Galleria mellonella* L., like elsewhere in the world. Nowadays, commercial production *in vitro* by the solid-state process using animal offal homogenate as a medium also occurs, and various studies have attempted to improve storage and application methods, summarized by Dolinski *et al.* (2017). Various application techniques are used to apply nematodes in the field, such as in aqueous suspensions with pressurized sprayers, mist blowers, electrostatic sprayers and irrigation systems. Like the cadaver method developed for rearing and release of the AgMNPV virus, release of nematodes within infected insect cadavers has been tested with good results. With this method, smallholder farmers can produce their own nematodes.

The nematode *D. siricidicola* applied for classical control of the *Sirex* wood wasp is produced by Embrapa Forest in Paraná.

Entomopathogenic fungi

In the 1970s, grower cooperatives in the Northeastern region of Brazil started the production of *M. anisopliae* for spittlebug *M. posticata* control in sugarcane. Currently, local sugarcane mills still mass produce this fungus, essentially as unformulated products for local on-farm use, but several registered commercial products are available as wettable powders, oil dispersions and suspension concentrates (Mascarin *et al.*, 2018). In addition to on-site production, about 26 commercial companies are producing pathogens for the Brazilian market and do so through solid-state fermentation, by growing them on a carrier (often enriched cereal grains), in plastic autoclavable bags or in trays. One of the largest producers of entomopathogenic fungi was Itaforte, now part of Koppert Biological Systems Brazil. The other much-used fungus, *B. bassiana*, is usually produced by solid-state fermentation (SSF), liquid-state fermentation (LSF) and di-phasic fermentation. Due to the increase in use, global and domestic companies are looking to expand markets and portfolios for entomopathogenic fungi in Brazil. Nowadays, there are 34 *M. anisopliae*-based products and 23 *B. bassiana*-based products registered in Brazil (Table 6.3) (Agrofit, 2019).

Fungi and bacteria for plant disease control

Products based on *Trichoderma* spp. and strains can be mass produced using different organic and inorganic carriers through either solid or liquid fermentation technologies. *Trichoderma* may be delivered in many different ways, such as through seed treatment, by bio-priming, as seedling dip, by soil application and as foliar spray. *Bacillus* spp. and strains for disease control can be mass produced preferably in liquid media based on a variety of industrial or agricultural organic waste products, such as the residue of the glutamic fermentation of molasses, cotton meal, maize bran, soybean bran and wheat bran (Bettiol *et al.*, 2014).

Entomopathogenic viruses

The *A. gemmatalis* multiple nucleopolyhedrosis virus (AgMNPV) to control soybean caterpillars was for a long time produced *in vivo* on their target host insect or on factitious hosts on farms, but is now also produced *in vivo* by companies and commercially available as a wettable powder formulation, like the *H. zea* virus (HzNPV). Viruses are provided on the Brazilian market by, among others, Certis, Andermatt and Ag BiTech (Sosa-Gómez, 2017).

Bacterial-based products

Most of the commercial Bt products can easily and cheaply be produced *in vitro* in large fermentors in liquid culture. For insect control in agricultural fields, powders or concentrated liquid suspensions containing a mixture of spores and toxin crystals are used. These formulations are applied directly to locations (plant or the environment) where the pest larvae feed. Important providers of Bt in Brazil are Valent Bioscience and Sumimoto.

Registration and the biocontrol market

Specific registration procedures for (micro) biocontrol agents were implemented in Brazil in 2006 and can be downloaded from ABCBio (2019). The information needed for registration of arthropods as biocontrol agents concerned: (i) biological characterization of the organism; (ii) description of potential health risks for humans and animals; (iii) identification of potential risks for the environment; (iv) explanation of the rearing methods and provision of a protocol for quality control; and (v) a manual for use, as well as a description of the beneficial potential of the organism. Much more information is demanded for registration of microbial organisms, including tests concerning toxicologicy and ecotoxicology, mutagenesis and teratogenesis, chronic toxicology and carcinogenesis, skin and eye irritation, and effects on birds, mammals, fish, invertebrates and plants.

Interestingly, in Brazil an easier way of registration exists for biocontrol agents: registration for use in organic agriculture. Several companies requested registration via this route for the following organisms: *Trichoderma virens* (Miller, Giddens & Foster) von Arx, *T. harzianum*, *Trichoderma asperellum* Samuels, Lieckf. & Nirenberg, *Trichoderma koningii* Oudem, *Trichoderma lignorum* (Tode) Harz, *Trichoderma gamsii* Samuels & Druzhin., *T. viride*, *C. rosea*, *B. amyloliquefaciens*, *B. subtilis*, *Bacillus megaterium* de Bary, *Bacillus pumilus* Meyer and Gottheil, *B. licheniformis*, *B. methylotrophicus*, *P. lilacinus*, *Coniothyrium minitans* Campb. and *Arthrobotrys oligospora* Fresen etc. (MAPA, 2016).

In 2007, the foundation of the Brazilian Association of Biological Control Companies (ABCBio) was fundamental for organization of this field of agribusiness. But even with the foundation of ABCBio and with the increase in the availability of properly registered bioagents, a problem continues to exist in Brazil, which concerns commercialization of non-registered biocontrol agents. Thus, commercial biocontrol in Brazil faces several challenges, like illegal production and distribution of non-registered products of unknown efficacy and without quality control, limited shelf life, and unreliable distribution and application methods.

6.3.7 Area under biological control in Brazil

A rough, incomplete and probably vastly underestimated approximation, taking into account that reliable data are not available for several biocontrol agents listed in Tables 6.1 and 6.3, as well as the use of unregistered agents, results in a total area of more than 24,774,000 ha (Table 6.4) under biocontrol with natural enemies and microbial agents against pests and diseases in 2017. According to I. Delaberia (Piracicaba, 2018, personal communication) on-farm production of non-registered microbial control agents is larger than that of registered microbials. Here we provide the crop and forest areas under biocontrol corrected for overlapping use of different types of biocontrol agents. The actual areas treated per year might be considerably higher if more than one cropping cycle is realized, which is, for example, the case for maize and sugarcane. If the use of *B. thuringiensis* products, which were applied on 5.5 million hectares, is included, the total area under biocontrol would increase by a few million hectares, but, strictly speaking, *B. thuringiensis* products are not considered biocontrol agents.

In addition to the use of augmentative biocontrol, Brazil has an area of at least 3,250,000 ha under classical biocontrol (Tables 6.1 and 6.4), but we expect this to be a considerable underestimate. This makes Brazil most likely the country with the largest area under augmentative biocontrol worldwide (van Lenteren *et al.*, 2018b). The area under biocontrol is expected to increase strongly during the coming decades and an indication of this growth is presented by the currently strong growth of Brazilian producers of biocontrol agents, which are illustrated below.

Readers are referred to three books about biocontrol in Brazil (Bueno, 2009; Bettiol and Morandi, 2009; Parra *et al.*, 2002) for more detailed information about several of the biocontrol projects summarized above.

6.4 New Developments of Biological Control in Brazil

Biocontrol in Brazil is used in systems that are quite different from those in, for example, Europe. Many biocontrol programmes in Europe are characterized by application of a number of species of biocontrol agents on greenhouse or farm plots of < 50 ha where high-value crops like vegetables, fruits and ornamental flowers are produced, while in Brazil biocontrol is applied on large farms of thousands of hectares of soybean, cotton, sugarcane, maize and bean, pastures and forests by applying one or two biocontrol agents. This is a completely different dimension in biocontrol, demanding specific mass production, shipping and release methods, as well as particular monitoring and impact evaluation approaches. Brazilian producers of biocontrol agents have succeeded in overcoming these specific conditions and biocontrol agents like *Trichoderma* spp., *Bacillus* spp.,

Metarhizium, *Beauveria* and *Paecilomyces* could relatively easily be adapted for application on such large-scale agricultural surfaces, as they can be applied with similar techniques as chemical pesticides. As a result, the biocontrol companies have developed and registered products mainly for this market. Some insects are also produced and released in these large-scale crops, such as the parasitoid *C. flavipes*, and biocontrol is used by small-scale farmers in Brazil as well. Application of biocontrol is increasing now at a rate of 15–20% annually, according to ABCBio, which is similar to developments worldwide (van Lenteren *et al.*, 2018b).

Another new development concerns conservation biocontrol. The recent increase in organic agriculture in Brazil has contributed to the development of diversified agroecosystems. In these systems, plants are present that are attractive for parasitoids and predators as refuge or for food. Also, banker plants are used to stimulate natural enemy development. For example, in a greenhouse rose crop, basil (*Ocimum basilicum* L.) and ornamental pepper (*Capsicum annuum* L.) were used as banker plants, and aided in the biocontrol of the spider mite *Tetranychus urticae* by the predatory mite *Neoseiulus californicus*.

Although not strictly a biocontrol programme, populations of the mosquito *Aedes aegypti* (L.) have recently been controlled by releases of genetically modified mosquitoes that are called 'Aedes do Bem' (i.e. good mosquitoes) in the municipality of Piracicaba, São Paulo State. Initial releases in an area of 51.5 ha in 2015 resulted in a decrease of 82% in the number of wild larvae of *A. aegypti* in the treated area compared with a control area without releases. Nowadays, the whole municipality of Piracicaba (some 400,000 inhabitants), as well as several other municipalities, receives releases of the 'good mosquitoes', keeping *A. aegypti* at very low populations and resulting in a reduction of 97% of cases of Dengue disease (Oxitec, 2019).

Growth of the area under biocontrol might be stimulated by taking away a number of factors that currently frustrate research and application. Some of these factors are as follows.

- Improvement of mass production, packaging, shipment and release/dispersal of biocontrol agents for use on very large areas, as well as development of effective monitoring programmes. Progress has been made with the use of *Cotesia* and *Trichogramma*, but automation of rearing and development of cheap aerial release and monitoring techniques are of high priority for these and other natural enemies used in the type of agriculture that is so typical for Brazil. Another priority is improvement of formulations for microbial control agents for optimal application and better survival of the agents in the field. If researchers together with the biocontrol industry succeed in removing these limitations, vast areas will be available for biocontrol in the coming decade.
- Overuse of pesticides. Overuse might be combated by aiming at a more rational use of pesticides in Brazil, in particular using them only when needed and not in the form of calendar sprays as often used today, thereby creating more possibilities for use of biocontrol agents. Since 2008, Brazil has been the largest global consumer of synthetic pesticides (Pignati *et al.*, 2017) and farmers are often convinced that these pesticides are the only solution for pest and disease control.
- Lack of technology transfer. Most research on biocontrol takes place at universities and research institutes and is published in peer-reviewed international journals, but lacks a practical follow-up. A better infrastructure linking research, extension service, practical field trials and the farming community might result in faster innovation and application of biocontrol results obtained at the research level. Useful research information is available about biocontrol agents that might potentially solve a number of pest and disease problems existing in Brazilian agriculture, were an efficient communication network between scientists and farmers to be put in place.
- Evaluation of biocontrol agents is still often an ad hoc affair. Critical use of criteria for evaluation of natural enemies would result in a quick separation of useless and potentially useful species and prevents spending money and time on irrelevant species.
- The quality of several biocontrol agents, including microbial agents, particularly those produced without official registration, is of serious concern. Adhering to strict registration and application of quality con-

trol methods would improve the reliability of biocontrol.

- Insufficient guidance and inspection of biocontrol in the field. Due to the increasing costs of biological products per hectare, several farmers started to release lower numbers, or to spray lower dosages or spray less frequently, with poor results, which led to a number of farmers giving up using biocontrol.
- Better networking and collaboration between research groups working on finding biocontrol agents for control of the same or similar pests might also result in quicker identification of candidates. On a national scale, the biannual congresses on biocontrol organized by the Brazilian Entomological Society could present a platform for regular work meetings discussing research progress.
- Finally, communication about successes obtained with biocontrol is often poor. The biocontrol community should be proud about the important successes obtained during recent decades and not only stress the excellent levels of pest and disease control that resulted, but also illustrate the benefits for the environment and health of plants, animals, farmers and consumers. Biocontrol strongly contributes to restoration of biodiversity and sustainable agriculture.

Next to factors currently limiting application of biocontrol in Brazil, there are also many elements that might stimulate its use, for example:

- Brazil has one of the richest sources of biodiversity and hundreds, if not thousands, of species of potential invertebrate and microbial biocontrol agents await discovery.
- Brazil is a very important exporter of agricultural products. International markets increasingly demand stricter control and lower levels of pesticide use, and also Brazilian consumers are more concerned about high levels of residues on food. These restrictions and concerns will lead to the growth of biocontrol.
- Biocontrol for large-scale farming has in some important cases been an initiative of the agricultural industry, such as projects in sugarcane, where the whole technology was developed in such a way that the farmers and the sugarcane industry itself produced the biocontrol agents; in 2018, *T. galloi* was produced on-farm for about 500,000 ha and *C. flavipes* for more than 2 million ha. Another example of on-farm production of a biocontrol agent was the large-scale use of the AgMNPV against soybean caterpillars on more than 2 million hectares during the 1980s and 1990s. A third example is large-scale on-farm production of *Trichoderma*. It is important to consider that this kind of biocontrol, i.e. the possibility of on-farm production of biocontrol agents, is still present in the minds of the farmers and might well be used in future projects.
- The number of commercial producers of biocontrol agents is still limited in Brazil and most are of a small size, though recently some producers merged, resulting in a large biocontrol company. Another evolution in this area is that in the State of São Paulo alone, 11 start-up companies were financed by the São Paulo Research Foundation (FAPESP) to develop production of natural enemies.
- Finally, a very important improvement of registration of arthropod natural enemies has been realized in Brazil. The methodology is now fine-tuned to arthropods, which speeds up the whole procedure. An example of improvement is that registration of a natural enemy is for control of a certain pest and no longer for use in a certain crop.

If scientists and biocontrol practitioners are able to overcome the limitations mentioned above, and make use of the positive factors in the setting of the impressive amount of agricultural production, microbial agents and arthropod natural enemies might become the 'modern minerals' of Brazil.

6.5 Acknowledgements

The following persons are thanked for helping us to find information about biocontrol projects and areas under biocontrol in Brazil: I. Delaberia (ESALQ/USP), E.T. Iede (Embrapa - Florestas), J.R. Salvadori (Universidade de Passo Fundo), V. Costa (Instituto Biológico), J. Virginio (Moscamed Brasil), D.R. Sosa-Gómes (Embrapa Soja), C.L. Hivizi (Coprodia, MT), J.E. Marcondes de Almeida (Instituto Biológico, Campinas), E. Morsoletto dos Santos (Biocontrol).

References

ABCBio (2019) Critérios e procedimentos para importação de organismos para controle biológico – IN. MAPA.522016 [Criteria and procedures for importing organisms for biological control]. Associação Brasileira das Empresas de Controle Biológico [Brazilian Association of Biological Control Companies]. Available at: https://www.abcbio.org.br/informativos/legislacao/ (accessed 27 October 2019).

Agrofit (2019) Sistema de Agrotóxicos Fitossanitários. Available at: http://agrofit.agricultura.gov.br/agrofit_cons/principal_agrofit_cons (accessed 27 October 2019).

Alves, S.B. and Lopes, R.B. (2008) *Controle Microbiano de pragas na América Latina* [*Microbial control of pests in Latin America*]. FAPESP/FEALQ, Piracicaba, São Paulo, Brazil.

Arrigoni, E.B., Dinardo, L.L., Conde, A.J. and Terán, F.O. (1986) Aplicação de *Neoplectana carpocapsea* Weiser, 1955 em condições de campo para o controle de *Migdolus* spp. (Coleoptera: Cerambycidae) [Application of *Neoplectana carpocapsea* at field conditions for control of *Migdolus* spp.]. *Nematologia Brasileira*, 8, 181–189.

Ausique, J.J.S., D'Alessandro, C.P., Conceschi, M.R., Mascarin, G.M. and Delalibera I. Jr (2017) Efficacy of entomopathogenic fungi against adult *Diaphorina citri* from laboratory to field applications. *Journal of Pesticide Science* 90, 947–960.

Barbosa, L.R., Rodrigues, A.P., Soler, L.S., Fernandes, B.V., Castro, B.M.C., Wilker, C.F. and Zanuncio, J.C. (2017a) Establishment in the field of *Cleruchoides noackae* (Hymenoptera: Mymaridae), an exotic egg parasitoid of *Thaumastocoris peregrinus* (Hemiptera: Thaumastocoridae). *Florida Entomologist*, 100, 372–374.

Barbosa, M.F.C., Demite, P.R., Moraes, G.J. and Poletti, M. (2017b) *Controle biológico com ácaros predadores e seu papel no manejo integrado de pragas* [*Biological control of predatory mites and their role in IPM*]. 1st edn. Promip Holding S.A., Limeira, São Paulo, Brazil.

Barreto, R.W. (2008) Latin American weed biological control science at the crossroads. In: Julien, M.H., Sforza, R., Bon, M.C., Evans, H.C., Hatcher, P.E., Hinz, H.L. and Rector, B.G. (eds) *Proceedings of the XII International Symposium on Biological Control of Weeds*, La Grande Motte, France, p.109. DOI: 10.1079/9781845935061.0109.

Batista Filho, A., Costa, V.A. and Hojo, H. (2018) *Neodusmetia sangwani* (Subba Rao) (Hymenoptera: Encyrtidae) to control *Antonina graminis* (Maskell) (Hemiptera: Pseudococcidae) in pastures in Brazil: a revision. *Arquivos do Instituto Biológico* 84, 1–8.

Bento, J.M.S., Moraes, G.J. de, Bellotti, A.C., Castillo, J.A., Warumby, J.F. and Lapointe, S.L. (1999) Introduction of parasitoids for the control of the cassava mealybug *Phenacoccus herreni* (Hemiptera: Pseudococcidae) in north-eastern Brazil. *Bulletin of Entomological Research* 89, 403–410.

Bento, J.M.S., Moraes, G.J. de, Matos, A.P. and Bellotti, A.C. (2000) Classical biological control of the mealybug *Phenacoccus herreni* (Hemiptera: Pseudococcidae) in northeastern Brazil. *Environmental Entomology* 29, 355–359.

Berti Filho, E., Costa, V.A. and Aagesen, T.L. (1989) Occurrence of natural enemies of *Musca domestica* L. (Diptera: Muscidae) in poultry areas of Bastos, State of São Paulo, Brazil. *Revista de Agricultura* 64, 98.

Bettiol, W. (1996) Biological control of plant pathogens in Brazil: application and current research. *World Journal of Microbiology & Biotechnology* 12, 505–510.

Bettiol, W. (2011) Biopesticide use and research in Brazil. *Outlooks Pest Management* 22, 280–284. DOI: 10.1564/22dec10.

Bettiol, W. and Morandi, M.A.B. (2009) *Biocontrole de doenças de plantas: uso e perspectivas* [*Biological control of plant disease: use and perspectives*]. Embrapa Meio Ambiente, Jaguariúna, São Paulo, Brazil.

Bettiol, W., Maffia, L.A. and Castro, M.L.M.P. (2014) Control biológico de enfermedades de plantas en Brasil [Biological control of plant diseases in Brazil]. In: Bettiol, W., Rivera, M.C., Mondino, P., Montealegre, J.R and Colmenárez, Y.C. (eds) *Control biológico de enfermedades de planta en América Latina y el Caribe* [*Biological control of plant diseases in Latin America and the Caribbean*]. Universidad de la República, Montevidéo, Uruguay, pp. 91–137. Available at: http://www.fagro.edu.uy/index.php/documentos/category/46-libros-fagro.html.

Bueno, V.H.P. (1999) Protected cultivation and research on biological control of pests in greenhouses in Brazil. *IOBC/WPRS Bulletin* 22, 21–24.

Bueno, V.H.P. (2000) *Controle biológico de pragas: produção massal e controle de qualidade* [*Biological pest control: mass production and quality control*]. 1st edn. Editora UFLA, Lavras, Minas Gerais, Brazil.

Bueno, V.H.P. (2009) *Controle biológico de pragas: produção massal e controle de qualidade* [*Biological pest control: mass production and quality control*]. 2nd edn. Editora UFLA, Lavras, Minas Gerais, Brazil.

Bueno, V.H.P. and Poletti, M. (2009) Progress with biological control and IPM strategies in protected cultivation in Brazil. *IOBC/WPRS Bulletin* 49, 31–36.

Bueno, V.H.P. and Sampaio, M.V. (2009) Desenvolvimento e multiplicação de parasitoides de pulgões. [Development and production of aphid parasitoids]. In: Bueno, V.H.P. (ed.) *Controle biológico de pragas: produção massal e controle de qualidade*. Editora UFLA, Lavras, Minas Gerais, Brazil, pp.117–167.

Bueno, V.H.P. and van Lenteren, J.C. (2011) Controle biológico de pragas em cultivos protegidos [Biological control of pests in protected cultivation]. *Ciencia & Ambiente* 43, 211–230.

Bueno, V.H.P., Silva, A.R., Carvalho L.M. and Moura, N. (2009) Control of thrips with *Orius insidiosus* in greenhouse cut roses: use of a banker plant improves the performance of the predator. *IOBC/WPRS Bulletin* 49, 183–187.

Bueno, V.H.P., van Lenteren, J.C., Lins, J.C., Calixto, A.M., Montes, F.C., Silva, D.B., Santiago, L.D. and Pérez, L.M. (2013) New records of *Tuta absoluta* (Meyrick) (Lepidoptera: Gelechiidae) predation by Brazilian Hemipteran predatory bugs. *Journal of Applied Entomology* 137, 29–34.

Cavalcante, A.C.C., Mandro, M.E.A., Paes, E.R. and Moraes, G.J. de (2017) *Amblyseius tamatavensis* Blommers (Acari: Phytoseiidae) a candidate for biological control of *Bemisia tabaci* (Gennadius) biotype B (Hemiptera: Aleyrodidae) in Brazil. *International Journal of Acarology* 43, 10–15. DOI: 10.1080/01647954.2016. 1225816.

Carvalho, R.S. and Nascimento, A.S. (2002) Criação e utilização de *Diachasmimorpha longicaudata* para controle biológico de moscas-das-frutas (Tephritidae) [Rearing and use of *Diachasmimorpha longicaudata* to biological control of fruit flies]. In: Parra, J.R.P., Botelho, P.S.M., Corrêa-Ferreira, B.S. and Bento, J.M.S. (eds) *Controle biológico no Brasil: parasitoides e predadores*. Manole, São Paulo, Brazil, pp. 165–179.

Chagas, M.C.M. and Parra, J.R.P. (2000) *Phyllocnistis citrella* Stainton, 1856 (Lepidoptera: Gracilariidae): técnica de criação e biologia em diferentes temperaturas [*Phyllocnistis citrella*: mass production technique and biology at different temperatures]. *Annais da Sociedade Entomológica do Brasil* 29, 227–235.

Chagas, M.C., Parra, J.R.P., Milano, P., Nascimento, A.M., Parra, A.L.G.C. and Yamamoto, P. (2002) *Ageniaspis citricola*: criação e estabelecimento no Brasil [*Ageniaspis citricola*: rearing and establishment in Brazil]. In: Parra, J.R.P., Botelho, P.S.M., Corrêa-Ferreira, B.S., and Bento, J.M.S. (eds) *Controle biológico no Brasil: parasitoides e predadores*. Manole, São Paulo, Brazil, pp. 377–391.

Charles, V.K. (1937) A fungus on lace bug. *Mycologia* 29, 216–221.

CIA (2017) The World Factbook: Brazil. Available at: https://www.cia.gov/library/publications/the-world-factbook/geos/br.html (accessed 7 July 2019).

Corrêa-Ferreira, B.S. (2002) *Trissolcus basalis* para controle de percevejos da soja [*Trissolcus basalis* for controlling soybean stinkbugs]. In: Parra, J.R.P., Botelho, P.S.M., Corrêa-Ferreira, B.S. and Bento, J.M.S. (eds) *Controle biológico no Brasil: parasitoides e predadores*. Manole, São Paulo, Brazil, pp. 449–476.

Costa, A.S. and Muller, G.W. (1980) Tristeza control by cross protection. *Plant Disease* 64, 538–541.

Costa, J.M. da, Williams, R.N. and Schuster, M.F. (1970) Cochonilhas dos capins, *Antonina graminis* no Brasil: II. Introdução de *Neodusmetia sangwani*, inimigo natural da cochonilha [Rhodesgrass scale, *Antonina graminis* in Brazil: II. Introduction of *Neodusmetia sangwani*, a natural enemy of the scale]. *Pesquisa Agropecuária Brasileira* 5, 339–343.

Costa, V.A., Berti Filho, E. and Silveira Neto, S. (2004) Parasitoides (Hymenoptera: Chalcidoidea) de moscas sinantrópicas (Diptera: Muscidae) em aviários de Echaporã, SP [Parasitoids of synanthropics flies in poultry farm in Echaporã, SP]. *Arquivos do Instituto Biológico* 71, 203–209.

Dolinski, C., Monteiro, C., Andaló, V. and Leite, L.G. (2017) Studies on entomopathogenic nematodes in Brazil: past and future. *Nematoda*, 4:e102017. DOI: 10.4322/nematoda.01017

Embrapa (2018) Pragas em plantios florestais produtivos [Pests in commercial forests]. Empresa Brasileira de Pesquisa Agropecuária [Brazilian Agricultural Research Corporation]. Available at: www.embrapa.br/florestas/transferencia-de-tecnologia/pragas-florestais (accessed 18 September 2018).

Fischbein, D. and Corley, J.C. (2015) Classical biological control of an invasive forest pest: a world perspective of the management of *Sirex noctilio* using the parasitoid *Ibalia leucospoides* (Hymenoptera: Ibaliidae). *Bulletin of Entomological Research* 105, 1–12.

Fiuza, L.M., Polanczyk, R.A. and Crickmore, N. (2017) Bacillus thuringiensis *and* Lysinibacillus sphaericus*: characterization and use in the field of biocontrol*. Springer, Dordrecht, Netherlands.

Freitas, G.D., Oliveira, A.C., Morais, E.J. and Barcelos, J.A.V. (1990) *Utilização do hemiptero predator* Podisus connexivus *Bergroth, 1891 (Hemiptera: Pentatomidae) para controle de lagartas desfolhadoras*

de Eucalyptus *spp.* [*Use of hemipteran predator* Podisus connexivus *Bergroth, 1891 to control defoliator caterpillars in* Eucalyptus *spp.*]. Mannesmann Fi-El Florestal, Paraopeba, Minas Gerais, Brazil.
Gabriel, D. (2017) Cochonilha dos capins *Antonina graminis* [Rhodesgrass scale *Antonina graminis*]. Instituto Biológico, Documento Técnico 30, 1–14.
Garcia, F.R.M. and Ricalde, M.P. (2013) Augmentative biological control using parasitoids for fruit fly management in Brazil. *Insects* 4, 55–70. DOI: 10.3390/insects4010055.
Garlet, J., Costa, E.C., Boscardin, J., Deponti, G., Shwengber, R. and Machado, L.M. (2013) *Leptocybe invasa* em *Eucalyptus* sp. no estado do Rio Grande do Sul, Brasil [*Leptocybe invasa* in *Eucalyptus* sp. in the state of Rio Grande do Sul, Brazil]. *Ciência Rural* 43, 2175–2177.
Gassen, D.N. and Tambasco, F.J. (1983) Controle biológico dos pulgões do trigo no Brasil [Biological control of wheat aphids in Brazil]. *Informe Agropecuário* 9, 49–51.
Gomez-Torres, M.L., Nava, D.E., Gravena, S., Costa, V.A. and Parra, J.R.P. (2006) Registro de *Tamarixia radiata* (Waterston) (Hymenoptera: Eulophidae) em *Diaphorina citri* Kuwayama (Hemiptera: Psyllidae) em São Paulo [Record of *Tamarixia radiata* from *Diaphorina citri* in São Paulo state]. *Revista de Agricultura* 81, 112–117.
Grafton-Cardwell, E.E., Stelinski, L.L. and Stansly, P.A. (2013) Biology and management of Asian citrus psyllid, vector of the Huanglongbing pathogens. *Annual Review of Entomology* 58, 413–432.
Gravena, S. (2005) *Manual prático de manejo ecológico de pragas de citros* [*Practical handbook on ecological management of citrus pests*]. UNESP, Jaboticabal, SP/Brasil.
Haji, F.N.P., Prezotti, L., Carneiro, J.S. and Alencar, J.A. (2002) *Trichogramma pretiosum* para o controle de pragas no tomateiro industrial [*Trichogramma pretiosum* to control pests of industrial tomato]. In: Parra, J.R.P., Botelho, P.S.M., Corrêa-Ferreira, B.S. and Bento, J.M.S. (eds) *Controle biológico no Brasil: parasitoides e predadores*. Manole, São Paulo, Brazil, pp. 477–494.
Hensley, S.D. and Hammond, A.M. (1968) Laboratory techniques for rearing the sugarcane borer on an artificial diet. *Journal of Economic Entomology* 61, 1742–1743.
IBGE (2018) Instituto Brasileiro de Geografia e Estatistica. Available at: https://seriesestatisticas.ibge.gov.br (accessed June 2018)
IBSP (2014) Instituto Biológico de São Paulo. Available at: www.biologico.sp.gov.br (accessed 30 September 2019).
Iede, E.T., Penteado, S.R.C. and Filho, W.R. (2012) The woodwasp *Sirex noctilio* in Brazil: monitoring and control. In: Slippers, B., de Groot, P. and Wingfield, M.J. (eds) *The Sirex woodwasp and its fungal symbiont: research and management of a worldwide invasive pest*. Springer, Dordrecht, Netherlands, pp. 217–228.
Indústria Brasileira de Árvores (Brazilian Tree Industry) (2014) Available at: http://www.iba.org/shared/iba_2014_pt.pdf (accessed 28 August 2015).
Kurylo, C., Garcia, M.S., Costa, V.A., Tibola, C, Ramrio, G.A. and Finkenauer, E. (2010) Occurrence of *Ctenarytaina eucalypti* (Maskell) (Hemiptera: Psyllidae) and its natural enemy *Psyllaephagus pilosus* Noyes (Hymenoptera: Encyrtidae) in *Eucalyptus globulus* in the State of Rio Grande do Sul, Brazil. *Neotropical Entomology* 39, 671–673.
Laboratório de quarentena 'Costa Lima' ['Costa Lima' quarantine laboratory] (2005) Embrapa Meio Ambiente, Jaguariúna, São Paulo, Brazil, 1 CD-ROM (Documentos, 45).
Li, Z., Alves, S.B., Roberts, D., Fan, M., Delalibera, I., Tang, J., Lopes, R.B., Faria, M. and Rangel, D.E.N. (2010) Biological control of insects in Brazil and China: history, current programs and reasons for their successes using entomopathogenic fungi. *Biocontrol Science and Technology* 20, 117–136.
Lin, N.-Q., Huber, J.T. and La Salle, J. (2007) The Australian Genera of Mymaridae (Hymenoptera: Chalcidoidea). *Zootaxa* 1596, 1–111.
MAPA (2016) Serviço de estudos normativos da produção orgânica/COAGRE/DEPROS/ SDC/MAPA Priorização das CPOrg-UFs dez. 2013/2014/2015 [Service of normative studies of organic production]. Available at: http://www.agricultura.gov.br/assuntos/sustentabilidade/organicos/produtos-fitossanitarios/arquivos-especificacao-de-referencia/priorizacao-nacional-cporgs.pdf (accessed 31 May 2018).
Mascarin, G.M., Lopes, R.B., Delalibera, I. Jr, Fernandes, E.K.K., Luz, C. and Faria, M. (2018) Current status and perspectives of fungal entomopathogens used for microbial control of arthropod pests in Brazil. *Journal of Invertebrate Pathology* 165, 46–53. DOI: 10.1016/j.jip.2018.01.001
Monteiro, L.B. (2002) Criação de ácaros fitófagos e predadores: um caso de produção de *Neoseiulus californicus* por produtores de maçã [Rearing of phytophagous and predatory mites: A case of production of *Neoseiulus californicus* by apple growers]. In: Parra, J.R.P., Botelho, P.S.M., Corrêa-Ferreira, B.S. and Bento, J.M.S. (eds) *Controle biológico no Brasil: parasitoides e predadores*. Manole, São Paulo, Brazil, pp. 351–362.

Moscamed Brasil (2018) Publicações. Available at: www.moscamed.org.br (accessed 18 September 2018).

Moscardi, F. (1999) Assessment of the application of Baculoviruses for control of Lepidoptera. *Annual Review of Entomolology* 44, 257–289.

Oxitec (2019) Aedes do BEM™, conheça a tecnologia inovadora que combate o *Aedes aegypti* selvagem, o mosquito transmissor da dengue, Zika e chikungunya [Aedes do Bem™, the innovative technology to combat *Aedes aegyptii*, the mosquito transferring dengue, Zika and chikungunya]. Available at: http://oxitec.com.br/ (accessed 5 July 2019)

Panizzi, A.R. (2015) Growing problems with stink bugs (Hemiptera: Heteroptera: Pentatomidae): species invasive to the US and potential neotropical invaders. *American Entomologist* 61, 223–233. DOI: 10.1093/ae/tmv068.

Parra J.R.P. (2014) Biological control in Brazil: an overview. *Scientia Agricola* 71, 345–355.

Parra, J.R.P., Botelho, P.S.M., Corrêa-Ferreira, B. and Bento, J.M.S. (2002) *Controle biológico no Brasil: parasitoides e predadores* [*Biological control in Brazil: parasitoids and predators*]. Manole, São Paulo, Brazil.

Parra, J.R.P., Zucchi, R.A., Coelho Jr., A., Geremias, L.D. and Consoli, F.L. (2015) *Trichogramma* as a tool for IPM in Brazil. In: Vinson, S.B., Greenberg, S.M., Liu, T.X., Rao, A. and Volosciuk, L.F. (eds) *Biological Control of Pests Using* Trichogramma. *Current Status and Perspectives.* Northwest A & F University Press, Yangling, Shaanxi, China, pp. 472–476.

Parra, J.R.P., Alves, G.R., Diniz, A.J.F. and Vieira, J.M. (2016) *Tamarixia radiata* (Hymenoptera: Eulophidae) × *Diaphorina citri* (Hemiptera: Liviidae): mass rearing and potential use of the parasitoid in Brazil. *Journal of Integrated Pest Management* 7, 1–11. DOI: 10.1093/jipm/pmw003.

Pereira, O.L., Barreto, R.W., Cavalazzi, J.R.P. and Braun, U. (2007) The mycobiota of the cactus weed *Pereskia aculeate* in Brazil, with comments on the life-cycle of *Uromyces pereskiae*. *Fungal Diversity* 25, 167–180.

Peronti, A.L.B.G., Martinelli, N.M., Alexandrino, J.G., Marsaro Júnior, A.L., Penteado-Dias A.M. and Almeida, L.M. (2016) Natural enemies associated with *Maconelliccoccus hirsutus* (Hemiptera: Pseudococcidae) in the state of São Paulo, Brazil. *Florida Entomologist* 99, 21–25.

Pignati, W.A., Souza e Lima, F.A.N., Lara, S.S., Correa, M.L.M., Barbosa, J.R., Leão, L.H.C. and Pignatti, M.G. (2017) Spatial distribution of pesticide use in Brazil: a strategy for health surveillance. *Ciência & Saúde Coletiva* 22, 3281–3293.

Pires, E.M., Soares, M.A., Nogueira, R.M., Zanuncio, J.C., Moreira, P.S.A. and Oliveira, M.A. (2015) Sete décadas de estudos com Asopinae predadores no Brasil (1933–2014). [Seven decades of studies with Asopinae predators in Brazil (1933–2014]. *Bioscience Journal* 31, 1530–1549.

Polanczyk, R.A., van Frankenhuyzen, K. and Pauli, G. (2017) The American *Bacillus thuringiensis* based biopesticides market In: Fiuza, L.M., Polanczyk, R.A. and Crickmore, N. (eds) *Bacillus thuringiensis* and *Lysinibacillus sphaericus: characterization and use in the field of Biocontrol*. Springer, Dordrecht, Netherlands, pp. 173–184.

Poletti, M. and Omoto, C. (2012) Susceptibility to deltamethrin in the predatory mites *Neoseiulus californicus* and *Phytoseiulus macropilis* (Acari: Phytoseiidae) populations in protected ornamental crops in Brazil. *Experimental and Applied Acarology* 58, 385–393.

Rodrigues, S.M.M., Bueno, V.H.P. and Bueno Filho, J.S. de S. (2001) Desenvolvimento e avaliação do sistema de criação aberta no controle de *Aphis gossypii* Glover (Hem.: Aphididae) por *Lysiphlebus testaceipes* (Cresson) (Hym.: Aphidiidae) em casa-de-vegetação [Development and evaluation of the open rearing system in the control of *Aphys gossypii* by *Lysiphlebus testaceipes* in greenhouse]. *Neotropical Entomology* 30, 433–436

Rodrigues, S.M.M., Bueno, V.H.P. and Sampaio, M.V. (2005) Efeito da liberação inoculativa sazonal de *Lysiphlebus testaceipes* (Hym.: Aphidiidae) na população de *Aphis gossypii* (Hem.: Aphididae) em cultivo de crisântemo em casa de vegetação comercial [Effect of seasonal inoculative release of *Lysiphlebus testaceipes* on the *Aphis gossypii* population in a chrysanthemum crop in a commercial greenhouse]. *Bolletin de Sanidad Vegetal Plagas* 31, 199–207.

Sá, L.A.N., Pessoa, M.C.P.Y., Moraes, G.J., Marinho-Prado, J.S., Prado, S.S. and Vasconcelos, R.M. (2016) Quarantine facilities and legal issues of the use of biocontrol agents in Brazil. *Pesquisa Agropecuária Brasileira* 51, 502–509. DOI: 10.1590/S0100-204X2016000500010.

Salvadori, J.R. and Salles, L.A.B. (2002) Controle biológico dos pulgões do trigo [Biological control of wheat aphids]. In: Parra, J.R.P., Botelho, P.S.M., Corrêa-Ferreira, B.S. and Bento, J.M.S. (eds) *Controle biológico no Brasil: parasitoides e predadores*. Manole, São Paulo, Brazil, pp. 427–447.

Sato, M.E., Silva, M.Z., Souza Filho, M.F., Matioli, A.L. and Raga, A. (2007) Management of *Tetranychus urticae* (Acari: Tetranychidae) in strawberry fields with *Neoseiulus californicus* (Acari: Phytoseiidae) and acaricides. *Experimental and Applied Acarology* 42, 107–120.

Schmitt, A.T., Gowen, S.R. and Hague, N.G.M. (1992) Baiting techniques for the control of *Cosmopolites sordidus* Germar (Coleoptera: Curculionidae) by *Steinernema carpocapsae* (Nematoda: Steinernematidae). *Nematropica* 22, 159–163.

Schühli, G.S., Penteado, S.C., Barbosa, L.R., Reis Filho, W. and Iede, E.T. (2016) A review of the introduced forest pests in Brazil. *Pesquisa Agropecuária Brasileira* 51, 397–406. DOI: 10.1590/S0100-204X2016000500001.

Schuster, M.F. and Boling, J.C. (1971) Biological control of rhodesgrass scale in Texas by *Neodusmetia sangwani* (Rao): effectiveness and colonization studies. *Bulletin Texas Agricultural Experiment Station* 1104, pp.5–15.

SEB (2018) Sociedade Entomológica do Brasil. Available at: http://www.seb.org.br/ (accessed 31 October 2018).

Silva, A.G.D.A., Gonçalves, C.R., Galvão, D.M., Gonçalves, A.J.L., Gomes, J., Silva, M.N. and Simoni, L. (1968) *Quarto catálogo dos insetos que vivem nas plantas do Brasil, seus parasitos e predadores. Parte II 1° Tomo: Insetos, hospedeiros e inimigos naturais* [*Fourth catalogue of insects living on plants in Brazil, their parasites and predators. Part II 1st issue: Insects, hosts and natural enemies*]. Ministério da Agricultura, Rio de Janeiro, Brazil.

Silveira, L.C.P., Bueno, V.H.P. and van Lenteren, J.C. (2004) *Orius insidiosus* as biological control agent of thrips in greenhouse chrysanthemums in the tropics. *Bulletin of Insectology* 57, 103–109.

SNIF (2018) Associação Brasileira de Produtores de Florestas Plantadas (ABRAF). Available at: http://snif.florestal.gov.br/pt-br/ (accessed June 2018).

Sosa-Gómez, D.R. (2017) Microbial control of soybean pest insects and mites, In: Lacey, L.A. (ed.) *Microbial Control of Insect and Mite Pests*. Elsevier, Amsterdam, Netherlands, pp. 199–208.

Souza, L. da S. and Fialho, J.F. (2003) *Sistema de produção de mandioca para a região do Cerrado* [*Cassava production system for the Cerrado region*]. Embrapa Mandioca e Fruticultura Cruz das Almas, BA, Brazil.

Souza-Pimentel, G.C., Reis, P.R., Silveira, E.C., Marafeli, P.P., Silva, E.A. and Andrade, H.B. (2014) Biological control of *Tetranychus urticae* (Tetranychidae) on rose bushes using *Neoseiulus californicus* (Phytoseiidae) and agrochemical selectivity. *Revista Colombiana de Entomología* 40, 80–84.

Torres, J.B., Zanuncio, J.C. and Moura, M. (2006) The predatory stinkbug *Podisus nigrispinus*: biology, ecology and augmentative releases for lepidopteran larval control in *Eucalyptus* forests in Brazil. *Biocontrol News and Information* 27, 1–18.

Valdebenito-Sanhueza, R.M. (1991) Possibilidades do controle biológico de *Phytophthora* em macieira [Possibilities of biological control of *Phytophthora* in apple orchards]. In: Bettiol, W. (ed.) *Controle biológico de doenças de plantas* [*Biological control of plant diseases*]. Embrapa-CNPMA, Jaguariúna, São Paulo, Brazil, pp. 303–305.

Valicente F.H., Tuelher, E.S., Pena, R.C., Andreazza, R. and Guimarães, M.R. (2013) Cannibalism and virus production in *Spodoptera frugiperda* (J.E. Smith) (Lepidoptera: Noctuidae) larvae fed with two leaf substrates inoculated with Baculovirus spodoptera. *Neotropical Entomology* 42,191-199. DOI: 10.1007/s13744-013-0108-6.

van Lenteren, J.C. and Bueno, V.H.P. (2019) Advances in augmentative biological control in IPM. In: Kogan, M. and Heinrichs, E. (eds) *Integrated Management of Insect Pests in Agriculture. Volume 2. Current and Future Developments in IPM*. Burleigh Dodds Scientific Publishing, Cambridge, UK. DOI: 10.19103/AS.2019.0047.15

van Lenteren, J.C., Bueno, V.H.P., Smit, J., Soares, M.A., Calixto, A.M., Montes, F.C. and Jong, P. (2017) Predation of *Tuta absoluta* eggs during the nymphal stages of three Neotropical mirid predators on tomato. *Bulletin of Insectology* 70, 69–74.

van Lenteren, J.C., Bueno, V.H.P., Calvo, F.J., Calixto, A.M. and Montes, F.C. (2018a) Comparative effectiveness and injury to tomato plants of three Neotropical mirid predators of *Tuta absoluta* (Lepidoptera: Gelechiidae). *Journal of Economic Entomology* 111, 1080–1086.

van Lenteren, J.C., Bolckmans, K., Köhl, J., Ravensberg, W. and Urbaneja, A. (2018b) Biological control using invertebrates and microorganisms: plenty of new opportunities. *BioControl* 63, 39–59. DOI: 10.1007/s10526-017-9801-4.

Vilela, E.F. and Zucchi, R.A. (2015) *Pragas introduzidas no Brasil: insetos e ácaros* [*Pests introduced in Brazil: insects and mites*]. FEALQ, Piracicaba, São Paulo, Brazil.

Vilela, E.F., Rech Filho, E.L., Bartha, G.B.M. Jr, Alves, E.R. de Andrade, Lopes, M.A., Guimarães, E.P., Cabral, P.R., Soares, C.O., Rosinha, G.M.S., Buainain, A.M., Nutti, M.R. and Callegaro, G.M. (2017) Food and nutrition security in Brazil. In: *Challenges and Opportunities for Food and Nutrition Security in the Americas. The View of the Academies of Sciences*. IANAS, IAP and BMBF, México DF, pp. x76–109. [Free public access of this publication in English and Spanish at www.ianas.org]

Walder, J.M.M., Costa, M.L.Z., and Mastrangelo, T.A. (2009) Produção massal do parasitoide *Diachasmimorpha longicaudata* para o controle biológico de moscas-das-frutas [Mass production of parasitoid *Diachasmimorpha longicaudata* for biological control of fruit flies]. In: Bueno, V.H.P. (ed.) *Controle biológico de pragas: produção massal e controle de qualidade*. Editora UFLA, Lavras, Minas Gerais, Brazil, pp. 221–231.

Watanabe, M.A., Moraes, G.J. de, Gastaldo, I. Jr and Nicolella, G. (1994) Controle biológico do ácaro rajado com ácaros predadores fitoseídeos (Acari: Tetranychidae, Phytoseiidae) em culturas de pepino e morango [Biological control of two spotted mite with predatory mites in cucumber and strawberry crops]. *Scientia Agricola* 51, 75–81.

Wilcken, C.F. and Oliveira, N.C. (2015) Gorgulho-do-eucalipto *Gonipterus platensis* Marelli [The eucalyptus weevil *Gonipterus platensis* Marelli]. In: Villela, E.F. and Zucchi, R.A. (eds.) *Pragas introduzidas no Brasil: insetos e ácaros*. FEALQ, Piracicaba, São Paulo, Brazil, pp. 779–791.

Wilcken, C.F., Firmino-Winckler, D.C., Dal Pogetto, M.H.F.A., Dias, T.K.R., Lima, A.C.V., Sá, L.A.N. and Ferreira-Filho, P.J. (2015) Psilídeo-de-concha-do-eucalipto, *Glycaspis brimblecombei* Moore [Red gum lerp psyllid, *Glycaspis brimblecombei*]. In: Vilela, E.F. and Zucchi, R.A. (eds) *Pragas introduzidas no Brasil: insetos e ácaros*. FEALQ, Piracicaba, São Paulo, Brazil, pp. 883–897.

Yorinori, J.T. and Gazziero, D.L.P. (1989) Control of wild poinsettia (*Euphorbia heterophylla*) with *Helminthosporium* sp. In: Delfosse, E.S. (ed.) *Proceedings of the VII International Symposium on Biological Control of Weeds*. Instituto Sperimentale per la Patologia Vegetale, Rome, pp. 571–576.

Zanuncio, J.C., Guedes, R.N.C., Oliveira, H.N. and Zanuncio,T.V. (2002) Uma década de estudos com percevejos predadores: conquistas e desafios [A decade of studies with predatory bugs: success and challenges]. In: Parra, J.R.P., Botelho, P.S.M., Corrêa-Ferreira, B.S. and Bento, J.M.S. (eds) *Controle biológico no Brasil: parasitoides e predadores*. Manole, São Paulo, Brazil, pp. 495–509.

7 Biological Control in Chile

Lorena Barra-Bucarei*, Luis Devotto Moreno and Andrés France Iglesias

Instituto de Investigaciones Agropecuarias, INIA-Quilamapu, Chile

* E-mail: lbarra@inia.cl

Abstract

The first introduction of a predator of olive black scale in 1903 marked the start of biocontrol in Chile. In 1915 the Ministry of Agriculture began a programme of introduction of beneficial organisms and since then approximately 200 species of beneficial insects have been introduced into the country. Many complete and permanent successes were obtained in the period 1900–1969 and examples are control of: olive black scale; scales in grapevine, citrus, berries, avocado and other fruit; mealybugs in citrus, grapevine, avocado and other fruit; and woolly apple aphid in apple. The use of microbial control agents started in the 1950s and numerous microorganisms have since been collected, identified and mass produced to control pests and diseases. Weed biocontrol began in Chile in the 1950s with successful classical biocontrol of St John's wort with a phytophagous weevil, followed in the 1970s by using a plant pathogenic fungus for control of *Rubus ulmifolius* (zarzamora) and in the 1980s–1990s by releasing phytophagous weevils, moths and mites for control of gorse weed. An important large project executed during the 1970s and 1980s resulted in classical biocontrol of several virus-transmitting aphid species on thousands of hectares of wheat and barley. In forestry, parasitoids have been introduced for classical biocontrol of European pine shoot moth and of the sirex wood wasp in pine, and the eucalyptus psyllid in eucalyptus on vast areas. Currently, there are many commercial and governmental research initiatives for predators, parasitoids, entomopathogenic fungi and nematodes, and with bacteria and fungi for pest and disease control.

7.1 Introduction

Chile has an estimated population of almost 17,800,000 (July 2017) and its main agricultural products are grapes, wine, apples, pears, onions, wheat, maize, oats, peaches, garlic, asparagus, beans, beef, poultry, wool, fish and timber (CIA, 2017). According to Schick *et al.* (2017, pp. 192, 193 and 198):

> ... the advent of export agriculture created two types of agriculture: one dedicated to supplying food for the local population, developed by Peasant Family Agriculture, in small, low-tech plots of land; and commercial agriculture, with a great deal of technology and significant foreign investment, designed to produce for world markets. ... mainland Chile has an area of 75.6 million hectares (ha), 51.7 million of which are suitable for silvo-agriculture and 35.5 million of which are used for agricultural livestock raising or forestry. However, due to geographical and economic factors, the area under cultivation currently stands at just 2.12 million ha. This area is distributed among 1,303,210 ha of annual and permanent crops, 401,018 ha of sown fields and 419,714 ha of fallow land. A total of 17,070,776 ha are covered by native forest and bushes; 12,549,478 by natural meadows; 2,707,461 by forest plantations, and 1,062,352 by improved pastures. Of the 1.3 million ha with annual and permanent crops, 704,575 ha are used for annual crops (wheat, corn, oats, potato and raps), 296,587 ha for fruit trees (table vines, apple, avocado, walnut, cherry), 137,593 ha for wine-grape vines and 78,072 ha for vegetables (corn, lettuce, tomato, onion, marrow). As for meat production, poultry accounts for the largest share, ... next pork ..., beef ..., lamb ... and horsemeat. ... The Chilean salmon aquaculture industry is the now second largest export sector in the country and the world's second largest salmon producer after Norway.

7.2 History of Biological Control in Chile

Many agricultural and forest pests are exotic to Chile and have been introduced into the country by different means, including tourism, machinery packing, ships, aircraft and as a result of self-dispersion of some species. Once introduced, they usually do not meet any natural enemies in the country and their populations can multiply easily.

7.2.1 Period 1880–1969

After the initiative of a farmer who imported a natural enemy of olive black scale *Saissetia oleae* (Olivier) into Chile in 1903, the Ministry of Agriculture began a programme to introduce beneficial insects in 1915. Since then, classical biocontrol has been the dominant practice, with approximately 200 species of beneficial insects introduced into Chile to control different pests. Many of the natural enemies have been successfully established and they controlled pests without the farmers realizing it. The main sources of the beneficial insects introduced into Chile are the USA, Peru, Canada, Germany,

England and Argentina (Rojas, 2005). From the 1930s, Chile also became an exporter of biocontrol agents, resulting in successes worldwide in controlling mosquitoes, aphids, scales and mites, among others. The use of microbial agents for biocontrol started in the 1950s. Numerous microorganisms have subsequently been used to control pests and diseases, some of which are commercialized for use in integrated pest management (IPM) or organic production.

Biological control of agricultural pests with arthropod natural enemies

Biocontrol of pests in Chile has a long history and started in 1903 with the introduction of *Rhyzobius ventralis* (Erichson), a coccinellid predator of eggs and nymphs of the olive black scale, by farmer T. Schneider. This pest had been present in Chile since 1868. Between 1931 and 1946, natural enemies of *S. oleae* of the genera *Coccophagus*, *Lecanobius*, *Metaphycus* and *Scutellista* were introduced from the USA and Peru. The parasitoid *Metaphycus helvolus* (Compere) was reintroduced in 1951 by the La Cruz National Insectarium (now La Cruz Regional Research Center of the Instituto de Investigaciones Agropecuarias (INIA)) and this time the parasitoid successfully established and dispersed (Zuñiga, 1986). To this day, *M. helvolus* is one of the most effective and important natural enemies of *S. oleae* (Rojas, 2005). The different natural enemies introduced for control of *S. oleae* in Chile have also controlled other pests with similar characteristics (Table 7.1).

Other important pests in Chile are mealybugs, because they are considered quarantine pests by many countries importing Chilean fruit. Species of the genera *Pseudococcus* and *Planococcus* establish colonies on fruits, leaves, stems, trunks and roots of many hosts. The weakening caused by these sap-feeding insects can be serious and in some cases mealybugs kill the plant. Other deleterious effects arise from toxin production, virus transmission and honeydew secretion. Failures in mealybug control cause rejection of fruit in the inspecting ports or importing countries and consequently important economic losses. Biocontrol of these species started in Chile in 1931 when the predator *Cryptolaemus montrouzieri* (Mulsant) and the parasitoid *Leptomastidea abnormis* (Girault) were introduced. The La Cruz National Insectarium was founded in 1939 and the first natural enemy it produced was *C. montrouzieri*. New introductions of *C. montrouzieri* from the USA occurred until 1996. Between 1951 and 1958, four other parasitoids were introduced. Parasitoids of the most common mealybug species are listed in Table 7.2. *Coccophagus gurneyi* (Compere) and *C. montrouzieri* have been the most effective white mealybug natural enemies.

Table 7.1. Pests controlled by natural enemies originally introduced to control *Saissetia oleae* (retrieved from González and Rojas, 1966; Rojas, 2005; Ripa and Droguett, 2008).

Natural enemies	Pests	Main host plants
Coccophagus caridei, *Metaphycus flavus*, *M. helvolus*	*Ceroplastes sinensis* (Del Guercio), Chinese wax scale	Grapevine
	Ceroplastes cirripediformis (Comstock), Florida wax scale	Orange, lemon, mandarin, grapefruit, cherry, plum, lucuma
	Parthenolecanium corni (Bouché), European fruit lecanium scale	Grapevines, chestnut, gooseberry, sarsaparrilla (*Smilax* sp.), cranberry
	Parthenolecanium persicae (Fabricius), European peach scale	Peach, grapevine
	Pulvinaria mesembryanthemi Vallot, Soft scale	Avocado
C. caridei, *M. flavus*, *M. helvolus*, *M. stanleyi*	*Coccus hesperidum* (Linnaeus), Florida soft scale	Orange, lemon, mandarin, grapefruit, raspberry, blueberry, guava
	Protopulvinaria pyriformis (Cockerell), Pyriform scale	Avocado, lucuma, guava, orange
	Saissetia coffeae (Walker), Hemispherical soft scale	Lemon, grapefruit, orange, olive, lucuma, mango, guave

Table 7.2. Parasitoids introduced into Chile to control mealybugs (retrieved from Ripa and Rodríguez, 1999; Rojas, 2005; Ripa *et al.*, 2008).

Parasitoids	Pests	Main host plants
Leptomastidea abnormis (Girault), *L. dactylopii* (Howard), *Pauridia peregrina* (Timberlake), *Allotropa citri* (Muesebeck), *Anagyrus pseudococci* (Girault), *Pseudophycus perdignus* (Compere)	*Planococcus citri* (Risso)	Lemon, mandarin, orange, grapefruit, persimmon, pomegranate, cherimoya, guava, mango
Coccophagus gurneyi (Compere), *Tetracnemus pretiosus* (Timberlake)	*Pseudococcus calceolariae,* (Maskell)	Lemon, mandarin, orange, pear, grapefruit, blueberry, persimmon, cherimoya, plum, peach, quince, avocado, raspberry, sarsaparilla
	Pseudococcus longispinus (Targioni and Tozzetti)	As above, plus guava, sour cherry, lucuma, mango, apple, passion fruit, loquat, olive, grapevine
Leptomastix epona (Noyes), *Pseudaphycus flavidulus* (Bréthes)	*Pseudococcus viburni* (Maskell)	Alfalfa, persimmon, cherry, plum, citrus, raspberry, chickpea, lentil, apple, blackberry, nectarine, loquat, potato, melon pear, pear, radish, grapevine, sarsaparilla

The woolly apple aphid *Eriosoma lanigerum* (Hausmann) is distributed throughout the apple orchard area in Chile and may cause serious damage. In the 1920s, this pest caused the death of a large number of trees, which led producers to abandon cultivation of apple. To control the pest, the parasitoid *Aphelinus mali* (Haldeman) was introduced from Uruguay in 1922 and had established by 1927 (Howard, 1929). The parasitoid adapted quickly to the country's conditions and re-energized the then weakened apple industry.

Microbial control of agricultural and forest pests

The first report about the use of entomopathogenic fungi (EPF) in Chile dates back to the 1950s when Dutky (1957) studied the effect of a pathogen of the beetle *Hylamorpha elegans* (Burmeister), which occurred in the southern regions of Chile and was a primary pest of wheat and pastures. The first report on biocontrol with entomopathogenic nematodes (EPN) also dates back to the 1950s, when *Steinernema* sp. (strain DD-136) was used for control of several soil pests, including larvae of coleopterans (*H. elegans, Pantomorus cervinus* (Boh.) (Curculionidae)) and lepidopterans (*Dalaca noctuides* Pfitzner, *Agrotys* sp.).

Biological control of weeds

Weed biocontrol began in Chile in the 1950s by introducing the beetles *Chrysolina hyperici* (Foster) and *Chrysolina quadrigemina* (Suffrian), which are specific to St John's wort (*Hypericum perforatum* L.). These species were successfully released and established in the country (Julien and Griffiths, 1998).

7.2.2 Period 1970–2000

Biological control of agricultural pests with arthropod natural enemies

One of the most successful cases of biocontrol in Chile was the programme to control cereal aphids. Until the beginning of the 1970s, wheat (*Triticum aestivum* L.) only showed low aphid densities not causing major problems. For still unknown reasons, several aphid species increased dramatically and began to cause serious yield losses to this Chilean staple food. In 1974 chemical insecticides were first applied on more than 120,000 ha of wheat (Zúñiga, 1986). In addition to the direct damage, the aphids transmitted viruses, such as the barley yellow dwarf virus (BYDV). Herrera and Quiroz (1988) showed that BYDV was able to reduce wheat yield by 10%, resulting in a yield loss of 80% in combination with direct damage caused by the aphids. The most damaging aphids were the pale green grass aphid *Metopolophium dirhodum* (Walker) and the grain aphid *Sitobion avenae* (Fabricius). The first attempts to control these aphids with beneficials go back to 1973 with introductions of parasitoids from Czechoslovakia and some predators, mainly coccinellids, from Canada, which had established in 1975.

In 1976, the Chilean government signed agreements with the Food and Agriculture Organization of the United Nations (FAO) and the United Nations Development Programme (UNDP) to bring expert consultants from California, through a National Biological Control of Aphids Program, focused on wheat and barley. Results of this programme were the implementation of new laboratory equipment at INIA, training of Chilean professionals and reliance on the support of international consultants, all of which contributed to the success achieved in biocontrol of wheat and barley aphids. Several parasitoids were brought to Chile, released and spread across the country; the aphid population was crushed and they never recovered for the next 40 years, saving a great amount of money for growers and consumers (Zúñiga, 1986). In addition, almost 300,000 ha per year were not sprayed with insecticides, avoiding environmental damage. In the mid-1980s, a new and dangerous aphid was accidentally introduced into Chile: the Russian wheat aphid *Diuraphis noxia* (Mordvilko). The same parasitoids introduced a decade earlier for control of other aphids also successfully controlled the Russian wheat aphid (Norambuena and Gerding, 1990; Rojas, 2005).

Apple orchards occupy more than 37,000 ha in Chile (ODEPA, 2016). *Cydia pomonella* (Linnaeus) causes important economic losses in apple, pear and nut orchards. Biocontrol of this pest was studied by using parasitoids of the genera *Trichogramma*, *Mastrus* and *Ascogaster* (Devotto *et al.*, 2010; Zaviezo and Mills, 2001). Torres and Gerding (2000) evaluated the parasitization efficiency of five species of *Trichogramma* and concluded that *T. cacoeciae* and *Trichogramma* sp. 'Cato' were efficient biocontrol agents of *C. pomonella*. Unfortunately, the apple industry in Chile still relies on chemical insecticides to control *C. pomonella*, instead of biocontrol.

Microbial control of agricultural pests

The Universidad Austral de Chile conducted studies in the 1970s to identify the different species of the genus *Entomophthora* existing in the country and the insects for which infection had been reported. They concluded that five species of this genus infected insects located between Biobío and Los Ríos regions (Aruta *et al.*, 1974). In the same decade, studies were carried out with EPF as antagonists of insect larvae of the family Scarabaeidae present in grasslands. In the 1980s Ripa and Rodríguez (1989) found that eight strains of *Metarhizium anisopliae* (Metschnikoff) Sorokin exhibited high virulence in *Naupactus xanthographus* (Germar) larvae. At that time the development of biocontrol with EPF did not yet meet the standards of other countries where commercial formulations were already being developed. In addition to the work conducted by the Universidad Austral de Chile, INIA established a collection of microbial agents as a result of sampling carried out throughout the country. The collection started in 1996 with native strains that were isolated by the Insect Pathology Program of the Regional Research Center INIA-Quilamapu, Chillán. More than 500 accessions of EPF from different latitudes were incorporated between 1997 and 2000. Among the genera that make up this collection, *Metarhizium* and *Beauveria* are the most important. At the end of this decade, strains of *M. anisopliae* were evaluated against *Otiorhynchus sulcatus* (Fabricius) and all strains were pathogenic to varying degrees and depending on fungal spore concentrations (Gerding *et al.*, 2000). Strains of *Beauveria bassiana* (Balsamo) Vuillemin and *M. anisopliae* from the INIA collection were evaluated in 2000 as pathogens of *Aegorhinus superciliosus* (Guérin), *Asynonychus (Naupactus) cervinus* (Boheman) and *O. sulcatus*, which are considered to this day as economically important pests, mainly of berry fruits. Some of the evaluated strains showed high pathogenicity levels on insects in laboratory trials. Two of these strains are commercial products and are used in berry crops.

The first tests of bacteria for insect control started in 1996, when the effectiveness of two

Bacillus thuringiensis (Berliner) formulations was evaluated. The concentrations recommended by the manufacturers of Dipel and Javelin were effective to control *Helicoverpa zea* (Boddie). Assays were also carried out to evaluate the control effect of four *B. thuringiensis* strains against the European pine shoot moth *Rhyacionia buoliana* (Denis & Schiffermüller) (Huerta and Cogollor, 1995).

In the 1980s, nematodes of the family Rhabditidae (not identified at species level) for control of *N. xanthographus* larvae were evaluated; later the presence of saprophytic nematodes of the genus *Caenorhabditis* was noted in larvae of this same pest (Ripa, 1992). At the end of the 1980s, Jiménez *et al.* (1989) found that six out of seven species of Lepidoptera were susceptible to *Steinernema carpocapsae* (Weiser). In the 1990s the EPN *Pellioditis pellio* (Schneider) was evaluated for control of the native horsefly *Scaptia lata* (Guérin-Méneville). Nematodes of the family Mermithidae were found in *Procalus mutans* (Blanchard) and *P. reduplicatus* (Bechyné) beetles. Further, native EPNs associated with their symbiont bacteria belonging to the INIA collection were evaluated in the laboratory against *Deroceras reticulatum* (Müller), *A. superciliosus* and *A. cervinus* and promising results were obtained. The INIA collection contained 34 accessions by 2000, with *Heterorhabditis* and *Steinernema* as the most common genera.

Biological control of forest pests

The forestry sector is very important to the Chilean economy and comprises 2,426,772 ha, of which 59% is *Pinus radiatea* D. Don and 34% *Eucalyptus* spp. (INFOR, 2016). One of the most important pests affecting pine nationwide was *R. buoliana*. The Chilean quarantine authority Servicio Agricola y Ganadero (SAG), INIA, the National Forest Corporation (Corporación Nacional Forestal) (CONAF) and several private forestry companies launched a programme to introduce natural enemies to control this pest. The Biological Control Center for the pine moth was implemented in the Regional Research Center INIA-Remehue in 1987 and the specific parasitoid *Orgilus obscurator* (Nees) was introduced from Europe that same year (Ide *et al.*, 2007). After a few years, this parasitoid reduced the moth outbreak below the economic damage threshold throughout the pine production area. The large forestry companies founded CPF S.A. (Forestry Pest Control Company) for *R. buoliana* control in the large private plantations, while CONAF provides consultancy, parasitized *R. buoliana* larvae and *O. obscurator* adults to small and medium-sized forest owners. The Regional Research Center INIA-Quilamapu started to mass produce and release the egg parasitoid *Trichogramma nerudai* (Pintureau & Gerding) in 1994 to further reduce moth outbreaks.

Another important forestry pest is the eucalyptus psyllid *Ctenarytaina eucalypti* (Maskell), which was detected in *Eucalyptus* spp. outbreaks in the Arica and Parinacota region in 1999. It spread rapidly over plantations throughout the country, causing damage as dehydration and death of primary shoots. To control this psyllid, INIA, SAG and CPF S.A. introduced the parasitoid *Psyllaephagus pilosus* (Noyes) from France and Peru in 2001. The parasitoid established and studies carried out by Rodríguez and Saiz (2006) demonstrated that parasitism of *C. eucalypti* was greater than 80%, especially during periods of maximum psyllid density.

Weed control with arthropod natural enemies and microbial agents

Ulex europaeus L. or gorse, known as *espinillo* in Chile, is an important weed in agriculture and forestry. *Apion ulicis* (Foster), a seed-consuming weevil of gorse, was introduced in 1976, but results were disappointing (Norambuena and Piper, 2000). In the 1980s, the gorse moth *Agonopterix ulicetella* (Stainton), a defoliator that specifically attacks this weed, was introduced. New populations of this defoliator were introduced from the USA and England in the 1990s. After an intensive selection process, introduction, breeding and determination of its specificity, SAG authorized the release of founding populations in Chile in 1997. In 1996, *A. ulicetella* and the mite *Tetranychus lintearius* (Dufour) were introduced from Hawaii, as well as another ecotype of the same mite from Portugal. Martínez *et al.* (2000) concluded that *A. ulicetella* was not a risk for plant species that are different from their natural host after field release.

Rubus ulmifolius (Schott.), known as *zarzamora*, is one of the most important weedy shrubs in Chile and in the 1970s it occupied approximately

5 million hectares. In 1974, the Universidad Austral de Chile conducted studies to introduce the fungus *Phragmidium violaceum* (Schultz) Winter. It was confirmed that this phytopathogen has the advantage of being a specific agent for this weed, without being a major threat to other species (Oehrens and González, 1974), and was then used for control of the weed.

Biological control of diseases

Biocontrol of diseases began in Chile with the evaluation and subsequent commercialization of a strain of *Agrobacterium radiobacter* (Beijerinck and van Delden) to inhibit the development of *Agrobacterium tumefaciens* Smith & Townsend in orchards. This product is registered and commercialized under the name Biobacter 84 G. The Universidad Austral de Chile conducted several studies to evaluate disease-controlling bacteria in potato (*Solanum tuberosum* L.). Isolates with antagonistic activity on potato bacterial wilt *Pseudomonas solanacearum* (Smith) (now *Ralstonia solanacearum*) were evaluated and those with high levels of antagonism against the pathogen were selected (Ciampi-Panno *et al.*, 1987). Among the bacteria that exhibited high antagonistic potential were some strains of *Pseudomonas fluorescens* (Migula) and an application method was later evaluated. The antagonistic strain FC8, associated with a modification, was more efficient in controlling *P. solanacearum* than its direct application on the tubers. The same university began research in the 1990s to develop formulations in capsules of *P. fluorescens* and *Bacillus subtilis* (Ehrenberg) to control *R. solanacearum*. These formulations were positively evaluated, because they did not influence the antagonistic ability of the strains of the microbial control agent.

A first course on biocontrol of diseases using microbial agents was held in the 1990s. Researchers from the Brazilian Agricultural Research Corporation (Embrapa) and the Department of Plant Health of the Faculty of Agricultural Sciences of the Universidad de Chile taught the course. This stimulated studies on microbial control agents. The research focused on control of diseases caused by *Botrytis*, *Rhizoctonia*, *Fusarium*, *Erwinia* and *Phytophthora*, among others. Studies with mycopathogenic fungi also began and the antagonistic ability of native *Trichoderma harzianum* (Rifa) strains against *Botrytis cinerea* Pers. in apple trees was evaluated, but they obtained only partial control. Also, the evaluation of solid and liquid fermentation of *Trichoderma* mass-production technologies started.

7.3 Current Situation of Biological Control in Chile

7.3.1 Introduction

The demand for biocontrol agents has increased significantly recently, which is partly due to the greater interest of agricultural companies and producers in complying with national and international regulations for pesticide use, as well as the need to incorporate these more sustainable technologies in their production systems. This situation has aroused interest among technology-based companies, universities, institutes and research centers in conducting research, technology transfer and commercialization of biocontrol agents.

In 2010, Chile signed the Accession Agreement to the Organization for Economic Cooperation and Development (OECD). By becoming a member, Chile acquired commitments associated with the normalization of agrochemical use and the assurance of a high level of protection for human, animal and environmental health, so that risk is reduced as the sale and use of pesticides decreases. The OECD reported in 2005 that Chile was among the countries with the highest values of agrochemical use worldwide with 0.46 t km^{-2} compared with the mean of 0.21 t km^{-2} in other member countries. The task was not easy and pesticide reduction goals were set for the end of 2014. Unfortunately, these goals were not met, because the consumption of chemical pesticides has increased in the past few years.

Pesticides intended for use in agricultural or forestry, whether they are imported, produced, commercialized or used in Chile, must be previously authorized by SAG. The 2017 SAG list of authorized pesticides has 1215 registered products. Less than 7% are classified in the ‘natural substance – biological pesticide – pheromone’ category, which includes 37 products for disease control, 22 for insect control and no products for weed control.

According to producers and importers, key factors that discourage registration of biocontrol agents in Chile are high cost, lengthy processing times and the large number of documents to submit. This situation has led companies to prefer registering the biopesticides under a series of names that allow them to avoid registration as pesticides, such as growth stimulants, resistance inducers, plant defence enhancers and biodiversity enhancers. Given the limited number of biocontrol agents registered by SAG, farmers have opted to use those agents that appear on the lists of products for organic production of certifying companies. Two companies in Chile, Ecocert Chile S.A and BioAudita-Eco Garantía (BioAudita), provide services to control, guarantee and certify organic products according to the regulations and standards in force in the country and in the main export markets. They also maintain lists of products for use in organic production systems. Meanwhile, SAG (2017) has a register of 56 biocontrol agents authorized for organic agriculture, but less than 4% of these are on the authorized pesticide list.

Important national companies providing biocontrol agents are Bionativa, Biogram, Biomycota, Biocaf, Biobichos, Biofuture, ControlBest and Xilema, among others. Other companies are spin-offs of universities, such as the Biopacific Company that currently relies on foreign capital to produce and commercialize biological agents. Also, international companies are active on the Chilean market (Koppert, Biobee). Still other companies, such as Natural Chile and Controlbest, have focused on providing all-inclusive services associated with biocontrol, such as pest monitoring and IPM. Many of the above-mentioned companies voluntarily participate in the National Biosupply Network (www.bioinsumos.cl), a technical entity created in 2014 by the Association of Chilean Producers of Biological Control Agents.

Although a lot of work has been done and is ongoing in the field of biocontrol in Chile, there are some issues that still need to be addressed. These concern quality control of commercialized biocontrol agents products, development of effective formulations enhancing the effect of agents, and implementation of technology transfer models allowing knowledge-generating entities, such as universities and research centres, to reach the market with their research developments.

INIA has developed an interesting approach in response to the national demand for microbial control agents. The model starts with the collection, identification and characterization of microorganisms that make up the Chilean Collection of Microbial Genetic Resources (CChRGM). This collection includes 1,857 accessions, of which 1,780 belong to the public collection (Table 7.3) and 77 to the private collection. From the CChRGM, 60% have potential for use as biocontrol agents, especially the entomopathogenic fungi (EPF) with 1,083 accessions, mycopathogenic fungi with 178 accessions and nematophagous fungi with 51 accessions. Furthermore, there is an important working collection of more than 100 EPF accessions. The CChRGM is conserved in the Microbial Genetic Resources Bank (Banco de Recursos Genéticos Microbianos, BRGM), the only public microbial bank in the country and the only South American bank recognized as an International Depositary Authority (IDA) under the Budapest Treaty, which allows it to receive microorganism deposits involved with patents (WIPO, 2012). Henceforth, the Biological Control Technological Center (CTCB) is responsible for assessing the germplasm conserved in the collections. This

Table 7.3. Microorganism collection of INIA's Microbial Genetic Resources Bank (BRGM).

Type of germplasm	Most common genera	Number of accessions
Bacteria	*Bacillus*, *Pseudomonas*, *Streptomyces*	29
Entomopathogenic fungi	*Beauveria*, *Lecanicillium*, *Metarhizium*, *Paecilomyces*	1,083
Phytopathogenic fungi	*Botrytis*, *Colletotrichum*, *Fusarium*, *Phytophthora*, *Rhizoctonia*, *Venturia*	414
Mycopathogenic fungi	*Clonostachys*, *Fusarium*, *Trichoderma*	178
Nematophagous fungi	*Arthrobotrys*, *Mortierella*, *Pochonia*	51

centre began operating in 2007 with the aim of strengthening the development of technologies to mass produce pest, disease and weed control agents and contribute to the integrated management of cleaner and more sustainable national agriculture and forestry. It has six lines of research and it had managed 17 projects by 2017 for an amount greater than US$6 million.

7.3.2 Use of predators and parasitoids

New parasitoids have recently been introduced, such as *Trioxys pallidus* (Haliday) from Iran to control *Chromaphis juglandicola* (Kaltenbach), an aphid affecting the walnut tree (*Juglans regia* L.). This parasitoid showed a high dispersion capacity and good level of aphid population reduction in walnuts three seasons after their release.

New natural enemies were also imported for an 'old pest', the codling moth *Cydia pomonella* L. in apples, because control was insufficient. However, release of the parasitoid *Ascogaster quadridentata* (Wesmael) did not result in improved control, but researchers from the Pontificia Universidad Católica de Chile identified a new parasitoid of codling moth larvae, *Goniozus legneri* (Gordh), which showed 50% larval parasitism in the laboratory (Zaviezo *et al.*, 2007).

In addition to the above-mentioned studies, companies supplying biocontrol agents are evaluating application methodologies under field conditions. Xilema Company, for example, studied a method to release *Eriopis chilensis* (Hofmann) against *Eriosoma lanigerum* Hausmann with the objective to incorporate it into IPM for apple. Their studies demonstrated that releasing 100 and 150 adults of *E. chilensis* per 0.5 ha is effective to reduce the pest.

In the past few years, attention has been centred on a pest that causes important damage in Chile's fruit sector: the grapevine moth *Lobesia botrana* (Denis and Schiffermüller). It was first detected in the central zone of the country. SAG declared the pest under mandatory control in 2014 and 2016 (SAG, 2016). This has motivated the interest of various research groups to study biocontrol alternatives for this important pest. Currently, INIA is evaluating the use of different species of the egg parasitoid *Trichogramma*. They will be used as alternatives for *G. legneri* larval parasitoids and predators of the genus *Chrysoperla*.

Pest control programmes in the forestry sector, developed jointly by public institutions and the private sector, resulted in controlling six pests by using predators and parasitoids. Two pests have been targeted in the implementation of new control programmes. One project concerns control of the woodwasp *Sirex noctilio* (Fabricius), which led SAG to declare mandatory control and the execution of a joint control programme with Argentina. The parasitoids *Megarhyssa nortoni* (Cresson) and *Ibalia leucospoides* (Hochenwarth) are being used to control the woodwasp (Beèche *et al.*, 2012). Another parasitoid, *Anaphes nitens* (Girault), is used to control the eucalyptus weevil *Gonipterus platensis* (Marelli) (Valente *et al.*, 2017).

7.3.3 Use of microbial agents to control pests and diseases

During the past decade, the use of microbial agents as an augmentative biocontrol strategy for pest and disease control has significantly increased. Different species of fungi, nematodes and bacteria that affect pests have been identified in Chile. Nowadays, several of the microorganisms that have successfully controlled pests are being mass produced and commercialized in Chile. Below we summarize the recent research efforts.

Entomopathogenic fungi

INIA is still the major institute for EPF studies. Evaluation of native strains of *B. bassiana* and *M. anisopliae* to control *Tuta absoluta* (Meyrick) eggs and larvae have recently been prioritized. *Metarhizium* spp. exhibit high control levels, with 80% egg mortality and 90% larval mortality, whereas *B. bassiana* causes 60% egg mortality and 50% larval mortality. Native strains of *M. anisopliae* were also evaluated against the black vine weevil *O. sulcatus* under laboratory conditions and high percentages of larval mortality were obtained, varying as a function of spore concentrations and fungal strain (Gerding *et al.*, 2000). Strains of *B. bassiana* and *M. anisopliae* have also been evaluated against the curculionids *A. cervinus* and *O. sulcatus* in raspberry.

Different levels of pathogenicity were found, the most aggressive being *B. bassiana* against *A. cervinus* and *M. anisopliae* against *O. sulcatus*. *Hylamorpha elegans* (Burmeister) is an important cereal and grassland pest in Chile. The Qu-M845 and Qu-M270 strains of *Metarhizium* were evaluated to control it and although both were effective, the second strain showed better results (Rodríguez *et al.*, 2004). Sepúlveda (2015) evaluated the enzymatic and insecticidal activity of six native strains of *Metarhizium* spp. to control *A. superciliosus*, an important minor fruit pest in Chile. All the evaluated strains produce destruxin A and concentrations of 100 mg l^{-1} generate 100% insect mortality.

Mass production of EPF can be based on liquid, solid and biphasic fermentation techniques. Liquid fermentation is habitually used for bacteria. INIA has optimized the mass production of *M. anisopliae* on different substrates (Barra-Bucarei *et al.*, 2016). Also formulation studies were done, in order to maintain the viability and pathogenicity of the strains under field conditions. This is achieved by using inert materials such as solvents, emulsifiers, gelling agents and other additives such as nutrients or stimulants. Since 2007, the CTCB has studied formulations for EPF. First, the use of chitin and its derivatives in formulations was studied; next, granular formulations on the basis of sodium alginate were tested (Gerding-González *et al.*, 2007); and a third study evaluated inverse and encapsulated emulsions.

From 2002 to 2017 INIA sold approximately 26,000 doses (1 dose is needed to treat 1 ha of crop) of different *M. anisopliae* strains. The principal target pests, for which 42% of the doses were sold, were curculionids like *Aegorhinus superciliosus* (Guérin), *A. nodipennis* (Hope) and *Naupactus xanthographus* (Germar), which occur mainly in blueberries.

Entomopathogenic nematodes

Research on EPNs has increased lately, stimulated by the existence of germplasm adapted to the wide variety of climates and soil types in Chile. INIA has conducted assays with strains of *Steinernema australe* Edgington, Buddie, Tymo, Hunt, Nguyen, France, Merino & Moore and *S. unicornum* Egington, Buddie Tymo, France, Merino & Hunt to control *A. nodipennis* (Maldonado *et al.*, 2012), which is a native insect that attacks blueberries at the neck, crown and roots, and *Dalaca pallens* (Blanchard), which causes important grassland losses. These insects spend the greater part of their life in the soil, which complicates their control. INIA is also evaluating strains of the native entomopathogenic nematodes *S. australe*, *S. unicornum* and *S. feltiae* (Fiipjev) Wouts, Mracek, Gerdin & Bedding to overwintering pupae of *L. botrana*, and several strains are able to cause 40–50% pupal mortality. Use of EPNs would allow lengthening of the control period of this quarantine pest. Further, EPNs have been evaluated to control forest pests. Successful inoculations were made by SAG in 2013 in forest plantations with baits containing the nematode *Deladenus siricidicola* (Bedding), aimed at reducing *S. noctilio* populations and preventing their dispersion.

Other work with EPNs concerns mass production technologies in *in vivo* systems (*Galleria mellonella* Linnaeus larvae). Since 2008, EPNs produced with INIA technology have been applied on 200 ha per year to control *A. superciliosus* and *A. nodipennis* in blueberries. Currently the CTCB is developing *in vitro* production protocols for EPNs by combining different liquid media and is evaluating the productivity of the nematode *S. unicornum* strain QU-N85 and its symbiotic bacteria. *In vitro* production appeared not to affect the parasitic and pathogenic ability of the nematode.

Bacteria for control of insects, nematodes and diseases

Several genera of bacteria have been evaluated to control insects. Native *B. thuringiensis* strains demonstrated high toxicity in *T. absoluta* larvae (Niedmann and Meza-Basso, 2006), *Plutella xylostella* (Linnaeus) and *Agrotis* spp. Commercial formulations of *B. thuringiensis* (Dipel, XenTari, and Turilav) also caused high percentages of *T. absoluta* larval mortality (Ramírez *et al.*, 2010).

Phytoparasitic nematodes are a major problem in many crops. Chemical control and resistant cultivars of fruit are used, but results are not always satisfactory. Rhizobacteria of the genera *Bacillus*, *Brevibacterium*, *Oerskovia* and *Pseudomonas* have been used to control *Globodera rostochiensis* (Wollenweber), a phytoparasitic nematode in potatoes, with *Oerskovia turbata* (Erikson) as the most effective species. Clay and liquid formulations of *Bacillus* and *Pseudomonas* native strains have also been evaluated against nematodes of the genera *Xiphinema* and *Meloidogyne* that affect grapevine.

Fungi and bacteria for control of diseases

One of the most studied agents to control diseases in Chile is the fungus *Trichoderma* against the following pathogens: *Alternaria alternata* (Keissl) (Roco and Pérez, 2001), *B. cinerea* (Lolas *et al.*, 2004), *Cladosporium echinulatum* (Berkeley) (Sandoval *et al.*, 2009), *Pyrenochaeta lycopersici* (Schneider and Gerlach) (Besoain *et al.*, 2007; Sánchez-Tellez *et al.*, 2013) and *Rhizoctonia solani* (Kühn) (Montealegre *et al.*, 2010).

In forests, one of the most important pathogens is *B. cinerea*, causing serious damage in nurseries. Seventy-one strains of the antagonistic fungi of *B. cinerea* were evaluated for pathogen colonization and sporulation in *in vitro* assays using leaf discs of *Eucalyptus globulus* Labill. (Molina *et al.*, 2006). Strains of the genera *Trichoderma*, *Clonostachys*, *Pencillium* and *Cladosporium* significantly reduced pathogen colonization and sporulation. In addition, strains of *Trichoderma* and *Clonostachys* were evaluated under nursery conditions and the latter achieved a better level of effectiveness against *Botrytis* in *E. globulus* (Zaldúa and Sanfuentes, 2011). Strains of these same genera were evaluated against *Fusarium circinatum* (Nirenberg and O'Donnell) and the *Clonostachys* UDC-222 strain significantly increased seedling survival of *Pinus radiata*. Currently, the biggest forest nurseries of Chile use *Trichoderma* for *B. cinerea* and *F. circinatum* control.

González (2001) evaluated the ability of *Bacillus lentimorbus* Dutky strains to inhibit *Fusarium solani* (Dutky) growth. The bacterium was able to inhibit pathogen growth between 27% and 77%, depending on the method used. Evaluated *Bacillus* spp. strains showed good results against *Erwinia carotovora* (Smith) (Toledo and Sandoval, 2004) and *Phytophthora infestans* (Mont.) de Bary (Rojas and Sandoval, 2009). Several products now found on the market for disease control are based on *Trichoderma* and *Bacillus* and some of them are already on the SAG list of authorized pesticides, while others are in the registration process. It is estimated that annually over 5,000 ha are using products based on microorganisms for control of plant diseases, mainly in orchards and vegetables.

Areas under biological control in Chile

Currently, according to INIA information (L. Devotto, Quilamapu, Chile, 2019, personal communication), 100% (225,042 ha) of the wheat area and 99% (1,434,085 ha) of the pine area are under classical biocontrol. Although it is complicated to obtain a complete picture of areas under biocontrol, an overview of data is given in Table 7.4. It is stressed that the estimated areas under classical biocontrol (7,726,465 ha) and augmentative biocontrol (62,197 ha) are underestimates.

7.4 New Developments of Biological Control in Chile

Until a few years ago, biocontrol of pests in Chile consisted almost exclusively of classical biocontrol. This approach will be continued in the future in spite of increasing restrictions that exist in all countries with respect to Access and Benefit regulations as formulated in the Nagoya protocol (Mason *et al.*, 2018).

The Centro de Entomología Aplicada Ltda. (Applied Entomology Center Ltd) is currently developing a toxic bait for use in agriculture to control the Argentine ant *Linepithema humile* Mayr in fruit orchards. Funding is provided by the Foundation for Agricultural Innovation (Fundación para la Innovación Agraria) (FIA). This ant is a limiting factor in the use of natural enemies, because it establishes a mutual relationship with sap-sucking and honeydew-secreting pests from whom it obtains its main food and repels its biocontrol agents.

The Bionativa Company, known in Chile for the production of biocontrol agents for diseases, is developing a hybrid formulation with FIA funding that includes natural extracts and microorganisms to control fruit and vegetable powdery mildew. The objective is to incorporate these formulations when there are no alternatives, or existing alternatives exhibit operational limitations and/or residues for its use. Favourable application times are at post-harvest, sprouting,

Table 7.4. Major biological control programmes in Chile.

Natural enemy/ antagonist	Pest and crop	Type of biocontrol[a]	Area (ha) under biocontrol
Metaphycus helvolus	Olive black scale in olives	CBC	21,904[b]
Parasitoids and predators	Scales in citrus, avocado and other fruit	CBC	~20,000[b]
Cryptolaemus spp., *Leptomastix* spp.	White mealybugs in citrus, other fruit, alfalfa, potato, olives	CBC	~130,000[b]
Aphelinus mali	*Eriosoma lanigerum* aphids in apple	CBC	37,000
Trichogramma spp.	*Cydia pomonella, Rhyacionia buoliana, Tuta absoluta,* and *Lobesia botrana* in forestry, fruit and vegetables	ABC	2,500
Metarhizium anisopliae	*Aegorhinus* spp., *Naupactus xanthographus*, *Asynonychus cervinus, Dalaca pallen, Otiorhynchus sulcatus,* and other curculionids	ABC	4,372
Chrysolina spp.	*Hypericum perforatum* weed, in pastures and nature	CBC	?, but large
Parasitoids and predators	Wheat aphids in wheat	CBC	225,042
Orgilus obscurator	*Rhyacionia buoliana* European pine shoot moth in pine forests	CBC	1,434,085
Psyllaephagus pilosus	*Ctenarytaina eucalypti* eucalyptus psyllid in eucalyptus forests	CBC	825,000
Agonopterix ulicetella	*Ulex europaeus* gorse weed in agriculture and forestry	CBC	?, but large
Phragmidium violaceum	*Rubus ulmifolius* weedy shrub in agriculture and nature	CBC	5,000,000
Agrobacterium radiobacter	*Agrobacterium tumefacien* in fruit orchards and berries	ABC	6,432
Trichoderma spp	*Botrytis cinerea* in fruit and vegetables	ABC	21,162
Chrysoperla spp.	Aphids in agriculture	ABC	100
Tupiocoris cucurbitaceus	*Tuta absoluta* and *Trialeurodes vaporariorum* in vegetables	ABC	20
Trioxys pallidus	*Chromaphis juglandicola* aphids in walnut orchards	CBC	33,434[b]
Megarhyssa nortoni, Ibalia leucospoide	*Sirex noctilio* woodwasps in pine forests	CBC	1,419,744
Anaphes nitens	*Gonipterus platens*, eucalyptus weevil in eucalyptus plantations	CBC	Part of 825,000
Bacillus spp.	*Pseudomonas, Xanthomonas* and *Clavibacter* in fruit, grape and vegetables	ABC	24,657
Microbials various	Soil diseases in fruit, grape and vegetables	ABC	30,548
Entomopathogenic nematodes	*Aegorhinus* spp. and *Naupactus xanthographus*	ABC	580

[a]Types of biocontrol: CBC = classical biological control, ABC = augmentative biological control
[b]Area of crop harvested in 2017 according to FAO (http://www.fao.org/faostat/en/#data/qc)

and close to harvest. As for formulations, INIA is developing two initiatives to prepare encapsulated and microencapsulated entomopathogenic fungi and nematodes to increase the field performance of these agents.

With FIA funding, INIA and the Natural Chile Company are executing an initiative to increase the efficiency and sustainability of pest management in agriculture through intensive monitoring and release of natural enemies by unmanned aerial vehicles (UAVs, i.e. drones). The project aims to use pest monitoring field data to prepare flight plans for the UAV, so that releases will spread the necessary amount of natural enemies at specific locations related to the degree of infestation. Another approach is biocontrol and education. Currently, those interested in these technologies have no access to courses at high school, college or university. Almost all the Chilean biocontrol companies were spin-offs from INIA, through former staff, students and/or short-term trainees. To improve this situation, an alliance between INIA and three secondary-level schools was established to offer theoretical and hands-on education on the production and use of biocontrol agents to students who will become agricultural technicians after their training.

The potential of native rhizobacteria to control phytoparasitic nematodes is being evaluated by collaboration between the companies Biorend and Biogram, and the Universidad de Chile along with funding from the Scientific and Technological Development Support Fund (Fondo de Desarrollo Científico y Tecnológico) (FONDEF). In addition, the development of a commercial formulation which is adapted to the production conditions of the central zone of Chile is being pursued.

The Universidad de Concepción, the Universidad de Talca, the Center for Advanced Studies in Arid Zones (Centro de Estudios Avanzados en Zonas Árida) (CEAZA) and INIA study endophytic microorganisms in native species, determine the endophytic colonization ability, persistence and localization of the microorganisms inside plants, and their effect as pest and disease agents, in the following projects:

- CEAZA works with an endophyte of a *Lecanicillium lecanii* (Zimm.) Zare & Gams strain isolated from native plants. Assays demonstrated that the fungus shows activity as a pathogen of adult *Bemisia tabaci* (Gennadius) and *Macrosiphum euphorbiae* (Thomas) and as a mycopathogen against the genera *Rhizoctonia*, *Sclerotinia*, *Botrytis* and *Mucor*.
- INIA established the first collection of endophytic microorganisms isolated in different regions of the country. This collection consists of more than 100 accessions of fungi and bacteria, some of which are being identified to be included in the CChRGM. Particularly the endophytic actinobacteria of native *S. tuberosum* lines to control bacterial diseases and promote potato growth are of interest. It has been possible to isolate endophytic actinobacteria of native potatoes and evaluate them for their antagonistic ability against *Pectobacterium carotovorum* (Jones) Waldee subsp. *carotovorum* and *Pectobacterium atrosepticum* (Van Hall), which cause soft rot and black foot rot in potatoes.
- Another area of research at INIA concerns the endophytic colonization ability exhibited by fungi of the genera *Beauveria* and *Metarhizium* and their potential use for pest and disease control. The colonization ability of native strains of nematophagous fungi of the genera *Beauveria*, *Trichoderma*, *Paecilomyces*, *Clonostachys*, *Fusarium* and *Metarhizium* has also been studied. Studies have been promising, because all the evaluated strains, except *Beauveria*, exhibited some degree of endophytic colonization in tomato seedling, thus implying that it might be an alternative for control of phytoparasitic nematodes of this species.

References

Aruta, C., Carrillo, R. and González, S. (1974) Determinación para Chile de hongos entomopatógenos del género *Entomophthora* [Determination of Chilean entomopathogenic fungi of the genus *Entomophthora*]. I. *Agro sur* 2(2), 62–70.

Barra-Bucarei, L., Vergara, P. and Cortes, A. (2016) Conditions to optimize mass production of *Metarhizium anisopliae* (Metschn.) Sorokin 1883 in different substrates. *Chilean Journal of Agricultural Research* 76(4), 448–454.

Beèche, M., Lanfranco, D., Zapata, M. and Ruiz, C. (2012) Surveillance and control of the Sirex woodwasp: the Chilean experience. In: Slippers, B., de Groot, P. and Wingfield, M.J. (eds) *The Sirex Woodwasp and its Fungal Symbiont*. Springer, Dordrecht, Netherlands, pp. 229–245.

Besoain, X.A., Pérez, L.M., Araya, A., Lefever, L. and Montealegre, J.R. (2007) New strains obtained after UV treatment and protoplast fusion of native *Trichoderma harzianum*: their biocontrol activity on *Pyrenochaeta lycopersici*. *Electronic Journal of Biotechnology* 10(4), 604–617.

CIA (2017) The World Factbook: Chile. Available at: https://www.cia.gov/ library/publications/the-world-factbook/geos/ci.html (accessed 12 July 2019).

Ciampi-Panno, L., Bustamante, P. and Polette, M. (1987) Isolation of soil bacteria with inhibitory activity to *Pseudomonas solanacearum*. In: Civetta, L., Collmer, A., Davis, R.E. and Gillaspie, A.G. (eds) *Plant Pathogenic Bacteria, Proceedings of the Sixth International Conference on Plant Pathogenic Bacteria, Maryland, June 2–7, 1985*. Springer, Dordrecht, Netherlands, pp. 733-739.

Devotto, L., Del Valle, C., Ceballos, R. and Gerding, M. (2010) Biology of *Mastrus ridibundus* (Gravenhorst), a potential biological control agent for area-wide management of *Cydia pomonella* (Linneaus) (Lepidoptera: Tortricidae). *Journal of Applied Entomology* 134(3), 243–250.

Dutky, S. (1957) Report on white grub control project in Chile. *Agricultura Técnica* 17(1), 92–105.

Gerding, M., France, I. and Cisternas, A. (2000) Evaluation of Chilean strains of *Metarhizium anisopliae* var. *anisopliae* against *Otiorhynchus sulcatus* Fab. (Coleoptera: Curculionidae). *Agricultura Técnica* 60(3), 216–223.

Gerding-González, M., France, A., Sepúlveda, M. and Campos, J. (2007) Use of chitin to improve a *Beauveria bassiana* alginate-pellet formulation. *Biocontrol Science and Technology* 17(1), 105–110.

González, R.F. (2001) Control biológico de *Fusarium solani* (Mart) Sacc. en *Lycopersicon esculentum* Mill mediante bacterias y *Trichoderma* spp. [Biological control of *Fusarium solani* in *Lycopersicon esculentum* by bacteria and *Trichoderma* spp.]. PhD thesis, Universidad de Chile, Facultad de Ciencias Agrarias y Forestales, Santiago, Chile.

González, R.H. and Rojas, S (1966) Estudio analítico del control biológico de plagas Agrícolas en Chile 1 [Analytical study of the biological control of agricultural pests in Chile 1]. *Agricultura Técnica* 26(4), N°4.

Herrera, G. and Quiroz, C. (1988) Pérdidas de rendimiento en trigo causadas por la infección natural del VEAC, en ensayos realizados desde 1976 a 1985 [Wheat yield losses caused by natural VEAC infection in trials conducted from 1976 to 1985]. *Agricultura Técnica* 48(2), 75–80.

Howard, L.O. (1929) *Aphelinus mali* and its travels. *Annals of the Entomological Society of America* 22(3), 341–368.

Huerta, A.and Cogollor, G., (1995) Control de la polilla del brote del pino (*Rhyacionia buoliana* Den. et Schiff.) mediantes cepas de del bacteria *Bacillus thuringiensis* var. Kurstaki [Control of the pine shoot moth (*Rhyacionia buoliana*) through strains of the bacterium *Bacillus thuringiensis* var. Kurstaki]. *Revista ciencias forestales Universidad de Chile* 10, N°1-2 a 12.

Ide, S., Lanfranco, D. and Ruiz, C. (2007) Detección de superparasitismo y multiparasitismo sobre larvas de *Rhyacionia buoliana* (Lepidoptera-Tortricidae) en las Regiones VIII y IX de Chile. Bosque (Valdivia) [Superparasitism and multiparasitism detection on larval *Rhyacionia buoliana* in Regions VIII and IX of Chile]. *Agricultura Técnica* 28(1), 57–64.

INFOR (Instituto de Investigaciones Forestal) (2016) Anuario forestal [Forest Yearbook] 2016. Chilian Statistical Yearbook of Forestry. Available at: http://wef.infor.cl/publicaciones/publicaciones.php#/12 (accessed 12 January 2019).

Jiménez, R.M., Gallo, D.P. and Silva, V.E. (1989) Susceptibility of various species of lepidopterous larvae to the entomopathogenic nematode *Steinernema carpocapsae*. *IDESIA* 11, 49-51.

Julien, M.H. and Griffiths, M.W. (1998) *Biological Control of Weeds: a World Catalogue of Agents and their Target Weeds*, 4th edn. CAB International, Wallingford, UK.

Lolas, M., Donoso, E., Gonzáles, V.and Carrasco, G. (2004) Use of a Chilean native strain 'Sherwood' of *Trichoderma virens* on the biocontrol of *Botrytis cinerea* in lettuces grown by a float system. *Acta Horticulturae* 697, 437–440.

Maldonado, A., Merino, L. and France, A. (2012) Selection of entomopathogenic nematodes against *Dalaca pallens* (Lepidoptera: hepialidae). *Chilean Journal of Agricultural Research* 72(2), 201.

Martínez, G.G., Norambuena, M.H., Carrillo, L.L, Neira, C.M. and Rodríguez, A.F. (2000) Estudio de especificidad de la polilla del espinillo *Agonopterix ulicetella* (Stainton) para el control biológico del espinillo (*Ulex europaeus*) [Host-specificity study of the gorse moth *Agonopterix ulicetella* for the biological control of gorse *Ulex europaeus*]. *Agro Sur* 28(1), 133–150.

Mason, P.G., Cock, M.J.W., Barratt, B.I.P., van Lenteren, J.C., Brodeur, J. *et al.*(2018) Best Practices for the use and exchange of biological control genetic resources relevant for food and agriculture. *BioControl* 63(1), 149–154. DOI: 10.1007/s10526-017-9810-3

Molina, G., Zaldúa, S., González, G. and Sanfuentes, E. (2006) Selección de hongos antagonistas para el control biológico de *Botrytis cinerea* en viveros forestales en Chile [Selection of antagonistic fungi for the biological control of *Botrytis cinerea* in forest nurseries in Chile]. *Bosque (Valdivia)* 27(2), 126–134.

Montealegre, J., Valderrama, L., Sánchez, S., Herrera, R., Besoain, X.and Pérez, L.M. (2010) Biological control of *Rhizoctonia solani* in tomatoes with *Trichoderma harzianum* mutants. *Electronic Journal of Biotechnology* 13(2), 1–2.

Niedmann, L. and Meza-Basso, L. (2006) Evaluación de cepas nativas de *Bacillus thuringiensis* como una alternativa de manejo integrado de la polilla del tomate (*Tuta absoluta* Meyrick; Lepidoptera: Gelechiidae) en Chile [Evaluation of native strains of *Bacillus thuringiensis* as an integrated management option for the tomato moth (*Tuta absoluta*) in Chile]. *Agricultura técnica* 66(3), 235–246.

Norambuena M.H. and Gerding, P.M. (1990) El pulgón ruso del trigo [The Russian wheat aphid]. *IPA La Platina* 59, 48–52.

Norambuena, H. and Piper, G.L. (2000) Impact of *Apion ulicis* Forster on *Ulex europaeus* L. seed dispersal. *Biological Control* 17(3), 267–271.

ODEPA (Oficina de Estudios y Políticas Agrarias) (2016) Superficie plantada nacional, regional, número de huertos e infraestructura frutícola [National, regional planted area, number of orchards and fruit infrastructure]. Available at: http://www.odepa.cl/estadisticas/productivas/ (accessed 12 January 2019).

OECD (2005) Evaluaciones del Desempeño Ambiental Chile [OECD Environmental Performance Reviews Chile 2005]. OECD, Chile. Available at: https://repositorio.cepal.org/bitstream/handle/11362/1288/1/S0500003_es.pdf (accessed 6 August 2019).

Oehrens, E. and González, S. (1974) Introducción de *Phragmidium violaceum* (Schulz) Winter como factor de control biológico de zarzamora (*Rubus constrictus* Lef. et *M.* y *R. ulmifolius* Schott) [Introduction of *Phragmidium violaceum* for biologicla control of *Rubus constrictus*]. *Agro Sur* 2(1), 30–33.

Ramírez, L., Ramírez, N., Fuentes, L.S., Jiménez, J. and Hernández-Fernández, J. (2010) Estandarización de un bioensayo y evaluación preliminar de tres formulaciones comerciales de *Bacillus thuringiensis* sobre *Tuta absoluta* (Meyrick) (Lepidoptera: Gelechiidae) [Standardization of a bioassay and preliminary evaluation of three commercial formulations of *Bacillus thuringiensis* on *Tuta absoluta*]. *Revista Colombiana de Biotecnología* 12(1), 12.

Ripa, R. (1992) Burrito de los frutales *Naupactus xanthographus* [Grape snout beetle *Naupactus xanthographus*]. *Boletín Técnico* 192, 9–23.

Ripa, R. and Droguett, I. (2008) Manejo de plagas en paltos y cítricos [Pest management in avocados and citrus fruits]. Instituto de Investigaciones Agropecuarias, La Cruz.

Ripa, S. and Rodríguez, A.F. (1989) Susceptibility of larvae of *Naupactus xanthographus* (Coleoptera: Curculionidae) to eight isolates of *Metarhizium anisopliae* (Deuteromycotina: Hyphomycetes). *Agricultura Técnica* 49(4), 336–340.

Ripa, R. and Rodríguez, F. (1999) Plagas de cítricos, sus enemigos naturales y manejo [Citrus pests, their natural enemies and management]. *Colección libros INIA* N°3. Instituto de Investigaciones Agropecuarias (INIA, Chile).

Ripa, R., Larral, P. and Rodríguez, S. (2008) Manejo Integrado de Plagas (MIP). Manejo de plagas en paltos y cítricos [IPM. Pest management in avocado and citrus fruits]. *Instituto de Investigaciones Agropecuarias, INIA, Chile* 399, 41–50.

Roco, A. and Pérez, L.M. (2001) *In vitro* biocontrol activity of *Trichoderma harzianum* on *Alternaria alternata* in the presence of growth regulators. *Electronic Journal of Biotechnology* 4(2), 1–2.

Rodríguez, F. and Sáiz, F. (2006) Parasitoidismo de *Psyllaephagus pilosus* Noyes (Hym.: Encyrtidae) sobre el psílido del eucalipto *Ctenarytaina eucalypti* (Maskell) (Hem.: Psyllidae) en plantaciones de eucaliptos en la V Región [Parasitism of *Psyllaephagus pilosus* Noyes on the eucalyptus psyllid *Ctenarytaina eucalypti* in eucalyptus plantations in the V Region]. *Agricultura Técnica* 66(4), 342–351.

Rodríguez, M., France, A. and Gerding, M. (2004) Evaluación de dos cepas del hongo *Metarhizium anisopliae* var. *anisopliae* (Metsh.) para el control de larvas de gusano blanco *Hylamorpha elegans* Burm. (Coleoptera: Scarabaeidae) [Evaluation of two strains of the fungus *Metarhizium anisopliae* var. *anisopliae* for the control of larvae of white grub *Hylamorpha elegans*]. *Agricultura Técnica* 64(1), 17–24.

Rojas, S. (2005) *Control biológico de plagas en Chile. Historia y avances* [*Biological control of pests in Chile. History and progress*]. INIA. Centro Regional de Investigación La Platina. Edit. Ograma, La Cruz, Chile.

Rojas, C. and Sandoval, C. (2009) Evaluacion *in vivo* de la efectividad biocontroladora de *Bacillus subtilis* y dos cepas nativas de *Trichoderma* spp. sobre la incidencia de tizon tardio (*Phytophthora infestans*) y otros patógenos foliares en plantas de tomate cv. Maria Italia [*In vivo* evaluation of the biocontrol effectiveness of *Bacillus subtilis* and two native strains of *Trichoderma* spp. on the late blight (*Phytophthora infestans*) incidence and other foliar pathogens in tomato plants cv. Maria Italy]. PhD thesis, Universidad de Talca (Chile), Escuela de Agronomía.

SAG (Servicio Agrícola y Ganadero) (2016) Estrategia 2016–2017 [Strategy 2016–2017]. Programa Nacional de *Lobesia botrana*. Available at: http://www.sag.cl/ambitos-de-accion/lobesia-botrana-o-polilla-del-racimo-de-la-vid (accessed 27 October 2019).

SAG (Servicio Agrícola y Ganadero) (2017) Lista de Plaguicidas Autorizados [Authorized Pesticides List]. Available at: http://www.sag.cl/ambitos-de-accion/plaguicidas-y-fertilizantes/78/registros (accessed 12 January 2019).

Sánchez-Téllez, S., Herrera-Cid, R., Besoain-Canales, X., Pérez-Roepke, L.M. and Montealegre-Andrade, J.R. (2013) *In vitro* and *in vivo* inhibitory effect of solid and liquid *Trichoderma harzianum* formulations on biocontrol of *Pyrenochaeta lycopersici*. *Interciencia* 38(6), 425.

Sandoval, C., Terreros, V. and Schiappacasse, F. (2009) Control de *Cladosporium echinulatum* en clavel mediante el uso de bicarbonatos y *Trichoderma* [*Cladosporium echinulatum* control in carnation using bicarbonates and *Trichoderma*]. *Ciencia e Investigación Agraria* 36(3), 487–498.

Schick, C.M., Mattar, C., Neira, R., Mora, M. and Espinoza, J. (2017) Sutainable agriculture and healthy food in Chile. In: *Challenges and Opportunities for Food and Nutrition Security in the Americas. The View of the Academies of Sciences*. IANAS, IAP and BMBF, México DF, pp. 190–211. [Free public access of this publication in English and Spanish at www.ianas.org]

Sepúlveda, M. (2015) Actividad enzimática e insecticida de seis cepas nativas de *Metarhizium* spp. para el control de *Aegorhinus superciliosus* (Coleoptera: Curculionidae) [Enzymatic and insecticidal activity of six native strains of *Metarhizium* spp. for the control of *Aegorhinus superciliosus* (Coleoptera: Curculionidae)]. PhD thesis, Universidad de Concepción Facultad de Agronomía.

Toledo Ramírez, D.B. and Sandoval Briones, C. (2004) Evaluación *in vitro* del efecto de cepas nativas de la bacteria *Bacillus* sp. en el biocontrol de la bacteria *Erwinia carotovora* [*In vitro* evaluation of the effect of native strains of *Bacillus* sp. in the biocontrol of the bacterium *Erwinia carotovora*]. PhD thesis, Escuela de Agronomia, Universidad de Talca, Chile,.

Torres, P. and Gerding, P. (2000) Evaluation of five species of *Trichogramma* as biological control agents for *Cydia pomonella* (L.) (Lepidoptera: Tortricidae). *Agricultura Tecnica* 60(3), 282–288.

Valente, C., Gonçalves, C.I., Reis, A. and Branco, M. (2017) Pre-selection and biological potential of the egg parasitoid *Anaphes inexpectatus* for the control of the *Eucalyptus* snout beetle, *Gonipterus platensis*. *Journal of Pest Science* 90, 911–923.

WIPO (World Intellectual Property Organization) (2012) Communication of the Republic of Chile Relating to the Acquisition of the Status of International Depositary Authority by the Colección Chilena de Recursos Genéticos Microbianos (CChRGM). Available at: http://www.wipo.int/treaties/en/notifications/budapest/treaty_budapest_283.html (accessed 1 October 2019).

Zaldúa, S. and Sanfuentes, E. (2011) Control of *Botrytis cinerea* in *Eucalyptus globulus* mini-cuttings using *Clonostachys* and *Trichoderma* strains. *Chilean Journal of Agricultural Research* 70(4), 576–582.

Zaviezo, T. and Mills, N. (2001) The response of *Hyssopus pallidus* to hosts previously parasitised by *Ascogaster quadridentata*: heterospecific discrimination and host quality. *Ecological Entomology* 26(1), 91–99.

Zaviezo, T., Romero, A., Castro, D. and Wagner, A. (2007) Primer registro de *Goniozus legneri* (Hymenoptera: Bethylidae) para Chile [First record of *Goniozus legneri* for Chile]. *Ciencia e Investigación Agraria* 34(1), 57–61.

Zúñiga, E. (1986) Control biológico de los áfidos (hom.; aphididae) de los cereales en Chile [Biological control of aphids of cereals in Chile]. I. Revisión histórica y líneas de trabajo. *Agricultura Técnica* 46(4), 475–477.

Please see the supplementary file "Addenda and Corrections" for names of natural enemies introduced into Chile for control of forest pests.

8 Biological Control in Colombia

Takumasa Kondo[1]*, Maria R. Manzano[2] and Alba Marina Cotes[3]

[1]*Corporación Colombiana de Investigación Agropecuaria (Agrosavia), Centro de Investigación Palmira, Valle del Cauca, Colombia;* [2]*Universidad Nacional de Colombia, sede Palmira, Departamento de Ciencias Agrícolas, Palmira, Valle del Cauca, Colombia;* [3]*Corporación Colombiana de Investigación Agropecuaria (Agrosavia), Centro de Investigación Tibaitata, Cundinamarca, Colombia*

* E-mail: tkondo@agrosavia.co

Abstract

The history of biological control of agricultural pests in Colombia dates back to 1880. The first documented cases include both classical biocontrol and the use of biopesticides against insect pests. The need to protect four national production systems of great economic importance (oil palm, sugarcane, coffee and greenhouse flowers and vegetables crops) from arthropod pests by reducing insecticide use has stimulated the development of research centres that support, together with local universities, the research and implementation of augmentative biocontrol programmes. Conservation biocontrol is a recent activity in Colombia. Although there are more than 52 officially registered companies in the country that produce microbial control agents (entomopathogenic fungi, bacteria, nematodes, baculoviruses and antagonistic microorganisms, including fungi and bacteria), predators and parasitoids, it is necessary to implement more rigorous quality control programmes in the production phase. Additionally, there is only a small amount of officially registered biocontrol agents, which is a limiting factor in their commercialization. Despite the quantity and diversity of agents released in Colombia, there are only partial records of the area under biocontrol, which is estimated at approximately 550,000 ha. It is necessary to implement post-release monitoring studies to determine the effectiveness of the biocontrol agent used. As an obstacle in Colombia and several other countries, the application of the principles of the Convention on Biological Diversity (CBD) is highlighted, which has made it very difficult or impossible to collect and export natural enemies which are essential for biocontrol programmes.

8.1 Introduction

Colombia has an estimated population of almost 47,700,000 (July 2017) and its agricultural, forestry and fishery products are coffee, cut flowers, bananas, rice, tobacco, maize, sugarcane, cocoa beans, oilseed, vegetables, shrimp and forest products (CIA, 2017). According to Hodson de Jaramillo *et al.* (2017), Colombian agriculture is concentrated in productive regions characterized by technified monocultures, including oil palm, banana, cassava, coffee, cotton, flowers, maize, potato, rice, sorghum and sugarcane; and the export of high-value crops such as oil palm, coffee and sugarcane. The agricultural potential of Colombia amounts to 26 million hectares, distributed as follows: land suitable for agriculture, 11 million hectares; livestock raising, 6 million hectares; agroforestry, 4 million hectares; forestry production, 3 million hectares; and in bodies of water, 2 million hectares. Food and nutritional security depend in great part on the production of cereals grown on small farms (smallholdings) that are supplied and traded in local markets, with the most important cereals being rice and maize; 41% of the agroindustrial employment is concentrated in the production of perennial crops for domestic consumption and export. The production of cacao is mainly implemented by small producers.

8.2 History of Biological Control in Colombia

8.2.1 Period 1880–1969

Classical biological control of woolly apple aphid and cottony cushion scale

An early case of successful classical biocontrol in Colombia is that of *Eriosoma lanigerum* (Hausmann), a pest of apples, with the use of *Aphelinus mali* (Haldeman) introduced from the USA in 1933 (Hagen and Franz, 1973). Another example of a successful classical biological programme in Colombia was documented in 1948 on the occasion of the IXth Pan American Conference held in Bogota: according to Valenzuela (1993), a number of acacias were introduced into the country with the purpose of beautifying the avenues of the capital city, but the imported plants were infested with a pest, *Icerya* sp., likely *I. purchasi* (Maskell) (Kondo, 2015), i.e. cottony cushion scale, which increased their population to such an extent that the people of Bogota soon began to call it the 'white pest'. Fortunately, with the initiative of the entomologist L.M. Murillo and the collaboration of entomologists F.J. Otoya, H. Osorno and C. Marín, a classical biocontrol programme was implemented by importing the vedalia lady beetle *Rodolia cardinalis* (Mulsant) from the USA, which ended up successfully controlling the pest (Valenzuela, 1993).

Microbial control of locusts

The first activities with microbial control agents also took place in this period. In his book, *A History of Applied Entomology*, Howard (1930) mentioned successful biocontrol of 'injurious locusts' with an entomopathogenic bacterium, *Coccobacillus acridiorum* D'Herelle, recounting the work in 1913 by L.Z. Uribe in Tocaima and by F. Lleras in Guaduas, Cundinamarca, Colombia.

8.2.2 Period 1970–2000

Augmentative biological control of pests in open field crops

Up to the early 1990s, the tobacco budworm *Heliothis virescens* (F.) was the main insect pest of cotton. Other pests of cotton include the Colombian pink bollworm *Sacadodes pyralis* (Dyar) and whiteflies and their control has generally been by extensive use of pesticides (Bellotti *et al.*, 2005). Cotton integrated pest management (IPM) for controlling *H. virescens* was based mainly on biocontrol through mass releases of the hymenopteran parasitoids *Trichogramma* sp. and *Apanteles* sp. (Bellotti *et al.*, 2005). As a result of these releases, insecticide applications were reduced to only two or three applications per cycle and this also resulted in higher yields (Bellotti *et al.*, 1990). According to van Lenteren and Bueno (2003), *Trichogramma* spp. were still applied on 30,000 ha of cotton in 1991, but by 2003 their use had decreased to about 5,000 ha. The use of pesticides to control lepidopteran pests was greatly reduced with the introduction of transgenic Bt cotton (Santos *et al.*, 2009). During the 1990s, mass releases of two parasitoids, *Bracon kirkpatricki* (Wilkinson) of African origin and imported from Mexico (Smith and Bellotti, 1996) and *Catolaccus grandis* (Burks) imported from Texas, were conducted for the control of the boll weevil in Colombia, but with little success (García and Sánchez, 1995). The entomopathogenic fungus *Lecanicillium lecanii* R. Zare & W. Gams (= *Verticillium lecanii*) is now used to control another important pest of cotton: the silverleaf whitefly *Bemisia tabaci* (Gennadius) (Gómez *et al.*, 2012).

The fall armyworm *Spodoptera frugiperda* (J.E. Smith) is a severe pest of maize causing at least 35% economic loss (Torres and Cotes, 2005) and is a secondary insect pest of cotton, rice and sorghum (Vélez-Arango *et al.*, 2008). In the 1990s inundative releases of the parasitoids *Trichogramma atopovirilia* Oatman & Platner and *Telenomus remus* Nixon were used to control the moth eggs on maize in combination with applications of *Nomuraea rileyi* (Farlow) and *Bacillus thuringiensis* Berliner in an IPM system in Valle del Cauca (García *et al.*, 1999). *T. remus* apparently became established in the region from specimens released in the Caribbean in 1976 (Yaseen *et al.*, 1982).

Augmentative biocontrol was intensively applied in the Valle del Cauca region, where periodic releases of various species of *Trichogramma* are conducted on ca. 200,000 ha cultivated with cassava, cotton, sorghum, soybean, sugarcane and tomato (van Lenteren and Bueno, 2003).

Augmentative biological control of pests in forestry

Augmentative biocontrol is applied in large forest areas to control defoliating caterpillars (van Lenteren and Bueno, 2003). An example is biocontrol of *Oxydia trychiata* (Guenee), a pest of conifers, particularly of *Pinus patula* Schltdl. & Cham. and *Cupressus lusitanica* Mill., which is controlled by the egg parasite *Telenomus alsophilae* Viereck imported from eastern North America (Bustillo and Drooz, 1977). Earlier studies with *T. alsophilae* targeted the geometrid *Glena bisulca* Rindge, another pest of *C. lusitanica*, but without success (Drooz and Bustillo, 1972).

Augmentative biological control of pests in greenhouse vegetables and ornamentals

Worldwide, Colombia is the second largest exporter of cut flowers, which are produced on more than 6,000 ha, of which about 98% are exported (Bueno, 2005). Many pests have been newly introduced into Colombia through worldwide exchange of plant propagation material, including *Spodoptera exigua* (Hübner), *Liriomyza trifolii* (Burgess), *Frankliniella occidentalis* (Pergande) and the B biotype of *B. tabaci* (*Bemisia argentifolii*) (Nicholls *et al.*, 1998). Several of these pests are managed through biocontrol and most of the following examples are from Nicholls *et al.* (1998). Multiple releases of *Encarsia formosa*

Gahan were conducted at the first sign of whiteflies in order to obtain overlapping generations of the parasitoid. Control of *Myzus persicae* (Sulzer) in greenhouses has been attempted with the parasitoid *Aphidius matricariae* Haliday to a limited extent in IPM programmes on chrysanthemums in Colombia. Another natural enemy used for aphid control is the predatory midge *Aphidoletes aphidimyza* (Rondani), which can kill more than 50 aphids during their development and is known to attack at least 60 different aphid species; it is available from international commercial suppliers.

Leaf miners *L. trifolii* and *L. huidobrensis* (Blanch.) are major pests of floricultural crops in Colombia. In *Gypsophila paniculata* L., leaf miners are controlled with the ectoparasitoid *Diglyphus begini* (Ashmead) (Cure and Cantor, 2003). The use of *D. begini* in inoculative and inundative release programmes is common in Colombia (Nicholls *et al.*, 1998). Finally, several IPM programmes for greenhouse pests are implemented to control tomato pests in Colombia (De Vis *et al.*, 1999).

Augmentative biological control of pests in sugarcane

Biocontrol with the use of various parasitoids in sugarcane mills has a long history in Colombia. Many private sugarcane plantations in Colombia have small insectaries for the mass rearing of sugarcane borer parasitoids (tachinid flies and *Trichogramma* spp.) (Drooz *et al.*, 1977). The egg parasitoid *Trichogramma pretiosum* Riley was introduced to Colombia by the end of the 1970s and from here it spread to other countries in South America, including Brazil, Costa Rica, Ecuador, Paraguay and Venezuela (Bueno and van Lenteren, 2002). Another parasitoid, a Peruvian race of the tachinid fly *Billaea claripalpis* (Wulp), which has a shorter life cycle than the native strain, was also introduced to control the sugarcane borer *Diatraea saccharalis* (Fabricius) (Hagen and Franz, 1973).

Classical and augmentative biological control of coffee berry borer in coffee

The coffee berry borer *Hipothenemus hampei* (Ferrari), one of the major pests of coffee worldwide, is a species of African origin and was introduced into Colombia in 1988 (Bustillo, 2006). Until the year 2000, around 2 billion individuals of *Cephalonomia stephanoderis* Betrem and around 500 million individuals of *Prorops nasuta* Waterston were released (Maldonado-Londoño and Benavides-Machado, 2007). The coffee borer is also controlled with native strains of *Beauveria bassiana* (Bals.-Criv.) Vuill., *Hirsutella eleutherathorum* (Nees ex Gray) Petch (Bustillo *et al.*, 1998) and *Metarhizium anisopliae* (Metsch.) (Bernal *et al.*, 1994). In the early 2000s, the entomopathogens *B. bassiana* and *M. anisopliae* were applied on 550,000 ha of coffee against the coffee berry borer (van Lenteren and Bueno, 2003). According to Bustillo (2007), entomopathogens and parasitoids can be used in an IPM programme for the coffee borer, where the time interval between applications of the entomopathogenic fungi and the release of the parasitoids is 8 days in order to reduce the risks of parasitoid infection by the fungi. The risk of infection in parasitoids is reduced when the parasitoids are released before the entomopathogenic fungi are applied (Reyes *et al.*, 1995; Mejía *et al.*, 2000).

Use of microbial control agents

In the 1970s, a baculovirus (Baculoviridae) was imported from the USA to control *Trichoplusia ni* (Hübner) and *Heliothis* spp. in cotton (Moscardi, 1999). Chet and Baker (1981) isolated a naturally suppressive fungus *Trichoderma hamatum* (Bonord.) Bainier from soil, which showed a control effect against different pathogens such as *Rhizoctonia solani* J.G. Kühn on radishes and beans, *Pythium* sp. on peas and *Athelia rolfsii* (Curzi) Tu & Kimbr on beans. After this discovery, R. Baker trained researchers from the Center for Microbiological Research (CIMIC) at the Andes University to produce and apply *Trichoderma* sp. and fluorescent *Pseudomonas* spp. to control *Fusarium oxysporum* f. sp. *dianthi* (Fod.) on carnation. In 1982, some flower companies, also advised by Baker, began the production of their own isolates of *Trichoderma* to control *Fusarium oxysporum* Schlecht. in carnation crops. Since then, a lot of research has followed with microbial control agents against insects. At first this was on entomopathogenic fungi such as *B. bassiana*, *M. anisopliae*, *Isaria fumosorosea* Wize (= *Paecilomyces fumosoroseus*) and *L. lecanii* (van Lenteren and Bueno, 2003; Bellotti *et al.*, 2005) and next on bacteria, primarily *B. thuringiensis*, and baculoviruses (Bellotti *et al.*, 2005).

Between 1994 and 1996, population outbreaks of the locust *Rhammatocerus schistocercoides* Rehn occurred in the Colombian Orinoco region, affecting pastures of native savannahs, particularly *Axonopus* sp. and *Trachypogon* sp. To mitigate the problems caused by this pest, Agrosavia (the Colombian Corporation for Agricultural Research, Corporación colombiana de investigación agropecuaria, formerly Corpoica) identified a strain of *M. anisopliae* that showed 70% of control on the locust nymphs (León *et al.*, 2018).

In the 1990s, the National Coffee Research Center (Cenicafé) promoted the production of the fungus *B. bassiana* to control the coffee berry borer *H. hampei* (Ferrari). This project significantly stimulated the development of microbial control and appeared to encourage many companies and research institutes to produce biopesticides (Cadena, 2005). As an example, Agrosavia started a research and development strategy in 1994, from the search for microorganisms to field evaluation of formulated products, accompanied by field or greenhouse demonstrations. This strategy included the isolation and conservation of potential biocontrol microorganisms, their molecular and ecophysiological characterization, determination of biocontrol activity and study of modes of action, culture medium development, pre-formulation studies, compatibility with chemicals, optimization, scale-up at pilot plant level, efficacy trials, registration and commercial production.

Use of macrobial control agents

Colombia was early in developing mass production technologies at commercial levels for parasitoids and predators, with 30 mass production facilities in 1990, but this number decreased to nine producers by 2000 (van Lenteren and Bueno, 2003). See next section for more information about mass production facilities and species that are produced in Colombia.

8.3 Current Situation of Biological Control in Colombia

8.3.1 Natural biological control of pests in cassava

Many pests in cassava are kept at low levels, without causing damage, by naturally occurring beneficial organisms. In the Neotropics there is an ample complex of natural enemies that exercise a certain level of natural biocontrol on the principal pests of cassava. Sixty-two species of natural enemies are associated with mite pests, 48 with the cassava hornworm *Erinnyis ello* (L.), 73 with mealybugs and 28 with whiteflies (Bellotti *et al.*, 2005). Eggs of *E. ello* are parasitized by various wasps, including species of the genera *Trichogramma*, *Telenomus* and *Ooencyrtus*; and larvae are parasitized by flies of the families Sarcophagidae and Tachinidae and wasps of the families Braconidae, Eulophidae and Ichneumonidae (Gómez-Jiménez, 2018). Populations of the cassava green mite *Mononychellus tanajoa* (Bondar) are also kept at low numbers by various species of naturally occurring predatory phytoseiid mites *Oligota minuta* Cameron, the coccinellid *Stethorus* sp. and the Chrysopid *Chrysopa* sp. (Bellotti *et al.*, 2005; Kondo *et al.*, 2018).

8.3.2 Classical biological control of the Colombian fluted scale

A recent case of a fortuitous classical biocontrol took place on San Andres and Old Providence islands in the Caribbean Sea, involving a scale insect closely related to *I. purchasi*. The Colombian fluted scale (CFS) *Crypticerya multicicatrices* Kondo & Unruh is a highly polyphagous scale insect that affects more than 140 species of plants (Kondo *et al.*, 2014, 2016). Between 2010 and 2013, outbreaks of this invasive pest were reported to be causing a lot of damage, resulting in the loss of self-sustainability of local food production and also dissatisfaction of tourists who were visiting the islands, due to the visual decay caused by high infestations on tropical trees and coconut palms near beaches (Kondo *et al.*, 2014). Damage by CFS includes stunted growth on soursop (Kondo, 2008), cosmetic damage to commercial parts of affected plants and reduction in the quality of the product due to sooty moulds that grow on their excreted honeydew. When it affects leaves, it can decrease the plant's photosynthesis. Very severe attacks may cause defoliation and death of the host plant (Kondo *et al.*, 2012). On 8 April 2013, a lady beetle, *Anovia* sp., was found feeding on the ovisacs of CFS and during a second visit to the island on 26 October 2013 it was found that the lady beetle had already spread over the entire island and that populations of the CFS were decimated to the point that it was very difficult to find any specimens (Kondo *et al.*, 2014). The lady beetle has now been identified as *Anovia punica* Gordon (González and Kondo, 2014). Besides *A. punica*, many other natural enemies have been reported in the literature, including *Isaria* sp. (Kondo *et al.*, 2012), *Delphastus quinculus* Gordon, *Diomus seminulus* (Mulsant) (González *et al.*, 2012), *Syneura cocciphila* (Coquillet) (Gaimari *et al.*, 2012), a hyperparasitoid *Cheiloneurus* sp. (Kondo *et al.*, 2014), a true parasitoid *Brethesiella* cf. *abnormicornis* (Girault) (Montealegre *et al.*, 2016) and two lacewings, *Chrysoperla* sp. and *Ceraeochrysa* sp. (Kondo *et al.*, 2014). The vedalia beetle *Rodolia cardinalis* has also been reported feeding on CFS together with *A. punica* in the city of Cali, Valle del Cauca, Colombia (Pinchao *et al.*, 2015). Until July 2017, Agrosavia maintained a breeding stock of *A. punica* and conducted periodical releases on San Andres Island, with more than 3,000 individuals released on the island due to resurgence of the pest in urban areas, probably caused by the application of pesticides to control mosquitoes. Notwithstanding, populations are under control by *A. punica* in other parts of the islands.

8.3.3 Conservation biological control of pests in sugarcane, chilli pepper, oil palm, coffee and ornamentals

In recent years, there has been an increasing interest in conservation biocontrol in Colombia (Wyckhuys *et al.*, 2013). Research is focused mainly on selecting flowering plants around crops (Díaz *et al.*, 2012) that offer sugar sources for parasitoids from flower nectar (Carrillo *et al.*, 2006) or extrafloral nectaries (Hernandez *et al.*, 2013). Studies have shown that the flower nectar of *Acmella oppositifoli* (Lam.) R.K. Jansen, *Bidens pilosa* L., *Mangifera indica* L. and *Lippia nodiflora* (L.) Michx. prolong the life of tachinids used as natural enemies of *Diatraea* spp. that affect sugarcane in the Department of Valle del Cauca (Vargas *et al.*, 2006). The role of weeds and other plants in the surroundings of chilli pepper crops in supporting natural enemies of aphids is being studied at present (Melo and Manzano, 2017). Ants have been used as biocontrol agents to control some oil palm pests (Aldana *et al.*, 2000) and exploration studies of natural enemies for the control of the coffee berry borer have been conducted (Armbrecht and Gallego, 2007). In flower production systems, corridors of wild vegetation where *Chrysoperla* spp. and *Trichogramma* spp. have been previously established help to reduce pest problems in flower crops (Arias & Arias Bioinsumos S.A.S., Palmira, 2018, personal communication).

8.3.4 Augmentative biological control

Pests in cassava

Most pests in cassava are kept at low levels by naturally occurring beneficial organisms, but some pests need occasional releases of mass-produced agents. The chrysopid species *Chrysoperla carnea* (Stephens) is available commercially for the control of various soft-bodied arthropod pests (aphids, mites, thrips, whiteflies, etc.) (see later, Table 8.1). The cassava hornworm *E. ello*, one of the most important cassava pests in Colombia, has been successfully controlled with *B. thuringiensis* applied to first-, second- and third-instar larvae; and a commercial product (Bio-virus) which is based on a baculovirus was developed in 2003 for the biocontrol of this lepidopterous pest (Bellotti *et al.*, 2005).

Pests in citrus

In Colombia, 160,408 ha are planted with citrus, with a total of 1,681,877 t of fresh fruit harvested in 2013 (DANE, 2016). A study conducted in the municipality of Caicedonia, Valle del Cauca, on a citrus orchard planted with Valencia orange (*Citrus sinensis* L.) showed that releases of native predatory phytoseiid mites, i.e. a combination of *Neoseiulus anonymus* (Chant & Baker), *Neoseiulus californicus* (McGregor), *Iphiseiodes zuluagai* Denmark & Muma and *Amblyseius herbicolus* (Chant) at 500 individuals per tree, and *Ch. carnea* (100 larvae per tree) were efficient in controlling the broad mite *Polyphagotarsonemus latus* (Banks), showing a similar control compared with that used by farmers who applied abamectine. However, for controlling the citrus rust mite *Phyllocoptruta oleivora* (Ashmead), phytoseiid mites were not efficient and control by *Ch. carnea* was only moderate (Imbachi *et al.*, 2012).

Biocontrol of the citrus weevil *Compsus viridivittatus* (Guérin-Méneville) has been carried out with applications of the entomopathogens *B. bassiana* on the foliage and *M. anisopliae* on the soil. The most frequently encountered parasitoid of this pest that has a potential for use in a biocontrol programme is the platygastrid parasitoid wasp *Fidiobia* sp. (ICA, 2012). A study conducted in the municipality of Caicedonia, Valle del Cauca, suggests that the release of 10,000 individuals per hectare, or 500 eggs of *C. viridivittatus* parasitized with *Fidiobia* sp. per 20 trees in the middle stratum of the tree, two or three times per productive period, may keep the citrus weevil populations under control (Carabalí, 2018).

A recent invasive species, the Asian citrus psyllid *Diaphorina citri* Kuwayama, has become an important new pest (García *et al.*, 2016). Damage caused by *D. citri* results from the removal of large amounts of sap from the leaves and because it can act as a vector of the bacterium that causes the catastrophic disease known as huanglongbing (HLB) or 'citrus greening' (Mead and Fasulo, 2010). In December 2015, the Colombian Agricultural Institute (ICA) declared a phytosanitary emergency when it detected adults of *D. citri* that tested positive for the bacterium '*Candidatus* Liberibacter asiaticus' (Las), the causative agent of HLB, in the department of Guajira (ICA, 2015). The disease spread to other northern departments of Colombia. Chemical control and eradication of infected trees have since been carried out by ICA and programmes have been aimed at controlling the insect vector, *D. citri*. In the department of Valle del Cauca, a total of 16 species of natural enemies of the psyllid were found, distributed in six families of five orders: *Azya orbigera* Mulsant, *Cheilomenes sexmaculata* (Fabricius), *Chilocorus* cf. *cacti* (L.), *Curinus colombianus* Chapin, *Cycloneda sanguinea* (L.), *Harmonia axyridis* (Pallas), *Hippodamia convergens* Guérin-Méneville, *Olla v-nigrum* (Mulsant), *Scymnus rubicundus* Erichson; *Allograpta* (*Fazia*) CR-2 aff. *hians* (Enderlein), *Leucopodella* sp.; *Zelus* cf. *nugax* Stål; *Polybia* sp.; *Tamarixia radiata* (Waterston); *Ceraeochrysa* sp. and *Ceraeochrysa* cf. *claveri* (Navás) (Kondo *et al.*, 2015). A mass rearing and release programme for *T. radiata*, the main parasitoid of *D. citri*, has been in progress at Agrosavia since 2013 for research purposes. Agrosavia's research-based mass release programme is aimed at maintaining populations of *D. citri* at low levels in urban areas, house gardens, hedge trees and shrubs of the Rutaceae family (e.g. *Citrus* spp., *Swinglea glutinosa* (Blanco) Merr. and *Murraya paniculata* (L.) Jack) because these populations can act as reservoirs for the bacterium (Kondo, 2018). Mass production of *T. radiata* is an expensive operation, with hand labour consuming 85% of the production costs; thus it is not viable commercially and generally needs to be subsidized or state-funded (Kondo, 2017).

Coffee berry borer and red mite in coffee

Currently, the coffee berry borer *H. hampei* affects some 800,000 ha of coffee plantations in Colombia and more than half a million families rely on the coffee industry (Bustillo, 2006). Biocontrol of *H. hampei* has been carried out with an IPM approach incorporating a combination of parasitoids and entomopathogens (Bellotti *et al.*, 2005). The National Center for Coffee Research (Cenicafé) produced and mass released the imported parasitoids *Cephalonomia stephanoderis* and *Prorops nasuta* (Bustillo *et al.*, 1995; Rivera *et al.*, 2010; Bustillo-Pardey, 2018) and *Phymastichus coffea* LaSalle (Aristizabal *et al.*, 2004; Jaramillo *et al.*, 2005, 2006). Fifteen years after their release, *P. nasuta* had established and contributed to the control of coffee berry borer populations in the field (Maldonado-Londoño and Benavides-Machado, 2007). At present, the company Biocafe commercially produces the parasitoids *P. nasuta*, *C. stephanoderis* and *P. coffea* for release, especially on organic farms. The efficiency of several native species of the nematode genera *Steinernema* and *Heterorhabditis* have been studied against *H. hampei*, which under controlled and field conditions have been shown to be able to reach the infested fruits and reduce coffee borer populations (Bustillo, 2006). Further, *Ch. carnea* is released to control the red mite *Olygonichus yothersi* (McGregor), an occasional and seasonal pest of coffee (L.M. Constantino, Chinchiná, 2018, personal communication).

Pests in cotton, sorghum and maize

In cotton, the entomopathogenic fungus *Lecanicillium lecanii* is now used to control the silverleaf whitefly *Bemisia tabaci* (Gennadius) (Gómez *et al.*, 2012). Additionally, antagonistic fungi *Trichoderma* spp. have shown good results in controlling *Athelia rolfsii* (Curzi) C.C. Tu & Kimbr. (= *Sclerotium rolfsii*) and *Rhizoctonia solani* in cotton (Hoyos *et al.*, 2008). A total of 21 isolates of *Trichoderma* spp. have been reported in Colombia based on the 5′ end of the translation elongation factor-1α (EF1-α1) gene and an RNA polymerase II gene encoding the second largest protein subunit (RPB2) by using mixed primers (Smith *et al.*, 2013).

Infestations of *S. frugiperda* have diminished because of the use of Bt technology in cotton and maize crops in Colombia since 2003 (Polanía *et al.*, 2008, 2009). However, *Trichogramma* spp. are still released to control eggs of this pest on maize and sorghum (Biocol, 2017). Studies conducted by Agrosavia demonstrated that the use of a native baculovirus (nucleopolyhedrovirus SfMNPV) is a promising alternative biocontrol agent of *S. frugiperda*. The effectiveness of geographical strains of SfMNPV has been evaluated under laboratory and field conditions, obtaining results in controlling the pest similar to that of chemical insecticides (Villamizar *et al.*, 2009; Gómez *et al.*, 2013). Naturally occurring SfMNPV showed a wide genetic diversity (Gómez *et al.*, 2010) and a biopesticidal formulation based on a selected virus was obtained (Villamizar *et al.*, 2010, 2012a). These studies allowed the development of an SfMNPV-based commercial biopesticide, which is in the process of registration.

Pine woolly aphid and hornworm in forestry

A recent problem in forestry is the pine woolly aphid *Pineus boerneri* Annand, which was recorded for the first time in Colombia in 2008, causing infestations on various pine species, namely *Pinus kesiya* (A. Chev.), *P. tecunumanii* Eguiluz & J. P. Perry, *P. maximinoi* H.E. Moore and *P. oocarpa* Schiede ex Schltdl. (Rodas *et al.*, 2014). After a search for local natural enemies of *P. boerneri*, it was shown that a single individual of *Ceraeochrysa* sp. can consume up to 140 adelgids per day (Rodas *et al.*, 2014).

The hornworm *E. ello* is an important pest of rubber *Hevea brasiliensis* L. in Colombia. Preliminary studies have shown that the granulovirus ErelGV, of the genus Betabaculovirus, family Baculoviridae, is highly pathogenic and virulent to the larvae of this lepidopteran pest (Cuartas *et al.*, 2018).

Pests in greenhouse vegetables and ornamentals

Many of the biocontrol programmes in the history of biological control in Colombia are still in use. Currently, the entomopathogenic fungus *L. lecanii* is used for the control of a wide range of greenhouse pests, including aphids, scale insects and other hemipterans. Indigenous strains of *L. lecanii* have been shown to be effective at controlling the silverleaf whitefly *B. tabaci* on eggplant (Gómez *et al.*, 2012) and melon (Cotes *et al.*, 2009). Also, *B. bassiana* is available for the control of aphids, thrips and whiteflies in floriculture. Predatory mites *Iphiseius degenerans* (Berlese), *Neoseiulus* (*Amblyseius*) *cucumeris* (Oudemans) and *Stratiolaelaps scimitus* Womersley (as *Hypoaspis miles*) and the minute pirate bugs *Orius insidiosus* (Say), *O. laevigatus* (Fieber) and *O. tristicolor* (White) are used to control western flower thrips *Frankliniella occidentalis* Pergande in different greenhouse crops (Andrade *et al.*, 1989). The two-spotted spider mite *Tetranychus urticae* Koch is a common pest of roses, gerberas, chrysanthemums and other floricultural crops. The predatory mite *Phytoseiulus persimilis* Athias-Henriot, commercially available in Colombia since the early 1970s, was registered for use as a natural enemy of spider mites in 2004 (Bueno, 2005) and is also used in rose crops (Daza *et al.*, 2010). It is now available from the company Bichopolis (Y. Martinez, Tabio, 2018, personal communication). *Neoseiulus californicus* is, among others, used to control *T. urticae* in commercial rose crops (Forero-Patiño *et al.*, 2010). The western predatory mite *Galendromus* (= *Typhlodromus*) *occidentalis* (Nesbitt), used to control spider mites, can tolerate lower humidities than *P. persimilis*. Further, *N. cucumeris* and *Neoseiulus* (*Amblyseius*) *barkeri* Hughes can be used for the control of the cyclamen mite and broad mite, respectively.

Recently, studies on whitefly control have been conducted with *Amitus fuscipennis* (MacGown & Nebeker) and *E. formosa* (De Vis and Fuentes 2001; De Vis and van Lenteren, 2008). For biocontrol of the greenhouse whitefly, *Trialeurodes vaporariorum* Westwood, good results have been obtained with the parasitoid *E. formosa* (Cantor *et al.*, 2011; Manzano, 2018). Currently, both *A. fuscipennis* and *E. formosa* are registered as natural enemies of whiteflies (UJTL, 2016). In Colombia, two predators, the coccinellid *Delphastus catalinae* (Horn) and the predatory mite *P. persimilis*, have shown good results in the control of the B biotype of *B. tabaci* (*B. argentifolii*) in greenhouse environments (Kondo *et al.*, 2018).

The tomato borer *Tuta absoluta* (Meyrick) damages tomato crops by feeding on leaves, stems, flowers and fruits (Bajonero *et al.*, 2008). To control the pest in greenhouse tomatoes, a baculovirus (PhopGV) is used with good results (Gómez *et al.*, 2014), as well as the egg parasitoid *T. pretiosum* (García, 1985), and currently the larval parasitoid *Apanteles gelechiidivoris* Marsh is commercially released in combination with sex pheromone traps with good control results (Aguilar *et al.*, 2010). The gall midge *Prodiplosis longifila* Gagné causes damage to solanaceous crops and is presently considered one of the most destructive pests of tomato (Hernandez *et al.*, 2015). Although several species of parasitoids of the genus *Synopeas* Förster parasitize *P. longifila* (Hernandez-Mahecha *et al.*, 2018), they have not been released as biocontrol agents as was done in Peru to control *P. longifila* on asparagus (Díaz-Silva, 2011). The antagonistic fungus *Trichoderma koningiopsis* Samuels, C. Suárez & H.C. Evans is currently used to control *Fusarium oxysporum* f. sp. *lycopersici* (Saccardo) Snyder & Hansen (Moreno *et al.*, 2009); and *Rhizoctonia solani* Kühn on greenhouse tomatoes with good results (Cotes *et al.*, 2001).

In 1996, the Association of Colombian Flower Exporters (Asocolflores) created a code of conduct for the flower sector aimed at a sustainable production of flowers, which led to the creation of the Florverde standards in 2002. Florverde aims to achieve high standards in flower production, encouraging improvement among member companies through a dynamic system that measures their social and environmental performance (Asocolflores *et al.*, 2006). The implementation of Florverde standards has resulted in cleaner and sustainable agricultural practices that rely heavily on biocontrol techniques. In the Colombian flower industry, *Trichoderma* spp. are now extensively used to control *F. oxysporum*, *R. solani*, *Sclerotinia sclerotiorum* (Lib.) de Bary and *Botrytis cinerea* Pers., among other pathogens. Companies have invested in laboratories to produce this fungus. The positive results obtained in floriculture have encouraged the production and adoption of additional biocontrol products registered for other crops.

Pests in oil palm

Oil palm plantations covered 404,104 ha in 2010 (Fedepalma, 2011), with biodiesel production as one of the main reasons for their recent expansion. Research into control of diseases and pests of palm has been carried out mainly by Cenipalma (Colombian Oil Palm Research Center). The parasitoid *T. pretiosum* is released to control *Stenoma impressella* (Busck) (Castillo *et al.*, 2000). The oil palm defoliator *Loxotoma elegans* Zeller is controlled with mass releases of parasitoids of *Trichogramma* spp., and initial foci of infestations are controlled with large-scale applications of *Bt* and *Beauveria* spp. (Aldana *et al.*, 2010). The egg parasitoid *Ooencyrtus* sp. was mass reared and released to successfully control the palm borer *Cyparissius daedalus* Cramer (Aldana *et al.*, 2004). The lace bug *Leptopharsa gibbicarina* Froeschner is involved in the development of pestalotiopsis disease in the *Pestalotiopsis–Leptopharsa* complex that can reduce the production of palm oil by up to 36%. This pest is controlled by predatory ants (Barrios-Trilleras *et al.*, 2015). The establishment of *Crematogaster* spp. is enhanced by planting nectariferous plants such as *Urena trilobata* Vell., *Croton trinitatis* (Millsp.) and *Cassia reticulata* Willd., the latter encouraging nesting by the ants. Ant colonies are removed from palms and redistributed to other palms with high pest density. Ant predation resulted in a reduction of pestalotiopsis disease, and lower production costs due to less use of insecticide sprays (Aldana *et al.*, 2000). Other predator species such as *Ch. carnea* and *Ch. rufilabris* (Burmeister) are released together to control eggs of *L. gibbicarina* and of some lepidopteran species (Arias & Arias Bioinsumos S.A.S., Palmira, 2018, personal communication). In the early 2000s, *B. bassiana* was used against *Opsiphanes cassina* C. & R. Felder on 130,000 ha of oil palm (van Lenteren and Bueno, 2003). Recently, the entomopathogenic nematode *Heterorhabditis* sp. has been studied for control of the palm root borer *Sagalassa valida* Walker (Bustillo, 2014) and it is at present produced commercially (see Table 8.1). *Heterorhabditis* sp. is traditionally reared on larvae of the greater wax moth *Galleria mellonela* L. (Bustillo, 2014), but studies are now done to multiply *Hetterorhabditis* sp. *in vitro* in order to improve production (Moreno-Salguero *et al.*, 2014).

Pests in potato

An important case of biocontrol in potato is the use of baculoviruses to control the tuber moths *Tecia solanivora* (Povolný) and *Phthorimaea operculella* (Zeller), which cause severe damage when their larvae mine the tubers in the field and during storage. Agrosavia sampled populations of *T. solanivora* and found five geographical granulovirus isolates; analysis by restriction endonuclease cleavage patterns revealed the presence of three different genotypic variants. Based on their DNA restriction patterns and biological activity, two isolates were selected for further analysis as potential biocontrol agents (Espinel *et al.*, 2010). At present, a formulation of Phop-GV, 'Baculovirus Corpoica' is the only baculovirus product registered in Colombia and is recommended for the control of *T. solanivora* in stored potatoes (Espinel *et al.*, 2012). The entomopathogenic fungus *B. bassiana* is commonly used to control the Andean potato weevil *Premnotrypes vorax* (Hustache) (Cotes *et al.*, 2004). For the control of the phytopathogenic fungus *R. solani* on potato, the antagonistic fungus *T. koningiopsis* has shown good control (Beltrán *et al.*, 2012).

Pests in sugarcane

In the early 2000s the parasitoids *Trichogramma* sp., *Lydella minense* (Townsend) and *Billaea claripalpis* were released to control the sugarcane borer *D. saccharalis* and other caterpillars in some 130,000 ha (van Lenteren and Bueno, 2003). Recently, a species complex of stem borers belonging to the genus *Diatraea* caused economic damage in sugarcane and showed a dynamic change in species composition involving the following species: *D. saccharalis*, *D. lineolata* Walker, *D. busckella* Dyar & Heinrich, *D. indigenella* Dyar & Heinrich, *D. tabernella* Dyar and *D. rosa* Heinrich (Pérez & Martínez, 2011; Vargas *et al.*, 2013; Vargas, 2018). The same three parasitoids as mentioned above are released to control these borers (Gómez and Vargas, 2014). The recommended release rate is 30 tachinid flies (using either *L. minense* or *B. claripalpis*) plus 50 square inches (323 cm^2) of *Trichogramma exiguum* Pinto and Platner cards (~ 85,000 adults) per hectare when 2.5% of internodes were bored in the previous crop (Vargas *et al.*, 2015). Due to the low parasitism level of *L. minense* on *D. tabernella* (Vargas *et al.*, 2013), control has been attempted recently with the larval parasitoid *Cotesia flavipes* (Cameron). This parasitoid was originally released in the 1970s, but on that occasion the parasitoid only established in certain areas of Colombia (Gaviria, 1990). However, recent biological studies support the use of *C. flavipes* to control *D. busckella*, *D. indiginella*, *D. sacharalis* and *D. tabernella* (Salamanca *et al.*, 2016). This parasitoid is now mass produced by several local companies (see Table 8.1) (Y. Gutierrez, Incauca, El Ortigal, 2018, personal communication). Data from two sugar mills of Valle del Cauca showed that 63% (= 127,254 ha) of the sugarcane-planted area (201,275 ha) was under biocontrol and that 1,813,729 tachinid flies, 1,004,594 square inches (6,481,239 cm^2) of *T. exiguum* and 25,350 g of *C. flavipes* cocoons were released during the year 2016 (Y. Gutierrez, Incauca, El Ortigal, 2018, personal communication). The common green lacewings *Ch. carnea* and *Ch. rufilabris* are mass produced and released to control outbreaks of the yellow sugarcane aphid *Sipha flava* (Forbes) (E. Arias, Palmira, 2018, personal communication).

Trichogramma pretiosum is known to control sugarcane leaf-feeding insects such as *Mocis latipes* Guenée (Bustillo, 2013). In Colombia, three species of the genus *Trichogramma* are currently mass reared and commercially released, namely *T. atopovirilia*, *T. exiguum* and *T. pretiosum* (Díaz-Niño, 2018).

Sugarcane borer in rice

Rice is damaged by the borer *D. sacharalis*. The wasp *T. exiguum* is periodically released to control its eggs (Fedearroz, 2010) and *C. flavipes* to control its larvae. Mixed releases of *Ch. carnea* and *Ch. rufilabris* are carried out to control mites and soft body insects (Arias & Arias Bioinsumos S.A.S., Palmira, 2018, personal communication).

Pests in various other crops

Black sigatoka or black leaf streak disease *Pseudocercospora fijiensis* (Morelet) is the most significant foliar disease of bananas (*Musa* spp.) worldwide (Arango *et al.*, 2016). In Colombia, good control has been achieved with *Bacillus subtilis* (Ehrenberg) Cohn (Gutierrez-Monsalve *et al.*, 2015).

The Andean blackberry *Rubus glaucus* Benth. is affected by grey mould disease *Botrytis cinerea* Pers. Good control of this disease was obtained with the antagonistic yeast *Rhodotorula glutinis* (Fresenius) Harrison (Zapata *et al.*, 2013) and the antagonistic fungus *T. koningiopsis* (Uribe-Gutiérrez *et al.*, 2013). *T. koningiopsis* is also effective in controlling *S. sclerotiorum* and *Sclerotinia minor* (Jagger) on lettuce (Moreno *et al.*, 2010).

Dried beans *Phaseolus vulgaris* L. are an important source of protein in many parts of Africa, South-east Asia and South America (EOL, 2016). The bean weevil *Acanthoscelides obtectus* (Say) is effectively controlled by the naturally occurring parasitoid *Dinarmus basalis* (Rondani), with a reduction of 88–97% of weevil populations on stored beans (Schmale *et al.*, 2006). On green and snap bean crops, the parasitoid wasp *A. fuscipennis* shows high parasitism rates on the greenhouse whitefly *T. vaporariorum* (Manzano, 2000), as a result of its good searching capacity (Manzano *et al.*, 2002a) and high intrinsic rate of growth (Manzano *et al.*, 2002b), but it only occurs abundantly in high tropical mountain regions (Manzano *et al.*, 2003). This parasitoid should be considered for IPM programmes to control *T. vaporariorum* in common bean crops (Cardona *et al.*, 2005).

In the department of Valle del Cauca, the lacewing *Ch. carnea* is widely used for the control of various species of aphids, e.g. *Aphis gossypii* Glover and *M. persicae* in hot peppers (*Capsicum* spp.). The passion fruit flower bud fly *Dasiops inedulis* Steyskal is an important pest of yellow passion fruit *Passiflora edulis* fo. *flavicarpa* O. Deg. in Colombia (Chacón and Rojas, 1984). As an IPM strategy for *D. inedulis*, Quintero *et al.* (2012) recommended the use of the pupal parasitoids *Pachycrepoideus vindemmiae* Rondani and *Spalangia* sp., which are commercially available for the control of the common housefly and fruit flies in the same region.

Flies in oil palm, poultry and livestock

In Colombia, pest flies in confinement areas in poultry and other livestock settings are controlled on a large scale by periodic releases of the parasitoid *Spalangia cameroni* Perkins reared on pupae of *Musca domestica* L. (Jiménez, 2001). These parasitoids are also released to control flies in landfills, compost from sugarcane residues and domestic residues. In the year 2000, around 300 million parasitoids were produced and commercialized in Colombia, and exported to other Latin America countries (Jiménez, 2001). On palm crops, the parasitoid *Spalangia* sp. was released to control the pupal stage of the stable fly *Stomoxys calcitrans* (L.), which reproduces on empty fruit bunches, sludge and fibres of oil palm by-products used to improve crop soil conditions (Bedoya, 2007). Some oil palm companies have their own mass rearing of *Spalangia*.

Control of vectors of human diseases

The mosquitoes *Aedes aegypti* (L.), *Culex* spp. and *Anopheles* spp., all transmit human diseases in Colombia. *A. aegypti* is a vector of yellow fever (Vasconcelos *et al.*, 1999), dengue (Ministry of Social Protection, 2017a), zika (Cetron, 2016) and mayaro viruses (Muñoz and Navarro, 2012). Because of resistance of mosquitoes to the insecticide Temephos, currently toxins of *Bacillus thuringiensis israelensis* (*Bti*H14) are widely used (Ministry of Social Protection, 2017a). This biopesticide is sold in a granular commercial preparation for the control of the larval stages of *A. aegypti* in outdoor breeding sites (sewers, puddles, etc.). For control inside homes, a water-diluted commercial preparation is sold. *Bti* is applied on water bodies every 7–10 weeks, with good control results (Ministry of Social Protection, 2017a). *Lysinibacillus sphaericus* (Neide, 1904) (as *Bacillus sphaericus*) in combination with an insect growth regulator is used to control *A. aegypti* and *Culex quinquefasciatus* Say in sewers in Cali (Giraldo-Calderón *et al.*, 2008). For malaria control, *L. sphaericus* is commercially produced for the control of *Anopheles* larvae and applied as granules (G, 2 g m^{-2}) and water diluted granules (WDG, 0.15 g m^{-2}) (Ministry of Social Protection, 2017b). The larvivorous fish *Gambusia affinis* Baird & Girard and *Poecilia reticulata* Rosen & Bailey are released in permanent water storage bodies in dengue endemic areas (Ministry of Social Protection, 2017a). The copepod *Mesocyclops longisetus* (Thiébaund) achieved good control of *A. aegypti* larvae that live in catch basins (cisterns that receive street gutter discharges) in the city of Cali (Suarez-Rubio and Suarez, 2004).

Use of microbial control agents

Agrosavia's activities were partly described in the historical section. Since 1994, this corporation has been equipped with advanced technological facilities and developed expertise in the development of biopesticides to control insect pests and plant pathogens, which has allowed the registration of domestic market products with high technical specifications, licensing production and formulation technologies for countries in South America and Europe. The locally developed technologies by Agrosavia have positioned Colombia as an outstanding country in South America for its development capabilities, production and marketing of biopesticides. The main developments in different crops are summarized in a recently published book that describes experiences on biocontrol of soil-borne (Moreno-Velandia *et al.*, 2018), foliar (Cotes *et al.*, 2018c) and post-harvest diseases (Zapata *et al.*, 2018), as well as the use of entomopathogens as viruses (Villamizar *et al.*, 2018), bacteria (Grijalba *et al.*, 2018) and fungi (Espinel-Correal *et al.*, 2018a, b). Also included are chapters on the different aspects concerning IPM (Cotes and Elad, 2018), biopesticide development (Cotes *et al.*, 2018a; Díaz *et al.*, 2018), registration (Santos *et al.*, 2018), commercialization (Gómez *et al.*, 2018) and future perspectives (Barrera *et al.*, 2018; Cotes *et al.*, 2018b). However, there is an increasing uncertainty in the commercial development of formulated products based on most of those biocontrol agents and their feasibility in a broad range of agroecosystems because of the small market size, short lifespan, inconsistency of efficacy and high specificity towards pests, which increases the cost factor for the growers. Additionally, many commercial biopesticides are based on the same active ingredient and are recommended for a reduced number of pests; and there is a lack of products for managing other limiting pests.

Currently, 104 microbial biopesticides have been registered by the Colombian Agricultural Institute (ICA, 2016b), and are commercially available. Isolation and selection of antagonistic microorganisms to control plant diseases was studied, mainly with *Trichoderma* spp. and *Bacillus* spp. (Bettiol *et al.*, 2008; Cotes, 2011, 2014). At present, 33 are recommended to control plant pathogens, with 21 products based on *Trichoderma harzianum* Rifai, three on *Trichoderma lignorum* (Tode) Harz, four on *Trichoderma viride* Pers., one on *Trichoderma atroviride* (Karsten) Bissett, one on *T. koningiopsis* and three based on mixtures of various species of *Trichoderma*. Bacterial biopesticides include eight products based on *Bacillus subtilis* (Ehrenberg) Cohn, one based on *Bacillus pumilus* Meyer & Gottheil and one based on *Streptomyces racemochromogenes* Sugai. Most of the above products are recommended for the control of diseases caused by soil fungi and only a few are recommended for the control of foliar pathogens (ICA, 2016b).

It seems obvious that the rich source of genetic diversity of Colombia (Smith *et al.*, 2013) represents a potential for finding novel biocontrol agents to control specific phytosanitary problems. Many of the currently registered biopesticides are based on microorganisms obtained from international collections, but a few are well known indigenous biocontrol agents. However, despite the rich genetic biodiversity of Colombia, novel biocontrol microorganisms have not been discovered. Furthermore, it is important to mention that there are also some biopesticides that were registered in the past and currently appear as unregistered; in some cases, developers of those products overspent on staffing, buildings and equipment and made unrealistic promises for financial returns and so local companies decided not to continue with the commercial production of such products.

Colombia has been a leader in Latin America in creating a regulatory system to register microbial control agents. Once registered, surveys to test the quality of microbial control agents are conducted to assure that manufacturers meet accepted standards. Table 8.1 provides a list of biopesticides and biocontrol agents that are produced in Colombia.

Use of macrobial control agents

Mass rearing of natural enemies in Colombia started from small private laboratories owned by sugarcane mills for the control of particular pests in sugarcane plantations. Initially, most of the production was done through small-scale cottage industries with little machinery, in small spaces, with little capital, and through non-systematic and usually non-automated activities (Kondo, 2011). Producers and manufacturers of biocontrol agents in Colombia are small and medium-sized companies, research institutes, cooperatives and agroindustries. Some of the producers are either too poor or too undeveloped for any significant sales. On the other hand, a number of economically competitive biocontrol products are now exported to different countries. Regarding arthropods sold in Colombia, the parasitoid *T. exiguum* is the only natural enemy registered by the ICA (Table 8.1). It seems that *T. exiguum* was registered in the 1980s, before the current ICA regulations were established. The new ICA registration process requires high-quality standardized production conditions in laboratories and companies, which cannot be afforded by most small producers. Currently, the Center for Biosystems at the University of Jorge Tadeo Lozano (Bogota), endorsed by the ICA, conducts quality control tests of various arthropods, including: (i) predatory mites such as *N. cucumeris*, *N. californicus*, *P. persimilis*, *S. scimitus*; (ii) parasitoids such as *Aphidius colemani* Viereck, *Dacnusa sibirica* Telenga, *E. formosa*, *Eretmocerus eremicus* Rose & Zolnerowich, *Diglyphus isaea* (Walker), *A. fuscipennis*, and *Trichogramma* sp.; and (iii) predators such as *Aphidoletes aphidimyza*, *Orius* sp., *Ceraeochrysa* sp. and *Chrysoperla* sp. The Center also carries out quality tests for various other natural enemies to control aphids, beetle grubs, collembolans, leaf beetles, mites, mosquitoes, whiteflies, and other insect pests (UJTL, 2016). We assume that the above arthropods are in the process of registration, since only one of them is officially registered by ICA at present.

Currently, there are 52 companies in Colombia that produce microbial control agents (entomopathogenic fungi, bacteria, baculoviruses and antagonistic microorganisms, including fungi and bacteria) and predators and parasitoids that are officially registered by the Colombian Agricultural Institute (ICA, 2016a, b). This number does not include private laboratories, for example those owned by sugar mills, which produce various species of natural enemies for release in their sugar plantations for the control of *Diatraea* spp., and newly created companies. To our knowledge, currently there are 60 national (registered and not registered) companies that sell biocontrol agents in Colombia (Table 8.2).

In Colombia, there are also several international companies that sell microbial control agents and natural enemies. For example, the Dutch company Koppert Biological Systems sells the predatory mite *Neoseiulus californicus* for the control of *Panonychus citri* (McGregor), *P. ulmi* (Koch), *T. urticae*, *Brevipalpus* spp., *Raoiella indica* Hirst, *P. latus* and *P. pallidus*; and the predatory mite *P. persimilis* for the control of *Tetranychus* spp., particularly *T. urticae*, mostly used by the flower industry. Bayer & Monsanto and Valent Biosciences Corporation (through Sumitomo Chemical Latin America) sell a product based on *Bacillus thuringiensis* var. *kurstaki* (*Btk*) for the control of the larval stages of a wide range of lepidopteran pests.

Information on the total area in which biocontrol is practised in Colombia is lacking. It is generally said to be about 5% of the total cultivated area, which translates to about 550,000 ha; however, no studies have been published that corroborate this estimation. An overview of various Colombian biocontrol programmes, including (when available) information on the hectarage on which they are used, is presented in Table 8.3.

8.4 Biological Control Hotspots in Colombia

Four major biocontrol hotspots and four other areas of high biocontrol activity are distinguished in Colombia, each related to a particular crop or production system (Fig. 8.1). The main hotspot of biocontrol is the Valle del Cauca region, with three main research centres (Cenicaña, CIAT, and Agrosavia, Palmira Research Center), two universities (Universidad Nacional

Table 8.1. Biopesticides and biocontrol agents produced in Colombia (data retrieved from ICA, 2016b).

Microorganism / natural enemy	Type of agent	Target pest	Company (and department[a] where located)
Amitus fuscipennis (MacGown & Nebeker)	Parasitoid	Whiteflies	**Unregistered:** *Bioterra (Cun)
Bacillus pumilus Meyer & Gottheil	Bacteria	Mildew *Podosphaera pannosa*	**Registered:** Bayer S.A.
Bacillus subtilis (Ehrenberg) Cohn	Bacteria	Nematodes and fungal pathogens (*Fusarium* sp., *Mycosphaerella fijiensis* Morelet, *Phoma* sp., *Podosphaera pannosa* and *Pyricularia* sp.)	**Registered:** Bayer S.A, Core Biotechnology S.A.S. (Val), Laverlam S.A. (Val), Mezfer de Colombia Ltda. (Cun), Semillas Valle S.A. (Val) **Unregistered:** *Biocontrol (Val), *Bio-crop S.A.S. (Val), *Bioquirama S.A.S. (Ant), *Mycotech Co. S.A.S. (Val), *Mycros International S.A.S. (Cun)
Bacillus thuringiensis var. *aizawai* (*Bta*)	Bacteria	Leafworm, armyworm	**Registered:** Valent Biosciences Corporation
Bacillus thuringiensis var. *israelensis* (*Bti*)	Bacteria	Fungus gnats, leaf miners, mosquitoes	**Unregistered:** *Bioquirama S.A.S. (Ant), *Mycros International S.A.S. (Cun)
Bacillus thuringiensis var. *kurstaki* (*Btk*)	Bacteria	Caterpillars (Lepidoptera)	**Registered:** Bayer S.A, Corporation Bio-crop S.A.S (Val), Laverlam S.A. (Val), Natural control (Ant), Vecol (Cun), Yaser (Val), Semillas Valle (Val), Valent Biosciences **Unregistered:** *Agroproductiva S.A. (Val), *Bioquirama S.A.S. (Ant)
Bacillus thuringiensis var. *thuringiensis* (*Bt*)	Bacteria	Various species of mites (e.g., rust mites, spider mites)	**Registered:** Plantador Colombia Ltda. (Cund) **Unregistered:** *Bioquirama S.A.S. (Ant), *Natural Control S.A. (Ant), *Organización Pajonales (Tol), *Safer *Agrobiológicos S.A.S. (Ant)
Beauveria bassiana (Bals.-Criv.) Vuill.	Fungus	Aphids, coffee borer, leaf beetles, leaf cutting ants, mites, thrips, weevils, whiteflies, stink bugs, scale insects, and lepidopterans. Attacks different growth stages	**Registered:** Bio-crop S.A.S (Val), Bioecológicos Ltda. (Cun), Bioprotección S.A.S. (Cal), Core biotechnology (Val), *Inproarroz (Cas), Laverlam (Val), Mycros International S.A.S. (Cun), Soluciones Microbianas del Trópico Ltda. (Cal) **Unregistered:** *Agroinsumos Biológicos (Cas), Agroproductiva S.A. (Val), *Biocontrol (Val), *Bioinsumos del Campo Ltda. (Met), *Biológicos y Ecológicos de Colombia (Cun), *Bionova Ltda. (Val), *Bioquirama S.A.S. (Ant), *Biotropical S.A. (Ant), Fundación Centro de Biotecnología "Mariano Ospina Perez" (Cun), *Funguscol (Val), *Laboratorio Biológico La Avispita (Met), * Live Systems Technology S.A. (Cun), *Mycol E.A.T. (Cas), *Mycotech Co S.A.S. (Val), *Natural Control S.A. (Ant), *Organización Pajonales (Tol), *Palmar del Oriente S.A. (Cun), *P.L.A. Biológicos (Tol), *Safer Agrobiológicos S.A.S. (Ant), *Sanoplant (Val), *Sociedad laboratorio de Suelos E.C.N. Ltda. (Mag)

Continued

Table 8.1. Continued.

Microorganism / natural enemy	Type of agent	Target pest	Company (and department[a] where located)
Beauveria bassiana Bals.-Criv.) Vuill.+ *Metarhizium anisopliae* Metchnikoff) Sorokin, *Purpureocillium lilacinum* (Thom) Luangsa-ard, Houbraken, Hywel-Jones & Samson (sold as *Paecilomyces lilacinus*)	Fungus	Coffee borer and bugs	**Registered:** Orius Biotecnología Ltda (Met)
Beauveria bassiana Bals.-Criv.) Vuill. + *Metarhizium anisopliae* Metchnikoff) Sorokin + *Bacillus thuringiensis* var. *kurstaki*	Fungus + Bacteria	Coffee borer and fruit flies	**Registered:** Safer Agrobiologicos S.A.S. (Ant)
Beauveria bassiana Bals.-Criv.) Vuill. + *Bacillus thuringiensis*	Fungus + Bacteria	*Copitarsia* sp.	**Registered:** Natural control (Ant)
Beauveria brongniartii (Saccardo) Petch	Fungus	Beetle grubs, scale insects (mealybugs), whiteflies	**Registered:** Sanitex S.A.S. **Unregistered:** *Biológicos y Ecológicos de Colombia (Cun), *Palmar del Oriente S.A. (Cun)
Billaea claripalpis (Wulp)	Parasitoid	*Diatraea* complex on rice, sugarcane, and maize	**Unregistered:** *Biocol (Val), *Productos Biológicos Perkins Ltda. (Val)
Cephalonomia stephanoderis Betrem	Parasitoid	Coffee berry borer	**Unregistered:** *Biocafé (Cal)
Ceraeochrysa sp.	Predator	Aphids, mites, scale insects, thrips, whiteflies, etc.	**Unregistered:** *Biológicos del Valle (Val)
Chrysoperla carnea (Stephens)	Predator	Aphids, mites, scale insects, thrips, whiteflies, etc.	**Unregistered:** *Arias & Arias Bioinsumos S.A.S. (Val), *BioAgro (Val), *Biocol (Val), *Biológicos del Valle (Val), *Productos Biológicos Perkins Ltda. (Val)
Chrysoperla rufilabris (Burmeister)	Predator	Aphids, mites, scale insects, thrips, whiteflies, etc.	**Unregistered:** *Biocol (Val), *Biológicos del Valle (Val)
Cotesia flavipes (Cameron)	Parasitoid	Lepidoptera (rice, sugarcane and maize)	**Unregistered:** *Arias & Arias Bioinsumos S.A.S. (Val), *Biocol (Val), *Bioenergy (Met), *Ingenio Mayagüez (Val)
Encarsia formosa Gahan	Parasitoid	Whiteflies	**Unregistered:** *Bioterra® (Cun)
Entomophthora virulenta I.M. Hall & P.H. Dunn	Fungus	Aphids, mites, thrips and whiteflies	**Unregistered:** *Agroproductiva S.A. (Val), *Bionova Ltda. (Val), *Laverlam S.A. (Val)

Granulovirus	Virus	Caterpillars, e.g., *Tecia solanivora*, *Phthorimaea operculella* (Lepidoptera)	**Registered:** Agrosavia (Cun) **Unregistered:** *Secretaria de Agricultura de Boyacá (Boy)
Heterorhabditis sp. (Nematoda: Heterorhabditidae)	Parasite	Lepidoptera (oil palm)	**Unregistered:** *Productos Biológicos Perkins Ltda. (Val)
Hirsutella thompsonii Fisher + *Akanthomyces* sp.	Fungus	Various species of mites (e.g., rust mites, spider mites)	**Unregistered:** *Bioquirama S.A.S. (Ant)
Isaria fumosorosea Wize (also referred to as *Paecilomyces fumosoroseus*)	Fungus	Mites, leaf miners, thrips, whiteflies	**Registered:** Bio-Crop S.A.S. (Val), Mycros International S.A.S. (Cun) **Unregistered:** *Bioecológicos Ltda. (Cun), *Biológicos y Ecológicos de Colombia (Cun), *Bioprotección S.A.S. (Cal), *Bioquirama S.A.S. (Ant), *Laboratorio Biológico La Avispita (Met), *Live Systems Technology S.A. (Cun), *Organización Pajonales (Tol), *Safer Agrobiológicos S.A.S. (Ant), *Sanoplant (Val)
Lecanicillium lecanii R. Zare & W. Gams	Fungus	Aphids, scale insects (mealybugs), whiteflies, thrips, root-knot nematodes	**Registered:** Agrosavia (Cun), Natural Control S.A. (Ant) **Unregistered:** *Biocontrol (Val), *Bioecológicos Ltda. (Cun), *Biológicos y Ecológicos de Colombia (Cun), *Bionova Ltda. (Val), *Bioprotección S.A.S. (Cal), *Bioquirama S.A.S. (Ant), *Biotropical S.A. (Ant), *Laboratorio Biológico La Avispita (Met), *Laverlam S.A. (Val), *Safer Agrobiológicos S.A.S. (Ant), *Sanoplant (Val), *Sociedad laboratorio de Suelos E.C.N. Ltda. (Mag)
Lydella minense (Townsend)	Parasitoid	*Diatraea* complex (rice, sugarcane, maize)	**Unregistered:** *Arias & Arias Bioinsumos S.A.S. (Val), *Biocol (Val), *Ingenio Maria Luisa (Val), *Ingenio Mayagüez (Val)
Lysinibacillus sphaericus (Meyer & Neide)	Bacteria	Fungus gnats, leaf miners, mosquitoes	**Unregistered:** *Bioquirama S.A.S. (Ant)
Metarhizium anisopliae (Metchnikoff) Sorokin	Fungus	Beetle grubs, leaf cutting ants, mosquito larvae, weevils, froghoppers	**Registered:** Bioprotección S.A.S. (Cal), Inproarroz S.A. (Met), Live Systems Technology S.A. (Cun), Sanitex S.A.S. (Cas), Soluciones Microbianas del Trópico Ltda (Cal) **Unregistered:** *Agroinsumos Biológicos (Cas), *Biocontrol (Val), *Bioecológicos Ltda. (Cun), *Bioinsumos del Campo Ltda. (Met), *Biológicos y Ecológicos de Colombia (Cun), *Bionova Ltda. (Val), *Laboratorio Biológico La Avispita (Met), *Laverlam S.A. (Val), *Bioquirama S.A.S. (Ant), *Funguscol (Val), *Mycol E.A.T. (Cas), *Natural Control S.A. (Ant), *Organización Pajonales (Tol), *Palmar del Oriente S.A. (Cun), *P.L.A. Biológicos (Tol), *Safer Agrobiologicos S.A.S. (Ant), *Sanoplant (Val), *Productos Biológicos Perkins Ltda. (Val), *Sociedad Laboratorio de Suelos E.C.N. Ltda. (Mag)

Continued

Table 8.1. Continued.

Microorganism / natural enemy	Type of agent	Target pest	Company (and department[a] where located)
Metarhizium anisopliae (Metchnikoff) Sorokin + *Paenibacillus popilliae* (Dutky) (also referred to as *Bacillus popilliae*)	Fungus + Bacteria	White grubs or beetles	**Registered:** Natural control (Ant)
Metarhizium sp.	Fungus	Beetle grubs, leaf cutting ants, mosquito larvae, weevils, froghoppers	**Unregistered:** *Orius biotecnologia Ltda. (Met), *Safer Agrobiológicos S.A.S. (Ant)
Nomuraea rileyi (Farl.) Samson	Fungus	Caterpillars (Lepidoptera)	**Unregistered:** *Bionova Ltda. (Val), *Laboratorio Biológico La Avispita (Met)
Orius insidiosus (Say)	Predator	Mites, eggs and early larvae of various insects, thrips, spider mites.	**Unregistered:** *Scientia Colombia S.A.S. (Val)
Pachycrepoideus vindemmiae (Róndani)	Parasitoid	Fruit flies (*Anastrepha* spp., *Toxotrypana curvicauda*), lance flies (*Dasiops* spp.), etc.	**Unregistered:** *Productos Biológicos Perkins Ltda. (Val)
Paenibacillus popilliae Dutky	Bacteria	Beetle grubs	**Registered:** Agrobiológicos de Colombia S.A. (Val)
Phymastichus coffea LaSalle	Parasitoid	Coffee berry borer	**Unregistered:** *Biocafé (Cal)
Phytoseiulus persimilis Athias-Henriot	Predator	Mites, thrips	**Registered:** Bichopolis (Cun)
Prorops nasuta Waterston	Parasitoid	Coffee berry borer	**Unregistered:** *Biocafé (Cal)
Purpureocillium lilacinum (Thom) Luangsaard, Houbraken, Hywel-Jones & Samson (including products sold as *Paecilomyces lilacinus*)	Fungus	Phytoparasitic nematodes, beetle grubs, ground pearls	**Registered:** Bio-crop S.A.S. (Val), Core Biotechnology S.A.S. (Val), Innovak Colombia Ltda. (Cun), Live Systems Technology S.A. (Cun), Mycros International S.A.S. (Cun), Natural Control S.A. (AntSoluciones Microbianas del Trópico Ltda. (Cal) **Unregistered:** *Agroinsumos Biológicos (Cas), *Agroproductiva S.A. (Val), *Biocontrol (Val), *Bioecológicos Ltda. (Cun), *Bioinsumos del Campo Ltda. (Met), *Biológicos y Ecológicos de Colombia (Cun), *Bionova Ltda. (Val), *Bioquirama S.A.S. (Ant), *Biotropical S.A. (Ant), *Funguscol (Val), *Laverlam S.A. (Val),), *Mycotech Co S.A.S. (Val),), *Orius Biotecnología Ltda (Met), *P.L.A. Biológicos (Tol), *Productos Biológicos Perkins Ltda. (Val), *Safer Agrobiologicos S.A.S. (Ant), Sanoplant (Val), *Sociedad laboratorio de Suelos E.C.N. Ltda. (Mag)

Spalangia cameroni Perkins	Parasitoid	Various flies (*Musca domestica*, *Stomoxys calcitrans*, *Haematobia irritans*, others)	**Unregistered:** *Productos Biológicos Perkins Ltda. (Val)
Trichoderma atroviride (Karsten) Bissett	Fungus	Various soil-borne pathogens (*Fusarium oxysporum*, *Rhizoctonia solani*)	**Registered:** Inproarroz Ltda. (Met)
Trichoderma koningiopsis (Samuels)	Fungus	Various soil-borne pathogens (*Fusarium oxysporum*, *Sclerotinia sclerotiorum*, *Rhizoctonia solani*)	**Registered:** Agrosavia (Cun)
Trichoderma viride (Pers) (including products sold as *Trichoderma lignorum*)	Fungus	Various soil-borne pathogens (*Fusarium oxysporum*, *Rhizoctonia solani, Helminthosporium oryzae, Gaeumannomyces graminis, Sarocladium oryzae*)	**Registered:** Biocontrol (Val), Biocultivos (Tol), Laverlam S.A. (Val), Yaser S.A.S. (Val)
Trichoderma spp. (mixtures of different species	Fungus	Various soil-borne and foliar pathogens (*Botrytis cinerea*, *Phytophtora* spp., *Fusarium* sp., *Rhizoctonia* spp.)	**Registered:** Natural Control S.A (Ant), Soluciones Microbianas del Trópico Ltda. (Cal)
Trichogramma atopovirilia Oatman & Platner	Parasitoid	Lepidoptera (eggs), especially *Diatraea* spp.	**Unregistered:** *BioAgro (Val), *Biodefensas Agrícolas Ltda. (Val), *Scientia Colombia S.A.S. (Val)
Trichogramma exiguum Pinto & Platner	Parasitoid	Lepidoptera (eggs), especially *Diatraea* spp. (*D. tabernella*, *D. busckella*, *D. indigenella*, *D. saccharalis*)	**Registered:** BioAgro (Val), Inbecol (Val), Productos biológicos Perkins Ltda (Val) **Unregistered:** *Arias & Arias Bioinsumos S.A.S. (Val), *Bioenergy (Met), *Biocol (Val), *Biodefensas Agrícolas Ltda. (Val), *Corporación AMA (Val), *Incauca (Cau), *Ingenio Maria Luisa (Val), *Orius Biotecnología Ltda (Met), *P.L.A. Biológicos (Tol), *Safer Agrobiologicos S.A.S. (Ant), * Sanoplant (Val), *Sociedad laboratorio de Suelos E.C.N. Ltda. (Mag)
Trichogramma pretiosum Riley	Parasitoid	Lepidopteran eggs (*Alabama* sp., *Anticarsia* sp., *Cydia* sp., *Diatraea* spp., *Elasmopalpus* sp., *Heliothis* spp., *Erinnyis* sp., *Spodoptera* spp., *Sitotroga* sp., others	**Unregistered:** *BioAgro (Val), *Biodefensas Agrícolas Ltda. (Val), *Ingenio Providencia (Val), *Scientia Colombia S.A.S. (Val)

[a]Department abbreviations: Ant: Antioquia, Boy: Boyacá, Cal: Caldas, Cas: Casanare, Cau: Cauca, Cun: Cundinamarca, Mag: Magdalena, Met: Meta, Tol: Tolima, Val: Valle del Cauca.
*Companies marked with asterisk (*) have no registry for the specific biocontrol agent. Any mention of unregistered biopesticides or beneficial insects reported in this document does not constitute a recommendation for that particular use by the authors or the authors' organizations. All applications must accord with the currently registered label for that particular biocontrol agent, crop, pest and region.

Table 8.2. Colombian companies that produce parasitoids, predators and pathogens.

Department	Company name (with location in parenthesis)
Antioquia	Bioquirama S.A.S. (Rionegro), Biotropical S.A. (Medellín), Natural control S.A. (La Ceja), Safer Agrobiológicos S.A.S. (Medellín)
Boyacá	Secretaria de Agricultura de Boyacá (Tunja)
Caldas	*Biocafé (Chinchiná), Bioprotección S.A.S. (Chinchiná), Soluciones Microbianas del Trópico Ltda. (Chinchiná)
Casanare	Agroinsumos Biológicos (Villanueva), Mycol E.A.T. (Villanueva), Sanitex S.A.S. (Villanueva)
Cauca	*Ingenio del Cauca - Incauca (El Ortigal)
Cundinamarca	Bichopolis (Tabio), Bioecológicos Ltda. (Sopo), Biológicos y Ecológicos de Colombia Ltda. (Bogotá), *Bioterra® (Bogotá), Agrosavia (Mosquera), Fundación Centro de Biotecnología "Mariano Ospina Perez" (Bogotá), Innovak Colombia Ltda. (Bogotá), Live Systems Technology S.A. (Bogotá), Mezfer de Colombia Ltda. (Bogotá), Mycros International S.A.S. (Cota), Palmar del Oriente S.A. (Bogotá), Plantador Colombia Ltda. (Facatativá), Vecol (Bogotá)
Magdalena	Sociedad Laboratorio de Suelos E.C.N. Ltda. (Santa Marta)
Meta	*Bioenergy (Villavicencio), Bioinsumos del Campo Ltda. (Villavicencio), Inproarroz Ltda. (Puerto López), Laboratorio Biológico La Avispita (Villavicencio), Orbitec S.A. (Villavicencio), Orius Biotecnología Ltda. (Villavicencio)
Tolima	Biocultivos (Ibagué), Organización Pajonales (Ibagué), P.L.A. Biológicos (Ibagué)
Valle del Cauca	*Acción Biocontrol (Palmira), Agrobiológicos de Colombia S.A. (Cali), Agroproductiva S.A. (Yumbo), *Arias & Arias Bioinsumos S.A.S. (Palmira), BioAgro (Cartago), Biocol (La Victoria), Biocontrol (Palmira), Bio-crop S.A.S. (Palmira), Biodefensas Agrícolas Ltda. (Andalucía), *Biológicos del Valle (Palmira), Bionova Ltda. (Cali), Core Biotechnology S.A.S. (Palmira), Corporación AMA (La Victoria), Fungicol (Palmira), Inbecol (Cali), *Ingenio María Luisa (Florida), *Ingenio Mayagüez (Candelaria), *Ingenio Providencia (Cali), Laverlam S.A. (Cali), Mycotech Co. S.A.S. (Cali), Productos Biológicos Perkins Ltda. (Palmira), Sanoplant (Palmira), Scientia Colombia S.A.S. (La Unión), Semillas Valle S.A. (Yumbo), Yaser S.A.S. (Cali)

*Companies marked with asterisk (*) have no registry for the specific biocontrol agent.

de Colombia, Palmira campus and Universidad del Valle) and some 25 companies that produce biopesticides and biocontrol agents, mainly to cover the demand of sugarcane plantations, various fruit crops, poultry and other livestock, and minor crops such as cotton, maize and sorghum. The second and third hotspots are in the departments of (i) Cundinamarca, with one research centre (Agrosavia, Tibaitata Research Center), four universities (Pontificia Univ. Javeriana, Univ. Jorge Tadeo Lozano, Univ. Militar Nueva Granada and Univ. Nacional de Colombia, Bogota campus) and 13 companies; and (ii) Antioquia with one research centre (Agrosavia, La Selva Research Center), two universities (Univ. Nacional de Colombia, campus Medellin and Univ. de Antioquia) and four companies. The latter two hotspots have mainly expanded to supply the demand for biocontrol agents to control pests of vegetables and ornamentals in greenhouses. The fourth hotspot is in the department of Caldas and neighbouring departments, with one research Centre (Cenicafé), one university (Univ. de Caldas) and three companies that specialize in biocontrol agents to control coffee pests, especially the coffee berry borer *H. hampei*. Apart from these four hotspots, biocontrol is commonly carried out in the departments of Magdalena, Nariño, Santander and part of Cundinamarca in large oil palm plantations (dotted-line circles in Fig. 8.1), which is promoted by Cenipalma. Research on biocontrol in research centres and institutions is partially supported by local universities, e.g., Univ. Nacional de Colombia (various campuses), Pontificia Univ. Javeriana, Univ. Jorge Tadeo Lozano, Univ. Militar Nueva Granada, Univ. del Tolima and Univ. del Valle.

Table 8.3. Overview of biological control activities in Colombia.

Plant / Animal	Pest	Natural enemy	Type of biocontrol[a] / area (ha) applied in 2016[d]	Reference
Apple	*Eriosoma lanigerum* (Hausmann)	*Aphelinus mali* (Haldeman)	CBC / ?	Altieri & Nicholls, 1999
Banana	*Pseudocercospora fijiensis* (M. Morelet) Deighton	*Bacillus subtilis* (Ehrenberg) Cohn	ABC / ?	Gutierrez-Monsalve *et al.*, 2015
Blackberry	*Botrytis cinerea* (Persoon)	*Rhodotorula glutinis* (Fresen.) F.C. Harrison	ABC / ?	Zapata *et al.*, 2013
		Trichoderma koningiopsis (Samuels)	ABC / ?	Mejía *et al.*, 2013
Cassava	*Erinnyis ello* (Linnaeus)	*Trichogramma perkinsi* Girault, *T. australicum* Girault	ABC / ?	Amaya, 1982
	Mononychellus tanajoa (Bondar)	*Chrysopa* sp., *Oligota minuta* Cameron, *Stethorus* sp., various species of predatory mites	NC / ?	Bellotti *et al.*, 2005
Chrysanthemum[b]	*Myzus persicae* (Sulzer)	*Aphidius matricariae* Haliday	ABC / ?	Nicholls *et al.*, 1998
Citrus	*Compsus viridivittatus* (Guérin-Méneville)	*Fidiobia* sp.	NC / ?	ICA, 2012
Citrus	*Diaphorina citri* Kuwayama	*Tamarixia radiata* (Waterston)	Test since 2013 /1,500	García *et al.*, 2016
Citrus	*Polyphagotarsonemus latus* (Banks)	Various species of predatory mites	NC / ?	Imbachi *et al.*, 2012
Coffee	*Hypothenemus hampei* (Ferrari)	*Cephalonomia stephanoderis* Betrem, *Phymastichus coffea* LaSalle, *Prorops nasuta* Waterston	CBC & ABC 1990–2000 / ?	Bustillo, 2006; Rivera *et al.*, 2010
		Heterospilus coffeicola Schmiedeknecht	CBC & ABC 1995–2000 / ?	Bellotti *et al.*, 2005
Coffee	*Olygonichus yothersi* (McGregor)	*Chrysoperla carnea* (Stephens)	ABC / 1,500	L.M. Constantino (Cenicafé, personal communication)
Common bean	*Acanthoscelides obtectus* (Say)	*Dinarmus basalis* (Rondani)	ABC / ?	Schmale *et al.*, 2006
Common bean, soy bean	*Anticarsia gemmatalis* Hübner, *Heliothis* spp.	*Trichogramma australicum* Girault, *T. pretiosum* Riley,	ABC / ?	Amaya, 1982
Cotton	*Alabama argillacea* (Hübner)	*Trichogramma* sp.	ABC / ?	Smith & Bellotti, 1996
Cotton	*Aphis gossypii* Glover	*Hippodamia convergens* Guérin-Méneville	CBC / ?	Smith & Bellotti, 1996
Cotton	*Anthonomus grandis* Boheman	*Bracon kirkpatricki* (Wilkinson)	CBC / ?	Smith & Bellotti, 1996
		Catolaccus grandis (Burks)	ABC / ?	García & Sánchez, 1995; Smith & Bellotti, 1996

Continued

Table 8.3. Continued.

Plant / Animal	Pest	Natural enemy	Type of biocontrol[a] / area (ha) applied in 2016[d]	Reference
Cotton	*Athelia rolfsii* (Curzi) C.C. Tu & Kimbr. (=*Sclerotium rolfsii*) and *Rhizoctonia solani* J.G. Kühn	*Trichoderma* spp.	ABC / ?	Hoyos *et al.*, 2008
Cotton	*Bemisia tabaci* (Gennadius)	*Lecanicillium lecanii* R. Zare & W. Gams	ABC / ?	Gómez *et al.*, 2012
Cotton	*Heliothis virescens* (Fabricius)	*Apanteles* sp., *Telenomus* sp.	ABC,1980s–1990s /?	Bellotti *et al.*, 2005
	H. virescens and other *Heliothis* spp.	*Trichogramma* sp	ABC,1980s–1990s /?	Bellotti *et al.*, 2005; Smith & Bellotti, 1996
Cotton	*Sacadodes pyralis* Dyar	*Telenomus* sp.	Experimental / ?	Smith & Bellotti, 1996
		Apanteles thurberiae Muesebeck	Native, released to reestablish NC / ?	
Cotton	*Spodoptera* spp.	*Telenomus remus* Nixon	ABC, 1993 / ?	Smith & Bellotti, 1996
Eggplant	*B. tabaci*	*L. lecanii*	ABC / ?	Gómez *et al.*, 2012
Flowers	Several species of aphids	*Chrysoperla carnea* Stephens	Corridors to prevent pest arrival / 1,000	Arias & Arias Bioinsumos[d]
Flowers, vegs[b]	Several species of aphids	*Aphidoletes aphidimyza* (Rondani)	ABC / ?	Nicholls *et al.*, 1998
	Several species of aphids and thrips	*C. carnea*	ABC / 2,000	Arias & Arias Bioinsumos[d]
Flowers, vegs[b]	*Frankliniella occidentalis* Pergande	*Hypoaspis miles* Berlese, *Iphiseius degenerans* (Berlese), *Neoseiulus cucumeris* (Oudemans), *Orius insidiosus* (Say), *Orius laevigatus* (Fieber), *Orius tristicolor* (White),	ABC / ?	Andrade *et al.*, 1989; Nicholls *et al.*, 1998
Flowers, vegs[b]	*Phytonemus pallidus* (Banks)	*Neoseiulus cucumeris* (Oudemans)	ABC / ?	Andrade *et al.*, 1989; Nicholls *et al.*, 1998
Flowers, vegs[b]	*Polyphagotarsonemus latus* (Banks)	*Neoseiulus barkeri* Hughes	ABC / ?	Andrade *et al.*, 1989; Nicholls *et al.*, 1998
Flowers, vegs[b]	*Tetranychus urticae* Koch	*Phytoseiulus persimilis* Athias-Henriot	ABC / 5,000	Andrade *et al.*, 1989; Nicholls *et al.*, 1998; Arias & Arias Bioinsumos[d]
Flowers, vegs[b]	*Trialeurodes vaporariorum* Westwood	*Encarsia formosa* Gahan	ABC / 500	Nicholls *et al.*, 1998
Flowers, vegs[b]	Several lepidopteran species	*Trichogramma* spp.	Corridors to prevent pest arrival / 500	Arias & Arias Bioinsumos[d]

Fruit crops, fabaceous street trees, palms	*Crypticerya multicicatrices* Kondo & Unruh	*Anovia punica* Gordon	FBC & CBC / 4,000	Kondo *et al*., 2014, 2016
Fruit crops	Lepidopteran complex: *Agraulis* sp., *Stenoma* sp., *Jocara* sp., *Oxydia* sp., *Cerconota annonella* Sepp.	*Trichogramma* spp.	ABC / 10,000	Arias & Arias Bioinsumos[d], Productos Biológicos Perkins
Fruit crops	Several aphid species	*C. carnea*	ABC / 35,000	Arias & Arias Bioinsumos[d]
Fruit crops	Several mite species, thrips species	*Phytoseiulus persimilis* Athias-Henriot	ABC / 1,500	Arias & Arias Bioinsumos[d]
Gypsophila paniculata L.	*Liriomyza huidobrensis* (Blanch.)	*Diglyphus begini* (Ashmead)	ABC / ?	Cure & Cantor, 2003; Nicholls *et al*., 1998
Hot peppers (*Capsicum* spp.) and vegs	*A. gossypii*, *M. persicae*	*C. carnea*	ABC / 2,000	BioCol (Biológicos de Colombia), Arias & Arias Bioinsumos[d]
Human health	*Aedes aegypti* (Linnaeus)	*Mesocyclops longisetus* (Thiébaud)	Experimental (1999–2000) / ?	Suarez-Rubio & Suarez, 2004
Lettuce	*Sclerotinia sclerotiorum* (Lib.) de Bary and *Sclerotinia minor* (Jagger)	*Trichoderma koningiopsis* (Samuels)	ABC / ?	Moreno *et al*., 2010
Maize	*Spodoptera frugiperda* (J.E. Smith)	*Baculovirus* (SfMNPV)	ABC / ?	Gómez *et al*., 2013; Villamizar *et al*., 2009, 2010, 2012a
		T. remus	ABC / ?	Yaseen *et al*., 1982
Maize, sorghum	*Spodoptera* spp.	*T. remus*	ABC / ?	Bennett & Street, 1984
Maize, sorghum, sugarcane	*Diatraea saccharalis* (Fabricius)	*Trichogramma perkinsi* Girault	ABC / ?	Amaya, 1982
Melon	*B. tabaci*	*L. lecanii*	ABC / ?	Cotes *et al*., 2009
Oil palm	*Cyparissius daedalus* Cramer	*Ooencyrtus* sp.	ABC / ?	Aldana *et al*., 2004

Continued

Table 8.3. Continued.

Plant / Animal	Pest	Natural enemy	Type of biocontrol[a] / area (ha) applied in 2016[d]	Reference
Oil palm	*Leptopharsa gibbicarina* Froesch.	*C. carnea*, with some *Chrysoperla rufilabris* (Burmeister)	ABC / ?	Arias & Arias Bioinsumos[d], Productos Biológicos Perkins Ltda (personal communication)
		Crematogaster spp.	ConsBC / ?	Barrios-Trilleras *et al.*, 2015
Oil palm	*Loxotoma elegans* Zeller	*Trichogramma* spp.	ABC / 50,000	Aldana *et al.*, 2005; Arias & Arias Bioinsumos[d]
Oil palm	*Stenoma impressella* Busck	*Trichogramma pretiosum*	ABC / ?	Castillo *et al.*, 2000
Oil Palm	*Sagalassa valida* Walker	*Heterorhabditis* sp.	ABC / ?	Ortiz *et al.*, 1994
Passion fruit	*Dasiops inedulis* Steyskal	*Pachycrepoideus vindemmiae* Rondani, *Spalangia* sp.	Component in IPM / ?	Quintero *et al.*, 2012
Pinus patula & *Cupressus lusitanica*	*Oxydia trychiata* Guenée	*Telenomus alsophilae* Viereck	ABC / ?	Bustillo & Drooz, 1977; Drooz *et al.*, 1977
Pinus spp.	*Pineus boerneri* Annand	*Ceraeochrysa* sp.	ABC / ?	Rodas *et al.*, 2014
Potato	*Premnotrypes vorax* (Hustache)	*Beauveria bassiana* (Bals.-Criv.) Vuill.	ABC / ?	Cotes *et al.*, 2004
Potato	*Rhizoctonia solani* J.G. Kühn	*Trichoderma koningiopsis* (Samuels)	ABC / ?	Beltrán *et al.*, 2012
Potato	*Tecia solanivora* (Povolný)	*Baculovirus* (PhopGV)	ABC / ?	Espinel *et al.*, 2012; Villamizar *et al.*, 2012b
Poultry, livestock	*Musca* spp.	*Pachycrepoideus* sp.	ABC / ?	Bueno & van Lenteren, 2002
Rice	Acari and soft body insects (aphids, scale insects, coccids)	*C. carnea* mixed with a low proportion of *C. rufilabris*	ABC / 5,000	Arias & Arias Bioinsumos[d]
Rice	*Diatraea saccharalis* (Fabricius)	*Cotesia flavipes* (Cameron)	ABC / 5,000	Arias & Arias Bioinsumos[d]
		Trichogramma exiguum Pinto & Platner	ABC / 38,000	Federación Nacional de Arroceros, 2010
Sugarcane	*Diatraea indigenella* Dyar	*Billaea claripalpis* (Wulp)	ABC / ?	Altieri & Nicholls, 1999; Vargas *et al.*, 2015
		Lydella minense (Townsend)	ABC / ?	Bueno & van Lenteren, 2002; Vargas *et al.*, 2015

Sugarcane	*D. saccharalis*	*B. claripalpis*	ABC / 56,896[c]	Bueno & van Lenteren, 2002; Vargas *et al.*, 2015
		Lixophaga diatraeae (Towns.)	ABC / ?	Bennett & Street, 1984
		L. minense[c]	ABC / ?	Bueno & van Lenteren, 2002; Vargas *et al.*, 2015
		Trichogramma sp.	ABC / ?	Bueno & van Lenteren, 2002
Sugarcane	*Diatraea* spp.	*C. flavipes*	ABC / 86,000	Arias & Arias Bioinsumos[d]; Smith & Bellotti, 1996
		Genea jaynesi (Aldrich)	ABC / ?	Smith & Bellotti, 1996
		Trichogramma exiguum Pinto & Platner	ABC / 98,000	Arias & Arias Bioinsumos[d]; Vargas *et al.*, 2015
		Trichogramma minutum Riley	ABC / ?	Bueno & van Lenteren, 2002; Gómez, 1995
Sugarcane	*Sipha flava* Forbes	*C. carnea* mixed with *Ch. rufilabris*	ABC / 5,000	Arias & Arias Bioinsumos[d], Biocol (Biológicos de Colombia)
Tomato[b]	*Fusarium oxysporum* f. sp. *lycopersici* W.C. Snyder & H.N. Hansen, *Rhizoctonia solani* J.G. Kühn	*T. koningiopsis*	ABC / ?	Cotes *et al.*, 2001; Moreno *et al.*, 2009
Tomato[b]	*Trialeurodes vaporariorum* Westwood	*Amitus fuscipennis* (MacGown & Nebeker)	ABC, used with *E. formosa* / ?	De Vis *et al.*, 1999; De Vis & Lenteren, 2008
		Encarsia formosa Gahan	ABC / ?	De Vis & Lenteren, 2008
		L. lecanii	ABC / ?	Cotes *et al.*, 2009
Tomato	*Tuta absoluta* (Meyrick)	*Apanteles gelechiidivoris* Marsh	ABC / ?	Bajonero *et al.*, 2008; Morales *et al.*, 2013
		T. exiguum	ABC / ?	García, 1996; Navarro, 1988
		Baculovirus (PhopGV)	ABC / ?	Gómez *et al.*, 2014
	T. absoluta and other lepidopteran pests	*Trichogramma* spp.	ABC / 1,000	Bajonero *et al.*, 2008; Arias & Arias Bioinsumos[d]
Tomato	*Scrobipalpupa* sp.	*T. australicum*	ABC / ?	Amaya (1982)

[a]Type of biocontrol: ABC = augmentative, CBC = classical, FBC = fortuitous, ConsBC = conservation biocontrol; NC = natural control
[b]Grown in greenhouses
[c]*Billaea claripalpis* and *Lydella minense* are being released currently by two sugarcane mills (Y. Gutierrez, Incauca, Colombia, personal communication)
[d]Personal communication of Marino Arias, Arias & Arias Bioinsumos company, Palmira, Colombia

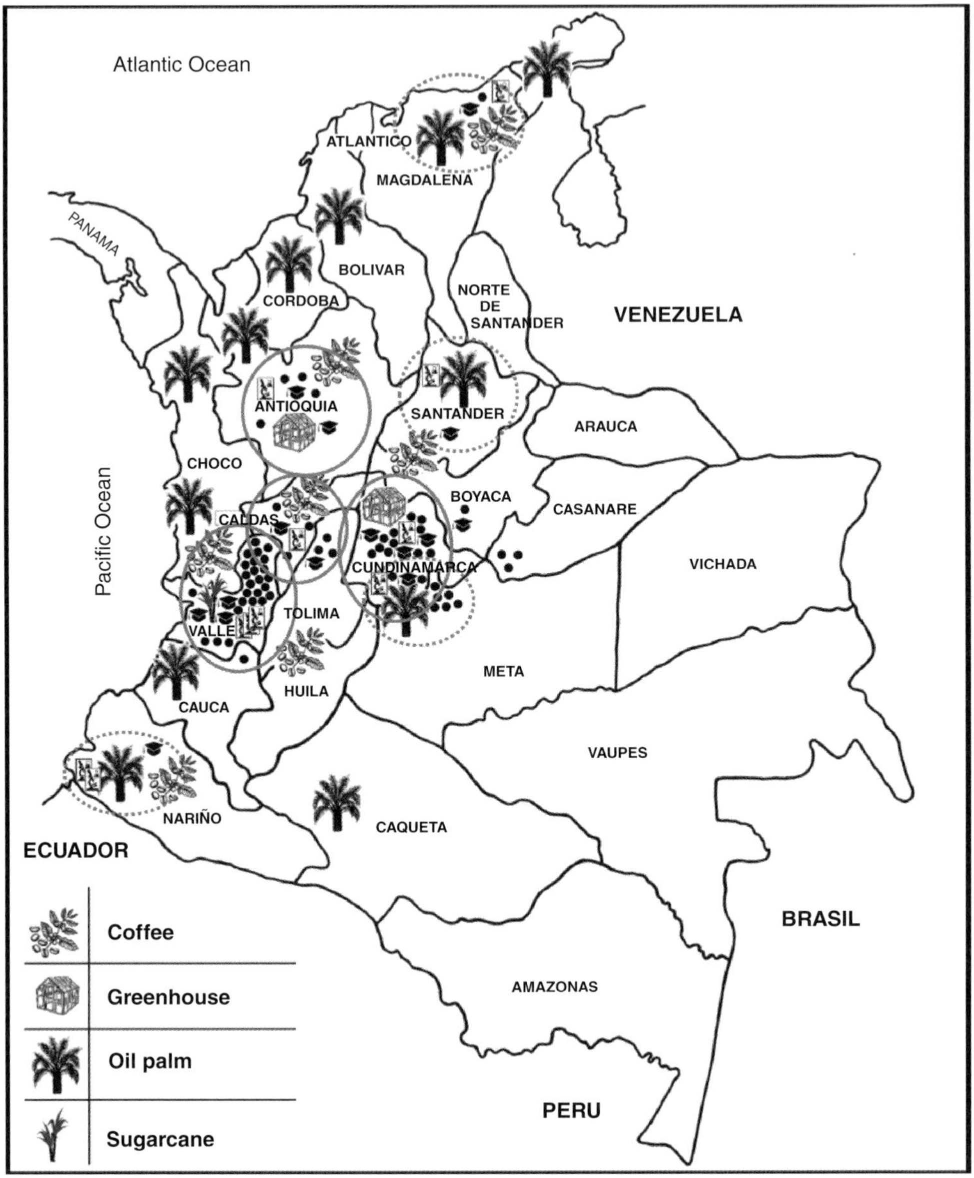

Fig. 8.1. Biological control hotspots in Colombia. Companies producing biocontrol agents are represented by dark dots. Research institutes associated with development and promotion of biocontrol are represented by a small box with a microscope symbol, and universities are represented by a square academic cap symbol. Full circles indicate the hotspots of biological control in Colombia. Dotted circles indicate oil palm growing areas in which biocontrol is frequently used.

8.5 New Developments of Biological Control in Colombia

Two rather new developments in Colombia deserve to be mentioned here.

1. The release of nematodes to control the oil palm root borer *Sagalassa valida* Walker in large areas planted with oil palm has stimulated research on the production of the entomopathogenic nematode *Heterorhabditis* sp. in the area of

Tumaco, department of Nariño. The newly developed mass production technique of symbiotic bacteria in a monoxenic culture is an important step to produce nematodes in liquid-culture bioreactors, as this is much more efficient than the currently used solid-media method (Moreno-Salguero *et al.*, 2014).

2. The second new development relates to biocontrol of plant diseases. Agrosavia developed and licensed to a Brazilian company an innovative formulation technology based on water-dispersible granules using a special isolate of *Trichoderma asperellum* Samuels, Lieckf. & Nirenberg. This successfully developed product, called Quality WG, was the first biofungicide registered in Brazil and is recommended for the control of *Fusarium* spp., *R. solani* and *S. sclerotiorum* in different crops. Recently, a Canadian biocontrol company acquired the Brazilian one and currently this formulation technology is massively applied since this company became the first worldwide *Trichoderma* producer.

Another new development concerns regulation of biocontrol in Colombia: a document developed by ICA demands the release of the biocontrol agents *B. claripalpis*, *C. flavipes*, *L. minense* and *T. exiguum*, and the implementation of a conservation biocontrol strategy through habitat manipulation, for the management of the *Diatraea* spp. complex in sugarcane plantations.

Biocontrol in Colombia was probably at its peak in the 1980s and 1990s, when various natural enemies were imported from abroad for augmentative biocontrol of invasive species affecting coffee, cotton, maize, sorghum and sugarcane. However, classical biocontrol is currently very little used, due to the complicated legislation in Colombia that acts as an impediment for import and use of exotic natural enemies (Bueno, 2005; B. Lohr, Agrosavia, Palmira, 2018, personal communication). In general, application of the Nagoya Protocol concerning Access and Benefit Sharing principles of the Convention on Biological Diversity (CBD) has made it very difficult or impossible to collect and export natural enemies for biocontrol (van Lenteren, 2012). Furthermore, there appears to be a deeply rooted culture of relying on chemical control among Colombian farmers (Wyckhuys *et al.*, 2010). For example, although biocontrol of tomato pests using the egg parasitoid *Trichogramma* sp. had a high impact in the 1980s in the Valle del Cauca region (García, 1985), recent surveys in tomato production areas showed that farmers do not incorporate biocontrol strategies (Hernandez *et al.*, 2015), despite natural enemies being commercially available in the area. Another reason for the lack of use of biocontrol horticulture might be lack of a research centre in the region, a situation quite different for crops that have research centres nearby, such as Cenicafé (coffee), Cenipalma (oil palm), Cenicaña (sugarcane) and the International Center of Tropical Agriculture (CIAT) (cassava and beans), with a high impact on biocontrol activities.

Still, Colombia is one of the pioneer countries in Central and South America in the area of biocontrol. From small insectaries for the mass rearing of parasitoids for controlling sugarcane borers about 40 years ago, today Colombia has become a leader in the region in the production of natural enemies and biopesticides, and in production systems such as cassava, citrus, coffee, cotton, sorghum and maize, forestry, greenhouse crops, human health, oil palm, potato, poultry and livestock, and sugarcane. There remains, however, one major concern, which is related to the quality of mass-produced biocontrol agents. Despite the success of the biocontrol industry in Colombia, many products are being sold unregistered and unregulated, and there exists a relatively large 'grey market' of unregistered, often ineffective products. The poor quality and unreliability of some unregistered products produced by new and inexperienced producers is a long-standing issue, which has resulted in a negative perception by farmers regarding biocontrol and, as a result, farmer demand. Undoubtedly, the registration process (cost and length of time) is the main reason for the use of unregistered products. Added to this is that several entrepreneurs have set up production units without any quality consciousness or the necessary facilities needed to produce quality products. Only a few of the entrepreneurs producing non-registered biopesticides commercialize quality products following regulatory procedures and contribute significantly to proper extension down to the end user. We hope that the quality control issue will be addressed in the near future, resulting in a brighter future for biocontrol in Colombia.

8.6 Acknowledgements

Many thanks to E. Torrado (Ethovision, Univ. Nac., Bogotá), E. and M. Arias (Arias & Arias Bioinsumos S.A.S.), J. Jiménez (Productos Biológicos Perkins Ltda.), L.N. Arenas (Bioagro), M.E. Cuellar (Unidad de Entomología, Laboratorio de Salud Pública Departamental del Valle del Cauca), R. Aldana (Cenipalma), O.A. Orjuela (Agroproductiva S.A.), Y. Martinez (Bichopolis), L.M. Constantino (Cenicafé), G. Vargas (Cenicaña) and Y. Gutierrez (Incauca) for providing information on the production and use of natural enemies; to B. L. Lohr (Agrosavia, Palmira Research Center) for providing information on the difficulties of importing natural enemies for biocontrol purposes. Special thanks to P.J. Gullan (The Australian National Univ.) for checking the English text, reviewing the content of the chapter and for useful comments.

References

(References with grey shading are available as supplementary electronic material)

[*] All chapters published in Cotes, A.M. (ed.) *Control biológico de fitopatógenos, insectos y ácaros. Volumes 1 and 2. Agentes de control biológico*. Editorial Agrosavia, Bogotá, Colombia, can be found at: http://hdl.handle.net/20.500.12324/33829 (volume 1) and http://hdl.handle.net/20.500.12324/33519 (volume 2).

Aguilar, M.L., Cantor, F., Cure, J.R., Rodríguez, D., Bajonero, J., Pérez, D. and Riaño, D. (2010) Integración del conocimiento y tecnología de la polinización y control biológico. [Integration of knowledge and technology of pollination and biological control]. Producción sostenible de tomate. Universidad Militar Nueva Granada, Bogota, Colombia.

Aldana, J.A., Calvache, H.H. and Arias, D.A. (2000) Programa comercial de manejo de *Leptopharsa gibbicarina* Froeschner (Hemiptera: Tingidae) con la hormiga *Crematogaster* spp. en una plantación de palma de aceite. [Commercial management program of *Leptopharsa gibbicarina* with the *Crematogaster* spp. ant in an oil palm plantation]. *Revista Palmas* 21 (Especial), 167–173.

Aldana, R.C., Calvache, H. and Higuera, O. (2004) Manejo integrado de *Cyparissius daedalus* (Cramer) en los Llanos Orientales. [Integrated Management of *Cyparissius daedalus* in the Llanos Orientales]. *Revista Palmas* 25 (Especial), 249–258.

Aldana, R.C., Aldana, J.A., Calvache, H. and Franco, P.N. (2005) *Plagas de la palma de aceite en Colombia*. [*Oil palm pests in Colombia*], 3a Edición. Cenipalma, Bogota, Colombia, pp. 65–68.

Aldana, R.C., Aldana, J.A., Calvache, H. and Franco, P.N. (2010) *Plagas de la palma de aceite en Colombia* [*Pests of the oil palm in Colombia*], 4a Edición. Cenipalma, Bogota, Colombia.

Altieri, M.A. and Nicholls, C.I. (1999) Classical biological control in Latin America. Past, present, and future. In: Bellows, T.S. and Fisher, T.W. (eds) *Handbook of Biological Control. Principles and Applications of Biological Control*. Academic Press, Cambridge, Massachusetts, pp. 975–991.

Amaya, N.M. (1982) Investigación, utilización y resultados obtenidos en diferentes cultivos con el uso de *Trichogramma* en Colombia Sur América. [Research, use and results obtained in different crops with the use of *Trichogramma* in Colombia South America]. In: *Les Trichogrammes. 1er Simposium Internacional*, Institute National de la Reserche Agronomique. Antibes, France, pp. 131–137.

Andrade, M., Briseño, J.A., Muñoz, P. and Jiménez, J. (1989). Búsqueda y reconocimiento de los enemigos naturales de las principales plagas en flores bajo invernadero en la sabana de Bogotá. [Search and recognition of natural enemies of the main pests in greenhouse flowers in the Bogotá savanna]. *Acta Biológica Colombiana* 1(5), 45–57.

Arango, R., Díaz-Trujillo, C., Dhillon, B., Aerts, A., Carlier, J., Crane, C.F. et al. (2016) Combating a global threat to a clonal crop: banana black *Sigatoka* pathogen *Pseudocercospora fijiensis* (synonym *Mycosphaerella fijiensis*) genomes reveal clues for disease control. PLoS Genetics 12(8), e1005876. DOI: 10.1371/journal.pgen.1005876

Aristizabal, A.L.F., Salazar, E.H.M., Mejía, M.C.G. and Bustillo, P.A.E. (2004) Introduction and evaluation of *Phymastichus coffea* (Hymenoptera: Eulophidae) in smallholder coffee farms, through participatory research. *Revista Colombiana de Entomologia* 30, 219–224.

Armbrecht, I. and Gallego, M.C. (2007). Testing ant predation on the coffee berry borer in shaded and sun coffee plantations in Colombia. *Entomologia Experimentalis et Applicata* 124, 261–267.

Asocolflores (Asociación Colombiana de Exportadores de Flores), SAC (Sociedad de agricultores de Colombia) and MAVDT (Ministerio de ambiente, vivienda y desarrollo territorial) (2006) Lineamientos y recomendaciones para el programa social y ambiental flor verde [Guidelines and recommendations for flowers]. Available at: http://www.minambiente.gob.co/images/AsuntosambientalesySectorialyUrbana/pdf/Evaluaci%C3%B3n_Ambiental_Estrategica/Lineamientos__recomendaciones_flor_verde.pdf (accessed 7 January 2018).

Bajonero, J., Córdoba, N., Cantor, F., Rodríguez, D. and Cure, J.R. (2008) Biología y ciclo reproductivo de *Apanteles gelechiidivoris* (Hymenoptera: Braconidae), parasitoide de *Tuta absoluta* (Lepidoptera: Gelechiidae). [Biology and life cycle of *Apanteles gelechiidivoris* parasitoid of *Tuta absoluta*]. *Agronomía Colombiana* 26, 417–426.

Barrera, G.P., Ghiringhelli, P.D., Mosher, S., Caro-Quintero, A., Massart, S. and Belaich, M.N. (2018) Omics in biological control. In: Cotes, A.M. (ed.) *Control biológico de fitopatógenos, insectos y ácaros. Volumen 2. Agentes de control biológico*. Editorial Agrosavia, Bogotá, Colombian, pp. 950–987.[*]

Barrios-Trilleras, C.E., Cuchimba-Triana, M.S. and Bustillo-Pardey, A.E. (2015) Parámetros poblacionales de *Leptopharsa gibbicarina* (Hemiptera: Tingidae) plaga de la palma de aceite. [Population parameters of *Leptopharsa gibbicarina* pest of oil palm]. *Revista Colombiana de Entomologia* 41, 1–5.

Bedoya, R. (2007) Manejo integrado de la mosca de los establos (*Stomoxys calcitrans*) en el Palmar del Oriente S.A. [Integrated management of the stable fly (*Stomoxys calcitrans*) in the Palmar del Oriente S.A.]. *Revista Palmas* 28, 383–388.

Bellotti, A.C., Cardona, C. and Lapointe, S.L. (1990) Trends in pesticide use in Colombia and Brazil. *Journal of Agricultural Entomology* 7, 191–201.

Bellotti, A.C., Melo, E.L., Arias, B., Herrera, C.J., Hernández, M.P., Holguín, C.M. Guerrero, J.M. and Trujillo, H. (2005) Biological control in the Neotropics: A selective review with emphasis on cassava. In: Hoddle, M.S. (ed.) *Second International Symposium on Biological Control of Arthropods*. Davos, Switzerland, pp. 206–227.

Beltrán, C., Moreno, C.A., Blanco, P., Villamizar, L. and Cotes, A.M. (2012) Biological control of *Rhizoctonia solani* and growth promotion activity of *Trichoderma koningiopsis* Th003 and *Trichoderma asperellum* Th034 formulations in potato (*Solanum tuberosum*). *IOBC WPRS Bulletin* 78, 223–227.

Bennett, D. and Street, G. (1984) The Commonwealth Institute of Biological Control in Integrated Pest Management Programs in Latin America. In: Allen, G. and Rada, A. (eds) *Proceedings of the International Symposium: The role of biological control in pest management*. IOBC/WPRS, Santiago, Chile. Ottawa University Press, Ottawa, Canada, pp. 41–53.

Bernal, M.G., Bustillo, A.E., and Posada, F.J. (1994) Virulencia de aislamientos de *Metarhizium anisopliae* y su eficacia en campo sobre *Hypothenemus hampei*. [Virulence of isolates of *Metarhizium anisopliae* and their field efficacy on *Hypothenemus hampei*]. *Revista Colombiana de Entomología* 20, 225–228.

Bettiol, W., Ghini, R., Morandi, M., Stadnik, M., Krauss, U., Stefanova, M. and Cotes, A.M. (2008) Controle biológico de doencas de plantas na América Latina. [Biological control of plant diseases in Latin America]. In: Batista, S. and Biaggioni, R. (eds) *Controle Microbiano de pragas na América Latina*. Fundação de Estudos Agrários Luiz de Queiroz- FEALQ, Piracicaba, Brazil, pp. 303–331.

Biocol (2017) Laboratorios Biocol, La Victoria, Valle, Colombia: Products. Available at: http://www.laboratoriosbiocol.com/?products=cotesia-flavipes-hymenoptera-braconidae (accessed 11 January 2018).

Bueno, V.H.P. (2005) Implementation of biological control in greenhouses in Latin America: how far are we? In: *2nd International symposium on Biological control of arthropods*. USDA Forest Service Publication FHTET-2005-08, USA, pp. 531–537.

Bueno, V.H.P. and van Lenteren, J.C. (2002) The popularity of augmentative biological control in Latin America: History and state of affairs. In: *Proceedings of the First International Symposium on Biological Control of Arthropods*, IOBC, Honolulu, USA, pp. 180–184.

Bustillo, A.E. (2006) Una revisión sobre la broca del café, *Hypothenemus hampei* (Coleoptera: Curculionidae: Scolytinae), en Colombia [A review of the coffee berry borer, *Hypothenemus hampei*, in Colombia]. *Revista Colombiana de Entomología* 32, 101–116.

Bustillo, A.E. (2007) Control biológico de la broca del café, *Hypothenemus hampei* (Ferrari) en Colombia. [Biological control of the coffee berry borer, *Hypothenemus hampei* in Colombia]. In: *Memorias XX Congreso Venezolano de Entomología*, July 22–26, 2007. Universidad Nacional Experimental del Táchira, San Cristóbal, Venezuela, pp. 11–17.

Bustillo, A.E. (2013) Insectos plaga y organismos benéficos del cultivo de la caña de azúcar en Colombia. [Insect pests and beneficial organisms of sugarcane crop in Colombia]. Cenicaña, Cali, Valle del Cauca, Colombia. Available at: https://repository.agrosavia.co/handle/20.500.12324/12701 (accessed 29 October 2019).

Bustillo, A.E. (2014) Nemátodos entomopatógenos y sus posibilidades para el control de plagas de la palma de aceite. [Entomopathogenic nematodes and their possibilities to control oil palm insect pests]. *Revista Palmas* 35, 53–58.

Bustillo, A.E. and Drooz, A. (1977) Cooperative Establishment of a Virginia (USA) Strain of *Telenomus alsopbilae* on *Oxydia trychiata* in Colombia. *Journal of Economic Entomology* 70, 767–770.

Bustillo, A.E., Villalba, G.D., Orozco, H.J., Benavides, M.P., Reyes, A.I.C. and Chaves, C.B. (1995) Biological control of pests in coffee in Colombia. In: *Proceedings of the Seizieme Colloque Scientifique International sur le Cafe*, 9–14 April, 1995. Kyoto, Japan, pp. 671–680.

Bustillo, A.E., Cardenas, E., Villalva, G.D.A., Benavides, P., Orozco, J. and Posada F.J. (1998) Manejo Integrado de la Broca del Café *Hypothenemus hampei* (Ferrari) en Colombia. [Integrated Management of the Coffee Berry Borer *Hypothenemus hampei* in Colombia]. CENICAFE, Chinchiná, Colombia. Available at: http://biblioteca.cenicafe.org/handle/10778/848 (accessed 11 January 2018).

Bustillo-Pardey, A.E. (2018) Parasitoides de origen africano para el control de la broca del café, *Hypothenemus* hampei (Ferrari) en Colombia [Parasitoids of African origin for the control of the coffee berry borer, *Hypothenemus hampei* in Colombia]. In: Cotes, A.M. (ed.) *Control biológico de fitopatógenos, insectos y ácaros. Volumen 1. Agentes de control biológico*. Editorial Agrosavia, Bogotá, Colombia, pp. 520–528. [*]

Cadena, G. (2005) Desarrollos científicos de Cenicafé en la última década [Scientific developments of Cenicafé in the last decade]. *Revista Academia Colombiana de Ciencias* 29, 89–99.

Cantor, F., Rodriguez, D. and Cure, J.R. (2011) Manejo de mosca blanca de los invernaderos *Trialeurodes vaporariorum* y la cría de su biocontrolador *Encarsia formosa* [Management of the greenhouse whitefly *Trialeurodes vaporariorum* and the breeding of its biocontrol agent *Encarsia formosa*]. Universidad Militar Nueva Granada. Colombia.

Carabalí, A. (2018) *Fidiobia* sp., parasitoide del picudo de los cítricos, *Compsus viridivittatus* (Coleoptera: Curculionidae) [*Fidiobia* sp., parasitoid of the citrus weevil, *Compsus viridivittatus*]. In: Cotes, A.M. (ed.) *Control biológico de fitopatógenos, insectos y ácaros. Volumen 1. Agentes de control biológico*. Editorial Agrosavia, Bogotá, Colombia, pp. 509–512.[*]

Cardona, C., Rodriguez, I., Bueno, J.M. and Tapia, X. (2005) Biología y manejo de la mosca blanca *Trialeurodes vaporariorum* en habichuela y fríjol [Biology and management of the whitefly *Trialeurodes vaporariorum* on snap and dry beans]. Centro Internacional de Agricultura Tropical (CIAT), Cali, Colombia.

Carrillo, D., Serrano, M.S. and Torrado-Leon, E. (2006) Effect of nectar producing on the reproduction of *Diadegma* aff *insulare* Cresson (Hymenoptera: Ichneumonidae), parasitoid of *Plutella xylostella* L. (Lepidoptera: Plutellidae). *Revista Colombiana de Entomología* 32, 18–23.

Castillo, J., Aldana, J., Calvache, H. and Grijalva, O. (2000) Evaluación de técnicas de liberación de *Trichogramma pretiosum* Riley para el manejo de *Stenoma cecropia* Meyrick [Evaluation of liberation techniques of *Trichogramma pretiosum* for the management of *Stenoma cecropia* in the oil palm crop]. *Palmas (Colombia)* 21, 203–211.

Cetron, M. (2016) Revision to CDC's Zika travel notices: minimal likelihood for mosquito-borne Zika virus transmission at elevations above 2,000 meters. *MMWR (Morbidity and Mortality Weekly) Report* 65.

Chacón, P. and Rojas, M. (1984) Entomofauna asociada a *Pasiflora mollisima*, *P. edulis* f. *flavicarpa* y *P. quadrangularis* en el departamento del Valle del Cauca [Entomofauna associated with *Pasiflora mollisima*, *P. edulis* f. *flavicarpa* and *P. quadrangularis* in the department of Valle del Cauca]. *Turrialba* 34, 297–311.

Chet, I. and Baker, R. (1981) Isolation and biocontrol potential of *Trichoderma hamatum* from soil naturally suppressive to *Rhizoctonia solani*. *Phytopathology* 71, 286–290.

CIA (Central Intelligence Agency) (2017) The World Factbook: Colombia. Available at: https://www.cia.gov/library/publications/the-world-factbook/geos/co.html (accessed 10 January 2018).

Cotes, A.M. (2011) Registry and regulation of biocontrol agents on food commodities in South America. Proc. IS on Biological control of postharvest diseases. *Acta Horticulturae* 905, 301–306.

Cotes, A.M. (2014) Control biológico de enfermedades de plantas en Colombia [Biological control of plant diseases in Colombia]. In: Bettiol, W., Rivera, M., Mondino, P., Montealegre, J. and Colmenárez, Y. (eds) *Control Biológico de enfermedades de plantas en América Latina y el Caribe*. Facultad de Agronomía. Montevideo. Uruguay, pp. 169–179.

Cotes, A.M. and Elad, Y. (2018) El Control Biológico en un contexto de Manejo Integrado de enfermedades [Biological control in the context of integrated plant diseases management]. In: Cotes, A.M. (ed.) *Control biológico de fitopatógenos, insectos y ácaros. Volumen 2. Agentes de control biológico*. Editorial Agrosavia, Bogotá, Colombia, pp. 794–821.[*]

Cotes, A.M., Cárdenas, A. and Pinzón, H. (2001) Effect of seed priming in the presence of *Trichoderma koningii* on seed and seedling disease induced in tomato by *Rhizoctonia solani* and *Fusarium oxysporum* f.sp. *lycopersici*. *IOBC WPRS Bulletin* 24, 259–263.

Cotes, A.M., Gil, J., Espinel, C., Villamizar, A.M., Gómez, M., Zuluaga, M.V., Torres, L. and Lopez-Ávila, A. (2004) Desarrollo de un insecticida microbiano para el control biológico del gusano blanco de la papa [Development of a microbial insecticide for the biological control of the potato white worm]. *Editorial Arte y Fotolito*, Bogotá, Colombia.

Cotes, A.M., Espinel, C., Villamizar, L., Garzón, I., Lozano, M. and López-Ávila, A. (2009) Biopesticide based on *Lecanicillium lecanii* developed for whiteflies control. *IOBC WPRS Bulletin* 45, 263–266.

Cotes, A.M., Fargetton, X. and Köhl, J. (2018a) Diseño conceptual, selección y prueba de concepto de microorganismos biocontroladores [Conceptual design, selection and proof of concept of biological control microorganisms]. In: Cotes, A.M. (ed.) *Control biológico de fitopatógenos, insectos y ácaros. Volumen 2. Agentes de control biológico*. Editorial Agrosavia, Bogotá, Colombia, pp. 594–627.[*]

Cotes, A.M., Mosher, S., Barrera, G., Kobayashi, S. and Elad, Y. (2018b) Nuevas estrategias para el control biológico de fitopatógenos [Novel strategies for plant pathogens biological control]. In: Cotes, A.M. (ed.) *Control biológico de fitopatógenos, insectos y ácaros. Volumen 2. Agentes de control biológico*. Editorial Agrosavia, Bogotá, Colombia, pp. 878–921.[*]

Cotes, A.M., Zapata, Y., Beltrán, C., Kobayashi, S., Uribe, L. and Elad, Y. (2018c) Control Biológico de patógenos foliares [Biological control of foliar pathogens]. In: Cotes, A.M. (ed.) *Control biológico de fitopatógenos, insectos y ácaros. Volumen 1. Agentes de control biológico*. Editorial Agrosavia, Bogotá, Colombia, pp. 58–143.[*]

Cuartas, P.E., Villamizar, L.F., Barrera, G.P, Ruiz, J.C., Campos, J.C., León-Martínez, G., and Gómez-Valderrama, J. (2018) Novel biopesticide based on *Erinnyis ello* betabaculovirus: characterization and preliminary field evaluation to control *Erinnyis ello* in rubber plantations. *Pest Management Science* 75(5), 1391–1399. DOI: 10.1002/ps.5259.

Cure, J.R. and Cantor, F. (2003) Atividade predadora e parasítica de *Digliphus begini* (Ashm.) (Hymenoptera: Eulophidae) sobre *Liriomyza huidobrensis* (Blanch.) (Diptera: Agromyziidae) em cultivos de *Gypsophila paniculata* L. [Predacious and parasitic activity of *Diglyphus begini* on *Liriomyza huidobrensis*, in *Gypsophila paniculata* plants]. *Neotropical Entomology* 32, 85–89.

DANE (Departamento Administrativo Nacional de Estadística) (2016) Resultados Encuesta Nacional Agropecuaria (ENA). [National Agricultural Survey Results (ENA)]. 3er Censo Nacional Agropecuario. Tomo 2. Bogotá, Colombia.

Daza, M., Cantor, F., Rodríguez, D., Bustos, A. and Cure, J.R. (2010) Criteria for the production of *Phytoseiulus persimilis* (Parasitiformes: Phytoseiidae) under greenhouse conditions. *Acta Biológica Colombiana* 15, 37–46.

De Vis, R.M. (2001) Biological control of whitefly on greenhouse tomato in Colômbia: *Encarsia formosa* or *Amitus fuscipennis*? PhD thesis. Wageningen University, Wageningen, Netherlands.

De Vis, R.M. and Fuentes, L.E. (2001) Manejo integrado de plagas y enfermedades [Integrated management of pests and diseases]. In: Lee, R. and Escobar, H. (eds) *Manual de producción de tomate bajo invernadero*. Ediciones Universidad Jorge Tadeo Lozano, Bogotá, Colombia, pp. 1–10.

De Vis, R.M. and van Lenteren, J.C. (2008) *Amitus fuscipennis*, an alternative to the biological control of *Trialeurodes vaporariorum* by *Encarsia formosa*? *Bulletin of Insectology* 61, 313–325.

De Vis, R.M., Fuentes, L.E. and van Lenteren, J.C. (1999) Development of biological control of *Trialeurodes vaporariorum* with *Encarsia formosa* and *Amitus fuscipennis* on greenhouse tomato in Colombia. *Bulletin of the International Organization for Biological Control (WPRS)* 22, 267–270.

Díaz, A., Gómez, M., Grijalba, E., Santos, A., Cruz, M., León, D., Alarcón, E. and Cotes, A.M. (2018) Desarrollo y escalamiento de bioplaguicidas [Development and scaling of biopesticides]. In: Cotes, A.M. (ed.) *Control biológico de fitopatógenos, insectos y ácaros. Volumen 2. Agentes de control biológico*. Editorial Agrosavia, Bogotá, Colombia, pp. 628–691.[*]

Díaz, M.F., Ramírez, A. and Poveda, K. (2012) Efficiency of different egg parasitoids and increased floral diversity for the biological control of noctuid pests. *Biological Control* 60, 182–191.

Díaz-Niño, M.F. (2018) Parasitoides del género *Trichogramma* (Hymenoptera: Trichogrammatidae) como controladores biológicos de plagas [Parasitoids of the genus *Trichogramma* as biological control agents of pests]. In: Cotes, A.M. (ed.) *Control biológico de fitopatógenos, insectos y ácaros. Volumen 1. Agentes de control biológico*. Editorial Agrosavia, Bogotá, Colombia, pp. 491–495.[*]

Díaz -Silva, F. (2011) Aspectos agroecológicos para el manejo integrado de *Prodiplosis longifila* Gagné en la irrigación de Chavimochic. [Agroecological aspects for the integrated management of *Prodiplosis longifila* in the irrigation of Chavimochic]. Escuela de Ciencias Biológicas' Universidad de Trujillo' Peru. Available at: http://www.academia.edu/15424714/Aspectos_Agroecol%C3%B3gicos_para_el_Manejo_Integrado_de_Prodiplosis_longifila_Gagn%C3%A9_en_la_Irrigaci%C3%B3n_Chavimochic (accessed 8 January 2018).

Drooz, A.T. and Bustillo, A.E. (1972) *Glena bisulca*, a serious defoliator of *Cupressus lusitanica* in Colombia. *Journal of Economic Entomology* 65, 89–93.

Drooz, A.T., Bustillo, A.E., Fedde, G.F. and Fedde, V.H. (1977) North American egg parasite successfully controls a different host genus in South America. *Science* 197, 390–391.

EOL (Encyclopedia of Life) (2016) *Phaseolus vulgaris* – Black Bean. Available at: http://eol.org/pages/645324/overview (accessed 18 January 2018).

Espinel, C., Lery, X., Villamizar, L., Gómez, J., Zeddam, J.L., Cotes, A.M. and López-Ferber, M. (2010) Genetic and biological analysis of Colombian *Phthorimaea operculella* granulovirus isolated from *Tecia solanivora* (Lepidoptera: Gelechiidae). *Applied and Environmental Microbiology* 76, 7617–7625.

Espinel, C., López-Ferber, M., Zeddam, J.L., Villamizar, L., Gómez, J., Cotes, A.M. and Léry, X. (2012) Experimental mixtures of *Phthorimaea operculella* granulovirus isolates provide high biological efficacy on both *Phthorimaea operculella* and *Tecia solanivora* (Lepidoptera: Gelechiidae). *Journal of Invertebrate Pathology* 110, 375–381.

Espinel-Correal, C., Barrera Cubillos, G.P., Torres Torres, L.A., Gómez Valderrama, J.A., Cuartas Otalora, P.E., Borrero Echeverry, F. and Villamizar Rivero, L.F. (2018a) Nuevas estrategias para el control biológico de insectos [Novel strategies for insect biological control]. In: Cotes, A.M. (ed.) *Control biológico de fitopatógenos, insectos y ácaros. Volumen 2. Agentes de control biológico*. Editorial Agrosavia, Bogotá, Colombia, pp. 922–949.[*]

Espinel-Correal, C., Torres, L., Villamizar, L., Bustillo, A., Zuluaga, M.V. and Cotes, A.M. (2018b) Hongos entomopatógenos en el control biológico de insectos plagas [Entomopathogenic fungi in insect pests biological control]. In: Cotes, A.M. (ed.) *Control biológico de fitopatógenos, insectos y ácaros. Volumen 1. Agentes de control biológico*. Editorial Agrosavia, Bogotá, Colombia, pp. 334–367.[*]

Fedearroz. (2010) Correo [Correspondence]. *Boletín informativo de la Federación Nacional de Arroceros* No. 239. Bogotá, Colombia.

Fedepalma (2011) La agroindustria de la palma de aceite en Colombia y en el mundo, Anuario Estadístico 2011. 182. [The oil palm agroindustry in Colombia and the world, Statistical Yearbook 2011. 182.]. Federación Nacional de Cultivadores de Palma de Aceite, Bogotá, Colombia.

Forero-Patiño, J.J., Argüelles, Z.P., Cantor, R. and Rodríguez, D. (2010) Bioensayo para evaluar la compatibilidad de adultos de *Neoseiulus californicus* (Parasitiformes: Phytoseiidae) con algunos pesticidas de uso comercial en condiciones semicontroladas [Bioassay for evaluating the compatibility of adults of *Neoseiulus californicus* with some commercial use pesticides under semicontrolled conditions]. *Agronomía Colombiana* 28, 273–280.

Gaimari, S.D., Quintero, E.M. and Kondo, T. (2012) First report of *Syneura cocciphila* (Coquillett, 1895) (Diptera: Phoridae), as a predator of the fluted scale *Crypticerya multicicatrices* Kondo & Unruh, 2009 (Hemiptera: Monophlebidae). *Boletín del Museo de Entomología de la Universidad del Valle* 13, 26–28.

García, F. (1985) Manejo del cogollero *Scrobipalpula absoluta* (Meyrick) en el Valle del Cauca. In: Manejo Integrado de Plagas, conferencias, Instituto Colombiano Agropecuario, Palmira, Colombia. pp. 329–334. Available at: https://books.google.com.co/books?id=K9cOAQAAIAAJ&pg=PP10&dq=Plagas+del+tomate+y+su+manejo.+Instituto+Colombiano+Agropecuario,+Palmira,+Colombia.&hl=es-419&sa=X&ved=0ahUKEwiggtWLoY_gAhVLMqwKHfFhDbkQ6wEIMjAC#v=onepage&q=Plagas%20del%20tomate%20y%20su%20manejo.%20Instituto%20Colombiano%20Agropecuario%2C%20Palmira%2C%20Colombia.&f=false (accessed 7 January 2018).

García, F. (1996) El control biológico aplicado en Colombia [Applied biological control in Colombia]. In: Zapater, M.C. (ed.) *El Control Biológico en América Latina*. International Organization for Biological Control, Buenos Aires, Argentina, pp. 31–33.

García, R. and Sánchez, M.J. (1995) Evaluación del comportamiento de *Catolaccus grandis* (Burks) parasitoide del picudo del algodonero [Evaluation of the behavior of *Catolaccus grandis*, parasitoid of the cotton boll weevil]. In: *Resúmenes. XXII Congreso Sociedad Colombiana de Entomología*, 26–28 July, 1995. Santafé de Bogotá, Colombia.

García, R., Mosquera, F.A., Vargas, A.T. and Rojas, C.A. (1999) Manejo integrado del gusano cogollero del maíz *Spodoptera frugiperda* (JE Smith) [Integrated management of the corn armyworm *Spodoptera frugiperda*]. (No. Doc. 19034)* CO-BAC. Santafé de Bogotá, Colombia.

García, Y., Ramos, Y.P., Sotelo, P.A. and Kondo, T. (2016) Biología de *Diaphorina citri* (Hemiptera: Liviidae) bajo condiciones de invernadero en Palmira, Colombia [Biology of *Diaphorina citri* under glass house conditions in Palmira, Colombia]. *Revista Colombiana de Entomología* 42, 36–42.

Gaviria, J. (1990) El control biológico de los insectos plaga de la caña de azúcar en Colombia [Biological control of insect pests of sugarcane in Colombia]. In: Buenaventura-O., C.E. (ed.) *III Congreso de la Sociedad de Técnicos de la Caña de Azúcar y I Congreso de la Asociación de Técnicos Azucareros de America Latina y el Caribe*. Tomo 1. Atalac, Tecnicaña, Cali, Colombia, pp. 201–227.

Giraldo-Calderón, G., Pérez, M., Morales, C.A. and Ocampo, C.B. (2008) Evaluación del triflumurón y la mezcla de *Bacillus thuringiensis* más *Bacillus sphaericus* para el control de las formas inmaduras de *Aedes aegypti* y *Culex quinquefasciatus* en sumideros de Cali, Colombia [Evaluation of triflumuron and the mixture of *Bacillus thuringiensis* plus *Bacillus sphaericus* for control of the immature stages of *Aedes aegypti* and *Culex quinquefasciatus* in catch basins]. *Biomédica* 28, 224–233.

Gómez, J.A., Guevara, J., Barrera, G.P., Cotes, A.M., and Villamizar, L. (2010) Aislamiento, identificación y caracterización de nucleopoliedrovirus nativos de *Spodoptera frugiperda* en Colombia [Isolation, identification and characterization of native nucleopolyhedrovirus of *Spodoptera frugiperda* in Colombia]. *Revista Facultad Nacional de Agronomía*, Medellín 63, 5511–5520.

Gómez. J.A., Guevara, J., Cuartas, P., Espinel, C. and Villamizar, L. (2013) Microencapsulated *Spodoptera frugiperda* nucleopolyhedrovirus: insecticidal activity and effect on arthropod populations in maize. *Biocontrol Science and Technology* 23, 829–846.

Gómez, J., Herrera, L., Uribe, D., López-Ferber, M. and Villamizar, L. (2014) An immunological method for granulovirus detection in larvae of *Tuta absoluta*: Searching for isolates with prospects for biological control of this pest in Colombia. *International Journal of Pest Management* 60, 136–143.

Gómez, L.L.A. (1995) Manipulación y aumento de enemigos naturales en el MIP. [Manipulation and increase of natural enemies in IPM]. In: Manejo Integrado de Plagas - Curso Internacional. Instituto Colombiano Agropecuario, Santafé de Bogotá, Colombia, pp. 75–78. Available at: https://repository.agrosavia.co/bitstream/handle/20.500.12324/1776/77044_47084.pdf?sequence=1&isAllowed=y (accessed 15 January 2018).

Gómez, L. and Vargas, G. (2014) Los barrenadores de la caña de azúcar, *Diatraea* spp., en el valle del río Cauca: Investigación participativa con énfasis en control biológico [Sugarcane borers, *Diatraea* spp., in the Cauca river valley: Participatory research with emphasis on biological control]. Cenicaña, Documento de trabajo No. 734. Cali. Colombia.

Gómez, M., Villamizar, L., Espinel, C., Varón, E., Jiménez, N. *et al.* (2012) Uso de *Lecanicillium lecanii* para el control de la mosca blanca *Bemisia tabaci* en algodón y berenjena. [Use of *Lecanicillium lecanii* for the control of the whitefly *Bemisia tabaci* in cotton and eggplant]. Ed. Produmedios, Bogotá, Colombia.

Gómez, M.I., Alarcón, A., León, M., Oehlschlager, C. and Solórzano, L. (2018) Comercialización de agentes de control biológico [Commercialization of biological control agents]. In: Cotes, A.M. (ed.) *Control biológico de fitopatógenos, insectos y ácaros. Volumen 2. Agentes de control biológico*. Editorial Agrosavia, Bogotá, Colombia, pp. 762–793.[*]

Gómez-Jiménez, M.I. (2018) Parasitoides como controladores de *Erinnyis ello* [Parasitoids as control agents of *Erinnyis ello*]. In: Cotes, A.M. (ed.) *Control biológico de fitopatógenos, insectos y ácaros. Volumen 1. Agentes de control biológico*. Editorial Agrosavia, Bogotá, Colombia, pp. 507–509.[*]

González, G. and Kondo, T. (2014) Geographical distribution and phenotypic variation of *Anovia punica* Gordon (Coleoptera: Coccinellidae: Noviini), a predatory ladybeetle of fluted scales (Hemiptera: Coccoidea: Monophlebidae). *Insecta Mundi* 0398, 1–6.

González, G., López, R. and Kondo, T. (2012) First report of *Delphastus quinculus* Gordon and *Diomus seminulus* (Mulsant) (Coleoptera: Coccinellidae) feeding on eggs and first-instar nymphs of *Crypticerya multicicatrices* Kondo & Unruh (Hemiptera: Monophlebidae). *Insecta Mundi* 0268, 1–6.

Grijalba, E., Hurst, M., Ibarra, J., Jurat-Fuentes, J.L. and Jackson, T. (2018) Bacterias entomopatógenas en el control biológico de insectos [Entomopathogenic bacteria in insect biological control]. In: Cotes, A.M. (ed.) *Control biológico de fitopatógenos, insectos y ácaros. Volumen 1. Agentes de control biológico*. Editorial Agrosavia, Bogotá, Colombia, pp. 296–333.[*]

Gutierrez-Monsalve, J., Mosquera, S., González, L., Mira, J. and Villegas, V. (2015) Effective control of black Sigatoka disease using a microbial fungicide based on *Bacillus subtilis* EA-CB0015 culture. *Biological Control* 87, 39–46.

Hagen, K.S. and Franz, J.M. (1973) A history of biological control. In: Smith, R.F., Mittler, T.E. and Smith, C.N. (eds) *History of Entomology*, Annual Reviews. Palo Alto, California.

Hernandez, L.M., Otero, J.T. and Manzano, M.R. (2013) Biological control of the greenhouse whitefly by *Amitus fuscipennis*: Understanding the role of extrafloral nectaries from crop and non-crop vegetation. *Biological Control* 67, 227–234.

Hernandez, L.M., Guzman, Y.C., Martínez-Arias, A., Manzano, M.R. and Selvaraj, J.J. (2015) The bud midge *Prodiplosis longifila*: damage characteristics, potential distribution and presence on a new crop host in Colombia. *SpringerPlus* 4, 205. DOI: 10.1186/s40064-015-0987-6

Hernandez-Mahecha, L.M., Manzano, M.R., Guzmán, Y.C. and Buhl, P.N. (2018) Parasitoids of *Prodiplosis longifila* Gagné (Diptera: Cecidomyiidae) and other Cecidomyiidae species in Colombia. *Acta Agronómica* 67, 184–191.

Hodson de Jaramillo, E., Castaño, J., Poveda, G., Roldán, G. and Chavarriaga, P. (2017) Food and Nutrition Security in Colombia. In: *Challenges and Opportunities for Food and Nutrition Security in the Americas. The View of the Academies of Sciences*. IANAS, IAP and BMBF, México DF, pp. 212–241. [Free public access of this publication in English and Spanish at www.ianas.org]

Howard, L.O. (1930) A history of Applied Entomology (Somewhat Anecdotal). *Smithsonian Miscellaneous Collections* 84, 1–564.

Hoyos, L., Chaparro, P., Abramsky, M., Chet, I. and Orduz, S. (2008) Evaluación de aislamientos de *Trichoderma* spp. contra *Rhizoctonia solani* y *Sclerotium rolfsii* bajo condiciones *in vitro* y de invernadero [Evaluation of *Trichoderma* spp. isolates against *Rhizoctonia solani* and *Sclerotium rolfsii* under *in vitro* and greenhouse conditions]. *Agronomía Colombiana* 26, 451–458.

ICA (Instituto Colombiano Agropecuario) (2012) Manejo fitosanitario del cultivo de cítricos (Citrus) – Medidas para la temporada invernal. [Phytosanitary management of citrus cultivation (Citrus) – Measures for the rainy season]. Available at: http://www.ica.gov.co/getattachment/89f7ca91-2820-4d06-9826-74964de55de6/Manejo-fitosanitario-del-ciltivo-de-Citricos.aspx (accessed 9 January 2018).

ICA (Instituto Colombiano Agropecuario) (2015) Declaración el estado de emergencia fitosanitaria en el territorio nacional por la presencia de adultos de *Diaphorina citri* infectados con la bacteria de la enfermedad del HLB [Declaration of the state of phytosanitary emergency in the national territory due to the presence of adults of *Diaphorina citri* infected with the bacterium of the HLB disease]. 10 de diciembre, 2015. Resolución No. 00002390. Bogota, Colombia. Available at: http://extwprlegs1.fao.org/docs/pdf/col151548.pdf (accessed 6 January 2018).

ICA (Instituto Colombiano Agropecuario) (2016a) Empresa bioinsumos, septiembre 22 de 2016 [Bio-inputs company, September 22, 2016]. Available at: http://www.ica.gov.co/getdoc/2ad9e987-8f69-4358-b8a9-e6ee6dcc8132/PRODUCTOSBIOINSUMOS-MAYO-13-DE-2008.aspx (accessed 6 January 2018).

ICA (Instituto Colombiano Agropecuario) (2016b) Productos registrados, Bioinsumos – septiembre 23 de 2016. [Registered products, Bioinputs – September 23, 2016]. Available at: http://www.ica.gov.co/Areas/Agricola/Servicios/Fertilizantes-y-Bio-insumos-Agricolas/Listado-de-Bioinsumos/2009/EMPRESAS-REGISTRADAS-BIOINSUMOS-JULIO-8-DE-2008.aspx (accessed 6 January 2018).

Imbachi, K., Mesa, C., Nora, C., Rodríguez, I.V., Gómez, I. *et al.* (2012) Evaluación de estrategias de control biológico de *Polyphagotarsonemus latus* (Banks) y *Phyllocoptruta oleivora* (Ashmead) en naranja Valencia [Evaluation of strategies of biological control of *Polyphagotarsonemus latus* and *Phyllocoptruta oleivora* on Valencia oranges]. *Acta Agronómica* 61, 364–370.

Jaramillo, J., Bustillo, A.E., Montoya, E.C. and Borgemeister, C. (2005) Biological control of the coffee berry borer *Hypothenemus hampei* (Coleoptera: Curculionidae) by *Phymastichus coffea* (Hymenoptera: Eulophidae) in Colombia. *Bulletin of Entomological Research* 95, 467–472.

Jaramillo, J., Borgemeister, C. and Baker, P. (2006) Coffee berry borer *Hypothenemus hampei* (Coleoptera: Curculionidae): searching for sustainable control strategies. *Bulletin of Entomological Research* 96, 223–233.

Jiménez, J. (2001) Cría, manejo y comercialización del parasitoide de moscas comunes, *Spalangia cameroni* Perkins [Breeding, management and commercialization of the parasitoid of common flies, *Spalangia cameroni*]. In: *Memorias. 28 Congreso de la Sociedad Colombiana de Entomología*. SOCOLEN, Pereira, Colombia, pp. 82–86.

Kondo, T. (2008) Las escamas de la guanábana: *Annona muricata* L. [The scale insects on soursop: *Annona muricata* L.]. Novedades Técnicas, Revista Regional, Corpoica, Centro de Investigación Palmira. Año 9/No. 10/Septiembre/2008, Palmira, Colombia, pp. 25–29.

Kondo, T. (2011) Notas sobre el uso correcto del término técnico para referirse a la cría masiva de insectos y otros artrópodos: cría masiva vs. cría masal y cría en masa [Notes on the correct use of the technical term to refer to the massive breeding of insects and other arthropods: massive breeding vs. mass breeding and breeding in mass]. *Boletín del Museo de Entomología de la Universidad del Valle*, 12, 26–28.

Kondo, T. (2015) Insectos plaga del árbol urbano con énfasis en los insectos escama (Hemiptera: Coccoidea) en Colombia [Insect pests of the urban tree with emphasis on scale insects in Colombia]. In: *Simposio: Entomología Forestal. 41 Congreso de la Sociedad Colombiana de Entomología (SOCOLEN)*. Julio 29–31, 2015. Universidad Nacional de Colombia, Sede Medellín, Medellín, Colombia.

Kondo, T. (2017) Protocolo de cría y liberación de *Tamarixia radiata* Waterston (Hymenoptera: Eulophidae) [Rearing and release protocol of *Tamarixia radiata* Waterston]. Corporación Colombiana de Investigación Agropecuaria (Corpoica), Mosquera, Colombia. Available at: https://repository.agrosavia.co/handle/20.500.12324/13147 (accessed 12 January 2018).

Kondo, T. (2018) *Tamarixia radiata* (Waterston) (Hymenoptera: Eulophidae), parasitoide del psílido asiático de los cítricos, *Diaphorina citri* Kuwayama (Hemiptera: Liviidae) [*Tamarixia radiata*, parasitoid of the Asian citrus psyllid, *Diaphorina citri*]. In: Cotes, A.M. (ed.) *Control biológico de fitopatógenos, insectos y ácaros. Volumen 1. Agentes de control biológico*. Editorial Agrosavia, Bogotá, Colombia, pp. 518–522.[*]

Kondo, T., Gullan, P. and Ramos-Portilla, A.A. (2012) Report of new invasive scale insects (Hemiptera: Coccoidea), *Crypticerya multicicatrices* Kondo & Unruh (Monophlebidae) and *Maconellicoccus hirsutus* (Green) (Pseudococcidae), on the islands of San Andres and Providencia, Colombia, with an updated taxonomic key to iceryine scale insects of South America. *Insecta Mundi* 0265, 1–17.

Kondo, T., Gullan, P. and González, G. (2014) An overview of a fortuitous and efficient biological control of the Colombian fluted scale, *Crypticerya multicicatrices* Kondo & Unruh (Hemiptera: Monophlebidae: Iceryini), on San Andres island, Colombia. *Acta Zoologica Bulgarica*, Supplement 6, 87–93.

Kondo, T., González, G., Tauber, C., Guzmán-Sarmiento, Y.C., Vinasco Mondragon, A.F. and Forero, D. (2015) A checklist of natural enemies of *Diaphorina citri* Kuwayama (Hemiptera: Liviidae) in the department of Valle del Cauca, Colombia and the World. *Insecta Mundi* 0457, 1–14.

Kondo T., Ramos-Portilla A.A., Peronti, A.L.B.G. and Gullan, P.J. (2016) Known distribution and pest status of fluted scale insects (Hemiptera: Monophlebidae: Iceryini) in South America. *Redia* XCIX, 187–195.

Kondo, T., Rincón, D.F., Pérez-Álvarez, R., Vásquez Ordóñez, A.A. and González, F.G. (2018) Uso de depredadores como agentes de control biológico para insectos plaga [Use of predators as biological control agents of insect pests]. In: Cotes, A.M. (ed.) *Control biológico de fitopatógenos, insectos y ácaros. Volumen 1. Agentes de control biológico*. Editorial Agrosavia, Bogotá, Colombia, pp. 486–543.[*]

León, M.G., Arcila, C.A.M., Pulido, L.A. and Kondo, T. (2018) Cambio climático y control biológico de insectos: visión y perspectiva de la situación [Climate change and biological control of insects: current situation and perspectives]. In: Cotes, A.M. (ed.) *Control biológico de fitopatógenos, insectos y ácaros. Volumen 2. Agentes de control biológico*. Editorial Agrosavia, Bogotá, Colombia, pp. 1034–1054.[*]

Maldonado-Londoño, C.E. and Benavides-Machado, P. (2007) Evaluación del establecimiento de *Cephalonomia stephanoderis* y *Prorops nasuta* controladores de *Hypothenemus hampei* en Colombia [Evaluation of the establishment of *Cephalonomia stephanoderis* and *Prorops nasuta* controllers of *Hypothenemus hampei* in Colombia]. *Cenicafé* 58, 333–339.

Manzano, M.R. (2000) Evaluation of *Amitus fuscipennis* as biological control agent of *Trialeurodes vaporariorum* on bean in Colombia. PhD thesis. Wageningen University, Wageningen, Netherlands.

Manzano, M.R. (2018) Parasitoides de mosca blanca [Parasitoids of whiteflies]. In: Cotes, A.M. (ed.) *Control biológico de fitopatógenos, insectos y ácaros. Volumen 1. Agentes de control biológico*. Editorial Agrosavia, Bogotá, Colombia, pp. 495–499.[*]

Manzano, M.R., van Lenteren, J.V. and Cardona, C. (2002a) Searching and oviposition behaviour of *Amitus fuscipennis*, a parasitoid of the greenhouse whitefly. *Journal of Applied Entomology* 12610, 528–533.

Manzano, M.R., van Lenteren, J. and Cardona, C. (2002b) Intrinsic rate of population increase of *Amitus fuscipennis* MacGown and Nebeker Hymenoptera: Platygasteridae according to climatic conditions and bean cultivar. *Journal of Applied Entomology* 1261, 34–39.

Manzano, M.R., van Lenteren, J. and Cardona, C. (2003) Influence of pesticide treatments on the dynamics of whiteflies and associated parasitoids in snap bean fields. *BioControl* 486, 685–693.

Mead, F.W. and Fasulo, T.R. (2010) Asian citrus psyllid, *Diaphorina citri* Kuwayama (Insecta: Hemiptera: Psyllidae). *FDACS/DPI Entomology. Circular 180*. University of Florida, Gainesville, Florida, USA. Available at: http://www.crec.ifas.ufl.edu/extension/greening/pdf/asiaticcitruspsyllid.pdf (accessed 8 January 2018).

Mejía, J.W., Bustillo, A.E., Orozco, J. and Cháves, B. (2000) Efecto de cuatro insecticidas y de *Beauveria bassiana* sobre *Prorops nasuta* (Hymenoptera: Bethylidae), parasitoide de la broca del café [Effect of four insecticides and *Beauveria bassiana* on *Prorops nasuta*, parasitoid of the coffee berry borer]. *Revista Colombiana de Entomología* 26, 117–123.

Mejía, C., Bolaños, A., Bermúdez, C.A., Acosta, M.A., Marcia, C., Gómez, M.I., Villamizar, L.F., Diaz, A., Uribe, L.I., Cotes, A.M. and Zapata, J.A. (2013) Desarrollo de prototipos de bioplaguicidas a base de *Rhodotorula glutinis* LvCo7 para el control de *Botrytis cinerea* en cultivos de mora. [Development of prototypes of biopesticides based on *Rhodotorula glutinis* LvCo7 for the control of *Botrytis cinerea* in blackberry cultivation]. CORPOICA, Bogotá, Colombia.

Melo, C. and Manzano, M. (2017) Diversidad de áfidos y sus enemigos naturales presentes en plantas acompañantes de cultivos de ají en el Valle del Cauca, Colombia. Coloquio Biodiversidad, ecología y agroecosistemas [Diversity of aphids and their natural enemies present in accompanying plants of pepper crops in Valle del Cauca, Colombia. Colloquium Biodiversity, ecology and agroecosystems]. Universidad del Valle, Universidad del Quindío, Universidad del Cauca, & Politécnico Colombiano (Jaime Isaza Cadavid), Cali. Available at: http://geahnacursos.correounivalle.edu.co/ (accessed 8 January 2018).

Ministry of Social Protection of Colombia (2017a) Guía de Vigilancia Entomológica y Control de Dengue [Guide to Entomological Surveillance and Dengue Control]. Available at: https://www.minsalud.gov.co/proteccionsocial/Paginas/Dengue.aspx (accessed 10 January 2018).

Ministry of Social Protection of Colombia (2017b) Guía de Vigilancia Entomológica y Control de Malaria [Guide for Entomological Surveillance and Malaria Control]. Available at: https://www.minsalud.gov.co/Documents/Salud%20P%C3%BAblica/Ola%20invernal/ENTOMOLOGICA%20Y%20%20CONTROL%20MALARIA%20.pdf (accessed 10 January 2018).

Montealegre, X., Sotelo, P. and Kondo, T. (2016) Development and reproduction of *Brethesiella* cf. *abnormicornis* (Girault) (Hymenoptera: Encyrtidae), a parasitoid of the Colombian fluted scale, *Crypticerya multicicatrices* Kondo & Unruh (Hemiptera: Monophlebidae). *Environmental Entomology* 45, 376–383.

Morales, J., Rodríguez, D. and Cantor, F. (2013) Estandarización de la cria masiva de *Apanteles gelechiidivoris* Marsh (Hymenoptera: Braconidae) para el control de *Tuta absoluta* Meyrick (Lepidoptera: Gelechiidae) [Standardization of a massive rearing system of *Apanteles gelechiidivoris* Marsh (Hymenoptera: Braconidae), biological control for *Tuta absoluta* Meyrick (Lepidoptera: Gelechiidae)]. *Revista Facultad de Ciencias Basicas (Universidad Militar)* 9, 20–37.

Moreno, C.A., Castillo, F., González, A., Bernal, D., Jaimes, Y. *et al.* (2009) Biological and molecular characterization of the response of tomato plants treated with *Trichoderma koningiopsis*. *Physiological and Molecular Plant Pathology* 74, 111–120.

Moreno, C.A., Cotes, A.M., Smith, A., Beltrán, C., Villamizar, L. *et al.* (2010) Desarrollo de un bioplaguicida a base de *Trichoderma koningiopsis* Th003 y uso en el cultivo de lechuga para el control del moho blanco *Sclerotinia sclerotiorum* y *Sclerotinia minor*. [Development of a biopesticide based on *Trichoderma koningiopsis* Th003 and use in the cultivation of lettuce for the control of white mold *Sclerotinia sclerotiorum* and *Sclerotinia minor*]. Editorial Produmedios, Bogotá, Colombia.

Moreno-Salguero, C.A., Chavarría-Hernández, N., Bustillo-Pardey A.E., Jiménez-Velásquez, J., Gérman, A., Vargas-Orozco, G., Daza-Merchan, T. and Serna-Cock, L. (2014) Potencial reproductivo del nematodo entomopatógeno *Heterorhabditis* sp. Tumaco (Rhabdita: Heteorhabditidae) en cultivo monoxénico. [Reproductive potential of the entomopathogenic nematode *Heterorhabditis* sp. in monoxenic culture]. In: *XXXVII Congreso Nacional de Control biológico*, noviembre 7 de 2014. Mérida, México.

Moreno-Velandia, C.A., Cotes, A.M., Beltrán, C, Bettiol, W. and Elad, Y. (2018) Control biológico de fitopatógenos del suelo [Biological control of soil-borne phytopathogens]. In: Cotes, A.M. (ed.) *Control biológico de fitopatógenos, insectos y ácaros. Volumen 1. Agentes de control biológico*. Editorial Agrosavia, Bogotá, Colombia, pp. 144–221.[*]

Moscardi, F. (1999) Assessment of the application of baculoviruses for control of Lepidoptera. *Annual Review of Entomology* 44, 257–289.

Muñoz, M. and Navarro, J.C. (2012) Virus Mayaro: un arbovirus reemergente en Venezuela y Latinoamérica [Mayaro virus: a reemerging arbovirus in Venezuela and Latin America]. *Biomédica* 32, 288–302.

Navarro, M.A. (1988) Biological control of *Scrobipalpula absoluta* (Meyrick) by *Trichogramma* sp. in the tomato (*Lycopersicon esculentum* Mill.). In: *International symposium on Trichogramma and other egg parasites*, 2, 1986, Guangzhou. Colloques de l'INRA 43, 453–458.

Nicholls, C.I., Parrella, M.P. and Altieri, M.A. (1998) Advances and perspectives in the biological control of greenhouse pests with special reference to Colombia. *Integrated Pest Management Reviews* 3, 99–109.

Ortiz, L.E., Calvache, H.H. and Luque, J.E. (1994) Control microbiano de *Sagalassa valida* Walker (Lepidóptera: Glyphipterigidae) con el nematodo *Steinernema carpocapsae* (Weiser) en Tumaco (Nar.) [Microbial control of *Sagalassa valida* Walker with the nematode *Steinernema carpocapsae* in Tumaco (Nar.)]. *Revista Palmas* 15, 29–37.

Pérez, E.F. and Martínez, K.S. (2011) Distribución espacial y ciclo de vida de *Diatraea* spp. en plantaciones de *Saccharum officinarum* (Caquetá, Colombia) [Spatial distribution and life cycle of *Diatraea* spp. in plantations of *Saccharum officinarum* (Caquetá, Colombia)]. *Ingenierías & Amazonia* 4, 122–130.

Pinchao, E.C., Kondo, T. and González, G. (2015) *Rodolia cardinalis* (Mulsant) (Coleoptera: Coccinellidae), a new predator of *Crypticerya multicicatrices* Kondo and Unruh (Hemiptera: Monophlebidae). *Insecta Mundi* 0431, 1–7.

Polanía, I.Z., Rodriguez, J.A.A., Maldonado, H.A.A., Cruz, R.M. and Bayona-R, M.A. (2008) Susceptibilidad de cuatro noctuidos plaga (Lepidoptera) al gene Cry1Ac del *Bacillus thuringiensis* incorporado al algodonero [Susceptibility of four noctuid pests (Lepidoptera) to the Cry1Ac gene of *Bacillus thuringiensis* incorporated into cotton]. *Revista Colombiana de Entomología* 34, 41–50.

Polanía, I.Z., Maldonado, A., Adrian, H., Mejía Cruz, R. and Díaz Sánchez, J.L. (2009) *Spodoptera frugiperda*: respuesta de distintas poblaciones a la toxina Cry1Ab. [*Spodoptera frugiperda*: response of different populations to the Cry1Ab toxin]. *Revista Colombiana de Entomología* 35, 34–41.

Quintero, E.M., López, I.C. and Kondo, T. (2012) Manejo integrado de plagas como estrategia para el control de la mosca del botón floral del maracuyá *Dasiops inedulis* Steyskal (Diptera: Lonchaeidae). [IPM as a strategy to control the passionfruit flower-bud fly, *Dasiops inedulis*]. *Corpoica – Ciencia y Tecnología Agropecuaria* 13, 31–40.

Reyes, I.C., Bustillo, A.E. and Chaves, B. (1995) Efecto de *Beauveria bassiana* y *Metarhizium anisopliae* sobre el parasitoide de la broca del café, *Cephalonomia stephanoderis* [Effect of *Beauveria bassiana* and *Metarhizium anisopliae* on the coffee berry borer parasitoid *Cephalonomia stephanoderis*]. *Revista Colombiana de Entomología* 21, 199–204.

Rivera, P.A., Montoya, E.C., and Benavides, P., (2010) Biología del parasitoide de *Prorops nasuta* (Hymenoptera: Bethylidae) en el campo y su tolerancia a insecticidas [Biology of the parasitoid of *Prorops nasuta* in the field and its tolerance to insecticides]. *Cenicafé* 61, 99–107.

Rodas, C.A., Serna, R., Bolaños, M.D., Granados, G.M., Wingfield, M.J. and Hurley, B.P. (2014) Biology, incidence and host susceptibility of *Pineus boerneri* (Hemiptera: Adelgidae) in Colombian pine plantations. *Southern Forests: a Journal of Forest Science* 77, 165–171.

Salamanca, L., Rondón, M., Vargas, G. and Manzano, M.R. (2016) Interacción *Cotesia flavipes* (Hymenoptera: Braconidae) con *Diatraea* spp. (Lepidoptera: Crambidae): influencia del hospedero en características biológicas del parasitoide [Interaction of *Cotesia flavipes* with *Diatraea* spp.: influence of the host on biological characteristics of the parasitoid]. In: *Resúmenes. XXXVIII Congreso Nacional de Entomología*, noviembre 30 a diciembre 2 de 2016. Talca, Chile, p. 63.

Santos, A.M., Uribe, L.A., Torres, L., Betancourt, R.A., Moreno, F., Alarcón, E.A. and Gómez, M.I. (2018). Marco regulatorio para el registro de bioplaguicidas [Regulatory framework for the registration of biopesticides]. In: Cotes, A.M. (ed.) *Control biológico de fitopatógenos, insectos y ácaros. Volumen 2. Agentes de control biológico*. Editorial Agrosavia, Bogotá, Colombia, pp. 692–715.[*]

Santos, O., Restrepo, O.D., Argüelles, J. and Garramuño, E.A. (2009) Evaluación del comportamiento del complejo *Spodoptera* con la introducción de algodón transgénico al Tolima, Colombia [Evaluation of the behaviour of the *Spodoptera* complex at the introduction of transgenic cotton in Tolima, Colombia]. *Revista Corpoica: Ciencia y Tecnología Agropecuaria* 10, 24–32.

Schmale, I., Wäckers, F.L., Cardona, C. and Dorn, S. (2006) Biological control of the bean weevil, *Acanthoscelides obtectus* (Say) (Col.: Bruchidae), by the native parasitoid *Dinarmus basalis* (Rondani) (Hym.: Pteromalidae) on small-scale farms in Colombia. *Journal of Stored Products Research* 42, 31–41.

Smith, A.M., Beltran, C., Kusunoki, M., Cotes, A.M., Motohashi, K., Kondo, T. and Deguchi, M. (2013) Diversity of soil dwelling *Trichoderma* in Colombia and their potential as biocontrol agents against the phytopathogenic fungus *Sclerotinia sclerotiorum* (Lib.) de Bary. *Journal of General Plant Pathology* 79, 74–85.

Smith, L. and Bellotti, A.C. (1996) Successful biocontrol projects with emphasis on the Neotropics. In: *Proceedings of Cornell Community Conferences on Biological Control*. Apr. 11-13, 1996. Cornell University, Ithaca, New York. Available at: http://web.entomology.cornell.edu/shelton/cornell-biocontrol-conf/talks/bellotti.html (accessed 12 January 2018).

Suarez-Rubio, M. and Suarez, M.F. (2004) The use of the copepod *Mesocyclops longisezus* as a biological control agent for *Aedes aegypti* in Cali, Colombia. *Journal of the American Mosquito Control Association* 20, 401–404.

Torres, L. and Cotes, A.M. (2005) Efecto de la crio-conservación sobre la viabilidad y actividad biocontroladora de *Nomuraea rileyi* contra *Spodoptera frugiperda* (Lepidoptera: Noctuidae) [Effect of cryopreservation on the viability and biocontrol activity of *Nomuraea rileyi* against *Spodoptera frugiperda* (Lepidoptera: Noctuidae)]. *Revista Colombiana de Entomología* 31, 133–138.

UJTL (2016) Centro de Biosistemas [Biosystem Center]. Available at: http://www.utadeo.edu.co/es/noticia/destacadas/home/1/centro-de-bio-sistemas-de-utadeo-lanza-dos-bioinsumos-para-el-control-de-plagas-agricolas-de-la (accessed 18 January 2018).

Uribe-Gutiérrez, L.A., Bolaños-Almeida, C.A., Zapata-Narváez, J.A., Gómez Álvarez, M.I. and Villamizar-Rivero, L.F. (2013) Desarrollo de prototipos de bioplaguicidas a base de *Rhodotorula glutinis* LvCo7 para el control de *Botrytis cinerea* en cultivos de mora. [Development of prototypes of biopesticides based on *Rhodotorula glutinis* LvCo7 for the control of *Botrytis cinerea* in blackberry cultivation]. In: Zapata-Narváez, J.I., Cotes-Prado, A.M., Uribe-Gutiérrez, L.A., Díaz-García, A., Villamizar-Rivero, L.F., Gómez-Álvarez, M.I., Saldarriaga-Cardona, A., Álvarez-Zambrano, R. and Gómez, E. (eds) Desarrollo de prototipos de bioplaguicida a base de *Rhodotorula glutinis* LvCo7 para el control de *Botrytis cinerea* en cultivos de mora. CORPOICA, Bogotá, Colombia, pp. 58–68.

Valenzuela, G. (1993) Aspectos históricos del control biológico [Historical aspects of biological control]. In: Palacios, F. (ed.) *Control biológico en Colombia: historia, avances y proyecciones*. Universidad Nacional de Colombia, Palmira, Colombia, pp. 1–8.

van Lenteren, J.C. (2012) The state of commercial augmentative biological control: plenty of natural enemies, but a frustrating lack of uptake. *BioControl* 57, 1–20.

van Lenteren, J.C. and Bueno, V.H. (2003) Augmentative biological control of arthropods in Latin America. *BioControl* 48, 123–139.

Vargas, G.A., Obando, V.P. and Gómez, L.A. (2006) *Jaynesleskia jaynesi*: otra alternativa para el manejo de *Diatraea* spp. [*Jaynesleskia jaynesi*: another alternative for the management of *Diatraea* spp.]. Cenicaña, Cali, Colombia. Carta Trimestral, 28, 3–5. Available at: https://www.cenicana.org/pdf_privado/carta_trimestral/ct2006/ct2_06/ct2_06_p3-5.pdf (accessed 29 October 2019).

Vargas, G. (2018) Los barrenadores del tallo *Diatraea* y su control biológico mediante parasitoides de huevos y larvas [The sugarcane stem borers *Diatraea* and their control by means of egg and larval parasitoids]. In: Cotes, A.M. (ed.) *Control biológico de fitopatógenos, insectos y ácaros. Volumen 1. Agentes de control biológico*. Editorial Agrosavia, Bogotá, Colombia, pp. 513–518.[*]

Vargas, G., Lastra, L. and Solis, A. (2013) First record of *Diatraea tabernella* (Lepidoptera: Crambidae) in the Cauca river Valley of Colombia. *Florida Entomologist* 96, 1198–1201.

Vargas, G., Gómez, L.A. and Michaud, J.P. (2015) Sugarcane stem borers of the Colombian Cauca River Valley: current pest status, biology, and control. *Florida Entomologist* 98, 728–735.

Vasconcelos, P.F., Travassos da Rosa, A.P.A., Pinheiro, F.P., Rodrigues, S.G., Travassos-da-Rosa, E.S., Cruz, A.C.R. and Travassos-da-Rosa, J.F.S. (1999) *Aedes aegypti*, dengue and re-urbanization of yellow fever in Brazil and other South American countries – past and present situation and future perspectives. *Dengue Bulletin* 23, 55–66.

Vélez-Arango, A.M., Arango-I., R.E., Villanueva-M.D., Aguilera-G.E. and Saldamando-B., C.I. (2008) Identificación de biotipos de *Spodoptera frugiperda* (Lepidoptera: Noctuidae) mediante marcadores mitocondriales y nucleares [Identification of *Spodoptera frugiperda* biotypes through using mitochondrial and nuclear markers]. *Revista Colombiana de Entomología*, 34, 145–150.

Villamizar, L., Espinel, C. and Cotes, A.M. (2009) Efecto de la radiación ultravioleta sobre la actividad insecticida de un nucleopoliedrovirus de *Spodoptera frugiperda* (Lepidoptera: Noctuidae) [Effect of ultraviolet radiation on the insecticidal activity of a *Spodoptera frugiperda* nucleopolyhedrovirus]. *Revista Colombiana de Entomología* 35, 116–121.

Villamizar, L., Barrera, G., Cotes, A.M. and Martínez, F. (2010) Eudragit S100 microparticles containing *Spodoptera frugiperda* nucleopolyehedrovirus: physicochemical characterization, photostability and in vitro virus release. *Journal of Microencapsulation* 27, 314–324.

Villamizar, L., Espinel, C.G., Gómez, M., Cuartas, J., Barrera, G. and Cotes, A.M. (2012a) Desarrollo de un bioplaguicida a base de nucleopoliedrovirus para el control del gusano cogollero del maíz *Spodoptera frugiperda*. [Development of a biopesticide based on nucleopoliedrovirus for the control of the corn leafworm *Spodoptera frugiperda*]. Produmedios, Bogotá, Colombia.

Villamizar, L., Espinel, C.G., Gómez, M.I., Moreno, C.A., Gómez, J. and Cotes, A.M. (2012b). Desarrollo de un bioplaguicida a base de granulovirus para el control de la polilla guatemalteca de la papa (*Tecia solanivora*) en el campo. [Development of a granulovirus-based biopesticide for the control of the Guatemalan potato moth (*Tecia solanivora*) in the field]. Produmedios, Bogotá, Colombia.

Villamizar, L., Cuartas, P., Gómez, J., Barrera, G.P., Espinel, C. and Lopez-Ferber, M. (2018) Virus entomopatógenos en el control biológico de insectos [Entomopathogenic viruses in the biological control of insects]. In: Cotes, A.M. (ed.) *Control biológico de fitopatógenos, insectos y ácaros. Volumen 1. Agentes de control biológico*. Editorial Agrosavia, Bogotá, Colombia, pp. 368–409.[*]

Wyckhuys, K.A., Acosta, F.L., Rojas, M. and Ocampo, J. (2010) The relationship of farm surroundings and local infestation pressure to pest management in cultivated *Passiflora* species in Colombia. *International Journal of Pest Management* 57, 1–10.

Wyckhuys, K.A., Lu, Y., Morales, H., Vazquez, L.L., Legaspi, J.C., Eliopoulos, P.A. and Hernandez, L.M. (2013) Current status and potential of conservation biological control for agriculture in the developing world. *Biological Control* 65, 152–167.

Yaseen, M., Bennet, F.D. and Barrow, R.M. (1982) Introduction of exotic parasites for control of *Spodoptera frugiperda* in Trinidad, the eastern Caribbean and Latin America. In: Brathwaite C.W. and Pollard, G.V. (eds) *Urgent Plant Pest and Disease Problems in the Caribbean*. IICA Miscellaneous Publication, Port-of-Spain, Trinidad & Tobago, pp. 161–171.

Zapata, J., Villamizar, L., Díaz, L., Uribe, L., Bolaños, C., Gómez, M. and Cotes, A.M. (2013) Development of a biopesticide prototype based on the yeast *Rhodotorula glutinis* Lv316 for controlling *Botrytis cinerea* in blackberry. *IOBC WPRS Bulletin* 86, 263–269.

Zapata, Y., Cotes, A.M., Jijakli, H. and Wisniewski, M. (2018) Control biológico de patógenos en poscosecha [Biological control of postharvest pathogens]. In: Cotes, A.M. (ed.), *Control biológico de fitopatógenos, insectos y ácaros. Volumen 1. Agentes de control biológico*. Editorial Agrosavia, Bogotá, Colombia, pp. 222–255.[*]

9

Biological Control in Costa Rica

Helga Blanco-Metzler[1]* and Rossy Morera-Montoya[2]

[1]*Crop Protection Research Centre (CIPROC), University of Costa Rica, San José, Costa Rica;* [2]*Graduate Programme in Agricultural Sciences and Natural Resources, University of Costa Rica, San José, Costa Rica*

* E-mail: helga.blanco@ucr.ac.cr

Abstract

The first biological control activities in Costa Rica date from around 1915, but it was only in the 1950s that studies were conducted on the introduction, mass rearing and release of biocontrol agents in crops such as fruit trees, coffee, bananas, sugarcane, rice and pineapple. Subsequently, the use of entomopathogens became important in the 1990s as a strategy to be used in integrated pest management (IPM) programmes. From the 1990s, biocontrol laboratories were established and alliances were developed with national and international institutions and universities to strengthen biocontrol activities in the country. Currently, new areas of study have been incorporated in biocontrol research, such as molecular biology, host-plant resistance, identification of new biocontrol agents, the use of X-rays to sterilize hosts of natural enemies and the use of volatile compounds produced by the plant after pest attack to attract natural enemies. All these activities help to develop sustainable pest management. Implementation of biocontrol in Costa Rica will keep increasing, since there is great pressure to reduce the use of chemical pesticides for all the export crops.

9.1 Introduction

Costa Rica has an estimated population of slightly more than 4,930,000 (July 2017) and its agricultural products are bananas, pineapples, coffee, melons, ornamental plants, sugar, maize, rice, beans, potatoes, beef, poultry, dairy and timber (CIA, 2017). According to Jiménez *et al.* (2017, pp. 246 and 248):

> In the past 50 years the largest area devoted to agricultural activities was achieved in 1984, with 53.8% of the national territory. This figure gradually declined year after year until 2000, and thereafter remained fairly stable until it reached 35.6% in 2013. Forest cover has increased since 2000, reaching 51% in 2010 ... Data from the 2014 Agricultural Census show that Costa Rica has a total of 557,888.6 hectares (ha) under perennial agricultural crop cultivation (excluding forest plantations) and 133,249.8 ha planted with annual crops. Smaller productive units (farms) tend to grow crops such as maize, beans, vegetables, palm trees, fruit trees, coffee and some livestock, whereas larger farms produce banana, sugarcane, rice, pineapple, orange, tilapia and milk. In 2014, the Costa Rican agricultural sector accounted for 22.8% of the value of the country's exports, while the livestock and fishing sectors contributed 3.2%.

9.2 History of Biological Control in Costa Rica

9.2.1 Period 1880–1969

The first case of biocontrol in Costa Rica dates from 1915, when an entomopathogenic bacterium was used to control locusts (Picado and Sanchez 1915, quoted by Hilje *et al.*, 1989). In the same period, observations were published about parasitism of the guava fruit fly *Anastrepha striata* Schiner by *Diachasmimorpha longicaudata* (Ashmead) (Hilje *et al.*, 1989). The citrus blackfly *Aleurocanthus woglumi* (Ashby) arrived in Costa Rica in 1920. In response, the parasitoids *Eretmocerus serius* (Silvestri) and *Encarsia opulenta* Silvestri were introduced in 1930 from Cuba, which controlled the pest (Hernández, 1996). Next, *Aphelinus mali* Haldeman was imported from the USA between 1933 and 1936 to control *Eriosoma lanigerum* Hausmann (Hernández, 1996). Below we summarize the situation of biocontrol in different crops during this period.

Pests in coffee

After the 1963–1964 eruption of the Irazú Volcano, several insects that earlier caused no problems in coffee plantations developed to pest status, such as the mealybug *Planococcus citri* (Risso), the leaf miner *Leucoptera coffeella* (Guérin-Meneville) and the red mite *Oligonychus yorthesi* McGregor (Hernández, 1996). To control *P. citri*, *Cryptolaemus montrouzieri* Mulsant was imported, mass reared and released. In 1968 *Leptomastidea abnormis* Girault was imported from Chile and released to improve control of the mealybug (Hernández, 1996; Hanson, 1991).

Mediterranean fruit fly in citrus

In 1955 the agronomist L.Á. Salas detected a specimen of the Mediterranean fruit fly *Ceratitis capitata* (Wiedemann) for the first time in the canton of Santa Ana. Previous infestations had only been found near the Juan Santamaría Airport (Padilla-Monge, 2012) and chemical control was used to avoid spreading of the pest to the rest of the country. Salas started importing the parasitoids *Biosteres longicaudatus* (Ashmead) and *Psyttalia concolor* (Szepligati) along with *Pachycrepoideus vindemmiae* (Rondani) in 1955 with the cooperation of the Agriculture and Livestock Ministry and the United States Department of Agriculture (USDA) (Jirón and Mexzón, 1989; Ovruski *et al.*, 2000). In the same year, eight species of parasitoids were introduced from Hawaii (*Aganaspis daci* (Weld); *Aceratoneuromyia indica* Silvestri; *Fopius arisanus* (Sonan); *F. vandenboschi* (Fullaway); *Diachasmimorpha* (formerly *Biosteres* or *Opius*) *longicaudata*; *D. tryoni* (Cameron); *Psyttalia incise* (Silvestri); *Dirhinus giffardii* (Silvestri)) and one from Italy (*P. concolor*). Rearing of five of these species was established in the laboratory and they were later released in the field. At the beginning of the 1960s, Costa Rica provided these species for field releases to 11 American countries (Ovruski *et al.*, 2000).

Because of the problematic fruit fly situation in Central America, the Regional International Organization for Plant Protection and Animal Health (OIRSA by its Spanish initials) approved the creation of the Mediterranean Fruit Fly Department in 1957 in Costa Rica, with the support of a group of Mexican technicians who specialized in fruit fly control. Next, the Biological Control Office was created to control Mediterranean fruit fly and a locust of the genus *Schistocerca* spp. (Padilla-Monge, 2012). Through this office, 11 species of hymenopteran parasitoids from Hawaii were released: *Biosteres formosanus* (Fullaway) (= *Opius formosanus* Fullaway); *Opius oophilus* Fullaway (= *B. arisanus* (Sonan); *B. tryoni* (Cameron) (= *O. tryoni* Cameron); *O. incise* Silvestri; *B. vandenboschi* (Fullaway) (= *O. vandenboschi* Fullaway); *B. compensans* (Silvestri) (= *O. compensans* Silvestri); *B. longicaudatus* Ashmead (= *O. longicaudatus* Ashmead), including the varieties of *taiensis* Fullaway and *novocaledonicus* Fullaway; *A. indica* (= *Syntomosphyrum indicum* Silvestri); *Trybliographa daci* Weld; and *D. giffardi* (Kuitert, 1962, and Salas, 1958, quoted by Jirón and Mexzón, 1989). However, the impact of these parasitoid introductions on fruit fly populations is unknown, because no monitoring programme was carried out.

Sugarcane borers in sugarcane

The sugar industry basically consisted of small farmers in rural areas, who formed an association of farmers and sugar mills in 1940: the Sugarcane Industry Association (LAICA by its Spanish initials). Biocontrol in sugarcane was started in the 1950s by the Ministry of Agriculture and Livestock (MAG) with the release of *Billaea* (*Paratheresia*) *claripalpis* Wulp for control of *Diatraea* spp. (Hernández, 1996).

9.2.2 Period 1970–2000

This period is characterized by prospecting and development of mass-rearing methods for natural enemies. Below we summarize the situation of biocontrol in different crops during this period.

Pests in avocado and pineapple

Four strains of *Metarhizium anisopliae* (Metsch.) and four strains of *Beauveria bassiana* (Bals.) were evaluated for the control of thrips in avocado. Disappointingly, none of the entomophathogenic fungi caused sufficient mortality of thrips, since percentages of mortality varied between 3% and 11% (Vargas *et al.*, 2011; Villalobos-Moya *et al.*, 2011). Many efforts have been made to find alternatives for chemical control in pineapple, mainly for *Dysmicoccus brevipes* Cockerell. Gratereaux-Baez (2009) studied the entomopathogenic fungi *M. anisopliae*, *B. bassiana* and *Paecilomyces* spp. to control this mealybug, including selection of the strains, application of a fungal mix, and different liquid and wettable powder formulations, which are still used in the field with good control results, but no data are available about the area treated.

Spiralling whitefly in banana

Following an outbreak of the spiralling whitefly *Aleurodicus dispersus* Russell in 1997, Blanco-Metzler and Laprade (1998a) and Blanco-Metzler *et al.* (2018) studied why some farms were severely attacked while other, surrounding farms were not. The difference in infestation of farms was linked to applications of nematicides during the summer, the vapours of which eliminated natural enemies. This finding marked the beginning of research into the relationships between pests and their natural enemies in banana. Four species of parasitoids were found (*Encarsiella noyesi* (Hayat), *E. aleurodici* (Girault), a new species of *Encarsiella* and *Encarsia guadalupae*) along with two species of predators (*Symnus* sp., *Nephaspis* sp.), two species of mites of the Phytoseiidae family and three spiders (*Crysso* sp., *Plesiometa argyra* Walckenaer and *Gasteracantha cancriformis* L.) (Blanco-Metzler and Laprade, 1998b; Blanco-Metzler and Laprade, 2000). Use of another, more selective nematicide during summer reduced the mortality of the natural enemies.

Fruit flies in citrus

During the 1980s, Jirón and Mexzón (1989) and others took stock of the parasitoids of common fruit flies. Parasitoids imported and released in fruit fly biocontrol programmes were found mainly in flies of the genus *Anastrepha*, rather than in *C. capitata*. In 1981, the Mediterranean Fruit Fly Biological Control Program introduced parasitoids such as *B. longicaudatus* and *A. indica* into mango plantations. In 1986, Elizondo-Solís and Quesada (1990) evaluated the distribution of *Aleurocanthus woglumi* (Ashby) and its parasitoids in four citrus areas of the country. Only *E. opulenta* was found parasitizing the pest, and levels of parasitism varied from 2% to 69%. Also, the fungus *Aschersonia aleyrodidis* (Webber) and the predators *Delphastus* sp. and *Chrysopa* sp. were found to be affecting citrus blackfly populations. In 1995, MAG'S Biological Control Laboratory was opened and started to produce biocontrol agents (Padilla-Monge, 2012). This laboratory merged with the Fruit Fly National Program to produce and distribute parasitoids such as *P. vindemmiae* for the control of *C. capitata* and *Anastrepha* spp., *Sphalangia* spp. for the control of stable fly *Stomoxys calcitrans* L. and *Trichogramma* spp. to control lepidopteran eggs. Also, the pest *Pseudococcus* spp. was reared with the aim of breeding the coccinellid *C. montrouzieri*. With the introduction of mass rearing of natural enemies of fruit flies, quality control programmes developed in Mexico have also been implemented during mass rearing in Costa Rica.

Pests in coffee

Sampling of parasitoids associated with coffee plantations was carried out in the area of Santa Bárbara in 1990 and more than 80 species were collected. The majority of species belonged to the Encyrtidae: *Coccidoxenoides peregrinus* (Timberlake), *Leptomastidea abnormis* (Girault), *Encyrtus infelix* (Embleton) and *Metaphycus helvolus* (Campere) (Hanson, 1991). This project constitutes one of the pioneer studies in biocontrol of tropical coffee plantations.

Lepidopterans in cotton

Trichogramma spp. have been applied for control of Lepidoptera in cotton (Hernández, 1996), but are no longer used.

Macadamia nut borer in macadamia (cashew)

From 1982 to 1993, cultivation of macadamia trees substantially increased due to both excellent ecological conditions and governmental incentives to small farmers to provide an alternative to coffee. This increase also caused an increase in phytosanitary problems, and the macadamia nut borer *Ecdytolopha (Gymnandrosoma) torticornis* became the main pest. Blanco-Metzler *et al.* (2007, 2009) studied the biology, behaviour and population regulation of this pest. They found six parasitoids: Trichogrammatidae, *Apanteles* I, *Apanteles* II, *Ascogaster* sp., *Pristomerus* sp. and a larval hyperparasitoid from the Perilampidae. The mortality of larvae and pupae in a macadamia orchard caused by predators was quantified during the period between nutfall and harvest. Predation by the fire ant *Solenopsis geminata* F. significantly reduced the abundance of nutborer larvae (Blanco-Metzler *et al.*, (2007). González *et al.* (1996) evaluated the effect of strains of *B. bassiana* on *E. torticornis*. At present, no classical or augmentative biocontrol is applied against the pest, but natural control by ants and parasitoids does result in decrease of pest populations.

Cycad aulacaspis in ornamentals

A. Zúñiga and H. Blanco-Metzler (San José, Costa Rica, 2019, personal communication) evaluated the establishment of the parasitoid

Coccobius fulvus (Compere and Annecke) on the cycad aulacaspis *Aulacaspis yasumatsui* Takagi in a commercial crop of *Cycas revoluta* Thun. These wasps were imported from a botanical garden located in Florida. The cycad crop environment was modified by planting the melifera species *Melanthera aspera* (L.) in one of the treatments. A higher percentage of parasitism was obtained in plots with the accompanying vegetation and parasitism was also higher than in plots with only a parasitoid treatment. Thus, it was recommended to implement the accompanying vegetation management strategy to protect and increase natural enemy populations. However, this form of conservation biocontrol is currently not applied.

Pests in oil palm

Sampling for natural enemies in palm oil plantations resulted in finding 33 parasitoid species, two species of Eulophinae wasps in *Stenoma cecropia* (Meyrick) pupae and a *Peleopoda* sp. As entomopathogens, *Metarhizium* sp., *Paecilomyces* sp. and *Microsporidian* sp. were found (Mexzón and Chinchilla, 1991). Mexzón (1997) studied which plants growing in palm oil plantations were most visited by natural enemies and found that these belonged to the families Asteraceae, Euphorbiaceae, Leguminosae and Malvaceae; the majority consisted of perennial plants and had extra floral nectaries. This form of conservation biocontrol is common practice is many perennial orchards.

Stemborers and spittlebugs in sugarcane

In the 1980s, damage to sugarcane crops due to *Diatraea* spp. stem borers became so high that a biocontrol programme was established. Monitoring of sugarcane crops showed low natural parasitism, so LAICA and the Entomology Program of the Research and Extension Area of Sugar Cane (DIECA) imported *Lydella (= Metagonistylum) minense* (Townsend), *Lixophaga diatraea* (Townsend) and *Cotesia* (= *Apanteles*) *flavipes*, which has a shorter life cycle than the tachinids (Badilla *et al.*, 1991). DIECA mass produced *C. flavipes* and was the first to become active at a commercial level in biocontrol in Costa Rica. The parasitoid attacks all three predominant sugarcane borers: *D. tabernella* Dyar (most widespread), *D. saccharalis* (the best host in the laboratory) and *D. guatemalella* Schaus (Badilla *et al.*, 1991; Chaves-Solera, 2008). The programme was a biocontrol and economic success. In 1991, for example, the cost of production of the parasitoid, transport and post-release field evaluations was US$60,761.60 (US$40.2 per hectare with three releases of 6,000 adults per hectare). Total cost of the programme was US$66,592.70, with a liquid payback of US$925,408.10, resulting in a cost–benefit ratio of 1:15 (Badilla *et al.*, 1991). The use of *Trichogramma* for control of Lepidoptera in sugarcane was reported by Hernández (1996).

As well as stem borers, spittlebugs cause damage to sugarcane, in particular *Aeneolamia*, *Prosapia* and *Zulia* spp. (Thompson and León, 2005). For the control of spittlebugs, isolates from the entomopathogenic fungus *M. anisopliae* are used, one from Brazil and another one produced in Costa Rica (Salazar and Badilla, 1997; Badilla, 2002). From 1989 to 1996, 3,246 kg (3.25×10^{16} conidia) were applied on 9,106 ha, obtaining a reduction in spittlebug populations ranging from 5% to 70%. Spittlebugs were also found to be attacked by spiders of the family Salticidae, by Dermaptera species of the genus *Doru* and by nymphs of the predatory fly *Salpingogaster nigra* Shiner (Badilla, 2002).

As a result of the successful use of parasitoids for control of sugarcane borers and fungi for spittlebugs, DIECA registered its products as COTEDIECA (*C. flavipes*), METADIECA (*M. anisopliae*) and BEAUVEDIECA (*B. bassiana*) in 1996 and commercialized these products in Costa Rica and other Central American countries.

M. Acuña and H. Blanco-Metzler (San José, Costa Rica, 2009, personal communication) selected multiple isolates of *M. anisopliae* for the control of the spittlebug *P. simulans* (Walker) under greenhouse conditions and afterwards they evaluated the virulence and pathogenicity of the most promising isolates in nymphs of *P. simulans* in the field and its potential as a microbial control of cercopids in forages.

Shootborers in timber trees

Hypsipyla grandella (Zeller) is a species that attacks forests of species from the Meliaceae family like mahogany (*Swietenia* spp.) and cedar (*Cedrela* spp.). Five *Trichogramma* species were detected parasitizing this pest in 1970: *T. beckeri* (Nagarkatti), *T. fasciatum* (Perkins), *T. pretiosum* (Riley), *T.* near *pretiosum* and *T. semifumatum* (Perkins) (Nagaraja and Nagarkatti, 1973; Grijpma, 1972, quoted by Blanco-Metzler *et al.*, 2001). A number of parasitoids were identified, including *Apanteles* sp., *Hypomicrogaster hypsipylae* De Santis, *Bassus* sp., *Bracon chontalensis* (Cameron) and *Brachymeria conica* (Ashmead), and pest and parasitoid population dynamics were studied (Blanco-Metzler *et al.*, 2001). Varón *et al.* (2005) sampled predatory ants present in mahogany and cedars found in coffee plantations in Turrialba. Six species were found in all and five of them were found in coffee and cedars; the most dominant and abundant species were *S. geminata* and *Pheidole radoszkowskii* (Forel), followed by *Crematogaster* spp. All of these natural enemies play a role in natural control of the pest, but they are not mass produced and applied in augmentative control programmes

Pests in vegetables

The diversity in vegetables grown on small areas and many locations makes it difficult to use biocontrol. The most common vegetable in which biocontrol has been applied is cabbage, for control of *Plutella xylostella* Linnaeus. Ochoa *et al.* (1989) initiated a study of the biology of the parasitoid *Diadegma insulare* (Cresson) and its relationship with the host. Fuentes and Carballo (1995) tested isolates of the entomophatogenic collection of DIECA and CATIE on *P. xylostella*. Isolate 447 was the most effective for control of *P. xylostella* in laboratory conditions and was, therefore, selected for use in the field. However, the IPM programme in which these biocontrol agents were used was terminated.

In the 1990s, *B. bassiana* isolates from different regions were tested for the pepper weevil *Anthonomus eugenii* Cano (Carballo *et al.*, 2001). All isolates were pathogenic for the pepper weevil in laboratory conditions, but only eight obtained a high mortality, a short lethal time (TL_{50}) and a high yield of conidia in rice. Mortality was higher with an oil formulation than with water. Several of these isolates were selected for use in the field, but the IPM programme was terminated.

The National Program for Organic Agriculture (Programa Nacional de Agricultura Organica (PNAO) at the National Learning Institute (Instituto Nacional de Aprendizaje) (INA) was created in Cartago in the 1990s. Its objective was to reduce the use of chemicals in one of the main horticultural zones (slopes of the Irazú Volcano) by training farmers in the production and use of entomopathogens in their crops. This programme had a moderate level of success, because some of the farmers started marketing entomopathogens without knowing the type of fungus they were selling, for example confusing *Metarhizium* with *Penicillium*, based only on the green colour of the spores. Also, due to a lack of quality control, the pathogens were not registered by the Ministry of Agriculture and Livestock (MAG) and commercialization was stopped.

White grubs in various crops

Rice, bean, maize, potato and sugarcane are the crops most affected by white grubs. Different species of *Phyllophaga*, *Anomala*, *Ataenius* and *Cyclocephala* occur in these crops (León, 1996). Fungi of the Deuteromycetes, such as *Paecilomyces*, *Hirsutela*, *Verticillium*, *Akanthomyces*, *Beauveria* and *Metarhizium* as well as strains of *B. popilliae* were evaluated for commercial use. Only *Beauveria* and *Metarhizium* appeared effective in controlling *Phyllophaga* (Hidalgo, 2001). Use of predators and parasitoids for control of white grubs is rare, because they are difficult to rear. Many invertebrate and vertebrate natural enemies are known to attack *Phyllophaga* in Costa Rica, including the toad *Bufo marinus* L. Grub larvae are attacked by Asilidae, Bombyliidae, Tachinidae, Pelecinidae, Tiphiidae, and Scoliidae; pupae are attacked by Asilidae and Bombyliidae; and adults by Tachinidae and Pyrgotidae (Hanson and Gauld, 1996). Currently, no augmentative biocontrol is used against white grubs.

9.3 Current Situation of Biological Control in Costa Rica

9.3.1 Introduction

Advances in biological knowledge and methods contributed to the discovery of new pest control options. Today, molecular biology, host-plant resistance, identification of new biocontrol agents, the use of X-rays to sterilize hosts of natural enemies and the use of plant-produced volatiles to attract natural enemies all help to develop sustainable pest management. In recent decades, important sampling programmes concerning Hymenoptera have also been carried out. Around 20,000 species are estimated to exist in Costa Rica (Gaston and Hanson, 1996). Sampling concentrated on the most abundant family, the Braconidae (van Der Ent and Shaw, 1998). Hanson and Gauld (2006) published the *Hymenoptera of the Neotropical Region*, an obligatory reference book for those who work on biocontrol with parasitoids in Central America and a reference book worldwide. Gaston and Gauld (1993) studied the diversity of pimplines in Costa Rica.

9.3.2 Overview of crops with biological control activities

Pests in banana

The Center for Biological Control and Molecular Biology (CCBBM) was created in 2007 within the National Banana Corporation (its acronym in Spanish is CORBANA), with the aim to reduce pesticide use in banana production. CCBBM focuses on prospecting and harvesting of native microorganisms, in developing natural pesticides and in production of microorganisms. Entomopathogens are now being used experimentally to control the banana root borer *Cosmopolites sordidus* (Germar) (Cubillo *et al.*, 2008). Rodriguez-Morales and Guillén (2009) have developed field application techniques, as well as mass production methods for entomopathogens (Rodríguez-Morales 2009, 2010, 2015). Guillén-Sánchez and Uribe Lorio (2010) evaluated the susceptibility of L_3 larvae of *C. sordidus* to the nematode *Heterorhabditis* sp. and they reported mortalities of 24% 24 h after applying the nematode, 88% after 48 h and 100% after 72 h, demonstrating the potential of this biocontrol agent on *C. sordidus*. The banana mealybug *Pseudococcus elisae* was experimentally treated with entomopathogens and its metabolic substances (Vargas-Castro *et al.*, 2010).

Pests in citrus

The Fruit Fly National Program continues to produce the parasitoid *D. longicaudata* on *C. capitata* to control the Mediterranean fly and other flies from the genus *Anastrepha*. Currently *Anastrepha ludens* (Loew) is also evaluated as a host for mass rearing (Morera-Montoya, 2016). Initially, parasitoids were released in paper bags, which were hung on branches or places in shadow, and they were protected from rain with a plastic cover. Currently a plastic device is used which was developed by the Mexican Mosca Med programme to protect parasitoids from rain and natural enemies and to encourage mating. The device consists of intertwined cardboard tapes with a mixture of honey and finely shredded toilet paper, so that the parasitoids have a resting place and food (Morera-Montoya, 2016). With the introduction of mass-breeding programmes such as the one of *D. longicaudata*, quality control methods for parasitoids were also put into practice.

Since the arrival of huanglongbing (HLB) with the Asian citrus psyllid, the commercial orange-production company Ticofruit has taken the initiative to produce and release the specific and efficient parasitoid *Tamarixia radiata* Watersen. León and Fallas (2010) studied the presence of *T. radiata* in the field. A problem is that the parasitoid is not compatible with pesticides applied during harvest. Ticofruit also found that larvae of chrysopids (Neuroptera) hid in citrus orchard waste, which protects them from pesticides. The chrysopids have a high predation capacity: one individual can eat more than 800 eggs and nymphs of *Diaphorina citri* Kuwayama and causes a mortality of 70% (Delgado, 2017).

Coffee berry borer in coffee

In 2002, the coffee berry borer *Hypothenemus hampei* (Ferrari) was detected in Costa Rica and severely complicated coffee production. The pest is hard to control and mainly depends on good practices carried out by the farmer (Borbón, 2001). IPM of coffee berry borer consists of biological, chemical, semiochemical, cultural and manual methods and includes proper hand harvesting of beans. Instituto del Café de Costa Rica (ICAFE) imported the following parasitoids to control the coffee berry borer: *Prorops nasuta* (Waterson), *Cephalonomia stephanoderis* (Betrem), *Heterospilus coffeicola* Schneideknecht and *Phymastichus coffea* (LaSalle). G. Alpízar, M. Blanco-Metzler and O. Borbón (San José, Costa Rica,

2004, personal communication) studied *P. coffea*; and M. Barrantes, H. Blanco-Metzler and O. Borbón (San José, Costa Rica, 2004, personal communication) evaluated *C. stephanoderis* in the laboratory. These species were used on 500 ha in the field for several years, but since ICAFE evaluated effectiveness and methods of application of *B. bassiana* and *M. anisopliae*, the parasitoids are no longer applied.

The predation potential of *S. geminata*, *P. radoskowskiiy* and *Crematogaster torosa* (Mayr) on various stages of the coffee berry borer was determined (Varón *et al.*, 2004) and laboratory results showed that the three species could reach levels of predation of 100%. However, in the field much lower levels of predation were found, probably because they are generalist predators.

Fruit flies in guava

Morera-Montoya *et al.* (2013) found that *Doryctobracon* sp. was parasitizing larvae of *Anastrepha* spp. during a study of guava packaging. In 2012, with the aid of the International Atomic Energy Agency (UN-IAEA), an X-ray machine was acquired and used to sterilize the Mediterranean fruit fly and to produce the parasitoid *D. longicaudata*. X-raying of the flies guaranteed that parasitoid releases were free of fertile fruit flies.

False codling moth in macadamia (cashew)

Mexzón (2001) studied occurrence of insects on weeds of macadamia plantations and found Sciaridae, Bibionidae, Drosophilidae, Muscidae, Mycetophilidae and Trichogrammatidae. A rearing of *Trichogrammatoidea cryptophlebiae* Nagaraja was started and levels of parasitism of 80% in the false codling moth *Thaumatotibia* (= *Cryptophlebia*) *leucotreta* Meyrich were obtained in the laboratory. Since the mass production costs were extremely high, the pest could not be controlled in a cost-effective manner with this parasitoid.

Oil palm defoliator in palm plantations

Chinchilla (2003) suggested using vegetation management to attract natural enemies of the oil palm defoliator *Opsiphanes cassina* C. and Felder, the most common pest of oil palms. Among the parasitoids of this pest are *Cotesia* sp., which was found to feed on the following plant species: *Ageratum conyzoides* L., *Amaranthus spinosus* L., *Baltimora recta* L., *Byttneria aculeate* Jacquin, *Cassia tora* L., *Solanum jamaicense* and *Vitis syciodes* L. *Horismenus* sp. fed on *B. aculeata*, *C. reticulata*, *C. tora*, *Melanthera aspera* L., *Scleria melaleuca* Miller and *V. sycioides*. *Conura* sp. fed on *A. spinosus*, *B. aculeata*, *C. tora*, *M. aspera*, *S. melaleuca*, *Urena lobata* L. and *V. sycioides* (Mexzón and Chinchilla, 1999). Egg parasitoids from the genera *Ooencyrtus* and *Telenomus*, larval parasitoids from the genera *Brachymeria*, *Cotesia*, *Conura* and *Horismenus*, and tachinid flies can all contribute to population reduction of *O. cassina* (Mexzón and Chinchilla, 1999) in this form of conservation biocontrol.

Aphids in sugarcane

In Costa Rica a wide range of predators have been found to feed on juice-sucking aphids, including *Ceraeochrysa* sp., *Chrysoperla carnea* (Stephens), *Olla v-nigrum* (Mulsant) and *Cycloneda sanguine* L. (Salazar, 2012). Also, conservation biocontrol to reduce rodents and damaging Lepidoptera became quite popular and consists of placing perches to encourage the presence of birds of prey (sparrow hawks, eagles, falcons and owls) and the great kiskadee bird. Currently, this is used by sugar mills in Guanacaste and San Carlos (Salazar *et al.*, 2016), but quantitative data about areas on which this type of conservation biocontrol is used are not available.

Pests in various crops

The National Institute of Innovation and Transfer of Agricultural Technology (Instituto Nacional de Innovación y Transferencia en Tecnología Agropecuaria) (INTA) has intensified the identification of pests and their natural enemies in many fruit trees, as in the naranjilla crops (*Solanum quitoense* var. Septentrional) (León and Flores, 2007).

Aphids create problems in several crops; thus in 2010, a sampling programme in cultivated and uncultivated plants for parasitoids of aphids was carried out. With the results an important list was generated, identifying species according to their distribution in the country.

Aphidiine represented 2,832 species, of which ten species belonging to six genera have not been identified yet. *Aphidius colemani* (Viereck) and *Lysiphlebus testaceipes* (Cresson) represented 90% of all specimens collected; a lot of these species are probably exotic, as they attack the majority of their exotic aphid hosts (Zamora *et al.*, 2010).

Although not a major pest in Costa Rica, a field study of the greenhouse whitefly *Trialeurodes vaporariorum* (Westwood) and its native parasitoid *Encarsia formosa* Gahan provided important insights into the natural distribution of whitefly and its parasitoid and helped in understanding the evolution of foraging behaviour of *E. formosa* (Burger *et al.*, 2004).

9.4 New Developments of Biological Control in Costa Rica

Currently, many biocontrol projects are being carried out to support development and production of entomopathogenic fungi by small providers. Further, the majority of exporting companies of fruit, flowers, cotton, coffee and others, have implemented IPM practices and have laboratories and trained professionals for production and use of entomopathogens, parasitoids and predators. Also some natural enemies are imported via Koppert Biological Systems (The Netherlands) and Biobest (Belgium). We were not able to obtain a complete set of data on areas on which biocontrol agents are applied. The data on areas under augmentative biocontrol that we could collect are presented in Table 9.1, showing that it is used on at least 15,650 ha. However, this is an underestimate of the use of biocontrol in Costa Rica, because the data in Table 9.1 do not include information from exporting companies and application of imported biocontrol agents. More importantly, the areas under classical and conservation biocontrol are not incorporated in the estimate.

In DIECA, methods for biocontrol of the main sugarcane pests have advanced greatly. In order to enhance efficiency and economics of these biocontrol agents, mass breeding and quality control of parasitoids and entomopathogens need to be further improved.

Research done to discover the diversity of entomopathogens in Costa Rica has made it possible to evaluate and develop the future use of native strains of fungal and bacterial entomopathogens.

The University of Costa Rica and the Phytosanitary Service of the State are now negotiating the establishment of a National Program for Biological Control (PROCOBI) to determine and regulate the quality of the biocontrol agents that are currently being marketed. PROCOBI will also initiate research of natural enemies of the most important pests

Table 9.1. Crops under augmentative biological control in Costa Rica.

Crop	Pest	Biocontrol agent	Area applied
Orange	*Diaphorina citri*	*Tamarixia radiata*	6,000 ha
		Ceraeochrysa valida	6,000 ha
Guava	*Anastrepha striata*	*Diaschasmimorpha longicaudata*	20 ha
Jocote (*Spondias*)			30 ha
Passion fruit			20 ha
Sugarcane	*Zulia vilion*	*Metarhizium anisopliae*	4,799.9 ha
	Aenolamia albofasciata	*Beauveria bassiana*	3,882.3 ha
	Leptodictya tabida		
	Saccharosydne saccharirora		
	Metamasius hemipterus		
	Diatraea spp.	*Cotesia flavipes*	
Pineapple	*Stomoxys calcitrans*	*Spalangia* sp.	780 ha
Palm tree	*Musca domestica*	*Spalangia* sp.; applied by volume of pupae; 10 ml m^{-2} per patch	?
	Stomoxys calcitrans		
Coffee	*Hypothenemus hampei*	*Beauveria bassiana*	4,000 ha

and will offer training to improve the use of biocontrol agents in the field.

In conclusion: implementation of biocontrol in Costa Rica will keep increasing, since there is great pressure to reduce the use of chemical pesticides for all the export crops. Negative factors that may affect an increase of biocontrol are: (i) the poor quality control of biocontrol products sold by small providers; (ii) the aggressive way in which pesticide salesmen approach farmers; and (iii) climate change.

9.5 Acknowledgements

Many thanks to C. Guillén (CORBANA), R. León (INTA), J.D. Salazar and A. Rodriguez (DIECA) who provided valuable material used in this chapter.

References

(References with grey shading are available as supplementary electronic material)

Badilla, F. (2002) Un programa exitoso de control biológico de insectos plaga de la caña de azúcar en Costa Rica [A successful programme of biological control of insect pests in sugarcane]. *Manejo Integrado de Plagas y Agroecología* 64, 77–87.

Badilla, F., Solis, A. and Alfaro, D. (1991) Control biológico del taladrador de la caña de azúcar *Diatraea* spp. (Lepidoptera: Pyralidae) en Costa Rica [Biological control of the sugar cane borer *Diatraea* spp.]. *Manejo Integrado de Plagas* 20–21, 39–44.

Blanco-Metzler, H. and Laprade, S. (1998a) *La mosca blanca en espiral* Aleurodicus dispersus *Russell (Homoptera: Aleyrodidae) en banano* [*The spiralling whitefly in banana*]. Informe de consultoría. UCR/CORBANA, San José, Costa Rica.

Blanco-Metzler, H. and Laprade, S. (1998b) Enemigos naturales de la mosca blanca *Aleurodicus dispersus* Russell (Homoptera:Aleyrodidae): parasitoides y depredadores. [Natural enemies of the spiraling whitefly *Aleurodicus dispersus* : parasitoids and predators]. *Agronomía Mesoamericana* 9(2), 41–44.

Blanco-Metzler, H. and Laprade, S. (2000) Variación espacial y estacional de la mosca blanca *Aleurodicus dispersus* Russell (Homoptera: Aleyrodidae) y sus parasitoides [Spatial and seasonal variation of the spiraling whitefly *Aleurodicus dispersus*]. *Revista Manejo Integrado de Plagas* 55, 43–48.

Blanco-Metzler, H., Vargas, C. and Hauxwell, C. (2001) Indigenous parasitoids for the control of *Hypsipyla grandella* (Lepidoptera: Pyralidae) in Latin America. In: *International Workshop for the control of the Meliaceae shootborers*, Kandy, Sri Lanka, 19–23 August 1996. Proceedings ACIAR, Australia, pp. 140–145.

Blanco-Metzler, H., Watt, A.D. and Cosens, D. (2007) The effect of predators on the population dynamics of *Ecdytolopha torticornis* Meyrick (Lepidoptera: Tortricidae). *Agronomía Costarricense* 31, 33–39.

Blanco-Metzler, H., Watt, A.D. and Cosens, D. (2009) The effect of parasitism on the population dynamics of *Gymnandrosoma torticornis* (Lepidoptera: Tortricidae). *Revista Biología Tropical* 57(4), 1245–1252.

Blanco-Metzler, H., Laprade-Coto, S. and Weintraub, P.G. (2018) Factors affecting the biological control of the spiralling whitefly *Aleurodicus dispersus* (Hemiptera: Aleyrodidae) in banana plantations in Costa Rica: a study case. *Biological Control Science and Technology* 28(10), 989–994.

Borbón, O. (2001) *Situación actual de la broca del fruto del cafeto en Costa Rica (*Hypothenemus hampei *Ferrari) [Current situation of the coffee berry borer in Costa Rica*]. CICAFE, Costa Rica.

Burger, J.M.S., Gort, G., van Lenteren, J.C. and Vet, L.E.M. (2004) Natural history of whitefly in Costa Rica: an evolutionary starting point. *Ecological Entomology* 29(2), 150–163.

Carballo, M., Rodriguez, L. and Durán, J. (2001) Evaluación de *Beauveria bassiana* para el control del picudo del chile en laboratorio [Evaluation of *Beauveria bassiana* for the control of the pepper weevil]. *Revista Manejo Integrado de Plagas* 62, 54–59.

Chaves-Solera, M. (2008) El Control Biológico de Plagas como Experiencia Empresarial de la Agroindustria Azucarera Costarricense [The biological control of pests as commercial experience of the Costa Rican agribusiness]. In: *Workshop 'Insumos para la Construcción de una Agenda Agrícola Ambiental'*, organized for the Universidad Nacional (UNA) and CATIE, 7–8 October 2008, San José, Costa Rica.

Chinchilla, C. (2003) Manejo integrado de problemas fitosanitarios en palma aceitera *Elaeis guinensis* en América Central [Integrated pest management of phytosanitary problems in oil palm in Central America]. *Revista Manejo Integrado de Plagas y Agroecología* 67, 69–82.

CIA (2017) The World Factbook: Costa Rica. Available at: https://www.cia.gov/library/publications/the-world-factbook/geos/cs.html (accessed 12 July 2019)

Cubillo, D., Laprade, S. and Obregón, M. (2008) Eficacia biológica de cepas de *Beauveria bassiana* en el control de adultos de *Cosmopolites sordidus* (Coleoptera: Curculionidae), en condiciones de laboratorio y campo [Biological efficacy of *Beauveria bassiana* strains in the adult control of *Cosmopolites sordidus* under laboratory and field conditions]. In: Sandoval, J. (ed.) *Informe Anual 2007*. Dirección de Investigaciones, CORBANA, San José, Costa Rica, pp. 153–156.

Delgado, C.Y. (2017) Control biológico Ticofruit – en línea, presentación [Biological control Ticofruit – online presentation]. Available at: https://prezi.com/jpp9n1iczf15/control-biologico-ticofrut/ (accessed 16 June 2017).

Elizondo-Solís, J. and Quesada, J. (1990) Identificación y evaluación de los enemigos naturales de la mosca prieta de los cítricos *Aleurocanthus woglumi* Ashby (Homoptera: Aleyrodidae) en cuatro zonas citrícolas de Costa Rica [Identification and evaluation of natural enemies of citrus fruit fly *Aleurocanthus woglumi* Ashby (Homoptera: Aleyrodidae) in four citrus areas of Costa Rica]. *Turrialba* 40(2), 190–197.

Fuentes, G. and Carballo, M. (1995) Evaluación de aislados de *Beauveria bassiana* (BALS.) Vuill para el control de *Plutella xyllostella* (Lepidoptera: Iponomeutidae) [Evaluation of *Beauveria bassiana* isolates (BALS.) Vuill for the control *of Plutella xyllostella* (Lepidoptera: Iponomeutidae)]. *Revista Manejo Integrado de Plagas* 35, 14–18.

Gaston, K. and Gauld, I. (1993) How many species of pimplines (Hymenoptera: Ichneumonidae) are there in Costa Rica? *Journal of Tropical Ecology* 9, 491–499.

Gaston, K. and Hanson, P. (1996) The size and composition of the hymenopteran fauna of Costa Rica. *Biogeography* 23, 105–113.

González, H.O., Carballo, M. and Blanco-Metzler, H. (1996) Efecto de cepas de *Beauveria bassiana* (Bals) Vuill sobre la mortalidad de *Ecdytolopha torticornis* Meyrick (Lepidoptera: Tortricidae) en macadamia [Effect of strains of *Beauveria bassiana* (Bals) Vuill on the mortality of *Ecdytolopha torticornis* Meyrick (Lepidoptera: Tortricidae) in macadamia]. *Revista Manejo Integrado de Plagas* 40, 17–23.

Gratereaux-Baez, W.V. (2009) Potencial del uso de entomopatógenos para el control de cochinilla (*Dysmicoccus brevipes*) en producción orgánica de piña (*Ananas comosus*) [Potential of the use of entomopathogens for mealybug control (*Dysmicoccus brevipes*) in organic pineapple production (*Ananas comosus*)]. MSc thesis. CATIE, Turrialba, Costa Rica.

Guillén-Sanchez, C. and Uribe Lorio, L. (2010) Potencial de los nematodos entomopatógenos para el control del picudo negro (*Cosmopolites sordidus*) (Germar) (Coleoptera: Dryopthoridae) [Potential of entomopathogenic nematodes for the control of the black weevil (*Cosmopolites sordidus*) (Germar) (Coleoptera: Dryopthoridae)]. In: Sandoval, J. (ed.) *Informe Anual 2009*. Dirección de Investigaciones, CORBANA, San José, Costa Rica, pp. 298–299.

Hanson, P. (1991) Los parasitoides asociados al cafeto en Costa Rica [The parasitoids associated with coffee in Costa Rica]. *Revista Manejo Integrado de Plagas* 20–21, 8–10.

Hanson, P. and Gauld, I. (1996) Control biológico de *Phyllophaga*: depredadores y parasitoides [The parasitoids associated with coffee in Costa Rica]. In*:* Shannon, P.J. and Carballo, M. (eds) *Biología y control de* Phyllophaga *spp*. CATIE serie Técnica, Informe Técnico No. 277. CATIE/PRIAG, Turrialba, Costa Rica.

Hanson, P. and Gauld, I. (2006) *Hymenoptera de la Región Neotropical* [*Hymenoptera of the Neotropical Region*]. Issue 77 of *Memoirs of the American Entomological Institute*. American Entomological Institute, Gainesville, Florida.

Hernández, J. (1996) El control biológico en Costa Rica [Biological control in Costa Rica]. In: Zapater, M.C. (ed.) *El control biológico en América Latina*. IOBC, Buenos Aires, Argentina, pp. 35–40.

Hidalgo, E. (2001) Uso de microorganismos para el control de *Phyllophaga* spp. [Use of microorganisms for the control of *Phyllophaga* spp.]. *Revista Manejo Integrado de Plagas* 60, 50–56.

Hilje, L., Cartín, L. and Víctor March, L.E. (1989) El combate de plagas agrícola dentro del contexto histórico Costarricense [The combat of agricultural pests within the Costa Rican historical context]. *Revista Manejo Integrado de Plagas* 14, 68–86.

Jiménez, V.M., Acuña-Gutiérrez, C., Ávila-Agüero, M.L., Alvarado, A., Blanco, M. *et al.* (2017) Costa Rica and its commitment to sustainability. In: *Challenges and Opportunities for Food and Nutrition Security in the Americas. The View of the Academies of Sciences*. IANAS, IAP and BMBF, México DF, pp. 245–263. [Free public access of this publication in English and Spanish at www.ianas.org]

Jirón, L. and Mexzón, R. (1989) Parasitoid hymenopterans of Costa Rica: geographical distribution of the species associated with fruit flies (Diptera: Tephritidae). *Entomophaga* 34, 53–60.

León, R. (1996) Problemática de *Phyllophaga* spp. en Costa Rica [Problems of *Phyllophaga* spp. in Costa Rica]. In*:* Shannon, P.J. and Carballo, M. (eds) *Biología y control de* Phyllophaga *spp*. CATIE/PRIAG, Turrialba, Costa Rica.

León, R. and Fallas, G. (2010) Presencia de *Tamarixia radiata* (Hymenoptera: Eulophidae) parasitoide de *Diaphorina citri* (Homoptera: Psyllidae) en cultivos cítricos en Costa Rica [Presence of *Tamarixia radiata* (Hymenoptera: Eulophidae) parasitoid of *Diaphorina citri* (Homoptera: Psyllidae) in citrus crops in Costa Rica]. *XIII Congreso Agropecuario Forestal (CONAGROF)*, Colegio Ing. Agr. De Costa Rica. 4–6 agosto 2010. San José, Costa Rica. p. 76.

León, R. and Flores, G. (2007) Identificación de plagas y sus enemigos naturales en el cultivo de Naranjilla (*Solanum quitoense* var. Septentrional) en la Región Central Oriental [Identification of pests and their natural enemies in the cultivation of Naranjilla (*Solanum quitoense* var. Septentrional) in the Eastern Central Region]. Informe Final. INTA, San José, Costa Rica.

Mexzón, R. (1997) Pautas de manejo de las malezas para incrementar las poblaciones de insectos benéficos en el cultivo de palma aceitera (*Elaeis guinensis* Jacquin) [Weed management guidelines to increase beneficial insect populations in oil palm (*Elaeis guinensis* Jacquin]. *Agronomía Mesoamericana* 8(2), 21–32.

Mexzón, R. (2001) *Reconocimiento de las especies vegetales en las que se refugian o alimentan las avispas parasitoides de los taladradores (Lepidoptera: Tortricidae) y de los chinches fitófagos (Hemiptera: Alydidae, Pentatomidae) de la nuez de la macadamia en Costa Rica* [*Recognition of the plant species in which the parasitoid wasps of the borers (Lepidoptera: Tortricidae) and of the phytophagous bugs (Hemiptera: Alydidae, Pentatomidae) of the macadamia nut in Costa Rica take refuge or feed*]. Proyecto de investigación Número 813-AO-146 [Research project Number 813-AO-146]. Centro de Investigaciones en Protección de cultivos (CIPROC), Universidad de Costa Rica, San José, Costa Rica.

Mexzón, R. and Chinchilla, C. (1991) Entomofauna perjudicial, enemigos naturales y malezas útiles en palma aceitera (*Elaeis guineensis*) J. en América Central [Detrimental entomofauna, natural enemies and useful weeds in oil palm (*Elaeis guineensis*) J. in Central America]. *Revista Manejo Integrado de Plagas* 20–21, 1–7.

Mexzón, R. and Chinchilla, C. (1999) Plant species attractive to beneficial entomofauna in oil palm plantations in Costa Rica. ASD Oil Palm Papers (CR) 19. *Palmas 24(1), 33–57*.

Morera-Montoya, R. (2016) Nueva alternativa para la liberación en campo del parasitoide de moscas de la fruta *Diachasmimorpha longicaudata* [New alternative for the field release of the parasitoid of fruit flies *Diachasmimorpha longicaudata*]. Programa Nacional de Moscas de la Fruta. Servicio Fitosanitario del Estado (SFE), Ministerio de Agricultura y Ganadería, San José, Costa Rica.

Morera-Montoya, R., Blanco-Metzler, H and Loría, C.L. (2013) Evaluación de materiales de embolsado para el control de las moscas de la fruta *Anastrepha* spp. en la guayaba taiwanesa (*Psidium guajava*) en Alajuela, Costa Rica [Evaluation of bagging materials for the control of fruit flies *Anastrepha* spp. in the Taiwanese guayaba (*Psidium guajava*) in Alajuela, Costa Rica]. *Entomología Mexicana* 12(2), 1206–1212.

Ochoa, R., Carballo, M. and Quezada, R. (1989) Algunos aspectos de la biología y comportamiento de *Plutella xylostella* (Lepidoptera: Plutellidae) y de su parasitoide *Diadegma insulare* (Hymenoptera: Ichneumonidae) [Some aspects of the biology and behavior of *Plutella xylostella* (Lepidoptera: Plutellidae) and its parasitoid *Diadegma insulare* (Hymenoptera: Ichneumonidae]. *Revista Manejo Integrado de Plagas* 11, 21–30.

Ovruski, S., Aluja, M., Sivinski, J.and Wharton, R. (2000) Hymenopteran parasitoids on fruit-infesting Tephritidae (Diptera) in Latin America and the southern United States: diversity, distribution, taxonomic status and their use in fruit fly biological control. *Integrated Pest Management Reviews* 5, 81–107.

Padilla-Monge, M.M. (2012) *Servicio Fitosanitario del Estado, nuestra huella en el tiempo*. [*Phytosanitary Service of the State, our footprint in time*]. MAG/SFE, San José, Costa Rica.

Rodríguez-Morales, A. (2009) Efecto de la adición de fuentes nitrogenadas, sobre el crecimiento del hongo *Beauveria bassiana* (CB-60) [Effect of the addition of nitrogenous sources, on the growth of the fungus *Beauveria bassiana* (CB-60)]. In: Sandoval, J. (ed.) *Informe Anual 2008*. Dirección de Investigaciones, CORBANA, San José, Costa Rica, pp. 142–145.

Rodríguez-Morales, A. (2010) Evaluación de sustratos alternativos para la producción de hongos antagonistas, entomopatógenos y nematófagos [Evaluation of alternative substrates for the production of antagonistic, entomopathogenic and nematophagous fungi]. In: Sandoval, J. (ed.) *Informe Anual 2009*. Dirección de Investigaciones, CORBANA, San José, Costa Rica, pp. 143–146.

Rodríguez-Morales, A. (2015) Bioprospección de nematodos entomopatógenos (NEPs) y su evaluación como posible agente de control biológico del picudo negro, *Cosmopolites sordidus* Germar (Coleoptera: Dryophthoridae) [Bioprospecting entomopathogenic nematodes (NEPs) and their evaluation as a possible biological control agent of the black weevil, *Cosmopolites sordidus* Germar (Coleoptera: Dryophthoridae)]. In: Sandoval, J. (ed.) *Informe Anual 2015*. Dirección de Investigaciones, CORBANA, San José, Costa Rica.

Rodríguez-Morales, A. and Guillén, C. (2009) Evaluación de la estabilidad del hongo *Beauveria bassiana* en trampas tipo cuña [Evaluation of the stability of *Beauveria bassiana* in wedge-type trap]. In: Sandoval, J. (ed.) *Informe Anual 2008*. Dirección de Investigaciones, CORBANA, San José, Costa Rica, pp. 146–150.

Salazar, J. (2012) Áfidos en el Cultivo de la Caña de Azúcar (disco compacto) [Aphids in the Sugar Cane Crop (compact disk)]. In: LAICA, DIECA (ed.) *Memoria Congreso Tecnológico de DIECA* (5, 2012) Grecia, Costa Rica.

Salazar, J. and Badilla, F. (1997) Evaluación de dos cepas del hongo entomopatógeno *Metarhizium anisopliae* y seis insecticidas granulados en el control de salivazo (*Aeneolamia postica*) (Hom: Cercopidae) en caña de azúcar en la Región de San Carlos, Costa Rica [Evaluation of two strains of the entomopathogenic fungus *Metarhizium anisopliae* and six granulated insecticides in spittlebug control (*Aeneolamia postica*) (Hom: Cercopidae) in sugarcane in the San Carlos Region, Costa Rica]. *X Congreso Nacional Agronómico Nacional y de Recursos Naturales, 8–12 July 1996*. San José, Costa Rica.

Salazar, J., Oviedo, R., Cadet, E. and Sáenz, C. (2016) Control biológico y otras estrategias de manejo de plagas implementadas en el cultivo de la caña de azúcar en Costa Rica [Biological control and other pest management strategies implemented in the cultivation of sugarcane in Costa Rica]. In: *Congreso Nacional Agropecuario, Forestal y Ambiental, 14, 27–29 October 2016*, Heredia, Costa Rica.

Thompson, V. and León, R. (2005) La identificación y distribución de los salivazos de la caña de azúcar y los pastos (Homoptera: Cercopidae) en Costa Rica [The identification and distribution of spittlebugs from sugarcane and pastures (Homoptera: Cercopidae) in Costa Rica]. *Revista Manejo Integrado de Plagas y Agroecología* 75, 43–51.

van der Ent, L. and Shaw, S.R. (1998) Species richness of Costa Rican Cenocoeliini (Hymenoptera: Braconidae), a latitudinal and altitudinal search for anomalous diversity. *Journal of Hymenopteran Research* 7, 15–24.

Vargas, A., Villalobos, K. and González, A. (2011) Evaluación de *Beauveria bassiana* y *Metarhizium anisopliae* en condiciones de campo para el combate de trips en el cultivo de aguacate (*Persea americana* Mill) en San Pablo de León Cortés, Costa Rica [Evaluation of *Beauveria bassiana* and *Metarhizium anisopliae* in field conditions for the combat of thrips in the avocado crop (*Persea americana* Mill) in San Pablo de León Cortés, Costa Rica]. *Métodos en Ecología y Sistemática* 6(3), 62–70.

Vargas-Castro, E., Rodríguez-Morales, A. and Guillén-Sánchez, C. (2010) Proyecto de evaluación *in vitro* de diferentes hongos entomopatógenos y sus metabolitos secundarios, para el control biológico de la cochinilla del banano, *Pseudococcus elisae* (Hemiptera: Pseudococcidae) [Project of *in vitro* evaluation of different entomopathogenic fungi and their secondary metabolites, for the biological control of the banana cochineal, *Pseudococcus elisae* (Hemiptera: Pseudococcidae)]. In: Sandoval, J. (ed.) *Informe Anual 2010*. Dirección de Investigaciones, CORBANA, San José, Costa Rica, pp. 95–99.

Varón, E., Hanson, P., Borbón, O., Carballo, M. and Hilje, L. (2004) Potencial de hormigas como depredadoras de la broca del café (*Hypothenemus hampei*) en Costa Rica [Potential of ants as predators of the coffee berry borer (*Hypothenemus hampei*) in Costa Rica]. *Revista Manejo Integrado de Plagas y Agroecología* 73, 42–50.

Varón, E., Barbera, N., Hanson, P., Carballo, M. and Hilje, L. (2005) Potencial de depredación de *Hypsipyla grandella* por hormigas en cafetales de Costa Rica [Predation potential of *Hypsipyla grandella* by ants in coffee plantations in Costa Rica]. *Revista Manejo Integrado de Plagas y Agroecología* 74, 17–23.

Villalobos-Moya, K., Vargas-Martinez, A., and González-Herrera, A. (2011) Evaluación de *Beauveria bassiana* y *Metarhizium anisopliae* en condiciones de campo para el combate de trips en el cultivo de aguacate (*Persea americana*) en San Pablo de León Cortés, Costa Rica [Evaluation of *Beauveria bassiana* and *Metarhizium anisopliae* in field conditions for the combat of thrips in the avocado crop (*Persea americana*) in San Pablo de León Cortés, Costa Rica]. *Métodos en Ecología y Sistemática* 6(3), 53–56.

Zamora, D., Hanson, P. and Stary, P. (2010) Survey of the aphid parasitoids (Hymenoptera: Braconidae: Aphidiinae) of Costa Rica with information on their aphid (Hemiptera: Aphidoidea): plant associations. *Psyche Journal of Entomology* 2010, 1–7. DOI:10.1155/2010/278643.

10 Biological Control in Cuba

María Elena Márquez[1*], Luis L. Vázquez[2], Mayra G. Rodríguez[3], Jorge L. Ayala Sifontes[4], Fermín Fuentes[5], Mayra Ramos[6], Leopoldo Hidalgo[3] and Lidcay Herrera[7]

[1]*Universidad de La Habana, Havana, Cuba;* [2]*Instituto de Investigaciones de Sanidad Vegetal-INISAV, Havana, Cuba;* [3]*Centro Nacional de Sanidad Agropecuaria-CENSA, apartado 10, Mayabeque, Cuba;* [4]*Dirección Provincial de Sanidad Vegetal, Sancti Spiritus, Cuba;* [5]*Laboratorio Provincial de Sanidad Vegetal, Havana, Cuba;* [6]*Instituto Superior de Tecnologías y Ciencias Aplicadas-InSTEC, Havana, Cuba;* [7]*Facultad de Ciencias Agropecuarias, Universidad Central Martha Abreu, Villa Clara, Cuba*

* E-mail: marquezmariaelena18@gmail.com

Abstract

The first biological control project in Cuba concerned the introduction of the parasitoid *Eretmocerus serius* in 1929, resulting in successful classical biocontrol of citrus blackfly in citrus. The subsequent biocontrol success that is still in use on large areas today was obtained in the 1940s by mass rearing and releasing the native dipteran parasitoid *Lixophaga diatraeae* for control of the sugarcane borer. Nowadays, many native and exotic *Trichogramma* species are successfully applied against lepidopteran defoliators in the field. Other current augmentative biocontrol programmes involve the use of microbial agents, nematodes, parasitoids and predators for pest and disease management in various crops. A network of 175 mass rearing centres for entomophages and entomopathogens (CREE) and four industrial plants belonging to the Enterprise System of the Ministry of Agriculture, guarantee the mass production of native strains of microbial control agents, such as *Beauveria bassiana*, *Metarhizium anisopliae*, *Lecanicillium lecanii*, *Bacillus thuringiensis*, *Trichoderma* spp. and *Heterorhabditis* spp. Each year these microbial control agents are applied on about 2,400,000 ha of field crops. Conservation biocontrol practices to increase natural enemy populations and the promotion of natural reservoirs of *Pheidole megacephala* predatory ants, along with capturing and re-release of the coccinellids *Cycloneda sanguinea*, *Coleomegilla cubensis*, *Hippodamia convergens* and *Chilocorus cactus* L. in urban agriculture, are widely applied by farmers in Cuba.

10.1 Introduction

Cuba has an estimated population of about 11,150,000 (July 2017) and its main agricultural products are sugar, tobacco, citrus, coffee, rice, potatoes, beans and livestock (CIA, 2017). According to Hernández *et al.* (2017, pp. 266, 277, 278):

> Cuba's total land area (10,988.4 million ha) includes 6,240.3 million ha for agricultural use. A total of 3,371.6 million ha is occupied by forests while the remainder comprises aqueous surfaces and other land unsuitable for agriculture ... Cuba is well known for ... technologies based on 'Conservation Agriculture' for soils and crops, regarding the farm as a basic management unit ... and ... Development of integrated pest management systems; design of bioproducts that support other methods of control.

10.2 History of Biological Control in Cuba

10.2.1 Period 1880–1969

Information about biocontrol projects during this period can be found in Bruner *et al.* (1945), Martínez *et al.* (2000) and Vázquez and Pérez (2016). Two major projects were as follows.

1. The introduction of the parasitoid *Eretmocerus serius* Silvestri from India for control of citrus blackfly, *Aleurocanthus woglumi* Ashby. From 1930 until today, citrus blackfly has been kept at densities below the economic threshold, making chemical control redundant.

2. The mass rearing and augmentative release of the native dipteran parasitoid, *Lixophaga diatraeae* (Towns) (Diptera: Tachinidae), for control of sugarcane borer *Diatraea saccharalis* Fabricius. L.C. Scaramuzza (Bruner *et al.*, 1945) started mass rearing and inundative releases of this native parasitoid in the antique sugarmill 'Mercedes' (now 'Seis de Agosto') in Matanzas province, for control of the borer. This parasitoid has been used since then in all sugarcane areas in Cuba and resulted in termination of chemical control.

10.2.2 Period 1970–2000

The information about the biocontrol situation in Cuba during this period is summarized without giving much detailed information, to prevent overlap with the section about the current situation. Cuba showed many activities in the field of augmentative releases, from 1982 in sugarcane fields and from 1988 in the rest of the agricultural crop (Fuentes *et al.*, 1998; Pérez and Vázquez, 2001; Vázquez and Pérez, 2016) (Table 10.1). *Trichogramma* species were applied in 1999 on 283,251 ha for control of Lepidoptera in pastures, cassava and vegetables (Pérez and Vázquez, 2001). Like in the pre-1970 period, sugarcane borers *D. saccharalis* were controlled with the native tachinid parasitoid *L. diatraea*. From 1960 to 1980, technological advances resulted in an increased production of this parasitoid in six laboratories. In 1982, a National

Table 10.1. Overview of major biological control projects and areas on which biological control is applied in Cuba.

Natural enemy	Origin	Started	Target pest and crop	Type of biocontrol[a]	Success	Area under biocontrol in 2016
Period 1880–1969						
Rodolia cardinalis	USA	1928	*Icerya purchasi* in citrus	CBC	Yes	
Eretmocerus serius	India	1930	*Aleurocanthus woglumi* in citrus	CBC	Yes	23,000 ha
Scymnus smithianus, Catana clauseni, Encarsia (= *Prosp altella*) *divergens* and *E. smithi*	Maleisia	1930	*A. woglumi* in citrus	CBC	No	
Paratheresia claripalpis	Trinidad & Antigua	1934–37	*Diatraea saccharalis* in sugarcane borer	CBC	No	
P. claripalpis	Brazil	1938–39	*D. saccharalis* in sugarcane	CBC	No	
Lydella minense	Brazil	1934/49	*D. saccharalis* in sugarcane	CBC	No	
Harmonia (= *Leis*) sp.	USA	1936	*Aphis spiraecola* in citrus	CBC	No	
Encarsia berlesei	Italy	1938	*Pseudaulacaspis pentagona* in *Morus alba*	CBC	No	
Dactylosternum abdominale, D. hydrophiloides	Jamaica	1949	*Cosmopolites sordidus* in banana and plantain	CBC	No	
Plaesius javanus	Jamaica	1949	*Cosmopolites sordidus* in banana and plantain	CBC	No	
Lixophaga diatraeae	Native	1950	*D. saccharalis* in sugarcane	ABC	Yes	1,600,000 ha
Period 1970–now						
Cryptolaemus montrouzieri	Russia / Trinidad & Tob	1970/2002	Pseudococcidae in differents crops	CBC	Yes	?
Lydella minense	Martinique	1971	*D. saccharalis* in sugarcane	CBC	No	
Trichogramma spp.	Exotic & Native	1982	Various lepidopterans in various crops	ABC	Yes	283,252 ha
Beauveria bassiana	France	1988	Various pests in various crops	ABC	Yes	132,201 ha
Heterorhabditis spp.	Native	1988	*Cylas formicarius* in sweet potato	ABC	Yes	1,618 ha
Lecanicillium lecanii	Native	1988	Hemipterans in various crops	ABC	Yes	76,733 ha
Metarhizium anisopliae	Native	1988	Various pests in various crops	ABC	Yes	67,470 ha
Pheidole megacephala	Native	1988	*C. formicarius* in sweet potato	ConsBC	Yes	140,000 ha
Trichoderma harzianum, *T. viride*	Native	1993	Soil-borne disaeases in various crops	ABC	Yes	60,032 ha
Cotesia flavipes	Brazil, Venezuela, Peru	1995–98	*D. saccharalis* in sugarcane	CBC	No	
Cephalonomia stephanoderis	Mexico	2003	*Hypothenemus hampei* in coffee	CBC	Yes	?
Phymastichus coffee	Mexico	2004	*H. hampei* in coffee			
Biopesticides						
Bacillus thuringiensis vars strains	Exotic & Native	1988	Various pests in various crops	ABC	Yes	238,309 ha

[a]Type of biological control: CBC = classical biological control, ABC = augmentative biological control, ConsBC = conservation biological control

Program for Biological Control (Programa Nacional de Lucha Biológica) was initiated and in 1995 the ca. 50 centres for reproduction of entomophagous organisms (Centros Reproductores de Entomófagos) (CRE) produced 78 million *L. diatraea* parasitoids for releases on a seasonal cumulative area of 1.6 million hectares (Fuentes *et al.*, 1998). Another interesting programme concerned the control of the sweet potato weevil *Cylas formicarius* (Fabricius) on 11.586 ha in 1999 with the predatory ant *Pheidole megacephala* (Fabricius) and, more recently, also by entomopathogenic nematodes *Heterorhabditis* spp. (Vázquez and Pérez, 2016). Further, the use of insect pathogenic fungi and strains of the bacterium *Bacillus thuringiensis* Berliner (Fernández-Larrea, 1999) was particularly impressive. In 1999, *B. thuringiensis* was applied on 238,309 ha (Vázquez and Pérez, 2016).

During this period, Cuba had 280 centres for reproduction of entomophagous organisms and entomopathogens (Centros Reproductores de Entomófagos y Entomopatógenos) (CREE) (Altieri and Nichols, 1999), where large amounts of insect pathogenic fungi and *B. thuringiensis* as well as *Trichogramma* spp. and sugarcane borer parasitoids were produced. *B. thuringiensis* was initially imported from the former USSR in the 1980s and used with success against the tobacco budworm *Heliothis virescens* (Fabricius) and the striped grass looper *Mocis latipes* Guenée. This stimulated interest in searching for native strains of *B. thuringiensis* and the development of technologies for mass production of several other entomopathogenic species (Fernández-Larrea, 1999). The Cuban production of biocontrol agents for sugarcane pests is carried out in 31 CREE of the Ministry of Agriculture's special branch for the Sugar Industry. Other biocontrol agents are produced in 175 CREE and four industrial plants that belong to the enterprise and cooperative system of the Ministry of Agriculture. These are applied in other crops, including coffee and forest nurseries.

The area under biocontrol in Cuba in 1988 was 313,802 ha, while in 1999 the total area of agricultural crops under biocontrol had increased to 874,430 ha (Pérez and Vázquez, 2001). Since 1974 biocontrol has been managed by the 76 Territorial Stations of Plant Health (ETPP), which introduce biocontrol within integrated pest management (IPM) and agroecological pest management (APM) programmes (Vázquez and Pérez, 2016).

10.3 Current Situation of Biological Control in Cuba

10.3.1 Introduction

During the first decades of application of biocontrol in Cuba a number of classical biocontrol programmes were implemented and later augmentative biocontrol became an important pest management method. Currently, the conservation of natural enemies is being strongly adopted. Biocontrol has been consolidated for a long time in Cuba, while new technological developments have been introduced regularly. Below, we present a synthesis of scientific results and adoption of those programmes and the current scope of biocontrol in the country's agricultural production systems.

10.3.2 Biological control agents used in Cuba

Parasitoids

The success obtained with classical biocontrol of citrus blackfly with *E. serius* since the 1930s has already been mentioned above. The parasitoid continues to provide good pest control as long as insecticides are not used indiscriminately against other pests in citrus. Attempts to introduce seven other parasitoid species in Cuba have failed because they did not establish or their population development was very low under Cuban climate conditions (Table 10.1) (Vázquez *et al.*, 2005)

Augmentative biocontrol of the sugarcane borer using *L. diatraeae* has been in use since 1950 and was supported by many studies (Alemán, 2000). In addition, a pupal parasitoid of the sugarcane borer, *Tetrastichus howardi* (Olliff), was more recently mass reared and released (Alvarez *et al.*, 2003). Further, studies to employ *Trichogramma* spp. have been carried out to control lepidopteran defoliator complexes (*Mocis* spp.

and *Leucania* spp.) (Rego *et al.*, 1990; De la Torre, 1993). The results of all these studies are that *Trichogramma fuentesi* Torre and *Trichogramma exiguum* Pinto & Platner are recommended for control of *D. saccharalis*, while *Trichogramma pintoi* Voegelé, *Trichogramma pretiosum* (Riley) and *Trichogramma rojasi* Nagaraja and Nagarkatti are used against lepidopteran defoliators. *Trichogramma* spp. inundative releases are also made against *M. latipes* in pastures with Bermuda grass (*Cynodon dactylon* (L.) Pers and Pangola grass (*Digitaria eriantha* Steud (= *D. decumbens*), *Erinnyis ello* (Linnaeus) in cassava (*Manihot esculenta* Crantz) and *H. virescens* in tobacco (*Nicotiana tabacum* L.) (Fuentes, 1994). Trichogrammatidae are further employed to control *Plutella xylostella* Linnaeus, *Ascia monuste eubotea* (Godart) and *Trichoplusia ni* (Hübner) in Cruciferae (*Brassica* spp.); *Diaphania hyalinata* (Linnaeus) and *Diaphania nitidalis* (Stoll) in *Cucurbita* spp.; *Spodoptera* spp. in tomato (*Solanum lycopersicum* L.), potato (*Solanum tuberosum* L.), sweet pepper (*Capsicum annuum* L.) and sweet potato (*Ipomoea batatas* L.); *Erinnyis alope* (Drury) in papaya (*Carica papaya* L.); and *Rhyacionia frustrana* (Comstock) and *Spodoptera sunia* (Guenée) in pine (*Pinus* spp.) plantations (Fuentes, 1994; Caballero *et al.*, 2000; Vázquez *et al.*, 2010). The production of these egg parasitoids is done by the Cuban network rearing system CREE, and can reach ca. 15 billion (15,000,000,000) individuals per year (CNSV, 2016).

Egg parasitoids of the genus *Telenomus* are used against *Spodoptera frugiperda* (Smith) in maize (*Zea mays* L.), sorghum (*Sorghum bicolor* (L.) Moensch) and rice (*Oryza sativa* L.) (Meneses, 2008), *Spodoptera* spp. in tomato and sweet pepper and *Spodoptera exigua* (Hübner) in onion (Armas *et al.*, 1997). The parasitoids *Chelonus insularis* (Cresson) and *Euplectrus platyhypenae* Howard are used for control of *S. frugiperda* in maize, sorghum and rice, and *Spodoptera* spp. in tomato, sweet pepper, potato and sweet potato (Vázquez *et al.*, 2010). The tachinid parasitoid *Archytas marmoratus* (Tns.) is used to control *S. frugiperda* in maize, sorghum and rice crops and also against *M. latipes* in pastures. The hymenopteran parasitoid *Lysiphlebus testaceipes* (Cresson) is used against aphids (Vázquez *et al.*, 2010) and the tachinid parasitoid *A. marmoratus* is used against *Leucania unipuncta* (Howard) (Gómez and Grillo, 1999).

Predators

Predators are used in five strategies:

1. Classical biocontrol programmes.
2. Augmentative biocontrol programmes.
3. Use of artificial reservoirs (they are allowed to reproduce in artificial reservoirs adjacent to field crops).
4. Trap–rear–release approach (they are trapped and reared in small-scale field insectaries).
5. Reservoir plants (they are reared on host plants that form a reservoir for establishment and direct action in the surrounding crops).

Strategies 3 and 5 are aspects of conservation biocontrol, which is widely applied in Cuba.

For strategy 2 (augmentative biocontrol), the following predators are mass reared in small insectaries: *Chrysopa* spp. for control of aphids and the tobacco whitefly *Bemisia tabaci* Gennadius in different crops (Rijo and Acosta, 1997); *Cycloneda sanguinea limbifer* (Casey) and *Coleomegilla cubensis* (Casey) to control aphids and mealybug populations (Alemán *et al.*, 2004; Caballero and Sánchez, 2007); and *Cryptolaemus montrouzieri* Mulsant against mealybugs in fruit orchards and ornamental plants (CNSV, 2016).

The use of artificial reservoirs (strategy 3) has been practised by breeding the ant lion *P. megacephala* to control the sweet potato weevil *C. formicarius* in sweet potato crops (Castiñeiras *et al.*, 1982). During the past few years, an area of ca. 140,000 ha has been treated with this ant species (CNSV, 2016).

Strategy 4, (the trapping–rearing–release approach) is used with the coccinelids *C. sanguinea limbifer*, *C. cubensis*, *Hippodamia convergens* Guérin-Méneville and *Chilocorus cacti* L. in urban agriculture to control *B. tabaci*, aphids of the genera *Myzus*, *Aphis* and *Rhopalosiphum*, scale insects, mealybugs and the Asian citrus psyllid *Diaphorina citri* Kuwayama in diverse crops (Milán *et al.*, 2006).

Reservoir plants (strategy 5) are used to provide shelter for predators. They are planted as living barriers surrounding field crops. Among them are: maize, sorghum and sunflower (*Helianthus annus* L.) for chrysopids, syrphids and coccinelids; fennel (*Foeniculum vulgare* Mill.) and whitetop weed (*Parthenium hysterophorus* L.) for coccinelids; and blackjack (*Bidens pilosa* L.) to rear *Orius* sp. bugs (Vázquez *et al.*, 2008).

Entomopathogenic nematodes

The species *Neoaplectana*, strain P2M, has been mass produced to control *Pachnaeus litus* (Germar) in citrus and specifically to protect nursery plants (Montes and Montejo, 1991); this strain was later identified as *Heterorhabditis indica* Poinar, Karunakar & David (Stack *et al.*, 2000). Entomopathogenic nematodes are produced exclusively *in vivo*, using late larval instars of *Galleria mellonella* L. as small 'biological reactors', a technology that CREE at Quivicán adapted during the 1990s by employing *Heterorhabditis bacteriophora* (Poinar) strain HC1 (Sánchez, 1997). This technology was later employed by other CREE units in Cuba. This biocontrol agent is applied to control *C. formicarius* in sweet potato; the rice water weevil *Lissorhoptrus brevirostris* (Suffrian) in rice; *Atta insularis* (Guérin-Menéville) in citrus and ornamental plants (Sánchez, 1997); white grubs (*Phyllophaga* spp.) in pineapple; *P. xylostella* in cabbage; seedling cutworms (*Agrotis* spp.) in several vegetables; *S. frugiperda* in maize; borer weevils (*Ips* spp.) in pine plantations; and *D. hyalinata* in cucurbits (Valdés *et al.*, 2005; Vázquez *et al.*, 2010). The nematode is also used against *Hypothenemus hampei* Ferrari (Rodríguez, 2015) and mealybugs (Rodríguez *et al.*, 1998) in coffee plantations. The production of nematodes now reaches 908 billion (908,000,000,000) individuals at CREE network units in all provinces of Cuba (CNSV, 2016).

Entomopathogenic fungi

Beauveria bassiana (Balsamo) Vuillemin cultures were initially developed and mass produced to control weevil populations of *Cosmopolites sordidus* (Germar) in plantains (Castiñeiras *et al.*, 1992), *C. formicarius* in sweet potato, *P. litus* in citrus (Jiménez and Fernández, 1980) and *Lissorhoptrus oryzophilus* Kuschel in rice (Meneses, 2008). Currently, *B. bassiana* is also used to control *Typophorus nigritus* F. in sweet potato; *A. insularis* in ornamental plants; *H. hampei* and *Xylosandrus compactus* (Eichhoff) in coffee; *Apate monachus* (Fabricius) in guava, mango and coffee; *Lagochirus dezayasi* Dillon in cassava; *Diabrotica balteata* LeConte in tomato and sweet pepper; *Phyllophaga* spp., *Ips* spp. and *A. insularis* in pine nurseries (*Pinus* spp.); *Phyllophaga* spp. in pineapple (*Ananas comosus* L.); *Pseudacysta perseae* (Heidemann) in avocado (*Persea americana* Mill.); *Corythucha gossypii* (Fabricius) in plantains (*Musa* spp.); and *B. tabaci* in tomato (Vázquez *et al.*, 2010).

Metarhizium anisopliae Metschnikoff (Sorokin) is mass produced (Lujan *et al.*, 1992) to control *L. oryzophilus* in rice (Meneses *et al.*, 1980), *C. sordidus* in plantains and bananas crops (Castiñeiras *et al.*, 1992), *C. formicarius* in sweet potato (Castiñeiras *et al.*, 1984), *P. litus* in citrus (Jiménez and Fernández, 1980) and *A. insularis* in ornamental plants. *M. anisopliae* has been used in combination with *B. bassiana* to control *Thrips palmi* Karny in potato, beans, cucumber, sweet pepper and other crops, *P. litus* in citrus, *S. frugiperda* in maize (Vázquez *et al.*, 2010) and against *Rhyacionia frustrana* (Comstock) in pine plantations (Duarte *et al.*, 1992).

Lecanicillium lecanii (Zimmermann) Zare & Gams is mass produced (Elósegui, 2006) to control *B. tabaci* in tomato, beans, cucumber and other crops (Murguido and Elizondo, 2007), *Aphis gossypii* Glover and *Myzus persicae* (Sulzer) in papaya, *Brevicoryne brassicae* L. and *Lipaphis erysimi* (Kaltenbach) in cabbage (*Brassica oleracea* L. var. *capitata*) (Vázquez *et al.*, 2010) and the coffee leaf rust *Hemileia vastatrix* Berk. E Br. (González and Martínez, 1998).

For specific regional needs, mass rearing of *Nomuraea rileyii* (Farlow) de Samson was developed to control *S. frugiperda* in maize and rice, and *Oiketicus* sp. in guava (Vázquez *et al.*, 2010) and *Anticarsia gemmatalis* (Hübner) in soybean (*Glycine max* L. Merr.) and beans. Also the fungus *Paecilomyces fumosoroseus* (Wize) is cultured to control *B. tabaci* in tomato and cabbage, *B. brassicae* in cucurbits, and *M. persicae* and *A. gossypii* in horticultural crops (Vázquez *et al.*, 2010).

The Cuban CREE network produces annually over 200 t of *B. bassiana*; 120 t of *L. lecanii* strain Y57, and 100 t of *M. anispoliae* strain 'Niña Bonita' (CNSV, 2016).

Microorganisms for the control of nematodes

A technology for local small-scale production of the fungus *Purpureollicium lilacinun* (Thom) Samson (= *Paecilomyces lilacinus*) was developed. *P. lilacinun* has the capacity to parasitize eggs of *Meloidogyne incognita* (Kofoid and White) Chitwood; it also has teratogenic effects on juvenile nematodes, due to the presence of secondary metabolites. The fungus was produced during

1992–1999 by the CREE network to control *M. incognita* in various crops and *Radopholus similis* (Cobb) Thorne in plantain, banana and coffee plantations, and its application covered 2,000 ha in 1999 (Fernández, 2007). However, development of commercial formulations based on the use of these native Cuban strains had to be interrupted until human-safety studies were completed.

Trichoderma spp. are used combined with other control measures for the management of nematodes important in horticultural production, under a National Program of Urban Agriculture and Protected Crops (Fernández, 2007). Best results for protected crops have been achieved with the use of *T. harzianum* Rifai strain A-34 in tomato by applying 8 kg ha^{-1}, a dose that reduced infestation values of 3–4 (values vary from 0 to 5) to 1 at the end of the cropping cycle of about 110 days (Méndez and Polanco, 2006). Urban agriculture in Matanzas province is practised using this strain (Stefanova, 2007). Positive results were also obtained in cucumber crops by Cuadra *et al.* (2008), who applied *Trichoderma viride* Persoon. ex Fr. strain 2684 at the beginning of the cropping cycle.

Gram-positive bacterium *Tsukamurella paurometabola* (Steinhaus) (ex *Corynebacterium paurometabola*), strain C924, has proven nematicidal activity on *Meloidogyne* spp., *R. similis* and *Pratylenchus* spp. and is used in greenhouse vegetable crops. It acts directly on chitinase activity and produces hydrogen sulfide. The combined action of both processes produce weakening of the outer chorion layer, as well as vacuole formation inside the egg affecting embryogenesis, and cuticle lysis and vacuole formation problems in larvae (Mena *et al.*, 2003). This bacterial strain also promotes plant growth, which can stimulate the susceptibility of plants for colonization by mycorrhiza.

In Cuba, several isolates of the fungus *Pochonia chlamydosporia* (Goddard) Zare & Gams were obtained from soil of the rhizosphere of coffee plantations infested with a complex of root-knot nematode (RKN) species (*Meloidogyne* spp.) and screened as a biocontrol agent (Hidalgo-Díaz *et al.*, 2000). The strain IMI SD187 (RRes 392, Cvc-108) was selected and a new management strategy for the control of RKN in organic vegetable production was developed (Atkins *et al.*, 2003). A technology for mass production of the selected strain was implemented on a pilot-plant scale at the National Center for Animal and Plant Health (Centro Nacional de Sanidad Agropecuaria) (CENSA) following Good Manufacturing Practices (Montes de Oca *et al.*, 2009). Efficacy of the product obtained (KlamiC) has been confirmed in field trials over 2 years, in six crops sequences with an annual application (Peteira *et al.*, 2005). The fungus is still producing at a volume of ca. 680 kg per year and is used on local farms. It must be applied to the seedling and at transplanting time to promote the rhizosphere colonization.

Bacteria-based pesticides

Several native strains of *B. thuringiensis* have been isolated and studied to control agricultural pests, mainly for control of lepidopterans (Fernández-Larrea, 1999, 2013). It was first employed for control of *H. virescens* in tobacco and *M. latipes* in pastures. Later, its use was extended to control *P. xylostella* and *T. ni* in Cruciferae; *S. frugiperda* and *S. exigua* in maize; *Spodoptera* spp. in tomato, potato and sweet potato; *Helicoverpa* (= *Heliothis*) *zea* in sweet pepper; *T. ni* and *Liriomyza trifolii* (Burguess) in potato; *D. hyalinata* and *D. nitidalis* in cucurbits; *E. ello* in cassava and *Herse* (= *Agrius*) *cingulata* Fabricius in sweet potato; *E. alope* and *Davara caricae* Dyar in papaya; *Hedylepta* (= *Omiodes*) *indicata* (Fabricius) in beans (*Phaseolus vulgaris* L.); *Herpetogramma bipunctalis* (Fabricius) in beet (*Beta vulgaris* L.); and *M. latipes* in pastures (Vázquez *et al.*, 2010). It was also used to control *Varroa destructor* Anderson & Trueman in beehives (Márquez, 2002).

Studies to optimize culture media and reproductive parameters in the fermentation process of the acaricide strain LBT 13 were started in 1993 (Márquez and Fernández-Larrea, 1999). Mortality of *Polyphagotarsonemus latus* Banks increased and fecundity was decreased at high concentrations of spores, with an $LD_{50} = 1.1 \times 10^7$ spores ml^{-1} and an $LD_{95} = 4.4 \times 10^9$ spores ml^{-1}; this strain is now also used against several species of mites of agricultural importance (Almaguel, 2006).

Annually, over 500 t of biopesticides based on different *B. thuringiensis* strains are mass produced at CREE and other biofactories (CNSV, 2016).

Antagonists for control of plant diseases

Research since the 1990s has resulted in the selection, small-scale and later large-scale industrial production of *Trichoderma harzianum* Rifai strains (Martínez *et al.*, 2008); *T. harzianum* is integrated in soil pathogen management programmes used as seed treatment (Stefanova, 2007), and as foliar application in several crops (Santana *et al.*, 2010; Reyes *et al.*, 2009; Pérez *et al.*, 2010). *T. harzianum* is hand-made at the CREE network in all provinces, reaching an annual volume of 400 tons (CNSV, 2016).

10.3.3 Adoption of biological control in agricultural production

Currently, mass production of biocontrol agents is carried out in a network that includes 206 local CREE laboratories and four industrial plants for production of microbial control agents, all located close to agricultural regions. The network is managed by companies and cooperative associations that cover annually an area of 1.7 million hectares of crops (CNSV, 2016). In addition, several Research Centers develop and sell their products direct to farmers.

The National Program for Biological Control is managed by the Plant Health Service (Dirección de Sanidad Vegetal, formerly Centro Nacional de Sanidad Vegetal) (CNSV), belonging to the Ministry of Agriculture. The Program involves 14 provincial laboratories (Laboratorios Provinciales de Sanidad Vegetal) (LAPROSAV) that provide certified strains and ecotypes of biocontrol agents, advise about production and carry out quality control of products. There is also a network of 72 ETPP that advise farmers how to use biocontrol agents as well as monitor their efficacy once they are applied to crops (Pérez and Vázquez, 2001).

The adoption of biocontrol was for a large part the result of collaboration between technicians and farmers, under the supervision of specialists from the Plant Health Service, and at the same time resulted in diversification of use and improvement in the techniques for release of entomophagous organisms, or spraying of entomopathogens (Vázquez *et al.*, 2010). Generally, in farms belonging to the agroecological movement and in urban agriculture, farmers have adopted natural enemy conservation practices and also use augmentative biocontrol on an area of 470,000 ha of crops (Vázquez *et al.*, 2008). Production systems of the agroecological movement have achieved a 50% reduction in chemical and biological pesticide applications. In addition, the levels of infestation of harmful organisms have decreased because of multiple and cumulative effects of practices aimed at conserving soils, promoting diversification and integration of crop and cattle systems, managing non-crop vegetation, and other tactics that favour ecological services that help to reduce pest populations and that increase the regulation of pests (Vázquez *et al.*, 2008).

Integration of biocontrol in IPM programmes has also been successfully achieved in conventional agriculture, especially in rice, plantain, banana, sweet potato, coffee, sugarcane, citrus, cabbage, bean, maize, potato, pastures, tobacco, tomato and cassava, among others (Pérez and Vázquez, 2001).

Biocontrol not only results in pest control, but is also much appreciated by farmers and technicians, by research centres and universities that have developed innovative technologies, and by the Ministry of Agriculture, which has prioritized its financial support since 1982 for sugarcane, then in 1988 for agriculture and livestock, and intensified its financial support again in 1993. Table 10.2 summarizes the contribution of biocontrol for sustainable production in Cuba. During the past few decades, biocontrol has become a leading component of pest management. A report by CNSV (2016) stated that chemical pesticides are employed along with biocontrol agents in only 25–30% of the cropped area; while the major proportion of land is cultivated by using biocontrol and other non-chemical alternatives. The high level of adoption of biocontrol in agricultural production, in addition to the contributions shown in Table 10.2, has created capabilities for the design of 'agriculture of the future', because of its potential to reduce pesticide use to a minimum and by enhancing self-sufficiency of local production of many biocontrol agents. However, as pointed out by Vázquez *et al.* (2008), for an even better result it is necessary

Table 10.2. Main contributions of biological control to sustainable agricultural production in Cuba.

Aspects	Contributions of biological control
Economical	Reduction of costs for the importation of chemical pesticides
	Territorial autonomy in managing the production according to the demands
	Leadership in biocontrol methods to face the economic crisis of the 1990s
Ecological–environmental	Reduction of toxicity in agroecosystems
	Adoption of conservation strategies using natural enemies
	Improvement of conservation of natural enemies, pollinators and rhizospheric biota, among other organisms
	Increment of knowledge by farmers of entomophagous arthropod species (e.g. insects, mites and spiders) with beneficial functions
	Use of endemic strains and ecotypes of natural enemies
	Recovery of the quality of ecotypes of natural enemies, by passing them through local hosts
	Use of vegetative barriers (e.g. maize, sorghum and sunflower) to favour the conservation of natural enemies
Technological	Facilitation and acceptance of the integrated pest management (IPM) approach in conventional agricultural production systems with reduction in the use of chemical pesticides
	Development of new technologies and procedures for the application of augmentative release
	Synergy between the policy of integration and diversification of agricultural production (reduction of monocultures and specialized production)
	Contribution to agroecological management of pests (AMP) to promote the epizootics and the establishment of biocontrol agents in crops that are agro-ecologically managed
	Technologies for farmers to multiply entomophagous arthropods *in situ* from local populations
Socio-cultural	Change in the perception of farmers who value the use of new methods in pest control
	New and more complex knowledge for technicians and specialists in agricultural production
	Local employment sources, mainly for women and youth that work at the CREE network
	Introduction of new technologies in the Cuban agriculture, in contrast with the conventional use of chemical pesticides, that contribute to the national food security
	Integration of the community, principally of children and youth, to the CREE network through circles of interest in the education system
	Active participation of farmers that are involved in the adoption and diversification of new technologies for biocontrol
	Empowerment of farmers for their self-sufficiency in pest management

to increase the use of conservation biocontrol by the implementation of an agroecological pest management programme. This would not only benefit the local populations of natural enemies inhabiting crops and adjacent areas, but also help survival of the biocontrol agents that are regularly released.

The estimated agricultural area currently under biocontrol in Cuba amounts to 2,400,000 ha (Table 10.1). When the use of *B. thuringiensis* is included, the total area under biocontrol is about 10% more.

10.4 New Developments of Biological Control in Cuba

10.4.1 Introduction

The development of biocontrol in Cuba has been a complex and time-consuming process requiring constant efforts, in combination with agrarian politics favouring biocontrol as a pest management method (Vázquez and Pérez, 2016).

A project connected to the National Science and Technology Program for Agricultural Biotechnology was prepared in 2010 with the following objectives: (i) optimize technological processes; (ii) develop more stable and efficient formulations; (iii) improve production; (iv) satisfy internal demands for bioproducts to replace imported products; and (v) produce products for export based on the growing demand of foreign markets (González, 2012). The result of this project was a portfolio that included 23 bioproducts (biopesticides, biofertilizers and biostimulants). The development of products and local self-sufficient technologies and the application of 2,000–2,500 t of microbial control agents per year (CNSV, 2016) represents one of the most significant goals achieved during this period. It has also been important to comply with various technical stages to ensure safe, effective and reliable biocontrol, which also facilitates product registration (Fernández-Larrea, 2013).

Basic and applied research promoted new knowledge for priority programmes through cooperation among scientific institutions, universities, specialists from Plant Health Service of Provincial Laboratories, technicians from agricultural production units, and farmers. Most researchers focused their studies on selection, mass reproduction, host range and use of bioproducts.

10.4.2 Conservation biological control

An important research approach of the past 20 years concerns conservation biocontrol and consists of protecting and stimulating the development of beneficial organisms that are spontaneously present in agroecosystems. The purpose is to increase the pest control activity of the most efficient naturally occurring species, or by a complex of beneficial species. Complementarily, released biocontrol agents may benefit from conservation biocontrol. Various natural-enemy conservation practices have been developed in Cuba and were integrated into the APM approach (Vázquez *et al.*, 2008). These APM programmes have been enhanced and generalized during the past 15 years. They can be classified as: (i) chemical compounds management (signalling system, replacement of chemical pesticides by biocontrol agents, use of ecologically selective pesticides); (ii) management of plant diversity (semi-natural habitat, living barriers, etc.); (iii) mixed-crop systems (polycultures, agroforestry, livestock–forestry, etc.); (iv) tolerance to weeds; (v) integration with green manure; (vi) crop rotation; (vii) cropping system mosaic; (viii) promotion and management of natural enemy reservoirs (artificial and natural); and (ix) management of epizootics (Vázquez *et al.*, 2008). In-field, small-scale insect rearing is a common practice used by growers and is part of APM (Milán *et al.*, 2007). Also plant management within and outside crops for conserving natural enemies is currently under research and aims to increase the pest regulatory capacity of natural enemies (Matienzo *et al.*, 2007; Vázquez *et al.*, 2008; Ceballos *et al.*, 2009). An extensive recent review by Wyckhuys *et al.* (2013) concerning conservation biocontrol concluded that Cuba ranked as the third country in Latin America, based on the number of scientific contributions about conserving natural enemies among developing countries.

10.4.3 Technologies for in-field arthropod rearing and survival

Predatory mites from the family Phytoseiidae have been studied during the past 45 years due to their high species diversity and potential pest regulation capacity. They are inoculatively released to manage a complex of mite pests in Cuba. Currently, 24 genera with 61 species have been reported (Ramos and Rodríguez, 2006). Cuban predatory mite species are mainly associated with phytophagous mites, in particular tetranichids, tarsonemids, tenuipalpids and to a lesser extent with small insects such as thrips (Rodríguez *et al.*, 2013). To promote their persistence in agroecosystems, use of diversified crops in polycultures, living barriers and green manure crops is practised.

A method has been developed to maintain founder colony quality and to assess quality of mass-reared predatory mites under laboratory conditions (Rodriguez *et al.*, 2013). Mass rearing of *Amblyseius largoensis* (Muma) and *Neoseiulus longispinosus* (Evans) was successfully realized on bean leaves infested with *Tetranychus tumidus* (Banks) (Pérez-Madruga *et al.*, 2014). This technology is now being studied to improve mass production in regional laboratories for massive reproduction.

Martínez *et al.* (2011) tested rearing of the parasitoid species *Diaeretiella rapae* McIntosh on *S. bicolor* as host plant and the aphid *Rhopalosiphum maidis* (Fitch) as host to be applied as banker plant technique, with the excellent result of > 90% *D. rapae* parasitism. This simple rearing system was adopted by local growers to control aphid populations in Brassicaceae crops in urban agriculture.

Predatory insects like coccinelids efficiently reduce populations of spider mites, aphids, scales and juvenile stages of lepidopteran pests in agricultural and ornamental crops (Milán *et al.*, 2007) and small-scale rearings have been developed to be applied in urban agriculture. Insectaries are constructed using recycled or waste materials and growers use this technology to rear and release beneficials in their own farms, without relying on specialized production and saving resources (Milán *et al.*, 2006).

10.4.4 Technologies for production of microbial control agents

Small-scale technologies to produce microbial control agents are successfully implemented by the CREE as part of the National Science Technology Program for Agricultural Biotechnology. Inoculation and incubation of microorganisms is carried out using solid substrates (crop subproducts such as split rice, rice straw) in trays, vials or bags for drying, with control of temperature and humidity by monophasic and biphasic methods (Elósegui, 2006; Márquez *et al.*, 2010). Active ingredients are spores, conidia, mycelia or toxins, which are prepared as colonized substrate. Once dried, they can be stored for 3 months at 10–20°C. Microbial control products can be used directly or suspended in water to eliminate solid substrates. Currently, fungal production is mainly by biphasic fermentation, obtaining a concentrated volume of spores in solid substrates, resulting in longer spore survival and resistance to desiccation, exposure to sunlight and high temperatures when applied. Several fungal species are produced by CREE using highly efficient native strains of *B. bassiana*, *T. harzianum*, *T. viride*, *M. anisopliae* and *L. lecanii*. Several new microorganisms are waiting to be developed, such as *Trichoderma asperellum* Samuels, Lieckfeldt & Nirenberg potential microbial control for several fungi (Infante *et al.*, 2013).

For insect management, at the beginning of this century CENSA attained modest results in the solid fermentation of entomopathogenic nematodes (Rodriguez, 2015) and nowadays works in liquid fermentation and formulation of nematodes for management of *Cosmopolites sordidus* in banana and plantain, under a project financed by the European Union (MUSA, 727624; topic: SFS-11-2016), working with partners from European, African and Latin American institutions.

Also four Bt-bioproducts (Thurisave 3, Thurisave 24, Thurisave 13 and Thurisave 26) are used to reduce lepidopteran, nematode and mite populations. They are produced by submerging cultures in fermenters, and later spores and toxic crystals are concentrated by sedimentation (Fernández-Larrea, 2013). The process takes 72–96 h and the product can be stored for 6 months at 28°C.

Progress in research for formulation of microbials is slow, with currently only one pilot-scale production project concerning *T. harzianum*. Elósegui *et al.* (2009) described collection of *T. harzianum* A34 strain spores cultured on rice grains and grain shells, using the separation method with a fluid-bed and dual cyclone, and separation by an electric vibrating sieve. Twenty minutes after harvesting with the first method, only 2.2% of the spores were recovered, while 29.7% of the spores were retrieved when using a vibrating sieve. Spore viability obtained showed that the contamination levels were acceptable and also that the size of particles required were under regular values for fungi spraying. Later Elósegui *et al.* (2015) showed the importance of the correct composition of the culture substrate for an efficient separation of spores in solid bioculturing.

Six bio-nematicides were developed from 2005 to 2010 by solid and submerged fermentation. To produce KlamiC, based on *P. chlamydosporia* var. *catenulata* strain IMI SD187, CENSA developed a technology of solid-state fermentation in polypropylene bags with filter, following a Good Manufacture Practice guidance which permitted the obtaining of a consistent and safe product (Montes de Oca *et al.*, 2009).

Using similar technology, *T. harzianum* strain A-34 and *T. viride* strain 2684 were multiplied in solid substrate, to obtain Tricosave and Trifesol, respectively (Stefanova, 2007; Cuadra *et al.*, 2008). A product named Nemacid was obtained from waste of submerged fermentation of *L. lecanii* strain 3166 by the Instituto de Investigaciones de Derivados de la Caña de Azúcar (ICIDCA) (Gómez *et al.*, 2001). This product is under registration process. Another product, HeberNem, registered by Centro de Ingeniería Genética y Biotecnología (CIGB) in Camagüey, succesfully controls *Meloidogyne* spp. and *Pratylenchus* spp. (Mena *et al.*, 2002). A novel nematicide obtained from *B. thuringiensis* strain LBT25 is being developed. It has a toxic effects on eggs and induces larval inmobilization and vacuolization of the gall-forming nematode *Meloidogyne* spp., and reduced soil crop infestation by juveniles by 48–66% (Márquez and Fernández, 2006; Márquez *et al.*, 2010).

Marín *et al.* (2013) found *in vitro* antagonistic activity of HeberNem against various phytopathogens, such as growth inhibition of development of *Alternaria longipes* (Ellis & Everh.), *Bipolaris oryzae* (Breda de Haan), *Colletotrichum gloeosporioides* (Penz.), *Fusarium oxysporum* f. sp. *cubense* (E.F. Sm.), *Pestalotia palmarum* (Cooke) and *Thielaviopsis paradoxa* (De Seynes). Other positive attributes found for HeberNem are promotion of plant growth in beans (*P. vulgaris*), maize (*Z. mays*), plantain (*Musa* spp.) and lettuce (*Lactuca sativa* L.), enhancing their height, root system, foliar development and dry biomass. Under experimental conditions, the strain produced indole acetic acid, solubilized phosphates and produced ammonium from organic matter. The strain also produced lytic exoenzymes, which might protect plants from pathogens (Marín *et al.*, 2014). A new formulation, HeberNem-S, is being studied to facilitate application (Hernández *et al.*, 2007) and it is indicated that this new formulation does not affect the bacterium *T. paurometabola* .

Combined use of KlamiC, HeberNem, *T. asperellum* and EcomiC (an arbuscular mycorrhyzic fungus), integrated with other pest management tactics such as biofumigation and 'trap-crops', decreased infestations by root knot nematodes in protected crops and resulted in similar crop yields as those treated with the chemical nematicide Agrocelhone®. Interestingly, the combined use of the biological agents was cheaper and resulted in better environmental conservation (Hernández *et al.*, 2015).

For the period 2017–2020, research at CENSA is evaluating new uses and new target organisms of the microbial control agent KlamiC to study its effects on seed germination and plant growth in crops such as plantain, banana, etc., and to assess its possibilities for controlling the mollusc *Praticolella griseola* Pfeiffer.

10.4.5 Registration of microbial control agents

Since 2007, Cuba has had a specific regulation for the registration of biological pesticides (biopesticides of microbial origin) with a separate procedure for chemical pesticides, although within the same regulatory board, the National Pesticide Registry (Hidalgo-Díaz, 2004). This procedure is based on the precepts of the UN's Food and Agriculture Organization (FAO), the Organization for Economic Cooperation and Development (OECD) and the US Environmental Protection Agency (EPA) with the aim of guaranteeing the efficacy and safety of biopesticides (Ceballos and Montes de Oca, 2016).

10.4.6 Final considerations

With the increase in use of augmentative biocontrol, mainly by means of microbial control agents, there is an increasing demand for studies that evaluate essential characteristics of agroecosystems contributing to more efficient pest control, and to promote epizootics and the establishment of beneficial microorganisms. Particularly in conventional agriculture, with numerous sprays with chemical pesticides, biocontrol cannot be used in a sustainable way.

The use of molecular tools is still challenging for species identification, for understanding modes of action, and for establishing interactions in multi-trophic species systems and the environment.

Regarding production of biocontrol agents, mass rearing levels must be warranted to satisfy a growing demand, and studies concerning up-scaling of production, product quality control, improvement of formulation and extended shelf life are of high priority. For microbial control agents, studies to approve application techniques are urgently needed, as current techniques have been adapted from chemical pesticides.

Cuban research has generated essential practical knowledge about biocontrol, positioned in a sustainable ecoagricultural setting. This initiative has influenced different societal sectors and resulted in new, innovative technologies for agriculture, with a strong commitment in agricultural extension service linked to environmental care and food autonomy.

10.5 Acknowledgements

M.G. Rodriguez and L. Hidalgo-Diaz were supported by project 'Microbial Uptakes for Sustainable management of major banana pests and diseases' (MUSA, 727624; topic: SFS-11-2016), with financial support through the European Union.

References

Alemán, J. (2000) Quality assessment in *Lixophaga diatraeae* Towsend (Diptera: Tachidae) mass-rearing. *Revista Protección Vegetal* 15 (1), 64–65.

Alemán, J., Martínez, M.A., Milán, O., Massó, E., and Rijo, E. (2004) Alternativas para la reproducción artificial de *Cryptolaemus montrouzieri* [Alternatives for the artificial production of *Cryptolaemus montrouzieri*]. *Revista Protección Vegetal* 19(2), 131–132.

Almaguel, L. (2006) Combate integral contra ácaros fitófagos [Integral control of phytophagous mites]. *Boletín fitosanitario* 11(1),71–100.

Altieri, M.A. and Nichols, C.I. (1999) Classical biological control in Latin America. In: Bellows, T.S. and Fisher, T.W. (eds) *Handbook of Biological Control*. Academic Press, London, pp. 975–991.

Alvarez, J.F., Naranjo, F. and Grillo, H. (2003) Estudios bioecológicos de *Tetrastichus howardi* (Olliff) (Hymenoptera: Eulophidae), parásito pupal de *Diatraea saccharalis (*F.) (Lepidoptera: Crambidae) en Cuba [Bioecological studies of *Tetrastichus howardi*, pupal parasite of *Diatraea saccharalis* in Cuba]. *Revista Centro Agrícola* 2, 63–67.

Armas, J.L., Ayala, J.L., Estévez, N. and Gómez, R.E. (1997) Manual para la reproducción y empleo de *Telenomus sp*. parasitoides de huevos de la palomilla del maíz, *Spodoptera frugiperda* (J. Smith) [Manual for the reproduction and use of *Telenomus* sp. parasitoids of eggs of the corn moth, *Spodoptera frugiperda*]. *Boletín técnico, INISAV* 2, 49–71.

Atkins, S., Hidalgo-Díaz, L., Kalisz, H., Mauchline, T., Hirsch P. and Kerry, B. (2003) Development of a new management strategy for the control of root-knot nematode (*Meloidogyne* spp.) in organic vegetable production. *Pest Management Science* 59, 183–189.

Bruner, S.C., Scaramuzza, L.C. and Otero, A.R. (1945) Catálogo de los insectos que atacan a las plantas económicas de Cuba [Catalogue of insects that attack economically important plants of Cuba]. *Boletín* 63, Estación Experimental Agronómica Santiago de la Vegas.

Caballero, S. and Sánchez, D. (2007) Reproducción del depredador *Coleomegilla cubensis* en laboratorio [Reproduction of the predator *Coleomegilla cubensis* in the laboratory]. *Fitosanidad* 11(2), 123.

Caballero, S., Fernández, S. and Rivero, A. (2000) *Trichogramma pintoi*, control biológico efectivo de *Plutella xylostella* (L.) en col (*Brassica oleracea* var. *capitata* L.) [*Trichogramma pintoi,* effective biological control of *Plutella xylostella* in cabbage (*Brassica oleracea* var. capitata)]. *Centro Agrícola* 27(1), 71–76.

Castiñeiras, A., Caballero, S., Rego, G., and González, M. (1982) Efectividad técnico-económica del empleo de la hormiga leona *Pheidole megacephala* en el control del Tetúan del Boniato *Cylas formicarius elegantulus* [Technical-economic effectiveness of the use of the lioness ant *Pheidole megacephala* in the control of the sweet potato weevil *Cylas formicarius elegantulus*]. *Ciencia y Técnica en la Agricultura. Proteccion de Plantas* 5, 103–109.

Castiñeiras, A., Pérez, M., Obregón, O. and Castañeda, I. (1984) Virulencia de tres cepas de *Metarhizium anisopliae* sobre adultos de *Cylas formicarius elegantulus* (Coleoptera: Curculionidae) [Virulence of three strains of *Metarhizium anisopliae* on adults *of Cylas formicarius elegantulus*]. *Ciencia y Técnica en la Agricultura. Proteccion de Plantas* 7(1), 129–135.

Castiñeiras, A., Calderón, A. and Ponce, E. (1992) Control biológico de *Cosmopolites sordidus* (Germ.) con *Beauveria bassiana* (Bals.) Vuill [Biological control of *Cosmopolites sordidus* with *Beauveria bassiana*]. *Protección Vegetal* 7(1), 13–16.

Ceballos, M., Martínez, M., de Los, A., Duarte, L., Baños, H.L. and Sánchez, A. (2009) Asociación áfidos-parasitoides en cultivos hortícolas [Associations of aphid parasitoids in horticultural crops]. *Revista Protección Vegetal* 24(3), 180–183.

Ceballos Vázquez, M. and Montes de Oca Martínez, N. (2016) Registro sanitario de bioplaguicidas microbianos en América Latina y Cuba. Caso de estudio: bionematicida cubano KlamiC [Health registration of microbial biopesticides in Latin America and Cuba. Case study: Cuban bionematicide KlamiC]. *Revista Protección Vegetal* 31(2), 120–133.

CIA (2017) The World Factbook: Cuba. Available at: https://www.cia.gov/library/publications/resources/the-world-factbook/geos/cu.html (accessed 7 July 2019).

CNSV (Centro Nacional de Sanidad Vegetal) (2016) *Análisis del cumplimiento del programa de producción de medios biológicos* [*Analysis of compliance with the production program of biological media*]. Ministerio de la Agricultural, Havana.

Cuadra, R., Ortega, J., Morfi, O.L., Soto, L. and Zayas, P. (2008) Efecto de los medios biológicos Trifesol y Nemacid sobre los nematodos de las agallas en la producción protegida de hortalizas [Effect of the biopesticides Trifesol and Nemacid on root knot nematodes in the protected production of vegetables]. *Revista Protección Vegetal* 23(1), 38–42.

De la Torre, S.L. (1993) Trichogramma. *Biología, Aplicación y Sistemática* [Trichogramma. *Biology, Application and Systematics*]. Editorial Científico Técnica, Havana, pp. 29–30.

Duarte, A., Menéndez, J.M., Fernández, A. and Martínez, J. (1992) Utilización del biopreparado *Metarhizium anisopliae* Cepa Niña Bonita en plantaciones de *Pinus caribaea* para el control de *Rhyacionia frustrana* [Use of the biopesticide *Metarhizium anisopliae* Niña Bonita strain in plantations of *Pinus caribaea* for the control of *Rhyacionia frustrana*]. *Revista Forestal Baracoa* 22(2), 17–23.

Elósegui, O. (2006) Producción de hongos para el control biológico [Production of fungi for biological control]. In: Vázquez, L.L. (ed.) *Memorias Del Curso Taller Manejo Agroecológico De Plagas En El Sistema De Producción* [*Proceedings of the Agro-ecological Pest Management Workshop in the Production System*]. Editorial Cidisav, Havana, pp.73–101.

Elósegui, O., Fernández-Larrea, O., Grijuela, P.E., Borges, M.G., Rovesti, L. and Jiménez, J. (2009) Colecta de esporas de *Trichoderma harzianum* Rifai cepa A34 por lecho fluidizado y ciclón dual y por tamizaje vibratorio [Collection of spores of *Trichoderma harzianum* Rifai strain A34 by fluid-bed and dual cyclone, and separation by electric vibratory sieving]. *Fitosanidad* 13(4), 265–269.

Elósegui, O., George, E., Fernández-Larrea, O., Pérez, C. and Laguardia, E. (2015) Evaluación de la colecta de esporas de *Trichoderma harzianum* Rifai desarrolladas sobre tres sustratos sólidos mediante lecho fluidizado y tamizaje vibratorio [Evaluation of the spores of *Trichoderma harzianum* Rifai developed on three solid substrates by fluid-bed and vibratory screening]. *Fitosanidad* 19(1), 23–27.

Fernández, E. (2007) Manejo de fitonematodos en la agricultura cubana [Management of phytonematodes in Cuban agriculture]. *Fitosanidad* 11(3), 57–60.

Fernandez-Larrea, O. (1999) A review of *Bacillus thuringiensis* (Bt) production and use in Cuba. *Biocontrol News and Information* 20(1), 47N–48N.

Fernández-Larrea, O. (2013) Programa para la recuperación de bioplaguicidas, biofertilizantes y bioestimulantes en Cuba [Program for the recovery of biopesticides, biofertilizers and biostimulants in Cuba]. *Agricultura Orgánica* 19(2), 2–5.

Fuentes, A., Llanes, V., Méndez, F. and González, R. (1998) El control biológico en la agricultura sostenible y su importancia en la protección de la caña de azúcar en Cuba [Biocontrol in sustainable agriculture and its importance in the protection of sugarcane in Cuba]. *Phytoma* 95, 24–26.

Fuentes, F. (1994) *Producción y uso de Trichogramma como regulador de plagas* [*Production and use of Trichogramma for pest control*]. Red de Acción em Alternativas Al uso de Agroquímicos (RAA), Lima, Peru.

Gómez, E., Álvarez, R.M., San Juan, A.N., Zayas, M.A., Croche, G. and Cruz, X. (2001) Nematicida a partir del hongo *Verticillium lecanii* [Nematicide from the fungus *Verticillium lecanii*]. *Revista Terralia* 24, 30–31.

Gómez, J. and Grillo, H. (1999) Especies de Diptera, Tachinidae parasitando larvas de *Leucania* spp. en caña de azúcar [Species of Diptera, Tachinidae parasitizing larvae of *Leucania* spp. in sugar cane]. *Centro Agrícola* 26(2), 33–34.

González, A.M. (2012) Bioproductos: ciencia o necesidad? [Bioproducts: science or need?] *Revista Institucional del Grupo Empresarial de producciones Biofarmacéuticas y Químicas. LABIOFAM* 2(2), 54–61.

González, E. and Martínez, B. (1998) Control biológico de la roya del cafeto (*Hemileia vastatrix* Berk. y Br.) con *Verticillium lecanii* (Zimm.) Viegas [Biological control of coffee tree rust (*Hemileia vastatrix*) with *Verticillium lecanii*]. *Revista Protección Vegetal* 13(2), 81–84.

Hernández, A., Weekers, F., Mena, J., Pimentel, E., Zamora, J., Borroto, C., and Thonart, P. (2007) Culture and spray-drying of *Tsukamurella paurometabola* C-924: stability of formulated powders. *Biotechnology Letters* 29(11), 1723–1728.

Hernández, M.A., Arévalo, J., Cerio, W.G.M., Rodríguez, G. and Hidalgo, L. (2015) Uso de KLAMIC en estrategias de manejo de *Meloidogyne incognita* en diferentes sistemas hortícolas de Cuba [Use of KLAMIC in management strategies of *Meloidogyne incognita* in different horticultural systems of Cuba]. *Nematropica* 45(2), 65.

Hernández, M.T.C, de la Noval, W.T., Madruga, R.P., López, R.P.C. and Rodríguez, A.C. (2017) Food and nutrition security: A Cuban perspective. In: *Challenges and Opportunities for Food and Nutrition Security in the Americas. The View of the Academies of Sciences*. IANAS, IAP & BMBF, México-DF, pp. 264–285.

Hidalgo-Díaz, L. (2004) Registration of biopesticides in Cuba. In: Wabule, P.N., Nguariva, F.K., Kimmins, F.K. and Siverside, P.J. (eds) *Registration for Biocontrol in Kenya. Proceeding of the pest control Products Board/Kenya*. Agricultural Research Institute, Nakuru, Kenya, pp. 117–132.

Hidalgo-Díaz, L., Bourne, J.M., Kerry, B.R. and Rodriguez, M.G. (2000) Nematophagous *Verticillium* spp. in soils infested with *Meloidogyne* spp. in Cuba: isolation and screening. *International Journal of Pest Management* 46, 277–284.

Infante, D., Martínez, B., Peteira, B., Reyes, Y. and Herrera, A. (2013) Molecular identification of thirteen isolates of *Trichoderma* spp. and evaluation of their pathogenicity towards *Rhizoctonia solani Kühn*. *Biotecnología Aplicada* 30, 23–28

Jiménez, J. and Fernández, R. (1980) Efectividad de entomopatógenos para el control del picudo verde azul de los cítricos [Effectiveness of entomopathogens for the control of citrus blue green weevil]. *Protección de Plantas* 3(1), 76–86.

Luján, M., Milán, O., Cabrera, T. and Vázquez, T. (1992) Propagación bifásica de diferentes cepas de *Metarhizium anisopliae*. Su eficacia y capacidad epizoótica sobre larvas de *Mocis latipes* [Biphasic propagation of different strains of *Metarhizium anisopliae*. Its efficiency and epizootic capacity on larvae of *Mocis latipes*]. *Protección de Plantas* 2(1), 29–40.

Marín, M., Wong, I., García, G., Morán, R., Basulto, R., Pimentel, E. and Mena, J. (2013) Actividad antagónica *in vitro* de *Tsukamurella paurometabola* C-924 frente a fitopatógenos [Antagonistic activity *in vitro* of *Tsukamurella paurometabola* C-924 against phytopathogens]. *Revista Protección Vegetal* 28(2), 132–133.

Marín, M., Mena, J., Wong, I., Morán, R., Franco, R., Chavelis, P., García, G., Basulto, R., Hernández, A., Veloz, L. *et al.* (2014) Determinación de la capacidad promotora del crecimiento vegetal de *Tsukamurella paurometabola* c-924 y caracterización de los principales mecanismos involucrados en el proceso [Determination of the plant growth promoting capacity of *Tsukamurella paurometabola* c-924 and characterization of the main mechanisms involved in the process]. *Revista Anales de la Academia de Ciencias de Cuba* 4(2), 8.

Márquez, M.E. (2002) Evaluación de un producto de *Bacillus thuringiensis* para el control de la varroasis [Evaluation of a *Bacillus thuringiensis* product for varroasis control]. *Fitosanidad* 7(1), 3–7.

Márquez, M.E. and Fernández-Larrea, O. (1999) Producción y evaluación de biopreparados de *B. thuringiensis* sobre *Polyphagotarsonemus latus* (Acarina) Tarsoneidae [Production and evaluation of biopreparations of *B. thuringiensis* on *Polyphagotarsonemus latus*]. *Fitosanidad* 3(3), 53–58.

Márquez, M.E. and Fernández, E. (2006) Selección de cepas de *Bacillus thuringiensis* con efecto nematicida [Selection of *Bacillus thuringiensis* strains with nematicidal effect]. *Manejo Integrado de Plagas y Agroecología* 78, 63–69.

Márquez, M.E., Fernández-Larrea, O., Jiménez Ramos, J., Elosegui Claro, O., Gómez González, R. *et al.* (2010) *Producción masiva de bioplaguicidas para el control biológico de plagas y enfermedades* [*Mass production of biopesticides for biological control of pests and diseases*]. Centro de Información y Documentación de Sanidad Vegetal (CIDISAV), Havana. Available at: http://www.inisav.cu (accessed 7 July 2019).

Martínez, R., Blanco, N. and Torre, C. de la (2000) Bosquejo histórico de los trabajos realizados para el establecimiento del control biológico de la mosca prieta de los cítricos en Cuba [Historical sketch of the studies carried out for the establishment of biological control of the citrus black fly in Cuba]. *Fitosanidad*, 4(3), 99–105.

Martínez, B., Reyes, Y., Infante, D., González, E., Baños, H. and Cruz, A. (2008) Selección de aislamientos de *Trichoderma* spp. candidatos a biofungicidas para el control de *Rhizoctonia* sp. en arroz [Selection of isolates of *Trichoderma* spp. candidates for biofungicides for the control of *Rhizoctonia* sp. in rice]. *Revista Protección Vegetal* 23(2), 118–125.

Martínez, M. de los A., Ceballos, M. Duarte, L., Baños, H. L., Sánchez, A. and Chico, R. (2011) Cría de *Diaeretiella rapae* McIntosh en un sistema de plantas banco [Rearing of *Diaeretiella rapae* McIntosh in a system of banker plants]. *Revista Protección Vegetal* 26 (2), 129–130.

Matienzo, Y., Rijo, E., Milán, O., Torres, N., Larrinaga, J. and Massó, E. (2007) Contribución al conocimiento de especies botánicas con potencialidad para el fomento de reservorios de insectos benéficos en la agricultura urbana [Contribution to the knowledge of botanical species with potential for the promotion of reservoirs of beneficial insects in urban agriculture]. *Fitosanidad* 11(2), 128–129.

Mena, J., Veloz, L., Vázquez, R.P., Expósito, M., Salazar, E., León, L., Coca, Y., Ramírez, Y., Rodríguez, G., Hernández, A.T and Pimentel, E. (2002) Population of *Corynebacterium paurometabolum* strain C-924 in soils treated with HeberNem. *Nematology* 4(2), 287.

Mena, J., Pimental, E., Veloz, L., Hernández, A.T., León, L. and Ramírez, Y. *et al.* (2003) Aislamiento y determinación de cepas bacterianas con actividad nematicida. Mecanismo de acción de *C. paurometabolum* C – 924 sobre nematodos [Isolation and determination of bacterial strains with nematicidal activity. Mechanism of action of *C. paurometabolum* C - 924 on nematodes]. *Biotecnología Aplicada*, 20(4), 248–252.

Méndez, M.R. and Polanco, A.A. (2006) Métodos de control de nematodos con *Trichoderma harzianum* en casas de cultivos protegidos [Control methods on nematodes with *Trichoderma harzianum* in protected cultivation]. *Fitosanidad* 10(2), 174.

Meneses, R. (2008) *Manejo integrado de los principales insectos y ácaros del arroz*. [*Integrated management of the main rice insects and mites*]. Instituto de Investigaciones Del Arroz, Cuba (IIArroz) – N° 716-2008.

Meneses, R., Echevarría, G. and Monzón (1980) Efectividad de *Beauveria bassiana* (Balsamo) Vuillemin y *Metarhizium anisoplieae* (Metchnikoff) Sorokin en el control de *Lissorhoptrus brevirostris* (Suffr) (Coleoptera: Curculionidae) [Effectiveness of *Beauveria bassiana* and *Metarhizium anisoplieae* in controlling *Lissorhoptrus brevirostris*]. *Centro agrícola* 7(1), 107–120.

Milán, O., Cueto, N., Larrinaga, J., Massó, E., Hernández, N., Pineda, M., Caballero, S., Peñas, M., Rodríguez, L., Esson, A.I. *et al.* (2006) Informe científico-técnico para la reproducción y uso de coccinélidos: insectos benéficos para el combate de fitófagos en los agroecosistemas sostenibles en Cuba. [Scientific-technical report for the reproduction and use of coccinellids: beneficial insects to control of phytophages in sustainable agroecosystems in Cuba]. Registro 2139-2006, Centro Nacional de Derecho de Autor (Cenda), Havana. Available at: http://www.cenda.cu (accessed 7 July 2019).

Milán, O., Cueto Zaldívar, N., Larrinaga Lewis, J., Matienzo Brito,Y., Massó Villalón E., Delís Hechavarría E., Ramos Torres, T., Pineda Duverge, M., Granda Sánchez, R., Peñas Rodríguez, M. *et al.* (2007) Reproducción rústica de los coccinélidos (Coleoptera: Coccinellidae) para su utilización contra fitófagos en agroecosistemas sostenibles [Small-scale rearing of coccinellids for use against phytophages in sustainable agroecosystems]. *Boletín Fitosanitario* 12, 15 pp.

Montes, M. and Montejo, R. (1991) Lucha biológica contra *Pachnaeus litus* Germar con la utilización de *Heterorhabditis heliothidis* (Khan, Brooks and Hirsch-mann) Poinar [Biological control of *Pachnaeus litus* with the use of *Heterorhabditis heliothidis*]. In: Pavis, C. and Kermarrec, A. (eds) *Les Colloques* 58: *Rencontres Caraïbes en lutte biologique*. Institut National de la Recherche Agronomique, Paris, pp. 157–160.

Montes de Oca, N., Arévalo, J., Nuñes, A., Riverón, Y., Villoch, A. and Hidalgo-Díaz, L. (2009) KlamiC: Experiencia Técnica-Productiva [KlamiC: Technical-Productive Experience]. *Revista Protección Vegetal* 24(1), 62–65.

Murguido, C.A. and Elizondo, A.I. (2007) El manejo integrado de plagas de insectos en Cuba [The integrated management of insect pests in Cuba]. *Fitosanidad* 11(3), 23–28.

Pérez, J., Martínez, B., Rivas, E., Moreno, M. and Díaz, M.E. (2010) Aplicación de *Trichoderma* para el control del tizón gomoso del tallo (*Didymella bryoniae* (Auersw) Rehm) en el cultivo de sandía (*Citrullus vulgaris*) Schrad [Application of *Trichoderma* for the control of gummy stem blight (*Didymella bryoniae*) in a watermelon crop (*Citrullus vulgaris*)]. *Fitosanidad* 14(1), 67, 51.

Pérez, N. and Vázquez, L.L. (2001) Manejo ecológico de plagas [Ecological management of pests]. In: ACTAF (ed.) *Transformando el campo cubano. Avances de la Agricultura Sostenible*. Ed. ACTAF, Havana, pp. 191–223.

Pérez-Madruga, Y., Alonso-Rodríguez, D., Chico, R. and Rodríguez Morell, H. (2014) Cría de *Neoseiulus longispinosus* (Evans) sobre *Tetranychus tumidus* Banks utilizando el método de las bandejas [Rearing of *Neoseiulus longispinosus* on *Tetranychus tumidus* using the tray method]. *Revista Protección Vegetal* 29(2), 141–144.

Peteira, B., Puertas, A., Hidalgo-Díaz, L., Hirsch, P.R., Kerry, B.R. and Atkins, S.D. (2005) Real-time PCR to monitor and assess the efficacy of two types of inoculum of the nematophagous fungus *Pochonia chlamydosporia* var. *catenulata* against root-knot nematode populations in the field. *Biotecnología Aplicada* 22, 261–266.

Ramos, M. and Rodríguez, H. (2006) Riqueza de los fitoseidos (Acari: Mesostigata) en agroecosistemas en Cuba [Richness of phytoseiids in agroecosystems in Cuba]. *Fitosanidad* 10(3), 1–6.

Rego, G., Collazo, D. and Borges, A. (1990) Eficacia técnico-económica de métodos alternos de liberación de *Lixophaga diatraeae* y *Trichogramma* spp. en el control biológico de *Diatraea saccharalis* en caña de azúcar [Technical-economic efficacy of alternative methods of release of *Lixophaga diatraeae* and *Trichogramma* spp. in the biological control of *Diatraea saccharalis* in sugarcane]. *Revista de Ciencia y Técnica de Agricultura. Serie Protección de Plantas* 13(2), 7–17.

Reyes, Y., Martínez D., B. C., Infante, D. M., Cruz, A. T. and Borrego, J. G. (2009) Selección de aislamientos de *Trichoderma* spp. para el biocontrol del Tizón de la Vaina en Arroz en condiciones de campo [selection of isolates of *Trichoderma* spp. for the biological control of the blight of the pod in rice in field conditions]. *Phytopathology* 99(6) (Supplement), 190.

Rijo Camacho, E. and Acosta Amador, N. (1997) Biología de cuatro especies de la familia Chrysopidae en Cuba [Biology of four species of the Chrysopidae family in Cuba]. In: *Resúmenes. Tercer Seminario Científico Internacional de Sanidad Vegetal*. Havana, Cuba, pp. 23–27.

Rodríguez, H., Montoya, A., Pérez-Madruga, Y. and Ramos, M. (2013) Reproducción masiva de ácaros depredadores *Phytoseiidae*: retos y perspectivas para Cuba [Mass rearing of predatory mites *Phytoseiidae*: challenges and perspectives for Cuba]. *Revista Protección Vegetal* 28(1), 12–22.

Rodríguez, I., Martínez, M.A., Sánchez, L. and Rodríguez, M.G. (1998) Comprobación en campo de la efectividad de *Heterorhabditis bacteriophora* cepa HC1 en el control de chinches harinosas (Homoptera: Pseudococcidae) del cafeto [Field test of the effectiveness of *Heterorhabditis bacteriophora* strain HC1 in the control of mealybugs (Homoptera: Pseudococcidae) of the coffee tree]. *Revista Protección Vegetal* 13(3), 195–198.

Rodríguez Hernández, M.G. (2015) Entomopathogenic nematodes in Cuba: from laboratories to popular biological control agents for pest management in a developing country. In: Campos–Herrera, R. (ed.) *Nematode Pathogenesis of Insects and Other Pests – Ecology and Applied Technologies for Sustainable Plant and Crop Protection*. Series Sustainability in Plant and Crop Protection. Springer, Dordrecht, Netherlands, pp. 343–364.

Sánchez, S. (1997) El CREE de 'Margarita', algo más que un sueño [The CREE of 'Margarita', something more than a dream]. *Agricultura Orgánica* 3(2), 5–8.

Santana, Y., Casola, C. and Cussy, P. (2010) Uso de *Trichoderma harzianum* R. como control biológico de *Cercospora beticola* (sacc.), en el cultivo de la remolacha (*Beta vulgaris* Lin.) en sistemas de organopónicos [Use of *Trichoderma harzianum* as a biological control agent of *Cercospora beticola*, in beet (*Beta vulgaris*) in organoponic systems]. *Fitosanidad* 14(1), 66–67.

Stack, C.M., Easwaramoorthy, S.G., Metha, U.K., Downes, M.J., Griffin, C.T. and Burnell, A.M. (2000) Molecular characterisation of *Heterorhabditis indica* isolates from India, Kenya, Indonesia and Cuba. *Nematology* 2(5), 477–487

Stefanova, M. (2007) Introducción y eficacia del biocontrol de fitopatógenos con *Trichoderma* spp. en Cuba [Introduction and efficacy of the biocontrol of phytopathogens with *Trichoderma* spp. in Cuba]. *Fitosanidad* 11(3), 75–80.

Valdés, R., Mora, E., Castro, M., Pozo, E. and Cárdenas M. (2005) Empleo de nematodos entomopatógenos (*Heterorhabditis* spp.) como contribución al Manejo Integrado de *Diaphania hyalinata* (L.) (Lepidoptera, Pyralidae) en el cultivo del pepino en sistemas de organopónicos [Use of entomopathogenic nematodes (*Heterorhabditis* spp.) as a contribution to the integrated management of *Diaphania hyalinata* in cucumber crop in organoponic systems]. *Centro Agrícola* 32(4), 47–53.

Vázquez, L.L. and Pérez, N. (2016) Control biológico [Biological control]. In: Funes, F. and Vásques, L.L. (eds) *Avances de la Agroecología en Cuba*. Estación experimental de Pastos y Forrajes Indio Hatuey, Matanzas, pp. 169–182.

Vázquez, L.L., Medina, H. and Castellanos, J.A. (2005) Notas sobre la introducción de insectos entomófagos en Cuba [Notes on the introduction of entomophagous insects in Cuba]. *Fitosanidad* 9(4), 63–65.

Vázquez, L.L., Matienzo, Y., Veitía, M. and Alfonso J. (2008) Conservación y manejo de enemigos naturales de insectos fitófagos en los sistemas agrícolas de Cuba [Conservation and management of natural enemies of phytophagous insects in the agricultural systems of Cuba]. Ed. Centro de Información y Documentación de Sanidad Vegetal (CIDISAV), Havana.

Vázquez, L.L., Caballero, S., Carr, A., Gil, J., Armas, J.L., Rodríguez, A., Becerra, M., Rodríguez, L.A., Granda, R., Corona, T. *et. al*. (2010) Diagnóstico de la utilización de entomófagos y entomopatógenos para el control biológico de insectos por los agricultores en Cuba [Diagnosis of the use of entomophages and entomopathogens for the biological control of insects by farmers in Cuba]. *Fitosanidad* 14(3), 159–169.

Wyckhuys, K.A.G., Lu,Y.H., Morales, H., Vazquez, L.L., Legaspi, J.C., Eliopoulos, P.A. and Hernandez, L.M. (2013) Current status and potential of conservation biological control for agriculture in the developing world. *Biological Control* 65, 152–167.

11 Biological Control in Dominica

Joop C. van Lenteren*

Laboratory of Entomology, Wageningen University, Wageningen, the Netherlands

* E-mail: joop.vanlenteren@wur.nl

Abstract

Introduction of two tachinid parasitoid species in 1951 for biocontrol of the sugarcane borer resulted in good control. Native and imported parasitoids were evaluated in the 1960s for control of fruit flies and noctuid moths on various fruits in Dominica, but none of them was effective. Several other imports of exotic natural enemies in this period for classical biocontrol of banana weevil, coffee leaf miner and diamondback moth were not successful. Complete island-wide control of the citrus blackfly was obtained after releases of two exotic parasitoids at the end of the 1990s.

11.1 Introduction

Dominica has an estimated population of about 73,900 (July 2017). Its economy was in the past dependent on agriculture, with banana, citrus and mango as main crops, and more recently on ecotourism (CIA, 2017). The material for this chapter mainly originates from Cock (1985) and articles therein.

11.2 History of Biological Control in Dominica

11.2.1 Period 1880–1969

Fruit flies

In the early 1960s during a visit to Dominica, F.D. Bennett reported damage to mangoes, citrus, guava and hog plums (*Spondias mombin* L.) by the fruit fly *Anastrepha obliqua* Macquart (reported as *A. mombinpraeoptans* Sein) and by fruit-piercing noctuid moths (*Gonodonta* spp.). He found the native *Opius anastrephae* Vier. to have parasitized *A. obliqua*. Parasitoids (*Aceratoneuromyia indica* Silvestri, *Doryctobracon crawfordi* (Vier.), *Biosteres longicaudatus* (Ashmead)) were imported from Mexico, via the CIBC station in Trinidad, mass reared and released. Later, in 1964 and 1965, other parasitoid species (*Doryctobracon trinidadensis* (Gah.), *D. cereus* (Gah), *Ganaspis* sp., *Opius bellus* (Gah.) and *Pachycrepoideus vindemiae* Rondani) were sent from Trinidad and released. None of these parasitoids exerted sufficient control (Cock, 1985).

Banana weevil

The most important pest of banana in the Caribbean is the banana weevil, or banana borer, *Cosmopolites sordidus* (Germ.). Biocontrol attempts with the predatory beetles *Hololepta* (= *Leionota*) *quadridentata* (F.) and *Plaesius javanus* Erichs., both sent in the 1950s from Trinidad, failed because the natural enemies did not establish, but a native *Dactylosternum* sp. was found during a survey in the 1970s (Cock, 1985).

Coffee leaf miner

The coffee leaf miner *Perileucoptera coffeella* Guer. is the principal pest of coffee in the Caribbean. An attempt in 1959 to introduce parasitoids failed (Cock, 1985).

Diamondback moth

The diamondback moth *Plutella xylostella* (L.) feeds on leaves of cruciferous plants. It is attacked by a number of local parasitoids, but control is insufficient. *Apanteles (Cotesia) plutellae* Kurd was obtained from India in 1970 and from the Netherlands in 1971; *Diadromus collaris* (Grav.) was obtained from India in 1972. After releases of *A. plutellae* in 1971–1972 and *D. collaris* in 1972 in Dominica, a few adults of *A. plutellae* were recovered (Cock, 1985).

Sugarcane moth borers

Sugarcane is the most important crop in the Caribbean, and sugarcane moth borers *Diatraea* spp. are its most widespread pests, with *Diatraea saccharalis* (F.) as the most important species. In the early 1950s, it was found that 20–30% of cane joints were damaged by *D. saccharalis* over extensive areas. None of the collected larvae were parasitized and two tachinid parasitoids, *Lixophaga diatraeae* (Tns.) and *Paratheresia claripalpis* (Wulp), were introduced from Trinidad and released at the end of 1951 and start of 1952. In 1954, infestation did not exceed 5% and parasitism was high; both parasitoids had apparently become established. The amazon fly *Lydella*

Table 11.1. Overview of biological control activities in Dominica.

Biocontrol agent / exotic (ex), native (na)	Pest / crop	Type of biocontrol[a] /since	Effect /area under biocontrol[b]	Reference
Opius anastrephae / na	Fruit flies, citrus etc.	NC	Insufficient control	Cock 1985
Aceratoneuromyia indica / ex		CBC / 1960s	Insufficient control	
Biosteres longicaudatus / ex		CBC / 1960s	Insufficient control	
Doryctobracon crawfordi / ex		CBC / 1960s	Insufficient control	
Doryctobracon trinidadensis / ex		CBC / 1960s	Insufficient control	
D. cereus / ex		CBC / 1964–1965	Insufficient control	
Ganaspis sp. / ex		CBC / 1964–1965	Insufficient control	
Opius bellus / ex		CBC / 1964–1965	Insufficient control	
Pachycrepoideus vindemiae / ex		CBC / 1964–1965	Insufficient control	
Dactylosternum sp. / na	Banana weevil, banana	NC	Insufficient control	Cock 1985
Hololepta quadridentata / ex		CBC / 1950s	No control / not established	
Plaesius javanus / ex		CBC / 1950s	No control / not established	
Cotesia plutellae / ex	Diamondback moth, vegetables	CBC / 1970–1971	No control / established	Cock 1985
Diadromus collaris / ex		CBC / 1972	No control / not established	
Lixophaga diatraeae / ex	Sugarcane borers, sugarcane	CBC / 1951–1952	Complete control / 2,500 ha	Cock 1985
Paratheresia claripalpis / ex		CBC / 1951–1952	Complete control / 2,500 ha	
Lydella minense / ex		CBC / 1951–1952	No control / not established	
Encarsia opulenta / ex	Citrus blackfly, citrus	CBC / 1997–1998	Complete control / 250 ha	Lopez *et al.* 2009
Amitus hesperidum / ex		CBC / 1997–1998	Complete control / 250 ha	
Eucelatoria bryani / ex	Armyworms, many crops	CBC / 1972	No control / not established	Cock 1985
Trichogrammatoidea armigera / ex		CBC / 1972	No control / not established	
Euplectrus platyhypenae / ex		CBC / 1982	No control / not established	
Telenomus remus / ex		CBC / 1982	Insufficient control / established	

[a]Type of biocontrol: ABC = augmentative, CBC = classical, NC = natural, ConsBC = conservation biological control
[b]Area of crop harvested in 2016 according to FAO (http://www.fao.org/faostat/en/#data/qc)

(= *Metagonistylum*) *minense* (Town.) was also introduced but did not become established. Up to the 1980s, biocontrol of *D. saccharalis* with the tachinids had been successful (Cock, 1985).

Armyworms

In the region, five species of *Spodoptera* and two species of *Heliothis* are of economic importance and attack a wide range of crops. The armyworm parasitoids *Eucelatoria bryani* Sabrosky and *Trichogrammatoidea armigera* Nagaraja had been imported from Trinidad and were released in 1972, followed by *Euplectrus platyhypenae* Howard and *Telenomus remus* Nixon in 1982. It seems that only *T. remus* established (Cock, 1985).

Dominica as provider of natural enemies

Dominica also served as a source of natural enemies during this period. The following species were sent to other Caribbean Islands and Kenya (Cock, 1985): *Tetrastichus gala* Walker to St Lucia(1938); *Mirax insularis* Mues. to St Lucia (1938) and Kenya (1962); and *Brachyufens* sp. to Jamaica (1958).

11.2.2 Period 1970–2000

Some of the introductions of natural enemies made during this period have been mentioned above, as they were related to biocontrol programmes started before 1970.

Citrus blackfly

Citrus blackfly *Aleurocanthus woglumi* (Ashby.), native to Asia, was first found in 1913 in Jamaica and subsequently spread over the Caribbean, including Dominica, where it was first recorded in 1969 but serious pest problems only began to appear in 1994 (FAO, 2000). Releases of two species of parasitoids, *Encarsia opulenta* Silvestri and *Amitus hesperidum* Silvestri, originating from the US Department of Agriculture Plant Protection Center in Mission, Texas, were made at a citrus orchard in May 1997 as part of a classical biocontrol programme that linked researchers from Texas and agricultural officials at the Ministry of Agriculture in Dominica. Parasitoids were recovered from parasitized blackfly pupae at the site of release. Because of this positive result, releases with material obtained from Texas were made at other sites up to May 1998. Later, locally collected parasitoids were released all over the island. Levels of parasitism, based on samples of 40 citrus leaves with parasitized citrus blackfly pupae per orchard, ranged between 7% and 99%. Substantial to complete control of blackfly was achieved within 1.5 years in all regions, currently with complete control islandwide. Lopez *et al.* (2009, p. 497), during post-release surveys in 2000 and 2001, concluded that citrus blackfly populations were 'low to non-existent in 50 of 51 field sites examined' and that '*E. perplexa* and *A. hesperidum* have kept citrus blackfly populations under effective biocontrol in Dominica'.

Dominica as provider of natural enemies

A shipment of parasitized citrus blackfly material was sent from Dominica to St Kitts in1998, where emerging *A. hesperidum* and *E. opulenta* were released.

11.3 Current Situation of Biological Control in Dominica

Natural enemies that were introduced, released and did establish in previous periods are still supposed to be present. In particular, classical biocontrol of pests in sugarcane and of citrus blackfly in citrus is successful (Table 11.1). No information could be obtained about current biocontrol activities in a recent literature search.

11.4 New Developments of Biological Control in Dominica

No information could be obtained in a recent literature search. Also no reliable estimate of areas under biocontrol could be obtained.

References

CIA (2017) The World Factbook: Dominica. Available at: https://www.cia.gov/library/publications/the-world-factbook/geos/do.html (accessed 7 July, 2019).

Cock, M.J.W. (ed.) (1985) *A Review of Biological Control of Pests in the Commonwealth Caribbean and Bermuda up to 1982*. Technical Communication No. 9, Commonwealth Institute of Biological Control. Commonwealth Agricultural Bureaux, Farnham Royal, UK.

FAO (2000) *Overview of FAO Project – TCP/DMI/8922: Management of Citrus Blackfly in Dominica*. National workshop on the Management of citrus blackfly in Dominica, Roseau, Dominica 12–13 December 2000.

Lopez, V.F., Kairo, M.T.K., Pollard, G.V., Pierre, C., Commodore, N. and Dominique, D. (2009) Post-release survey to assess impact and potential host range expansion by *Amitus hesperidum* and *Encarsia perplexa*, two parasitoids introduced for the biological control of the citrus blackfly, *Aleurocanthus woglumi* in Dominica. *BioControl* 54(4), 497–503.

12 Biological Control in the Dominican Republic

Colmar Serra[1]* and Joop C. van Lenteren[2]

[1]Instituto Dominicano de Investigaciones Agropecuarias y Forestales (IDIAF), Centro de Tecnologías Agrícolas (CENTA), Santo Domingo, Dominican Republic; [2]Laboratory of Entomology, Wageningen University, Wageningen, The Netherlands

* E-mail: colmar.serra@gmx.net

Abstract

During the first part of the 20th century several introductions were made, resulting in successful classical biocontrol of coconut scale and cottony cushion scale. After 1970, biocontrol activities increased and prospecting for biocontrol agents of pests, diseases and weeds took place. Other successful classical biocontrol programmes were implemented after 1980, resulting in control of citrus blackfly, whiteflies, papaya mealybug, pink hibiscus mealybug and Anastrepha fruit flies. A number of cases of natural control by predatory bugs, chrysopids, syrphids, coccinellids, entomopathogenic fungi and nematodes were documented for arthropod pests, such as whiteflies, thrips, mites, lepidopterans, and dipteran and lepidopteran leaf miners. Conservation biocontrol enhancing the action of parasitoids, predators or pathogens of lepidopterans, whiteflies, leaf miners, thrips, aphids and mites in vegetables was developed, as well as for citrus psyllids and fruit flies. Augmentative biocontrol of the coffee berry borer, using an exotic parasitoid and a pathogen, of the ello sphinx in cassava, of thrips and other vegetable pests in greenhouses and open fields, and of bilharzia-transmitting snails, was implemented. Also, prospecting for native predatory mites has been executed recently. Conservation biocontrol based on integrated approaches using cultural control measures, monitoring systems, selective pesticides and native natural enemies is step by step reaching larger groups of agricultural producers, above all those who produce crops destined for export markets, which are subject to severe regulations concerning types and residues of synthetic pesticides.

12.1 Introduction

The Dominican Republic has an estimated population of about 10,735,000 (July 2017) (CIA, 2017) and according to Bethancourt *et al.* (2017) its main agricultural products are rice, maize, sugarcane, coffee, cacao, tobacco, coconut, beans, pigeon peas, cassava, taro, sweet potato, yam, potato, plantain, banana, onion, eggplant, squash, tomato, avocado, papaya, pineapple, sweet orange, passion fruit, cucumber and chayote. Beef cattle, pigs, poultry and egg production are also very important. Aquaculture and fishing activities have expanded in recent years, accounting for 11.3% of livestock production units. The country has 1.9 million hectares dedicated to agricultural production, including over 0.69 million hectares for agricultural use, 0.56 million hectares for animal husbandry, 0.46 million hectares for the cultivation of agricultural products combined with animal husbandry, 0.07 million hectares for planting of forest and timber trees, 19,000 ha for a combination of tree planting and animal husbandry and 17,000 ha for growing flowers and/or ornamental plants. According to the same report (Bethancourt *et al.*, 2017, pp. 287 and 289):

> Twenty-three percent of Dominican soils are considered suitable for agricultural crops, with specific use and management practices, while another 16% can be used for grazing and rice production with modern mechanization methods and intensive management methods ... A third of the Dominican population is rural, earning its livelihood from agricultural and forestry activities.

12.2 History of Biological Control in the Dominican Republic

12.2.1 Period 1880–1969

Natural biological control: fungi attacking weeds

Ciferri and Fragoso (1927) mentioned plant fungal pathogens found in the Dominican Republic on weeds: *Petrakina mirabilis* Ciferri, which infects cattail *Typha domingensis* (Pers); and *Uredo eichhorniae* Fragoso and Ciferri, and *Doassansia eichhorniae* Ciferri, which were both infecting water hyacinth *Eichhornia crassipes* (Mart.) Solms.

Classical biological control of coconut scale and cottony cushion scale

Coconut scale *Aspidiotus destructor* Sign. is an important pest in the Caribbean. Numerous inter-island introductions have been made for the control of this scale, including introductions into the Dominican Republic, where the coccinellids *Chilocorus cacti* (L.), *Cryptognatha nodiceps* Marshall and *Pentilia castanea* Mulsant were released (Cock, 1985). During extensive recent sampling in the Dominican Republic, neither *P. castanea* nor *C. nodiceps* were found, so they might have failed to establish (Pérez-Gelabert, 2008), while *C. cacti* apparently did establish. Gómez-Menor (1941) reported that *Cryptolaemus mountrouzieri* Mulsant was introduced into the Dominican Republic in the 1930s, probably for biocontrol of *Icerya purchasi* Maskell in citrus, and did establish. The same author mentioned the introduction of *Rodolia cardinalis* (Mulsant).

Introduction of vertebrates for classical biocontrol of rats and insects

The small Indian mongoose *Herpestes javanicus* (É. Geoffroy Saint-Hilaire) was brought to the Dominican Republic after an earlier introduction to Jamaica in 1872 (see Chapter 20: Jamaica) to control rats that had been accidentally introduced into the island by Spanish conquerors in the 15th century. However, due to its polyphagous predation behaviour, the introduction is no longer considered beneficial, as mongooses also eat amphibians, reptiles, birds and small mammals; they are supposed to have caused the extinction of at least five species in Jamaica (MIMARENA, 2012). Later, the cane toad *Rhinella marina* (L.) was introduced to control pest insects in sugarcane, but this introduction is now also considered non-beneficial as it consumes other organisms (MIMARENA, 2012). Although no data were found in the literature concerning the introduction of this species into the Dominican Republic, the local herpetologist S. Inchaustegui (Santo Domingo, 2018, personal communication) assumes it happened probably at the beginning of the 20th century and that is was known as the Bogaert toad, maybe as reference to the introduction of this species in the Bogaert family's rice fields.

12.2.2 Period 1970–2000

During this period, the number of accidentally introduced invasive pest species dramatically increased, due to intensified international exchange of agricultural produce as well as tourism. An overview of introduced plant pathogens and arthropod pests can be found in Serra *et al.* (2003), who mentioned some 50 damaging species introduced between 1975 and 2003. Around 1978, a biocontrol laboratory was established at the Centro Sur de Desarrollo Agropecuario (CESDA) of the Secretaría de Estado de Agricultura (SEA) in San Cristóbal, as part of support by the US Agency for International Development (USAID) and the Interamerican Institute of Cooperation for Agriculture (IICA). Initially, the laboratory reared the egg parasitoids *Trichogramma* sp., *Telenomus remus* Nixon, two *Chelonus* spp. (including *C. insularis* Cresson), as well as other natural enemies (Serra, 2006; F. Díaz, Santo Domingo, 2018, personal communication). An inventory was made of natural enemies of the ello sphinx moth *Erinnyis ello* (L.), which is a severe pest in cassava. Based on the inventory, monitoring of the pest was recommended and, when necessary, spraying with a microbial agent based on *Bacillus thuringiensis* Berliner. In this period, the Consejo Estatal del Azucar (CEA) mass reared *Lixophaga diatraeae* (Townsend) in its Duquesa laboratory and released it in sugarcane fields to control the sugarcane borer *Diatraea saccharalis* (F.). *Lydella* (= *Metagonistylum*) *minense* (Townsend) was also introduced and liberated, but it could not be recovered (F. Díaz, Santo Domingo, 2018, personal communication).

Natural, classical and augmentative biological control of citrus pests

Citrus production was of growing importance before the detection of the huanglongbing disease (HLB) in 2006 (Matos *et al.*, 2009) and fruit is particularly used to produce juice and concentrates. In 1993, citrus was produced on about 10,000 ha. The citrus root weevil *Diaprepes abbreviatus* (L.), the citrus rust mite *Phyllocoptruta oleivora* (Ashmead) and the brown citrus aphid *Toxoptera citricidus* Kirkaldy, the vector of citrus tristeza virus, were the most important citrus pests at that time. In 1994, the citrus leaf miner *Phyllocnistis citrella* (Stainton) invaded the country and became another important citrus pest. The citrus blackfly *Aleurocanthus woglumi* Ashby, another invasive pest, is no longer a problem, as it was brought under classical biocontrol. A brief summary of the studies carried out for biocontrol of citrus pests in this country is presented below.

CITRUS ROOT WEEVIL. Because of the feeding habits of the pest, chemical control of this weevil is unprofitable. Therefore, since 1990, citrus agribusinesses together with the Fundación de Desarrollo Agropecuario (Center for Agricultural Development FDA, now Centro de Desarrollo Agropecuario y Forestal, Center for Agricultural and Forestry Development) (CEDAF) have supported studies for biocontrol on parasitoids, including *Ceratogramma etiennei* Delvare (Etienne *et al.*, 1992), *Quadrastichus* (= *Tetrastichus*) *haitiensis* (Gahan) and *Fidiobia* sp. (Serra, 2006); fungal entomopathogens *Metarhizium anisopliae* Metschnikoff and *Beauveria bassiana* (Balsamo) Vuillemin; and entomopathogenic nematodes (*Heterorhabditis* sp.) (Serra, 2006). Within the Programa Nacional de Manejo Integrado de Plagas (PNMIP) programme, a laboratory mass rearing for natural enemies of *D. abbreviatus* was established in the area of the Citrícola del Este Consortium. In 1993, the root weevil was successfully controlled by augmentative releases of the egg parasitoid *Q. haitiensis* (Serra, 2006). Mass production of *B. bassiana* was initiated at the Dominican Citrus Consortium and this microbial control agent was used for augmentative biocontrol of the root weevil on 1,400 ha in 1994, sufficient for one-third of the company's demand (Serra, 2006). Also, in Central Romana, a commercial *B. bassiana* product was applied to the soil to control *D. abbreviatus*. *B. bassiana* and *Heterorhabditis* sp. (on larvae of *Galleria melonella* L.) were also reared at the Laboratorio de Control Biológico (LABOCOBI) of the Universidad Autónoma de Santo Domingo (UASD) (Project UASD/SEA/ Institut National de la Recherche Agronomique (INRA)) and in the Citrícola del Este Consortium for control of citrus root weevil.

BLACK CITRUS APHID The parasitoid *Aphidius colemanii* Viereck was imported from Argentina through Florida to control *Toxoptera aurantii* (Boyer de Fonscolombe), but did not establish (Serra *et al.*, 2003). A survey of the main natural enemies of the black aphid was made in 1992 and the predators *Cycloneda sanguinea* (L.) and *Chrysopa* spp. and several entomopathogenic fungi were found (O.G. Jiménez, Santo Domingo, 2006, personal communication).

BROWN CITRUS APHID. Serra (2006) collected parasitoids of the brown citrus aphid *T. citricidus* (Kirk.), which were identified by G.A. Evans (Gainesville, Florida, 2003, personal communication) as belonging to the genus *Pachyneuron*, often hyperparasitic on primary parasitoids (braconids and others) of aphids and other sucking pests.

CLOUDY-WINGED WHITEFLY. The whitefly *Dialeurodes citrifolii* (Morgan) occasionally attacks citrus trees, but the naturally occurring fungus *Aschersonia aleyrodii* Webber causes epidemics that drastically reduce whitefly populations (Schmutterer, 1990).

CITRUS LEAF MINER. The citrus leaf miner *P. citrella*, native to Asia, was found in the country in 1994 and caused severe defoliation of new shoots, reducing plant growth. Due to its mining habits, it is difficult to control with chemical pesticides and the latter interfere with naturally occurring biocontrol agents. The Fundación de Desarrollo Agropecurio (FDA, now Centro de Desarrollo Agropecuario y Forestal (CEDAF)) and citrus companies, with the support of INRA and the Laboratory of Biological Control of the US Department of Agriculture (USDA), studied natural enemies of the leaf miner. Six species of parasitoids of the genera *Elasmus* and *Horismenus*, *Cirrospillus quadristriatus* (Subba Rao and Ramamani) and the genus *Aprostocetus* (families Eulophidae and Elasmidae), were found, with *Horismenus* sp. occurring most frequently. Also many predators (coccinellids, chrysopids, reduviid bugs and ants) were found (Taveras, 2000; Serra, 2006). Due to the native natural enemies that attack this leaf miner, pest problems are now less problematic than shortly after its introduction. In addition, application of a *B. thuringiensis* product provides good leaf miner control (Serra, 2006).

CITRUS BLACKFLY. *A. woglumi* of Asian origin is known to have occurred in the country since the 1920s (Pérez-Gelabert, 2008) and initially caused serious damage to citrus. With the introduction in 1996 and establishment of its parasitoid *Encarsia perplexa* Huang & Polaszek, initially misidentified as *E. opulenta* (Silvestri), the blackfly was brought under complete classical biocontrol, and citrus production costs considerably decreased as insecticides were no longer needed (Evans and Serra, 2002; Serra, 2006).

Biological control of whiteflies and other pests in tomato and aubergine

Since the 1970s, the sweet potato whitefly *Bemisa tabaci* (Gennadius) 'A' biotype has been present in the Dominican Republic. In 1978, the greenhouse whitefly *Trialeurodes vaporariorum* Westwood was found in the Constanza mountain valley (1,000 m above sea level (asl)) in vegetables and potato crops (Serra *et al.*, 2003). In 1987, the PNMIP started as a collaborative project of several agricultural institutes sponsored by USAID and dealt initially mainly with control of the *B. tabaci*-transmitted begomovirus (Serra, 2006). Since the 1980s, the whitefly parasitoid *Encarsia formosa* Gahan has been accidentally and/or intentionally imported and did establish, but high spray frequencies against other pests have prevented effective biocontrol (Serra *et al.*, 2003). As part of the Whitefly Research & Extension Project (WREP) at the Instituto Superior de Agricultura (ISA) in La Herradura, Santiago, the biology, host-stage and host-plant preferences, reproduction techniques, efficiency and interspecific competition of the whitefly parasitoids *Encarsia sophia* (Girault & Dodd) (= *E. transvena*), *E. formosa* and *E. telemachusi* Evans (= *E.* sp. nr. *pergandiella* or *Encarsia* sp. *parvella* group), were studied by Ramirez (1995). Serra *et al.* (1996) found that *E. formosa* was not able to compete with *E. telemachusi* on *B. tabaci* under lowland conditions. From 1993, as a result of very heavy losses in tomato crops due to the whitefly-begomovirus complex, plans were made to start a mass rearing of whitefly parasitoids in the Azua valley, which, however, never materialized. Interestingly, *Amitus fuscipennis* Mac Gown & Nebeker was found in the country, which reached very high percentages of parasitism in *T. vaporariorum* in undisturbed areas at higher altitudes (Serra *et al.*, 1996; Evans and Serra, 2002).

Control of greenhouse whitefly on snap beans and various other crops in the Constanza valley with a predatory lacewing, *Chrysopa* sp., and a predatory ladybird beetle, *Delphastus pusillus* (LeConte), and using sorghum as refugee crop for the predators, was studied with support of the Japan International Cooperation Agency (JICA) (Serra, 2006).

Other pests of tomato and their natural enemies were studied in a Dominican–German Project (IPL-GTZ: German Society for Technical Cooperation) project, which started as the 'Neem Project' in 1987. Twenty main pest species from 11 families and 25 secondary pest species were reported on tomatoes in the South-western part of the country. Twenty-eight parasitoid species of two orders and 16 families were obtained from the seven most important pests (e.g. leaf miners, leaf-eating and fruit-boring worms, whiteflies, bugs). Among these were 11 species, eight genera and four families of parasitoids of the vegetable leaf miner *Liriomyza sativae* Blanchard, including three eulophid species (*Diglyphus beginii* (Ashmead), *D. isaea* Wlk. and a *Neochrysocharis* sp.), the pteromalid *Halticoptera circulus* Walker, two eucoilids (*Ganaspidium utilis* Beardsley and *Disorygma pacifica* (Yoshimoto)) and two braconid *Opius* spp. (including *O. forticornis* Cameron). In addition, eight species of eight genera and four families of parasitoids of the tomato pinworm *Keiferia lycopersicella* (Walsingham) were found, among which were the eulophids *Euderus* sp. and *Zagrammosoma* sp., as well as the four braconids *Pseudapanteles dignus* (Muesebeck), *Glyptapanteles* sp., *Parahormius pallidipes* Ashmead and *Chelonus* sp., the ichneumonid *Temelucha* sp. and the formicid *Conomyrma* sp. Noctuid *Spodoptera* spp. were attacked by a platygastrid egg parasitoid and the complex *Pseudoplusia includens* (Walker)/ *Trichoplusia ni* (Hübner), the encyrtid *Copidosoma* sp., the eulophid *Euplectrus* sp. and the braconid *Glyptapanteles* sp. The stink bug *Nezara viridula* (L.) was attacked by ceraphronid egg parasitoids and a tachinid fly.

Furthermore, 23 species of predators belonging to eight orders and 22 families were reported, including the chrysopid *Chrysopa* sp., two mirid bugs (*Nesidiocoris* (= *Cyrtopeltis*) *tenuis* (Reuter) and *Engytatus modestus* (Distant)), the reduvid *Zelus* sp., as well as the coccinellids *Hippodamia convergens* (Guérin-Méneville), *Coleomegilla cubensis* (Casey) and *C. sanguinea* and nine spider species (Serra, 1992).

Mirid predators were mainly found on solanaceous crops like tomatoes and tobacco, and included, besides the two already mentioned species, *Macrolophus praeclarus* (Distant) and *Tupiocoris notatus* (Distant) (Marcano, 1964; Schmutterer, 1990; Pérez-Gelabert, 2008). They are important predators of whiteflies, thrips, aphids, mites, leaf miners, lepidopteran eggs and small larvae, but as they are zoophytophagous they may cause plant and fruit damage in the absence of prey. With the massive appearance of *B. tabaci* in tomatoes after the end of the 1980s, high populations of predatory mirids, mainly *N. tenuis*, were observed in this crop (Serra, 1992). A field trial with these predators in tomato was done in Las Cabuyas, Azua, by M. Reyes and A. Abud in 1989 (Azua, 1989, personal communication), who concluded that the high mirid densities were partly responsible for the poor production of the heavily whitefly-infested field, due to their phytophagous activity. But poor yields due to phytophagy were not found in other field trials done in the same region (Serra, 1992). Later, in laboratory tests, Serra *et al.* (1995) compared premature drop of flower buds and flowers caused by *E. modestus* and *N. tenuis* on tomato plants with and without whitefly prey. The highest drop of flower buds occurred on plants with only whiteflies, *E. modestus* caused a higher loss of flowers than *N. tenuis* when no prey was available, and drop of flowers was similar for both mirid species when prey was present.

Tappertzhofen (1996) studied the population dynamics of *B. tabaci* on aubergine *Solanum melongena* L. and found primary parasitoids of the genera *Encarsia* and *Eretmocerus*, as well as the hyperparasitoid *Signiphora aleyrodis* Ashmead (Evans and Serra, 2002).

Studies on the efficiency of different imported (commercial and non-commercial) and native strains of entomopathogenic fungi were undertaken between 1993 and 1997 at ISA in La Herradura, Santiago, in order to evaluate their role in integrated pest management (IPM) programmes in tomatoes, ornamentals and other crops infected by *B. tabaci* and/or *T. vaporariorum*. A collection of fungal strains was built, obtained from the UK, Trinidad and Tobago (TRI), Mexico (*Hirsutella thompsonii* (Fisher) and *H. nodulosa* Petch, from *Aceria guerreronis* Keifer) and, above all, through national bioprospecting. Native and imported strains of *Lecanicillium* (= *Verticillium*) *lecanii* R. Zare & W. Gams (UK, TRI), *Isaria* (= *Paecilomyces*) *fumosorosea* Wize (TRI), *B. bassiana*, *Metarhizium anisopliae* (Metschnikoff) Sorokin were tested successfully against *B. tabaci*, *in vitro* and in a screenhouse under lowland conditions, and against *T. vaporariorum* in protected ornamental crops and a tomato field at higher altitudes.

In laboratory tests, TRI strains of *I. fumosorosea*, *L. lecanii*, *B. bassiana* and *M. anisopliae* were compared with locally collected strains of *L. lecanii* and *I. fumosorosea*, as well as with commercial products such as PreFeRal (*I. fumosorosea*, W.R. Grace & Co.-Conn., Maryland, USA), Mycotal (*L. lecanii*, Koppert BV, Berkel en Rodenrijs, the Netherlands), Vertisol (*L. lecanii*) and Vektor (*Entomophthora virulenta* I.M. Hall & P.H. Dunn, both Laverlam, Cali, Colombia). Among the tested non-commercial strains, *I. fumosorosea* TRI and some native *I. fumosorosea* strains gave the most consistent results concerning the infectiousness of whitefly and spore viability, followed by *L. lecanii* (TRI) and native strains. When whiteflies parasitized by *E. sophia* were sprayed with *I. fumosorosea* TRI, PreFeRal and Mycotal, the number of emerged parasitoids was similar to that emerging from a water-sprayed check, so apparently the fungal treatment did not negatively influence parasitism (Serra *et al.*, 1997).

Naturally occurring fungi (*I. fumosorosea* and *L. lecanii*) showed a limited effect on *T. vaporariorum* in ornamental crops depending on local climate (altitude 400 m asl, cold season), as well as the crop and its pest and disease management programme, in particular the use of fungicides. In the Constanza valley, use of *L. lecanii* resulted in high mortality rates in *T. vaporariorum* in non-sprayed ornamentals and vegetables. In the Jarabacoa valley, the centre of ornamental production at 500 m asl, both whiteflies coexist, but *T. vaporariorum* dominates. Here, naturally occurring *I. fumosorosea* showed an efficiency of < 50% in reducing mainly *T. vaporariorum* in gerbera, aster and other ornamentals. However, in lowlands under rather hot and dry climate conditions (annual rainfall 500–1,000 mm, average temperatures 24–28°C, and mainly with *B. tabaci*) practically no natural infection with fungi was found for both whiteflies. An exception was a single field with sour gourd *Momordica charantia* L. and aubergine *S. melongena* L., in La Vega, with an untypical *T. vaporariorum* infestation during a cold period, where *I. fumosorosea* reached medium to high levels of biocontrol. Under semi-field conditions in the cold season and at 500 m asl, the commercial products based on *I. fumosorosea* and *L. lecanii* increased whitefly mortality in ornamentals. Good perspectives were expected for the use of *L. lecanii* for whitefly control in mountain valleys of the Cordillera Central, but unfortunately the collection at ISA of about 50 fungal strains got lost after the end of the project (Serra, 2006).

Biological control of the coffee berry borer

The coffee berry borer *Hypothenemus hampei* Ferrari, was found on the island in 1995 near Cotuí, resulting in heavy yield losses in coffee plantations at low altitudes (< 1,000 m asl). According to unpublished data of the Programa de Manejo Integrado de la Broca del Cafe (PROMIB) in 1999, 62,500 ha were infested; fruit damage was 9.2% in 1999, 15.4% in 2002 and 6.7% in 2006 (Contreras and Camilo, 2007). Together with the rapid spread of coffee rust *Hemileia vastatrix* Berk. & Broome, extensive areas of mostly old and extensively managed coffee plantations by small farmers have lost their profitability. Initially 20,000 l of endosulfan were sprayed in coffee plantations in 1997, but in 1998 a biocontrol programme was established. A combination of measures consisting of mass production and release of the parasitoid *Cephalonomyia stephanoderis* Betrem, application of the commercial or artisanal production microbial control agent *B. bassiana*, manual removal of infested fruit and chemical control was implemented together with a trapping system for the borer (Brocap-CIRAD) (Serra, 2006). Parasitoids were multiplied first in four and later in eight regional facilities, and annually between 500,000 and 2.6 million wasps were released in 70% of the coffee plantations in a joint project of the UASD, Instituto Dominicano de Investigaciónes Agropecuarias y Forestales (IDIAF), Consejo Dominicano del Café (CODOCAFE) and the Ministry of Agriculture (MA) (Serra, 2006; Contreras and Camilo, 2007). Today, *C. stephanoderis* is still produced at LABOCOBI, UASD. Natural infection of the coffee berry borer with the entomopathogenic fungus *B. bassiana* was 2–7%. Within PROMIB the use of combined sanitary cultural measures (collection of fruits left on the plant or the ground) with the application of *B. bassiana* biocontrol measures, achieved a reduction of 27.3% of infestation compared with the untreated check (Contreras and Camilo, 2007). In addition, 18 isolates of *B. bassiana* obtained from beetles of different localities were screened in the laboratory and resulted in borer mortalities varying from 22% to 37% (Pérez and Caro, 2002).

Biological control of rice stalk stink bug

The rice stalk stink bug *Tibraca limbativentris* (Stål.) might have been present since the 1980s and was confirmed in 1997. A survey of natural enemies was conducted and as a result a native egg parasitoid (*Telenomus* sp.) was mass reared, its best release rate determined (Cordero, 2004) and released by the PNMIP. Due to the success, additional releases were no longer needed (Serra *et al.*, 2003).

Biological control of weeds

R. Charudattan (USA) collected 82 fungal and bacterial isolates from diseased water hyacinths (*Eichhornia crassipes* (Mart.) Solms) in 1974 in the Dominican Republic (Freeman *et al.*, 1976). The pathogenicity of 35 of these isolates was tested. Fungal species of the genera *Phoma* and *Helminthosporium* were found to be virulent. The species *Cochliobolus stenospilus* T. Matsumoto & W. Yamam, until then an unreported pathogen of water hyacinth, appeared to be extremely pathogenic. Barreto *et al.* (2000) mentioned the collection of still another fungus on water hyacinth in the country in 1998: *Myrothecium roridum* Tode ex Fr.

Biological control of bilharzia-transmitting snails

In the Dominican Republic, accidentally introduced *Biomphalaria glabrata* (Say) snails had since 1947 been transmitting *Schistosoma mansoni* Sambon, which causes bilharzia (also known as schistosomiasis) in humans. In 1970, a centre for the eradication of bilharzia was established, which, among other snail control methods, imported, reared and released *Marisa cornuareitis* (L.) snails for biocontrol of *B. glabrata*. Sampling showed that *M. cornuareitis*, together with later-released imported snails *Tarebia* (= *Thiara*) *granifera* (Lamarck) and *Melanoides* (= *Thiara*) *tuberculata* (Müller), were outcompeting *B. glabrata*. The percentage of infected humans showed a marked decrease after 1970 (Schwartz, 2015).

Rearing and augmentative releases of natural enemies for control of various pests

Laboratory rearing of natural enemies was started mainly for scientific purposes and in a few cases for pest control by augmentative releases. Since 1988, LABOCOBI, co-financed by CEDAF and GTZ, has developed mass production of beneficial organisms, among which are egg parasitoids of the genus *Trichogramma*, reared in eggs of the cereal moth *Sitotroga cerealella* (Olivier). *Trichogramma* spp. parasitize eggs of the corn earworm *Helicoverpa zea* (Boddie), tobacco budworm *Heliothis virescens* (F.), armyworms *Spodoptera* spp., cabbage looper *Trichoplusia ni* (Hübner), hawkmoth *Manduca* spp. and cassava hornworm *E. ello*. From 1984, the melon and tomato growing and exporting company Dominican Export (DOMEX) imported and released *Trichogramma* sp. and a chrysopid species, probably from Israel. In 1992–1993, DOMEX provided egg parasitoids to three other Dominican tomato-growing companies (L. Zoquier, Santo Domingo, 2018, personal communication).

Mass releases of *Trichogramma* species have been made in cassava since 1990 (R. Taveras, Santo Domingo, 2018, personal communication) and tomatoes (in 1989–1990 and later) (Serra, 1992). Peña *et al.* (1992) published the mass production method used at LABOCOBI, as well as results of experimental releases in a field trial in cassava in 1991 and 1992. For several years, a mixture of *Trichogramma* species was released, but nowadays only *T. pretiosum* Riley is reared. An annual average of 25 million *T. pretiosum* individuals as parasitized *S. cerealella* eggs have been released to control cassava hornworms in up to 1,070 ha per year in the North-west region of the country (Taveras and Guzmán, 2018). The agrobusiness consortium Transagrícola CxA reared *Trichogramma* in 1997 and 1998, but terminated the rearing due to technical problems and contamination of the colonies.

Until around 2005, the Centro de Tecnologías Agrícolas (CENTA) of IDIAF mass produced *Telenomus* sp. on *Spodoptera frugiperda* (J.E. Smith), reared on sorghum grains, and entomopathogenic nematodes in larvae of the wax moth *G. melonella*.

IPM of arthropod pests with biocontrol measures

Development of IPM started in the Dominican Republic in the 1970s by using pest monitoring in sugarcane crops and application of a *B. thuringiensis* (Bt) product by the State Sugar Council (CEA) (F. Taveras, San Cristóbal, 2000, personal communication).

In 1976, an IPM project was started for the control of the ello sphinx or cassava hornworm *E. ello*, with the participation of the University of California, IICA, the Dominican–German Project (PDA) of the Department of Plant Health (DSV) and GTZ. An alert system was established to recommend the best moment to take control measures with, among others, a selective Bt-based insecticide (Serra, 2006). For biocontrol of *E. ello*, the conservation of egg parasitoids of the genera *Trichogramma* and *Telenomus* was encouraged (Schmutterer, 1990; Taveras and Guzmán, 2018).

In the second half of the 1970s, DSV with PDA carried out taxonomic work related to pest management in the areas of entomology, nematology, weed biology and phytopathology for various crops. In 1979, PDA with the Cotton Institute used a monitoring system in this crop based on empirical thresholds to determine the best moment for control measures in the South-west area of the island. Bt-based insecticides against lepidopteran caterpillars and the preservation of natural enemies were included in this IPM programme.

In 1981, studies were done to establish tentative management thresholds for tobacco pests by the Biological Control Laboratory, which was created with support of the PDA and DSV of CESDA in San Cristóbal (Serra, 2006).

12.3 Current Situation of Biological Control in the Dominican Republic

12.3.1 Natural, classical and augmentative biological control of arthropod pests

Papaya mealybug

The Dominican Republic participated in a US-financed project on classical biocontrol of the papaya mealybug *Paracoccus marginatus* Williams and Granara de Willink, which is supposed to be native to Mexico and/or Central America and has invaded several Caribbean countries where it attacks more than 55 plant species, including a number of crops and ornamentals. The aim of this project was to reduce papaya mealybug populations in the Caribbean and Central America to limit the risk of introduction of the pest into the USA. However, the mealybug has already been found in Florida (Kaufmann *et al.*, 2001). In 2000, individuals of four species of parasitoids, *Anagyrus* (= *Apoanagyrus*) *californicus* (Compere), *Anagyrus loecki* Noyes, *Acerophagus papaya* Noyes and Schauff, and *Pseudleptomastix mexicana* Noyes and Schauff, all originally collected in Mexico, were released in the Dominican Republic. Within a year, the releases resulted in a more than 99% reduction in papaya mealybug densities, concurrent with an increase of parasitism from less than 1% to almost 60% (Kaufmann *et al.*, 2001; Meyerdirk and De Chi, 2003).

Pink hibiscus mealybug

Since its arrival in Grenada in 1995, the pink hibiscus mealybug *Maconellicoccus hirsutus* (Green) has spread over the Caribbean area, Central America and the USA (Florida and California, 2002–2003), causing high economic losses to various crops. The invasions resulted in the establishment of a region-wide programme for classical biocontrol. In 2002, mealybug outbreaks were confirmed in the capital cities of Haiti and the Dominican Republic. A team of researchers of DSV/Min.Agric., Junta Agroempresarial Dominicana (JAD), IDIAF, Universidad Nacional Pedro Henríquez Ureña (UNPHU) and UASD released the parasitoids *Gyranosoidea indica* Shaffee, Alam & Agarwal and *Anagyrus kamali* Moursi obtained from Puerto Rico through USDA/APHIS (Animal and Plant Health Inspection Service) and monitored the spread of the pest, its population dynamics, its range of host plants, percentage parasitism, the presence of associated predators, and mealybug-tending ants. The pest was found on *Hibiscus* spp. and *Calliandra surinamensis* Benth., together with the parasitoid *G. indica*, and to a lesser extent on other plant species such as the probably native *Allotropa* sp.. Heavy outbreaks of the mealybug were reduced in most cases by the coccinellid *Cryptolaemus montrouzieri* Mulsant that has been present in the country since the 1930s (Gómez-Menor, 1941) and the voracity of which interferes with the parasitoids. At low densities of the pink hibiscus mealybug, the predator moves to heavily infested plants, resulting in recovery of the mealybug during the hot months. The coccinellids *Scymnus* sp. and occasionally *C. sanguinea* as well as the syrphid *Baccha clavata* (Fabricius) and larvae of a non-identified lepidopterous species were also preying on the mealybug. Associated ants (e.g. *Solenopsis* spp.) moved mealybugs to new sites and defended them against natural enemies. Interestingly, at sites with and without releases of parasitoids, parasitism increased and low or non-existent levels of mealybugs were the result, with concurrent recovery of affected plants. During sampling, low levels of hyperparasitism of the released parasitoid species by the encyrtid wasps *Procheiloneurus* sp. and *Acerophagus* sp., in addition to the signiphorid *Chartocerus* sp., were observed. Since the implementation of this classical biocontrol project, only very limited population increases of the mealybug on several host plants were registered (Serra *et al.*, 2007a).

Anastrepha fruit flies

The West Indian fruit fly *Anastrepha obliqua* (Macquart) infests several fruit species, in particular those of the family Anacardiaceae, and most importantly mango *Mangifera indica* L., while the Caribbean fruit fly *A. suspensa* (Loew) attacks fruits of guava, tropical almond and others. It was first reported in Hispaniola more than 75 years ago (Stone, 1942). In South America, their populations are attacked by the parasitoids *Utetes anastrephae* (Viereck) and *Doryctobracon areolatus* (Szépligeti), which are largely sympatric and coexist through microhabitat specialization based on different ovipositor lengths and asymmetries in larval competitive abilities during multi-parasitism. In Hispaniola only *U. anastrepha* was found, which is supposed to be native, as preliminary studies on host plants of *Anastrepha* spp. suggested (Serra *et al.*, 2011a). A classical biocontrol programme was started in 2005 with participation of SEA/DSV and USDA/APHIS–IDIAF–UASD, by releasing *D. areolatus*. The introduction was considered safe, as the two parasitoids share an evolutionary history over a substantial portion of their distribution area, and thus it was expected that there would be no negative interactions when the two species would be reunited in the

Dominican Republic, and that overall parasitism would increase. IDIAF checked its establishment in two release zones and found that, immediately following releases, *D. areolatus* averaged 9% of total adult insects recovered and 2 years after release constituted a mean of 13% (Serra *et al.*, 2011a). There was no evidence of competitive exclusion of *U. anastrephae* by *D. areolatus.* After 8 years, Forchue (2014) found that the parasitoid had spread only up to 21.7 km from the closest release sites. Recently, *D. areolatus* was detected in Jarabacoa, almost 50 km from the closest site where it was recorded before (C. Serra, unpublished data). In Piedra Blanca, Monseñor Nouel province, *D. areolatus* was found to parasitize up to 60% of *A. suspensa* in the period from 18 October 2007 to May 2008 in a commercial guava plot at a distance of 50 km from the release site. It is unknown how *D. areolatus* reached this location. Parasitoids alone are unlikely to provide economic levels of control, but they can serve as components of an IPM programme to maintain 'fly-free' or 'low prevalence' fruit export zones.

Diamondback moth

The diamondback moth *Plutella xylostella* (L.) is by far the most important pest in cabbage (*Brassica oleracea* L.) in the Dominican Republic, needing numerous insecticide sprays. Santos *et al.* (2000) provided lists of primary larval–pupal parasitoids found on the island, among which were *Diadegma insulare* (Cresson) on higher altitudes, as well as *Conura petioliventris* Cameron, *C. hirtifemora* Ashmead and *C. pseudofulvovariegata* (Becker) in lowlands. Also several hyperparasitods were found in cabbage, belonging to the genera *Brachymeria, Ceraphron, Aprostocetus, Trichomalopsis, Conura* and *Pachyneuron*, according to R.D. Cave (USA). The status of other species found during the study (*Ooencyrtus submetallicus* (Howard) and species of the genera *Baryscapus, Zaglyptus* and *Temelucha*) is still unclear (Santos *et al.*, 2000). López (2016) studied pest and natural enemy populations at the Constanza Experimental Station of IDIAF, in the province of La Vega, which is the largest cabbage production area. Parasitism by *D. insulare* varied between 14% and 44% and total parasitism ranged between 21% and 59%. The high percentages of parasitism suggest that natural biocontrol of the pest can play an important role when the right pest control methods are selected in an IPM programme.

Pigeon pea pod fly

The pigeon pea *Cajanus cajan* (L.) Millspaugh, native to India, is grown in the Dominican Republic on more than 25,000 ha and eaten locally as well as processed as canned or frozen beans for export (Taveras and Hansson, 2015). The Asian pigeon pea pod fly *Melanagromyza obtusa* (Malloch) was first reported in the Dominican Republic in 2002 (Serra *et al.*, 2003), 2 years after being found in Puerto Rico. It feeds on and develops in seed pods, where it is well protected from natural enemies and insecticides. Sampling done by IDIAF between 2002 and 2003 in more than 20 localities in different production zones showed crop infestation levels of 10–80% (average 50%) in 95% of the localities sampled and no parasitoids were found. However, in 2004 a few parasitoids from the genera *Habrobracon* and *Elasmus* were recovered on pigeon pea and on alternative hosts, the associated weed species *Rhynchosia minima* (L.) DC. and *R. reticulata* (Sw.) DC. (Fabaceae) (Segura *et al.*, 2005). Due to low natural parasitism, the introduction and release of parasitoids originating from Southern Asia and Australia were recommended (D. Meyerdirk, Santo Domingo, 2003, personal communication). A project of the Ministry of Agriculture of Dominican Republic, IDIAF, the University of Puerto Rico, the Ministry of Agriculture of Puerto Rico and USDA/APHIS was initiated in 2005, and *Ormyrus orientalis* Walker and two other species, supposedly the eulophid *Callitula* sp. and eventually a scelionid, were shipped to the Dominican Republic, but they did not survive transport and no further shipments were carried out.

Several years later, a new species of parasitoid as well as *O. orientalis* were found attacking pigeon pod fly. Their origin is unclear but probably they were introduced fortuitously. In 2008, Serra (unpublished data) collected parasitoids obtained from pod flies originating from Alto Yaque, Santiago province, which was described in 2015 as a new species, *Pediobius cajanus* Taveras & Hansson. This parasitoid may have switched to the pigeon pea pod fly from another native host. There are at least six to ten *Melanagromyza* spp. present on Hispaniola island, and other plants of the family Fabaceae may provide hosts for the development of this parasitoid (Pérez-Gelabert, 2008). This parasitoid caused 25.8% parasitism of the pod fly (Taveras and Hansson, 2015). The other parasitoid found, *O. orientalis*, showed parasitism of 2%; it is supposed to have entered the country with pod-fly infested peas from elsewhere in the Caribbean (Guzmán *et al.*, 2018).

Red palm mite

The red palm mite *Raoiella indica* Hirst, native to Asia, was detected in the Dominican Republic in 2006. Its distribution, host plants (e.g. coconut, ornamental palms and bananas) and natural enemies were studied by Serra (2007). The most important natural enemy found was the phytoseiid *Amblyseius largoensis* (Muma), besides other non-identified mites and tiny coccinellids of the genus *Stethorus*. Pérez-Gelabert (2008) reported the species *Stethorus caribus* Gordon & Chapin for the Dominican Republic. Serra (2007) concluded that under dry climate conditions, which are excellent for development of *R. indica*, biocontrol by native natural enemies might be inefficient and that the importation of exotic Asian natural enemies should be considered. After population explosions of *R. indica* during dry seasons, populations suddenly dropped to low numbers, supposedly due to mechanical effects of the rain, but above all to the presence of fungal pathogens, including *Hirsutella* sp., among others.

Various other pest mites

Many injurious mite species are present in the Dominican Republic, among which are several tetranychid species (*Tetranychus urticae* Koch, *T. evansi* Banks & Pritchard and *T. cinnabarinus* (Boisduval)) on various crops, *T. gloveri* Banks and *Mononychellus caribbeanae* (McGregor) on cassava, and *Eutetranychus banksi* (McGregor) and *Panonychus citri* (McGregor) on citrus. Further, the following accidentally introduced mites are present: the tarsonemids *Polyphagotarsonemus latus* (Banks) on pepper, aubergine and numerous other plants, and *Steneotarsonemus spinki* Smiley on rice; and the eriophyids *Aculops lycopersici* Massee on tomatoes, and *Aceria sheldoni* Ewing and *Phyllocoptruta oleivoa* (Ashm.) on citrus (Schmutterer, 1990; Pérez-Gelabert, 2008). Pellerano (2014) made an inventory of mites and their natural enemies on numerous wild and cultivated crops. He found species of Araneidae, Coccinellidae, Syrphidae, *Scolothrips* sp. and the pathogen *Hirsutella* sp. on the phytophagous mite *Steneotarsonemus konoi* Smiley & Enmanouel and on another unidentified tarsonemid mite. The broad mite *P. latus* was observed to be attacked by the coccinellids *Scymnus* sp. and *H. convergens* in hot peppers (Medrano and Serra, 2018).

Pests of oriental vegetables

Production of oriental vegetables is of great socio-economic importance for the Dominican Republic, because of large exports to North America and Europe. Due to high rejection rates of exported vegetables caused by pesticide residues, IDIAF started to develop IPM programmes for a project funded by the National Council of Agriculture and Forestry Research (CONIAF) and with support from UASD and producers of the Asociación Dominicana de Exportadores de Vegetales Orientales (ADEXVO). One aspect of this project concerned natural control of the main pests (aphids, thrips, whiteflies, leafhoppers, leaf miners, beetles and moths) of yard-long beans *Vigna sesquipedalis* (L.), Chinese aubergines *S. melongena* L. and bitter gourd *M. charantia* (Baltensperger and Serra, 2007; Serra *et al.*, 2007b). The following natural enemies were found: the heteropterans *Orius insidiosus* (Say), *N. tenuis* and other mirids; the coccinellids *H. convergens*, *C. sanguinea*, *C. cubensis* and a *Scymnus* sp.; the syrphid *Pseudodorus clavatus* F.; a dolichopodid fly *Condylostylus* sp.; a chrysopid *Chrysoperla* sp.; a predatory thrips *Franklinothrips vespiformis* (Crawford); vespids *Polistes crinitus* (Felton) and others; the ants *Solenopsis geminata* (Fab.) and *S. invicta* Buren; *Tapinoma melanocephalum* Fab.; and not yet identified parasitoids (Medrano and Serra, 2018). Integration of natural control by native biocontrol agents in the production of oriental vegetables is currently limited, due to high quality requirements but also because of the lack of awareness of the role that biocontrol may play and the limited availability of mass-reared biocontrol agents.

Pests in organic production of fruit and coffee

According to FAO (2018, p. 1):

> Contemporary ideas of organic production were introduced into the Dominican Republic in the early 1980s, but it was not until the mid-1990s that production expanded. Organic production is now an important component of the agricultural sector. In 1999 organic produce was estimated to contribute about 20 percent of total agricultural exports with a value of US$9.6 million. In 2000, the value of organic exports had doubled to US$20.9 million. The Dominican Republic is currently one of the leading exporters of tropical organic products globally. Much of the information on organic production is either documented in grey literature or is still undocumented. Organic production in the country is clearly dominated by banana crops which account for about 80 percent of all organic exports.

FAO (2018) reported that, in addition to banana, other organic fruits, cocoa and coffee are exported. As organic farmers cannot use conventional synthetic pesticides, pest control is based on a holistic approach consisting of cultural measures, including natural and conservation biocontrol.

Use of exotic natural enemies

Vegetable production in protected structures like screenhouses has become an important subsector of Dominican agriculture at higher altitudes and amounted to 900 ha in 2015. Production is mainly for the export market and consists of bell peppers, followed by tomatoes, cucurbits, other peppers and herbs. The main pest problems in these cultures are viruses transmitted by thrips, whiteflies and aphids, fruit-boring worms, mites, nematodes, soil-borne bacterial and foliar fungal diseases. Control is mainly by synthetic chemical pesticides, but inappropriate application may cause pesticide residues, pest resurgence and, as a consequence, economic losses, in particular when prohibited residues are found on harvested products that are rejected by the receiving countries. Inundative biocontrol of insect vectors seems to be a suitable alternative. Some farmers obtained natural enemies from abroad through a local supplier, such as the anthororid *Orius laevigatus* (Fieber) and the phytoseiid *Typhlodromips* (= *Amblyseius*) *swirskii* Athias-Henriot for control of whiteflies, thrips and spider mites. The programme 'Exportando Calidad e Inocuidad' (ECI), with funding of USAID and local producers of peppers, tomatoes and other vegetables under screenhouse cropping conditions in mountain valleys like in Rancho Arriba, Ocoa province, showed interest in the importation of the above-mentioned natural enemies. There is also evidence of secretly imported predatory mite *Phytoseiulus persimilis* Athias-Henriot for spider-mite control (Pellerano, 2014). However, several Dominican Republic specialists, including C. Serra, do not approve the importation of exotic species without a risk analysis, and prefer that, first, studies are undertaken to get a better knowledge of the native species of natural enemies. Thus it is proposed to perform local prospecting for natural enemies before importing exotic species.

Natural control of recently introduced exotic pests

Several exotic insect species with invasive potential have entered the island during the past 10 years. For example, on *Ficus* spp., mainly ornamental *F. benjamina* (L.), the cuban laurel thrips *Gynaikothrips ficorum* (Marshal) and the galling thrips *G. uzeli* (Zimmermann) have been defoliating plants for several years. Their populations have declined probably due to competition with other ficus pests, as well as some predation by the occasionally occurring predatory thrips *F. vespiformis* and coccinellids.

Another recent invader is the ficus whitefly *Singhiella simplex* (Singh), which causes defoliation mainly in *F. benjamina*. Populations of the two thrips species have declined recently, probably due to competition with the ficus whitefly and to predation by the occasionally occurring predatory thrips *F. vespiformis* and coccinellids. The aphelinid parasitoid *Encarsia protransvena* Viggiani and several tiny coccinellids have been observed to attack the ficus whitefly. Parasitism by *E. protransvena* has steadily increased, reaching 40% parasitism of *S. simplex* in shrubs of *F. retusa* in Santo Domingo, and it is thought that the parasitoid will result in sufficient control of this whitefly on other plants like *F. benjamina* in the future (Serra *et al.*, 2011b).

The invasive eulophid erythrina gall wasp *Quadrastichus erythrinae* Kim has infested *Erythrina* spp., mainly the ornamental *E. variegata* L., in several regions of the country during the past 5 years, causing defoliation and death to numerous trees. No natural enemies have yet been detected, but a reduction of damage in the affected species has been observed recently.

The invasive margaroid scale *Crypticerya genistae* Hempel entered the country about 5 years ago from Haití. It infests plants from the Fabaceae family, but also numerous species from other families. Occasionally, coccinellids such as *R. cardinalis* have been found on infested pigeonpea plants. Predators of this scale reported from other islands have not yet been found in the Dominican Republic (Pérez-Gelabert, 2008).

Finally, the invasive cycad aulacapsis scale *Aulacapsis yasumatsui* Takagi has spread widely over the Santo Domingo province since 2015, and heavily infests and kills plants of *Cycas revoluta* Thumb. Recently, a minute beetle, *Cybocephalus nipponicus* Endrödy-Younga, and larvae of the cosmopterigid pink scavenger caterpillar moth *Anatrachyntis* (= *Pyroderces*) *rileyi* (Wals.) were found to prey on the scale. Also a number of ceraphronid parasitoids *Aphanogmus* sp. were recovered from *A. yasumatsui*, which is known as a parasitoid of *C. nipponicus* (Evans *et al.*, 2005). The beetle is of Asian origin and attacks at least 14 species of armoured scales worldwide. It has been introduced into North America since 1989 and was released in Florida in 1999 (Smith and Cave, 2006). It has been reported from the Cayman Islands, St Kitts/Nevis and Barbados, attacking *A. yatsumatsui* and the coconut scale *Aspidiotus destructor* Signoret (Smith and Cave, 2007).

Classical biocontrol might be successful for control of several new invasive pests, but biocontrol is impeded by the lack of economic importance of some species, funding, and a legal framework, as well as the current opposition of the Ministry of Environment to introduction of exotic beneficial species, even after a risk analysis.

Use of native Anthocoridae

Orius insidiosus has been studied for control of several pests. Different methods and media to rear this anthocorid have been tested (Pinales and Serra, 2018). The predator was tested for control of whitefly on nine pepper varieties (Pinales, 2015). Recently mass production of *O. insidiosus* and the exotic *O. laevigatus* has started at the Bioterio de Control Biológico (former LABOCOBI) of the UASD on eggs of the grain moth *S. cerealella*, and up to 70,000 individuals are produced weekly. Many predators were released in 2018, benefiting greenhouse producers of Rancho Arriba, Jarabacoa and El Seybo, as well as farmers using open-field production of oriental vegetables at La Vega, Santo Domingo, and Monte Plata.

Inventory of native predatory mites

Only two phytoseiid predatory mites had been reported from the island by Pérez-Gelabert (2008): *Proprioseiopsis sandersi* (Chant) and *Amblyseius obtusus* (Koch), earlier named *Amblyseius perlongisetus* Berlese and *Amblyseiopsis musae* Garman, respectively. During a recent survey conducted in natural vegetation areas, 23 species belonging to 11 genera were collected, including 21 native species already described and reported from other countries, and two new to science: *Phytoseius dominicensis* Ferragut & Moraes sp. nov. and *Typhloseiopsis adventitius* Ferragut & Moraes sp. nov. (Ferragut *et al.*, 2011).

Also, a survey of plant-inhabiting mesostigmatid mites was conducted by Abo-Shnaf *et al.* (2016). Phytoseiidae was by far the most numerous family represented, with 12 species. *Amblyseius tamatavensis* Blommers was the most common species, reported from four provinces on five plant species. This species is now considered as a serious candidate for biocontrol of *B. tabaci*, biotype 'B', in Brazil (Cavalcante *et al.*, 2017). In addition, two species of Ascidae, four of Blattisociidae (all four were new species: *Lasioseius dominicensis* n. sp., *L. oryzae* n. sp., *L. prorsoperitrematus* n. sp. and *L. sanchezensis* n. sp. (Abo-Shnaf *et al.*, 2016)); and three of Melicharidae were found. Moraes *et al.* (2015) had earlier collected *L. chaudhrii* (Wu & Wang) on rice *Oryza sativa* L. at El Pozo, María Trinidad Sánchez province, and this species is now considered as a potential biocontrol agent of tarsonomid pest mites on rice in Asia.

Natural enemies and the effect of pesticides

During the past three decades, several studies have shown the negative effects of insecticides on natural enemies (e.g. Serra, 1992). Results of these studies can be used in IPM programmes to select those pesticides with the least negative effects on beneficial organisms. Examples are the IPM programmes developed for oriental vegetables (Serra *et al.*, 2007b).

12.3.2 Augmentative microbial control of arthropod pests

Sweet potato weevil

The sweet potato weevil, *Cylas formicarius elegantulus* (Summers), is the key pest of sweet potato, and J.M. Romain (Santo Domingo, 2000, personal communication) tested the effect of foliar sprays with *B. bassiana* in combination with pheromone treatments. The idea was that male weevils would be attracted by the pheromone and become infected by the fungus. Then, after removing the pheromone capsules, the male weevils would move to the galleries in the soil and infested tubers, where they would contaminate male, female and inmature weevils. Unfortunately, the quantity and quality of harvested tubers did not reflect a significant decrease in weevil numbers. In another experiment, Taveras and Caro (2018) tested a liquid and solid treatment with *B. bassiana*, and they found that both treatments significantly reduced the amount of damaged tubers and increased the amount of infected insects.

Banana weevils and orchid thrips in banana

Several collections were made and initial tests were done at LABOCOBI with entomopathogenic fungi (*B. bassiana* and *M. anisopliae*) and nematodes (*Heterorhabditis* sp.) in order to control the banana weevil *Cosmopolites sordidus* (Germar) (Del Orbe *et al.*, 1998; Matos *et al.*, 1998). On-farm trials were done to control the banana weevils *C. sordidus* and *Metamasius hemipterus* (L.) in an organic banana field by spraying plants with *B. bassiana* and as a solid paste (conidia in rice) between and surrounding two discs of pseudostems as a trap. While the untreated control showed 16% of mortality, the spraying and solid treatment resulted in 31% and 46% mortality, respectively (Taveras and Cuevas, 2006). In organic banana plantations in the North-west, *F. vespiformis* and predatory mites of the genera *Typhlodromina*, *Amblyseius*, *Lasioseius* and *Neoseiulus* were detected feeding on nymphs and adults of the orchid thrips *Chaetanaphothrips orchidii* (Moulton), which cause aesthetic damage to bananas (L. Sánchez and M. Arias, Santo Domingo, 2017, personal communication).

Asian citrus psyllid

The Asian citrus psyllid *Diaphorina citri* Kuwayama is the vector of bacteria (*Candidatus* Liberibacter spp.) causing huanglongbing (HLB) or citrus greening, currently one of the most serious diseases affecting citrus and found in the country in 2001 (Serra *et al.*, 2003). In 2006, the disease was found at Luperón, in Puerto Plata province (Matos *et al.*, 2009) and quickly spread later to commercial citrus orchards and backyard citrus trees. After a very heavy impact of the disease from 2008, the citrus area strongly decreased and in three years 100,000 trees were eliminated on an area of 71,266 ha. The psyllid was parasitized by *Tamarixia radiata* (Waterston), but seldom more than 10% (Serra *et al.*, 2011b). Prospecting for entomopathogenic fungi was done by collecting infected insects in citrus orchards in Monte Plata, Palenque (San Cristóbal) and San José de las Matas (Santiago). The 37 isolates obtained resulted after purification in five strains of four species: *M. anisopliae*, *I.* (= *Paecilomyces*) *fumosorosea*, *B. bassiana* and *Fusarium* sp. (strains Ma-8, Pf-10, Bb-1, Bb-6 and F-6). In the IDIAF-CENTA laboratory, several *in vitro* bioassays were conducted, resulting in mortalities of 50–100% of *D. citri* (Serra *et al.*, 2016).

12.4 New Developments of Biological Control in the Dominican Republic

Until recently, the resources invested in research have been insufficient to evaluate the potential role of documented biocontrol agents in the country (Pérez-Gelabert, 2008), or to introduce exotic species to develop effective biocontrol. However, the interest in investing in studies on biocontrol with national funds from the Ministry of Higher Education, Science

Table 12.1. Overview of biological control in the Dominican Republic.

Biocontrol agent	Pest and crop	Type of biocontrol[a]	Area (ha) under biocontrol	References
Beauveria bassiana, *Metarhizium anisoplae*, *Heterorhabditis* sp.	*Cosmopolites sordidus*, banana	ABC	Testing phase	Del Orbe *et al*., 1998, Matos *et al*., 1998, Taveras and Cuevas, 2006
Trichogramma spp., *T. pretiosum*	*Erinnyis ello*, cassava	ABC	1,070	R. Taveras, pers. com., Santo Domingo, 2018
Encarsia perplexa	*Aleurocanthus woglumii*, citrus	CBC	17,588[b]	Data Min.Agr.; Evans and Serra, 2002; Serra, 2006
Rodolia cardinalis	*Icerya purchasi*, citrus	CBC	17,588[b]	Gómez-Menor, 1941
Cycloneda sanguinea (L.), Chrysopid *species*, entomopathogenic fungi	*Toxoptera citricida*, citrus	FBC	17,588[b]	O.G. Jiménez, pers. comm., Santo Domingo, 2006
Ceratogramma etiennei, Quadrastichus haitiensis, *Fidiobia* sp.	*Diaprepes abbreviatus*, citrus	CBC	17,588[b]	Etienne *et al*., 1992, Méndez and Labarre, 1996, Serra, 2006
Beauveria bassiana, *Metarhizium anisopliae*, *Heterorhabditis* sp.	*Diaprepes abbreviatus*, citrus	ABC	> 1,400 ha	Serra, 2006
Horismenus sp., *Elasmus* sp., *Cirrospilus quadristiatus*	*Phyllocnistis citrella*, citrus	CBC	> 1,400 ha	Taveras, 2000, Serra, 2006
Tamarixia radiata	*Diaphorina citri*, citrus	FBC	17,588[b]	Data Min.Agr.; Serra *et al*., 2011b
Chilocorus cacti	*Aspidiotus destructor*, coconut	CBC	50,116[b]	Schmutterer, 1990
Amblyseius largoensis	*Raoiella indica*, coconut, banana, palm	NC	50,116[b]	Serra, 2007
Cephalonomia stephanoderis	*Hypothenemus hampei*, coffee	CBC	62,500	Contreras and Camilo, 2007
Beauveria bassiana	*Hypothenemus hampei*, coffee	ABC	62,500	Contreras and Camilo, 2007; Pérez and Caro, 2002
Orius insidiosus	Thrips, whiteflies, mites, lepidopterans, egg plant, pepper, tomato	ABC	250	Data CEIRD; R. Taveras, C., Martínez, pers. com., Santo Domingo, 2018
Anagyrus californicus, *A. loecki*, *Acerophagus papaya*, *Pseudleptomastix mexicana*	*Paracoccus marginatus*, ornamentals, vars crops	CBC	3,087[b]	Kaufmann *et al*., 2001

Continued

Table 12.1. Continued.

Biocontrol agent	Pest and crop	Type of biocontrol[a]	Area (ha) under biocontrol	References
Trichogramma spp.	*Diaphania hyalinata*, melon	ABC	Part of 1,500[b]	L. Zoquier, pers. com., Santo Domingo, 2018
Cryptolaemus mountrouzieri, Anagyrus kamali, Gyranosoidea indica, Allotropa sp.	*Maconellicoccus hirsutus*, ornamentals, vars crops	NC, FBC	Areas with malvaceous hosts	Serra *et al*., 2007a
Telenomus sp.	*Tibraca limbativentris*, rice	ABC	174,389[b]	Serra *et al*., 2003
Lixophaga diatraeae	*Diatraea saccharalis*, sugarcane	ABC	114,742[b]	F. Díaz, pers. com., Santo Domingo, 2018
Trichogramma spp.	*Spodoptera* spp., tomato	ABC	Small part of 4,543[b]	Data CEIRD; L. Zoquier and R. Taveras, pers. com., Santo Domingo, 2018
Doryctobracon areolatus	*Anastrepha* spp. and other fruit flies, fruit trees	CBC	> 25,000	Serra *et al*., 2011a
Utetes anastrephae	*Anastrepha* spp.	FBC	?	

[a]Type of biocontrol: ABC = augmentative, CBC = classical, FBC=fortuitous biological control; NC = natural control
[b]Area of crop harvested in 2017 according to FAO (http://www.fao.org/faostat/en/#data/qc)

and Technology (MESCyT) and CONIAF has been positive and has resulted in funding programmes by the Ministry of Agriculture and farmer associations and cooperatives. For several crops and pests, such as tomatoes, oriental vegetables and citrus, inventories of natural enemies are now available. Also, a number of important natural enemy species are still reared in the laboratory, though there is no commercial production. The current lack of resources and trained professionals, among which are taxonomists, prevents significant advances in the field of biocontrol. For biocontrol of many key pests, only laboratory experiments have been done and critical evaluation of practical application in the field is lacking. Still, in the past few decades aspects of augmentative biocontrol for pests in cassava, tomato, banana and coffee have been developed, as well as in the field of classical biocontrol against exotic invasive insect pests, like biocontrol of the pink hibiscus mealybug. Table 12.1 gives an overview of the current situation.

Based on information obtained from different sources mentioned in Table 12.1, the estimated area under classical biocontrol is at least 170,000 ha, while it is at least 356,500 for augmentative biocontrol.

Pest control in the Dominican Republic is heavily dominated by the chemical pesticide industry and farmers are used to periodically applying pesticide cocktails, usually not based on monitoring of pest populations and not incorporating selective and/or microbiological pesticides. However, the increasing request by foreign markets and local consumers for products that are free of pesticide residues creates opportunities for farmers who seek to produce within certification systems for organic food and for 'Good Agricultural Practices' (GAP) labels, which are partly based on pest management by biocontrol.

12.5 Acknowledgements

The following are thanked for their contributions to biocontrol in the Dominican Republic, and for taxonomic support: P. Alvarez (Min. Agr.), D. Torres & M. Reyes (Univ. Autónoma de Santo Domingo), R. Taveras, M. Castillo, L. López & R.E. Guzmán (LABOCOBI-UASD), M. Ortíz, J. Nuñez, A. Schulz, (former ISA-WREP team, Santiago), H. Schmutterer (Univ. of Giessen, Germany), D. Gerling (Tel Aviv Univ., Israel), A. Abud-Antún (UASD, Junta Agroempresarial Dominicana), M. Pellerano (consultant), L. Sánchez, Y. Segura (IDIAF), F. Taveras, N. Ferreira, J. Cicero & E. de Jesus Marcano (former 'Neem project' IPL-GTZ or IPL, San Cristóbal), F. Díaz (Fertilizantes Santo Domingo), D.E. Meyerdirk, M. Ciomperlik, A. Roda, J. Sivinski, (USDA/APHIS), G.A. Evans, R.D. Cave, Ph. Stansly (Univ. of Florida, USA), R. Ochoa (USDA, Beltsville, Maryland, USA), D. Pérez-Gelabert (Smithsonian Institute, Washington, D.C., USA), F. Méndez (Consorcio de Cítricos Dominicanos), C. Gómez Moya (IIBI, Univ. Tecnológica del Cibao Oriental, UTECO).

References

(References with grey shading are available as supplementary electronic material)

Abo-Shnaf, R.I.A., Sánchez, L. and De Moraes, G.J. (2016) Plant inhabiting Gamasina mites (Acari: Mesostigmata) from the Dominican Republic, with descriptions of four new species of *Lasioseius* (Blattisociidae) and complementary descriptions of other species. *Systematic and Applied Acarology* 21(5), 607–646. DOI: 10.11158/saa.21.5.5.

Baltensperger, S. and Serra, C.A. (2007) Fluctuaciones poblacionales de plagas artrópodas en el cultivo de berenjena china (*Solanum melongena* L.) bajo la influencia de tres diferentes tipos de manejo de plagas y el clima en La Vega, D.R. [Fluctuations of populations of Chinese aubergine pests under the influence of three types of pest management and the climate]. In*: Resultados de Investigación I. Instituto Dominicano de Investigaciones Agropecuarias y Forestales. 1 Edición*. Santo Domingo, Dominican Republic, p. 20.

Barreto, R., Charudattan, R., Pomella, A. and Hanada, A.R. (2000) Biological control of Neotropical aquatic weeds with fungi. *Crop Protection* 19, 697–703.

Bethancourt, H., Tallaj, J.A., Tapia B.L., Peña, P.P., Sánchez, A.R. *et al.* (2017) Food and nutrition security in the Dominican Republic. A vision for the next 50 years. In: *Challenges and Opportunities for Food and Nutrition Security in the Americas. The View of the Academies of Sciences*. IANAS, IAP and BMBF, México DF, pp. 286–315. [Free public access of this publication in English and Spanish at www.ianas.org]

Cavalcante, A.C.C., Mandro, M.E.A., Paes, E.R. and Moraes, G.J. de (2017) *Amblyseius tamatavensis* Blommers (Acari: Phytoseiidae) a candidate for biological control of *Bemisia tabaci* (Gennadius) biotype B (Hemiptera: Aleyrodidae) in Brazil. *International Journal of Acarology* 43,1,10–15. DOI: 10.1080/01647954.2016.1225816

CIA (2017) The World Factbook: Dominican Republic. Available at: https://www.cia.gov/library/publications/the-world-factbook/geos/dr.html (accessed 12 February 2019)..

Ciferri, C. and Fragoso, R.G. (1927) Hongos parásitos y saprofitos de la República Dominicana [Parasitic fungi and saprophytes from the Dominican Republic]. *Boletín de la Sociedad Española de Historia Natural* 27, 68–81.

Cock, M.J.W. (ed.) (1985) *A Review of Biological Control of Pests in the Commonwealth Caribbean and Bermuda up to 1982*. Technical Communication No. 9, Commonwealth Institute of Biological Control. Commonwealth Agricultural Bureaux, Farnham Royal, UK.

Contreras, T. and Camilo, J.E. (2007) Situacion de la broca del café (*Hypothenemus hampei*) en República Dominicana [Situation of the coffee berry borer in the Dominican Republic]. In: Barrera, J.F., García, A., Domínguez, V. and Luna León, C. (eds) *La Broca del Café en América Tropical: Hallazgos y Enfoques. Simposio Internacional Broca del Café*. Sociedad Mexicana de Entomología: El Colegio de la Frontera Sur, Tapachula, Mexico, pp. 43–55.

Cordero, R.D. (2004) Efecto del Parasitoide (*Telenomus* sp.) Sobre el control biológico de poblaciones de *Tibraca limbativentris* Stål en el cultivo de arroz (*Oryza sativa* L.) en Cotuí, República Dominicana [Effect of the parasitoid (*Telenomus* sp.) on the biological control of populations of *Tibraca limbativentris* in the cultivation of rice in Cotuí, Dominican Republic]. BSc thesis. Instituto Superior de Agricultura. Santiago de Los Caballeros, Dominican Republic.

Del Orbe, J., Díaz, C.A. and Garrido Jansen, L. (1998) Control Biológico del Picudo Negro (*Cosmopolites sordidus)* del Plátano (*Musa* AAB) Mediante la Aplicación de Nematodos Entomopatógenos [Biological Control of Banana Weevil (*Cosmopolites sordidus*) in Plantains Using Pathogenic Nematodes]. Resúmenes 3[era] Jornada Científica SODIAF, Santo Domingo, Dominican Republic.

Etienne, J., Abud-Antún, A., Reyes, M. and Díaz, F. (1992) *Toxoptera citricidus* (Homoptera: Aphididae) en la República Dominicana: Una amenaza para las plantaciones de cítricos [*Toxoptera citricidus* in the Dominican Republic: a threat to citrus plantations]. *Proceedings 28th Annual Meeting of the Caribbean Food Crops Society* 28, 22–30. Available at: https://ideas.repec.org/p/ags/cfcs92/258938.html (accessed 13 April 2019).

Evans, G.A. and Serra, C.A. (2002) Parasitoids associated with aleyrodids (Homoptera: Aleyrodidae) in Hispaniola and descriptions of two new species of *Encarsia* Förster (Hymenoptera: Aphelinidae). *Journal of Hymenoptera Research* 11(2), 197–212.

Evans, G.A., Dessart, P. and Glenn, H. (2005) Two new species of *Aphanogmus* (Hymenoptera: Ceraphronidae) of economic importance reared from *Cybocephalus nipponicus* (Coleoptera: Cybocephalidae). *Zootaxa* 1018, 47–54.

FAO (2018) Report on Organic Production in the Dominican Republic. Available at: http://www.fao.org/docrep/004/Y1669E/y1669e0k.htm (accessed 26 August 2018).

Ferragut, F., Moraes, G.J de. and Navia, D. (2011) Phytoseiid mites (Acari: Phytoseiidae) of the Dominican Republic, with a re-definition of the genera *Typhloseiopsis* De Leon. *Zootaxa* 2997, 37–53.

Forchue, A. (2014) Actualización de la distribución a nivel regional de parasitoides exóticos *Doryctobracon areolatus* (Hymenoptera: Braconidae) liberados en 2006 [Actualization of the regional distribution of exotic parasitoids liberated in 2006]. BSc thesis, Pontificia Universidad Católica Madre y Maestra, Santo Domingo.

Freeman, T.E., Charudattan, R., Conway, K.E. and Zettler, F.W. (1976) *Biological Control of Aquatic Weeds with Plant Pathogens*. Contract Report A-76-2 of the US Army Corps Engineers Waterways Experiment Station. Vicksburg, Mississippi.

Gómez-Menor, J. (1941) Cóccidos de la República Dominicana (Hem. Cócc.) [Coccids of the Dominican Republic]. *EOS* 16, 125–143.

Guzmán, R., López, L.D. and Taveras, R. (2018) Mosca asiática, *Melanagromyza obtusa*, (Malloch) (Diptera: Agromyzidae), sus enemigos naturales y otras plagas asociadas al cultivo del guandul en R.D. [Pigeon pod fly, its natural enemies and other pests associated with the pigeon pea crop in the

Dominican Republic]. In: Cepeda Ureña., J. (ed.) *Consejo Nacional de Investigaciones Agropecuarias y Forestales (CONIAF). Socialización de Resultados de Investigación en Manejo Integrado de Plagas*. Santo Domingo, Dominican Republic, pp. 53–66.

Kauffman, W.C., Meyerdirk, D.E., Miller, D., Schauff, M., González Hernández, H. and Villaneuva Jiménez, J.A. (2001) Papaya mealybug biological control in Puerto Rico and Dominican Republic. Abstract of the ESA meeting 9–12 December 2001, San Diego, California. Available at: https://esa.confex.com/esa/2001/techprogram/paper_3261.htm (accessed 12 February 2019).

López, L.D. (2016) Estudio poblacional de *Plutella xylostella* (L.) (Lepidoptera:Plutellidae) y *Diadegma insulare* Cresson (Hymenoptera: Ichneumonidae) bajo la influencia de tratamientos comerciales y selectivos en el cultivo de repollo (*Brassica oleracea* L.). [Population studies of the diamondback moth and *Diadegma insulare* under the influence of commercial and selective treatments in the cabbage crop]. MSc thesis, Universidad Nacional Pedro Henríquez Ureña, Santo Domingo, Dominican Republic.

Marcano, E. de Js. (1964) Apuntes para el estudio de los insectos dañinos a nuestra agricultura [Notes on the study of insects harmful to our agriculture]. Notas mimeográficas, Santo Domingo, Dominican Republic.

Matos, L.A., Julio C. Borbón, J.C. and Cuevas Moronta, A. (1998) Patogenicidad de *Beauveria bassiana* (Bals.) Vuill. y *Metarhizium anisopliae* (Metsch) Sorokin en el control de *Cosmopolites sordidus* Germar en plátano (*Musa* AAB) [Pathogenic potential of *Beauveria bassiana* and *Metarhizium anisopliae* controlling *Cosmopolites sordidus* in plantain]. *Resúmenes 3era Jornada Científica SODIAF*, Santo Domingo, Dominican Republic, p. 22.

Matos, L., Hilf, M.E. and Camejo, J. (2009) First report of 'Candidatus Liberibacter asiaticus' associated with citrus huanglongbing in the Dominican Republic. *Plant Disease* 93(6), 668.

Medrano C.S. and Serra, C.A. (2018) Inventario de artrópodos asociados a cultivos de vegetales orientales en la provincia de La Vega, Dominican Republic [Inventory of arthropods associated with oriental vegetables in the province of La Vega, Dominican Republic]. In: Cepeda Ureña., J. (ed.) *Consejo Nacional de Investigaciones Agropecuarias y Forestales (CONIAF). Socialización de Resultados de Investigación en Manejo Integrado de Plagas*. Santo Domingo. Dominican Republic. pp. 34–52 [see text in pdf Guzman *et al.*, 2018]

Méndez, F. and Labarre, E. (1996) Ensayo de introducción de un himenóptero oofago en el marco del control biológico de *Diaprepes abbreviatus* (Col., Curculionidae) en República Dominicana [Intent of introduction of an oophagous hymenopteran for biological control of *Diaprepes abbreviatus* in the Dominican Republic]. INRA Guadeloupe/Consorcio Cítricos Dominicanos/UASD, Villa Altagracia, 1–5.

Meyerdirk, D.E. and De Chi, L.W. (2003) Models for minimizing risks of dangerous pests: the pink hibiscus mealybug and papaya mealybug. Symposium: Challenges and Opportunities in Protecting the Caribbean, Latin America, and the United States from Invasive Species. *Proceedings 39th Annual Meeting of the Caribbean Food Crops Society* 39, 11.

MIMARENA (2012). *Especies Exóticas Invasoras: una amenaza a la biodiversidad. Orientaciones para un manejo adecuado* [*Invasive alien species: a threat to biodiversity. Guidance for proper handling*]. Ministerio de Medio Ambiente y Recursos Naturales, Santo Domingo, Dominican Republic.

Moraes, G.J. de, Abo-Shnaf, R.I.A., Pérez-Madruga, Y., Sánchez, L., Karmakar, K. and Ho, C.C. (2015) The Lasioseius Phytoseioides species group (Acari: Blattisociidae): new characterisation, description of a new species, complementary notes on seven described species and a taxonomic key for the group. *Zootaxa* 3980(1), 1–41.

Pellerano, M.A. (2014) Ácaros fitófagos de la República Dominicana [Phytophagous mites of the Dominican Republic]. MSc thesis, Univ. Autónoma de Santo Domingo, Santo Domingo, Dominican Republic.

Peña, A.A., Guzman, R.E. and Guzman R. (1992) Liberación experimental de *Trichogramma* spp. en un campo de yuca en Moca, Rep. Dom [Experimental field release of *Trichogramma* spp. in a yuca field in Moca]. *Proceedings 28th Annual Meeting of the Caribbean Food Crops Society* 28, 95–103.

Pérez, Q. and Caro, A. (2002) Control de la broca del café (*Hypothenemus hampei* Ferr.) con *Beauveria bassiana* bajo condiciones de laboratorio [Control of the coffee berry borer (*Hypothenemus hampei*]. Resúmenes XLVIII Reunión Anual PCCMCA 2002, 14–20 de Abril del 2002 Boca Chica, República Dominicana, p. 119.

Pérez-Gelabert, D. (2008) Arthropods of Hispaniola (Dominican Republic and Haiti): a checklist and bibliography. *Zootaxa* 1831, 1–530.

Pinales de Soriano, C. (2015) Compatibilidad del enemigo natural *Orius insidiosus* en variedades de ají 'Morrón' (*Capsicum annuum* L. var. *annuum*) en cultivos protegidos [Compatibility of the natural enemy *Orius insidiosus* in varieties of bell pepper in protected crops]. MSc thesis, Universidad Autónoma de Santo Domingo, Dominican Republic, 72 pp.

Pinales de Soriano, C. and Serra, C. (2018) Estudios biológicos del chinche depredador *Orius insidiosus* (Say) (Hemiptera: Anthocoridae) para la adaptación de la metodología de cría masiva [Biological studies of the predatory bug *Orius insidiosus* for the adaptation of the methodology of mass rearing]. In: *Memorias XIV Jornada de Investigación Científica*, Univ. Autónoma de Santo Domingo, 19-20/11/2015, Santo Domingo, Dominican Republic, pp. 57–68.

Ramírez, W. (1995) La biología y preferencias de dos especies del género *Encarsia* (Hymenoptera: Aphelinidae) como parasitoides de la 'Mosca Blanca', *Bemisia tabaci* (Gennadius) (Homoptera: Aleyrodidae) [The biology and preferences of two species of the genus *Encarsia* as parasitoids of the 'whitefly', *Bemisia tabaci*]. BSc thesis, Pontificia Universidad Católica Madre y Maestra & Instituto Superior de Agricultura, Santiago de los Caballeros, Dominican Republic.

Santos, B.M., Sánchez, L., Reyes, M., Pérez, Q. and Castillo, M. (2000) Parasitoides de *Plutella xylostella* en República Dominicana [Parasitoids of *Plutella xylostella* in the Dominican Republic]. *Manejo Integrado de Plagas (Costa Rica)* 58, 76–77.

Schmutterer, H. (1990) *Crop Pests in the Caribbean, with Particular Reference to the Dominican Republic*. Deutsche Gesellschaft für Technische Zusammenarbeit (GTZ), Eschborn.

Schwartz, S. (2015) Schistosomiasis in the Dominican Republic. Upstream Alliance, Stanford University. Available at: http://schisto.stanford.edu/pdf/Dominican%20Republic.pdf (accessed 12 February 2019).

Segura, Y., Cedano, J., Godoy de Lutz, G. and Serra, C.A. (2005) Estudios para un manejo integrado de la Mosca Asiática (*Melanagromyza obtusa* (Malloch), Diptera: Agromyzidae) del Guandul (*Cajanus cajan* L. Millsp.) en la República Dominicana [Studies for IPM of the pigeon-pea pod fly in the Dominican Republic]. *Abstracts of the 51st Annual Meeting, Inter-American Society for Tropical Horticulture*, 10–14/10/2005, Boca Chica, Dominican Republic, p. 157.

Serra, C.A. (1992) Untersuchungen zum Einsatz von Niemsamenextrakten im Rahmen integrierter Ansätze zur Bekämpfung von Tomatenschädlingen in der Dominikanischen Republik [Research on the use of neem-seed extracts with integrated concepts for the control of tomato pests in the Dominican Republic]. PhD thesis, Univ. of Giessen, Wissenschaftlicher Fachverlag, Giessen, Germany.

Serra, C.A. (2006) *Manejo Integrado de Plagas de Cultivos – Estado Actual y Perspectivas para la República Dominicana* [*Integrated Pest Management in Crops – Actual Status and Perspectives for the Dominican Republic*]. Serie Proyecto AGORA, Kellogg Foundation/Centro de Desarrollo Agropecuario y Forestal, Santo Domingo, Dominican Republic. Available at: http://190.167.99.25/digital/CEDAF_MIP.pdf (accessed 12 February 2019).

Serra, C.A. (2007) Distribución e impacto preliminar del ácaro rojo de las palmeras (*Raoiella indica* Hirst, Prostigmata: Tenuipalpidae) en la República Dominicana [Distribution and preliminary impact of the red palm mite (*Raoiella indica*) in the Dominican Republic]. *Memorias 3er. Congreso Bianual SODIAF*, Santo Domingo, Dominican Republic.

Serra, C.A., Nuñez, J.B.and Ortíz, M. (1995) Perspectivas de *Cyrtopeltis* spp. (Hemiptera: Miridae) como depredadores de *Bemisia tabaci* (Genn.) (Homoptera: Aleyrodidae) para el manejo integrado de plagas en tomate [Perspectives of *Cyrtopeltis* spp. mirids as predators of *Bemisia tabaci* in IPM in tomatoes]. II. *Jornada Científica, Sociedad Dominicana de Investigadores Agropecuarios y Forestales, Inc*. Santo Domingo, Dominican Republic, 13 October 1995, p 3.

Serra, C.A., Ortíz, M., Nuñez, J.B., Schulz, A., and Benoit, P.F. (1996) Natural enemies of whiteflies in the Dominican Republic – their prospects for biocontrol in tomato and ornamental crops. In: *Proceedings of the XX International Congress of Entomology*, Firenze, Italy, August 25–31, 1996, p. 654.

Serra, C.A., Ortíz, M. and Ferreras, R. (1997) Evaluación de los efectos secundarios de insecticidas químicos y biológicos sobre *Encarsia transvena* (Timberlake) [Evaluation of secondary effects of chemical and biological insecticides on *Encarsia transvena*]. *Proceedings of the Caribbean Food Crops Society* 33, 475-479.

Serra, C.A., Jorge, P.E., Abud-Antún, A.J., Alvarez, P. and Peguero, B. (2003) Invasive alien species in the Dominican Republic: their impact and strategies to manage introduced pests,. *Proceedings 39th Annual Meeting of the Caribbean Food Crops Society* 39(1), 102–118.

Serra, C.A., Nuñez, C.A. and García, S. (2007a) El control natural y biológico clásico de la cochinilla rosada de los hibiscus (CRH), *Maconellicoccus hirsutus* (Green) [Natural and classical biological control of the pink hibiscus mealybug]. *Resúmenes de Investigación en Protección Vegetal*. Instituto Dominicano de Investigaciones Agropecuarias y Forestales, Santo Domingo, Dominican Republic.

Serra, C.A., Medrano, S., Ayala, C.A. and Galicia, J.A. (2007b) Alternativas para el manejo de artrópodos en vegetales orientales en la República Dominicana [Alternatives for the management of arthropods in oriental vegetables in the Dominican Republic]. *Proceedings of the 43rd Annual Meeting of the Caribbean Food Crops Society* 43, 125–132.

Serra, C.A., Ferreira, M., García, S., Santana, L., Castillo, M. *et al*. (2011a) Establishment of the West Indian Fruit Fly (Diptera: Tephritidae) Parasitoid *Doryctobracon areolatus* (Hymenoptera: Braconidae) in the Dominican Republic. *Florida Entomologist* 94(4), 809–816.

Serra, C.A., Cayetano, X., Féliz, A., Ferreira, M., García, S. *et al*. (2011b) Impacts of recently emerged IAS and major threats to the Dominican Agriculture. *Proceedings 47th Annual Meeting of the Caribbean Food Crops Society* 47(1), 146–156.

Serra, C., Díaz, C. and González G., F.J. (2016) Avances en el control microbiológico *in vitro* de la *Diaphorina citri* (Hemiptera: Psyllidae), vector del Huanglongbing, en cítricos dominicanos [Advances in microbiological control *in vitro* of *Diaphorina citri*, vector of Huanglongbing, in Dominican citrus]. *Resumenes 7 Congreso SODIAF*, 10–12 de noviembre 2016, Bávaro, Punta Cana, Dominican Republic.

Smith, T.R. and Cave, R.D. (2006) The life history of *Cybocephalus nipponicus*, a predator of the cycad aulacaspis scale, *Aulacaspis yasumatsui* (Homoptera: Diaspididae). *Proceedings of the Entomological Society of Washington* 108, 905–916.

Smith, T.R. and Cave, R.D. (2007) The Cybocephalidae (Coleoptera) of the West Indies and Trinidad. *Annals of the Entomological Society of America* 100, 164–172.

Stone, A. (1942) *Fruit Flies of the Genera* Anastrepha. USDA Miscellaneous Publication, 439. United States Department of Agriculture, Washington, DC.

Tappertzhofen, S. (1996) Zur Populationsentwicklung von *Bemisia argentifolii* im Südwesten der Dominikanischen Republik [Population dynamics of *Bemisia argentifolii* in the southwest of the Dominican Republic]. *Anzeiger für Schädlingskunde Pflanzenschutz Umweltschutz* 69, 153–156. DOI: org/10.1007/BF01906805.

Taveras, R. (2000) Fluctuation del minador de la hoja de los citricos (*Phyllocnistis citrella* Stainton) y la identification de sus parasitoides [Fluctuation of the citrus leaf miner (*Phyllocnistis citrella*) and identification of its parasitoids]. *Proceedings 36th Annual Meeting of the Caribbean Food Crops Society* 36, 111.

Taveras, R. and Caro, A. (2018) Eficiencia del hongo *Beauveria bassiana* en el control del piogán, *Cylas formicarius* (Coleoptera: Brentidae) en batata, *Ipomoea batata* [Efficiency of *Beauveria bassiana* in the control of the sweet potato weevil in sweet potato]. In: Cepeda Ureña, J. (ed.) *Consejo Nacional de Investigaciones Agropecuarias y Forestales (CONIAF). Socialización de Resultados de Investigación en Manejo Integrado de Plagas*. Santo Domingo, Dominican Republic. pp. 87–104. [see text in pdf Guzman *et al*., 2018]

Taveras, R. and Cuevas, A. (2006) Mortalidad causada por la aplicación del hongo *Beauveria bassiana* sobre los picudos del plátano *Cosmopolites sordidus* y *Metamasius hemipterus* en banano [Mortality caused by the application of the fungus *Beauveria bassiana* on banana weevils *Cosmopolites sordidus* and *Metamasius hemipterus* in banana]. *Proceedings 42nd Annual Meeting of the Caribbean Food Crops Society* 42, 121–125.

Taveras, R. and Guzmán, R. (2018) Control biológico del gusano de la flota de la yuca, *Erinnyis ello* (Lepidoptera: Sphingidae) en la zona norte de la República Dominicana [Biocontrol of the ello sphinx of cassava in the Northern zone of the Dominican Republic]. In: Cepeda Ureña., J. (ed.) *Consejo Nacional de Investigaciones Agropecuarias y Forestales (CONIAF). Socialización de Resultados de Investigación en Manejo Integrado de Plagas*. Santo Domingo, Dominican Republic. pp. 22–32. [see text in pdf Guzman *et al*., 2018]

Taveras, R., and Hansson, C. (2015) *Pediobius cajanus* sp. n. (Hymenoptera, Eulophidae), an important natural enemy of the Asian fly [*Melanagromyza obtusa* (Malloch)] (Diptera, Agromyzidae) in the Dominican Republic. *Journal of Hymenoptera Research* 4, 41–54. DOI: 10.3897/JHR.45.4964

Please see the supplementary file "Addenda and Corrections" for additional infomation.

13 Biological Control in Continental Ecuador and the Galapagos Islands

Carmen C. Castillo[1]*, Patricio G. Gallegos[2] and Charlotte E. Causton[3]

[1]*Plant Protection Department, Santa Catalina Research Station, INIAP (National Institute for Agriculture Research), Quito, Ecuador;* [2]*Entomologist, Independent Consultant, Conocoto, Quito, Ecuador;* [3]*Charles Darwin Foundation, Puerto Ayora, Santa Cruz Island, Galapagos, Ecuador*

* E-mail: carmen.castillo@iniap.gob.ec

Abstract

Biological control has been used against agricultural pests and diseases on mainland Ecuador for over 80 years and its use as a pest management tool is increasing. Classical and augmentative biocontrol techniques are now commonly used for a wide range of crops, fruits and flowers that are grown using traditional or organic farming approaches. The Ecuadorian government, through the Ministry of Agriculture and the National Institute of Agricultural Research, strongly encourages the use of biocontrol, conducts research on potential biocontrol agents, and has helped to establish laboratories for the production of biocontrol agents in different provinces. There are also now at least four Ecuadorian companies that produce entomopathogens, insect predators and parasitoids as well as cooperation agreements set up with international suppliers. It is estimated that augmentative biocontrol was used on over 65,000 ha of farmland in 2017, and conservation biocontrol on 150,000 ha. The area under classical biocontrol was difficult to reliably estimate. The aim of Ecuador is to offer more high-quality commodities that are produced using methods with low negative ecological impact. On the Galapagos Islands, classical biocontrol has been used once to control the invasive cottony cushion scale, which was seriously affecting threatened endemic plant species. Following the success of this programme, it is now being considered as a management tool for invasive plant and insect species, as well for the most damaging agricultural pests.

13.1 Introduction

Ecuador has an estimated population of almost 16,300,000 (July 2017) (CIA, 2017); According to Aguirre *et al.* (2017, pp. 317, 318, 319, 326 and 327):

> Ecuadorian workforce is largely based in agriculture ... In 2014, agriculture accounted for 24.5% of the country's employment ... There are various forms of agriculture, depending on the crop and the region where it is grown. Fruit, for example, is produced for export on large industrial plantations in the western tropical lowlands, whereas in the highlands, there are small farms for the local market, and family farming in the Amazon region. Maize, palm oil, cocoa and quinoa saw a significant increase in tons harvested in the past 20 years, whereas green coffee, barley and wheat suffered the largest declines in area and tons harvested ... Ecuador is the third largest producer of bananas by weight, but the leading producer in terms of the amount of US dollars generated by exports ... Ecuador is one of the world's most biodiverse countries and has nearly 25,000 species of vascular plants ... The various geographic regions are extremely rich in wild relatives related to cultivated species such as potato, bean, tomato and tropical and subtropical fruit trees. The country's natural forests are also home to the wild relatives of species such as avocado (*Persea* spp.) and papaya (*Carica* spp.). The high diversity of medicinal species is used on an everyday basis for the treatment of innumerable diseases, thanks to the traditional knowledge developed over thousands of years and advances in ethnobotany ...

According to the most recent information available at FAOSTAT (2018), the most commonly grown crops in Ecuador (by production tonnage in 2014) are fruit, sugarcane, plantain, cereals, oil palm, ground rice, rice, hard grains, maize and banana. Other important crops grown in Ecuador are potato, cocoa and citrus. Livestock production mainly concerns cattle, pork, sheep, goats and poultry. Fishing and aquaculture are an important economic activity in Ecuador. Based on monetary value (2014), Ecuador's largest exports after bananas are crustaceans and processed fish. Next are flowers and cacao (Aguirre *et al.*, 2017). In 2017, 12,355,146 ha were used for agricultural production, 4,816,821 ha in the coastal region, 3,765,969 ha in the highlands and 3,753,923 ha in the Amazon region (ESPAC, 2017).

On the Galapagos Islands, until the 1970s, residents relied on agriculture and fishing to make a living (Lundh, 2006), but now tourist visits to the islands provide an alternative economy. It is estimated that in 2015 there were about 224,755 visitors and 25,000 people living in the archipelago (Toral-Granda *et al.*, 2017), resulting in increased imports of food from mainland Ecuador. Less than 3% of the land area of the archipelago is made up of human settlements and agricultural lands; the remaining 97% is protected as a national park (DPNG, 2014).

In the last 15 years, considerable effort has been put into increasing local agricultural production in order to reduce the reliance on imports and the risks of introduction of non-native species transported accidentally on food

(Toral-Granda *et al.*, 2017). Emphasis is placed on promoting agroecological practices that will help ensure the long-term conservation of the Galapagos Islands (Guzmán and Poma, 2015). Agricultural production is focused on horticultural crops, coffee and pasture grass for cattle (Salvador Ayala, 2015).

13.2 History of Biological Control in Ecuador

Biological control has been used against agricultural pests for over 80 years in continental Ecuador. More recently, it was used for the first time in the Galapagos Islands to control an invasive scale insect that was affecting endemic plants.

13.2.1 Period 1880–1969

Most of the early results from applied biocontrol programmes in Ecuador have not been published and are only referred to in brief summaries of local meetings or found in internal reports (Klein-Koch, 1989). Biological agents historically used in Ecuador can be found in Tables 13.1 and 13.2 and the most important cases of classical biocontrol of the period 1880–1969 are summarized below.

One of the first available records of applied biocontrol dates back to 1937, when L. Rodriguez López of the Ecuadorian Department of Agriculture introduced and studied the parasitoid *Aphelinus mali* (Hald) from Chile for control of the woolly apple aphid *Eriosoma lanigerum* (Hausmann) in Tungurahua province (Merino, 1984). Next, in 1941, the predator *Rodolia cardinalis* (Mulsant) was imported from the USA, which established and controlled the scale *Icerya montserratensis* Riley and Howard (Merino, 1984). In 1955, the wasp *Amitus hesperidium* Silvestri was introduced from Mexico to control the citrus blackfly *Aleurocanthus woglumi* Ashby (Clausen, 1967; Merino, 1984) and was apparently successful, because the pest was not reported in a pest inventory made in 1986 in Ecuador (Klein-Koch, 1989). The ectoparasitic wasp *Aphytis lepidosaphes* Compere was introduced from the USA in 1958 for the biocontrol of *Lepidosaphes beckii* (Newman) in citrus. The wasp was mass reared and released in citrus plantations in several provinces and successfully controlled the pest two years after its release (Merino and Vásquez, 1962; Merino, 1984). A decade later, a study evaluated 54 citrus orchards and concluded that more than half of the orchards were not using insecticides against *L. beckii* and the other half used insecticides only for other minor pests (Merino and Vásquez, 1962). In 1992, *L. beckii* was no longer found in Pichincha province, one of the locations where the wasp was released in 1958 (Benzing and Goetz, 1993).

To control the sugarcane borer *Diatraea saccharalis* (Fabr.), the tachinid *Paratheresia claripalpis* Wulp was introduced in 1964 from Peru with promising results. To complement control by *P. claripalpis*, *Lydella* (= *Metagonistylum*) *minense* (Townsend) was introduced in 1990, but was not successful (Benzing and Goetz, 1993).

13.2.2 Period 1970–2000

During this period, a number of new classical biocontrol programmes were started. In addition to this, inundative biocontrol was used for the first time.

One of the first cases of inundative biocontrol occurred in 1977, when the mirid *Tytthus mundulus* (Breddin) was introduced from Hawaii as an egg predator of the leafhopper *Perkinsiella saccharicida* Kirkaldy, an important pest in sugarcane. It was mass reared in the laboratory and about one million mirids were released, but unfortunately this species did not establish (Morey, personal communication in Klein-Koch, 1989). In maize on the Ecuadorian coast, releases of *Trichogramma pretiosum* Riley were made during the 1980s to control *D. saccharalis*, with satisfactory results. However, this programme was discontinued when maize–soybean rotation was established and the pest population was drastically reduced due to the positive effect of crop rotation. During the same period, *L. minense* and *Lixophaga diatraeae* (Townsend) were introduced from Venezuela for control of sugarcane borer in Los Ríos province, but these species did not establish (J. Mendoza, Quevedo, 2018, personal communication).

In 1978, Ecuador received a donation of 24 million individuals of *Hippodamia convergens* Guerin from the USA to control the cottony cushion scale *I. purchasi* Maskell in fruit trees and ornamental plants located in the urban area

Table 13.1. Overview of biological control programmes in Ecuador.

Crop	Pest	Natural enemy	Type of biocontrol[a] / Efficacy	Origin	Year	Reference
Apple	*Eriosoma lanigerum*	*Aphelinus mali*	CBC / Complete	Chile	1937	Merino, 1984
Chirimoya (*Annona cherimola*)	*Anastrepha* spp.	*Biosteres longicaudatus* *Opius concolor* *Trybliographa daci*	CBC / Unknown	USA	1988–1989	Klein-Koch, 1989
Citrus	*Icerya montserratensis*	*Rodolia cardinalis*	CBC / Complete	USA	1941	Rodriguez-Lopez, 1942 cited in Merino, 1984
	Aleurocanthus woglumi	*Amitus hesperidium*	CBC / Partial	Mexico	1955	Merino, 1984; Clausen, 1967
	Lephidosaphes beckii	*Aphytis lepidosaphes*, *Encarsia* sp.	CBC / Complete Substantial	Mexico	1958	Merino and Vásquez, 1962 in Benzing and Goetz, 1993
	Icerya purchasi	*R. cardinalis*	CBC / Complete	USA	1978	Molineros, 1984; Barragán *et al.*, 2009
		Hippodamia convergens	Partial	USA	1978	
	Phyllocnistis citrella	*Ageniaspis citricola*	CBC / Substantial	Peru	1996	Cañarte *et al.*, 2005
Coffee	*Hypothenemus hampei*	*Cephalonomia stephanoderis*,	CBC / Substantial	Togo	1988	Klein-Koch, 1988; Cisneros and Tandazo, 1989; Delgado *et al.*, 1989; Mendoza *et al.*, 1994
		Prorops nasuta		Tanzania, Kenya, Togo	1987, 1988	
		Phymastichus coffea	CBC / Substantial	Colombia	1999	ANECAFE, 2014
Cotton	*Pectinophora gossypiella*	*Bracon kirkpatrick*	CBC / Unknown	Colombia	1988	Klein-Koch, 1989 in Benzing and Goetz, 1993
Maize	*Diatraea saccharalis* *D. lineolata*	*Lydella minense*	CBC / Unknown	Venezuela	1983	Klein-Koch, 1989
		Cotesia flavipes	CBC / Unknown	Colombia Pakistan/USA	1983 1987	Klein-Koch, 1989
		Paratheresia claripalpis	CBC / Unknown	Colombia Pakistan/USA	1983 1987	Klein-Koch, 1989
		Trichogramma semifumatum, *T. minutum*, *T. pretiosum*	ABC / Substantial	Colombia	1983	Klein-Koch, 1989
		Lydella (=*Metagonistylum*) *minense*, *Lixophaga diatraeae*	CBC / Not established	Venezuela	1980s	J. Mendoza personal communication

Continued

Table 13.1. Continued.

Crop	Pest	Natural enemy	Type of biocontrol[a] / Efficacy	Origin	Year	Reference
Mango	*Aulacaspis tubercularis*	*Cybocephalus nipponicus*	CBC / Complete	USA	2004	Arias *et al.*, 2004
Rice	*Rupela albinella*	*Telenomus* sp.	ABC / Substantial	Colombia	1983	Klein-Koch, 1989 in Benzing and Goetz, 1993
Several	*Tetranychus* spp.	*Phytoseiulus persimilis*	ABC / Substantial	Italy	1986	Klein-Koch, 1989
	Musca domestica	*Spalangia* spp. *Muscidifurax raptor*	ABC / Unknown	Colombia	1986	Klein-Koch, 1989
		Neoplectana carpocapsae (= *Steinernema feltiae*)	ABC / Unknown	USA	1983	Klein-Koch, 1989
	Anastrepha fraterculus, A. serpentina, A. striata, A. obliqua	*Diachasmimorpha longicaudata*	CBC / Substantial	Peru	2005	M. Arias personal communication
Sugarcane	*D. saccharalis*	*Paratheresia claripalpis*	CBC / Complete	Peru	1964	Morey pers. com. in Benzing and Goetz, 1993
		Cotesia (= *Apanteles*) *flavipes*	CBC / Unknown	Colombia	1978	Morey pers. com. in Klein-Koch, 1989
	Perkinsiella saccharicida	*Tytthus mundulus*	CBC / Not established	Hawaii	1977	Morey pers. com. in Klein-Koch, 1989

[a]Type of biocontrol: ABC = augmentative, CBC = classical biological control

Table 13.2. Use of native beneficial agents in Ecuador.

Crop	Pest	Natural enemy	Type of biocontrol[a] / Efficacy	Reference/remarks
Maize, soybean	*Spodoptera frugiperda*	*Trichogramma* spp.	ABC / Substantial	Klein-Koch, 1989
Several	*Aleurothrixus floccosus*	*Cales noacki*	CBC / Substantial	Klein-Koch, 1989, moved from South to North Ecuador
Sugarcane	*Perkinsiella saccharicida*	*Anagrus optabilis*	CBC / Substantial	Klein-Koch, 1989
		Tytthus parviceps	CBC / ABC / Substantial	Klein-Koch, 1989, the exotic *Tytthus mundulus* performed better
	Aphis sorghi, *Sipha flava*	*Cycloneda sanguinea*, *Coleomegilla maculata*	ABC / Complete	Klein-Koch, 1989; Morey, 1984 (pers. com. in Klein-Koch, 1988)
	Spodoptera frugiperda	*Polistes infuscatus ecuadorius*	ConsBC / Substantial	Klein-Koch, 1989, provide appropriate nesting places

[a]Type of biocontrol: ABC=augmentative, CBC = classical, ConsBC = conservation biological control

of the Ecuadorian capital, Quito (Molineros, 1984). The donation also included 40 individuals of the predator *R. cardinalis* and 10 of *Cryptochaetum iceryae* (Willston), which were multiplied and released with the same purpose as the releases of *H. convergens*. *C. iceryae* did not survive but *R. cardinalis* did. The control of *I. purchasi* was successful as a result of the introduction of these predators (Molineros, 1984; Barragán *et al.*, 2009), though it should be mentioned that the presence of *H. convergens* in Ecuador had already been recorded in 1957 (Merino, 1987, in Klein-Koch, 1989). More releases of *R. cardinalis* were done in the South of the country in 1980, which resulted in excellent control of *I. purchasi* in several crops (Franklin Santillán, Riobamba, 2018, personal communication).

In 1983, the parasitoid *Telenomus* sp. was introduced to Ecuador from Colombia to control the white rice borer *Rupela albinella* (Cr.). The parasitoid established and successfully controlled the rice borer (Klein-Koch, 1989; Benzing and Goetz, 1993). In the same year and also from Colombia, three species of *Trichogramma* (*T. semifumatum*, *T. minutum* and *T. pretiosum*) were imported to control *D. saccharalis* and *D. lineolata* (Walker) (Klein-Koch, 1989), but it is not known if these parasitoids were used in the field.

The entomopathogenic nematode *Neoaplectana carpocapsae* Weiser was introduced in Ecuador in 1983 from the USA and after multiplication was released in the field against soil-borne pests in vegetables, with success (Klein-Koch, 1989).

For a period of 10 years, the company Biológicos Ecuatorianos SA (Bioesa), located in Santo Domingo de los Tsachilas province, produced annually three billion parasitoids of *Trichogramma* sp. reared on *Sitotroga cerealella* Oliver eggs for releases in maize and soybean (75,000–120,000 parasitoids per hectare) and in sugarcane, banana and cotton (210,000– 240,000 parasitoids per hectare) against *D. sacharalis* and *Spodoptera frugiperda* (Smith) (Benzing and Goetz, 1993). Sadly, the company went out of business because the demand went down (J. Mendoza, Quevedo, 2018, personal communication).

In 1987–1988, under an agreement between the National Institute for Agriculture Research (Instituto Nacional de Investigaciones Agropecuarias) (INIAP), the German Cooperation Agency (GTZ) and Centre for Agricultural Bioscience International (CABI), parasitoids of the coffee berry borer *Hypothenemus hampei* (Ferrari), *Prorops nasuta* Waterston (270 individuals) and *Cephalonomia stephanoderis* Betrem (57 individuals) were imported from Kenya and Togo (Klein-Koch, 1988). After multiplication at INIAP, *P. nasuta* was released in the field on nine occasions (total of 2,808 parasitoids) in the South-eastern region of Ecuador, on 24 occasions in the coastal region (total of 1,328 wasps), and once in the Amazonian region (45 adults). Also, a total of 400 *C. stephanoderis* wasps were released. After a period of approximately three years, wasps were recovered. With the successful mass-rearing methodology developed at INIAP, field releases of both parasitoids were continued (Cisneros and Tandazo, 1989; Delgado *et al.*, 1989, Mendoza *et al.*, 1994), with parasitism rates reaching 83% with *C. stephanoderis* and 23% with *P. nasuta*. Thus, *C. stephanoderis* was the best adapted and most efficient in controlling the coffee berry borer (Mendoza *et al.*, 1994).

The Association of Coffee Exporters of Ecuador (ANECAFE), in collaboration with the International Institute for Biological Control (IIBC; now CABI) implemented an integrated management programme for the coffee berry borer. In 1999, the parasitoid *Phymastichus coffea* La Salle was introduced from Colombia so that studies could be carried out on its biology and in order to develop a mass-rearing methodology. Over three years a total of 210,000 wasps were released in six provinces of Ecuador, resulting in parasitism rates of 33%. Up to three generations of wasps were recovered in the field during this study (ANECAFE, 2014).

In 1988, the braconid *Bracon kirkpatricki* (Wilkinson) was introduced from Colombia to control the pink bollworm *Pectinophora gossypiella* (Saunders), but results were unsatisfactory (Benzing and Goetz, 1993).

For the control of fruit flies (*Anastrepha* spp.) in cherimoya, *Annona cherimola* (Mill.), the species *Biosteres longicaudatus* Ashmead, *Opius concolor* Szépligeti and *Trybliographa daci* Weld were introduced from the USA in 1988 and 1989. Releases were made in the low-elevation valleys of Pichincha and Azuay provinces (Klein-Koch, 1989), but the effects of these releases are not known.

Between 1985 and 1990, the Polytechnic University of Chimborazo Province (ESPOCH) created a biocontrol centre to study, rear and release predators and parasitoids as well as entomopathogenic fungi in Chimborazo province. After several studies, promising species were multiplied and released, e.g. *Incamya chilensis* Aldrich, *Trichogramma* sp. and *Podisus connexivus* Bergroth. The latter species was released in maize to control the borer *Agrotis deprivata* Walker in the Ecuadorian highlands (Andrade, 1990). Another biocontrol project, developed by ESPOCH between 2001 and 2003, was an inventory of natural enemies in the province of Chimborazo to identify the best species for mass production. *Encarsia formosa* Gahan was multiplied and released to control *Bemisia tabaci* (Genn.) and *Trialeurodes vaporariorum* (Westwood) in tomato *Solanum lycopersicum* L. in greenhouses. Also *Trichogramma pretiosum* Riley was multiplied and released to control *Tuta absoluta* (Meyrick) and *Scrobipalpula* sp.. Significant quantities were sold to farmers, but parasitoid releases were not well planned and were erratic, resulting in poor performance (A. Espinoza, Riobamba, 2018, personal communication). The contribution of ESPOCH to biocontrol in Ecuador has been important in terms of academic training and research using a field-based approach. Several of its graduates have developed companies that produce biocontrol agents, some of which have now been running for over 30 years (e.g. Biosagro and AgroVerde). These companies supply biocontrol agents to Chimborazo province and other provinces in the highlands and to the coastal region.

The parasitoid *Ageniaspis citricola* Logvinovskaya was introduced in Peru in 1996 to control *Phyllocnistis citrella* Stainton. Later, the parasitoid spontaneously immigrated into Ecuador and in 2001 it had already dispersed throughout the Ecuadorian coast and had reached the Amazonian province of Napo (Cañarte *et al.*, 2005). In 2004, 270 specimens of predatory beetle *Cybocephalus nipponicus* Endrody-Younga were imported from the USA for biocontrol studies of the white mango scale *Aulacaspis tubercularis* Newstead. After multiplication in the laboratory, trials in mango orchards in Guayas province were performed with 1,000 specimens released per hectare. In the field, the parasitoid populations increased and a successful reduction of the scale was obtained (Arias *et al.*, 2004).

13.3 Current Situation of Biocontrol in Ecuador

Information about many of the biocontrol programmes that have been carried out against crop pests and diseases in Ecuador is not publicly available. Because of this, the cases presented below are likely to be an underestimate of biocontrol activities in Ecuador and do not include all of the companies or farms involved in the production of natural enemies or currently implementing biocontrol. Table 13.3 gives an overview of information provided by farmers and institutions about crops, natural enemies and areas under biocontrol. Table 13.4 presents information provided by producers of biocontrol agents.

13.3.1 Banana

In 2014, 196,673 ha of bananas were planted in Ecuador (ANCUPA, 2014a). Approximately 150,000 ha of them are currently using *Bacillus thuringiensis* (Berliner) to control the banana moth *Antichloris* (= *Ceramidia*) *viridis* Druce with a high percentage of efficacy (> 80%), a practice that has been successfully applied for many years with excellent results (J. Mendoza, Quevedo, 2018, personal communication). For example, the Primobanano ranch has been using *B. thuringiensis* since 1989 with very good results for the control of *A. viridis*, *Sibine* sp. and other banana pests. The results depend on good monitoring of the pest. Only in a few cases, when the pest surpasses the economic threshold, do farmers apply insecticides. *B. thuringiensis* is available commercially under several names: Dipel, Thuricide, DiBeta, Javelin, among others. The company Estuardo Quirola Lojas, located in the province of Guayas, has started to apply *Lecanicillium lecanii* (Zimmerman) against thrips (*Chaetanaphothrips brevicaulis* Hood) in 10 ha of banana.

A recent development is the successful application of *Amblyseius swirskii* Athias-Henriot on 300 ha of banana for control of banana rust thrips *Chaetanaphothrips signipennis* (Bagnall), causing red spot in banana, as well as another species, the banana flower thrips, *Frankliniella parvula* Hood in 2018 (Cristóbal Fábrega Castellón, Koppert, Berkel and Rodenrijs, the Netherlands, 2019, personal communication).

13.3.2 Broccoli

Agroindustry Argentina, located in the province of Cotopaxi, has been growing broccoli on 62 ha for 14 years. By continuously applying *Trichoderma harzianum* Rifai through a sprinkler irrigation system and by drenching, this company has managed to eradicate cabbage club root *Plasmodiophora brassicae* Woronin from the soil and is also able to control *Sclerotinia* sp. and *Rhizoctonia* sp. (90% efficacy), *Alternaria* and *Botrytis* (50% efficacy) and damping-off disease (85% efficacy). With applications of *B. bassiana* they reduce aphid populations by 70% and *Plutella* sp. by 15%. The microbial control agents are produced locally.

13.3.3 Cacao

The company Estuardo Quirola Lojas controls frosty pod rot caused by *Moniliophthora roreri* (Cif.) H.C. Evans, Stalpers, Samson & Benny with products based on *T. harzianum*, *T. viride* and *T. pseudokoningii* Rifai in an area of 224 ha. Control is effective and will be extended by an additional 143 ha this year.

13.3.4 Coffee

Currently, technical advisers of ANECAFE do not recommend the use of insecticides to control the coffee bean borer *H. hampei* in coastal areas, where approximately 150,000 ha of traditional varieties are planted and production is seasonal. Also, fungicides are not used, in order to preserve native entomopathogenic fungi. In the Amazon region, farmers do not spray coffee with insecticides. However, in provinces such as Guayas and Santa Elena (coastal area), a new variety of coffee has been introduced recently that needs pesticide applications against insects, fungi and spider mites (Luis Duicela, Porteviejo, 2018, personal communication).

13.3.5 Oil palm

Since 2012, *Trichoderma* sp. and *Metarhizium* sp. have been extensively used in oil palm cultivations to control pests and diseases on around 500 ha in Quevedo (Los Ríos province) and 2,000 ha in Quinindé (Esmeraldas province) (ANCUPA, 2014b).

Table 13.3. Use of augmentative and conservation biological control and biopesticides in Ecuador, based on information given by farms/enterprises.

Crop	Pest/disease	Biocontrol agent	Efficacy (%)	Estimated area (ha)	In use since	Source, enterprise
Banana	*Antichloris viridis*	*Bacillus thuringiensis*, biopesticide	> 80	150,000	2014	J. Mendoza (personal communication)
	Chaetanaphothrips brevicaulis	*Lecanicillium lecanii*		10	2016	Enterprise Estuardo Quirola Lojas
Cacao	*Moniliophthora roreri*	*Trichoderma harzianum*, *T. viride*, *T. pseudokoningii*	100	224		Enterprise Estuardo Quirola Lojas
Coffee	*Hypothenemus hampei*	ConsBC with parasitoids and ABC with native entomopathogens		150,000	2017	ANECAFE
Oil palm	Pest and diseases	*Trichoderma* sp., *Metarhizium* sp.		2,500	2012	ANCUPA
	Lincus sp.	*Paecilomyces tenuipes*		500		Enterprise Palmar del Río
	Demotispa sp.	*Trichoderma* sp.		5		
	Stenoma cecropia, *Opsiphanes cassina*	*B. bassiana*, *Metarhizium* sp. and *B. thuringiensis*		2,000		Energy & Palma
	Alurnus humeralis	*Metarhizium* sp.		80	2015	Organic Crops
				120	2009	
Papaya	*Phytophthora palmivora*	*Metarhizium* sp.		160		Terrasol
Pineapple	*Thecla basilides*, *Spodoptera exigua*, *Mythimna unipuncta*	Dipel, biopesticide		300		Frugalp
	Rhynchophorus palmarum, *Metamasius* sp.	*B. bassiana*, *Metarhizium* sp.				
	P. palmivora	*Trichoderma* sp.		300		Terrasol
Rice	*Pyricularia oryzae*	*Trichoderma* sp., *B. subtilis*	100	360	2011	Agroreal S.A.,
	Hydrellia sp., *S. frugiperda*, *Diatraea* sp., *Sogatodes oryzicola*	*B. bassiana*, *Metarhizium* sp., *Lecanicillium* sp.				
Roses/Flowers	*Tetranychus urticae*	*Neoseiulus* (= *Amblyseius*) *californicus*	100	2	2014	Flower Farm A
		Paecilomyces sp., *Arthrobotrys* sp., *Trichoderma* sp., *Bacillus* sp., *Gliocladium* sp.		40		

	T. urticae	*N. californicus*	Variable	50	2013	Flower Farm B
	Botrytis sp., *Oidium* sp.,	*T. harzianum, B. subtilis, B. licheniformis, Paecilomyces* sp., *B. bassiana, Metarizium* sp., *Verticillium* sp.	100			
	Trialeurodes vaporariorum/ Bemisia tabasi					
	Plant nematodes					
	Botrytis sp.	*Trichoderma* sp.		80		Flower Farm C
	Pests	*B. thuringiensis,* biopesticide				
	T. urticae	*Phytoseiulus* sp., *Amblyseius* sp.		8	2013	Florisol
	Plant nematodes and soil-borne fungi	*Trichoderma* sp., *Paecilomyces* sp.		20		
	Thrips *Frankliniella occidentalis* and leaf miner	*Lecanicillium* sp., *Beauveria* sp.		12		
	Leaf miner, *Liriomyza* sp.	*Coenosia attenuata*		7	2014	
	Plant nematodes and soil-borne fungi	*Trichoderma* sp.	100	13	2009	FlorMachachi
Sugarcane	*Diatraea saccharalis*	*Billaea* (=*Paratheresia*) *claripalpis*	50-60	40,000		CINCAE
		Cotesia flavipes	30			
Vegetables	Several plant pathogens	*Trichoderma* sp., *Beauveria* sp., *Mycorrhiza* sp.		18	1998	A. Sancho (personal communication)
	Whitefly	*Encarsia formosa*	100		2002–2003	
		Trichogramma sp.	100, seasonal		1998	
	Armyworm, aphids, Whitefly	*B. thurigiensis B. bassiana, Azotobacter* sp., *Metarhizium* sp.		30	2007	Municipality of Quito, CONQUITO
Bell pepper, passion fruit, banana and cocoa	Soil-dwelling plant pathogens	*Trichoderma* sp.		45	2016	MAG. Santo Domingo de los Tsachilas province
Broccoli	*Plasmodiophora brassicae*	*T. harzianum*	100	62	2003	Agroindustry La Argentina
	Esclerotinia sp., *Rhizoctonia* sp.		90			
	Damping off		85			
	Aphids	*B. bassiana*	70			
	Plutella sp.		15			

Continued

Table 13.3. Continued.

Crop	Plest/disease	Biocontrol agent	Efficacy (%)	Estimated area (ha)	In use since	Source, enterprise
Broccoli, onion, potato, tamarillo, tomato, strawberry, blackberry	Soil-dwelling plant pathogens	*Trichoderma* sp.		5	2015	MAG. Tungurahua province
Fruit trees, tomatoes, bell peppers, onions, coffee	Soil-dwelling plant pathogens	*Trichoderma* sp.		100	2016	MAG. Azuay province
Onion, maize, cassava, veget-ables	Soil-dwelling plant pathogens	*Trichoderma* sp.		15	2015	MAG. Loja province
Tomato	White fly Diseases	*E. formosa* *Trichoderma* sp.		5	2002	AgroVerde
Subtotal				347,071		

Table 13.4. Use of augmentative biological control in Ecuador, based on information given by producers of biological control agents.

Crop	Pest/ disease	Biocontrol agent	Estimated number of hectares	Source, enterprise
Banana	Nematodes	*Trichoderma* spp., *Paecilomyces lilacinus*	5,000	Agrodiagnostic, since 2004
Cacao	*Moniliophthora roreri*	*Trichoderma* spp.	1,000	
Flowers	*Botrytis* sp.	*Trichoderma* spp.	280	
	Spider mites	*B. bassiana, Lecanicillium lecanii*	80	
	Thrips	*L. lecanii*	10	
Mango	Spider mites	*B. bassiana*	1,400	
	Soil-borne pathogens	*Trichoderma* sp.		
	Colletotrichum sp. in postharvest fruit	*Trichoderma* sp.		
Roses	Several	*P. lilacinus, P. fumosoroseus, Metarhizium anisopliae,*	350	Equabiológica, since 2005
Vegetables in open field		*B. bassiana, Arthrobotrys oligospora, L. lecannii,*	250	
Banana		*Gliocadium* sp.	150	
Palmito peach palm (*Bactris gasipaes*)			400	
Summer flowers			300	
Flowers	Several	*B. bassiana, P. lilacinus, Arthrobotrys* sp., *T. asperellum, Bacillus subtilis*, bacterial fermentations and protein extracts	1,000	Microtech, since 2013
Roses	Several	*T. harzianum, T. koningii, T. viride, T. hamatum,*	350	BioControlScience, since 1997
Summer flowers		*P. lilacinus,*	120	
Tropical flowers		*P. fumosoroseus, A.*	50	
Banana		*irregularis, Dactylella* sp., *M. anisopliae*	10,000	
Flowers	Spider mites	*Neoseiulus californicus* (previously *Amblyseius*) and *Phytoseiulus persimilis*	72	Koppert
Subtotal			20,812	
Total ha of tables 3+4			**367,883**	

The company Palmar del Río applies a native isolate of *Paecilomyces tenuipes* (Peck) Samson on 500 ha of oil palm located in Orellana province in the Amazon region against *Lincus* sp., which is a vector of a pathogen causing wilting. On a five ha trial area, *Trichoderma* sp. has recently been tested for control of *Demostipa* sp. (Palmar del Río, 2017).

Energy & Palma, located in San Lorenzo (Esmeraldas province), uses a mixture of *B. bassiana*, *Metarhizium* sp. and *B. thuringiensis* Berliner to control *Stenoma cecropia* Meyrick and

Opsiphanes cassina C. & R. Felder on 2,000 of their 6,000 ha of palm oil.

Organic Crops has controlled *Alurnus humeralis* Rosenberg using *Metarhizium* sp. since 2014 on 80–100 ha of palm in the area of Quinindé (Esmeraldas province), La Concordia and Santo Domingo (province of St Domingo de los Tsáchilas). Other palm growers with a total area of 120 ha have controlled this pest using *Metarhizium* sp. for the past 8 years.

13.3.6 Papaya

On approximately 160 ha of the Terrasol farm, *Trichoderma* sp. is applied to plant material after pruning to prevent infestation by the pathogen *Phytophthora palmivora* Butler.

13.3.7 Pineapple

The Frugalp farm has been applying Dipel since 2015 on 300 ha of pineapple to control *Thecla basilides* (Geyer), *S. exigua* (Hubner) and *Mythimna unipuncta* (Haworth). Sometimes *Rhynchophorus palmarum* (L.) attacks young pineapple plants and damages the fruit. *Metamasius* sp. (also a pest of palm) is another serious pest in pineapple. For control of these two pests a mix of *B. bassiana* and *Metarhizium* sp. is applied. After the pineapple harvest, the plant is cut and incorporated into the soil with an addition of *Trichoderma* sp. to avoid *P. palmivora* infestations. This is also the common control method on approximately 300 ha at the Terrasol farm.

13.3.8 Rice

The 'Sitio Nuevo' farm (Agroreal SA) has produced 360 ha of rice under biocontrol since 2011 and obtains 100% control of the rice blast *Pyricularia oryzae* (Cavara) with a mix of *Trichoderma* sp., *B. subtilis* (Ehrenberg) and other bacteria. Other pests such as *Hydrellia* sp., *S. frugiperda*, *Diatraea* sp. and *Sogatodes oryzicola* (Muir) are controlled by using a cocktail based on *B. bassiana*, *Metarhizium* sp. and *Lecanicillium* sp.

13.3.9 Roses, Flowers

Three farms in the Pichincha province that prefer to remain anonymous grow roses and other flowers and apply biocontrol. For the past two years, Farm A has been using *Amblyseius californicus* (McGregor) to control spider mites *Tetranychus urticae* Koch on two ha of roses. It also has 40 ha of summer flowers (*Gypsophila* sp. and other species) and roses that are treated with *Paecilomyces* sp., *Arthrobotrys* sp., *Trichoderma* sp., *Bacillus* sp. and *Gliocladium* sp. Farm B has 50 ha of roses and has controlled *T. urticae* with *A. californicus* since 2013. Also, diseases (*Botrytis* sp., *Oidium* sp. and *Peronospora* sp.) and pests like whitefly (*T. vaporariorum* / *B. tabaci*), nematodes and other pests are controlled using *T. harzianum*, *B. subtilis*, *B. licheniformis* (Weigmann), *Paecilomyces* sp., *B. bassiana*, *Metarhizium* sp. and *Verticillium* spp. This farm has managed to ensure quality flower production without using any synthetic pesticides. Farm C maintains 80 ha of roses and controls *Botrytis* sp. and spider mites using *B. thuringiensis* and *Trichoderma* spp.

The company Florisol has one farm located in Lasso (Cotopaxi province) with roses and another in Chavezpamba (Pichincha province) with summer flowers. Since 2013, eight hectares of roses have been controlled for spider mites using *Phytoseiulus* sp. and *Amblyseius* sp. On 20 ha of summer flowers and roses, *Trichoderma* sp. and *Paecilomyces* sp. are used against nematodes and soil fungi, and on 12 ha of summer flowers *Lecanicillium* sp. and *Beauveria* sp. are used against thrips *Frankliniella occidentalis* (Pergande) and leaf miners *Liriomyza* spp. Since 2014, workers have collected and released the hunter fly or killer fly *Coenosia attenuata* Stein on seven ha of flowers. The company FlorMachachi, located in Pichincha province, has used *Trichoderma* sp. and a mixture of beneficial microorganisms for the control of nematodes on 13 ha of roses since 2009.

13.3.10 Sugarcane

Since the 1960s, sugarcane plantations have been using the parasitic tachinid *Billaea* (= *Paratheresia*) *claripalpis* (Wulp), a native species, against the borer *D. saccharalis*. Also a Peruvian population of *B. claripalpis* was introduced, which has a shorter life cycle and reaches higher levels of parasitism. The parasitoids are used on about 40,000 ha of sugarcane, resulting in up to 50–60% larval parasitism. *Cotesia flavipes* (Cameron), introduced from Colombia during the 1980s is also used for borer control; its parasitism varies between regional zones and can reach up to 30%. During the same period, *L. minense* was also introduced from Colombia and, after successful mass production, was released for several years, but not recovered from the field and its production was therefore terminated (J. Mendoza, Quevedo, 2018, personal communication). CINCAE (Sugar Cane Research Center) has identified 26 parasitoids, 32 predators and 10 entomopathogenic fungal species as natural enemies of sugarcane pests.

13.3.11 Vegetables

For more than 15 years, the farm Rivera-Heredia–AgroVerde has produced tomato (*Solanum lycopersicum* L.) in greenhouses under a combined system of rational use of chemicals and organic alternatives such as pheromones to attract *T. absoluta*, *Trichoderma* sp. and other microorganisms for disease control, and *E. formosa* against whiteflies. *E. formosa* is periodically released on five ha of tomato. In the eastern part of Riobamba city, 10–40% of the farmers grow vegetables with rational use of pesticides in combination with biocontrol on approximately 200 ha. Another association of 64 farmers in Chimborazo province has been growing vegetables under organic certification for 19 years. Greenhouse tomato, cape gooseberry or goldenberry (*Physalis peruviana* L.), lettuce, cabbage, broccoli, onion and romanesco (*Brassica oleracea* L. var. Italica), among others, are produced on 18 ha. They apply locally produced microorganisms such as *Trichoderma* sp., *Beauveria* sp. and *Mycorrhiza*. *E. formosa* was introduced to control whiteflies over two years (2002–2003), after which they started to produce the parasitoid themselves. Annually, farmers also release *Trichogramma* sp. for control of lepidopteran pests. When pest populations become too large, farmers use light traps, pheromones (imported from Israel, Canada and Chile) and manual control (A. Sancho, Riobamba, 2018, personal communication).

13.3.12 Governmental and non-governmental research on biological control in Ecuador

For 58 years, INIAP has given special importance to research on biocontrol, which is one of its research mandates to ensure the well-being of farmers and consumers and to protect the environment (Delgado and Játiva, 2010; Domínguez and Zambrano, 2017). Research projects have been carried out at its seven stations (one in the Amazon region, four in the coastal area and two in the highlands) and four farms (two in the highlands and two in the Amazon region). INIAP has provided training in biocontrol to technicians, universities, companies and farmers in all the Ecuadorian regions and produced more than 40 local publications on biocontrol (INIAP, 2019).

Major research projects at INIAP include biocontrol of the coffee berry borer and coffee rust *Hemileia vastatrix* Berk and Br., biocontrol of the fall armyworm *S. frugiperda* in maize, biocontrol of the Andean potato weevil *Premnotrypes vorax* (Hustache) and Andean potato moths, a project about the identification of pests and natural enemies in *S. quitoense* (naranjilla, or *lulo* in Colombia) and biocontrol of oil palm pests. INIAP has collections of natural enemies, of entomopathogenic fungi and of insect parasitic nematodes (Gallegos *et al.*, 2011) and has conducted inventories of native beneficial insects (Sosa, 2009; Báez and Gallegos, 2013). Below are short summaries of the most important projects developed by INIAP.

Biological control of pests in potatoes and other crops

INIAP joined with other partners such as the International Potato Center, IPM-CRSP and Virginia Polytechnic Institute and State University, Ohio State University, Universidad San Francisco de Quito, Universidad Central del Ecuador and Universidad de la Fuerzas Armadas, to study entomopathogenic fungi and nematodes for pest control. In 2003, surveys of entomopathogenic fungi in different ecosystems were started for control of *P. vorax* in potato (Barriga, 2003; Landázuri, 2003; Guapi *et al.*, 2013) and of *Macrodactylus pulchripes* Blancharé in maize (Rueda, 2005; Ayala, 2006). More than 30 isolates of

B. bassiana, 20 of *Metarhizium* sp. and four of other species were obtained (Gallegos *et al.*, 2011). Next, their growth rate, sporulation and pathogenicity were determined in the laboratory and the field and formulations were developed for isolates that significantly reduced both pest species (Barriga, 2003; Landázuri, 2003; Rueda, 2005; Ayala, 2006; Guapi *et al.*, 2013). From this collection, the genetic diversity of 41 isolates of entomopathogenic fungi (Arahana *et al.*, 2013) was determined. In 2005, soil sampling for entomopathogenic nematodes took place in potato-grazing systems, potato–other crop systems, places where crops were stored, natural vegetation and fruit production systems in the provinces of Carchi, Cotopaxi, Chimborazo and Tungurahua. A total of 357 samples were evaluated, of which 7.8% were positive for entomopathogenic nematodes. Fifteen isolates of the genus *Steinernema* and 13 of the genus *Heterorhabditis* were identified (Argotti *et al.*, 2011); eight isolates were efficient in controlling *P. vorax* in greenhouse and field trials (Chacón, 2011). In addition, *Copidosoma koehleri* Blanchard was found parasitizing eggs of potato tuber moth *Phthorimaea operculella* (Zeller), while *Apanteles* sp. parasitized larvae of Andean potato tuber moth *Symmetrischema tangolias* (Gyer.) in potatoes (Báez and Gallegos, 2013).

In 2005 and 2006, Universidad de las Fuerzas Armadas and its Andean Institute of Agriculture (IASA, Spanish acronym) taught a Master of Science programme in biocontrol of agricultural pests. Fifteen of these projects were executed in this period, some in cooperation with INIAP, described in the previous paragraph.

Formulation of a baculovirus for potato moth control

The McKnight Foundation, the Catholic University of Ecuador, IRD (Institut de Recherche pour le Développement, France) and INIAP worked from 2006 to 2012 on the development of a commercial biopesticide based on the JLZ9F (Baculovirus) virus in combinations with use of *B. thuringiensis kurstaki* (Bacu-Turin) to control the Andean potato moths *Tecia solanivora* Povolny and *P. operculella* in potato tubers for consumption and for seed potatoes in the Andes (Suquillo *et al.*, 2012). Unfortunately, no agreement has been made with producers to scale up the microbial control agent.

Improvement of formulations for microorganisms

Since 2009, INIAP has been working in partnership with AgResearch, New Zealand, for production and formulation of high-quality microbial control agents and bioinoculants. The project has three components: (i) assisting in the development of a regulatory framework for microbial control agents; (ii) training of technicians in biocontrol research; and (iii) development and implementation of biocontrol systems for key crops in pilot projects. The project stresses the importance of developing a strong and effective partnership between researchers, producers of microbial control agents and the farmers/users for good implementation of biocontrol (Jackson, 2016).

Identification of natural enemies of pests in citrus, banana and cacao in the coastal region

INIAP, in a project supported by PROMSA (Program for Modernization of Agricultural Services), studied biocontrol of the citrus leaf miner *Phyllocnistis citrella* Stainton in Manabí province from 2000 to 2003. This pest was detected in 1995 and is considered the main pest of citrus. Eleven species of native parasitoids of *P. citrella* were found and the presence of the exotic parasitoid *Ageniaspis citricola* Logvinovskaya was reported for the first time. *A. citricola* probably spontaneously entered Ecuador from Peru, after its release in 1996 in Peru (Cañarte *et al.*, 2005). An IPM-CRSP project studied biocontrol of the banana black weevil *Cosmopolites sordidus* (Germar) by using traps with local isolates of *B. bassiana* in the coastal region (Solís *et al.*, 2001). Two native isolates of *Trichoderma* (*T. koningiopsis* and *T. stromaticum* Samuels & Pardo-Schulth) were tested in the field against the pathogens of cacao *Moniliophthora roreri* (Cif and Par) and *Crinipellis perniciosa* (Stahel) Aime and Phillips-Mora. Lowest damage and highest yield increase were obtained when treated with *T. koningiopsis* (Solís and Suárez, 2006).

Biological control of fruit flies and scale insects of tropical fruit

In the coastal region small, medium and large farms with fruit trees occur, but only large farms have the economic resources to apply pesticides against the attack of fruit flies and scale insects. Today, international market requirements for fruit are very demanding. The fruit flies *Anastrepha* spp. and *Ceratitis capitata* (Wiedemann) attack more than 200 species of fruit trees; and armed scales such as *Aulacaspis tubercularis* (Newstead), *Aspidiotus destructor* Signoret and *Diaspis boisduvalii* Signoret attack, among others, mango and banana. In 2005, INIAP received 1,000 parasitoids of *Diachasmimorpha longicaudata* (Ashmead) from Peru to evaluate its efficacy in controlling several fruit fly species (*C. capitata*, *Anastrepha fraterculus* (Wiedemann), *A. serpentine* (Wiedemann), *A. striata* (Schiner) and *A. obliqua* Macquart) along the Ecuadorian coast. After releases in urban areas, parasitism rates increased from 20% to 85% (Myriam Arias, 2018, personal communication). INIAP, supported by PROMSA and in partnership with the Ecuadorian Mango Foundation, introduced, reared, released and studied the control effect of *D. longicaudata* (Ashmead) in the laboratory and the field. Mass-rearing and release technologies were successful and fruit fly control in the field was encouraging, with parasitism of 85% in *Anastrepha* and 45% in *Ceratitis* (Arias *et al.*, 2009). The parasitoid *C. nipponicus* was studied for control of scales after first being imported from the USA in 2004 and mass rearing on *A. tubercularis*. Releases of 1,000 parasitoids per hectare controlled scale insects in mango, as mentioned in section 13.2.2 (Arias *et al.*, 2004).

Companies producing biological control agents

The company Agrodiagnostic, which started in 2004, estimates that its biological products based on *Trichoderma* spp. and *P. lilacinus* (Thom) Samson are now used on about 5,000 ha of banana to control nematodes and on about 1,000 ha of cacao with *Trichoderma* spp. against *Moniliophthora roreri* (Cif and Par). About 280 ha of flowers are treated with their *Trichoderma* spp. products against *Botrytis* sp., 80 ha with *B. bassiana* and *L. lecanii* against spider mites, and 10 ha with *L. lecanii* against thrips. Also 1,400 ha of mango are treated with *B. bassiana* against spider mites and with *Trichoderma* spp. against soil-dwelling pathogens and to avoid *Coletotrichum* rot in post-harvest fruit. It also covers several hectares of teak (*Tectona grandis* L.) and vegetables with *Trichoderma* sp. against *Fusarium* sp.

The company Equabiológica began the production of *T. harzianum*, *T. viridae* and *B. bassiana* in dry and granulated formulations in 2005. Currently, its products are applied on an area of 350 ha of roses, 250 ha of vegetables in open fields, 150 ha of banana, 400 ha of 'palmito' (peach palm, heart of palm) (*Bactris gasipaes* Kunth) and 300 ha of summer flowers. The company produces formulations based on *P. lilacinus*, *P. fumosoroseus* (Wize) Brown & Smith, *M. anisopliae* (Metsch.), *B. bassiana*, *Arthrobotrys oligospora* Fres., *V. lecannii* (Zimm.) and *Gliocadium* sp., among others, to deal with soil pathogens, nematodes and insects.

The company Microtech began in 2013. At present its products are applied in 1,000 ha of flowers. Its products are based on *B. bassiana*, *Streptomyces lydicus* De Boer, *P. lilacinus*, *Arthrobotrys* sp. and *T. asperellum* Samuels, Lieckf. & Nirenberg, *B. subtilis* (Ehrenberg), bacterial fermentations and protein extracts.

BioControlScience estimates that its products based on the microorganisms *T. harzianum*, *T. koningii* Oudem., *T. viride* Pers., *T. hamatum* (Bon.), *P. lilacinus*, *P. fumosoroseus* (Wise), *A. irregularis* (Matr.), *Dactylella* sp. and *M. anisopliae*, among others, were applied in 350 ha of roses, 120 ha of summer flowers, 50 ha of tropical flowers and 10,000 ha of banana in 2016.

International companies also play an important role in biological pest control. For example, Koppert Biological Systems estimated that in 2010 spider mites were controlled on 28 ha of flowers with *Neoseiulus californicus* (Mcgregor) (previously *Amblyseius*) and *Phytoseiulus persimilis* Athias-Henriot, and on more than 72 ha in 2016.

13.3.13 Governmental programmes for production of biological control agents

At a local level, the Municipality of the Metropolitan District of Quito and CONQUITO (Economy

Promotion Agency, Quito Municipality) run participatory urban agriculture projects, stimulating development of urban organic gardens cultivated under integrated pest and disease management practices. The project provides training and technical assistance in organic production of vegetables, animal husbandry and food. Since 2007, *B. thuringiensis* has been used for armyworm control. Currently, *B. bassiana*, *Azotobacter* sp. and *Metarhizium* sp. are also used in about 1,600 vegetable gardens covering 30 ha.

At the national level, the Ministry of Agriculture (MAGAP) helped to establish laboratories for the production of biocontrol agents in 13 Ecuadorian provinces in 2010, mainly to produce *Trichoderma* sp. For example, in 2016 the laboratory located in the province of Santo Domingo de los Tsachilas produced *Trichoderma* sp. for an area of 45 ha of crops such as bell pepper, passion fruit, banana and cacao. This production laboratory is managed by four farmer organizations with 15 owners and they are motivated to use *Trichoderma* sp. because of the positive effect in controlling soil pathogens on their farms. A similar laboratory is located in Azuay province, South Ecuador, which also produced *Trichoderma* sp. to be combined with 'bokashi'. Bokashi is usually made by combining vegetable and animal organic matter with soil, yeast, ashes, charcoal and molasses in certain proportions. Then this is piled up, watered and stirred two times per day during the first 10 days and once a day for 10 more days. After 20 days, it is placed next to the plants to alter soil composition. In 2016, about 20 organizations in Azuay, with an average of 20 farmers each, used *Trichoderma* sp. in crops such as fruit trees, kidney tomatoes, peppers, onions and coffee on a total of approximately 100 ha against soil pathogens. *Trichoderma* sp. is also used in seedlings against damping off. Farmers are satisfied and enthusiastic about the results obtained with *Trichoderma* sp. In Tungurahua province, *Trichoderma* sp. has been multiplied since 2015 for use on approximately five ha with broccoli, onion, potato, tamarillo (*Solanum betacea* Cav.), tomato, strawberry and blackberry against soil-dwelling plant pathogens. The laboratory of this province is managed by a corporation formed by five associations with a total of 94 farmers. Farmers are motivated to use *Trichoderma* sp. because of positive disease control results. In the southern province of Loja, a *Trichoderma* sp. laboratory has been maintained by the Tabloncillo community (made up of 20 farmers) since 2015. They use *Trichoderma* sp. on around 15–20 irrigated hectares with crops such as onion, maize, cassava and vegetables. Farmers are motivated by the results, but have not obtained extra market benefits for their products.

13.3.14 Legislation

IICA (Interamerican Institute for Agriculture Cooperation), AGROCALIDAD (Ecuadorian Agency to assure quality in Agriculture), MAGAP, INIAP (AgResearch-NZ project) and producers of biocontrol agents have been participating in workshops to develop regulations for the registration of biocontrol agents and bio-inoculants in Ecuador. Workshop participants validated national standards, regulations for registration and post-registration control of biocontrol agents, plant extracts, mineral preparations, semiochemicals and organic growth regulators for agricultural use. They also validated the operating manuals of the regulations (Viera *et al.*, 2016).

13.3.15 Area under augmentative biological control in Ecuador

Based on the figures presented in Tables 13.3 and 13.4, we estimate that in 2017 augmentative biocontrol was applied on more than 65,000 ha. This area amounts to more than 215,000 ha when the use of the biopesticide *B. thuringiensis*, which is strictly speaking not a biocontrol agent, is included. The area under conservation biocontrol is difficult to estimate, but is approximately 150,000 ha. Reliable estimates of the areas under classical biocontrol are not available, but considering that classical biocontrol is permanent and by using data from FAO (http://www.fao.org/faostat/en/#data/qc) for areas harvested in 2017 for apple (1211 ha), citrus (36,679 ha), green coffee (37,260 ha), mango (15,000 ha), rice (358,100 ha) and sugarcane (110,603 ha), exotic natural enemies might still control pests on at least 558,853 ha.

13.4 Current Situation on the Galapagos Islands

Classical biocontrol has been used once in the Galapagos islands; *Rodolia cardinalis* was released to control the highly invasive cottony cushion scale *I. purchasi*, that was threatening survival of endemic plant species (Causton *et al.*, 2017). *R. cardinalis* was released in 2002 after 6 years of rigorous studies and careful evaluation. In 1996, conservation scientists at the Charles Darwin Foundation (CDF) and Galapagos National Park Service managers formed a Technical Advisory Committee to evaluate the threats of the *I. purchasi* invasion and to develop a management strategy for control of the scale. Experimental and field studies revealed that *I. purchasi* was restricting plant growth (and possibly killing individuals) of many native and endemic plant species, including 16 IUCN Red Listed species (Causton *et al.*, 2004, 2017; Calderón Alvaréz *et al.*, 2012). This impact was not just restricted to the plant species themselves, but there were also potential or actual indirect impacts on the communities that depended on them as either refuges (Fessl *et al.*, 2010; Causton *et al.*, 2017) or for food (Roque-Albelo, 2003). The Technical Advisory Committee concluded that there was enough evidence to show that *I. purchasi* was having a significant impact on Galapagos ecosystems and that the best option for management in the long term, and with minimal risk to Galapagos species, was classical biocontrol using the coccinellid beetle *R. cardinalis*.

As this was the first time that biocontrol had been considered for use in the archipelago, there was concern that the deliberate importation of a self-dispersing alien species could have an impact on this World Heritage Protected Area (Causton, 2009). Studies in agricultural systems where *R. cardinalis* had been deliberately introduced (Quezada and DeBach, 1973) and in its native range (Prasad, 1989) suggested that the beetle was highly host specific and would be very effective at suppressing *I. purchasi* populations. Nevertheless, additional host-specificity trials were considered necessary to determine whether the introduction of this beetle would pose risks to native and endemic insects. At this time, there were no biosecurity protocols for importing biocontrol agents and our studies were based on experiences elsewhere (Causton *et al.*, 2004). These studies were conducted in a purpose-built biosecure unit at the Charles Darwin Research Station on Santa Cruz Island. In addition to testing *R. cardinalis* on non-target insects, studies were also carried out to eliminate the possibility of toxic effects of the beetles on birds that might feed on them. Ornithologists were concerned that insectivorous passerines could be at risk if they should consume *R. cardinalis*, because some species of coccinellids produce a defensive fluid that contains an alkaloid that can be toxic to some vertebrates if ingested (e.g. Pasteels *et al.*, 1973). The results of these studies and a risk analysis suggested that *R. cardinalis* would not present a significant threat to either non-target insects or insectivorous birds (Causton *et al.*, 2004, 2017; Lincango *et al.*, 2011).

Between 2002 and 2005, *R. cardinalis* adults and larvae (2,206 in total) were released on ten different islands in the archipelago. The Galapagos community was actively involved in the release programme and an intensive education and communication campaign was run to raise awareness about it. This was followed by an evaluation of the programme's success in reducing *I. purchasi* numbers. Follow-up studies were conducted directly after the release of *R. cardinalis* and involved the public (Calderon Alvarez *et al.*, 2012). Seven years later, the biocontrol programme was evaluated more fully to determine if there was successful suppression of *I. purchasi* and whether there had been any impacts on non-targets (Hoddle *et al.*, 2013). The evaluation was carried out over 2 years. Results from this evaluation and more recent surveys indicated that *R. cardinalis* had survived and spread and was present in many habitats on at least nine islands (an area of at least 400,000 ha) (Calderón Alvaréz *et al.*, 2012; Causton *et al.*, 2017). The evaluation found that overall populations were low and *I. purchasi* had been suppressed to non-damaging levels on several important plant hosts (Hoddle *et al.*, 2013). The level of control, however, was dependent on the plant species, habitat, season, and possibly on the presence of invasive ants that tend and defend the scale insect colonies. The feeding preferences and predation behavior of *R. cardinalis* were also evaluated. *R. cardinalis* did not attack non-target species, even when they were on the same branch as

I. purchasi (Hoddle *et al.*, 2013). These results suggest that this programme is unlikely to have had any significant adverse impacts on the non-target insect fauna in the Galapagos. Continued monitoring will allow us to understand the benefits of this biocontrol programme better, but in the meantime, evidence suggests that biocontrol, when conducted responsibly, can be an effective and safe tool for managing invasive species in natural ecosystems of high conservation value.

At the same time as the biocontrol programme was being developed, authorities in Galapagos were developing the first biosecurity protocols for the islands. In 1999, the Galapagos Inspection and Quarantine System (SICGAL) was inaugurated. This programme was later managed under the auspices of the Ecuadorian Service for Agricultural Health (SESA). In compliance with the objectives of the 2003 Regulation of the Total Control of Species Introduced to Galapagos (of the Special Law for Galapagos), protocols were developed to ensure that the introduction of biocontrol agents would be of minimum risk to Galapagos fauna and flora. These protocols were based on the experiences of the *R. cardinalis* biocontrol programme and followed the FAO standards for biocontrol (FAO, 1996; Zapata Erazo, 2005) and are currently under review. Since 2012, the Galapagos Biosecurity Agency has coordinated all biosecurity efforts in the archipelago.

13.5 New Developments of Biological Control in Ecuador and on the Galapagos Islands

13.5.1 Continental Ecuador

Ecuador has earned a good reputation around the world as an exporter of products such as banana, cacao, coffee, roses/summer flowers, broccoli and other vegetables. Nevertheless, to widen national and international markets, the aim of Ecuador is to offer more high-quality commodities produced using methods with low negative ecological impact. This is also in response to the increased demand for less-pesticide or pesticide-free daily-use products by farmers and consumers and strong pressure by agroecologist–social movements in Ecuador. To achieve this, more research, training and technology transfer for applied biocontrol have to be developed and executed in Ecuador. Ecuadorian producers of biocontrol agents have to guarantee quality and build confidence through tried and tested products in the field. Ecuadorian farmers need to be more receptive to adopting new or existing crop production technologies that have been shown to be successful and need to consider biocontrol as a key tool in IPM programmes. Finally, as more environmentally friendly products become available to consumers, this will increase market demands and raise awareness about food quality and the need to use products produced with a small ecological footprint.

13.5.2 Galapagos Islands

Classical biological control to manage invasive species in natural ecosystems

Classical biocontrol is now being evaluated as a tool for the management of other invasive species that are severely affecting Galapagos ecosystems: wild blackberry *Rubus niveus* Thunb. and the red quinine tree *Cinchona pubescens* Vahl, the tropical fire ant *Solenopsis geminata* (Fabricius) and the avian nest parasitic fly *Philornis downsi* Dodge and Aitken. These programmes are at different stages of development.

Several potential biocontrol agents have been found in the native range of *Philornis downsi*, a species that is significantly threatening the future of Darwin's finches and other small passerines in Galapagos (Causton *et al.*, 2013; Fessl *et al.*, 2017). The most promising is the chalcidid parasitoid wasp *Conura anullifera* (Walker) and studies are under way in the field on mainland Ecuador and in quarantine facilities at the University of Minnesota to evaluate the safety of this and other parasitoid species (Bulgarella *et al.*, 2015, 2017). Results to date suggest that *C. annulifera* is a specialist of *Philornis* species. Studies will continue to evaluate host specificity; additionally studies are being carried out to determine whether there are cues or habitat preferences that may restrict this wasp to ovipositing in *Philornis* pupae in birds' nests. This project is particularly innovative because some of the host testing is being conducted in the field in the native range of the candidate biocontrol agent.

Ant-decapitating phorid flies are being evaluated for use against the tropical fire ant *Solenopsis geminata*, an environmental and agricultural pest. This work is being undertaken by USDA-ARS (Gainesville, Florida), CDF, University of Texas, and ESPOCH. Flies from the genus *Pseudacteon* are known natural enemies of *Solenopsis geminata* (Plowes *et al.*, 2009). One species in particular, *P. bifidus*, showed promise in laboratory studies (Porter and Plowes, 2018) because it was highly host specific (Porter *et al.*, 2018), however, studies revealed that this species is too much of a specialist and not suitable for the biotype of *S. geminata* found in Galapagos: ants from Galapagos are smaller than those in the USA and flies were unable to parasitize enough ants in the laboratory to expect that they could become established in the field. Exploratory surveys are now under way (in collaboration with ESPOCH) to study fire-ant decapitating flies on mainland Ecuador, believed to be the source area of the fire ants found in Galapagos.

A fungus has been isolated from the native range of the Himalayan blackberry *Rubus niveus* and it is currently being tested for its potential as a biocontrol agent on plants of *R. niveus* from mainland Ecuador and Galapagos at the CABI greenhouses in the UK. On the other hand, dieback of the invasive quinine *Cinchona pubescens* in Galapagos prompted surveys for pathogens. Several fungus species have been identified on tree trunks, but it does not seem likely that any of these are causing the dieback or have potential as biocontrol agents. The search for promising control agents for these two species continues.

Augmentative biological control of agricultural pests

Local production of crops and fruits is a priority for Galapagos authorities to reduce reliance on imports from mainland Ecuador, decrease risks of introduced species introductions, and ensure the well-being of Galapagos residents. Searches for natural enemies of agricultural pests for use in augmentative biocontrol programmes is under way. On the Galapagos Islands, a student of the Central University of Ecuador, supported by IPADE (now the non-governmental organization (NGO) Alianza por la Solidaridad) and FUNDAR (an Ecuadorian NGO) within a support project of sustainable agricultural production in San Cristobal Island, searched for entomopathogenic fungi associated with *C. sordidus* and *Metamasius hemipterus* L., both curculionid pests in banana crops. From 28 isolates, seven were *Beauveria* sp., eight were *Paecilomyces* and one was a *Metarhizium* sp. Pathogenicity trials in the laboratory were performed and some isolates caused 100% mortality (Valverde, 2005). Currently, MAG (Ministry of Agriculture) with the Galapagos Biosecurity Agency and INIAP are investigating native entomopathogenic fungi associated with other agricultural pests including the coffee berry borer, *H. hampei*, and a coffee rust.

13.6 Acknowledgements

The following persons and organizations are thanked for providing information: J. Mendoza, CINCAE; M. Arias, ESPOL (Polytechnic University of the Coastal Region); C. Ruales and Antonio L. Reyes, USFQ (Universidad San Francisco de Quito), Microtech; W. Bilbao; K. Garcés, Agrodiagnostic; R. Felix, Florisol; G. Montero, Equabiologica; M. Chacón; B. Montero and C. Núñez, Koppert Ecuador; C. Troya, D. Narváez, E. Encalada, P. Lara and P. López, MAG; P. Garófalo, IMQ; A. Espinoza, ESPOCH; H. Rivera, Agroverde; V. Santillán, Biosagro; J. Hurtado, Empresa Estuardo Quirola Lojas; A.Sancho; R. Asipuela and J. Barba, Palmar del Río; F. Orellana, Energy & Palma; V. Bravo, ANCUPA; G. Vera, Organic Crops; J. Loor, Primobanano; L. Duicela, ANECAFE; G. Moreno, AgroArgentina; J. Mancheno, FlorMachachi; C. Falconí, BioControlScience; X. Cuesta, J. Ochoa, C. Asaquibay and F. Báez, INIAP; J. Cueva, Terrasol; G. Garzón, Frugalp; J.C. Camacho, Agroreal S.A.; and P. Landázuri, ESPE (Universidad de las Fuerzas Armadas).

On the Galapagos islands, the success of the biocontrol programme against *I. purchasi* was due to a large collaborative effort (see participants listed in Calderón Alvaréz *et al.*, 2012 and Hoddle *et al.*, 2013). Our special appreciation and thanks to the Galapagos National Park Directorate who worked with CDF on the *Rodolia* biocontrol project and who are our partners on the projects currently underway. Many thanks to G. Heimpel, S. Porter, H. Jäger, C. Ellison, and P. Couenberg for providing information and help with reviewing parts of the manuscript. This publication is contribution number 2226 of the Charles Darwin Foundation for the Galapagos Islands.

References

(References with grey shading are available as supplementary electronic material)

Aguirre, B., Barnes, C.W., Ordóñez, M.E. and Ruales, J. (2017) Food and nutrition security in Ecuador. In: *Challenges and Opportunities for Food and Nutrition Security in the Americas. The View of the Academies of Sciences*. IANAS, IAP and BMBF, México DF, pp. 316–341. [Free public access of this publication in English and Spanish at www.ianas.org]

ANCUPA (2014a) Boletín situacional, banano [Current situation concerning banana]. Ministerio de Agricultura, Ganaderia, Acuacultura y Pesca, Ecuador.

ANCUPA (2014b) Mejore las raíces de sus palmas usando *Trichoderma* [Improving roots of oil palm by using *Trichoderma*]. Boletín Técnico No. 11. Ministerio de Agricultura, Ganaderia, Acuacultura y Pesca, Ecuador.

Andrade, J. (1990) Perspectivas de control biológico en la provincia de Chimborazo (zona central) del Ecuador [Perspectives for biocontrol in Chimborazo province, central zone of Ecuador]. In: *Memorias del VI Seminario Nacional de Sanidad Vegetal*, 28–30 August 1990. Guayaquil, Ecuador, pp. 87–90.

ANECAFE (2014) Manejo Integrado de la Broca del Café [IPM for coffee berry borer]. Available at: http://www.anecafe.org.ec/noticias/manejo-integrado-de-la-broca-del-cafe-1 (accessed 8 August 2018).

Arahana, V.S., Bastidas, N., Torres, M. and González, P. (2013) Diversidad genética de una colección de hongos entomopatógenos de Ecuador utilizando un enfoque AFLP modificado [Genetic diversity of the Ecuadorian entomopathogenic fungal collection based on modified AFLP]. *ACI Avances en Ciencias e Ingenierías* 5(1), B49–B57.

Argotti, E., Hernández, C., Cazar, M., Gallegos, P., Alcazar, J. and Kaya, H. (2011) Aislamiento de nematodos entomopatógenos del género *Steinernema* y *Heterorhabditis* en sistemas de producción asociados a cultivos de *Solanum tuberosum* [Isolation of entomopathogenic nematodes of the genera *Steinernema* and *Heterorhabditis* from crop systems associated with *Solanum tuberosum*]. *ESPE Revista Ciencia y Tecnología* 3(1), 1–11.

Arias, M., Ramos, P., Jines, A. and Maldonado, E. (2004) Biología, comportamiento y uso de *Cybocephalus nipponicus* Endrody-Younga (Coleoptera: Nitidulidae) para el control biológico de la escama blanca del mango (*Aulacaspis tubercularis* Newstead) [Biology, behavior and use of *Cybocephalus nipponicus* to control white mango scale (*Aulacaspis tubercularis*)]. *Boletín técnico* No 124. EEB, INIAP, Ecuador.

Arias, M., Jines, A., and Pluas, M. (2009) Control biológico de larvas de moscas de la fruta con el parasitoide *Diachasmimorpha longicaudata* (Ashmead) Hymenoptera: Braconidae [Biocontrol of the fruit fly maggot with the parasitoid *Diachasmimorpha longicaudata*]. Leaflet Nº302, INIAP-SENACYT-MAGAP-FME-CEDEGE. Ecuador.

Ayala, O. (2006) Determinación de Agresividad de Hongos Entomopatógenos Para *Macrodactylus* sp. (Catzo del Maíz) en Chillanes, Bolívar [Assessment of the Effectiveness of Entomopathogenic Fungi to Control *Macrodactylus* sp. in Chillanes, Bolivar]. BSc thesis. Facultad de Ciencias Agrícolas, Universidad Central del Ecuador.

Báez, F. and Gallegos, P. (2013) Prospección y eficiencia de parasitoides nativos de las polillas de la papa *Tecia solanivora* Povolny, *Symmetrischema tangolias* Gyen y *Phthorimaea operculella* Zeller (Lepidoptera: Gelechiidae) en el Ecuador. [Prospecting and efficiency assessment of native parasitoids of potato tuber moths *Tecia solanivora*, *Symmetrischema tangolias* and *Phthorimaea operculella* in Ecuador]. *Memorias del V Congreso de la Papa, INIAP/CIP*. Guaranda, Ecuador, pp. 54–55.

Barragán, A., Dangles, O., Cardenas, R. and Onore, G. (2009) The history of entomology in Ecuador. *Annales de la Société Entomologique de France (N.S.)* 45(4), 410–423. DOI: 10.1080/00379271.2009.10697626.

Barriga, E. (2003) Evaluación de la Patogenicidad y Multiplicación en Sustratos de Aislamientos de *Beauveria brogniartii* y *Metarhizum anisopliae* Para el Control de *Premnotrypes vorax* en el Laboratorio y Campo. [Laboratory and Field Assessment of Pathogenicity and Multiplication of *Beauveria brogniartii* and *Metarhizum anisopliae* Isolates to Control *Premnotrypes vorax*]. MSc thesis. Facultad de Ciencias Agrícolas, Universidad Central del Ecuador.

Benzing, A. and Goetz, H. (1993) *Control Biológico de Plagas y Enfermedades de los Cultivos en Ecuador, una Recopilación de Experiencias, Estudios y Recomendaciones* [*Biocontrol of Pests and Diseases of Ecuadorian Crops, Compilation of Experiences, Studies and Recommendations*]. Escuelas radiofónicas populares del Ecuador (ERPE), Riobamba, Ecuador.

Bulgarella, M., Quiroga, M., Brito Vera, G.A., Dregni, J.S., Cunninghame, F., Mosquera, D.A., Muñoz Monje, L.D., Causton, C.E. and Heimpel, G.E. (2015) *Philornis downsi*, an avian nest parasite invasive to the Galápagos Islands, in mainland Ecuador. *Annals of the Entomological Society America* 108, 242–250.

Bulgarella, M., Quiroga, M.A., Boulton, R.A., Ramírez, I.E., Moon, R.D., Causton, C.E. and Heimpel, G.E. (2017) Life cycle and host specificity of the parasitoid *Conura annulifera* (Hymenoptera: Chalcididae), a potential biological control agent of *Philornis downsi* (Diptera: Muscidae) in the Galápagos Islands. *Annals of the Entomological Society of America* 110, 317–328. DOI: 10.1093/aesa/saw102

Calderón Alvaréz, C., Causton, C.E., Hoddle, M.S., Hoddle, C.D., Van Driesche, R. and Stanek, E.J. (2012) Monitoring the effects of *Rodolia cardinalis* on *Icerya purchasi* populations on the Galápagos Islands. *BioControl* 57, 167–179.

Cañarte, E., Valarezo, O., Navarrete, B., and Bautista, N. (2005) Control biológico del minador de la hoja de los cítricos *Phyllocnistis citrella*: estudio de caso *Ageniaspis citricola* en Ecuador [Biocontrol of the citrus leaf miner *Phyllocnistis citrella*: a case study of *Ageniaspis citricola* in Ecuador]. *Manual* No. 64. INIAP, Estación Experimental Portoviejo, Ecuador.

Causton, C.E. (2009) Success in biological control: the scale and the ladybird. In: De Roy, T. (ed.) *Galapagos: Preserving Darwin's Legacy*. Bloomsbury Publishing, London, pp. 184–190.

Causton, C.E., Lincango, M.P. and Poulsom, T.G.A. (2004) Feeding range studies of *Rodolia cardinalis* (Mulsant), candidate biological control agent of *Icerya purchasi* Maskell in the Galápagos Islands. *Biological Control* 29, 315–325.

Causton, C.E., Cunninghame, F. and Tapia, W. (2013) Management of the avian parasite *Philornis downsi* in the Galápagos Islands: a collaborative and strategic action plan. In: *Galápagos Report 2011–2012*. GNPS, GCREG, CDF and GC. Puerto Ayora, Galápagos, Ecuador, pp. 167–173.

Causton, C.E., Calderón Alvaréz, C., Hoddle, C., Hoddle, M.S., Lincango, M.P., Poulsom, T.G.A. and Van Driesche, R.G. (2017) Improving health of native Galápagos plants by introducing a specialized predator of the invasive cottony cushion scale. In: Van Driesche. R.G. and Reardon, R. (eds) *Suppressing Over-abundant Invasive Plants and Insects in Natural Areas by Use of their Specialized Natural Enemies*. FHTET, USDA Forest Service Morgantown, West Virginia, pp. 16–22.

Chacón, A. (2011) Evaluación de la Efectividad de Nematodos Entomopatógenos Para el Control Biológico del Gusano Blanco de la Papa (*Premnotrypes vorax* Hustache), (Coleoptera: Curculionidae) en San Gabriel, Carchi [Assessment of Effectiveness of Entomopathogenic Nematodes for Biocontrol of Andean Potato Weevil (*Premnotrypes vorax*), in San Gabriel, Carchi]. BSc thesis. Universidad Técnica de Babahoyo. Ecuador.

CIA (2017) The World Factbook: Ecuador. Available at: https://www.cia.gov/ library/publications/the-world-factbook/geos/ec.html (accessed 8 August 2018)

Cisneros, P. and Tandazo, A. (1989) Estudios preliminares del establecimiento del parasitoide *Prorops nasuta* W. en el sur del Ecuador [Preliminary assays for the establishment of the parasitoid *Prorops nasuta* in the South of Ecuador]. Ministerio de Agricultura y GTZ, Quito, Ecuador. *Sanidad Vegetal* 4 (4), 43–50.

Clausen, C.P. (1967) Biological control of citrus insects. In: Reuther, W. (ed.) *The Citrus Industry: Crop Protection, Postharvest Technology, and Early History of Citrus Research in California. Vol. IV: Diseases and Injuries; Viruses; Registration, Certification, Indexing; Regulatory Measures; Vertebrate Pests; Biological Control of Insects; Nematodes*. UCANR, Berkeley, California, pp. 276–317.

Delgado, A. and Játiva, P. (2010) *Políticas Institucionales de Investigación, Transferencia de Innovaciones y Prestación de Servicios Tecnológicos* [*Institutional politics for research, innovation transfer and presentation of technological services*]. Dirección General, Dirección de Planificiación y Economía Agrícola. Publicación Miscelánea No. 154. INIAP, Quito, Ecuador.

Delgado, D., Sotomayor, I., Páliz, V. and Mendoza, J. (1989) Cría, colonización y parasitismo de los entomófagos *Cephalonomia stephanoderis* Betrem y *Prorops nasuta* Waterston. [Rearing, colonization and parasitism of the entomophagous *Cephalonomia stephanoderis* and *Prorops nasuta*]. Ministerio de Agricultura y GTZ, Quito, Ecuador. *Sanidad Vegetal* 4 (4), 51–66.

Domínguez, J.M. and Zambrano, J.L. (2017) Plan Estratégico de Investigación, Desarrollo e Innovación (PE I+D+i) 2014 – 2017 del Instituto Nacional de Investigaciones Agropecuarias. [Strategic plan for research, development and innovation from 2014 to 2017]. INIAP, Quito, Ecuador.

DPNG (2014) Plan de manejo de las áreas protegidas de Galápagos para el BUEN VIVIR [Management plan for protected areas of the Galápagos for BUEN VIVER]. Dirección del Parque Nacional Galápagos, Puerto Ayora, Galapagos. Available at: http://www.galapagos.gob.ec/wp-content/uploads/downloads/2016/07/DPNG_Plan_de_Manejo_2014.pdf (accessed 28 October 2019).

ESPAC (2017) Encuesta de Superficie y Producción Agropecuaria Continua [Report about surfaces under agriculture and production]. Available at: www.ecuadorencifras.gob.ec. (accessed 19 October 2018).

FAO (1996) Code of conduct on the import and release of biological control agents. FAO, Rome. Available at: www.fao.org.

FAOSTAT (2018) Available at: http://www.fao.org/faostat/es (accessed 8 August 2018).

Fessl, B., Young, H.G., Young, R.P., Rodríguez-Matamoros, J., Dvorak, M., Tebbich, S. and Fa, J.E. (2010) How to save the rarest Darwin's finch from extinction: the mangrove finch on Isabela Island. *Philosophical Transactions Royal Society B* 365, 1019–1030.

Fessl, B., Anchundia, D., Carrión, J., Cimadom, A., Cotin, J. *et al.* (2017) Galapagos land birds (passerines, cuckoos, and doves): status, threats, and knowledge gaps. In: *Galapagos Report 2015–2016*. GNPS, GCREG, CDF and GC, Puerto Ayora, Galapagos, Ecuador, pp. 149–160.

Gallegos, P., Castillo, C., Asaquibay, C. and Oña, M. (2011) Diversidad de microorganismos y su potencial en el control biológico de plagas de cultivos de seguridad alimentaria de la sierra ecuatoriana [Diversity of microorganisms and their potential for biocontrol of pests of staple crops in the Ecuadorian Highlands]. *Memorias del VIII Simposio Internacional de Recursos Genéticos de América Latina y el Caribe, SINGEALC. 21–23 de noviembre del 2011*. INIAP, Ecuador. Publicación miscelánea No. 191. Available at: http://repositorio.iniap.gob.ec/bitstream/41000/3303/1/iniapscCD72di.pdf (accessed 8 August 2018).

Guapi, A., Gallegos, P. and Asaquibay, C. (2013) Evaluación de la eficacia del bioformulado de *Beauveria bassiana*, y tipos de aplicación para el control del gusano blanco de la papa *Premnotrypes vorax*, en dos localidades de la provincia de Chimborazo, Ecuador. [Assessment of the efficacy of a *Beauveria bassiana* formulation and application methods to control the Andean weevil *Premnotrypes vorax*, in two localities of Chimborazo province, Ecuador]. *Memorias V Congreso Ecuatoriano de la Papa*. Guaranda, Ecuador.

Guzmán, J.C. and Poma, J.E. (2015) Bioagriculture: an opportunity for island good living. In: *Galapagos Report* 2013-2014. GNPD, GCREG, CDF and GC, Puerto Ayora, Galapagos, Ecuador, pp. 25–29.

Hoddle, M.S., Crespo Ramírez, C., Hoddle, C.D., Loayza, J., Lincango, M.P., Van Driesche, R.G. and Causton, C.E. (2013) Post release evaluation of *Rodolia cardinalis* (Coleoptera: Coccinellidae) for control of *Icerya purchasi* (Hemiptera: Monophlebidae) in the Galápagos Islands. *Biological Control* 67, 262–274

INIAP (2019) Repositorio digitales del Ecuador [Local publications on biocontrol, Ecuador]. Instituto Nacional de Investigaciones Agropecuaries [National Institute for Agriculture Research], Quito, Ecuador. Available at: http://repositorio.iniap.gob.ec (accessed 31 July 2019).

Jackson, T. (2016) The New Zealand/Ecuador biocontrol project – a strategy for sustainable agriculture. International Conference of Sustainable Production of Agricultural Crops May 05, S9. *International Journal of Clinical and Biological Sciences*, 1 (Suppl. 2), July–Dec 2016.

Klein-Koch, C. (1988) Factores naturales de regulaciones y control biológico de la broca del café *Hypothenemus hampei* Ferr. [Natural factors for regulation and biocontrol of the coffee berry borer *Hypothenemus hampei*]. *Memorias del V Seminario Nacional de Sanidad Vegetal*. Cuenca, Ecuador, pp. 5–30

Klein-Koch, C. (1989) El control biológico de plagas en Ecuador [Biological control of pests in Ecuador]. Ministerio de Agricultura y GTZ, Quito, Ecuador. *Sanidad Vegetal* 4 (4), 5–20.

Landázuri, P. (2003) Prospección y evaluación de cepas de *Beauveria* spp. y *Metarhizum* sp. en las Provincias de Carchi, Cotopaxi, Chimborazo y Cañar, para el control biológico de gusano blanco de la Papa (*Premnotrypes vorax*) [Prospecting and assessment of *Beauveria* spp. and *Metarhizium* spp. isolates in Carchi, Cotopaxi, Chimborazo and Cañar Provinces for biological control of the Andean weevil (*Premnotrypes vorax*)]. MSc thesis. Instituto de Postgrado-Escuela Politécnica del Ejército. Sangolquí, Ecuador.

Lincango, P.M., Causton, C.E., Alvaréz Calderon, C. and Jiménez Uzcátegui, G. (2011) Evaluating the safety of *Rodolia cardinalis* to two species of Galápagos finch, *Camarhynchus parvulus* and *Geospiza fuliginosa*. *Biological Control* 56, 145–149.

Lundh, J.P. (2006) The farm area and cultivated plants on Santa Cruz, 1932–1965, with remarks on other parts of Galapagos. *Galapagos Research* 64, 12–25.

Mendoza, J., Quijije, R., Patiño, M., and Delgado, D. (1994) Resultados de varios estudios efectuados con *Prorops nasuta* y *Cephalonomia stephanoderis* para el control biológico de la broca del café *Hypothenemus hampei*, en Ecuador [Results of several studies of *Prorops nasuta* and *Cephalonomia stephanoderis* for the biocontrol of the coffee berry borer *Hypothenemus hampei* in Ecuador]. *Informe técnico* para los donantes Sociedad Alemana de Cooperación Técnica (GTZ) y a la Junta del Acuerdo de Cartagena (JUNAC).

Merino, G. (1984) Secuencia de la actividad entomológica en el país y apreciación de su desarrollo desde mediados de la década de 1940 [History of the entomological activity in the country and appreciation of its development since the middle of the 1940s]. In: *Memorias del Encuentro Entomológico Ecuatoriano*. Museo Ecuatoriano de Ciencias Naturales. Serie Misceláneas Núm. 3. Ed. Gráficas Robles. Quito, Ecuador, pp. 15–16.

Merino, G. and Vásquez, V. (1962) Historia y procedimiento de la implantación del control biológico del 'coma de los cítrus' *Lepidosaphes beckii* mediante la avispa *Aphytis lepidosaphes* en el Ecuador [History and methods to implementation of biocontrol of *Lepidosaphes beckii* with the parasitoid *Aphytis lepidosaphes* in Ecuador]. *Folleto informativo*. INIAP, Quito, Ecuador.

Molineros, J. (1984) Control de la escama algodonosa *Icerya purchasi* Maskell en Quito y sus alrededores, 1977–1979 [Control of the cottony cushion scale *Icerya purchasi* in Quito and its surroundings]. In: *Memorias del Encuentro Entomológico Ecuatoriano*. Museo Ecuatoriano de Ciencias Naturales. Serie Misceláneas Núm. 3. Ed. Gráficas Robles. Quito, Ecuador, pp. 29–30.

Palmar del Río (2017) Evaluación de hongos entomopatógenos e identificación de especies vegetales hospedantes de *Lincus* spp. [Assessment of entomopathogenic fungi and identification of plant hosts for *Lincus* spp.]. Available at: http://www.palmardelrio.com/sitio/files/Identificacion_de_Especies_Vegetales_Hospedantes.pdf and http://www.palmardelrio.com/sitio/files/Evaluacion_de_Hongos_Etomopatogenos.pdf (accessed 23 October 2018).

Pasteels, J.M., Deroe, C., Tursch, B. *et al*. (1973) Distribution and activities of the defensive alkaloids of the Coccinellidae. *Journal of Insect Physiology* 19, 1771–1784.

Plowes, R.M., LeBrun, E.G., Brown, B.V. and Gilbert, L.E. (2009) A review of *Pseudacteon* (Diptera: Phoridae) that parasitize ants of the *Solenopsis geminata* complex (Hymenoptera: Formicidae). *Annals of the Entomological Society of America* 102, 937–958.

Porter, S.D. and Plowes, R.M. (2018) Biology and rearing of the decapitating fly *Pseudacteon bifidus* (Diptera: Phoridae) a parasitoid of tropical fire ants. *Florida Entomologist* 101, 265–72.

Porter, S.D., Plowes, R.M. and Causton, C.E. (2018) The fire ant decapitating fly *Pseudacteon bifidus* (Diptera: Phoridae): host specificity and attraction to potential food items. *Florida Entomologist* 101, 55–60.

Prasad, Y.K. (1989) The role of natural enemies in controlling *Icerya purchasi* in South Australia. *Entomophaga* 34, 391–395.

Quezada, J.R. and DeBach, P. (1973) Bioecological and populations studies of the cottony- cushion scale, *Icerya purchasi* Mask., and its natural enemies, *Rodolia cardinalis* Mul. and *Cryptochaetum iceryae* Will. in southern California. *Hilgardia* 41, 631–688.

Roque-Albelo, L. (2003) Population decline of Galápagos endemic Lepidoptera on Volcan Alcedo (Isabela island, Galápagos Islands, Ecuador): an effect of the introduction of the cottony cushion scale? *Bulletin de l'Institut Royal des Sciences Naturelles de Belgique* 73, 1–4.

Rueda, M. (2005) Prospección de hongos entomopatógenos para *Macrodactylus pulchripes* Blanchard (Aguacuro del Maíz), y su comportamiento de ovipostura: San José de Minas, Pichincha. [Prospecting of entomopathogenic nematodes to control *Macrodactylus pulchripes* and its behavior in San Jose de Minas, Pichincha]. BSc thesis. Facultad de Ciencias Agrícolas. Universidad Central del Ecuador.

Salvador Ayala, G.M. (2015) Análisis del sistema de producción y abastecimiento de alimentos en Galápagos [Analysis of the food production and supply system in Galápagos]. MSc thesis. Latinoamerican Postgraduate University Leader in Social Sciences, Quito, Ecuador.

Solís, Z.K., Suárez, C., Vera, D., Williams, R., Carranza, I. and Flowers, W. (2001) Mass production of local strains of entomopathogenic fungi to control black weevil in plantain. In: *Eighth annual report of IPM CRSP, 2000-2001*. Integrated Pest Management, Collaborative Research Support Program, Office of International Research and Development (OIRD), Virginia Tech, Blackburg, Virginia, pp. 390–392.

Solís, H.K. and Suárez, C.C. (2006) Uso de *Trichoderma* spp. para control del complejo Moniliasis-Escoba de Bruja del cacao en Ecuador [Use of *Trichoderma* spp. to control the complex Moniliasis-cacao witches' broom in Ecuador]. Internal report. INIAP, EEP. Available at: http://repositorio.iniap.gob.ec/handle/41000/3368 (accessed 23 October 2018).

Sosa, C. (2009) Prospección de enemigos naturales del barrenador del fruto (*Neoleucinodes elegantalis*) de la naranjilla (*Solanum quitoense)* y evaluación de la incidencia de las plagas en el cultivo [Prospecting of natural enemies of the fruit borer (*Neoleucinodes elegantal*is) of naranjilla (*Solanum quitoense*) and assessment of pest incidence in the crop]. BSc thesis. Facultad de Ciencias Agrícolas. Universidad Central del Ecuador.

Suquillo, J., Rodríguez, P., Gallegos, P., Orbe, K. and Zeddam, J.L. (2012) Manual para la elaboración del bioinsecticida Bacu-Turin a través de premezclas concentradas para el control de las polillas de la papa: *Tecia solanivora*, *Phthorimaea operculella* y *Symmetrischema tangolias*. [Manual for the preparation of the biopesticide Bacu-Turin with concentrated mixes to control the Andean potato moths: *Tecia solanivora*, *Phthorimaea operculella* and *Symmetrischema tangolias*]. Manual no. 94. INIAP, Carchi, Ecuador.

Toral-Granda, M.V., Causton, C.E., Jäger, H., Trueman, M., Izurieta, J.C., Araujo, E., Cruz, M., Izurieta, A. and Garnett, S. (2017) Alien species pathways to the Galapagos Islands, Ecuador. *PLoS ONE* 12, e0184379. DOI: 10.1371/journal.pone. 0184379.

Valverde, I. (2005) Aislamiento e identificación de hongos entomopatógenos y pruebas de patogenicidad en *Cosmopolites sordidus* y *Metamasius hemipterus* en condiciones de laboratorio, San Cristóbal, Galápagos [Isolation and identification of entomopathogenic fungi and pathogenicity trials of *Cosmopolites sordidus* and *Metamasius hemipterus* in laboratory conditions]. BSc thesis. Facultad de Ciencias Agrícolas, Universidad Central del Ecuador.

Viera, W., Báez, F. and Jackson, T. (2016) Proyecto 'Biocontrol for sustainable farming systems'. Informe anual interno [Internal annual report of the project Biocontrol for Sustainable Farming Systems]. INIAP, Quito, Ecuador.

Zapata Erazo, C. (2005) *Manual de Procedimientos para Inspectores y Técnicos del Sistema de Inspección y Cuarentena para Galápagos* [*Operation Manual for Inspectors and Technicians of the Quarantine and Inspection System in Galapagos*]. Servicio Ecuatoriana de Sanidad Agropecuario (SESA), Quito, Ecuador.

14 Biological Control in El Salvador

Joop C. van Lenteren*

Laboratory of Entomology, Wageningen University, Wageningen, The Netherlands

*E-mail: joop.vanlenteren@wur.nl

Abstract

In the 1960s, natural control was shown to play an important role in the reduction of saturniid populations in *Spondias* fruit trees by dipteran and hymenopteran parasitoids and in the reduction of coconut weevil populations by a staphelinid. Later, other cases of high degrees of natural control were documented, such as that of the cotton leafworm by a native egg parasitoid and of the Mexican beetle by an ectoparasitic mite as well as by a pupal parasitoid. Classical biocontrol of citrus blackfly and purple scale in citrus was initiated in the 1970s, as well as augmentative biocontrol of lepidopteran pests with egg parasitoids. Another, very successful, augmentative biocontrol project during the 1970s concerned the killing of two mosquito species by an entomopathogenic nematode. Prospecting for nematophagous fungi in 1977 in El Salvador resulted in finding 18 species. Recently, several microbial control agents have been produced by the Escuela Agrícola Panamericana, Zamorano in Honduras for control of soil-borne diseases; nematodes and arthropod pests have been registered and are applied in El Salvador. *Tilipia* fish are used for biocontrol of *Aedes* mosquitoes, the vector of, among others, Zika, dengue and chikungunya viral diseases.

14.1 Introduction

El Salvador has an estimated population of more than 6 million (July 2017) and its agricultural activities concern coffee, sugar, maize, rice, beans, oilseed, cotton, sorghum, beef and dairy products (CIA, 2017).

14.2 History of Biological Control in El Salvador

The editors have not been able to find a contact in El Salvador who was able to write a chapter about biocontrol in this country. An extensive literature search resulted in a few interesting papers, which are summarized below.

14.2.1 Period 1880–1969

Natural control of native pests

The saturniid *Rothschildia aroma* Schaus is considered a pest of *Spondias* sp. (Anacardiacea), a fruit tree species of which several varieties are widely grown in backyards or along fences in the field in El Salvador. Quezada (1967) field-collected more than 1200 cocoons of the saturniid and found that 70% of these were parasitized: 91.3% by the tachinid *Belvosia nigrifrons* Aldrich and 6% by the tachinid *Lespesia* n. sp., 2.7% by the ichneumonid *Enicospilus americanus* Christ and a single individual of a hyperparasitoid of *B. nigrifrons*, the eulophid *Euplectrus comstockii* Howard. Another 19.5% of the cocoons died due to factors other than parasitism. Based on the fact that he never saw host plants heavily defoliated by the saturniid, Quezada (1967) concluded that the parasitoids might be responsible for a drastic reduction of the pest population.

When looking for natural enemies of the coconut weevil *Rhynchophorus palmarum* L., which, besides causing direct damage, is also a vector of red ring disease of palms, Quezada *et al.* (1969) observed that larvae and adults of the staphelinid *Xanthopygus cognatus* Sharp were eating eggs and larvae of the pest weevil. They mentioned that the staphelinid could be mass reared and prefererd to eat the coconut weevils, but could also survive on other prey.

14.2.2 Period 1970–2000

According to the BIOCAT database (Cock *et al.*, 2016; and Chapter 32, this volume), three introductions for classical biocontrol programmes have been made into El Salvador: one in the 1970s, one in the 1990s and one undated.

Classical biological control of citrus pests

Citrus blackfly *Aleurocanthus woglumi* Ashby, native to Asia, invaded El Salvador in 1965 and dispersed to most citrus orchards in the country by 1972. Native natural enemies such as the predators *Delphastus* sp. and *Chrysopa* sp. and the pathogenic fungus *Aschersonia aleyrodis* Web. were unable to control the pest sufficiently (Quezada, 1974). Therefore, the parasitoid *Encarsia* (= *Prospaltella*) *opulenta* Silvestri was introduced from Mexico in 1971 and complete biocontrol of the blackfly was obtained.

Purple scale *Lepidosaphes beckii* Compere was brought under substantial-complete biocontrol by *Aphytis lepidosaphes* Compere (Laing and Hamai, 1976) and apparently the parasitoid had itself established fortuitously (Rosen and DeBach, 1979).

Citrus snow scale *Unaspis citri* Comstock was the most severe armoured scale pest of citrus in El Salvador, but attempts to introduce the lady-beetle *Telsimia* sp. from Fiji to control it failed (Quezada *et al.*, 1973).

Natural and classical biological control of pests in cotton, maize and bean

In cotton, high percentages of the eggs of cotton leafworm *Alabama argillaceae* Hubner were parasitized by the native egg parasitoid *Trichogramma semifumatum* Perkins (Catareda *et al.*, 1976).

In 1977, the Ministry of Agriculture's National Center for Agricultural Technology (Centro Nacional de Tecnologia Agropecuaria) (CENTA) imported the egg parasitoid *Telenomus remus* Nixon from Trinidad for classical biocontrol of the fall armyworm *Spodoptera frugiperda* (Smith) in maize and other crops, but results are unknown (Quezada, 1989).

The Mexican bean beetle *Epilachna varivestris* Mulsant (Coleoptera: Coccinellidae) is a common pest in lima bean *Phaseolus lunatus* Linnaeus in El Savador. During field observations, up to 75% of the beetle pupae appeared to be parasitized by *Tetrastichus* sp. Beetles were also often infected by the ectoparasitic mite *Coccipolipus epilachnae* Smiley, resulting in the beetle's drastically reduced oviposition and increased mortality (MAD, 1984).

Augmentative biological control of lepidopteran pests

Smith and Bellotti (1996) mentioned that Colombian laboratories were producing *Trichogramma* spp. as well as one of its hosts, eggs of *Sitotroga cerealella* Oliver, for exportation to El Salvador and other countries for control of lepidopteran pests.

Augmentative biological control of mosquitoes

Petersen *et al.* (1978) reported about biocontrol of two mosquito species, *Anopheles albimanus* C.R.G. Wiedemann and *An. pseudopunctipennis* Theobald, in lakes in El Salvador by treating the water with the entomopathogenic nematode *Romanomermis culicivorax* Ross & Smith. This resulted in high rates of mosquito infection and the *Anopheles* populations had decreased by 94% at the end of the release period.

Classical biological control of weeds

Two species of curculionids, *Neochetina bruchi* Hustache and *Neochetina eichhorniae* Warner, were imported into Honduras by Escuela Agrícola Panamericana, Zamorano (see Chapter 19: Honduras) from Florida in 1990 for control of *Eichhornia crassipes* (Martius) Solms-Laubac, a water weed originating from the Amazon. Later, curculionids were field collected and released in several other Central American countries. They were introduced into El Salvador in 1994, but results of this introduction are unknown (Cave *et al.*, 2011).

Nematophagous fungi present in El Salvador

As a result of sampling at 38 locations in 1977 all over El Salvador, Búcaro (1983) found 18 species of nematophagous fungi, 11 species were predators and seven were parasites. *Stylopage hadra* Drechsler was the dominant species and might be an interesting candidate for biocontrol of nematodes, due to its aggressiveness. It was established for the first time that *Helicosporina veronae* Rambeli was able to kill nematodes.

14.3 Current Situation of Biological Control in El Salvador

14.3.1 Microbial control of pests and diseases

In Honduras, the Escuela Agrícola Panamericana, Zamorano, started a large project on production of microbial agents for pest and disease control (see Chapter 19: Honduras). One of the spin-offs of this project was the introduction into El Salvador and registration of several microbial agents, such as: (i) the fungi *Trichoderma harzianum* Rifai for control of soil diseases and *Beauveria bassiana* (Bals.-Criv.) Vuill. for control of lepidopteran and

Table 14.1. Overview of major biocontrol activities in El Salvador.

Biocontrol agent / exotic (ex), native (na)	Pest / crop	Type of biocontrol[a] / since	Area under biocontrol[b]	Reference
Belvosia nigrifrons / na	Saturniid *Rothschildia aroma*, fruit trees	NC	Control / ?	Quezada, 1967
Lespesia n. sp / na		NC	Control / ?	
Enicospilus americanus / na		NC	Control / ?	
Xanthopygus cognatus / na	Coconut weevil, coconut	NC	Partial control / 250 ha	Quezada *et al.*, 1969
Delphastus sp. / na	Citrus blackfly, citrus	NC	Insufficient control / ?	Quezada, 1974
Chrysopa sp. / na		NC	insufficient control / ?	
Aschersonia aleyrodis / na		NC	Insufficient control / ?	
Encarsia opulenta / ex		CBC / 1971	Complete control / 1,500 ha	
Aphytis lepidosaphes / ex	Purple scale, citrus	FBC / 1975s	Substantial control / 1,500 ha	Laing and Hamai ,1976
Telsimia sp. / ex	Citrus snow scale, citrus	CBC / 1970s	No control / not established	Quezada *et al.*, 1973
Trichogramma semifumatum / na	Cotton leafworm, cotton	NC	Partial control / 325 ha	Catareda *et al.*, 1976
Telenomus remus / ex	Fall armyworm, corn	CBC / 1977	? / ?	Quezada 1989
Tetrastichus sp. / na	Mexican bean beetle, lima bean	NC	Substantial control / 180 ha	MAD, 1984
Coccipolipus epilachnae / na		NC	Substantial control / 180 ha	
Romanomermis culicivorax / na	Mosquitoes, lakes	ABC	Substantial control / ?	Petersen *et al.*, 1978
Neochetina bruchi / ex	*Eichhornia crassipes* weed, lakes	CBC / 1994	? / ?	Cave *et al.*, 2011
N. eichhorniae / ex		CBC / 1994	? / ?	
Trichoderma harzianum / ex	Soil diseases various crops	ABC / 2008	Control / ?	Cave *et al.*, 2011
Beauveria bassiana / ex	Lepidopterans and coleopterans, various crops	ABC / 2008	Control / ?	
Metarhizium anisopliae / ex	Leafhoppers, cotton	ABC / 2008	Control / ?	
Purpureocillium lilacinum / ex	Nematodes, horticulture and fruit	ABC / 2008	Control / ?	
Tilapia sp. / ex	Mosquitoes, lakes	ABC / 2018	? / ?	PLANUSA 2018

[a]Type of biocontrol: ABC = augmentative biocontrol, CBC = classical biocontrol, ConsBC = conservation biocontrol, FBC = fortuitous biocontrol; NC = natural control
[b]Area of crop harvested in 2016 according to FAO (http://www.fao.org/faostat/en/#data/qc)

coleopteran pests in various crops in 2008; (ii) *Metarhizium anisopliae* (Metchnikoff) Sorokin, among others, for leafhopper control in sugarcane; and (iii) *Purpureocillium lilacinum* (Thom) Luangsaard, Houbraken, Hywel-Jones and Samson for nematode control in horticulture and fruit (Cave *et al.*, 2011)

14.3.2 Biological control of mosquitoes

Plan International USA (PLANUSA) is a non-governmental organization (NGO) working in more than 50 developing countries to end the cycle of poverty. In El Salvador they use small tilapia fish to control *Aedes aegyti* L. mosquitoes that spread viral diseases like Zika, dengue and chikungunya, by introducing the fish into water barrels, ponds and drinking-water tanks in homes.

In conclusion, natural enemies that were introduced, released and did establish in previous periods for classical biocontrol of pests in El Salvador are still supposed to be present, and particular successes in citrus have been obtained. Recently, microbial control agents are being used in various crops. Very limited quantitative data are available about areas under control (Table 14.1). Classical biocontrol is used on about 1,500 ha of citrus and natural control on more than 750 ha.

References

(References with grey shading are available as supplementary electronic material)

Búcaro, R.D. (1983) Hongos nematófagos de El Salvador [Nematophagous fungi of El Salvador]. *Revista de Biología Tropical* 31, 25–28.

Catareda, S.L., Mancia, J.E. and Quezada, J.R. (1976) *Trichogramma semifumatum* (Perkins) una especie nativa de El Salvador parásito de *Alabama argillacea* Hubner [*Trichogramma semifumatum*, a native parasitoid of *Alabama argillacea* in El Salvador]. *SIADES Comunicaciones Cientificas El Salvador* 8, 94.

Cave, R.D., Trabanino, R. and Pitty, A. (2011) Zamorano y sus contribuciones a la agricultura sostenible a través del control biológico de plagas [Zamorano and its contributions to sustainable agriculture through biological control of pests]. *Ceiba* 52, 26–38. DOI: 10.5377/ceiba.v52i1.966

CIA (2017) The World Factbook: El Salvador. Available at: https://www.cia.gov/library/publications/the-world-factbook/geos/es.html (accessed 22 August 2018).

Cock, M.J.W. (ed.) (1985) *A Review of Biological Control of Pests in the Commonwealth Caribbean and Bermuda up to 1982*. Technical Communication No. 9, Commonwealth Institute of Biological Control. Commonwealth Agricultural Bureaux, Farnham Royal, UK.

Cock, M.J.W., Murphy, S.T., Kairo, M.T.K., Thompson, E., Murphy, R.J. and Francis, A.W. (2016) Trends in the classical biological control of insect pests by insects: an update of the BIOCAT database. *Biocontrol* 61, 349–363. DOI: 10.1007/s10526-016-9726-3

Laing, J.E. and Hamai, J. (1976) Biological control of insect pests and weeds by imported parasites, predators and pathogens. In: Huffaker, C.B. and Messenger, P.S. (eds) *Theory and Practice of Biological Control*. Academic Press, New York, pp. 685–743.

MAD (Medio Ambiente y Desarrollo) (1984) Enemigos naturales de *Epilachna varivestris* Mulsant [Natural enemies of *Epilachna varivestris*]. *Boletín Informativo, El Salvador* 28, 1–12.

Petersen, J.J., Chapman, H.C., Willis, O.R. and Fukuda, T. (1978) Release of *Romanomermis culicivorax* for the control of *Anopheles albimanus* in El Salvador. II. Application of the nematode. *American Journal of Tropical and Medical Hygiene* 27, 1268–1273.

PLANUSA (Plan International USA) (2018) What we do. Available at: https://www.planusa.org/what-we-do (accessed 22 August 2018).

Quezada, J.R. (1967) Notes on the biology of *Rothschildia ?aroma* (Lepidotera: Saturniidae), with special reference to its control by pupal parasites in El Salvador. *Annals of the Entomological Society of America* 60, 595–599.

Quezada, J.R. (1974) Biological control of *Aleurocanthus woglumi* (Homoptera: Aleyrodidae) in El Salvador. *Entomophaga* 19, 243–254.

Quezada, J.R. (1989) Utilización del control biológico clássico [Use of classical biological control]. In: Andrews, K.L. and Quezada, J.R. (eds) *Manejo Integrado de Plagas Insectiles en la Agricultura: Estado Actual y Future* [*Integrated Pest Management of Insects in Agriculture: Current State and Future*]. Escuela Agrícola Panamericana, Departamento de Protección Vegetal, El Zamorano, Honduras, pp. 16–31.

Quezada, J.R., Amaya, C.C. and Herman Jr., L.H. (1969) *Xanthopygus cognatus* Sharp (Coleoptera: Staphylinidae), an enemy of the coconut weevil, *Rhynchophorus palmarum* L. (Coleoptera: Curculionidae) in El Salvador. *Journal of the New York Entomological Society* 77, 264–269.

Quezada, J.R., Cornejo, C., De Mira, A.D. and Hidalgo, F. (1973) *Principales Especies de Insectos Asociados a los Cultivos de Citricos en El Salvador* [*Main Species of Insects Associated with Citrus Culture in El Salvador*]. Ministry of Agriculture, San Salvador, El Salvador.

Rosen, D. and DeBach, P. (1979) *Species of Aphytis of the World (Hymenoptera: Aphelinidae)*. Junk Publishers, The Hague, Netherlands.

Smith, L. and Bellotti, A.C. (1996) Successful biocontrol projects with emphasis on the Neotropics. In: *Proceedings of the Cornell Community Conference on Biological Control, April 11–13, Cornell University, USA*. Available at: http://web.entomology.cornell.edu/shelton/cornell-biocontrol-conf/talks/bellotti.html (accessed 22 August 2018).

15 Biological Control in French Guiana, Guadeloupe and Martinique

Philippe Ryckewaert[1]* and Jean-François Vayssières[2]

[1]CIRAD, UR Hortsys, Campus Agro-environnemental Caraïbes, Le Lamentin, Martinique (French West Indies); [2]CIRAD, UR Hortsys, Campus International de Baillarguet, 34398–Montpellier, France

* E-mail: philippe.ryckewaert@cirad.fr

Abstract

Several biological control agents have been introduced successfully in French Guiana, Guadeloupe and Martinique: three tachinid dipterans and one hymenopteran for control of sugarcane borers, a ladybird and a hymenopteran parasitoid against the pink hibiscus mealybug, a hymenopteran parasitoid to control Asian citrus psyllid, another hymenopteran parasitoid against citrus blackfly and a hymenopteran parasitoid for control of fruit flies. Mass rearings of a lacewing and a *Trichogramma* egg parasitoid are being implemented in Martinique for augmentative biocontrol. Use of native natural enemies in conservation biocontrol projects is being initiated in several crops, after a period of intensive prospecting for natural enemies. A project started recently in French Guiana aims to control the mango mealybug by introducing two exotic parasitoids.

15.1 Introduction

French Guiana has an estimated population of about 290,000 (UN, 2018) and its main agricultural products are rice, vegetables, perennial fruit trees, pineapple, manioc, sugarcane, cocoa, bananas, flowers, cattle, pigs, poultry, goats, shrimps and forestry products (Wikipedia, 2019a).

Guadeloupe has an estimated population of almost 450,000 (UN, 2018) and its main agricultural products are sugarcane, bananas, fruits, vegetables, pineapple, root crops, coffee, flowers, cattle, goats, pigs, poultry and fish (Wikipedia, 2019b).

Martinique has an estimated population of slightly more than 385,000 (UN, 2018) and its main agricultural products are bananas, sugarcane, pineapples, avocados, vegetables, root crops, flowers, cattle, goats, pigs and poultry (Wikipedia, 2019c).

Other islands belonging to the French Antilles (Saint Martin, Saint Barthelemy, Desirade and Les Saintes) have hardly any agriculture; data for the Marie Galante island are included in the sections about Guadeloupe.

15.2 History of Biological Control in French Guiana, Guadeloupe and Martinique

15.2.1 Period 1800–1969

Use of giant toad

Prior to 1850, the giant toad *Bufo marinus* (L.) was introduced from French Guiana into Martinique to kill rats, *Rattus rattus* (L.), one of the principal pests of sugarcane. It is doubtful whether the toads had any effect upon the rat populations (Cock, 1985).

Use of introduced parasitoids against sugarcane borers in Martinique and Guadeloupe

The sugarcane borers *Diatraea saccharalis* L., *D. impersonatella* Walker and *D. centrella* (Möschler) have long caused extensive damage to the sugarcane crop in the West Indies (Stehlé, 1956). In 1938 the tachinid *Lydella* (= *Metagonistylum*) *minense* Townsend was introduced. Then, in 1947, another tachinid, *Lixophaga diatraeae* Townsend, was imported and finally *Paratheresia claripalpis* (Van der Wulp) was introduced in 1954 (Cochereau, 1990). In 1970 in Guadeloupe and in 1976 in Martinique, the hymenopteran parasitoid *Cotesia flavipes* Cameron was introduced from Barbados. In 1986, the rate of cane infestation by borers was less than 6%, with no significant economic consequences (Boulet, 1986). Today, biocontrol of sugarcane borers is considered very satisfactory on these islands, especially as no insecticide treatment is applied.

15.2.2 Period 1970–2000

Classical biological control of the pink hibiscus mealybug in Martinique and Guadeloupe

The pink hibiscus mealybug *Maconellicoccus hirsutus* Green was accidentally introduced into the island of Grenada in 1994 and mainly attacked ornamental plants (Kairo *et al.*, 2000). It then invaded the northern Caribbean, including Martinique and Guadeloupe in 1998 (Etienne *et al.*, 1998b). Soon, research was conducted to introduce natural enemies against this pest and two species were introduced into Guadeloupe and Martinique: the ladybird *Cryptolaemus montrouzieri* Mulsant and the parasitoid *Anagyrus kamali* Mursi. Mealybug populations have declined rapidly following the release of these two natural enemies in all countries where this biocontrol has been implemented (Kairo *et al.*, 2000). Today, this mealybug has become very rare in Martinique and Guadeloupe.

Classical biological control of the Asian citrus psyllid in Martinique and Guadeloupe

The Asian citrus psyllid *Diaphorina citri* Kuwayama is one of the vectors of the most serious citrus disease: huanglongbing (HLB), caused by the bacterium *Candidatus liberibacter asiaticus*. The psyllid was detected in Guadeloupe in 1998 (Etienne *et al.*, 1998a) and in Martinique in 2012 (Cellier *et al.*, 2014). In 1999, its main parasitoid *Tamarixia radiata* (Waterston) was introduced into Guadeloupe from a population of Réunion (Indian Ocean) and quickly dispersed over the island (Etienne *et al.*, 2001). Monitoring of parasitism on Guadeloupe was done in 2014 in several orchards and its parasitism levels varied from 40% to 70% (unpublished data). In Martinique, *T. radiata* was found shortly after the discovery of the psyllid, probably introduced on citrus plants imported from Guadeloupe. Only low densities of the psyllid were observed in Martinique, probably due to effective parasitism by *T. radiata* in orchards, which are not treated with insecticides today. Parasitism of *D. citri* on another host plant, *Murraya paniculata* (Rutaceae), sometimes exceeded 90% (unpublished data).

Classical biological control of the citrus blackfly in French Guiana

The citrus blackfly *Aleurocanthus woglumi* Ashby was a major pest of citrus fruit in the 1990s in French Guiana and chemical control proved ineffective. A pest-specific parasitoid, *Encarsia opulenta* (Silvestri), was introduced from Florida and adapted well locally (Janelle *et al.*, 2000). The orchards where the parasitoid was released soon showed good rates of parasitism, but it was found necessary to introduce parasitoids into each orchard, because distances between orchards are large and they are separated by the Amazonian forest. During the past 10 years, this whitefly has no longer been found in citrus orchards (C. Gourmel, French Guiana, 2018, personal communication).

Classical biological control of the carambola fruit fly in French Guiana

Although biocontrol was not included as an element of the eradication programme carried out on the carambola fruit fly *Bactrocera carambolae* Drew & Hancock, some biocontrol activities were implemented in French Guiana along the border with Brazil in collaboration with Embrapa (Empresa Brasileira de Pesquisa Agropecuária, Brazil). At the end of 2000, *Diachasmimorpha longicaudata* (Ashmead) was released along both sides of the Oyapock River (= border), from Taparabu to Clevelandia, including St Georges (Vayssières *et al.*, 2013). About 2 million *Ceratitis capitata* Wied. pupae parasitized by the braconid *D. longicaudata* were transported by plane from the Centro de Energia Nuclear na Agricultura (CENA) laboratory in Piracicaba (Brazil). Between 2001 and 2003, emergence of *D. longicaudata* was regularly recorded from parasitized *B. carambolae* and also from *Anastrepha* spp. in fruit sampled from along the French side of the river Oyapock and in the areas of St Georges and Regina, so the parasitoid had become well established after its release in 2000 (Vayssières *et al.*, 2013). Future biocontrol activities against the fruit fly include the introduction of other braconid parasitoids into French Guiana, such as *Fopius arisanus* (Sonan).

French Guiana, Guadeloupe and Martinique as providers of natural enemies

For control of the coffee leaf miner *Perileucoptera coffeella* Guérin-Meneville, a local parasitic braconid, *Mirax insularis* Muesebeck, was introduced into Puerto Rico from Guadeloupe. Although it initially became established, it had negligible effects on the miner and may have died out later (Cock, 1985).

15.3 Current Situation of Biological Control in French Guiana, Guadeloupe and Martinique

15.3.1 Introduction

At present, no new classical or augmentative controls are applied in these territories, but projects are under way. It is mainly conservation biocontrol that is used in crops, while limiting the use of chemicals as much as possible. However, the effectiveness of conservation biocontrol varies greatly depending on the situation, the environment, the time of year and the target crop. For example, citrus orchards today hardly require insecticides or miticides, or can be limited to spot treatments. On the other hand, in the absence of treatments, cucurbit crops are usually damaged by the melonworm *Diaphania hyalinata* (L.) and pickleworm *D. nitidalis* (Stoll), because these caterpillars have very few natural enemies.

15.3.2 Augmentative biological control

Since 2017, Martinique's Regional Federation of Protection Against Damaging Organisms (Fédération Régionale de Défense contre les Organismes Nuisibles) (FREDON) has undertaken mass rearing of two beneficials: (i) a polyphagous predator, the lacewing *Chrysoperla externa* (Hagen); and (ii) a parasitoid of lepidopteran eggs, *Trichogramma pretiosum* Riley. These natural enemies are particularly intended for releases in vegetable crops.

15.3.3 Conservation biological control

Inventories of natural enemies of crop pests have been made for individual crops or for several related crops. General inventories have been made for vegetables (Ryckewaert and Rhino, 2017) and fruit crops (Leblanc, 2000). Other inventories concern particular pests such as the whitefly *Bemisia tabaci* Gennadius (Ryckewaert and Alauzet, 2002; Pavis *et al.*, 2003), the citrus weevil *Diaprepes abbreviatus* L. (Etienne and Delvare, 1991) or a specific group of natural enemies such as ladybirds (Lucas, 2012; Nicolas, 2012), thrips (Etienne *et al.*, 2015), the genus *Coccophagus* (Panis, 2013) or predatory mites (Kreiter and de Moraes, 1997; de Moraes *et al.*, 1999; Kreiter *et al.*, 2013; Kreiter *et al.*, 2018). However, and particularly in relation to conservation biocontrol, some groups of arthropods have been poorly studied, such as spiders and predators present on the soil. Tables 15.1 and 15.2 list the main species and genera found on Martinique and Guadeloupe. There is little knowledge of French Guiana, but many species mentioned in these tables are present there as well (Gourmel, 2014).

15.4 Conclusions

Biocontrol brings many benefits to agriculture in Martinique, Guadeloupe and French Guiana, while avoiding numerous chemical treatments. However, current biocontrol in these territories is not always sufficient and might be improved by importation of new exotic species or by augmentative releases of natural enemies that are already present. However, the profitability of mass releases is not always obvious, while the introduction of exotic species is subject to very strict regulations in relation to environmental risks. Table 15.3 provides a summary of the biocontrol projects in French Guiana, Guadeloupe and Martinique. Based on the areas with a certain crop and natural enemies used (Table 15.3), we estimate that at least 20,000 ha are under classical biocontrol. In the near future, augmentative biocontrol may also be applied.

15.5 New Developments of Biological Control in French Guiana, Guadeloupe and Martinique

15.5.1 Augmentative biological control with *Tamarixia radiata* in Guadeloupe

Parasitism rates of the psyllid *D. citri* are often insufficient after the planting of young citrus plants. Therefore, a project is underway at FREDON Guadeloupe to start a mass rearing of the parasitoid *T. radiata* with as host plant the orange jasmine *M. paniculata*.

15.5.2 Classical biological control of the mango mealybug in French Guiana

The mango mealybug *Rastrococcus invadens* Williams, native to Asia, was discovered in French Guiana in 2014 (Germain *et al.*, 2015). It attacks

Table 15.1. Predators cited from Martinique and Guadeloupe (retrieved from de Moraes *et al.*, 1999; Gourmel, 2014; Kreiter and de Moraes, 1997; Kreiter *et al.*, 2013, 2018; Lucas, 2012; Nicolas, 2012; Ryckewaert and Rhino, 2017).

Predators	Main prey species
Predatory mites	
Many species, mainly from the Phytoseiidae family	Mites
Spiders	
Theridiidae, Araneidae, Thomisidae, Salticidae, Tetragnathidae, Oxypidae	Polyphagous
Predatory bugs	
Orius insidiosus Say. *O. pumilio* (Champion)	Thrips, aphids, whiteflies, mites
Macrolophus nr praeclarus (Distant), *Nesidiocoris tenuis* Reuter, *Cyrtopeltis* sp.	Aphids, whiteflies, larvae, caterpillars
Zelus longipes (L.), *Nabis capsiformis* Germar	Polyphagous
Lacewings	
Chrysoperla externa (Hagen), *Ceraeochrysa cubana* (Hagen), *Leucochrysa floridana* (Banks), *Chrysopa, Chrysocerca, Chrysopodes*	Aphids, psyllids, caterpillars
Hoverflies	
Pseudodorus clavatus (F.), *Syrphus*, *Allograpta*, *Ocyptamus*, *Toxomerus*, *Baccha*	Aphids
Ladybirds (main species)	
Coleomegilla (= *Megilla*) *maculata* (De Geer)	Aphids, caterpillars, worms
Cycloneda sanguínea (L.)	Aphids, psyllids
Coccinella septempunctata L., *Coelophora inaequalis* F.	Aphids
Zagreus (= *Exochomus*) *bimaculosus* Mulsant	Scales
Chilocorus nigritus (F.), *Chilocorus cacti* (L.)	Scales
Cladis nitidula F.	Aphids, scales
Delphastus pusillus Le Conte, *D. pallidus* Le Conte	Whiteflies
Cryptolaemus montrouzieri Mulsant	Mealybugs
Rodolia cardinalis Mulsant	*Icerya purchasi*
Ants (main species)	
Solenopsis geminata (Fabricius), *Pheidole fallax* Mayr, *Wasmannia rochai* Forel, *Nylanderia fulva* (Mayr), *Odontomachus brunneus* (Patton), *Camponotus sexguttatus* (Fabricius)	Polyphagous
Wasps	
Polistes spp.	Caterpillars, larvae
Predatory thrips	
Franklinothrips vespiformis (Crawford)	Thrips, mites
Carabids	
Species not determined	Polyphagous
Staphylinids	
Oligota sp.	Mites

at least 26 fruit (mango, citrus, bananas, etc.) and ornamental plant species, and this invasive pest may eventually invade neighbouring countries and spread over the Caribbean (Vayssières, 2017). Without effective biocontrol agents the mango mealybug population is increasing in size every year in all of French Guiana. Thirty years ago this was well controlled in West Africa after the introduction of *Gyranusoidea tebygi* Noyes and *Anagyrus mangicola* Noyes and often recorded in the field (Neuenschwander *et al.*, 1994; Bokonon-Ganta *et al.*, 2002; Neuenschwander, 2003). A donor-supported biocontrol project in French Guiana will be based on the introduction

Table 15.2. Parasitoids cited from Martinique and Guadeloupe (retrieved from Boulet, 1986; Etienne and Delvare, 1991; Gourmel, 2014; Janelle *et al*., 2000; Kairo *et al*., 2000; Leblanc, 2000; Panis, 2013; Pavis *et al*., 2003; Ryckewaert and Rhino, 2017; Stehlé, 1956; Vayssières *et al*., 2013).

Parasitoids	Hosts
Encarsia nigricephala (Dozier), *E. sophia* (= *transvena*) (Girault & Dodd), *E. luteola* (Howard) / *E. formosa* Gahan, *E. hispida* De Santis, *E. meritoria* Gahan, *E. tabacivora* (= *pergandiella*) Viggiani	*Bemisia tabaci* (Gennadius) and / or *Trialeurodes vaporariorum* Westwood
Eretmocerus tejanus Rose & Zolnerow, *Amitus bennetti* Viggiani & Evans, *A. fuscipennis* McGown & Nebeker, *Signiphora* sp.	
Encarsia cubensis Gahan	*Aleurotrachelus trachoides* Back
E. sophia (Girault & Dodd)	*Aleyrodes proletella* (L.)
Encarsia dispersa Polazeck, *E. guadeloupae* Viggiani, *Encarsiella noyesi* Hayat, *Aleuroctonus vittatus* (Dozier)	*Aleurodicus dispersus* Russell
Encarsia basicincta (Gahan), *E. nigricephala* (Dozier) *Eretmocerus portoricensis* (Dozier)	*Aleurothrixus floccosus* (Maskell)
Encarsia opulenta (Silvestri)	*Aleurocanthus woglumi*(Ashby)
Aphelinus gossypii Timberlake, *Diaeretiella rapae* (McIntosh), *Aphidius colemani* Viereck, *Lysiphlebus testaceipes* (Cresson), *Syrphophagus aphidivorus* (Mayr), *Pachyneuron aphidis* (Bouché)	Aphids
Aphytis sp., *Encarsia lounsburyi* (Berlèse & Paoli), *Coccophagus pulvinariae* Compere, *C. basalis* Compere, *Aprostocetus* sp., *Anagyrus kamali* Moursi, *Gyranusoidea* sp.	Scales, mealybugs
Cotesia (= *Apanteles*) *plutellae* (Kurdj.), *Conura hirtifemora* (Ashmead), *Oomyzus sokolowski* (Kurdjumov), *Trichogramma chilonis* Ishii	*Plutella xylostella* (L.)
Ageniaspis citrícola Logvinovskaya, *Galeopsomyia fausta* LaSalle & Pena, *Horismenus* spp., *Cirrospilus* sp., *Elasmus* sp., *Zagrammosoma* sp.	*Phyllocnistis citrella* Stainton
Cotesia sp., *Apanteles* sp., *Pseudapanteles* sp., *Trichogramma pretiosum* Riley	*Diaphania* spp.
Cotesia flavipes Cameron, *T. cretiosum, Lydella minense* Townsend, *Lixophaga diatraeae* Townsend, *Paratheresia claripalpis* (Van der Wulp)	*Diatraea saccharalis* (F.) *D. impersonatella* (Walker) *D. centrella* (Moschl)
Copidosoma floridanum (Ashmead), *Euplectrus* sp., *Telenomus remus* Nixon, *Trichogramma nubilale* Ertie & Davis	Noctuids (*Spodoptera, Helicoverpa*…)
Telenomus sp.	*Manduca sexta* (L.)
Pteromalus puparum (Linné), *Brachymeria* sp.	*Ascia monuste* (L.)
Dacnusa sp., *Opius* sp., *Chrysocharis caribea* Boucek, *Ch. vovones* (Walker), *Closterocerus purpureus* (Howard), *Diaulinopsis callichroma* Crawford, *Diglyphus begini* (Asmead), *Halticoptera circulus* (Walker)	*Liriomyza* spp.
Zaeucoila sp.	*Amauromyza maculosa* (Malloch)
Aprostocetus gala (Walker), *A. haitiensis* (Gahan), *Aprostocetus* sp., *Baryscapus fennahi* (Schauff), *Ceratogramma etiennei* Delvare	*Diaprepes abbreviatus* L.
Goetheana parvipennis (Gahan), *Thripastichus gentilei* (Del Guercio), *Megaphragma* sp., *Cerasinus sp.*	Thrips

Table 15.3. Overview of biocontrol activities in French Guiana, Guadeloupe and Martinique.

Biocontrol agent / exotic (ex), native (na)	Pest / crop	Type of biocontrol[a] / since	Effect /area (ha) under biocontrol	Reference
French Guiana				
Encarsia opulenta / ex	Citrus blackfly / citrus	CBC / 1995s	Control / established / 1,650[1]	Janelle *et al.*, 2000
Diachasmimorpha longicaudata / ex	Carambola fruit fly / various fruit	CBC / 2000	Partial control / established	Vayssières *et al.*, 2013
Fopius arisanus / ex		CBC / testing	Testing	
Gyranusoidea tebygi / ex	Mango mealybug / mango and various fruit	CBC / testing	Testing	Vayssières, 2017
Anagyrus mangicola / ex		CBC / testing	Testing	
Guadeloupe				
Lydella minense / ex	Sugarcane borers / sugarcane	CBC / 1938	Partial control / established	Cochereau, 1990
Lixophaga diatraeae / ex		CBC / 1947	Partial control / established	
Paratheresia claripalpis / ex		CBC / 1954	Partial control / established	
Cotesia flavipes / ex		CBC / 1970	Control / established / 14,173[2]	Boulet, 1986
Anagyrus kamali / ex	Pink hibiscus mealybug / ornamentals	CBC / 1999	Control / established	Kairo *et al.*, 2000
Cryptolaemus montrouzieri / ex		CBC / 1999	Control / established	
Tamarixia radiata / ex	Asian citrus psyllid / citrus	CBC / 1999	Control / established 694[3]	Etienne *et al.* 2001
Tamarixia radiata / ex		ABC / 2018	Boost biocontrol on young citrus	Ryckewaert (pers. com.)
Martinique				
Bufo marinus / ex	Rats / sugarcane	CBC / 1850	No control / established	Cock, 1985
Lydella minense / ex	Sugarcane borers / sugarcane	CBC / 1938	Partial control / established	Cochereau, 1990
Lixophaga diatraeae / ex		CBC / 1947	Partial control / established	
Paratheresia claripalpis / ex		CBC / 1954	Partial control / established	
Cotesia flavipes / ex		CBC / 1976	Control / established / 4,046[4]	Boulet, 1986
Anagyrus kamali / ex	Pink hibiscus mealybug / ornamentals	CBC / 1999	Control / established	Kairo *et al.*, 2000
Cryptolaemus montrouzieri / ex		CBC / 1999	Control / established	
Tamarixia radiata / ex	Asian citrus psyllid / citrus	CBC / 2012	Control / established / 440[4]	Ryckewaert (pers. com.)
Chrysoperla externa / na	Pests in vegetables	ABC / 2017	Testing phase	Ryckewaert (pers. com.)
Trichogramma pretiosum / na		ABC / 2017	Testing phase	

[a]Type of biocontrol: ABC = augmentative, CBC = classical
[1]DAAF/INSEE, 2018
[2]DAAF, 2018a
[3]FAO (http://www.fao.org/faostat/en/#data/qc)
[4]DAAF, 2018b

of these exotic parasitoids in controlled conditions at first, in order to test their behaviour on locally available potential hosts. Field release will then be considered as a second step, together with monitoring and efficiency records.

15.6 Acknowledgements

We thank our colleagues C. Gourmel (Biosavane, French Guiana) and J. Etienne (Guadeloupe) for providing information.

References

Bokonon-Ganta, A. H., de Groote, H. and Neuenschwander, P. (2002) Socio-economic impact of biological control of mango mealybug in Benin. *Agriculture, Ecosystems & Environment* 93(1–3), 367–378. DOI : 10.1016/S0167-8809(01)00337-1

Boulet, A. (1986) Lutte biologique contre les «borers» de la canne à sucre en Martinique, exemple d'une intervention bien menée [Biological control against sugarcane borers in Martinique, an example of a well-conducted intervention]. *Bulletin technique d'information Ministère d'Agriculture* 409 (411), 363–374.

Cellier, G., Moreau, A., Cassam, N., Hostachy, B., Ryckewaert, P., Aurela, L. and Rioualec, A.L. (2014) First report of '*Candidatus* Liberibacter asiaticus' associated with huanglongbing on *Citrus latifolia* in Martinique and Guadeloupe, French West Indies. *Plant Disease* 98, 683–683. DOI: 10.1094/PDIS-08-13-0879-PDN

Cochereau, P. (1990) Installation en Guadeloupe de *Cotesia flavipes* Cameron (Hymenoptera: Braconidae), un parasite larvaire du foreur de la canne à sucre *Diatraea saccharalis* L. [Establishment in Guadeloupe of *Cotesia flavipes*, a larval parasite of the sugarcane borer *Diatraea saccharalis*]. In: *Rencontres Caraïbes en Lutte Biologique*. INRA, Guadeloupe, pp. 443–450.

Cock, M.J.W. (ed.) (1985) *A Review of Biological Control of Pests in the Commonwealth Caribbean and Bermuda up to 1982*. Technical Communication No. 9, Commonwealth Institute of Biological Control. Commonwealth Agricultural Bureaux, Farnham Royal, UK.

DAAF (2018a) Mémento de la statistique agricole. AGRESTE Guadeloupe, Nov 2018. Available at: http://daaf.guadeloupe.agriculture.gouv.fr/IMG/pdf/Memento_2018_Guadeloupe_cle8b13a1.pdf (accessed 28 October 2019).

DAAF (2018b) Mémento de la statistique agricole. AGRESTE Martinique, Nov 2018. Available at: http://daaf.martinique.agriculture.gouv.fr/IMG/pdf/MEMENTO_2018_cle0edf11.pdf (accessed 28 October 2019).

DAAF/INSEE (2018) Agriculture: en 2017 la culture d'agrumes poursuit sa croissance Bilan économique. Direction de l'Alimentation, de l'Agriculture et de la Forêt / Institut nationale de la statistique et des études économiques. Available at: https://www.societe.com/societe/france-daaf-490125564.html (accessed 6 August 2019).

de Moraes, G.J., Kreiter, S. and Lofego, A.C. (1999) Plant mites (Acari) of the French Antilles. 3. Phytoseiidae (Gamasida). *Acarologia* 40(3), 237–264.

Etienne, J. and Delvare, G. (1991) The parasites of *Diaprepes abbreviatus* (Coleoptera: Curculionidae) of the French Antilles. *Bulletin de la Société Entomologique de France* 96(3), 295–299.

Etienne, J., Burckhardt, D. and Grapin, C. (1998a) *Diaphorina citri* (Kuwayama) in Guadeloupe, first record for the Caribbean (Hemiptera: Psyllidae). *Bulletin de la Société Entomologique de France* 103(1), 32–33.

Etienne, J., Matile-Ferrero, D., Leblanc, F. and Marival, D. (1998b) Premier signalement de la cochenille *Maconellicoccus hirsutus* (Green) en Guadeloupe; situation actuelle de ce ravageur des cultures dans les Antilles françaises [First report of the mealybug *Maconellicoccus hirsutus* in Guadeloupe; current situation of this pest of crops in the French West Indies]. *Bulletin de la Société Entomologique de France* 103(2), 173–174.

Etienne, J., Quilici S., Marival, D. and Franck, A. (2001) Biological control of *Diaphorina citri* (Hemiptera: Psyllidae) in Guadeloupe by imported *Tamarixia radiata* (Hymenoptera: Eulophidae). *Fruits* 56(5), 307–315.

Etienne, J., Ryckewaert, P. and Michel, B. (2015) Thrips (Insecta: Thysanoptera) of Guadeloupe and Martinique: updated check-list with new information on their ecology and natural enemies. *Florida Entomologist* 98, 298–304.

Germain, J. F., Laplace, D., Devarieux, A. and Boavida, C. (2015) First records of the mealybug *Rastrococcus invadens* Williams (Hemiptera: Pseudococcidae) in French Guiana and the Americas. *Zootaxa* 3905(3), 447–450. DOI: 10.11646/zootaxa.3905.3.11

Gourmel, C. (2014) *Catalogue Illustré des Principaux Insectes Ravageurs et Auxiliaires des Cultures de Guyane* [*Illustrated Catalogue of the Main Insects Pests and Beneficials of Crops in French Guiana*]. Coopérative Bio Savane, Macouria, French Guiana.

Janelle, J., Séguret, J., Etienne, J., Vaillant, F. and Didelot, D. (2000) Citrus blackfly. *Phytoma* 532, 60–63.

Kairo, M.T., Pollard, G.V., Peterkin, D.D. and Lopez, V.F. (2000) Biological control of the hibiscus mealybug, *Maconellicoccus hirsutus* Green (Hemiptera: Pseudococcidae) in the Caribbean. *Integrated Pest Management Reviews* 5(4), 241–254. DOI: 10.1023/A:1012997619132

Kreiter, S. and de Moraes, G.J. (1997) Phytoseiid mites (Acari: Phytoseiidae) from Guadeloupe and Martinique. *Florida Entomologist* 80, 376–382.

Kreiter, S., Mailloux, J., Tixier, M.S., Le Bellec, F., Douin, M., Guichou, S. and Etienne, J. (2013) New phytoseiid mites of the French West Indies, with description of a new species, and new records (Acari: Mesostigmata). *Acarologia* 53, 285–303. DOI: 10.1051/acarologia/20132095

Kreiter, S., Zriki, G., Ryckewaert, P., Pancarte, C., Douin, M. and Tixier, M.S. (2018) Phytoseiid mites of Martinique, with redescription of four species and new records (Acari: Mesostigmata). *Acarologia* 58, 366–407. DOI: 10.24349/acarologia/20184248

Leblanc, F. (2000) Rapport d'activité du laboratoire d'entomologie pour la période allant de septembre 1999 à juin 2000: synthèse des travaux entrepris et perspectives à court terme [Activity report of the laboratory of entomology for the period from September 1999 to June 2000: summary of work undertaken and short-term perspectives]. CIRAD-FLHOR, Le Lamentin.

Lucas, P.D. (2012) Les coccinelles de la Martinique: une ressource biologique méconnue pour la protection durable des cultures [Ladybirds from Martinique: an underestimated biological resource for sustainable crop protection]. *Coléoptères des Petites Antilles* 1, 86–94.

Neuenschwander, P. (2003) Biological control of cassava and mango mealybugs in Africa. In: Neuenschwander, P., Borgemeister, C. and Langewald, J. (eds) *Biological Control in Integrated Pest Management Systems in Africa*. CAB International, Wallingford, UK, pp. 45–59.

Neuenschwander, P., Boavida, C., Bokonon-Ganta, A., Gado, A. and Herren, H. (1994) Establishment and spread of *Gyranusoidea tebygi* and *Anagyrus mangicola* (Hymenoptera: Encyrtidae), two biological control agents released against the mango mealybug *Rastrococcus invadens* Williams (Hemiptera: Pseudococcidae) in Africa. *Biocontrol Science and Technology*, 4, 61–69. DOI: 10.1080/09583159409355313

Nicolas, V. (2012) Etude préliminaire des Coccinelles des Petites Antilles: Chilocorini et Coccinellini [Preliminary study of ladybirds of the Lesser Antilles: Chilocorini and Coccinellini)]. *Harmonia* 9, 10–20.

Panis, A. (2013) Les *Coccophagus* (Hymenoptera: Aphelinidae) de Guadeloupe [*Coccophagus* from Guadeloupe]. *Bulletin Société Entomologique de Mulhouse* 69, 37–43.

Pavis, C., Huc, J.A., Delvare, G. and Boissot, N. (2003) Diversity of the parasitoids of *Bemisia tabaci* B-biotype (Hemiptera: Aleyrodidae) in Guadeloupe Island (West indies). *Environmental Entomology* 32, 608–613. DOI: 10.1603/0046-225X-32.3.608

Ryckewaert, P. and Alauzet, C. (2002) The natural enemies of *Bemisia argentifolii* in Martinique. *BioControl* 47, 115–126. DOI: 10.1023/A:1014439715271

Ryckewaert, P. and Rhino, B. (2017) *Insectes et acariens des cultures maraîchères en milieu tropical humide: reconnaissance, bio-écologie et gestion agro-écologique* [*Insects and mites of vegetable crops in humid tropical environment: recognition, bio-ecology and agro-ecological management*]. Ed. Quae, Versailles, France.

Stehlé, H. (1956) Les insectes nuisibles à la Canne à sucre. Leurs parasites naturels et la lutte biologique aux Antilles françaises [Insect pests in sugarcane. Their natural parasites and biological control in the French West Indies]. *Journal d'Agriculture Traditionnelle et de Botanique Appliquée* 3(1), 60–81.

UN (United Nations) (2018) World Population Prospects. Available at: https://population.un.org/wpp (accessed 31 July 2018).

Vayssières, J.F. (2017) Rapport de mission en Guyane sur *Rastrococcus invadens* [Mission Report in French Guiana on *Rastrococcus invadens*]. CIRAD, Montpellier.

Vayssières, J.F., Cayol, J.-P., Caplong, P., Séguret, J., Midgarden, D., Van Sauers-Muller, A., Zucchi, R., Uramoto, K. and Malavasi, A. (2013) Diversity of fruit fly (Diptera Tephritidae) species from French Guiana: their main host plants with associated parasitoids during the period 1994–2003 and prospects for fly management. *Fruits* 68, 219–243. DOI: 10.1051/fruits/2013070

Wikipedia (2019a) Economy of French Guiana. Available at: https://en.wikipedia.org/wiki/Economy_of_French_Guiana) (accessed 31 July 2019).
Wikipedia (2019b) Economy of Guadeloupe. Available at: https://en.wikipedia.org/wiki/Economy_of_Guadeloupe) (accessed 31 July 2019).
Wikipedia (2019c) Economy of Martinique. Available at: https://en.wikipedia.org/wiki/Economy_of_Martinique) (accessed 31 July 2019).

16 Biological Control in Guatemala

Joop C. van Lenteren*
Laboratory of Entomology, Wageningen University, Wageningen, The Netherlands

* E-mail: joop.vanlenteren@wur.nl

Abstract

Biological control was initiated in Guatemala in the 1990s, after costs for chemicals had increased substantially and had resulted in a drastic decrease in production of, for example, cotton and tomato. Most often, augmentative biocontrol is used in Guatemala. Examples of successful programmes are control of: (i) lepidopterans in cotton and vegetables by microbial agents and egg parasitoids; (ii) cotton leafworm by egg parasitoids and a predator in cotton; (iii) coffee berry borer with a parasitoid and a microbial agent; (iv) sugarcane borer with a parasitoid and a microbial agent; (v) diamondback moth with parasitoids in cruciferous crops; and (vi) white grub in maize with an entomopathogenic nematode. In 1998, populations of *Anopheles* mosquitoes, which are vectors of malaria, were successfully reduced by killing their larvae with microbial agents, resulting in a 50% lower malaria prevalence. A recently started project concerns classical biocontrol of the Asian citrus psyllid with a parasitoid. Biocontrol is expected to grow because of, inter alia, demands for residue-free food by countries to which Guatemala exports.

16.1 Introduction

Guatemala has an estimated population of almost 15.5 million (July, 2017) and its major agricultural products are sugarcane, maize, bananas, coffee, beans, cardamom, cattle, sheep, pigs and chickens (CIA, 2017).

16.2 History of Biological Control in Guatemala

Latin American books and research papers listed in Chapter 1 (this volume) provide very little information about the history of biocontrol in Guatemala and nothing could be found at all for the period 1880–1969. In fact, only Estrada Hurtarte (1996) gives data for biocontrol projects taking place in the 1990s. This author wrote that pest control in Guatemala had been based mainly on chemical pesticides and that high costs of chemical control caused a drastic decrease in cotton production in the 1970s and 1980s. Later, tomato production faced similar problems after *Bemisia tabaci* (Gennadius) appeared. In the 1990s, the major biocontrol projects concerned pests in coffee, cotton, vegetables, basic grains, pastures and sugarcane. An overview of biocontrol agents studied and/or applied in this period is given in Table 16.1.

Biever (1996) mentioned that a parasitoid release programme utilizing *Cotesia plutellae* (Kurdjumov) for control of the diamondback moth *Plutella xylostella* (L.) in broccoli and cabbage had been in operation in Guatemala for 3 years, since October 1993, on about 250 ha. Releases were limited in 1996 because the logistics of transporting the parasitoids from the USA to the other countries became difficult.

16.3 Current Situation of Biological Control in Guatemala

16.3.1 Natural and augmentative biological control of pests in coffee

The National Coffee Association (Asociación Nacional del Café) provides information about biological and integrated pest management (IPM) on its website (Anacafé, 2018). The website first presents an overview of IPM and then lists control measures for the most important coffee pests. The major pest of coffee in Guatemala is the coffee berry borer *Hypthonemus hampei* (Ferrari) and the parasitoid *Cephalonomia stephanoderis* Betrem is used for its control, for which on-farm production technology is available for coffee farmers. Two genera of nematodes are problematic in coffee plantations: *Pratylenchus* spp. and *Meloidogyne* spp. Experiments have been done with the pathogenic fungus *Paecilomyces lilacinus* (Thom) Samson, but it seems not be used at the moment. Several scale species cause problems in coffee, including *Dysmicoccus cryptus* Williams (= *bispinosus*) Beardsley, *Geococcus coffeae* Green, *Planacoccus citri* (Risso), *Coccus viridis* (Green.), *Coccus hesperidum* L., *Saissetia oleae* Olivier and *Saissetia nigra* (Nietner), but currently no natural enemies are used for their control. The coffee leaf miner *Leucoptera coffeella* (Guérin-Meneville) is often kept under natural control by ten species of hymenopteran parasitoids, including *Bracon* spp. and *Zagrammosoma* spp.

Table 16.1. Biological control agents studied and/or used in Guatemala since the 1990s, with their target pests and crops (retrieved from Estrada Hurtarte, 1996, with updates from other sources).

Natural enemy, exotic (ex), native (na)	Pest / crop	Type of biocontrol[a]	Effect / area under biocontrol[b]
Bacillus thuringiensis[c] / ex	Lepidopterans in cotton and vegetables	ABC / 1990s	Control / > 50,000 ha
Baculovirus VPN 80 & 82 / na	Lepidopterans in cotton and vegetables	ABC / 1990s	Control / 3,600 ha
Trichogramma pretiosum / ex	Cotton leafworm in cotton	ABC / 1990s	Control / 14,000 ha
Chrysopa sp. / ex		ABC / 1990s	Control
Cephalonomia stephanoderis / ex	Coffee berry borer in coffee	ABC / 1990s	Control / 100 ha
Metarhizium anisopliae / na		ABC / 1990s	Testing / < 5 ha
Paecilomyces lilacinus / na	Nematodes in coffee	ABC / 1990s	Testing / < 5 ha
Cotesia flavipes / ex	Sugarcane borer in sugarcane	ABC / 1990s	Control / 156 ha
Metarhizium anisopliae / na		ABC / 1990s	Control / 156 ha
Cotesia plutellae / ex	Diamondback moth in crucifers	ABC / 1990s	/ 250 ha, terminated in 1996
Plutella xylostella / ex		ABC / 1990s	Control / 100 ha
Diplogasterid nematodes / na	White grub in corn	ABC / 1990s	Control / < 20 ha
Encarsia opulenta / ex	Citrus blackfly in citrus	CBC / 1990s	Control / 1,500 ha
Metarhizium anisopliae / na	Spittlebugs in pastures	ABC / 1990s–2006	Testing

[a]Type of biocontrol: ABC = augmentative biocontrol, CBC = classical biocontrol
[b]Area of crop harvested in 2016 according to FAO (http://www.fao.org/faostat/en/#data/qc)
[c]Strictly speaking, use of the biopesticide *B. thuringiensis* is not considered biocontrol

Biocontrol of the coffee tree root-attacking *Phyllophaga* spp., as well as the coffee berry borer, has been tested with *Metarhizium anisopliae* (Metchnikoff) Sorokin, resulting in larval infection of up to 65%.

16.3.2 Augmentative biological control of spittlebugs in grasslands

Cattle production is an important economic activity in some parts of Guatemala. Spittlebugs reduce biomass, protein content and digestibility of grass (in this case *Brachiaria decumbens* Stapf.). Zeno (2006) evaluated three commercial strains of *M. anisopliae* with regard to their effect on the mortality of two species of spittlebugs: *Aeneolamia albofasciata* (Lallemand) and *Prosapia simulans* (Walker). The results were disappointing, as treatments with *M. anisopliae* did not reduce spittlebug populations. However, near the study location, a native entomopathogenic fungus of the genus *Batkoa* was found, which caused epizootics in adult spittlebugs.

16.3.3 Classical biological control of the Mediterranean fruit fly

In a project of the US Department of Agriculture (USDA) Center for Plant Health Science and Technology (CPHST, 2009), experiments were conducted in Guatemala for control of the Mediterranean fruit fly *Ceratitis capitata* (Wied.) by releases of the exotic parasitoid *Fopius ceratitivorus* Wharton. Parasitoids were reared at USDA's Animal and Plant Health Inspection Service

(APHIS)/Moscamed (*mosca de la fruta del Mediterráneo*, or 'medfly', *C. capitata*) quarantine facility in San Miguel Petapa in Guatemala. During 2009, field releases of *F. ceratitivorus* in combination with releases of sterile fruit fly males were initiated in several areas in South-west Guatemala. According to CPHST (2009): 'circumstantial evidence obtained was that no recurrent infestation has been present in the area where releases were conducted'.

16.3.4 Classical biological control of Asian citrus psyllid

The Asian citrus psyllid *Diaphorina citri* Kuwayama, which is a vector of huanglongbing disease (caused by *Candidatus* Liberibacter *asiaticus*), is present in Guatemala. The country collaborates in several regional programmes to develop biocontrol of the psyllid (FAO, 2013).

16.3.5 Augmentative biological control of malaria vectors

Bacillus sphaericus Meyer and Neide was tested in1998 in 46 localities with the highest epidemiological malaria risk to reduce population of *Anopheles albimanus* Wiedemann. Larval mortality was high (> 90%) and the rate of malaria prevalence went down by 50% (Blanco *et al.*, 2000). Based on the positive results, use of *B. sphaericus* was recommended for control of this malaria-transmitting mosquito in Central America.

16.3.6 Areas under biological control in Guatemala

Based on the information provided above and in Table 16.1, it is estimated that Guatemala has at least 18,000 ha under augmentative and 1,500 ha under classical biocontrol.

16.4 New Developments of Biological Control in Guatemala

Estrada Hurtarte (1996) mentioned that the use of biocontrol in Guatemala was promising, because of: (i) demands for residue-free food or by prohibiting the use of certain pesticides by countries to which Guatemala exports; (ii) resistance development to pesticides; (iii) growing resistance by the public and farm labourers to the use of pesticides that may cause health risks and have a negative effect on the environment; and (iv) positive experiences of farmers using biocontrol, the availability of biocontrol agents and local production of some beneficial organisms. However, he also reported that there were a number of factors that frustrate application of biocontrol in Guatemala, including: (i) the very limited number of professionals to teach and apply biocontrol and to rear and sell natural enemies; (ii) lack of support for research staff and limited financial support of biocontrol programmes; (iii) strong pressure to use chemical control by those who sold pesticides; (iv) difficult and expensive registration procedures for biocontrol agents; and (v) lack of extension personnel in the field of biocontrol.

References

Anacafé (2018) Asociación Nacional del Café [National Coffee Association]. Available at: https://www.anacafe.org/caficultura/manuales/ (accessed 28 October 2019).

Biever, K.D. (1996) Development and use of a biological control – IPM system for insect pests of crucifers. *Proceedings of the 3rd International Workshop on the Management of Diamondback Moth and Other Crucifer Pests*, 29 October – 1 November 1996, 257–261.

Blanco, C.S.D., Martinez, A.A., Cano, V.O.R., Tello, G.R. and Mendoza, I. (2000) Introduction of *Bacillus spaericus* strain-2362 (GRISELESF) for biological control of malaria vectors in Guatemala [in Spanish]. *Revista Cubana de Medicina Tropical* 52, 37–43.

CIA (2017) The World Factbook: Guatemala. Available at: https://www.cia.gov/library/publications/the-world-factbook/geos/gt.html (accessed 18 April 2019).

CPHST (2009) Annual Report. Biological Control Unit, Center for Plant Health Science and Technology, APHIS, USDA, pp. 39–40. Available at: https://www.aphis.usda.gov/plant_health/cphst/downloads/2009BiologicalControlUnitAnnualReport.pdf (accessed 19 April 2019).

Estrada Hurtarte, R.E. (1996) Situación del control biológico de plagas en Guatemala [Biological control of pests in Guatemala]. In: Zapater, M.C. (ed.) *El Control Biológico en América Latina*. IOBC/NTRS, Buenos Aires, Argentina, pp. 43–53.

FAO (2013) *Managing huanglongbing/citrus greening disease in the Caribbean*. Issue Brief No. 4, Subregional Office for the Caribbean. FAO, Christchurch, Barbados. Available at: http://www.fao.org/3/a-ax739e.pdf (accessed 18 January 2019).

Zeno, S.C. (2006) Uso de *Metarhizium anisopliae* para el control biológico del salivazo (*Aeneolamia* spp. y *Prosapia* spp.) en pastizales de *Brachiaria decumbens* en El Petén, Guatemala [Use of *Metarhizium anisopliae* for control of spittlebugs (*Aeneolamia* spp. and *Prosapia* spp.) in pastures of *Brachiaria decumbens* in El Petén, Guatemala]. MSc thesis, Centro Agronómico Tropical de Investigación y Enseñanza (CATIE), Turrialba, Costa Rica. Available at: http://repositorio.bibliotecaorton.catie.ac.cr/handle/11554/4543 (accessed 18 January 2019).

17 Biological Control in Guyana

Joop C. van Lenteren*

Laboratory of Entomology, Wageningen University, Wageningen, The Netherlands

* E-mail: joop.vanlenteren@wur.nl

Abstract

Conservation biological control, implemented in the 1910s by erecting perches in rice fields for insectivorous birds, resulted in effective control of the fall armyworm. Guyana is supposed to be the first country where the tachinid parasitoid *Lydella minense*, which had been collected in 1932 in Brazil, was successfully introduced for sugarcane borer control in the same year. During this period, several native natural enemies of this borer, as well as of other pests, were also identified as a result of prospecting projects. Guyana provided some of these species to other countries in the region. Classical biocontrol projects during the period 1970–2000 were attempted and the pink hibiscus mealybug was brought under complete control in 1998 by a parasitoid and a predator. Current projects concern classical biocontrol of the carambola fruit fly by parasitoids, augmentative control of the red palm mite by predatory mites, augmentative control of the palm castniid by a native entomopathogenic bacterium and conservation biocontrol of pests in rice.

17.1 Introduction

Guyana has an estimated population of almost 740,000 (July, 2017) and its main agricultural products are sugarcane, rice, edible oils, beef, pork, poultry, shrimp and fish (CIA, 2017).

17.2 History of Biological Control in Guyana

17.2.1 Period 1875–1969

Conservation biological control of lepidopterans in rice

G.E. Bodkin, during the 1910s, found that *Spodoptera frugiperda* Smith was a serious defoliator of rice and suggested erecting perches in the fields for insectivorous birds. This appeared to be most effective, particularly in rice nurseries (Cock, 1985).

Classical biological control of sugarcane borer with the Amazon fly

In 1933, Myers reported that a tachinid parasitoid, which he had collected near Santarem in Brazil and called the Amazon fly *Lydella (Metagonistylum) minense* Tns, had established and become widespread in Guyana after he had introduced it in 1932. Thereafter, it was considered a valuable parasitoid of *Diatraea saccharalis* F. (Cock and Bennett, 2011). Cock (1985) referred to a number of papers related to sugarcane borer and its natural enemies in Guyana by J.F. Bates, H.E. Box and L.D. Cleare, who worked on mass rearing of the egg parasitoids *Trichogramma* sp. and *Telenomus* sp. (Myers, 1935), as well as the Amazon fly. As a result of introduction and releases of *L. minense*, populations of *D. saccharalis* went down, as well as the damage caused by the borer (Dasrat *et al.*, 1997).

Guyana as provider of natural enemies

N. Vesey took the giant toad *Bufo marinus* (L.) from Guyana to introduce it into Bermuda in 1875 in the hope that it would control insects, particularly roaches. The toad became extremely abundant and undoubtedly had some value as a predator of insects, slugs and snails, according to Cock (1985).

H.E. Box, engaged by a Puerto Rican sugar company, successfully introduced a braconid, *Alabagrus stigma* Brullé, from Guyana to Puerto Rico (Cock and Bennett, 2011).

J.C. Meyers and his wife (stationed at the Trinidad CABI station) went for several natural enemy exploratory trips to Guyana. Cock and Bennett (2011) provided a very interesting story of the couple's rather spectacular journeys. When travelling privately in Guyana in 1930, Mrs Myers discovered that balisier *Heliconia bihai* (L.) is a wild food plant of the sugarcane giant moth borer *Telchin licus* (Drury). J.C. Myers found a tachinid subsequently identified as *Palpozenillia palpalis* (Aldrich) parasitizing larvae of this borer. Later, puparia of this species were sent to Trinidad but attempts to breed it in the laboratory failed.

In 1933, during a trip on the Brazilian frontier of Guyana, Meyers collected froghoppers and their natural enemies. He found *Tomaspis* aff. *pubescens* Fabr. and another, large red *Tomaspis* sp. On the large red *Tomaspis* no parasitoids were found, but two syrphid predators were common,

Salpingogaster nigra (Schiner) and a pale yellow one, apparently new to science (Cock and Bennett, 2011). Dissection of many hundreds of nymphs of *T.* aff. *pubescens*, did not result in finding any parasitoids, but one adult froghopper had an egg of a new parasitoid, a phasiine tachinid fly that specialized on insects of the order Hemiptera, to which froghoppers belong. Meyers went back in 1934 and found, dissected and observed many *T.* aff. *pubescens*, but could not detect parasitoids. He did find large numbers of nymphs and a few adults of another, small froghopper (*Clastoptera* sp.) infesting a shrub, but found neither parasitoids nor predators.

17.2.2 Period 1970–2000

Biological control of pests of coconut and oil palms

The palm castniid *Lapaeumides dedalus* (Cram.) is a pest of coconuts in Guyana. Its larvae are gregarious, living in communal webs, from which they emerge to feed at night and bore in the crown of the tree. They are able to totally defoliate coconuts and other palms and repeated defoliations will kill the palms. Several parasitoids have been recorded from Guyana that attack eggs, larvae and pupae, but they do not provide adequate control. Possibilities for biocontrol by introduction of exotic natural enemies are limited, since the family Castniidae is restricted to the Americas. According to Cock (1985), the best prospect for biocontrol might be the introduction of tachinid parasites known from this pest in Brazil. In Guyana, R. Bhim and colleagues (mentioned in Cock, 1985) have used a bacterium (species as yet undetermined) against *L. dedalus* attacking oil palms and reported that this castniid moth is no longer a major pest problem on that crop.

Biological control of sugarcane borers

The sugarcane borer *D. saccharalis* was brought under classical biocontrol by introducing *L. minense* in 1932, but another borer species, *D. centrella* (Möschler), was occasionally causing serious damage. The parasitoid *Allorhogas pyralophagus* (Marsh), initially found in Mexico in the early 1980s, was mass reared and released by the Guyana Sugar Industry from 1989 to 1994. The parasitoid easily attacked borers in the laboratory, but parasitism in the field was below 1% (Dasrat *et al.*, 1997).

Classical biological control of the pink hibiscus mealybug

The pink hibiscus mealybug *Maconellicoccus hirsutus* Green has been present in Guyana since 1997 and attacks many different plant species. A project with involvement of the UN's Food and Agriculture Organization (FAO), CABI and the Caribbean Agricultural Research and Development Institute (CARDI) developed biocontrol of the mealybug based on the introduction of the parasitoid *Anagyrus kamali* Moursi and the predator *Cryptolaemus montrouzieri* Mulsant in the Caribbean region. These two natural enemies were introduced and released in Guyana in 1997 and resulted in complete control of the pest (Kairo *et al.*, 2001).

17.3 Current situation of biological control in Guyana

17.3.1 Biological control of the carambola fruit fly

Much work was recently done on biocontrol and the male annihilation technique (MAT) of the carambola fruit fly (*Bactrocera carambolae* Drew and Hancock) in Suriname, Guyana, French Guiana and Brazil. The carambola fruit fly supposedly originates from South-east Asia. An eradication programme based on MAT was developed and funded by the International Fund for Agricultural Development (IFAD) (Netherlands, France and the USA); it began officially in 1998 and covered Guyana, Suriname, French Guiana and Brazil (Midgarden *et al.*, 2016). As a result of MAT the distribution of *B. carambolae* was reduced to limited areas of Suriname and French Guiana by 2001, and Guyana was declared free of this pest. However, in 2002 funding for the eradication programme was reduced and eventually terminated. This resulted in expansion of the distribution of this fruit fly, with detections as far South-east as Curralinho, in the Para State of Brazil, and as far North as Orlando, Florida, and there was re-infestation of over 50% of the regions in Guyana. Currently, natural enemies of the pest are being evaluated, e.g. in Suriname (see Chapter 28: Suriname for more detail).

17.3.2 Augmentative biological control of red palm mite

Coconut palm is the third most important economic crop in Guyana. The red palm mite *Raoiella indica* Hirst was first found in Guyana in 2013 and is now one of the most serious pests of palm. Two predators, lacewings and the predatory mite *Amblyseius largoenesis* (Muma), are currently being studied for control of this pest. Also, the National Agricultural Research and Extension Institute (NAREI, 2016) is testing the effectiveness of the entomopathogenic fungi against red palm mite.

17.3.3 Conservation biological control of pests in rice

The Guyana Rice Development Board (GRDB, 2018) has several projects related to biocontrol. An example is conservation biocontrol of the paddy bug *Oebalus* sp. by ladybirds, spiders, damsel flies and dragon flies, *Telenomus* sp. egg parasitoids and entomopathogenic fungi.

Based on data presented in this chapter and areas harvested according to FAO (2016), an area of 44,000 ha might be under classical biocontrol and 12,000 ha under augmentative biocontrol (Table 17.1), but it was difficult to verify these data.

17.4 New Developments of Biological Control in Guyana

No information has been found about future developments of biocontrol in Guyana. However, in the light of the long history of the application of biocontrol in Guyana, new programmes might eventually be implemented.

Table 17.1. Overview of major biocontrol activities in Guyana.

Biocontrol agent / exotic (ex), native (na)	Pest / crop	Type of biocontrol[a] / since	Effect / area under biocontrol[b]	Reference
Insectivorous birds / na	Defoliating insects, rice	ConsBC, 1910	Control / ? ha	Cock, 1985
Trichogramma sp. / na	*Diatraea saccharalis*, sugarcane	ABC	Insufficient	Myers, 1935
Telenomus sp. / na	*D. saccharalis*, sugarcane	ABC	Insufficient	Myers, 1935
Lydella (Metagonistylum) minense / ex	*D. saccharalis* sugarcane	CBC, 1932	Control / 44,000 ha	Dasrat *et al.*, 1997
Allorhogas pyralophagus / ex	*D. centrella*, sugarcane	CBC, 1989–1994	Insufficient control, established	Dasrat *et al.*, 1997
Entomopathogenic bacterium / na	Palm castniid, coconut and oil palm	ABC	Control / 12,000 ha	Cock, 1985
Amblyseius largoenesis / ?	Red palm mite, coconut palm	ABC / 2015s	Testing phase	NAREI, 2016
Entomopathogenic fungi / ?	Red palm mite	ABC / 2015s	Testing phase	
Anagyrus kamali / ex	Pink hibiscus mealybug, various plants	CBC / 1997	Complete control / ? ha	Kairo *et al.*, 2001
Cryptolaemus montrouzieri / ex	Pink hibiscus mealybug	CBC / 1997	Complete control / ? ha	Myers, 1935
Suite of natural enemies / na	Paddy bug, rice	ConsBC /	Partial control / ? ha	GRDB, 2018

[a]Type of biocontrol: ABC = augmentative, CBC = classical, ConsBC= conservation biological control; NC = natural control
[b]Area of crop harvested in 2016 according to FAO (http://www.fao.org/faostat/en/#data/qc)

References

CIA (2017) The World Factbook: Guyana. Available at: https://www.cia.gov./library/publications/the-world-factbook/geos/gy.html (accessed 19 April 2019)

Cock, M.J.W. (ed.) (1985) *A Review of Biological Control of Pests in the Commonwealth Caribbean and Bermuda up to 1982*. Technical Communication No. 9, Commonwealth Institute of Biological Control. Commonwealth Agricultural Bureaux, Farnham Royal, UK.

Cock, M.J.W. and Bennett, F.D. (2011) John Golding Myers (1897–1942), an extraordinary exploratory entomologist. *CAB Reviews: Perspectives in Agriculture, Veterinary Science, Nutrition and Natural Resources* 6 (No. 008), 1–18.

Dasrat, B., Rajkumar, A., Richards-Haynes, C., Quashie-Williams, C. and Eastwood, D. (1997) Evaluation of *Allorhogas pyralophagus* Marsh (Hymenoptera: Braconidae) for the biological control of *Diatraea* spp. (Lepidoptera: Pyralidae) in sugar-cane in Guyana. *Crop Protection* 16, 723–726.

FAO (2016) Data available at http://www.fao.org/faostat/en/#data/qc (accessed 27 July 2019).

GRDB (2018) Guyana Rice Development Board. Available at: http://grdb.gy/entomology (accessed 19 April 2019).

Kairo, M.T.K., Pollard, G.V., Peterkin, D.D. and Lopez, V.F. (2001) Biological control of the hibiscus mealy bug, *Maconellicoccus hirsutus* Green (Hemiptera: Pseudococcidae) in the Caribbean. *Integrated Pest Management Reviews* 5, 241–254.

Myers, J.C. (1935) Second report on an investigation into the biological control of West Indian insect pests. *Bulletin of Entomological Research* 26, 181–252.

Midgarden, D., van Sauers-Muller, A., Signoretti Godoy, M.J. and Vayssières, J.-F. (2016) Overview of the Program to Eradicate *Bactrocera carambolae* in South America. In: Ekesi, S., Mohamed, S. and De Meyer, M. (eds) *Fruit Fly Research and Development in Africa – Towards a Sustainable Management Strategy to Improve Horticulture*. Springer International, Switzerland, pp. 705–736.

NAREI (2016) Annual Report 2016. National Agricultural Research and Extension Institute. Available at: http://www.narei.org.gy/wp-content/uploads/2016/06/ANNUAL-REPORT-2016-Final-draft.pdf (accessed 19 April 2019).

18 Biological Control in Haiti

Philippe Ryckewaert[1]* and Joop C. van Lenteren[2]

[1]CIRAD, UR Hortsys, Campus Agro-environnemental Caraïbes, Petit Morne, Martinique, French West Indies; [2]Laboratory of Entomology, Wageningen University, Wageningen, The Netherlands

* E-mail: philippe.ryckewaert@cirad.fr

Abstract

Biocontrol on Haiti is mainly in the form of classical biological control. Successful early projects are control of citrus blackfly by a hymenopteran parasitoid since the 1930s and control of the sugarcane borer by an introduced tachinid parasitoid since 1953. After 1970 three hymenopteran parasitoids were introduced for control of the coffee berry borer. A new project concerns the introduction of predators against the fluted scale *Crypticerya genistae*. Most Haitian farmers have insufficient financial resources to apply chemical control, which creates possibilities for use of conservation biocontrol and also allows the ecosystem service of biocontrol to exhibit its role in pest and disease reduction.

18.1 Introduction

Haiti has an estimated population of almost 10,900,000 and its main agricultural products are rice, mangoes, cacao, coffee, maize, beans, cassava, sweet potato, peanuts, bananas, pigeon peas, sugarcane and sorghum (Wikipedia, 2019). Information about the history and current use of biocontrol in Haiti is very limited.

18.2 History of Biological Control in Haiti

18.2.1 Period 1880–1969

Classical biological control of the citrus blackfly

Citrus blackfly *Aleurocanthus woglumi* Ashby, native to Asia, spread over the Caribbean during the 1910s and created serious damage to citrus crops. Of the various parasitoids and predators introduced into Caribbean islands, *Eretmocerus serius* Silv. appeared to be the most effective natural enemy of citrus blackfly. This parasitoid was introduced from Cuba into Haiti in the 1930s and successfully controlled the blackfly (Cock, 1985).

Classical biological control of the sugarcane borer

The parasitoid *Lixophaga diatraeae* Townsend was imported and released in Haiti and has resulted in a 50% reduction in infestation of *Diatraea saccharalis* L. since 1953 (Box, 1960).

18.2.2 Period 1970–2000

Classical biological control of the coffee berry borer

The coffee berry borer *Hypothenemus hampei* (Ferrari) is considered the main pest of coffee worldwide. Three species of specific parasitoids of this pest, *Cephalonomia stephanoderis* Betrem, *Prorops nasuta* Waterston and *Phymastichus coffea* LaSalle, were introduced into Haiti during this period (Azar, 2018).

18.3 Current Situation of Biological Control in Haiti

Hardly any information could be found about current use of augmentative and classical biocontrol in Haiti. However, as most Haitian farmers have very limited financial resources, they use little chemical control. Thus, the ecosystem service of biocontrol (natural biocontrol) can exhibit its role and also conservation biocontrol can be developed. Inventories of biocontrol agents that could be used in conservation biocontrol are, however, rare. We found two papers concerning arthropod inventories in Hispaniola (Haiti + Dominican Republic): Evans and Serra (2002) reported parasitoids associated with whiteflies and Pérez-Gelabert (2008) provided an extensive checklist with literature on the arthropods of Hispaniola. Based on information provided above and data from FAO (2019) about areas harvested in 2017, we estimate that more than 30,000 ha are under complete classical biocontrol (Table 8.1).

18.3.1 Classical biological control of the pink hibiscus mealybug

The pink hibiscus mealybug *Maconellicoccus hirsutus* (Green) is supposed to have been imported into Haiti in 2002. The parasitoid *Anagyrus kamali* Moursi was imported from the USA and its release resulted in a reduction in pest populations of 97.2% within the first year (Meyerdirk, 2006).

Table 18.1. Overview of major biocontrol activities in Haiti.

Biocontrol agent	Pest / crop	Type of biocontrol[a] / since / success	Area under biocontrol (ha)
Eretmocerus serius	Citrus blackfly in citrus	CBC / 1930s / complete	33,186[b]
Lixophaga diatraeae	Sugarcane borer in sugarcane	CBC / 1953 / partial	23,781[b]
Anagyrus kamali	Pink hibiscus mealybug	CBC / 2002 / complete	? but large
3 spp. of parasitoids	Coffee berry borer in coffee	CBC / 2018 / testing	

[a]Type of biocontrol: CBC = classical biological control
[b]Area of crop harvested in 2017 according to FAO (2019)

18.4 New Developments of Biological Control in Haiti

18.4.1 Classical biological control of the fluted scale

The fluted scale *Crypticerya genistae* (Hampel) is a recent pest in the Caribbean, causing significant damage to peanut crops in Haiti, as well as to other crops such as pigeon pea, beans and cassava. Surveys carried out in Haiti have failed to find effective natural enemies of this pest. As chemical control is not considered feasible on a large scale, the Ministry of Agriculture of Haiti (MARNDR) wants to implement classical biocontrol by introducing one or more effective exotic natural enemies. The fluted scale was also detected a few years ago on other islands in the Caribbean (Puerto Rico, Guadeloupe, Martinique and Barbados) and more recently on the South American continent in Colombia and French Guiana (Etienne and Matile-Ferrero, 2008; Ciomperlik, 2010; Kondo *et al.*, 2016). Observations in these fluted scale-infested areas have shown the presence of ladybirds of the genus *Anovia* (*A. circumclusa* (Gorham) and *A. punica* Gordon) which consumed this scale and related species, and strongly reduced their populations (Kondo *et al.*, 2014; Silva-Gomez *et al.*, 2017). Thus, the MARNDR would like to introduce first *A. circumclusa*. However, their populations are now very rare in their native areas, because the scale has also become rare, perhaps due to the effectiveness of these predators, and so the operation is currently on stand-by.

Acknowledgement

J. Frisner Clerveus (Ministry of Agriculture, Haiti) is thanked for providing information.

References

Azar, M.S. (2018) Plan de lutte antiparasitaire et de gestion des pestes et des pesticides (plagp) (Pest and pesticide management plan). Projet RESEPAG-II, Ministère de lAgriculture, des Ressources Naturelles et du Développement Rural (MARNDR), Haiti. Available at : http://documents.worldbank.org/curated/en/221021525851918328/pdf/resepag-fa-plagp-version-final-apr4.pdf (accessed 19 April 2019).

Box, H.E. (1960) Status of the moth-borer *Diatraea saccharalis* (F.) and its parasites in St Kitts, Antigua and St Lucia; with observations on Guadeloupe and an account of the situation in Haiti. *Proceedings of the International Society of Sugar Cane Technologists* 10, 901–914.

Ciomperlik, M. (2010) *Crypticerya genistae* scale, an invasive pest in Puerto Rico. In: *CPHST Biological Control Unit 2010 Annual Report*. US Department of Agriculture, 33–34. Available at: https://www.aphis.usda.gov/plant_health/cphst/downloads/2010BiologicalControlUnitAnnualReport.pdf (accessed 19 April 2019).

Cock, M.J.W. (ed.) (1985) *A Review of Biological Control of Pests in the Commonwealth Caribbean and Bermuda up to 1982*. Technical Communication No. 9, Commonwealth Institute of Biological Control. Commonwealth Agricultural Bureaux, Farnham Royal, UK.

Etienne, J. and Matile-Ferrero, D. (2008) *Crypticerya genistae* (Hempel), a new danger to Guadeloupe (Hemiptera, Coccoidea, Monophlebidae). *Bulletin de la Société Entomologique de France* 113, 517–520.

Evans, G.A. and Serra, C.A. (2002) Parasitoids associated with aleyrodids (Homoptera: Aleyrodidae) in Hispaniola and descriptions of two new species of *Encarsia* Förster (Hymenoptera: Aphelinidae). *Journal of Hymenopteran Research* 11(2), 197–212.

FAO (2019) FAOSTAT data. Available at: http://www.fao.org/faostat/en/#data/qc (accessed 19 April 2019).

Kondo, T., Gullan, P. and Gonzalez, G. (2014) An overview of a fortuitous and efficient biological control of the Colombian fluted scale, *Crypticerya multicicatrices* Kondo & Unruh (Hemiptera: Monophlebidae: Iceryini), on San Andres island, Colombia. *Acta Zoologica Bulgarica* 66, 87–93.

Kondo, T., Ramos-Portilla, A.A., Peronti, A.L.B. and Gullan, P.J. (2016) Known distribution and pest status of fluted scale insects (Hemiptera: Monophlebidae Iceryini) in South America. *Redia* 99, 187–195. DOI: 10.19263/REDIA-99.16.24

Meyerdirk, D.E. (2006) Offshore Biological Control Strategy Applied to Pink Hibiscus Mealybug. In: *5th National IPM Symposium 'Delivering on a Promise'*, Presentation 57.2. Available at: https://ipmsymposium.org/2006/sessions/57-2.pdf (accessed 12 September 2018).

Pérez-Gelabert, D. (2008) Arthropods of Hispaniola (Dominican Republic and Haiti): a checklist and bibliography. *Zootaxa* 1831, 1–530.

Silva-Gómez, M., Quiroz-Gamboa, J.A., Hoyos-Carvajal, L.M., Yepes, F.C., Maya, M.F. and Santos, A. (2017) Coccinellidae predator of *Crypticerya multicicatrices* (Hemiptera: Monophlebidae) in San Andrés Island, Colombia. *Boletín Científico. Centro de Museos. Museo de Historia Natural* 21(1), 165–173. DOI: 10.17151/bccm.2017.21.1.13

Wikipedia (2019) Haiti. Available at: https://en.wikipedia.org/wiki/Haiti (accessed 19 April 2019).

19 Biological Control in Honduras

Rogelio Trabanino[1]*, Abelino Pitty[1] and Ronald D. Cave[2]
[1]*Escuela Agrícola Panamericana, Zamorano, Honduras;*
[2]*University of Florida, Fort Pierce, Florida, USA*

* E-mail: rtrabanino@zamorano.edu

Abstract

Prospecting for natural enemies and microbial control agents has been an important activity in Honduras since 1968, with as major projects the inventories of parasitoids of diamondback moth, striped grass looper, fall armyworm, Mexican cabbage butterfly, dipteran leaf miners, diaspids, coccids and aleurodids. The creation of the Center for Biological Control in Central America within the Zamorano University in 1989 was the starting point of teaching, research and the implementation of classical, augmentative and conservation biocontrol in the region. The Center trains hundreds of students from 21 countries each year and thesis work by its students results in a lot of knowledge about natural enemies, microbial agents and their mass-production methods. In 1990, introduction of phytophagous weevils for classical control of water hyacinth was a success. Microbial control of diamondback moth with a virus was tested in the 1990s, followed by the development and production of several fungus-, virus- and nematode-based formulations for control of pests and diseases. *Trichoderma*, *Metarhizium*, *Isaria*, *Lecanicillium* and *Beauveria* are now registered as commercial products and are also used in other countries in the region. In 2008 a successful project was started to reduce agrochemical use in vegetables, consisting of local mass rearing of several predators for augmentative control of thrips, aphids, mites and whitefly, and training of hundreds of farmers. A similar successful project concerned training courses on biocontrol of the sweet potato weevil for more than 5,000 farmers in the poorest area of the country.

19.1 Introduction

Honduras has an estimated population of slightly more than 9 million (July 2017) (CIA, 2017). According to Alduvin *et al.* (2017, pp. 368–372):

> Twenty-four percent of the national territory has agricultural soils, while 76% are for forest use. ... The agricultural sector accounts for nearly 40% of overall employment and the majority of rural employment. ... crops such as banana, plantain, sugarcane, African palm and pineapple are grown on more than 30,000 farms, with a cultivated area of 227,326 ha ... Citrus fruits, coconut, papaya, mango and other fruit trees are common in family orchards, even in urban areas. Annual crops such as melon, watermelon, tomato, potato, chili and onion are produced at 8,840 farms covering 19,580 ha ... Basic grains such as maize, beans and rice are planted on 117,647 ha ... Coffee is grown in almost 300,000 ha in 15 of 18 of Honduras departments...and is now the third largest producer in Latin America. Approximately 110,000 registered producers depend directly on coffee ... An ecologically friendly agroforestry system (known as Quesungual) has improved the living conditions of small farmers with scant resources. In Honduras, nearly 78% of the land used in agriculture is on slopes ... Two decades ago, peasants in the southwestern part of the country – with the help of FAO – began to develop the system. Quesungual consists of: 1) No slashing and burning ... 2) Permanently covering the floor ... 3) Minimum soil disturbance: zero tillage and direct planting; 4) Efficient fertilizer use: accurate application regarding time, amount and form. This system is used to cultivate maize, beans, sorghum, vegetables and soybeans. Farmers now clear the vegetation by hand. Trees that were previously cut and burned constitute a source of fruit, firewood and furniture wood ... the system increases production, requires less labor, retains moisture better, is low-cost and has reduced greenhouse gas emissions and retained more carbon. Extensive livestock production is carried out on 9.7% of the territory ... Preference is given to dairy and beef cattle, pigs and poultry and to a lesser extent, goats, sheep and bees. Most fishing is carried out using small artisanal boats, which catch shrimp and lobster, as well as fish, while aquaculture produces tilapia and shrimp ...

19.2 History of Biological Control in Honduras

19.2.1 Period 1880–1969

The first prospecting for natural enemies in Honduras was done by F.D. Bennett (CABI, Trinidad and Tobago) in 1968. He searched for natural enemies of the mahogany shoot borer *Hypsipyla grandella* (Zeller) and in 1969 for fruit flies and their natural enemies. Also, he studied the possibilities for biocontrol of some sugarcane pests in Honduras. Further, he released parasitoids of the Mexican fruit fly *Anastrepha ludens* (Loew) in 1969 (Cock, 1985).

19.2.2 Period 1970–2000

Center for Biological Control in Central America

A Center for Biological Control in Central America (Centro para el Control Biológico en Centroamérica) (CCBCA) was created in 1989 as a part of the Integrated Pest Management (IPM) Program in Honduras (Manejo Integrado de Plagas en Honduras) (MIPH) of the Plant Protection Department (Departamento de Protección Vegetal) (DPV) of the Escuela Agrícola Panamericana, Zamorano, Honduras, also known as Zamorano University. The US Agency for International Development (USAID) in Honduras funded the MIPH programme to promote the development of teaching, research and the implementation of classical, augmentative and conservation biocontrol in the region. To achieve this goal, the CCBCA collaborated with many international institutions, including the International Institute of Biological Control (IIBC, now CABI), the US universities of Florida, Purdue, Iowa State and Cornell, and regional ones such as the University of Costa Rica, the National Autonomous University of Nicaragua, the Agronomy Center for Research and Higher Education Center (Centro Agronómico Tropical de Investigación y Enseñanza) (CATIE) and the International Institute of Tropical Agriculture (IITA). The work of the CCBCA, together with these organizations, has contributed to maintaining or even increasing crop yields and reducing the agronomic, economic, environmental and public health issues associated with the indiscriminate use of pesticides. Teachers and students of Escuela Agrícola Panamericana and graduate students from Latin American, North American and European universities drew upon the CCBCA for research in biocontrol and for giving (short) courses to train stakeholders working in various fields of agriculture.

Inventory of natural enemies of pests

Inventories of natural enemies of pests are important for two purposes. Firstly, they provide records of natural enemy species present in crops/agroecosystems, which contribute to pest mortality, and provide knowledge about the factors that play a significant role in the regulation of the pest populations. This information can then be used for developing technologies to conserve natural enemies, as well as for the selection of natural enemies for specific biocontrol projects. The second purpose is to find out the causes for absence of natural enemies of certain pests in particular crops, as these might have to be considered before importation.

Inventories of parasitoids were made for four lepidopteran pests. The diamondback moth *Plutella xylostella* (L.) acts as a host of three primary and two facultative parasitoids. In the South-central Honduran region, the striped grass looper *Mocis latipes* Guenée is parasitized by 31 species of parasitoids (Cave, 1992). Cave (1993) listed 43 species of parasitoids in Central America that attack larvae and pupae of the fall armyworm *Spodoptera frugiperda* (Smith) and designed a taxonomic key to identify the 26 known species occurring in Honduras. Cave and Cordero (1999) reported 12 species of flies and wasps that attack the Mexican cabbage butterfly *Leptophobia aripa* Boisduval.

Twenty-five species of hymenopteran parasitoids attack three species of *Liriomyza* leaf miners on several crops and weeds in the southern region of Honduras (Acosta and Cave, 1994). *Chrysonotomyia diastatae* (Howard), *Opius dissitus* Muesebeck and *Ganaspidium utilis* Beardsley were the most abundant species. The authors also provided a key for identification of parasitoids of *Liriomyza* in Central America.

Cave and Márquez (1994) found 25 parasitoids attacking species of Diaspididae, Coccidae and Aleyrodidae in citrus in Honduras and discussed the introduction of candidates for classical biocontrol programmes. Cave (1994) listed ten species of Hymenoptera parasitizing sweet potato whiteflies *Bemisia tabaci* (Gennadius) in Central America. Bográn *et al.* (1998) reported for common bean crops that the percentage of natural parasitism of *B. tabaci* varied between 21% and 32% in the first growing season and between 10% and 37% in the last growing season. The most frequently collected parasitoids were *Encarsia pergandiella* Howard and *E. nigricephala* Dozier.

Cave (1997) discovered the only currently known parasitoid attacking the bromeliad weevil *Metamasius quadrilineatus* Champion. Later, Wood and Cave (2006) described and named the parasitoid *Lixadmontia franki* Wood and Cave. The parasitic fly was released in Florida in 2007–2010 for biocontrol of the Mexican bromeliad weevil *Metamasius callizona* (Chevrolat) (Cave, 2008).

These inventories were used in the *Manual of Parasitoid Identification of Agricultural Pests in Central America* by Cave (1995a), which provided information on the rearing, mounting methods and morphology of parasitoids, a key to the families of parasitic Hymenoptera and a list of biological and taxonomic publications for identification of parasitoids of agricultural pests in Central America. The book helps in identifying 112 primary parasitoids and three hyperparasitoids of the Diptera and Hymenoptera, their known hosts, distribution and biology and provides one or more drawings of each species.

Introduction of natural enemies for classical biological control of aquatic weeds

Water hyacinth *Eichhornia crassipes* (Martius) Solms-Laubach is native to the Amazon region and one of the most problematic weeds worldwide. It accidentally established from the beginning of the 20th century in Honduras. In Zamorano, it had established possibly since the creation of the Escuela Agrícola Panamericana in 1942 and covered at least half of the surface of the Monte Redondo Lake in 1990. Students working in aquaculture practised manual removal of the weed for years without success. Two natural enemies of the weed, the weevils *Neochetina bruchi* Hustache (24 individuals) and *N. eichhorniae* Warner (123 individuals), were introduced into Honduras in February 1990 by Zamorano University and followed Honduran government protocols for the introduction of living organisms. The insects arrived from Florida with the cooperation of F.D. Bennett (University of Florida, USA). The introduction was directed by R.D. Cave, director of the CCBCA, and A. Pitty (weed management specialist) was in charge of weevil establishment in Monte Redondo Lake and follow-up studies. Because *N. bruchi* has a shorter life cycle and a higher oviposition rate, the proportion of this weevil increased from 16% after release to 37% by August 1991. The weevils cleaned Lake Monte Redondo permanently, without using any other control (see Figures 2 and 3 in Cave *et al.*, 2011). The weevils feed on the leaves and reduce photosynthesis, and the larvae bore into the stolons, which causes the plant to putrefy and sink (Pitty *et al.*, 1993).

In February 1992, another batch of weevils was introduced into the Yojoa Lake (an area of 79 km^2) by R.D. Cave, I. Dejud and A. Pitty, with the collaboration of the National Electrical Company (ENEE). The insects established at a slow rate and could be found on several occasions later. Both *N. eichornia* and *N. bruchi* are found at very low densities and did not control the weed. On 28 April 1994, weevils were released in the reservoir of Los Laureles Dam, Tegucigalpa, by A. Pitty, H. Mero and I. Zelaya, as well as in La Ceiba in collaboration with the Regional University Center of the Atlantic Coast (CURLA). Weevils were sent from Zamorano to El Salvador by Leopoldo Serrano (University of El Salvador). D. Greathead (CABI), through FAO, introduced *N. bruchi* into Cuba, where *N. eichhorniae* was already established. Results of the introductions into El Salvador and Cuba are unknown.

The fungus *Cercospora piaropi* (Andalusia) (= *Cercospora rodmanii* (Conway)) was introduced and released for weed control in Honduras by Zamorano researchers L. del Río and A. Pitty in February 1991. The fungus was not recovered and it was assumed that it did not establish due to unsuitable rainfall and relative humidity regimes present during the months of February, March and April of 1991.

The weevil *Neohydronomous affinis* (Hustache), native to South America, was introduced in 1995 and released in the Ticamaya Lagoon in the department of Cortes by A. Pitty and G. Godoy to control water lettuce *Pistia stratiotes* L. The weevils originated from Florida, where they had been introduced in 1987. It is not known if the weevils established in Honduras.

Introduction of natural enemies for classical and augmentative biological control of insect pests

Four species of parasitoids and an exotic baculovirus were introduced for classical biocontrol. Releases of 10,445 individuals of *Cotesia plutellae* (Kurdjumov), an Asian parasitoid of *P. xylostella*, were made at several sites between December 1988 and November 1989. However, only 33 individuals of the next generation were recovered and there was no evidence later that *C. plutellae* had established. About 200 adults from the pupal parasitoid *Diadromus collaris* (Gravenhorst) were released in 1990, but there was no evidence of establishment.

Telenomus remus (Nixon) is a parasitoid of Asian origin that attacks the eggs of ca. 30 species of Lepidoptera (Cave, 2000). It was released in maize and sorghum from 1991 to 1994 (Cave and Acosta, 1999). Parasitism rates varied from 20% to 92% and good cropping conditions and flowering non-crop plants favoured control of lepidopteran pests, because in these fields parasitism was higher than 65%. Today, Zamorano still produces *T. remus* in small quantities, mainly for control of maize pests on about 150 ha per year.

From August to December 1994, 2,787 adults of *Eretmocerus* sp. from Thirumala, India, were released in Comayagua (Gómez Cabrera, 1995) in order to increase biocontrol of *B. tabaci*. In samples taken from September 1994 until February 1995, parasitism ranged from 0 to 3% on 13 species of host plants where *B. tabaci* was present, and from 0 to 52% in *Trialeurodes* spp. on eight host plants (Gómez Cabrera, 1995). However, parasitism levels by *Eretmocerus* sp. were very low in comparison with those produced by the native *E. pergandiella*. Cave (1994) argued that biocontrol of *B. tabaci* would be more effective within an integrated programme that considered the use of non-chemical control methods.

Microbial control of diamondback moth

The relationship between the applied dose of the nuclear polyhedrosis virus (NPV) *Galleria mellonela* (L.) and the size of the larvae of *P. xylostella* was studied in the laboratory by Espinoza *et al.* (1994) and field experiments were done to determine the best conditions for application of the virus (Espinoza and Cherry, 1994). Escribano *et al.* (1999) compared biological, genetic and structural aspects of four isolated NPV strains of *S. frugiperda* from the USA, Nicaragua and Argentina, to select the best candidate for field experiments in Honduras and Mexico. The strains were not genetically identical. The Nicaraguan and US strains caused the greatest infectivity. Field tests in Honduras and Mexico produced similar *S. frugiperda* mortality after applying *S. frugiperda* NPV or spraying with chlorpyrifos (Williams *et al.*, 1999). The highest dose of the virus used caused 40% larval mortality. Parasitism by wasps and flies contributed another 40% mortality rate in plots applied with the virus. However, pesticide applications caused a reduction of the natural enemies, resulting in a resurgence of *S. frugiperda*. Morjan Erazo (1993) demonstrated that parasitism by *T. remus* in the field is not reduced by applications of *S. frugiperda* NPV. Williams *et al.* (1999) noted that the use of *S. frugiperda* NPV was more expensive than chemical control.

Cultural measures to improve natural enemy effectiveness

Cultural control measures may increase predator and parasitoid populations, an approach that is also addressed as conservation biocontrol. Zero-tillage cropping, for example, can help the presence and action of predators in agroecosystems, as in maize crops, where Morjan Erazo (1993) found more diversity of nocturnal predators, while González and Cave (1997a) reported 1.2 times more daytime foliar spiders on beans. These authors also found 2–3 times higher parasitism rates in eggs of the leafhopper *Empoasca kraemeri* Ross and Moore by the parasitoid *Anagrus gonzalezae* Triapitsyn in beans under zero tillage (González and Cave, 1997b).

Cañas Castro (1996) determined the natural enemy abundance and the dynamics of fall armyworm populations in maize plots where sugar was applied as food for natural enemies. The fire ant *Solenopsis geminata* (F.) and the parasitic fly *Lespesia archippivora* (Riley) were more abundant in maize with sugar than in the control. Damage to the leaves by the fall armyworm was reduced by 35% in sugar-treated maize compared with the control and infestation of plants by the fall armyworm was 19% lower in sugar-applied plots.

Non-crop resources are essential for both parasitoids and predators as source for alternative plant food (pollen, nectar, etc.), prey/hosts, sites for oviposition and shelter. Cordero *et al.* (2000) showed that weed management is important for lepidopteran pest control in sorghum and maize. Crops under traditional weed management, consisting of elimination of weeds at the time of planting, had the highest pest density per plant compared with crops with elimination of weeds 3 weeks after planting. However, the larvae in both weed management systems showed the same levels of parasitism (11–16%), indicating that the 12 species of parasitoids were not affected by the weed management system.

Training of teaching and research in biological control

CCBCA provides fourth-year students of Escuela Agrícola Panamericana with the opportunity to conduct undergraduate and graduate thesis research, which has produced over a hundred theses during the existence of this Center. The thesis topics covered a diversity of issues, contributing significantly to biocontrol knowledge in Central America.

In 1993, CCBCA organized a workshop about biocontrol in Latin America with attendants of six Central and South American countries. A valuable product of this workshop was the production of the *Teaching Handbook of Biological Control in Latin America* by Cave (1995b). Another workshop was organized in 1990 in Zamorano as a result of the successes obtained with the entomopathogenic fungus *Metarhizium anisopliae* (Metchnikoff) Sorokin to control spittlebugs (*Aeneolamia* spp. and *Prosapia* spp.) in sugarcane in Costa Rica and Brazil. During the workshop, mass production and application techniques were presented. Later, CCBCA produced the fungus and applied it in sugarcane. Unfortunately, results in the field did not yield the expected success and production was discontinued.

19.3 Current Situation of Biological Control in Honduras

19.3.1 Development and production of microbial control agents

In 2000, CCBCA changed its approach as well as its name into Biological Control Laboratory (BCL), although it continued R.D. Cave's biocontrol teaching legacy. Following the learning-by-doing philosophy of Zamorano, BCL produced *T. remus* and *S. frugiperda* NPV that served as teaching practice for students and to supply some customers with these organisms. Later on, farmers asked BCL for help with pest control in vegetable crops. Visits to Cuba, Bolivia and the USA taught the staff of BCL how to rear and commercialize biocontrol agents. Interactions with Fintrac, a private US-based non-governmental organization (NGO) funded by USAID, were initiated with the goal of increasing farmers' income by teaching them how to grow vegetables without synthetic pesticides. Farmers in this project were no longer allowed to use agrochemicals. Fintrac requested Zamorano to produce biopesticides to reduce the 40–50% seedlings mortality in jalapeño pepper. This alliance helped BCL to get experience with the extension service.

In 2000, BCL started to produce the antagonistic fungus *Trichoderma harzianum* Rifai to combat soil-borne fungi. Efficacy, speed of growth and production yield at the laboratory were evaluated to select appropriate strains with the help of P. Arneson (Cornell University, USA, and professor at Zamorano University) and a group of students. Next, field evaluations were done in horticultural crops showing the effectiveness of the product. In 2003, Honduras recorded its first biopesticide in Zamorano with the commercial name of Trichozam. Registration data were obtained by Méndez Martínez (2003) and Medina Terán (2003). After six months of Trichozam use in jalapeño pepper, damping-off mortality declined to 5% and the majority of the farmers started using the product. For 2003, it was estimated that 300 ha of jalapeño pepper, 1,300 ha of tomato and 1,800 ha of sweet pepper had been treated with Trichozam.

In 2004 the biopesticides Bazam (*Beauveria bassiana* (Bals.-Criv.) Vuill) and Verzam (*Lecanicillium lecanii* (Zimm.) Zare and W. Gams) were developed. Bazam controls lepidopteran larvae and coleopteran adults and soil-living larvae of the coffee berry borer. The studies for its development and registration were based on thesis works by Romero Villagra (2004), Rivera Jerez (2004), Toapanta Valencia (2005), and Espinoza Silva (2005). Bazam is also used in onion against thrips and mites. Verzam, which was registered for the control of aphids and whitefly, could not be used because of interference with the intense application of fungicides in vegetable production.

In 2004, *M. anisopliae* strains were collected in the southern region of Honduras and *Purpureocillium lilacinum* (Thom) Luangsaard, Hou-braken, Hywl-Jones and Samson was collected in Zamorano. Production of *M. anisopliae* was resumed and the new strain is now used on 3,000 ha of vegetable crops and coffee each year. Products containing these microorganisms were registered in 2006 with the trade names of Metazam and Pazam. Metazam controls spittlebugs in sugarcane. Some 8,000 ha are treated annually with Metazam in three sugar mills, reducing use of agrochemicals. Metazam is used against May beetles (*Phyllophaga* spp.) in sweet potato, instead of the common practice of applying insecticides 30 days before harvest. It is expected that from 2018 onwards a strong increase in use of Metazam will take place, because of the low production cost and the interest of sugarcane producers to reduce pesticide application in the light of certification for export. Pazam is a fungus that controls nematodes and can be used in horticultural crops, basic grains and fruit trees. Baseline studies of Vaquedano Gámez (2006), Fernandez Tondelli (2006) and Juárez Figueroa (2007) were used for its registration.

The products registered in Honduras were later used in other countries of the region. Trichozam was registered in 2005 in Jamaica and since then Jamaica has ordered annual shipments. El Salvador registered Bazam and Trichozam in 2008, and Pazam and Metazam in 2009. In 2010, these four products were registered in Nicaragua as well, and registration in the Dominican Republic, Guatemala and Panama is under way. Thesis work by Morán Quintero (2007), Castillo Samudio (2007), Morán Ruiz (2007), Parada López (2007), Mantilla Compte (2007) and Rosero Asqui (2008) assisted in the foreign registration of these products.

The fungus *B. bassiana* and the nematode *Steinernema carpocapsae* have been tested for coffee borer control by Bauer Stillman (2016) and Ávila Sosa (2010). The problem of rust on coffee was studied by Welchez Arita and Guerra Burgos (2013), Roldán Salazar (2016) and Yánguez Quintero (2016), resulting in the production of *L. lecanii* fungus, and efforts to get the fungus registered have been started.

During 2014, Zamorano obtained *Isaria fumosorosea* strain IF3581 from the Agriculture Research Service (USDA-ARS, USA) and also tested a native strain with good results. Zamorano is in the process of registering this fungus for control of *B. tabaci*. Prieto Navas (2016) evaluated three *I. fumosorosea* concentrations for the control of *B. tabaci* in pepper. Sanchez and Carlos (2015) evaluated *I. fumosorosea* and its compatibility with the ladybird *Thalassa montesumae* Mulsant for the control of green scale *Phalacrococcus howertoni* Hodges and Hodgson in croton and showed better control results when both biocontrol agents were applied than using *I. fumosorosea* alone.

Zamorano has registered all its products in Nicaragua and El Salvador and has subsidiaries in those countries. In Nicaragua, Zamorano is treating over 4,000 ha of rice with *Trichoderma* for the control of the black sheath rot (*Gaeumannomyces* sp.).

Finally, in the area of veterinary problems, biocontrol of the cattle tick *Boophilus microplus* (Canestrini) has been the subject of much interest. Bendeck Meléndez (2012) and Pineda Bonilla (2014) evaluated application of *B. bassiana*, *M. anisopliae* and an insecticide and they found that *M. anisopliae* was more effective than *B. bassiana* and similarly effective as the insecticide.

19.3.2 Development and use of invertebrate natural enemies for pest control

In 2008, a Millennium Challenge Account donation was obtained in collaboration with A. Rueda to develop a mass-rearing and release programme for three natural enemies to reduce the use of agrochemicals in vegetables. The pirate bug *Orius insidiosus* (Say) was selected for the control of thrips, aphids and whitefly; the predatory mite *Neoseiulus longispinosus* (Evans) for the control of spider mites *Tetranychus* spp.; and the nematode *Heterorhabditis bacteriophora* Poinar to control soil insects, especially *Phyllophaga* spp., *Cosmopolites sordidus* (Germar), larvae of several lepidopteran species, the sweet potato weevil *Cylas formicarius* F. and soil termites. This project convinced farmers that sustainable agriculture is the best way to stay in business. Five Zamorano graduates were recruited for the project, to work with nematodes, mites and pirate bugs and to train farmers in the use of biocontrol. A laboratory and greenhouses were built to produce *H. bacteriophora*, *N. longispinosus* and *O. insidiosus*. Training in the use and production of these organisms involved 800 farmers, 150 technicians and persons interested to start a business in production and commercialization of biocontrol agents. In addition, 300 students learned the production technology, biology and ecology of these natural enemies. A total of 15 research projects with students and technical staff took place. The results reinforced knowledge of Zamorano staff to be able to train farmers and technicians. The projects included several theses works (Suazo López 2008, Pantoja Guamán 2009, Perrera Viamil 2009, Turcios Rivas 2009, Turcios Rivera 2009, Villarroel Basantes 2009, Ávila Sosa 2010 and Diéguez Gonzales 2010). An example of the success of this project is that two sweet pepper exporting companies received technical assistance for biocontrol of spider mite *Tetranychus urticae* Koch. Their practice was to spray acaricides 1.5 times per week at a cost of US$3,200 ha^{-1} per year on a total of 69 ha without sufficient control. By the end of 2009, *N. longispinosus* was released and chemical applications were reduced to two applications during the whole cropping cycle. In 2012, *N. longispinosus* was no longer needed, due to the amount already present in greenhouses resulting in successful control of spider mite, and chemical control was completely terminated. Since 2015 mass rearing of two other predatory mites, *Amblyseius cucumeris* (Oudemans) and *A. swirskii* (Athias-Henriot), has been studied and in 2017 commercial production of *A. cucumeris* started.

Another successful project involved control of sweet potato *Ipomoea batatas* L. in 2010. The sweet potato weevil *C. formicarius* pierces the potato tubers, reducing their quality and market value. Chemical control is not effective, because the weevil is in the ground and very mobile. The effect of *H. bacteriophora* on larvae and adults was studied, and after a cycle of application, damage by the weevil was reduced by 95%. After 2010, production of *H. bacteriophora* (NemaPower) was increased and used in sweet potato. This crop is of increasing importance and the production area of 1,200 ha is mainly used for export to the European market. Farmers cannot apply any chemical pesticides in the last 30 days before harvest. Several tests were carried out in their fields with the product NemaPower at the recommended rate of 200 million nematodes per hectare applied through the drip irrigation system. Producers are now using NemaPower in combination with pheromone traps to reduce adult populations. Studies carried out by Nuñez Flores (2011) and Mata Marcillo (2013) provided technical assistance to sweet potato producers. During this project, 5,100 small producers in the poorest area of the country, known as the dry corridor, were trained. Most of them were producers of potato, carrot, pineapple, banana, coffee, passion fruit, sweet potato and pepper. Thirty-six demonstration plots were established in six departments that make up the dry corridor, to determine the effect of biocontrol compared with the conventional chemical control that they used. During the training courses, farmers also recognized the importance of cultural practices such as cleaning the plots, the toxicity problems caused by pesticide use and, most important, the effectiveness of biocontrol. These farmers' experiences are one of the most stimulating examples of Zamorano involvement, because of its importance for an economically weak sector that did not have any knowledge of biocontrol.

The nematode *H. bacteriophora* was also evaluated for control of the coffee berry borer *Hypothenemus hampei* Ferrari as an alternative for borer control since the prohibition of use of the pesticide endosulfan. Funds were obtained through the programme Feed the Future to increase the production of *H. bacteriophora* to an industrial level in 2015 and 2016. Originally, *G. mellonella* larvae were used as host and with this *in vivo* production some 600 doses could be produced (equivalent to 600 ha treated per year), but grower demands were for 2,400 doses per year. In 2017 and 2018, production by liquid fermentation was developed. Currently, the laboratory has two bioreactors of 150 l and 500 l capacity. Studies by Grijalva Teran (2014), Duron Alvarado (2016) and Ucan Yam (2016) showed that with the new technology a 10-day production cycle equals the number of doses that in the past took 1 year to rear. The bioreactors are also used to produce fungi by liquid fermentation, which is more efficient than the biphasic process used before. In 2017, we began to evaluate the feasibility of the production of blastospores and microsclerotia of some of our fungal products mentioned in the previous section.

The main pest of plantain, an important crop in Honduras, is the weevil *C. sordidus*. Control of the weevil by *H. bacteriophora*, *S. carpocapsae*, *B. bassiana* and *M. anisopliae* was studied by Chicas Nolasco (2016), Mojica Espinoza (2016), Morales

Ramírez (2012), Ramírez Herrera (2016) and Torres Estrada (2016).

19.3.3 Research and training

Many of the laboratory studies at Zamorano University are executed by fourth-year students as a graduation requirement to obtain the degree of Bachelor of Science and Production. At least 300 students from 21 countries in Latin America attend this University each year to learn about and get practice in the production and commercialization of biocontrol agents. They are exposed to the newest technologies that subsequently will allow them to implement these in their own countries, contributing to cheaper, more efficient and safer pest control, as well as helping to improve the environment. Zamorano University also trained about 6,000 farmers in the use of biocontrol and its advantages over chemical control, so that by now the small farmers in the poorest areas of Honduras do know about biocontrol. Thanks to FINTRAC, BCL can continue the production of biocontrol agents and Zamorano could start another round of training new farmers.

In conclusion, Honduras has a rich history in biocontrol. Table 19.1 provides a chronological overview of biocontrol activities in Honduras, showing several attempts to establish classical biocontrol with success in control of aquatic weeds. In-depth studies have been performed on occurrence of native natural enemies and many results have been obtained with augmentative biocontrol of pests and diseases with predators, parasitoids and microbial control agents. According to Table 19.2, the estimated area under biocontrol in Honduras is more than 25,000 ha.

Honduras has aided development of biocontrol in other countries. For control of water hyacinth, *N. bruchi* and *N. eichhorniae* beetles were sent to El Salvador and Cuba in the mid-1990s, and the parasitoid *L. franki* was shipped to Florida (USA) from 2007 to 2010 for control of the Mexican bromeliad weevil *M. callizona*. Microbial control agents produced in Honduras like Bazam, Trichozam, 2008, Pazam and Metazam are now registered or pending registration in several Central American countries. Currently, Zamorano is treating over 4,000 ha of rice with Trichozam for control of black sheath rot (*Gaeumannomyces* sp.).

19.4 New Developments of Biological Control in Honduras

We are convinced that research and application of biocontrol in Honduras is here to stay, particularly at Zamorano University, where the recent acquisition of new bioreactors will result in larger-scale production of entomopathogens. Increased production of microbial control agents can then be used in crops that are produced on large areas, such as coffee, a crop with great importance for Honduras. For example, *B. bassiana* will be used in great volume by small and large producers for control of the coffee berry borer *H. hampei*. Further, *I. fumosorosea* will be used in the near future to control *B. tabaci* in horticultural crops. We also expect that entomopathogenic nematodes will be a very strong tool for pest control in Honduras.

The use of predators and parasitoids in conservation, augmentative and classical biocontrol is increasing every year, because it is effective and even less expensive than current chemical control methods. It is important that Zamorano University continues its work in securing biocontrol programmes in the country, by studying natural enemies and producing biocontrol agents, and through training of farmers. This work is made possible with support from the Ministry of Agriculture, several NGOs and private organizations of coffee producers, the sugarcane association and the oil palm organization.

19.5 Acknowledgements

We thank Zamorano University for allowing us write this chapter, FINTRAC for supporting the projects in the early days of our work, the Millennium Challenge Account for funding research work resulting in biocontrol in Central America, Partnering for Innovation (Feed the Future) for financing equipment and for giving the opportunity to teach small producers about biocontrol and its benefits.

Table 19.1. Chronological overview of biocontrol activities in Honduras.

Biocontrol agent	Pest, disease or weed, and crop	Type of biocontrol[a] / used since	Area (ha) under biocontrol
Parasitoids	*Anastrepha ludens*	CBC/1969	?
Parasitoids 5 spp.	*Plutella xylostella*	NC	?
Parasitoids 31 spp.	*Mocis latipes*	NC	?
Parasitoids 26 spp.	*Spodoptera frugiperda*	NC	?
Parasitoids 12 spp.	*Leptophobia aripa*	NC	?
Parasitoids 25 spp.	Diaspididae, Coccidae, Aleyrodidae	NC	?
Parasitoids 10 spp.	*Bemisia tabaci*	NC	?
Lixadmontia franki	*Metamasius quadrilineatus*	NC	?
Neochetina bruchi + *N. eichhorniae*	*Eichhornia crassipes,* aquatic weed	CBC / 1990	1
Cotesia plutellae	*P. xylostella*	CBC / 1988	? Established
Diadromus collaris	*P. xylostella*	CBC / 1990	Not established
Telenomus remus	Lepidopterans in maize/ sorghum	ABC / 1991	150
Eretmocerus sp.	*Bemisia tabaci*	CBC / 1994	Not effective
Neohydronomous affinis	*Pistia stratiotes* aquatic weed	CBC / 1995	Not established
Metarhizium anisopliae	*Aeneolamia* spp. and *Prosapia* spp.	ABC / 1996	Not effective
Lespesia archippivora	*S. frugiperda*	ConsBC / 1996	?
Anagrus gonzalezae	*Empoasca kraemeri*	ConsBC /1997	?
S. frugiperda NPV	*S. frugiperda*	ABC / 1999	Effective
Trichoderma harzianum	Soil-borne fungi	ABC / 2000	8,000
Beauveria bassiana	Lepidopterans, coleopterans	ABC / 2004	2,800
Lecanicillium lecanii	Aphids, whitefly	ABC / 2004	Testing
Metarhizium anisopliae	Spittlebugs, beetles	ABC / 2004	11,000
Purpureocillium lilacinum	Nematodes	ABC / 2004	1,000
Orius insidiosus	Thrips, aphids, whitefly	ABC / 2008	62
Neoseiulus longispinosus	*Tetranychus* spp.	ABC / 2008	37
Heterorhabditis bacteriophora	Insects in soil	ABC / 2008	1,000
	Cylas formicarius, sweet potato weevil	ABC / 2010	800
Lecanicillium lecanii	*Hemileia vastatrix,* coffee rust	ABC / 2013	500
Isaria fumosorosea	*B. tabaci*	ABC / 2014	1,000
B. bassiana *M. anisopliae*	*Boophilus microplus*, cattle tick	ABC / 2012	+ Testing
Amblyseius cucumeris, *A. swirskii*	Mites and insects	ABC / 2015	50
H. bacteriophora + *Hypothenemus hampei*, *B. bassiana*	Coffee berry borer	ABC / 2015	2,400

[a]Type of biocontrol: ABC = augmentative, CBC = classical, ConsBC= conservation biological control; NC = natural control

Table 19.2. Area under augmentative biological control by agents produced at Zamorano University in Honduras; only aquatic weed concerns classical biological control.

Biocontrol agent	Pest/disease/weed	Crop	Area (ha) under biocontrol
Microbial control agent			
Trichoderma harzianum	*Fusarium*, *Pythium*, *Phytophthora*, *Esclerotinia*	Pepper, tomato, coffee, lettuce, cucumber, banana, bean, sweet potato, potato, maize, rice	8,000
Metarhizium anisopliae	*Phyllophaga*, *Symphyla*, *Aeneolamia*, *Aphididae*	Sugarcane, pepper, tomato, coffee, lettuce, cucumber, banana, sweet potato, potato, bean, maize	11,000
Purpureocillium lilacinum	*Meloidogyne*, *Radopholus*, *Pratylenchus*	Carrot, banana, lettuce, pepper, tomato, coffee, maize, potato, sweet potato, okra	1,000
Beauveria bassiana	*Hypothenemus hampei*, *Agriotes*, *Cosmopolites sordidus*	Coffee, banana, pepper, tomato, maize, sweet potato, potato	2,800
Lecanicillium lecanii	*Hemileia vastatrix*	Coffee	500
Isaria fumosorosea	*Bemisia tabaci*	Tomato, pepper,sweet potato	1,000
Entomopathogenic nematode			
Heterorhabditis bacteriophora	*Phyllophaga*, *Agriotes*, *Cylas formicarius*	Pepper, tomato, coffee, lettuce, cucumber, banana, sweet potato, potato, bean, maize	800
Arthropod natural enemy			
Telenomus remus	*Spodoptera frugiperda*	Maize	150
Neoseiulus longispinosus	*Tetranychus urticae*	Tomato, pepper, ornamentals, citrus	37
Orius insidiosus	*Thrip* spp.	Tomato, pepper	62
Amblyseius cucumeris	*Thrip* spp.	Tomato, pepper	50
Neochetina spp.	*Eichhornia crassipes*	Aquatic weed	1

References

[Thesis Zamorano*] = BSc thesis to obtain the degree of Agricultural Engineer at the Escuela Agrícola Panamericana, Zamorano, Honduras. Available at: https://bdigital.zamorano.edu/ (accessed 18 April 2019).

Acosta, N.M. and Cave, R.D. (1994) Inventario de los parasitoides de *Liriomyza* spp. (Diptera: Agromyzidae) en la región sur de Honduras [Inventory of parasitoids of *Liriomyza* spp. in the southern region of Honduras]. *Revista de Biología Tropical* 42, 201–216.

Alduvin, C., Molina, N., Espinal, D., Ordonez, C.A. and Moncada, M.M. (2017) Honduras: the green heart of Central America. In: *Challenges and Opportunities for Food and Nutrition Security in the Americas. The View of the Academies of Sciences*. IANAS, IAP and BMBF, México DF, pp. 366–385. [Free public access of this publication in English and Spanish at www.ianas.org]

Ávila Sosa, O.A. (2010) Control de la broca del café (*Hypothenemus hampei*) utilizando once cepas de *Beauveria bassiana* y el nemátodo *Heterorhabditis bacteriophora* [Control of coffee borer beetle (*Hypothenemus hampei*) using eleven strains of *Beauveria bassiana* and the nematode *Heterorhabditis bacteriophora*]. [Thesis Zamorano*]

Bauer Stillman, A. (2016) Control de la broca del café *Hypothenemus hampei* utilizando dos nematodos entomopatógenos, *Beauveria bassiana* y el insecticida clorpirifos durante la maduración del fruto [Control of coffee borer beetle *(Hypothenemus hampei)* using two entomopathogenic nematodes, *Beauveria bassiana* and the insecticide clorpirifos during fruit ripening]. [Thesis Zamorano*]

Bendeck Meléndez, N.V. (2012) Evaluación de *Beauveria bassiana* y *Metarhizium anisopliae* para el control de *Boophilus microplus* en vacas del hato lechero de Zamorano [Evaluation of *Beauveria bassiana* and *Metarhizium anisopliae* to control *Boophilus microplus* in cows from the Zamorano dairy herd]. [Thesis Zamorano*]

Bográn, C., Obrycki, J.J. and Cave, R.D. (1998) Assessment of biological control of *Bemisia tabaci* (Homoptera: Aleyrodidae) on common bean in Honduras. *Florida Entomologist* 81, 384–395.

Cañas Castro, L.A. (1996) Response of natural enemies to sugar applications in maize and their effect on fall armyworm *Spodoptera frugiperda* (J.E. Smith) populations in Honduras. MSc thesis, Purdue University, West Lafayette, Indiana.

Castillo Samudio, R.R. (2007) Efecto de la aplicación de (*Trichoderma harzianum*) en la producción de maíz dulce (*Zea mays*) variedad Golden Baby [Effect of the application of *Trichoderma harzianum* in the production of sweet corn (*Zea mays*) Golden Baby variety]. [Thesis Zamorano*]

Cave, R.D. (1992) Inventory of parasitic organisms of the striped grass looper, *Mocis latipes* (Lepidoptera: Noctuidae), in Honduras. *Florida Entomologist* 75, 592–598.

Cave, R.D. (1993) Parasitoides larvales y pupales de *Spodoptera frugiperda* (Smith) (Lepidoptera: Noctuidae) en Centro América con una clave para las especies encontradas en Honduras [Larval and pupal parasitoids of *Spodoptera frugiperda* in Central America with a key for species found in Honduras]. *Revista científica y tecnológica de la Escuela Agrícola Panamericana Zamorano* 4, 33–56.

Cave, R.D. (1994) ¿Es viable el control biológico de un vector de geminivirus, como *Bemisia tabaci*? [Is the biological control of a gemini virus vector, such as *Bemisia tabaci* viable?] *Manejo Integrado de Plagas (Costa Rica)* 34, 18–22.

Cave, R.D. (1995a) *Manual para el Reconocimiento de Parasitoides de Plagas Agrícolas en América Central* [*Agricultural Pests Parasitoids Recognition Manual for Central America*]. Zamorano Academic Press, Zamorano, Honduras.

Cave, R.D. (ed.) (1995b) *Manual para la Enseñanza del Control Biológico en América Latina* [*Biological Control Teaching Manual for Latin America*]. Zamorano Academic Press, Zamorano, Honduras.

Cave, R.D. (1997) *Admontia* sp., a potential biological control agent of *Metamasius callizona* in Florida. *Journal of the Bromeliad Society* 47, 244–249.

Cave, R.D. (2000) Biology, ecology and use in pest management of *Telenomus remus*. *Biocontrol News & Information* 21(1), 21N–26N.

Cave, R.D. (2008) *Lixadmontia franki* Wood and Cave released to control the Mexican bromeliad weevil. *Tachinid Times* 21, 12–14.

Cave, R.D. and Acosta, N.M. (1999) *Telenomus remus*: un parasitoide para el control biológico de *Spodoptera frugiperda* [*Telenomus remus*: a parasitoid for the biological control of *Spodoptera frugiperda*]. *Revista científica y tecnológica de la Escuela Agrícola Panamericana Zamorano* 40(2), 215–227.

Cave, R.D. and Cordero, R.J. (1999) *Leptophobia aripa* parasitoids (Lepidoptera: Pieridae) in Honduras. *Revista científica y tecnológica de la Escuela Agrícola Panamericana Zamorana* 40(1), 51–55.

Cave, R.D. and Márquez, C.G. (1994) Parasitoides de Diaspididae, Coccidae y Aleyrodidae atacando cítricos en Honduras [Diaspididae, Coccidae and Aleyrodidae parasitoids attacking citrus in Honduras]. *Revista científica y tecnológica de la Escuela Agrícola Panamericana, Zamorano* 35(1), 3–8.

Cave, R.D., Trabanino, R. and Pitty, A. (2011) Zamorano y sus Contribuciones a la Agricultura Sostenible a Través del Control Biológico de Plagas [Zamorano and its contributions to sustainable agriculture through biological control of pests]. *Ceiba* 52(1), 26–38. DOI: 10.5377/ceiba.v52i1.966.

Chicas Nolasco, A.A. (2016) Evaluación de *Heterorhabditis bacteriophora* y *Steinernema carpocapsae* mediante dos métodos de aplicación para el control de *Cosmopolites sordidus* (German) en plátano [Evaluation of *Heterorhabditis bacteriophora* and *Steinernema carpocapsae* through two application methods for controlling *Cosmopolites sordidus* in banana]. [Thesis Zamorano*]

CIA (2017) The World Handbook: Honduras. Available at: https://www.cia.gov/library/publications/the-world-factbook/geos/ho.html (accessed 18 April 2019).

Cock, M.J.W. (ed.) (1985) *A Review of Biological Control of Pests in the Commonwealth Caribbean and Bermuda up to 1982*. Technical Communication No. 9, Commonwealth Institute of Biological Control. Commonwealth Agricultural Bureaux, Farnham Royal, UK.

Cordero, R.J., Pitre, H.N. and Cave, R.D. (2000) Influence of weed management and parasitoids on lepidopterous pests in intercropped sorghum and maize in southern Honduras. *Revista científica y tecnológica de la Escuela Agrícola Panamericana, Zamorano* 41(1), 41–52.

Diéguez Gonzales, V.H. (2010) Susceptibilidad del nematodo entomopatógeno *Heterorhabditis bacteriophora* a nueve plaguicidas [Susceptibility of the entomopathogenic nematode *Heterorhabditis bacteriophora* to nine pesticides]. [Thesis Zamorano*]

Duron Alvarado, M.Y. (2016) Estimación de la concentración y tiempo letal de los nematodos entomopatogenos *Heterorhabditis bacteriophora*, *Steinernema carpocapsae* para el control de broca de café (*Hypothenemus hampei*) [Estimation of concentration and lethal time of the entomopathogenic nematodes *Heterorhabditis bacteriophora*, *Steinernema carpocapsae* for controlling *Hypothenemus hampei*]. [Thesis Zamorano*]

Escribano, A., Williams, T., Goulson, D., Cave, R.D., Chapman, J.W. and Caballero, P. (1999) Selection of a nucleopolyhedrovirus for control of *Spodoptera frugiperda* (Lepidoptera: Noctuidae): structural, genetic and biological comparison of four isolates from the Americas. *Journal of Economic Entomology* 92, 1079–1085.

Espinoza, A. and Cherry, A. (1994) Efecto de la insolación sobre la persistencia del virus de poliedrosis nuclear *Galleria mellonella* (L.) (Effect of sunlight on the nuclear polyhedrosis virus (*Galleria mellonella*) persistence). *Revista científica y tecnológica de la Escuela Agrícola Panamericana, Zamorano* 35, 53–56.

Espinoza, A., Cherry, A. and Cave, R.D. (1994) Efecto del VPN *Galleria mellonella* (L.) sobre larvas de *Plutella xylostella* (L.) (Effect of NPV *Galleria mellonella* on *Plutella xylostella* larvae). *Revista científica y tecnológica de la Escuela Agrícola Panamericana, Zamorano* 35, 57–61.

Espinoza Silva, L.R. (2005) Evaluación de cepas de *Beauveria bassiana* y de *Metarhizium anisopliae* en control biológico de *Boophilus microplus* [Evaluation of *Beauveria bassiana* and *Metarhizium anisopliae* strains on the biological control of *Boophilus microplus*]. [Thesis Zamorano*]

Fernández Tondelli, J.A. (2006) Evaluación de la eficiencia del control de garrapatas (*Boophilus microplus*) con tres frecuencias de aplicación de Bazam (*Beauveria bassiana*) [Efficiency evaluation on the control of tick (*Boophilus microplus*) with three application frequencies of Bazan (*Beauveria bassiana*)]. [Thesis Zamorano*]

Gómez Cabrera, L.E. (1995) Control biológico clásico de *Bemisia tabaci* (Gennadius) en Honduras [Classical biological control of *Bemisia tabaci* (Gennadius) in Honduras]. [Thesis Zamorano*]

González, A. and Cave, R.D. (1997a) Comparación de las poblaciones de arañas foliares diurnas en frijol común bajo dos sistemas de labranza [Population comparison of daytime leaf spiders on bean plants under two tillage systems]. *Revista científica y tecnológica de la Escuela Agrícola Panamericana, Zamorano* 38, 47–50.

González, A. and Cave, R.D. (1997b) Comparación del parasitismo de huevos de *Empoasca kraemeri* Ross & Moore (Homoptera: Cicadellidae) por *Anagrus* spp. (Hymenoptera: Mymaridae) en frijol común en labranza cero y labranza convencional [Comparison of egg parasitism of *Empoasca kraemeri* by *Anagrus* spp. in common bean plants under zero and conventional tillage]. *Revista científica y tecnológica de la Escuela Agrícola Panamericana, Zamorano* 38, 51–56.

Grijalva Teran, I.A. (2014) Evaluación de cuatro medios sólidos de crecimiento para la producción *in vitro* del nematodo entomopatógeno *Heterorhabditis bacteriophora* [Evaluation of four solid culture media for the *in vitro* production of the entomopathogenic nematode *Heterorhabditis bacteriophora*]. [Thesis Zamorano*]

Juárez Figueroa, R.H. (2007) Control del escarabajo del estiércol *Alphitobius diaperinus* con *Heterorhabditis bacteriophora*, *Beauveria bassiana* y *Metarhizium anisopliae* [Control of *Alphitobius diaperinus* with *Heterorhabditis bacteriophora*, *Beauveria bassiana* and *Metarhizium anisopliae*)] [Thesis Zamorano*]

Mantilla Compte, J. (2007) Movimiento de *Trichoderma harzianum* en un suelo de textura media cultivado con pepino (*Cucumis sativa*), suministrado a una y dos horas en el sistema de riego por goteo en El Zamorano, Honduras [Movement of *Trichoderma harzianum* in medium texture soils cultivated with cucumber (*Cucumis sativa*), providing one and two hours of drip irrigation in El Zamorano, Honduras]. [Thesis Zamorano*]

Mata Marcillo, J.E. (2013) Evaluación de la distribución del nematodo entomopatógeno *Heterorhabditis bacteriophora* a través de un sistema de riego por goteo y su desplazamiento en el suelo [Evaluation of the distribution of the entomopathogen *Heterorhabditis bacteriophora* through dripping irrigation systems and its movement on soils]. [Thesis Zamorano*]

Medina Terán, G.M. (2003) Eficacia de *Beauveria bassiana* y *Verticillium lecanii* aplicados a tres concentraciones en dos formulaciones para el control de *Spodoptera frugiperda* en jilote y *Aphis* sp. en pepino [The efficiency of *Beauveria bassiana* and *Verticillium lecanii* applied in three concentrations and two formulations for controlling *Spodoptera frugiperda* and *Aphis* sp. in cucumber]. [Thesis Zamorano*]

Méndez Martínez, J.A. (2003) Efectos de aplicación de *Trichoderma harzianum* y *Paecilomyces lilacinus* en el rendimiento de lechuga orgánica en Zamorano [Effects of the application of *Trichoderma harzianum* and *Paecilomyces lilacinus* in the production of organic lettuce in Zamorano]. [Thesis Zamorano*]

Mojica Espinoza, K.R. (2016) Evaluación de *Heterorhabditis bacteriophora* y *Steinernema carpocapsae* mediante dos métodos de aplicación para el control de *Cosmopolites sordidus* (German) en plátano [Evaluation of *Heterorhabditis bacteriophora* and *Steinernema carpocapsae* through two application methods to control *Cosmopolites sordidus* in banana crops]. [Thesis Zamorano*]

Morales Ramírez, F. (2012) Estimación de la concentración y tiempo letal del nematodo entomopatógeno *Heterorhabditis bacteriophora* (Nematoda: Heterorhabditidae) para el control de *Cosmopolites sordidus* (Coleoptera: Curculionidae) [Estimation of concentration and lethal time of the entomopathogenic *Heterorhabditis bacteriophora* to control *Cosmopolites sordidus*]. [Thesis Zamorano*]

Morán Quintero, N.R. (2007) Evaluación de cuatro cepas de *Trichoderma harzianum* para el control de *Rhizoctonia solani* en plántulas de pepino (*Cucumis sativa*) [Evaluation of four strains of *Trichoderma harzianum* to control *Rhizoctonia solani* on cucumber plants (*Cucumis sativa*)]. [Thesis Zamorano*]

Morán Ruiz, F.S. (2007) Efectividad del fraccionamiento de la dosis comercial 3×10^{11} UFC/ha de Trichozam (*Trichoderma harzianum*) en el crecimiento de las plántulas de siete cultivos hortícolas [Effectiveness of Trichozam (*Trichoderma harzianum*) at a commercial dose of 3×10^{11} UFC/ha on the growth of seven horticultural crops]. [Thesis Zamorano*]

Morjan Erazo, W.E. (1993) Depredadores nocturnos de plagas de maíz y de frijol en dos sistemas de labranza [Nocturnal pest predators on corn and bean crops under two tillage systems]. [Thesis Zamorano*]

Nuñez Flores, M.A. (2011) Efectividad del nematodo entomopatógeno *Heterorhabditis bacteriophora* y plan de manejo químico para el control de larvas de *Phyllophaga* sp. (Coleoptera: Scarabaeidae) en el cultivo de camote (*Ipomoea batatas*) [Effectiveness of the entomopathogenic *Heterorhabditis bacteriophora* and chemical control plan to control *Phyllophaga* sp. larvae in sweet potato (*Ipomoea batatas*) crops]. [Thesis Zamorano*]

Pantoja Guamán, D.O. (2009) Capacidad depredadora de *Orius insidiosus* (Say) sobre *Thrips tabaci* (Lindeman) en laboratorio y en cultivo de pepino [Predatory capacity of *Orius insidiosus* on *Thrips tabaci* under lab conditions in cucumber crops]. [Thesis Zamorano*]

Parada López, J.E. (2007) Estudio de factibilidad para la comercialización a través de E-commerce de bioplaguicidad marca Zamorano en Honduras [Feasibility study in Honduras for Zamorano branded biopesticides through E-commerce]. [Thesis Zamorano*]

Perrera Viamil, A.A. (2009) Efectividad del nematodo *Heterorhabditis bacteriophora* para el control de larvas de *Phyllophaga* sp. [Effectiveness of the nematode *Heterorhabditis bacteriophora* in controlling larvae of *Phyllophaga* sp.]. [Thesis Zamorano*]

Pineda Bonilla, M. (2014) Efecto del corte de pelo y el uso de hongos entomopatogenos (*Beauveria bassiana* y *Metarhizium anisopliae*) en el control de garrapata (*Boophilus microplus*) en ganado bovino [Effect of hair length and the use of *Beauveria bassiana* and *Metarhizium anisopliae* to control tick (*Boophilus microplus*) in bovines]. [Thesis Zamorano*]

Pitty, A., Dejud, I., Zelaya, I. and Barletta, H. (1993) Control biológico del lirio acuático (Biological control of water lily). *Agricultura de las Américas* 42(6), 18–20.

Prieto Navas, H.J. (2016) Evaluación de tres concentraciones de *Isaria fumosorosea* para el control de *Bemisia tabaci* en cultivo de chile dulce bajo macro túnel [Evaluation of three concentrations of *Isaria fumosorosea* to control *Bemisia tabaci* on sweet pepper crops in macro tunnels]. [Thesis Zamorano*]

Ramírez Herrera, J.M. (2016) Control del picudo (*Cosmopolites sordidus*) en el cultivo de plátano (*Musa paradisiaca*) usando tres agentes biológicos, *Heterorhabditis bacteriophora*, *Beauveria bassiana* y *Metarhizium anisopliae* [Control of *Cosmopolites sordidus* in banana plantations (*Musa paradisiaca*) using three biological agents, *Heterorhabditis bacteriophora*, *Beauveria bassiana* and *Metarhizium anisopliae*]. [Thesis Zamorano*]

Rivera Jerez, C.G. (2004) Evaluación de sensibilidad *in vitro* de Trichozam (*Trichoderma harzianum*) a nueve fungicidas [*In vitro* evaluation of sensitivity of Trichozam (*Trichoderma harzianum*) to nine fungicides]. [Thesis Zamorano*]

Roldán Salazar, E.L. (2016) Evaluación del efecto hiperparásito de tres concentraciones de *Lecanicillium lecanii* sobre la roya del café (*Hemileia vastatrix*) [Evaluation of the hyperparasitic effect of three concentrations of *Lecanicillium lecanii* on coffee rust (*Hemileia vastatrix*)]. [Thesis Zamorano*]

Romero Villagra, D.C. (2004) Efectos de la aplicación de *Paecilomyces lilacinus* en el control de *Meloidogyne* spp. en pepino [Effects of the application of *Paecilomyces lilacinus* on *Meloidogyne* sp. in cucumber]. [Thesis Zamorano*]

Rosero Asqui, A. (2008) Evaluación de cuatro cepas de *Trichoderma* sp. y sus combinaciones para el control de *Fusarium* sp. en sandía [Evaluation of four strains of *Trichoderma* sp. and their combinations to control *Fusarium* sp. in watermelon]. [Thesis Zamorano*]

Sanchez, B. and Carlos, F. (2015) Compatibilidad del hongo entomopatógeno *Isaria fumosorosea* con la mariquita *Thalassa montezumae* para control de la escama verde del croton, *Phalacrococcus howertoni* (Hemiptera: Coccidae), Zamorano, Honduras [Compatibility of the entomopathogenic fungus *Isaria fumosorosea* with the ladybug *Thalassa montezumae* for control of the green croton scale (*Phalacrococcus howertoni*), Zamorano, Honduras]. [Thesis Zamorano*]

Suazo López, O.O. (2008) Optimización de los parámetros de aplicación de *Heterorhabditis bacteriophora* para el control de *Spodoptera frugiperda* en el cultivo de maíz en condiciones de campo, Zamorano, Honduras [Optimization of application parameters of *Heterorhabditis bacteriophora* for control of *Spodoptera frugiperda* in maize under field conditions]. [Thesis Zamorano*]

Toapanta Valencia, L.D. (2005) Efecto de la aplicación de *Beauveria bassiana* y *Metarhizium anisopliae* para el control de *Phyllophaga* spp. y *Aeolus* spp. en cultivo de camote (*Ipomoea batatas*) [Effect of application of *Beauveria bassiana* and *Metarhizium anisopliae* to control *Phyllophaga* sp. and *Aeolus* sp. in sweet potato (*Ipomoea batatas*) cultures]. [Thesis Zamorano*]

Torres Estrada, H. (2016) Control del picudo (*Cosmopolites sordidus*) en el cultivo de plátano (*Musa paradisiaca*) usando tres agentes biológicos, *Heterorhabditis bacteriophora*, *Beauveria bassiana* y *Metarhizium anisopliae* [Control of banana weevil (*Cosmopolites sordidus*) in plantain (*Musa paradisiaca*) cultures by using three biocontrol agents, *Heterorhabditis bacteriophora*, *Beauveria bassiana* and *Metarhizium anisopliae*]. [Thesis Zamorano*]

Turcios Rivas, E.E. (2009) Evaluación del movimiento del nematodo *Heterorhabditis bacteriophora* y su capacidad infectiva bajo condiciones controladas de humedad y tres texturas de suelo [Evaluation of the movement of the nematode *Heterorhabditis bacteriophora* and its infective capacity in three soil textures under controlled humidity conditions]. [Thesis Zamorano*]

Turcios Rivera, C.W. (2009) Preferencia de depredación de *Neoseiulus californicus* (McGregor) (Acari: Phytoseiidae) sobre huevos, ninfas y adultos de *Tetranychus gloveri* (Banks) y *T. ludeni* (Zacher) (Acari: Tetranychidae) en Zamorano, Honduras [Prey preference of *Neoseiulus californicus* when exposed to eggs, nymphs and adults of *Tetranychus gloveri* and *T. ludeni* (Zacher) at Zamorano, Honduras]. [Thesis Zamorano*]

Ucan Yam, I. (2016) Evaluación de cuatro formulaciones para el almacenamiento del nematodo entomopatógeno *Heterorhabditis bacteriophora* en dos estados de actividad [Evaluation of four formulations for the storage of the entomopathogenic nematode *Heterorhabditis bacteriophora* in two activity stages]. [Thesis Zamorano*]

Vaquedano Gámez, L.C. (2006) Efecto de la aplicación de hongos entomopatógenos para el control de plagas en el cultivo de pepino, en el valle de Comayagua, Honduras [Effect of application of entomophagogenic fungi for control of pests in cucumber in the valley of Comayagua, Honduras]. [Thesis Zamorano*]

Villarroel Basantes, R.I. (2009) Respuesta funcional, respuesta numérica e interferencia de *Neoseiulus californicus* (McGregor) (Acari: Phytoseiidae) sobre *Tetranychus ludeni* (Zacher) y *Tetranychus gloveri* (Banks) (Acari: Tetranychidae) en Zamorano, Honduras [Functional and numerical response, and effect of *Neoseiulus californicus* on *Tetranychus ludeni* and *Tetranychus gloveri* in Zamorano, Honduras]. [Thesis Zamorano*]

Welchez Arita, J.A. and Guerra Burgos, J.O. (2013) Evaluación de la efectividad de cuatro fungicidas biológicos en el control del hongo de la roya de café *Hemileia vastatrix* [Evaluation of effectiveness of four biofungicides to control the coffee leaf fungus *Hemileia vastatrix*]. [Thesis Zamorano*]

Williams, T., Goulson, D., Caballero, P., Cisneros, J., Martínez, A.M., Chapman, J.W. *et al.* (1999) Evaluation of a baculovirus bioinsecticide for small-scale maize growers in Latin America. *Biological Control* 14, 67–75.

Wood, D.M. and Cave, R.D. (2006) Description of a new genus and species of weevil parasitoid from Honduras (Diptera: Tachinidae). *Florida Entomologist* 89, 239–244.

Yánguez Quintero, L.E. (2016) Evaluación del efecto hiperparásito de tres concentraciones de *Lecanicillium lecanii* sobre la roya del café (*Hemileia vastatrix*) [Evaluation of the hyperparasitic effect under three concentrations of *Lecanicillium lecanii* against coffee rust (*Hemileia vastatrix*)]. [Thesis Zamorano*]

20 Biological Control in Jamaica

Michelle A. Sherwood[1]* and Joop C. van Lenteren[2]

[1]Crop and Plant Protection Unit, Research and Development Division, Ministry of Industry of Agriculture and Fisheries, Old Harbour, St Catherine, Jamaica; [2]Laboratory of Entomology, Wageningen University, Wageningen, The Netherlands

* E-mail: masherwood@micaf.gov.jm

Abstract

Jamaica has a rich history of successful biological control of more than 25 pests of economic and quarantine importance. Approximately 14 classical, 13 natural and six augmentative biocontrol programmes were recorded, as well as two fortuitous introductions. The programmes concerned control of key pests on: (i) citrus, such as citrus blackfly, citrus red scale, cottony cushion scale, citrus root weevils, brown citrus aphid, lime swallowtail butterfly and Asian citrus psyllid; (ii) sugarcane, mainly sugarcane borers; (iii) banana, the banana weevil; (iv) cocoa, the cocoa thrips; (v) coconut, with coconut scale, two aphid species and red palm mite; (vi) sweet potato, sweet potato weevil; (vii) crucifers, with diamondback moth and cabbage looper; (viii) coffee, with coffee berry borer and coffee leaf miner; (ix) residential fruit tree crops and ornamentals, with pink hibiscus mealybug and ensign scale; (x) papaya, with papaya mealybug and citrus root weevil; and (xi) onion and scallion, with beet armyworm. Biocontrol agents used were species of parasitoids, predatory beetles and mites, nucleopolyhedroviruses, and entomopathogenic fungi and nematodes. Jamaica served as a provider of biocontrol agents for the Caribbean and Hawaii. A biocontrol facility is currently being built to develop control methods for endemic and invasive pests.

20.1 Introduction

Jamaica is approximately1 million hectares in size (82 km wide by 234 km long). Its varied topography and climate allow for a diversity of habitats and growing conditions. The population is currently approximately 2.7 million (PIOJ, 2009). The main traditional agricultural products are sugarcane, bananas, coffee and citrus, yams, ackees, vegetables, poultry, goats, milk and shellfish (CIA, 2017). One-fifth of the employed labour force works in agriculture. The rural population makes up about half of the total population, though not all family members are employed in farming, and more than 85% of rural people live on farms of less than 5 ha. The agricultural sector is extremely vulnerable to disturbance such as weather conditions, pest infestations, impact of natural disasters and changes in export market prices and trading regimes. Revitalization of the sector is of priority in order to achieve greater efficiency and competitiveness (PIOJ, 2009).

20.2 History of Biological Control in Jamaica

20.2.1 Period 1870–1969

The majority of the information summarized for this period comes from Cock (1985). Almost all projects during this period concerned classical biocontrol.

Classical biological control of rats

The earliest known record of biocontrol in Jamaica dates back to 1872 when four male and five female mongooses, *Herpestes auropunctatus* (Hodgson), were introduced from India by W.B. Espeut, a sugarcane producer in Jamaica (Espeut, 1882). The idea originated from Mrs Espeut, who had a pet mongoose in Ceylon (now Sri Lanka), and ship owners who had successfully controlled rat (*Rattus* sp.) infestations with mongooses. It was so successful that Mr Espeut sold many mongooses to other Caribbean islands, including Puerto Rico where it became established by 1877 (Long, 2003). Initially the mongooses appeared to have controlled the rat populations, as declines were observed by sugarcane growers (Pimentel, 1955). However, the mongoose, a diurnal and omnivorous animal, proved to be a nuisance itself, damaging crops and poultry production, serving as a rabies reservoir (Johnson *et al.*, 2015) and destroying ground-nesting birds and lizards (Brathwaite *et al.*, 1974).

Classical biological control of citrus pests: citrus blackfly, citrus red scale, cottony cushion scale and citrus weevils

CITRUS BLACKFLY. In 1913, the citrus blackfly *Aleurocanthus woglumi* Ashby, native to Asia, was found in Jamaica after first being reported in Cuba that year (Brathwaite *et al.*, 1974). By 1914 the blackfly was widespread on citrus, coffee and several other plants. It was attacked by the fungus *Aschersonia aleyrodis* Webber, but remained a serious pest. Attempts to control it by establishing nests of the ant *Crematogaster brevispinosa* Mayr failed. Introduction of the coccinellid predator *Delphastus catalinae* Horn was suggested in 1917, but was probably not undertaken. Further control attempts using *A. aleyrodis* proved unsuccessful. During the period 1929–1931 the US Department of Agriculture (USA) conducted explorations in Malaysia and Indonesia (on Java and Sumatra) which led to the introduction into Jamaica, via Cuba, of the aphelinids *Eretmocerus serius* Silv. and *Encarsia divergens* (Silv.) and the coccinellids *Catana clauseni* Chapin and *Scymnus smithianus* Silv. Of these, *C. clauseni* became temporarily established and *E. serius* (origin peninsular Malaysia) built up rapidly and, since it became widespread in 1933, has consistently kept this blackfly in check (Cock, 1985). Following reports during the 1950s and early 1960s that occasional heavy attacks still occurred, the Commonwealth Institute of Biological Control (CIBC, now CABI) recommended the introduction of another aphelinid, *Encarsia opulenta* (Silv.) from Mexico, where this species provided effective control. Consignments of field-collected adults were released in September–October 1964. Subsequent surveys indicated that *E. opulenta* was well established and within 2 years had largely supplanted *E. serius*. Since its establishment, there have been no reports of any major upsurges of citrus blackfly (Cock, 1985).

CITRUS RED SCALE. The aphelinid *Aphytis diaspidis* (How.) was introduced in 1934 and tried against citrus red scale *Aonidiella aurantii* (Mask.) and established at the liberation site (Cock, 1985).

COTTONY CUSHION SCALE. An attempt to import the coccinellid *Rodolia cardinalis* from Bermuda for control of cottony cushion scale *Icerya purchasi* Mask. in 1964 failed, but field collections at that time indicated its presence in Jamaica. In 1970 *R. cardinalis* was imported from Barbados, but it is not known whether it established (Cock, 1985).

CITRUS WEEVILS. Two genera of large weevils, *Diaprepes* spp. and *Exophthalmus* spp., damage citrus: the larvae cause considerable damage to roots; the adults cause comparatively less damage by feeding on the leaves, though damage to seedlings can be severe. While *Diaprepes* spp. are reported from other islands, *Exophthalmus* spp. are a problem only in Jamaica. Surveys in Jamaica resulted in finding four species of parasitoids: three eulophids, *Tetrastichus haitiensis* Gah., *T. gala* Wlk. (= *T. marylandensis* in lit.) and *Horismenus* sp.nr. *cupreus* (Ashm.), and a platygasterid, *Platystasius citri* Nixon. A consignment of *Brachyufens* sp. was sent from Dominica to Jamaica in 1958, but none was received alive. In 1959, W. Dixon of the Jamaica Department of Agriculture brought *T. 'marylandensis'* from Puerto Rico and released it, and another unrecorded introduction was made in 1961 (Cock, 1985). Cultures of *Beauveria bassiana* and *Metarhyzium anisopliae* were received from the University of California. Spores of *Bacillus papillae* were also imported. An assessment of these spores was conducted with adults and first-instar larvae *Prepodes* and *Pachneus* spp. *Beauveria* and *Metarhyzium* proved to be pathogenic to the first-instar larvae under precise conditions in the laboratory, but only with great difficulty could these conditions be duplicated in the field. *Bacillus papillae* exhibited little pathogenicity to fiddler beetle larvae. Fungal spores were kept for evaluation against scarabid larvae (MOA, 1957). No subsequent record has been found that reported the outcome of this evaluation.

Classical biological control of sugarcane moth borers

Sugarcane was first imported by Columbus in 1493 into the Dominican Republic. For approximately 150 years (from mid 17th century to start of 19th century) the Caribbean was the world's leading exporter of sugarcane and in 1805 Jamaica was the worlds' largest exporter, recording up to 100,000 t (Falloon, 2005). In most of the West Indies, sugarcane was the dominant agricultural crop during the 19th century and first half of the 20th century and it is still the most widely grown crop in the region as a whole (Cock, 1985). Sugarcane is attacked by a number of pests and biocontrol programmes have been successfully developed for several of these.

Sugarcane moth borers *Diatraea* spp. are the most widespread and in many places the most important pests in the Caribbean and several programmes have achieved satisfactory control. In Jamaica, *Diatraea saccharalis* (F.) causes damage to sugarcane. Some native parasitoids of *Diatraea* spp. occurring on wild grasses have adapted to the sugarcane ecosystem. Initially, during the first half of the 20th century, native *Diatraea* parasitoids with limited ranges within the Neotropical Region were redistributed, often leading to a degree of control. Several tachinid parasitoids were tested in the 1930s. In Jamaica, natural control of *D. saccharalis* by the native tachinid *Lixophaga diatraeae* (Tns.) has always played an important role in the reduction of the pest.

WEST INDIAN CANE FLY. The West Indian cane fly *Saccharosydne saccharivora* (Westw.) is usually a minor pest, but irregular intense outbreaks may occur over large areas in Jamaica. Although it has a well developed natural enemy complex, efforts were made to supplement this in Jamaica in the 1950s and 1960s, but several of the introduced species were already present (predators: *Coleomegilla maculata* (Deg.); parasitoids: *Anagrus flaveolus* Waterh.) and none of the additional species established (*Pseudogonatopus saccharivorae* Richards, *Tetrastichus* sp. nr. *vaquitarum* Wolc., *Tytthus mundulus* (Bredd.)). The native parasitoids *Tetrastichus* sp. and *Stenocranophilus quadratus* Pierce were having an effect in the reduction of cane fly populations (Cock, 1985).

Classical biological control of banana weevil

The first attempt to achieve biocontrol of the banana weevil *Cosmopolites sordidus* (Germ.) was made in Jamaica in 1918–1919 with the introduction of the beetles *Plaesius javanus* Erichs. from Indonesia (Java) and *Dactylosternum hydrophiloides* (Macleay) and *D. abdominale* (F.) from peninsular Malaysia. These efforts were unsuccessful and the same species were again imported in 1937–1938 from Fiji. Releases of *P. javanus* and *D. hydrophiloides* were continued until 1942: both species became established and provided a considerable measure of control (Cock, 1985). In 1948–1949, *P. javanus*, *D. hydrophiloides* and *D. abdominale* were imported from the Commonwealth Bureau of Biological Control (CBBC, now CABI) station in Trinidad and released in new banana plantations. For many years, it became standard practice to collect and release these predators in newly developed plantations and replanted fields. However, at the end of the 1950s chemical pesticides were recommended to treat the soil for weevil control. Both beetle species were still present in the 1980s (Cock, 1985).

Classical biological control of cocoa thrips

Small amounts of cocoa are produced in Jamaica. Cocoa thrips *Selenothrips rubrocinctus* (Giard) is a pest of cocoa, but also of mango and cashew. When conditions are improved by providing shade and windbreaks, an outbreak usually subsides. In 1937, adults of the eulophid parasitoid *Goetheana parvipennis* (Gahan) sent from Trinidad were released on infested mango trees. Small numbers were recovered at the end of the year and it has since been collected repeatedly from thrips on mango trees (Cock, 1985). In 1948 and 1950, *G. parvipennis* was introduced once more for control of *S. rubrocinctus* on mango and cashew. The wasp species was reared in large numbers in the laboratory and mass released in mango groves at several locations (Department of Agriculture Jamaica, 1949, 1951).

Natural and classical biological control of coconut scale and two aphid species

In 1948, the coconut scale *Aspidiotus destructor* Sign., though not yet recorded from Jamaica at that time, was a potential danger because of proximity to the Cayman Islands, where this pest was firmly established. Specimens of four species of coccinelids (*Pseudoazya* (*Azya*) *trinitatis* Gordon, *Cryptognatha nodiceps* Mshl., *Chnoodes* spp. and *Prodilis (Neoporia)* sp., new for Jamaica) were sent from the Trinidad branch of the Imperial Bureau Parasite Service of Belleville, Ontario. Releases were made in coconut growing areas, citrus and bamboo groves, in the flower garden at Hope Botanic Gardens and elsewhere. Shipments of coccinellids from Trinidad were discontinued in 1949 (Department of Agriculture Jamaica, 1949, 1950). The coconut scale was first recorded in Jamaica in 1957, and predatory coccinellids were requested from the West Indian Station of CIBC. Although no preliminary surveys could be made, *C. nodiceps* and *P. trinitatis* and miscellaneous coccinellids associated with scales on coconuts in Trinidad were sent in 1958. In 1960, *C. nodiceps* and *P. trinitatis* were both present, but their economic value was not assessed (Cock, 1985). *Chilocorus cacti* L. was observed feeding on the star scale *Vinsona stellifera* Westw., the Florida scale *Chrysomphalus aonidium* L. and the Boisduval's scale *Diaspis boisduvalii* Sign. The white entomophagous fungus *Cephalosporum lecanii* assisted in the control of the green scale *Coccus viridis* G. (Department of Agriculture Jamaica, 1951).

The ladybird beetle *Cycloneda sanguinea* L. was observed feeding in very large numbers in all the Irish potato-growing areas in Christinana (Manchester), controlling outbreaks of aphids. The predator and *C. cacti* were also found feeding on the coconut aphid (*Cerataphis lataniae*, Boisduval, 1867) and on *Myzus persicae* Sulz. affecting citrus (Department of Agriculture Jamaica, 1951).

Classical biological control of pineapple mealybug

The pineapple mealybug *Dysmicoccus brevipes* (Ckll.) is a common pest of pineapples. In the 1930s, a cecidomyiid predator of the mealybug, probably *Vincentodiplosis pseudococci* (Felt), was found in Jamaica, but no parasitoids. In 1939, the encyrtid parasitoid *Hambletonia pseudococcinna* Comp. as well as two coccinellids, *Hyperaspis* sp. and *Diomus* sp., were imported from Hawaii. In 1965, a specimen of *H. pseudococcinna* from a citrus tree at Montego Bay, Jamaica, was found, but since this encyrtid has been reared or collected from several of the Lesser Antilles where no introductions have been made, it is not impossible that it was already present in Jamaica before the introduction from Hawaii (Cock, 1985).

Classical biological control of various scales on trees and ornamentals

The parasitic aphelinid *Aphytis diaspidis* (How.) was introduced in 1934 and tried against a number of coccids, including citrus red scale *Aonidiella aurantii* Maskell on citrus, and white peach scale *Pseudaulacaspis pentagona* (Targioni), which is a pest of papaya, cotton, oleander and mulberry. The parasitoid was recorded as established at the liberation site (Cock, 1985). This parasitoid was later reared in Jamaica on the scale *Pentagona* (*Aulacaspis*) *pentagona* on oleander and released on Mahoe trees for control *of A. pentagona* (Department of Agriculture Jamaica, 1949).

Classical biological control of puncture vine

Puncture vine *Tribulus cistoides* L. is an annual herb, with small yellow flowers and barbed seed cases, which has become disseminated from the Old World throughout the tropics. Biocontrol is based on the use of the weevils *Microlarinus lareynii* (Duv.), which attacks seeds, and *M. lypriformis* (Woll.), which attacks the stems of puncture vine. These two weevils were successfully introduced from Italy to California. Both weevils have high dispersive powers and *M. lypriformis* has made its way unaided from California to Mexico, Florida, Jamaica and the Bahamas (Cock, 1985).

Jamaica as provider of biological control agents

During this period, Jamaica sent several species of biocontrol agents to other countries in the region and these shipments are summarized in the chapters of the receiving countries. It also provided a plant-killing fungus, *Cercosporella ageratinae (nomen nudum)* Hyphomycetes, to Hawaii for control of the weed *Ageratina riparia* (Regel) K. & R. (= *Eupatorium riparium* Regel) in Hawaii. It is credited with a 40–60% reduction of the weed in the high-moisture Kona ranchland (Cock, 1985).

20.2.2 Period 1970–2000

Classical biological control of fruit flies

The fruit fly *Anastrepha* spp. includes several important pests of soft fruits including citrus, guava and mango. In some areas, no sound fruit can be obtained. The species of greatest importance varies with the crop and with the location. Four consignments of the parasitoid *Pachycrepoideus vindemiae* were sent from the West Indian Station CIBC in 1970 to Jamaica, but whether they established is unknown (Cock, 1985).

Classical biological control of the sugarcane borer

Sugarcane, its pests and their biocontrol up to the 1970s has been summarized above. Developments during the period 1970–2000 are described in this section. Chemical control of sugarcane moth borer *D. saccharalis* is not recommended in Jamaica. Native natural enemies together with the imported parasitoid *Cotesia (Apanteles) flavipes* Cameron result in important reductions of moth borer populations (Falloon, 2005). Two other tachinid parasitoids, *Lydella* (= *Metagonistylum*) *minense* (Tns.) (in 1974) and *Billaea* (= *Paratheresia*) *claripalpis* (Wulp) (in 1973 and 1980) were introduced to Jamaica, but did not establish. The hymenopteran parasitoid *C. flavipes* was introduced into Jamaica in 1974 but did not establish. However, after a second introduction in 1980 it did establish (Cock, 1985).

A study was conducted which monitored the success of *C. flavipes* in parasitizing the borer and in reducing stalk damage. Borers were collected from the field and reared under observation in the laboratory and a record was kept of parasitoid emergence. Borer damage levels were monitored by conducting periodic stalk damage surveys, random selection of stalks from fields, splitting and examination of internal stalk damage. In 1983, the first field recovery of *Cotesia* was made. In 1987, the parasitoid was established and declared permanent, after which rearing and release were phased out. Prior to release of *Cotesia*, total parasitism by other parasitoids was 22%; and after *Cotesia* releases, total parasitism increased to 37% in 1987. However, in 1999 very low levels of *Cotesia* parasitism were detected. In 2000, *Cotesia* was imported from Barbados and in 2003–2004 from Guatemala for rearing and release in infested areas, but the project was discontinued after hurricane Ivan (Falloon, 2005). Seasonality in parasitism levels has been shown in the field since the introduction of *Cotesia* and interference between local and imported parasitoids is thought to occur. Parasitism did not show a net increase and monitoring showed no sustained decrease in stalk damage; therefore questions have been raised about the efficacy of *Cotesia* in Jamaica (Falloon, 2005).

Augmentative biocontrol of the sweet potato weevil

Six heterorhabditid entomopathogenic nematode isolates from the Caribbean (*Heterorhabditis* sp. D1 Strain (Group I; [local variant of D1 strain] JAM34 and Luquilla), *Heterorhabditis* sp. D1 strain (Group II; JAM23, SJo02, and SC12), a previously unrecorded species *Heterorhabditis* sp. (Group III; El Yunque) and two isolates from North America, *Heterorhabditis bacteriophora* Poinar HP88 and NJ70 were compared for their potential as biocontrol agents for the sweet potato weevil *Cylas formicarius* (Fabricius). Nematodes differed in their ability to invade and kill late instars of the weevil, but did not differ in the number of progeny produced within the weevil cadavers. Virulence of the nematodes to the weevil was not related to climatic origin, but differed among species groups. Three species (*Heterorhabditis*, D1 (JAM34), *Heterorhabditis* sp. (El Yunique), and *H. bacteriophora* HP88) were compared for their ability to survive, infect and reproduce in various soil temperature regimes. The tolerance of the three nematode species to soil temperature appeared to be related to their climatic origin. One species from the Caribbean, *Heterorhabditis* sp. D1 strain (JAM34), was more tolerant of higher temperatures than the temperate *H. bacteriophora* HP88 strain. HP88 was also better adapted to lower temperatures than JAM34. This research helps in determining the potential for a nematode-based biocontrol programme against sweet potato weevil (Lawrence, 1994) in Jamaica.

Natural and classical biological control of pests of cruciferous crops: diamondback moth and cabbage looper

DIAMONDBACK MOTH. The diamondback moth *Plutella xylostella* (L.) is a major pest of cruciferous crops in the island. Alam (1996), during monthly sampling of *Plutella* populations on three locations in Jamaica from 1988 to 1993, found 34 species of natural enemies. Five species were primary parasitoids (*Trichogramma* sp., *Diadegma insulare* (Cresson), *Cotesia* sp., *Oomyzus sokolowskii* (Kurdj.) and *Trichospilus diatraeae* (C. and M.); 11 were insect predators (three species each of carabids, coccinellids and syrphids and one species each of staphylinid and chrysopid); 15 species were spiders; and three species were entomophagous fungi. The parasitoids *D. insulare*, *Cotesia* sp. and *O. sokolowskii* were the most common ones, parasitizing significantly ($p = 0.01$) more (35–60%) in unsprayed than in sprayed (24–27%) fields. Four species of hyperparasitoids (*Spilochalcis* sp., *Horismenus* sp., *Cotaloccus* sp. and *A. fijensis*) were also recorded, attacking pupae of *D. insulare*,

C. plutellae and *Cotesia* sp. Studies on the mortality of *Plutella* due to parasitoids, predators and rainfall carried out at two sampling locations revealed that biotic and abiotic factors caused an average mortality of 84–92%. A larval parasitoid, *Cotesia plutellae* (Kurd.), was successfully introduced from Trinidad in the 1970s and again in the 1980s, after which it established, as 48% and 54% of parasitism were achieved within 60 days of its first release (Alam, 1996).

CABBAGE LOOPER. The cabbage looper *Trichoplusia ni* (Hb.) is a pest of many crops, including crucifers. During the monthly sampling described in the previous paragraph, Alam (1996) found several parasitoids as well as entomopathogenic fungi attacking the cabbage looper. Four species of parasitoids (*Cotesia* sp., *Euplectrus platyhpenae* (How), *Brachymeria* sp. and *Winthemia* sp.) were found attacking up to 38% of the larvae and pupae of the looper. Two entomopathogenic fungi (*Entomophthoralis* sp. and *Beauveria bassiana* (Balsmo)) were found infecting up to 34% of the larvae of *T. ni* (Alam,1996).

Fifteen species of spiders were found feeding on diamondback moth and other crucifer pests in Jamaica. Among these *Lycosa* spp., *Theridula. gonygaster* (Simon), *Argiope trifasciata* (Forskal) and *Habronathus* sp. were the most abundant, and these spiders could consume 0.13–0.17 g of food or 14–18 larvae per day. However, the use of chemical pesticides reduced the spider populations by 43–100% (Alam, 1996).

Natural and classical biological control of pine mite

Pine mite *Oligonychus milleri* (McG.) is a pest of conifers native to North America, but was found in the 1970s on *Pinus caribaea* Morelet (Pinaceae), which is an important forestry crop in Jamaica. Initial surveys for natural enemies in Jamaica resulted in finding 32 species of predacious mites and insects in association with this pest, but they did not provide adequate control. High level of attack by the pine mite on young trees reduced the rate of growth, affecting the economics of the crop. In 1974, stocks of mite predators were sent to Jamaica from the University of California at Riverside (USA) and releases were made of *Stethorus salutaris* Kapur and phytoseiid mites *Neoseiulus* (= *Amblyseius*) *californicus* (McG.), *Typhlodromus occidentalis* Nesbitt and *T. citri* Garman & McG. No recoveries of the released predators have been reported. The phytoseiid *Fundiseius cesi* (Muma), collected in Florida in 1975, was reared in Jamaica but further details have not been recorded (Cock, 1985).

Natural biological control of whiteflies

The whitefly *Bemisia tabaci* (Gennadius) is well distributed in Jamaica and was found feeding on 29 plant species belonging to 11 different families, with the order of preference being Leguminosae > Solanacae > Cruciferae > Labiatae. The pest transmitted viruses like golden mosaic and yellow mosaic to various host plants, causing 30–100% crop losses. Twenty-one species of natural enemies of *B. tabaci* were recorded, of which four were encyrtid parasitoids and one an aphelinid. Thirteen were predators (seven coccinellids, one chrysopid, three syrphids and two spiders) and three entomopathogens were found causing degrees of whitefly infection from 2% to 95% (Alam,1996).

Natural and augmentative biological control of citrus root weevils

Citrus root weevils *Exophthalmus* spp. are polyphagous insects with a preference for *Citrus* spp. The citrus industry developed in the 1920s in Jamaica and had increased to 11,412 ha by 1996 (Anonymous, 1997). Major pests are three species of citrus root weevils: *Exophthalmus vittatus* (L), *E. similis* (Drury) and *Pachneus citri* Marshall. Research work conducted by Clarke-Harris at the Caribbean Agricultural Research and Development Institute (CARDI) from1990 to 1993 resulted, inter alia, in finding five species of hymenopteran parasitoids from root weevil egg masses: *Fidiobia citri* (Nixon), *Aprostocetus haitiensis* (Gahan), *A. gala* (Walker) emerged both from E. *vittatus* and *E. similis* (Clarke-Harris, 1998) and *Euttetrastichus fennahi* and an unidentified species emerged only from *E. similis* eggs. The egg parasitoids were reared and released at several locations on the island. A total of 205,500 parasitoids with an average of 4,000 per week were released between April 1991 and July 1992. From 1991 to 1993, *F. citri* and *A. haitiensis* caused the highest levels of parasitism (between 25.6% and 31.4%) at one location, while at another location *A. gala* and *A. haitiensis* reached levels of parasitism between 18.9% and 27.5% (Clarke-Harris, 1998).

Natural biological control of coffee leaf miner

A study into control of the coffee leaf miner *Perileucoptera coffeella* Guérin-Méneville was conducted from 1992 to 1994 by Dalip (1998). The integrated pest management (IPM) strategy for management of the leaf miner included tree pruning, irrigation, fertilization, drought-resistant rootstock, parasitoids and plant extracts. Fourteen species of parasitoids – two braconids, eight eulophids, two pteromalids and two other unclassified species – emerged from coffee leaf miner-infested field-collected leaves. The level of parasitism ranged between 2.4% and 62.89%, and varied significantly ($p = 0.05$) among plantations (Dalip, 1998).

Augmentative biological control of coffee berry borer

Biocontrol of the coffee berry borer *Hypothenemus hampeii* Ferrari was initiated in Jamaica by the Coffee Industry Board (CIB), which contracted CARDI in 1999. Two parasitoids were sourced from the Regional Cooperative Programme for the Technological Development and Modernization of the Coffee Industry (PROMECAFE): *Cephalonomia stephanoderis* Betrem and *Prorops nasuta* Waterston, which prey on and parasitize immature stages of the borer. During the period of this 3-year project, over 550,000 *C. stephanoderis* were produced. The rearing of *P. nasuta* was not successful. The parasitoid *Phymastichus coffea* La Salle, native to Africa, was imported into Jamaica from Honduras in 2001, successfully reared and released, and data about borer infestation levels before and after release of parasitoids were collected, showing significantly reduced borer populations in fields where parasitoids were released than in control plots without parasitoid releases. Also, the mean marketable yield per tree was higher in fields with parasitoid releases than in fields without releases (CARDI, 2002).

20.3 Current Situation of Biological Control in Jamaica

20.3.1 Natural and classical biological control of the brown citrus aphid

The brown citrus aphid *Toxoptera citricida* (Kirkaldy) is one of the most serious pests of citrus and, as well as causing direct damage, it is an efficient transmitter of citrus tristeza closterovirus (CTV). In a project with CARDI (2006), a sampling project for parasitoids was executed, but also parasitoids were imported from abroad, mass reared and released.

DISTRIBUTION OF BROWN CITRUS APHID PARASITOIDS, INCLUDING *LIPOLEXIS OREGMAE* (GAHAN), IN JAMAICA

Aphid colonies with mummified individuals were sampled during an island-wide survey in 2003. However, the mummies failed to yield emerging adult parasitoids, a feature characteristic of parasitism by *Lysiphlebus testaceipes* (Cresson) in the brown citrus aphid. In 2005, another survey was conducted to determine the presence of two specific parasitoids, *L. testaceipes* and *L. oregmae (scutellaris)*. Both specific parasitoids were found: *L. oregmae* was found in most locations, whereas *L. testaceipes* was found in only a few. On farms where both species were found, *L. oregmae* was also the dominant parasitoid. Confirmation of the wide distribution of *L. oregmae* in Jamaica is an important finding, as in the Western Hemisphere this parasitoid was only known to occur in Florida and Bermuda, where it was introduced as a classical biocontrol agent against the brown citrus aphid (CARDI, 2006). How and when this natural enemy entered Jamaica is still unknown.

IMPORTATION AND LABORATORY REARING OF *LIPOLEXIS OREGMAE*. In view of the low incidence of parasitoids found in a 2003 survey and the very low densities of brown citrus populations during most of 2004, it was decided to import *L. oregmae* from Florida during 2005 to maintain a nucleus colony of this parasitoid. M. Hoy (University of Florida, Gainesville, USA) visited Jamaica and presented to various stakeholders (MINAG, CARDI and regulatory agencies concerning importation of biocontrol agents) aspects of the biology of pest and parasitoids and the benefits of classical biocontrol of brown citrus aphid and the Asian citrus psyllid. Next, parasitoids were imported from Florida and successfully reared at the Bodles Research Station in Jamaica at the end of 2005.

FIELD RELEASES OF *L. OREGMAE*. A field visit was made to Montpelier Citrus, a site where parasitism had not been found in earlier surveys. The aphid population at Montpelier Citrus was very low and clumped. A few infested flushes showed the presence of mummies, indicating that parasitoids were present. Given the very low aphid population and evidence of parasitism, it was decided that conditions were not favourable to release parasitoids during that visit. At another site, Clarendon Park, a release of *L. oregmae* was made in December 2005. Before release, samples of infested flushes were removed and taken to the laboratory to determine parasitism levels. From the pre-release sample, 183 mummies formed from which 116 *L. oregmae* adults emerged. One week after release, the site was visited again to take samples to assess post-release parasitism: no brown citrus aphids were observed and all flushes were newly hardened.

20.3.2 Natural biological control of susumba beetle or false Colorado potato beetle

Susumba beetle *Leptinotarsa undecemlineata* Stal. was found on susumba/gully bean *Solanum torvum* Sw. in Jamaica, where it commonly grows wild on abandoned lands, backyards and in gullies and is traditionally used in many dishes, including soups, chicken stew and seasoned rice. *L. undecemlineata* is common in Central America and Colombia. Several natural enemies were found attacking *L. undecemlineata*, including a tachinid fly, *Myiopharus doryphorae* (Riley), and a predatory reduvid bug. In Colombia, two predatory ladybird beetles, *Cycloneda sanguinea* L. and *Chilocorus cacti* L., were observed to attack the susumba beetle and as these species are also known to exist in Jamaica, they may aid in reducing beetle populations (MoAF, 2006).

20.3.3 Natural biological control of ensign scale

The ensign scale *Orthezia praelonga* Douglas is widely distributed across Jamaica and was found to be associated with at least 22 plant species across the island, including many ornamentals, fruit trees and other food crops (Plant Protection Unit, 2006). In a survey conducted in 2006, the youngest instars of the predator *Scymnus* spp. were found in the egg sacs of the ensign scale, while older instars were found feeding on nymphs on the leaves. Levels of egg predation ranged from 25% to 100%. Also, two species of entomopathogenic fungi, *Colletotrichum gloeosporoides* (Penz.) Penz. & Sacc and *Metarhizium anosipliae* (Metchnikoff) Sorokin, were isolated affecting adult female *O. praelonga* (Sherwood and Myers, 2006).

20.3.4 Natural biological control of lime swallowtail butterfly

The lime swallowtail butterfly *Papilio demoleus* L., considered a quarantine pest in Jamaica, was found in Jamaica in September 2006 affecting citrus nurseries and young citrus orchards. A population dynamics study was carried out from April 2007 to February 2008 to determine the role of local natural enemies. An egg parasitoid *Ooyncyrtus* sp. and a pupal parasitoid *Brachymeria* sp. were discovered. The population of lime swallowtail larvae remained < 1 per tree, while damage levels remained low. In a quick survey in February 2008, the pest was observed to be widespread in major citrus-growing areas at very low levels with little or no impact in nurseries or on citrus production. Seventeen months after introduction to Jamaica, *P. demoleus* was relegated to a non-quarantine pest (Sherwood *et al.*, 2008).

20.3.5 Classical biological control of the pink hibiscus mealybug

The pink hibiscus mealybug *Maconellicoccus hirsutus* (Green) was first reported in Jamaica in June 2007. A classical biocontrol programme was implemented by releasing the parasitoids *Anagyrus kamali* Moursi and *Gyranusoidea indica* Shafee, Alam & Agarwal, which were sourced through the US Department of Agriculture (USDA) in August 2007 and September 2008. Pre-release and monthly post-release monitoring was conducted at 13 locations on the island. Four months after the introduction, *A. kamali* had established and the mealybug population was reduced by 95.98% (Fig. 20.1) (Sherwood *et al.*, 2009). The level of reduction in infestation recorded for Jamaica after 20 months was 95.97%, which is comparable to similar programmes in Florida, Haiti, Belize and Bahamas, where reductions of 98.7%, 97.2%, 86.6% and 82% were recorded, respectively, within the first year of releasing *A. kamali* for control of the pink hibiscus mealybug (Meyerdirk, 2006).

The parasitoid *G. indica* was not recovered, though several releases were made. Twelve species of local natural enemies were observed in association with the mealybug, among which were three species of ladybird beetles, six parasitoids (one chalcid and an unknown species) and one species of reduvid bug. Resurgence in mealybug populations was observed at several sites, but not to pre-release levels. The successful implementation of this biological programme has reduced the risk of this pest spreading to other sites in the island parishes and minimized the impact on agriculture and natural areas (Sherwood *et al.*, 2009).

In 2008, two shipments of adult *Cryptolaemus montrouzieri* totalling 142 were imported from Trinidad and released in infested areas located in Kingston and Portland. Subsequent monitoring showed that pink hibiscus mealybug populations were reduced at both locations (Sherwood, 2008).

20.3.6 Natural biological control of red palm mite

The red palm mite *Raoiella indica* Hirst was first detected in Jamaica in April 2007 affecting coconut and ornamental palms, some severely. Sustainable approaches to manage this pest included a search for local natural enemies. Two predators were found feeding on the mite: the phytoseiid mite *Amblyseius largoensis* (Muma) and a ladybird beetle. *A. largoensis* was the most abundant predator (Goldsmith and Myers, 2008).

20.3.7 Natural and augmentative biological control of the citrus root weevil

The citrus root weevil *Exophthalmus vittatus*, a fiddler beetle, is one of the main insect pests affecting citrus plants and causes considerable economic losses in Jamaica (Alleyne, 1970). Between 1990 and 1997, the parasitoids *Fidioba citri* and *Aprostocetus haitiensis* were reared and released in Jamaican citrus orchards and

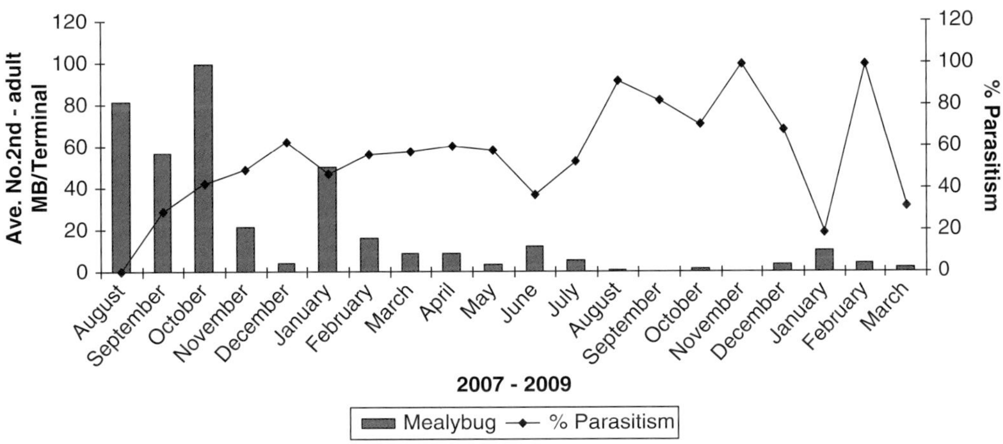

Fig. 20.1. Impact of parasitoids on 2nd nymphal stage to adult pink hibiscus mealybug in Jamaica from 15 August 2007 (moment of release of parasitoids) to 31 March 2009 (retrieved from Sherwood *et al.*, 2009).

were found to be effective against the eggs of the fiddler beetle by giving 50–80% parasitism. Since 2008, diagnostic field visits have been made to citrus growers experiencing severe infestation of citrus root weevils, some with 100% infestation level of *E. vittatus*, which potentially can destroy the entire production. An IPM programme, consisting of mechanical and cultural methods and selective chemicals, was proposed to manage the pest and also to reduce the risk of pesticide residues on fruits destined for the European export market, which has stringent requirements for agricultural market access (Sherwood and Diedrick, 2010). Explorations were also conducted in several orange orchards infested with citrus root weevils for parasitized egg masses. From the collected egg masses only *A. haitiensis* was obtained. Parasitism levels between 75% and 84% were recorded. A laboratory rearing of this parasitoid was started at the Bodles Research Station, resulting in a production of about 12,000 parasitoids, which were released in papaya orchards where parasitoids were not present. Also, at one of the papaya farms, an on-farm rearing facility was established in 2009 and staff were trained to rear the parasitoids. Since the start of this production facility, 19,500 wasps have been produced and released (Sherwood and Diedrick, 2010). Subsequent monitoring showed levels of egg parasitism between 45 and 98% (Sherwood and Diedrick, 2010).

20.3.8 Natural and fortuitous biological control of the papaya mealybug

In September 2010, the papaya mealybug *Paracoccus marginatus* Williams and Granara de Willink, native to Mexico, was detected in Jamaica, 17 years after it was first found in the Caribbean on a sample of teak (*Tectona grandis*). A delimiting survey subsequently determined that the infestation had spread beyond 10 km on residential properties around the location where it was first found. Host plants being impacted include papaya, cassava, gungo peas, West Indian cherry, hibiscus and frangipani. Though the infestation so far has been confined to residential areas, the risk of spread into agricultural, tourism and natural areas is high and will have a serious negative economic and environmental impact. Later surveys showed the presence of papaya mealybug at other locations and it was found on fruit trees (papaya, cherry), forest trees (wild tamarind, blue mahoe), ornamentals (hibiscus, frangipani and lantana) and field crops (gungo peas and cassava). In November 2010, a parasitoid, tentatively identified as an *Apoanagyrus* sp., caused between 5.6% and 9.1% parasitism at three of the surveyed locations (Sherwood, 2011).

Elsewhere, the papaya mealybug has been the object of classical biocontrol using the species-specific parasitoids *Anagyrus loecki* Noyes and Menezes, *Acerophagous papaya* Noyes and Schauff and *Pseudleptomastrix mexicana* Noyes and Schauff, which were often obtained from the USDA Animal and Plant Health Inspection Service (APHIS) parasitoid-rearing facility in Puerto Rico (see Chapter 26: Puerto Rico). In an example of releases made in India in 2009, control of the mealybug was reported within 5 months of initiating releases (Muniappan *et al.*, 2009). It is possible that one of the parasitoid species mentioned above has been fortuitously introduced into Jamaica, since parasitism has been detected at all sites surveyed so far. Native and/or fortuitously introduced natural enemies seem to be effecting some control, but augmentation seems still to be required to increase levels of control.

20.3.9 Fortuitous and augmentative biological control of Asian citrus psyllid

The Asian citrus psyllid *Diaphorina citri* Kuwayama, native to southern Asia, is a vector of the most serious citrus disease worldwide, referred to as citrus greening or huanglongbing (HLB). Plant yield and fruit quality are greatly reduced and citrus production is severely reduced within 5–10 years. In order to mitigate the spread and impact of this disease, management of its insect vector is critical. In Jamaica, the citrus psyllid is being parasitized by its natural enemy *Tamarixia radiata* (Waterston), which was apparently fortuitously introduced, as well as by native generalist predators such as ladybird beetles and lacewings. Although levels of parasitism and predation are high on *Murraya* (a preferred host of the psyllid) systems, it is unknown how much they contribute to psyllid mortality on citrus. Hence, knowledge of the population dynamics of *D. citri* Kuwayama is important in understanding the seasonal trends of its population and the role of mortality factors if it is to form the basis for a reliable IPM programme (Lee and Sherwood, 2012).

Studies were conducted in three agroecological zones: Good Hope farm in Trelawny (a dry, cool area in West Jamaica); Orange River Research Station in St Mary (a cool, wet area in North-east Jamaica); and Cambria Farms in Bogwalk, St Catherine (a hot, wet area in South Central Jamaica), which has the largest acreage of citrus in Jamaica (Lee and Sherwood, 2012). During the period 2011–2012, over 9,000 adult *T. radiata* parasitoids were reared at the Bodles Research Station and released into citrus orchards at the above locations (Lee and Sherwood, 2012). Samples of young flush infested with 4th to 5th instars of psyllids were collected at the three locations to assess for *T. radiata* parasitism, and *D. citri* nymphs were found to be parasitized at all locations (Lee and Sherwood, 2012). Populations of *D. citri* were lower during periods of high parasitism levels at all sites. Parasitism levels ranged between 8% and 38%. In St Mary, parasitoids were present from February to July in 2010 and in January 2011, in Trelawny from October 2010 to January 2011, and in St. Catherine they were present for most of the year (Lee and Sherwood, 2012). Parasitism levels were also determined from 15 June to 20 September 2011, during summer and the beginning of autumn, on seven farms in six districts with levels ranging from 0 to 100%. Parasitism was not detected at Rosemount and Redwood, while over 80% parasitism was recorded at Knollis, Rose Hall, Bog Walk and Wallens (Sherwood *et al.*, 2012).

20.3.10 Augmentative biological control of the coffee berry borer

Biocontrol of the coffee berry borer (CBB) *Hypothenemus hampeii* Ferrari was initiated in Jamaica by the CIB and initially it was based on the use of two species of parasitoids, as summarized above. In 2011, CIB, in collaboration with the University of the West Indies (Mona), undertook a major *Beauveria bassiana* project to combat the coffee berry borer, with detailed studies on its biology and mass production methods (Williams, 2012). Also, in 2012, CIB explored the use of *B. bassiana* by assessing the levels of borer infestation, the rates of infection by *B. bassiana*, as well as the amount of bean damage in CBB-infested beans with or without *B. bassiana* at two coffee farms that were agroecologically managed in a different way.

Borer infestation was the same (about 42%) at both farms, but *B. bassiana* infection differed significantly and was 11% at one and 24% at the other farm. The farm with higher *B. bassiana* infection had significantly lower bean damage (21%) than the farm with lower fungal infection (50% bean damage). Borer mortality without *B. bassiana* infection was 20.3%, while mortality was 100% with infection (Bryan and Robinson, 2012). In a laboratory assessment in 2013, borers treated with *B. bassiana* showed mortalities ranging from 33% to 77%, and this study assisted in determining the lethal concentration of *B. bassiana* for borer control in Jamaica (Brown and Robinson, 2013). In 2015, a new control approach was attempted with *B. bassiana* cultured on rice at CIB. The final dried product is packaged in 350 g bags, which are taken to the farms where the rice is removed by washing in water at a ratio of 5 g of the fungus to 1 litre. Using a spray pan or mist blower, the solution is then applied to the coffee trees. Development of the borer biocontrol project was critical to the 'Good Agriculture Practice' approach started by CIB in 2001, which is built around cost-effective agroecologically friendly management strategies (CIB, 2015).

20.3.11 Augmentative biological control of the sweet potato weevil

An important pest of sweet potato *Ipomoea batatas* in Jamaica is the sweet potato weevil *Cylas formicarius*. The entomopathogenic fungus *B. bassiana* was tested in a field trial in Jamaica for control of the weevil during 2017 and the trial would be repeated in 2018 (Sybron *et al.*, 2017).

20.3.12 Augmentative biological control of beet armyworm

Beet armyworm, *Spodoptera exigua* (Hubner), a polyphagous pest native to South Asia, has been reported from Jamaica since the 1970s and started to cause damage to *Allium* spp. in 2009.

After two outbreaks in 2009, a search was made for natural enemies in St Elizabeth in fields infested with beet armyworm. One parasitoid pupa was found, but no adult emerged. A parasitized larva of the fall armyworm *Spodoptera frugiperda* (J.E. Smith), tentatively identified as *Euplectrus* sp., was found on sweet pepper *Capsicum annum* at Thetford St Catherine, but attempts to rear the parasitoids failed at the third generation. *Euplectrus* sp. is a generalist parasitoid of noctuid moths and attacks mostly young larvae. New attempts to rear this parasitoid on larvae of *S. frugiperda* and *S. exigua* failed, because the wasps did not lay eggs in the larvae (Diedrick *et al.*, 2011). Another fall armyworm larva parasitized by *Euplectrus plathypenae* (Howard) was brought to the Entomology Laboratory in 2010 and successfully reared. This species, *E. plathypenae*, is a gregarious ectoparasitoid that can develop on noctuid and geometrid larvae. In the laboratory, parasitism of fall armyworm is significantly higher than that of the beet armyworm. Parasitoid performance in field cages containing beet armyworm on scallion was poor and no wasps were recovered after release (Diedrick *et al.*, 2011).

A microbial agent, SPEXIT SC (Andermatt BioControl, Switzerland), the nucleopolyhedrovirus (SeNPV) specific for *S. exigua*, was obtained from Andermant Biocontrol in Switzerland and field tested in 2017. The armyworm population was reduced by 96.7–99% after 2 weeks in plots with the virus and *Bacillus thuringiensis* (Bt) treatments were compared with insecticide treatment in the control plots (Haughton *et al.*, 2017). Also, in 2017 a few adult wasps were submitted to the Plant Protection Laboratory of the Research and Development Division, which were suspected to be parasitizing beet armyworm. Later, additional field specimens were collected and the parasitoid was confirmed to be *Chelonus insularis* (Cresson). Use of this parasitoid in local biocontrol programmes still needs to be evaluated (C. Haughton, Old Harbour, St Catherine, 2017, personal communication).

In conclusion, Jamaica has a long and rich history of successful biocontrol. Approximately 62,000 ha of commercial orchards continue to benefit from augmentative, natural, fortuitous and classical biocontrol. Island wide, classical and fortuitous biocontrol in ornamentals continue to have an impact (Table 20.1).

20.4 New Developments of Biological Control in Jamaica

In the 2030 National Vision Plan (PIOJ, 2009), agriculture is one of the major sectors mentioned for providing growth to the economy of Jamaica. Programmes like import substitution of locally produced crops and improved production programmes were designed for selected crops including Irish potato, onion, ginger, cassava and sweet potato. Managing the key pests of these root crops is critical to meet increased production goals. Currently there is still a high dependence on costly pesticides that also pose human and environmental health risks. The Pesticide Control Authority has increased efforts to reduce the use of pesticides from the highest risk classes, resulting in increased registration of pesticides of lower risk classes, as well as biopesticides.

In 2017, the Government of Jamaica strengthened research at its agricultural research station at Bodles, including the construction of a biocontrol facility, which is expected to result in an increase of bioharvesting and augmentation of local natural enemies. Agroecologically based field systems including intercrops, strip crops, shelterbelts and banker plants are being proposed to develop conservation biocontrol programmes in Jamaica.

New invasive pests have entered the country and some endemic pests caused more damage with the expansion of production of some crops; several of these pests offer good opportunities for using biocontrol (Table 20.2). Funding for these biocontrol programmes will require support from regional and international partners of the Ministry of Industry, Commerce, Agriculture and Fisheries.

20.5 Acknowledgements

Staff from the following entities are thanked for their contribution to biocontrol programmes in Jamaica over the years: the Ministry

Table 20.1. Overview of application of biological control in Jamaica.

Crop	Pest	Natural enemy	Area (ha) under biocontrol	References (for areas)
Classical biological control				
Banana	*Cosmopolites sordidus*	*Plaesius javanus, Dactylosternum hydrophiloides, D. abdominale*	3,000	AMID, 2018
Citrus	*Aleurocanthus woglumii*	*Eretmocerus serius*	7,846	FAO, 2016
Citrus	*Icerya purchasi*	*Rodolia cardinalis*	7,846	FAO, 2016
Citrus	*Exophthalmus* spp.	*Aprostocetus haitiensis*	7,846	FAO, 2016
Citrus	*Aonidiella aurantii*	*Aphytis diaspidis*	7,846	FAO, 2016
Citrus	*Toxoptera citricida*	*Lipolexis oregmae*	7,846	FAO, 2016
Cocoa	*Selenothrips rubrocinctus*	*Goetheana parvipennis*	2,335	FAO, 2016
Coconut	*Aspidiotus destructor*	*Cryptognatha nodiceps, Pseudoazya trinitatis*	16,077	Coconut Industry Board, 2016
Coffee	*Hypothenemus hampei*	*Cephalenomia stephanoderis, Phymastichus coffea*	6,498	FAO, 2016
Ornamentals	*Maconellicoccus hirsutus*	*Anagyrus kamali,*	1,000,000, whole island	PIOJ, 2009
Sugarcane	*Diatrea saccharalis*	*Cotesia flavipes*	26,255	FAO, 2016
Natural and fortuitous biological control[a]				
Citrus	*Papilio demoleus*	*Ooyncyrtus* sp. *Brachymeria* sp.	7,846	FAO, 2016
Citrus	*Diaphorina citri*	*Tamarixia radiata*	7,846	FAO, 2016
Coconut	*Raoiella indica*	*Amblyseius largoensis*	16,077	Coconut Industry Board, 2016
Ornamentals and crops	*Paracoccus marginatus*	*Apoanagyrus* sp	1,000,000, whole island	PIOJ, 2009
Augmentative biological control				
Citrus	*Exophthalmus* spp.	*Aprostocetus haitiensis*	7,846	FAO, 2016
Citrus	*Diaphorina citri*	*Tamarixia radiata*	7,846	FAO, 2016

[a]For these crops, the total area in Jamaica is given, because the natural enemies introduced for control of pests are commonly found throughout the crop areas

Table 20.2. Recent endemic and invasive pests in Jamaica with crops they damage, their potential natural enemies and information about possibilities for biological control.

Pest or disease	Crops infested	Biological control agent	Reference about biocontrol possibilities
Endemic			
Spodoptera exigua, beet armyworm	Onion, scallion (*Allium* spp.), callaloo (*Amaranthus* sp.)	*Beauveria bassiana*	Barbercheck and Kaya, 1991
		Steinernema feltiae	
		Heterorhabditis marelata	Kaya, 1985; Dalip, 2014
		Baculovirus formulations	Kaya, 1985; Dalip, 2014
		Cotesia marginiventris	Elvira *et al.*, 2013; Haughton *et al.*, 2017
			Ruberson *et al.*, 1994
Exophthalmus spp., citrus root weevil or fiddler beetle	Citrus (*Citrus* spp.), papaya (*Carica papayae*)	*Fidiobia citri* *Aprostocetus haitiensis*	Clarke-Harris, 1998; Sherwood and Diedrick, 2010

Continued

Table 20.2. Continued.

Pest or disease	Crops infested	Biological control agent	Reference about biocontrol possibilities
Invasive			
Diaphorina citri, Asian citrus psyllid	Citrus	*Tamarixia radiata*	Flores and Ciomperlik, 2017; Lee and Sherwood, 2012
Tarophagous esculenta, taro leafhopper	Dasheen (*Colocasia esculenta)*	*Cyrtorhinus fulvus*	Fatuesi and Vargo, 1986
Orthezia praelonga praelonga, ensign scale	Many host plants	*Colletotrichum gloesporoides*, *Metarhizium anosiplae, Scymnus spp.*	Plant Protection Unit, 2006; Sherwood and Myers, 2006; Kondo and Szita, 2013
Franklinella occidentallis, western flower thrips	Sweet pepper under protected cultivation	*Orius laevigatus*, predatory mite	Elimem *et al.*, 2018
Moniliophthora roreri, frosty pod rot	Cocoa (*Theobroma cacao*)	*Trichoderma ovalisporum*	Crozier *et al.*, 2015

Commerce, Agriculture and Fisheries (Research and Development Division and Rural Agricultural Development Authority-RADA, Plant Quarantine), the Caribbean Agricultural Research and Development Institute (CARDI), the Sugar Industry Research Institute (SIRI), the Coffee Industry Board and the University of the West Indies, Mona Campus Jamaica.

References

(References with grey shading are available as supplementary electronic material)

Alam, M.M. (1996) Fluctuation in the populations of *Plutella xylostella*, *Trichoplusia ni*, *Bemisia tabaci* and their natural enemies in Jamaica. PhD thesis, University of the West Indies, Mona, Jamaica.

Alleyne, E.H. (1970) A study of the morphology and biology of the immature stages of some Jamaican fiddler beetles (Coleoptera; Curculionidae). MSc thesis, University of the West Indies, Mona Campus, Jamaica.

AMID (Agricultural Marketing Information Division) (2018) *Estimates of Domestic Crop Production 2018*. Internal Database, Ministry of Industry Commerce Agriculture and Fisheries, Jamaica.

Anonymous (1997) *Citrus Review Study*. Vakakis and Associates, Athens, Greece.

Barbercheck, M. and Kaya, H.K. (1991) Competitive interactions between entomopathogenic nematodes and *Beauveria bassiana* (Deuteromycotina: Hyphomycetes) in soilborne larvae of *Spodoptera exigua* (Lepidoptera: Noctuidae). *Environmental Entomology* 20(2), 707–712.

Brathwaite, C.W.D., Phelps, R.H. and Bennett, F.D. (1974) Early examples of biological control in the Commonwealth Caribbean. In: *Proceedings of a Symposium on the Protections of Horticultural Crops in the Caribbean*. University of the West Indies, St Augustine, Trinidad, pp. 218–227

Brown, R. and Robinson, D.E. (2013) Efficacy of *Beauveria bassisana* against *Hypothenemus hampei*. University of the West Indies, Mona Campus, Jamaica.

Bryan, G. and Robinson, D.E. (2012) An evaluation of *Beauveria bassiana* as a natural control agent of the coffee berry borer, *Hypothenemus hampei* (Ferrari), in Jamaica. Poster presented at 24th International Conference on Coffee Science, Costa Rica, Nov. 11–16, 2012.

CARDI (2002) *Biological Control of the Coffee Berry Borer*. End of Project Report, 1999–2002. Caribbean Agricultural Research and Development Institute, Mona, Jamaica.

CARDI (2006) *Final report of the Citrus Replanting Project*. Caribbean Agricultural Research and Development Institute. Research Services – Entomology, Ministry of Agriculture, Kingston, Jamaica.

CIA (2017) The World Factbook: Jamaica. Available at: https://www.cia.gov/library/publications/the-world-factbook/geos/jm.html) (accessed 7 January 2019)

CIB (Coffee Industry Board) (2015) Fighting the berry borer infestation. Available at: http://www.ciboj.org/content/fighting-berry-borer-infestation-mon-11232015-1352 (accessed 10 August 2018).

Clarke-Harris, D.O. (1998) Investigations on the potentials of biological and cultural methods in the integrated management of citrus root weevils in Jamaica. MSc thesis, University of the West Indies, Mona Campus, Jamaica.

Cock, M.J.W. (ed.) (1985) *A Review of Biological Control of Pests in the Commonwealth Caribbean and Bermuda up to 1982*. Technical Communication No. 9, Commonwealth Institute of Biological Control. Commonwealth Agricultural Bureaux, Farnham Royal, UK.

Coconut Industry Board (2016) 75th Annual Report of the Coconut Industry Board – for year ended December 31, 2016. Available at: http://1ljzi2315iz11syhi3x3fdsh.wpengine.netdna-cdn.com/files/2018/01/75th-Annual-Report-for-year-ended-December-31-2016.pdf. (accessed 7 January 2019).

Crozier, J., Arroyo, C., Morales, H., Melnick, R.K., Strem, M.D, Vinyard, B.T., Collins, R., Holmes, K.A. and Bailey, B.A. (2015) The influence of formulation on *Trichoderma* biological activity and frosty pod rot management in *Theobroma cacao*. *Plant Pathology* 64, 1385–1395.

Dalip, K.M. (1998) Management of the coffee leaf miner *Perileucoptera coffeella* (Guér-Menv,. 1842), in Jamaica: epidemiology, economic importance, parasitoid complex and effect of selected insecticides and their ecological impact. PhD thesis, University of the West Indies, Mona Campus, Jamaica.

Dalip, K.M. (2014) *Final Report on Biological Control Programme: Strengthening a National Beet Armyworm Management Programme in Jamaica*. TCP/JAM/3402, Food and Agriculture Organisation of the United Nations, Rome.

Department of Agriculture Jamaica (1949) Investigations 1948 to 1949. *Bulletin* 45, 36–37.

Department of Agriculture Jamaica (1950) Investigations 1949 to 1950. *Bulletin* 47, 38–39.

Department of Agriculture Jamaica (1951) Investigations 1950 to 1951. *Bulletin* 49, 4–41.

Diedrick, W., Sherwood, M. and Myers Morgan, L. (2011) Development of an integrated pest management programme for beet armyworm (*Spodoptera exigua*) on escallion (*Allium fistulosum*) – Biological control component. Internal report, Research and Development Division, Ministry of Agriculture and Fisheries, Kingston, Jamaica.

Elimem, M,. Harbi, A., Limem-Sellemi, E., Othmen, S.B. and Chermiti, B. (2018) *Orius laevigatus* (Insecta; Heteroptera) local strain, a promising agent in biological control of *Frankliniella occidentalis* (Insecta; Thysanoptra) in protected pepper crops in Tunisia. *Euro-Mediterranean Journal for Environmental Integration* 3:5. Available at: https://www.researchgate.net/profile/Chermiti_B/publication/320912571 (accessed 23 March 2019).

Elvira, S., Ibargutxi, M.A., Gorria,N., Oz, D.M., Caballero, P. and Williams, T. (2013) Insecticidal characteristics of two commercial *Spodoptera exigua* nucleopolyhedrovirus strains produced on different host colonies. *Journal of Economic Entomology* 106, 50–56

Espeut, W.B. (1882) On the acclimatization of the Indian mungoos [mongoose] in Jamaica. *Proceedings of the Zoological Society of London* 1882, 712–714.

Falloon, T. (2005) Biological control of the sugarcane stalk borer (*Diatraea saccharalis* (Fabricius)) Lepidoptera: Pyralidae in the Caribbean. In: *Proceedings of The Jamaica Association of Sugar Technologies – 68th Annual Conference*. Ocho Rios, St Ann and Jamaica.

FAO (2016) FAOSTAT. Available at: http://www.fao.org/faostat/en/#home (accessed 23 February 2019).

Fatuesi, S. and Vargo, A.M. (1986) *An initial evaluation of biological and cultural control of taro pests in American Samoa*. Land Grant Technical Report No. 26. Available at: https://www.ctahr.hawaii.edu:443/adap/ASCC_LandGrant/Dr_Brooks/Report26.pdf; (accessed 23 March 2019).

Flores, D. and Ciomperlik, M. (2017) Biological control using the ectoparasitoid, *Tamarixia radiata*, against the Asian citrus psyllid, *Diaphorina citri*, in the Lower Rio Grande Valley of Texas. *Southwestern Entomologist* 42, 49–59.

Goldsmith, J.V. and Myers, L.R. (2008) Population dynamics of the red palm mite (*Raoiella Indica* Herst) and the search for sustainable management practices in Jamaica. Internal report, Research and Development Division, Ministry of Agriculture, Kingston, Jamaica.

Haughton, C., Sherwood, M., Webb, F. and Henry, D. (2017) The efficacy of Baculovirus bio-pesticides in the management of beet armyworm, *Spodoptera exigua* (Hübner) in scallion, *Allium fistulosum* L. in Jamaica. In: *Proceedings from the 30th Caribbean Academy of Sciences Conference*. Kingston, Jamaica, pp. 108–109.

Johnson, S.R., Berentsen, A.R., Ellis, C., Davis, A. and Vercauteren,K.C. (2015) Estimates of small Indian mongoose densities: implications for rabies management. *Journal of Wildlife Management* 80, 37–47.

Kaya, H.K. (1985) Susceptibility of early larval stages of *Pseudaletia unipuncta* and *Spodoptera exigua* (Lepidoptera: Noctuidae) to the entomogenous nematode *Steinernema feltiae* (Rhabditida: Steinernematidae). *Journal of Invertebrate Pathology* 46, 58–62. DOI: 10.1016/0022-2011(85)90129-6

Kondo, T. and Szita, E. (2013) The citrus orthezia, *Praelongorthezia praelonga* (Douglas) (Hemiptera: Ortheziidae), a potential invasive species. In: Pena, J. (ed.) *Potential Invasive Pests of Agricultural Crops*. CAB International, Wallingford, UK, pp. 301–319. Available at: https://www.researchgate.net/publication/246601458 (accessed 23 March 2019).

Lawrence, J.L. (1994) Potential of neotropical and temperate heterorhabditid nematodes as biological control agents for the sweet potato weevil, *Cylas formicarius* (Fabricius) (Coleoptera: Apionidae). MSc thesis. The University of Florida, USA.

Lee, G. and Sherwood, M.A. (2012) Seasonal abundance of the Asian citrus psyllid in three agro-ecological zones in Jamaica. Internal report, Research and Development Division, Ministry of Agriculture, Kingston, Jamaica.

Long, J.L. (2003) *Introduced Mammals of the World. Their History, Distribution and Influence*. CSIRO Publishing, Melbourne, Australia.

Meyerdirk, D.E. (2006) Offshore biological control strategy applied to pink hibiscus mealybug. In: *5th National IPM Symposium 'Delivering on a Promise'*, Presentation 57.2. Available at: https://ipmsymposium.org/2006/sessions/57-2.pdf (accessed 12 September 2018).

MOA (Ministry of Agriculture and Lands) (1957) Biological control of citrus blackfly (*Aleurocanthus woglumi*), banana weevil and fiddler beetles (*Prepodes* spp. and *Pachneus* spp.) on citrus. Investigations 1956–1957. *Bulletin No.* 57, 25–27.

MoAF (Ministry of Industry, Agriculture and Fisheries) (2006) An insect pest of Susumba, *Leptinotarsa undecemlineata* Stahl (Coleoptera; Chrysomelidae). Available at: http://www.micaf.gov.jm/sites/default/files/CIRCULAR%20SUSUMBER%20LEAF%20BEETLE_1.pdf (accessed 21 July 2018).

Muniappan, R., Shepard, B.M., Watson, G.W., Carner, G.R., Sartiami, D., Rauf, A. and Hammig, M.D. (2009) First report of the papaya mealybug, *Paracoccus marginatus* (Hemiptera: Pseudococcidae), in Indonesia and India. *Journal of Agriculture and Urban Entomology* 25(1), 37–40. Available at: https://www.researchgate.net/publication/232675839 (accessed 12 September 2018).

Pimentel, D. (1955) The control of the mongoose in Puerto Rico. *American Journal of Tropical Medicine and Hygiene* 41, 147–151.

PIOJ (Planning Institute of Jamaica) (2009) *Vision 2030 Jamaica National Development Plan*. Available at: http://www.vision2030.gov.jm/portals/0/ndp/vision%202030%20jamaica%20ndp%20full%20no%20cover%20(web).pdf (accessed 21 August 2018).

Plant Protection Unit (2006) Ensign scale (*Orthezia praelonga*). Plant Protection Flyer, Ministry of Agriculture and Fisheries, Research and Development Division, Bodles Research station, Jamaica.

Ruberson, J.R., Herzog, G.A., Lambert, W.R. and Lewis, W.J. (1994) Management of the beet armyworm (Lepidoptera: Noctuidae) in cotton: role of natural enemies. *Florida Entomologist* 77, 440–453.

Sherwood, M.A. (2008) Integrated pest management of *Maconellicoccus hirsutus*, pink hibiscus mealybug in Jamaica. Presented at the Plant Health Meeting August 8, 2008. Research and Development Division, Ministry of Agriculture and Lands, Kingston, Jamaica.

Sherwood, M.A. (2011) Management programme for the papaya mealybug, *Paracoccus marginatus* Williams and Granara de Willink (Hemiptera: Pseudococcus) in Jamaica (September, 2009 – March, 2011). Internal report, Research and Development Division, Ministry of Agriculture, Kingston, Jamaica.

Sherwood, M.A. and Diedrick, W. (2010) Management of the citrus root weevil, *Exophthalmus vittatus* on papaya (*Carica papaya*) at Advanced Farms, Trelawny. Internal report, Research and Development Division, Ministry of Agriculture, Kingston, Jamaica.

Sherwood, M.A. and Myers, L.R. (2006) Survey for natural enemies of ensign scale (*Orthezia praelonga*) in Jamaica. Plant Protection Annual report. Internal report, Research and Development Division, Ministry of Agriculture and Fisheries, Kingston, Jamaica.

Sherwood, M.A., Myers, L.R. Young, M. and Allen, G. (2008) Management of the lime swallowtail butterfly, *Papilio demoleus* (Lepidoptera, Papilionidae) in Jamaica. Internal report, Research and Development Division, Ministry of Agriculture and Fisheries, Kingston, Jamaica.

Sherwood, M.A., Myers, L.R., Robinson, D. and Lawrence, J. (2009) Management of pink hibiscus mealybug (PHMB), *Maconellicoccus hirsutus* Green (Hemiptera; Pseudococcidae) in Jamaica. Internal report. Plant Protection Unit, Research and Development Division, Ministry of Agriculture; Life Sciences Department, Entomology, University of the West Indies, (Mona); Caribbean Agricultural Research and Development Institute (CARDI, Jamaica).

Sherwood, M., Lee, G., Diedrick, W., Elliott, S., Williams, J. and Chang, P. (2012) Citrus greening survey in Clarendon and St Catherine. Internal report, Research and Development Division, Ministry of Agriculture and Fisheries, Jamaica.

Sybron, A., Johnson, E. and Sherwood, M. (2017) The efficacy of the fungus, *Beauveria bassiana*, as a biological control agent against the sweet potato weevil (*Cylas Formicarius*) on sweet potato (*Ipomoea batatas*) in Jamaica. Paper presented at 55th Annual Meeting and Convention of the US Sweet Potato Council, San Diego, California, 22–24 January 2017

Williams, M.P. (2012) Investigation of the shelf-life and viability of *Beauveria bassiana* conidia: under various temperatures using Sabourad and potato dextrose medium. BSc thesis, University West Indies, Mona Campus, Jamaica.

21

Biological Control in Mexico

Hugo César Arredondo-Bernal* and Beatriz Rodríguez-Vélez

Centro Nacional de Referencia de Control Biológico, Dirección General de Sanidad Vegetal. Km 1.5 Carretera Tecomán-Estación, Tecomán, Colima, México

* E-mail: hugo.arredondo@senasica.gob.mx

Abstract

During the 1940s and 1950s, the number of native and exotic species of biocontrol agents used in Mexico amounted to 59 and most of these were parasitoids. Classical biocontrol was dominant and included control of hemipteran pests such as the woolly apple aphid, scales, citrus mealybugs, spittlebug, and the rhodesgrass mealybug. One of the globally recognized successes was classical biocontrol of the citrus blackfly with an imported parasitoid. In the 1960s, construction of mass-production centres for biocontrol organisms took place, for rearing of *Trichogramma* spp., among others. More recent cases of classical biocontrol, sometimes in combination with augmentative biocontrol, are the control of grasshoppers, pink hibiscus mealybug, velvet soybean worm, eucalyptus psyllid, Asian citrus psyllid, brown citrus aphid, and fruit flies. Examples of augmentative control are control of the Mexican bollworm with a native parasitoid, of diamondback moth with native and exotic parasitoids and of aphids with predators and an entomopathogenic fungus. Augmentative biocontrol is particularly popular for control of aphids, thrips, leaf miners, mites and whiteflies in vegetables and ornamentals, where several species of predators, parasitoids and microbial agents are used. Examples of today's large-scale augmentative programmes are control of sugarcane borers with *Trichogramma* spp., of sugarcane aphids with predators in soybean and of several species of locusts and grasshoppers with an entomopathogenic fungus. Currently, there are 65 companies producing and marketing 40 species of beneficial organisms, and more than 50 species of arthropods produced outside Mexico are authorized for importation to be used in specific pest control programmes.

21.1 Introduction

Mexico is one of the main producers and exporters of food in the world. It has 124.1 million inhabitants, of which 9.1 million generate and transform agricultural and fishing goods. The heterogeneity of the territory, natural resources, diversity of climate, biodiversity and infrastructure results in an agricultural and fishing production that places Mexico in 11th position as producer of food, agricultural crops and primary livestock and 17th in world fisheries and aquaculture production (SADER, 2018).

The total land area of Mexico is 1,964,375 km^2. Cropland accounts for 56% of Mexico's total surface and the total agricultural area is 32.4 million hectares, equivalent to 29.4% of a total of 110.3 million hectares. Of the 32.4 million hectares of agricultural land, 79% correspond to rain-fed agriculture and the remaining 21% to irrigation agriculture. The remaining 77.9 million hectares consist of summer pastures (for cattle) or fallow land (INEGI, 2017).

Mexico, like other countries in the region, has enormous cultural wealth due to its indigenous people, who have interacted for thousands of years with the country's vast biological diversity. This interaction has resulted in the description of 5,500 species of useful plants and the selection and modification (domestication) of over 200 species (Ortíz-García *et al.*, 2017). Mexico's most important crop is maize: in 2017, the area under maize was 8.4 million hectares with a production of 32.2 million t (INEGI, 2017).

The main export products, in which the country is self-sufficient, are vegetables and fruits with, respectively, 28% and 25% of the total export value (mainly tomato, cucumber, lime, avocado, chilli, berries, banana and watermelon) (Ortíz-García *et al.*, 2017). However, the country has a deficit of cereals, meat, seeds and oilseeds, which are imported mainly from the USA (Ortíz-García *et al.*, 2017). In addition to the products mentioned before, other important agricultural products are soybeans, rice, cotton, coffee and wood products.

Mexico produces cattle and goats (for milk and meat), pigs and sheep (for meat), poultry (for meat and eggs) and bees (for honey). Livestock production also includes aquaculture (fish farming) and rabbit breeding (Ortíz-García *et al.*, 2017). Mexico has positioned itself as a major producer of animal protein in the world, occupying 7th place (SAGARPA, 2016).

The forest area of Mexico is 65.6 million hectares and represents 32.8% of the country's surface. Around 21.6 million hectares of forests (32.9%) have the potential for sustainable commercial production, of which approximately 8 million hectares are currently under management (CONAFOR-FAO, 2009).

21.2 History of Biological Control in Mexico

21.2.1 Period 1900–1969

A number of classical and augmentative biocontrol programmes were implemented during this period, mainly by the Mexican government and often based on earlier successes obtained in the USA. The first case of classical biocontrol was the introduction in 1902 of the myoktanine virus from Europe to control the field rat (Herrera *et al.*, 1902). In the 1940s and 1950s, the number of species of biocontrol agents used rose to 59 and most of these were parasitoids. Those years represent one of the most important phases of biocontrol in Mexico, resulting in global recognition of the successful control of the citrus blackfly. As in many other countries, import of biocontrol agents from abroad decreased after the 1950s due to the use of chemical pesticides, a situation that was dominant for the next 30 years. In the 1960s, construction of centres for mass production of beneficial organisms took place, including for the reproduction of *Trichogramma* species. Arredondo-Bernal and Rodríguez-del-Bosque (2008, 2015) and Willams *et al.* (2013) summarized the development of biocontrol in Mexico. The major pests for which classical biocontrol was developed during this period are mentioned in Table 21.1.

21.2.2 Period 1970–2000

Classical biological control of pests and weeds

Several new classical biocontrol programmes were carried out during this period, including the introduction of *Cephalonomia stephanoderis* Betrem and *Prorops nasuta* Waterston for the control of the coffee berry borer *Hypothenemus hampei* (Ferrari) (Barrera-Gaytán *et al.*, 2008). *Neochetina bruchi* Hustache and *N. eichhorniae* Warner were imported for control of the aquatic weed *Eichhornia crassipes* (Mart.) (Martínez-Jiménez, 2008). *Ageniaspis citricola* Logvinovskaya was introduced from Florida to control the citrus leaf miner *Phyllocnistis citrella* (Bautista-Martínez *et al.*, 2008). *Bracon kirkpatricki* (Wilkinson) was used against pink bollworm *Pectinophora gossypiella* Saunders. *Cotesia flavipes* (Cameron) was used for control of *Diatraea saccharalis* (Fabricius) and *D. lineolata* (Dyar) (Rodríguez-del-Bosque and Vejar-Cota, 2008; Williams *et al.*, 2013). *Copidosoma desantisi* Annecke & Mynhardt was imported for control of the potato tuber moth *Phthorimaea opercullela* (Zeller) (Bahena-Juárez, 2008) and *Aphytis melinus* DeBach was introduced to control *Aonidiella aurantii* populations (González-Hernández *et al.*, 2015).

Augmentative biological control of pests

During this period, several augmentative biocontrol projects were implemented in Mexico, such as the production of *Jaliscoa* (= *Catolaccus*) *grandis* Burks for release against *Anthonomus grandis* Boheman in northern Mexico (Loera-Gallardo *et al.*, 2008). Introduction, production and release of *Trichogrammatoidea bactrae* Nagaraja and *Trichogramma pintoi* Voegele and the rearing of *Diadegma insulare* (Cresson) and *Cotesia plutellae* (Kurdjumov) were developed for control of *Plutella xylostella* (L.) (Salazar-Solís and Salas-Araiza, 2008). Further, the mass production and release of *Ceraeochrysa claveri* (Navás), *Cycloneda sanguinea* L., *Harmonia axyridis* Pallas and *Olla v-nigrum* (Mulsant) and mass production and application of *Isaria fumosorosea* (strain CHE-CNRCB 305) were started for control of brown citrus aphid *Toxoptera citricida* Kirkaldy, a vector of citrus tristeza virus (López *et al.*, 2008). *Euseius* (= *Amblyseius*) *victoriensis* Womersly of Australian origin was introduced for laboratory tests against *Phyllocoptruta oleivora* (Ashmead) (Ruíz-Cancino *et al.*, 2008). A recurrent problem in the 1980s was the brown bug *Oebalus mexicana* (Sailer) on sorghum, which was controlled through applications of the entomopathogenic fungus *Beauveria bassiana* Balsamo (Vuillemin) (Salazar-Solís *et al.*, 2015). Biocontrol of the whitefly *Bemisia tabaci* (Gennadius) was developed through importation and release of two exotic parasitoids, *Eretmocerus mundus* Mercet and *E. eremicus* Rose & Zolnerowich, as well as with releases of the native parasitoid *Encarsia formosa* (Gahan) (Martínez-Carrillo *et al.*, 2015).

Also in this period, 20 insectaries (Regional Centers for Reproduction of Beneficial Organisms) were founded by the Federal Government until 1991, and these were later transferred to agricultural producers. Together with 65 private insectaries, they mass produced, introduced and commercialized the following species of parasitoids: *Trichogramma* spp., *Spalangia* spp, *Muscidifurax* spp., *Nasonia* spp., *Habrobracon* sp., *Aphelinus abdominalis* (Dalman), *Aphidius colemani* Viereck, *Aphytis lingnanensis* Compere, *B. kirkpatricki*, *C. flavipes*, *C. plutellae*, *Dacnusa sibirica* Telenga,

Table 21.1. Classical and augmentative biological control projects started in Mexico during period 1900–1969.

Crop	Pest	Natural enemy/pathogen	Origin	Type of biocontrol[a]	Area (ha) under biocontrol[b]	Year
Sugarcane, maize, sorghum, wheat	*Apodemus sylvaticus* (L.), *Microtus arvalis* (Pallas) *Mus musculus* (L.); field rats, mouse	Myoktanine virus	Europe	CBC	?	1902
Cotton	*Anthonomus grandis* (Boheman); Mexican cotton boll weevil	*Pyemotes ventricosus* (Newport)	Mexico	ABC	?	1904
Sugarcane	*Diatraea* spp.; sugarcane borers	*Lixophaga diatraeae* Townsend	Cuba	CBC	?	1922
		Billaea (=*Paratheresia*) *claripalpis* (Wulp)	Peru	CBC	?	1922
			Mexico	ABC	?	1950
		Lydella minense Townsend	?	CBC	?	1922
		Trichogramma minutum Riley	USA	ABC	1,400	1929
Citrus	*Aleurocanthus woglumi* (Ashby); citrus blackfly	*Eretmocerus serius* Silvestri	Panama	CBC	Established	1938
			Panama	CBC	2,636	1943
		Amitus hesperidum Silvestri	India and Pakistan	CBC	51,052[c]	1949/50
		Encarsia clypealis Silvestri	India and Pakistan	CBC	51,052[c]	1949/50
		E. perplexa Huang & Polaszek	India and Pakistan	CBC	51,052[c]	1949/50
		E. smithi (Silvestri)	India and Pakistan	CBC	51,052[c]	1949/50
		E. divergens (Silvestri)	Malaya	CBC	Established	1948
		E. sp. *nr. citrofila*	India and Pakistan	CBC	?	1949/50
		E. merceti Silvestri	India and Pakistan	CBC	Established	1949/50
		Serangium parcesetosum Sicard	India and Pakistan	CBC	?	1949/50
		Acletoxenus indicus Malloch	India and Pakistan	CBC	?	1949/50
Citrus	*Icerya purchasi* Maskell; cottony cushion scale	*Rodolia cardinalis* (Mulsant)	USA	CBC	Established on Leguminosae in urban areas of Mexico City	1939
Apple	*Eriosoma lanigerum* Hausmann; woolly apple aphid	*Aphelinus mali* (Haldeman)	USA	CBC	5,678	1940

Continued

Table 21.1. Continued.

Crop	Pest	Natural enemy/ pathogen	Origin	Type of biocontrol[a]	Area (ha) under biocontrol[b]	Year
Citrus	*Aonidiella aurantii* Maskell; California red scale	*Aphytis lingnanensis* Compere	USA	CBC	21,646[c]	1940s–50s
		A. chrysomphali (Mercet)	USA	CBC	21,646[c]	1940s–50s
		Encarsia perniciosi (Tower)	USA	CBC	33,656[c]	1940s–50s
		Comperiella bifasciata Howard	USA	CBC	36,656[c]	1940s–50s
		Chilocorus sp.	Mexico	?	?	1940s–50s
Citrus	*Planococcus citri* Risso; citrus mealybug	*Cryptolaemus montrouzieri* Mulsant	USA	CBC	?	1942
Grasslands	*Aeneolamia* and *Prosapia*; spittlebugs	*Castolus* sp.	Mexico	ABC	187,454	1950s
		Sinea sp.	Mexico	ABC	187,454	1950s
		Zelus janus Stål	Mexico	ABC	187,454	1950s
		Salpingogaster nigra Schiner	Trinidad	CBC	187,454	1956
Cotton	*Pectinophora gossypiella* Saunders; pink bollworm	*Bracon brevicornis* Wesmael	USA	CBC	? (present)[d]	1932/34 1953/55
		Exeristes roborator (Fabricius) (European strain)	USA	CBC	? (present)[d]	1933/35
		Bracon kirkpatricki (Walker)	USA	CBC	? (present)[d]	1935/36
		B. nigrorufum (Cushman)	USA	CBC	? (present)[d]	1937/44
		Chelonus blackburni Cameron	USA	CBC	? (present)[d]	1937/44
		Bracon platynotae (Cushman)	USA	CBC	?	1953/55
		B. gelechiae Ashmead	USA	CBC	? (present)[d]	1953/55
		Chelonus heliopae Gupta	USA	CBC	?	1953/55
		Ch. narayani Subba Rao	USA	CBC	? (present)[d]	1953/55
		Apanteles angaleti Muesebeck	USA	CBC	? (present)[d]	1953/55
Mangoes, citrus, chicozapote	*Anastrepha* spp.; Mexican fruit fly	*Diachasmimorpha longicaudata* (Ashmead)	Hawaii	CBC	18,872[e]	1954
		D. tryoni (Cameron)	Hawaii	CBC	18,872[e]	1954
		Fopius arisanus (Sonan)	Hawaii	CBC	18,872[e]	1954
		F. vandenboschi (Fullaway)	Hawaii	CBC	18,872[e]	1954
		Psyttalia incisi (Silvestri)	Hawaii	CBC	18,872[e]	1954
		Aceratoneuromyia indica (Silvestri)	Hawaii	CBC	18,372[e]	1955
		Aganaspis daci (Weld)	Hawaii	CBC	18,872[e]	1955
		Dirhinus giffardii (Silvestri)	Hawaii	CBC	18,872[e]	1955
		Pachycrepoideus vindemmiae (Rondani)	Hawaii	CBC	18,872[e]	1955

Citrus	*Lepidosaphes beckii* (Compere); purple scale	*Aphytis lepidosaphes* Compere	USA	CBC	27,621[c]	1954
Banana	*Cosmopolites sordidus* (Germar); banana root borer	*Plaesius javanus* Erichson	Fiji Islands	CBC	4,019	1955
Citrus	*Chrysomphalus aonidum* (L.); Florida red scale	*Encarsia aurantii* (Howard)	Hong Kong	CBC	277[c]	1949/50
		Pseudhomalopoda prima Girault	USA	CBC	277[c]	1955
		Pteroptrix (=*Casca*) *smithi* (Compere)	Hong Kong	CBC	?	1956
		Aphytis lingnanensis Compere	Hong Kong	CBC	277[c]	1957
		Aphytis holoxanthus DeBach	USA	CBC	4,404[c]	1957
						1960
Bean	*Epilachna varivestis* Mulsant; Mexican bean beetle	*Pediobius foveolatus* Crawford	Guam	CBC	3,981	1956
		Aplomyiopsis epilachnae (Aldrich)	Mexico	ABC	Experimental releases	1973
Grass	*Antonina graminis* (Maskell); rhodesgrass mealybug	*Anagyrus antoninae* Timberlake	USA	CBC	45,540	1957
		Neodusmetia sangwanii Subba Rao	India	CBC	45,540	1957
Alfalfa	*Therioaphis trifolii* Monell; spotted alfalfa aphid	*Aphelinus asychis* Walker	USA	CBC	25,290	1957/58
		Praon exsoletum (Ness)	USA	CBC	25,290	1957/58
		Trioxys complanatus Quilis Pérez	USA	CBC	25,290	1957/58
		Aphidius matricariae (Haliday)	USA	CBC	?	1957/58
		Hippodamia convergens Guerin	Mexico	ABC	2,961	1973
Citrus	*Aleurothrixus floccosus* (Maskell), woolly whitefly	*Amitus spiniferus* (Brèthes)	USA	CBC	? (present)[d]	1969
		Encarsia dominicana Evans	?	CBC	?	1950s

[a]Type of biocontrol: ABC = augmentative biocontrol, CBC = classical biocontrol

[b]Area on which natural enemies contributed to pest control (data estimated on information from scientific literature and internal reports); areas mentioned in this table do not include current area protected by these biological control agents

[c]Only Mexican lime and orange

[d]Source: https://www.cabi.org/isc/datasheet

[e]Only orange

Diglyphus isaea (Walker), *E. formosa*, *Eretmocerus califonicus* Howard, *Leptomastix dactylopii* Howard, *Macrocentrus prolificus* Wharton and *Telenomus remus* Nixon. The following predators were also imported and mass produced: *Neoseiulus cucumeris* (Oudemans), *Neoseiulus barkeri* Hughes, *Neoseiulus californicus* (McGregor), *Iphiseius degenerans* Berlese, *Aphidoletes aphidimyza* (Rondani), *Delphastus pusillus* (LeConte), *Galendromus occidentalis* (Nesbitt), *Chrysoperla* spp., *Orius insidiosus* (Say), *Phytoseiulus persimilis* Athias-Henriot and *Podisus maculiventris* (Say). In addition, species of entomopathogenic organisms of the genera *Beauveria*, *Metarhizium* and *Isaria* (= *Paecilomyces*), and *Bacillus thuringienisis*, *Steinernema carpocapsae*, *S. feltiae*, *S. glaseri* and *Heterorhabditis* sp. were mass produced and other species were imported (Rodríguez-del-Bosque and Arredondo-Bernal, 1999). Releases were made in crops such as maize, cotton, sugarcane, sunflower, coffee, tobacco, soybean, sorghum, vegetables, ornamental plants, beans, wheat, citrus and forests on about 1.5 million hectares annually (Reyes-Domínguez, 1996). As an example of activities in Mexico, production data of the Regional Centers is shown in Table 21.2.

21.3 Current Situation of Biological Control in Mexico

21.3.1 Overview of classical and augmentative biological control programmes

Exotic and native pests for which biocontrol was or is currently being implemented are presented in Table 21.3. The majority of the pests are hemipterans of the families Aleyrodidae, Aphididae, Coccidae, Diaspididae, Liviidae, Pseudococcidae and Triozidae, underlining that sucking insects tend to be the most problematic in Mexico. Some of these pests are controlled by natural enemies through inundative releases: *Aphis gossypii* Glover (Lomelí-Flores *et al.*, 2008), *B. tabaci* (Martínez-Carrillo *et al.*, 2015), *H. hampei* (Barrera-Gaytán *et al.*, 2008) and *Trialeurodes vaporariorum* Westwood (García-Valente and Ortega-Arenas, 2008). A case of successful biocontrol of weeds in Mexico is that of the exotic water hyacinth *E. crassipes* with phytophagous insects and plant pathogens (Martínez-Jiménez, 2008). Based on information presented in this chapter, it is estimated that classical biocontrol is functioning on about 11,810,400 ha, whereas augmentative control is applied on about 763,800 ha.

21.3.2 Major recent cases of biological control

Recent cases of biocontrol in Mexico are summarized below; for more detail about each of these programmes, refer to Arredondo-Bernal and Rodríguez-del-Bosque (2008, 2015).

Pink hibiscus mealybug

The pink hibiscus mealybug *Maconellicoccus hirsutus* (Green) is a polyphagous pest of Asian origin. It was detected in September of 1999 in Mexicali, Baja California. Considering the possible dispersion of *M. hirsutus* to the rest of the country and experience gained in the Caribbean, the Government of Mexico decided to apply various control methods consisting of the use of insecticides, cutting and burning of infested trees, as well as releasing 1,600 *Anagyrus kamali* Moursi and 600 *Gyranusoidea indica* Shafee, Alam & Argarwal parasitoids in blackberries, obelisk, grapefruit and carob trees, which are all common in urban areas. The parasitoids were imported from Puerto Rico and established (Santiago-Islas *et al.*, 2008). In the same year, the pink hibiscus mealybug was also found in Belize and the same parasitoids were released at the border of Quintana Roo to establish an ecological barrier and slow down the introduction of this pest from Central America to Mexico. In 2004, the hibiscus mealybug infested teak and *Acacia* spp. in the State of Nayarit (Santiago-Islas *et al.*, 2008). Therefore, from April 2004, imports of *A. kamali* from the International Regional Organization of Agricultural Health (OIRSA, located in Belize) and the Animal and Plant Health Inspection Service of USDA (located in Puerto Rico) began and 386,000 parasitoid individuals were imported. Another 82,000 *G. indica* wasps from Puerto Rico were introduced, but did not establish. The predator *Cryptolaemus montrouzieri* Mulsant was imported from the USA and Canada, and individuals from mass production centres located in Mexico were released as well.

Table 21.2. Production and release of biological control agents by the Regional Centers for Reproduction of Beneficial Organisms during 1994, 1995 and 1996 (retrieved from CNRCB, 1997).

Year	Biological control agent	Production[a]	Released[a]	Area treated (ha)
1994	*Trichogramma* spp.	17,893,000,000	13,340,500,000	708,901
1995	*Trichogramma* spp.	20,543,000,000	14,530,600,000	892,070
	Chrysoperla carnea (Stephens) (*sensu lato*)	193,700,000	159,800,000	29,222
	Diachasmimorpha longicaudata (Ashmead)	2,2000,000	–	–
	Aceratoneuromyia indica Silvestri	1,600,000	600,000	26
	Muscidifurax spp.	41,580,000	33,880,000	406.5
	Entomopathogenic fungi	11,417 L	7,660 L	10,930
1996	*Trichogramma* spp.	19,222,388,500	14,202,105,900	917,947
	Chrysoperla carnea (*sensu lato*)	344,496,000	291,009,750	105,741
	Habrobracon spp.	40,999,000	35,337,000	54,180
	Diachasmimorpha longicaudata	224,235	–	–
	Aceratoneuromyia indica	693,000	–	–
	Muscidifurax spp.	27,895,000	21,910,000	248
	Entomopathogenic fungi	22,807 L	17,611 L	20,004

[a]Numbers of individuals, except when followed by L (= litres)

Table 21.3. Exotic and native pests under biological control in Mexico since 1970.

Crop	Pest	Natural enemy	Origin	Type of biocontrol[a]	Area under biocontrol[b]	Project started[c]
Broccoli, cauliflower	*Plutella xylostella* (L.)	*Bacillus thuringiensis*	Exotic	ABC	9,613 ha	1987
		Diadegma insulare	Native	ABC	24,335 ha	1993
		Cotesia plutellae	Exotic	ABC	24,335 ha	1994
		Trichogrammatoidea bactrae	Exotic	ABC	?	1989
		Trichogramma pintoi	Exotic	ABC	15,570 ha	2000
		Trichogramma pretiosum	Native	ABC	15,570 ha	2000
Citrus	*Aonidiella aurantii* Maskell	*Aphytis melinus*	Exotic	ABC	6,715 ha	1993
Citrus	*Diaphorina citri* Kuwayama[d]	*Tamarixia radiata*	Exotic	ABC/CBC	266,474 ha	2010
		Isaria javanica (strain CHE-CNRCB 303 & 307) and *Metarhizium anisopliae* (strain CHE-CNRCB 224)	Native	ABC	28,356 ha	2012
Citrus	*Phyllocnistis citrella* Stainton	*Ageniaspis citricola*	Exotic	CBC	19,420 ha	1997
Citrus	*Phyllocoptruta oleivora* (Ashmead)	*Amblyseius victoriensis*	Exotic	CBC	?	1994

Continued

Table 21.3. Continued.

Crop	Pest	Natural enemy	Origin	Type of biocontrol[a]	Area under biocontrol[b]	Project started[c]
Citrus	*Toxoptera citricida* (Kirkaldy)[d]	*Ceraeochrysa claveri*	Native	ABC	39,678 ha	1998
		Cycloneda sanguinea	Native	ABC	39,678 ha	1998
		Harmonia axyridis	Exotic	ABC	39,678 ha	1998
		Isaria javanica (strain CHE-CNRCB 305)	Native	ABC	12,336 ha	2009
		Olla v-nigrum	Native	ABC	39,678 ha	1998
Coffee	*Hypothenemus hampei* (Ferrari)	*Cephalonomia stephanoderis*	Exotic	CBC/ABC	Established	1988
		Prorops nasuta	Exotic	CBC	?	1988/89
		Phymastichus coffea	Exotic	CBC	?	2000
		Beauveria bassiana	Native	ABC	Established	2000
Contaminated natural aquatic systems	*Chironomus plumosus* (L.)	*Buenoa scimitra*	Native	ABC	?	1999
		Gambusia affinis	Native	ABC	45 km long (river)	2010
Cotton	*Anthonomus grandis* Boheman	*Jaliscoa* (=*Catolaccus*) *grandis*	Native	ABC	?	1995
Cotton	*Pectinophora gossypiella* Saunders	*Bracon kirpatricki*	Exotic	CBC	?	1970
Eucalyptus	*Glycaspis brimblecombei* Moore	*Psyllaephagus bliteus*	Exotic	CBC	Urban / forest areas with eucalyptus	2001
Fruit, forest, ornamentals, wild plants, vegetables	*Maconellicoccus hirsutus* (Green)[d]	*Anagyrus kamali*	Exotic	ABC/CBC	11,300,000 ha	1999
		Gyranusoidea indica	Exotic	CBC	Unestablished	1999
		Cryptolaemus montrouzieri	Exotic	ABC	11,300,000 ha	2004
Grape	*Planococcus ficus* (Signoret)[d]	*Anagyrus pseudococci*	Exotic	ABC	28,365 ha	2001
		Cryptolaemus montrouzieri	Exotic	ABC	?	2001
		Chrysoperla carnea sensu lato	Native	ABC	28,365 ha	2001
Mango	*Aulacaspis tubercularis* Newstead	*Cybocephalus nipponicus*	Exotic	CBC	?	2007
Fruits	*Anastrepha* spp.[d]	*Diachasmimorpha longicaudata*	Exotic	ABC	10,685 ha per week	1992
Potato	*Phthorimaea opercullela* (Zeller)	*Copidosoma desantisi*	Exotic	CBC	?	1990

Continued

Table 21.3. Continued.

Crop	Pest	Natural enemy	Origin	Type of biocontrol[a]	Area under biocontrol[b]	Project started[c]
Rivers, lakes, irrigation systems	*Eichhornia crassipes* (Mart.) Solms.	*Acremonium zonatum*	Native	ABC	1,516 ha and 3 km long (Río Santiago, State of Jalisco)	1998
		Cercospora piaropi	Native	ABC		1998
		Neochetina bruchi	Exotic	ABC		1994
		Neochetina eichhorniae	Exotic	CBC		1976
Sorghum	*Melanaphis sacchari* (Zehntner)	*Chrysoperla carnea sensu lato*	Native	ABC	225,925 ha	2014
		Coleomegilla maculata	Native	ABC	81,394 ha	2017
Sorghum	*Oebalus mexicana* (Sailer)	*Beauveria bassiana*	Native	ABC	600–800 ha (hibernation areas)	1996
Soybean	*Anticarsia gemmatalis* Hübner[d]	AgNPV	Exotic	ABC	10,000 ha	2001
Soybean, beans, maize and sorghum	*Schistocerca piceifrons piceifrons* Walker[d]	*Metarhizium acridum*	Native	ABC	1,586 ha	2009
	Brachystola magna (Girard)[d]				13,564 ha	2011
	B. mexicana Bruner[d]				13,564 ha	2011
	Melanoplus differentialis (Thomas)[d],				13,564 ha	2011
	Sphenarium purpurascens Charpentier[d]				13,564 ha	2011
Strawberry	*Lygus hesperus* Knight	*Peristenus relictus*	Exotic	CBC	?	2014
Sugarcane	*Aeneolamia* and *Prosapia*[d]	*Metarhizium anisopliae,*	Native	ABC	35,000 ha	1994
Maize	*Diatraea saccharalis* (F.) *D. lineolata* (Walker)	*Cotesia flavipes*	Exotic	CBC	?	1985
Sugarcane	*Diatraea* spp.	*Trichogramma* spp.	Native	ABC	241,608 ha	1970
Tomato, chilli, pepper, cucumber, aubergine, strawberry, ornamentals	*Macrosiphum euphorbiae, M. rosae* (L.), *Aulacorthum solani* (Kaltenbach)	*Aphidius matricariae*	Exotic	ABC	38 ha	1996
		Aphidius ervi	Exotic	ABC	378 ha	2000
		Aphelinus abdominalis	Exotic	ABC	158 ha	1995
		Chrysoperla rufilabris	Native	ABC	38 ha	2000
		Chrysoperla carnea sensu lato	Native	ABC	39 ha	2000

Continued

Table 21.3. Continued.

Crop	Pest	Natural enemy	Origin	Type of biocontrol[a]	Area under biocontrol[b]	Project started[c]
Tomato, chilli, pepper, cucumber, aubergine, strawberry, ornamentals	*Myzus persicae* (Sulzer), and *Aphis gossypii* Glover	*Chrysoperla carnea sensu lato*	Native	ABC	58 ha	1990
		Aphidius colemani	Exotic	ABC	3,161 ha	1995
		Aphidoletes aphidimyza	Exotic	ABC	15 ha	1995
Tomato, cucumber, ornamentals	*Frankliniella occidentalis* (Pergande), and *Thrips tabaci* Lindeman	*Neoseiulus cucumeris*	Native	ABC	2,366 ha	1996
		Orius insidiosus	Native	ABC	825 ha	1996
		Amblydromalus limonicus	Native	ABC	159 ha	2011
		Orius laevigatus	Exotic	ABC	5 ha	2016
		Typhlodromips swirskii	Exotic	ABC	9,478 ha	2016
Tomato, aubergine, pepper, cucumber, ornamentals	*Trialeurodes vaporariorum* (Westwood), and *Bemisia tabaci* (Gennadius)	*Encarsia formosa*	Native	ABC	235 ha	1994
		Typhlodromips swirskii	Exotic	ABC	159 ha	2007
Vegetables, cotton, ornamentals	*Bemisia tabaci* (Gennadius)	*Chrysoperla carnea sensu lato*	Native	ABC	11,000 ha	1991
		Eretmocerus mundus	Exotic	CBC	?	1996
		Eretmocerus hayati	Exotic	CBC	?	1997
		Eretmocerus eremicus	Exotic	ABC	1,918 ha	2000
Tomato, ornamentals	*Liriomyza* spp.	*Diglyphus isaea*	Exotic	ABC	2,907 ha	1995
Tomato, pepper, cucumber, lettuce, ornamentals	*Bradysia* sp.	*Heterorhabditis bacteriophora*	Native	ABC	199 ha	1994
		Steinernema carpocapsae	Native	ABC	9 ha	1994
		Steinernema feltiae	Native	ABC	310 ha	1995
Urban areas and stabled livestock	*Musca domestica* L.	*Spalangia endius*	Native	ABC	?	1970
		Muscidifurax raptor	Native	ABC	?	1991
Tomato, aubergine, chilli, pepper	*Tetranychus urticae* Koch	*Neoseiulus californicus*	Native	ABC	2,340 ha	1994
		Phytoseiulus persimilis	Exotic	ABC	3,210 ha	1991
		Feltiella acarisuga	Exotic	ABC	1,887 ha	2003
Wetlands	*Arundo donax* L.	*Tetramesa romana*	Exotic	ABC	100 km long Río Bravo, Tamaulipas	2009
		Rhizaspidiotus donacis	Exotic	ABC		2010

[a]Type of biological control: ABC = augmentative biological control, CBC = classical biological control
[b]Current area on which natural enemies contributed to pest control (data estimated from scientific literature and internal reports)
[c]Year when biocontrol project started; some projects have been terminated
[d]See text, special biocontrol programme

The number of natural enemies released in all hibiscus mealybug infested zones since 2004 are presented in Table 21.4. To date 13% of the specimens are released in agricultural areas; of the rest, 1.5%, 9% and 76.5% are released in forest, marginal and urban areas, respectively, all over the country. A laboratory for the production of *A. kamali* and *C. montrouzieri* was built in Bahía de Banderas, Nayarit, with a capacity to produce up to 7 million *A. kamali* wasps per month. The predator and the parasitoid have established everywhere after release and cause more than 90% decrease in pink hibiscus mealybug in the majority of cases (Santiago-Islas *et al.*, 2008; Williams *et al.*, 2013). At present, populations of pink hibiscus mealybug are less than 0.5 individuals per branch on plants like carambolier, soursop, mango, Chinese guava, obelisk, teak and jackfruit in marginal and urban areas (Santiago-Islas *et al.*, 2008). Phytosanitary actions that include cultural practices, chemical control, use of soaps and biocontrol (as the main strategy carried out by the Federal Government) protect at least 843,000 ha of crops susceptible to the pest (jamaica, soursop, guava, mango and citrus fruits), with a production value of US$1 billion (Salcedo-Baca *et al.*, 2012). The total area of other commercial crops that might potentially be affected is 11.3 million hectares and these are now also protected by the above-mentioned control programme (SIAP, 2017).

Table 21.4. Biological control agents (number of individuals) released against pink hibiscus mealybug in Mexico since 2004.

	Species		
Year	*Anagyrus kamali*	*Gyranusoidea indica*	*Cryptolaemus montrouzieri*
2004	237,900	8,200	931,500
2005	216,752[a]	73,400	2,558,900
2006	2,079,707[a]	—	1,600,400
2007	14,334,473[a]	—	1,390,120
2008	8,340,250[a]	—	42,200
2009	21,304,400[a]	—	100,000[a]
2010	14,527,700[a]	—	150,000[a]
2011	21,622,000[a]	—	100,000[a]
2012	29,632,000[a]	—	200,000[a]
2013	30,743,200[a]	—	200,000[a]
2014	21,982,000[a]	—	131,000[a]
2015	17,381,000[a]	—	125,275[a]
2016	28,941,750[qa]	—	223,500[a]
2017	34,047,900[a]	—	330,900[a]
2018	38,504,000[a]		333,400[a]

[a]Produced in Mexico

Asian citrus psyllid

The Asian citrus psyllid *Diaphorina citri* (Kuwayama) is the vector of huanglongbing (HLB), a devastating citrus disease worldwide caused by the bacteria *Candidatus* Liberibacter spp.. The HLB has been detected in 498 municipalities in 24 states in Mexico, of which 403 are considered citrus growers (Sanidad Vegetal, 2018). Biocontrol of *D. citri* in Mexico is carried out through two laboratories that produce the parasitoid *Tamarixia radiata* (Waterston): one in Colima, the other in Yucatán (Sánchez-González *et al.*, 2015). The National Center of Reference of Biological Control (Centro Nacional de Referencia de Control Biológico) (CNRCB) developed mass production of *T. radiata* for release in specific areas. Additionally, control of *D. citri* with entomopathogenic fungi was investigated. The two mass-rearing laboratories have produced 56 million individuals of *T. radiata* since 2010. Releases totalling more than 51.3 million parasitoids have been made in 18 States that produce citrus, especially at locations without specific pest management regimes, like in urban zones, backyards, areas that are difficult to reach and organic farms. *T. radiata* is established all over the country. The releases have resulted in increased parasitism and control of psyllids compared with locations where *T. radiata* was not released (Moreno-Carrillo *et al.*, 2013; Cicero-Jurado *et al.*, 2013). At the same time application of the entomopathogenic fungi *Isaria javanica* (= *Paecilomyces fumosorosea*) (Friederichs & Bally) Samson & Hywel-Jones (strain CHE-CNRCB 303 & 307) and *Metarhizium anisopliae* (Metchnikoff) Sorokin (strain CHE-CNRCB 224) to control psyllid populations was developed, resulting in 60–81.8% mortality caused by mycosis. Between 2012 and 2015 a cumulative total of 34,644 ha of citrus was treated with these entomopathogenic fungi in the States of Colima, Hidalgo, Jalisco, Nayarit, Oaxaca, San Luis Potosí, Tamaulipas and Veracruz (Sánchez-González *et al.*, 2015). During 2016–2017, 28,356 ha were treated (M.A. Mellín, Colima, Mexico, 2018, personal communication), yielding an average of about 10,000 ha treated annually by entomopathogens.

Vine mealybug

Vine mealybug *Planococcus ficus* (Signoret) is an exotic mealybug that feeds on the sap of vine plants and infests roots, leafs, bunches, trunks and cords. It was detected in March 2001 at the coast of Hermosillo, Sonora, where it infested 150 ha of grapes (Fu-Castillo *et al.*, 2002). In the same year, *C. montrouzieri* and *Anagyrus pseudococci* (Girault) were imported from California, achieving the establishment of *A. pseudococci* with parasitism between 5% and 30% in 2001 and up to 36% in 2002 (Fu-Castillo, 2008). The relatively low percentage of parasitism was affected by the cryptic habits of the mealybug and the use of insecticides. Releases of *A. pseudococci* were realized on 500 ha in 2002 and 450 ha in 2003 (Fu-Castillo, 2008). Extreme temperatures (< 10°C in winter and > 40°C in summer) also had a negative impact on the life cycle and fertility of *A. pseudococci*. Additionally, *Chrysoperla carnea sensu lato* and *C. montrouzieri* were released, resulting in 40–80% population decreases of *P. ficus* 52 days after release (Fu-Castillo and Grageda-Grageda, 2002).

Brown citrus aphid

Brown citrus aphid *Toxoptera citricida* (Kirkaldy) is the vector of the citrus tristeza virus (CTV), which originated from Asia. It was detected in February 2000 in the north of Quintana Roo and Yucatán (Michaud and Álvarez, 2000). A classical biocontrol programme was initiated with releases of the predator *Harmonia axyridis*, also of Asian origin, to create an ecological barrier on the southern border, since the aphid had been present in Belize for years. Rearing of *H. axyridis* started in 1998 in Mexico, while releases started in 1999 in the State of Quintana Roo, and 18 million individuals were released in citrus plantations of the Yucatan peninsula until the end of mass production of the predator in 2002. The direct impact of these releases is not well known, but *H. axyridis* did not establish at release sites. Mass breeding of *Ceraeochrysa claveri*, a native predator of the Yucatan peninsula, started in 1998 and the first releases took place in Quintana Roo in 1999, reaching a total number of 3 million released individuals until the third trimester of 2002 (Munguía-Rosales, 2002). Also, about 800,000 *C. sanguinea* and *Olla v-nigrum* ladybird beetles were reared and released, but the effect on brown citrus aphid populations is unknown (Munguía-Rosales, 2002). As part of this biocontrol programme, the CNRCB created a technology package for the use of the entomopathogenic fungus *I. javanica* (strain CHE-CNRCB 305 [AMBAS1]), resulting in mortalities greater than 90% (Williams *et al.*, 2013). From 2009 until 2015, a cumulative total of 12,336 ha of citrus plantations was treated with *I. javanica* in Mexico, supported by a federal programme. Since then, those producers interested in biocontrol have obtained entomopathogenic fungi and applied them.

Acrididae locusts

Acrididae are species of gregarious and swarm-forming locusts and non-swarming grasshoppers (Hernández-Velázquez and Toriello, 2008). The Central American locust *Schistocerca piceifrons piceifrons* Walker is found every year affecting around 20,000 ha, with a risk of damaging 50 times this surface area when checks and control activities are neglected (Macías and Ramírez, 1998). It is polyphagous and feeds on over 400 plant species, though it has a preference for maize, soybean, beans, sesame, cotton, sorghum, banana, sugarcane and fruits (Barrientos-Lozano *et al.*, 1992). In 1993, the CNRCB and the Plant Health General Directorate-National Service for Agri-Food Health, Safety and Quality (Spanish acronym DGSV-SENASICA: Dirección General de Sanidad Vegetal (Secretaría de Agricultura y Desarrollo Rural) – Servicio Nacional de Sanidad, Inocuidad y Calidad Agroalimentaria) started the development of a microbial control programme of this locust with the entomopathogenic fungus *Metarhizium acridum* (= *M. anisopliae* var. *acridum*). The programme was accomplished by the General Directorate of Plant Health of the Mexican Government in the States of Yucatán, Quintana Roo, Campeche, Veracruz, Tamaulipas, Chiapas, Oaxaca, Colima and Nayarit. Evaluations carried out in collaboration with the Australian Plague Locust Commission indicate a reduction in locust populations of 86% 11 days after treatment with the strain CHE-CNRCB 213 and more than 90% with the strain CHE-CNRCB 206 (Hernández-Velázquez *et al.*, 2003; Barrientos-Lozano *et al.*, 2004). Since 2009, microbial control of *S. p. piceifrons* has been realized in different States of the country (Campeche, Chiapas, Hidalgo, Oaxaca, San Luis Potosí, Tabasco, Tamaulipas, Veracruz and Yucatan), reaching a cumulative treatment of 11,107 ha during the seven years to 2016, with an average of 1,586 ha treated per year. Considering that 20 t of methylated parathion in a concentration of 3% would be needed per hectare (SENASICA, 2009), the use of this entomopathogenic fungus replaced the application of 222.14 t of this insecticide.

The Federal Government also coordinates control of the grasshopper complex including *Brachystola magna* Girard, *Brachystola mexicana* Bruner, *Melanoplus differentialis* (Thomas) and *Sphenarium purpurascens* Charpentier. Soybeans, pumpkin, maize and sorghum are the most affected crops (SENASICA, 2012). In 2011, 4,000 ha of cultivated land in Guanajuato and 8,000 ha in Tlaxcala were treated with *M. acridum* (Williams *et al.*, 2013). During 2012–2016, the cumulatively treated surface was 54,256 ha in Chihuahua, Guanajuato and Tlaxcala, with an average of 13,564 ha treated per year (M. Sánchez, Mexico City, Mexico, 2017, personal communication).

Soybean caterpillar

The soybean caterpillar *Anticarsia gemmatalis* Hübner is a pest occurring in many American countries and is the most important defoliator of soybean plants in Mexico, causing yield losses up to 40% (Ávila-Valdez and Rodríguez-del-Bosque, 2008). In the region of Huasteca in Tamaulipas, a commercial programme of validation and transfer of technology was implemented in 2001 by using the *A. gemmatalis* nuclear polyhedrosis virus (AgNPV), imported from Brazil. Larval mortalities of 100% were obtained in the field (Ávila-Valdez and Rodríguez-del-Bosque, 2008). Until 2010, the accumulated area treated with AgNPV in the Huastecas was 92,000 ha, averaging about 10,000 ha per year. Research and transfer of technology costs to develop and apply AgNPV in Mexico were US$191,000, while the benefits were US$5.14 million, resulting in a benefit/cost ratio of 27:1. Treatment with AgNPV avoided the use of 92,000 l of insecticides and prevented the elimination of more than 10 billion predators, such as lacewings, ladybirds, bedbugs and spiders, and avoided the development of secondary pests like *Chrysodeixis* (= *Pseudoplusia*) *includens* (Walker) and *Trichoplusia ni* Hübner (Ávila-Valdez and Rodríguez-del-Bosque, 2015).

Red gum lerp psyllid

The red gum lerp psyllid *Glycaspis brimblecombei* Moore, originally from Australia, was first detected in Mexico in 1999 and found in 24 federal areas in 2002. It reduces growth and causes defoliation and death of hundreds of thousands of eucalyptus trees all over the country. In 2001 the Australian parasitoid *Psyllaephagus bliteus* Riek was imported from California to be released in different parts of Guadalajara, Jalisco, Mexico State, Michoacan and Mexico City. The parasitoid established and reduced the pest population to very low levels, where it no longer caused the death of the trees, and the parasitoid is now present all over the country where the pest occurs (Cibrián-Tovar, 2015). Romo-Lozano *et al.* (2007) made a benefit/cost analysis in Mexico City and found that it was 137:1 at locations with 15% in tree mortality, 391:1 with 35% tree mortality and 593:1 with 64% tree mortality, taking into account the number of trees that might have died between 2001 and 2003. Thus, this biocontrol method is very cost-effective even where tree mortality was lowest. The establishment and dispersal of the parasitoid have continued and follow the same course as that of *G. brimblecombei*, as it is now present all over the country where red eucalyptus occurs (Cibrián-Tovar, 2015).

Spittlebugs

Spittlebugs are widely present in the Americas. The most abundant species in sugar production areas in Mexico, which are located between the watershed of the Gulf of Mexico and the Pacific Ocean, are of the genera *Aeneolamia* and *Prosapia* (Alatorre-Rosas and Hernández-Rosas, 2015). During the past three decades, spittlebug control in Mexico has been focused on reduced use of insecticides and the application of *M. anisopliae*. Fungal treatments resulted in up to 70% mortality (Bautista and González, 2005; Alatorre-Rosas and Hernández-Rosas, 2015). From 1999 until 2011, the use of this entomopathogenic fungus has strongly increased (Fig. 21.1). A study conducted in 49 sugar mills (CONADESUCA, 2011) indicated that 114,230.64 ha are affected by the spittlebug, of which 26% are treated with

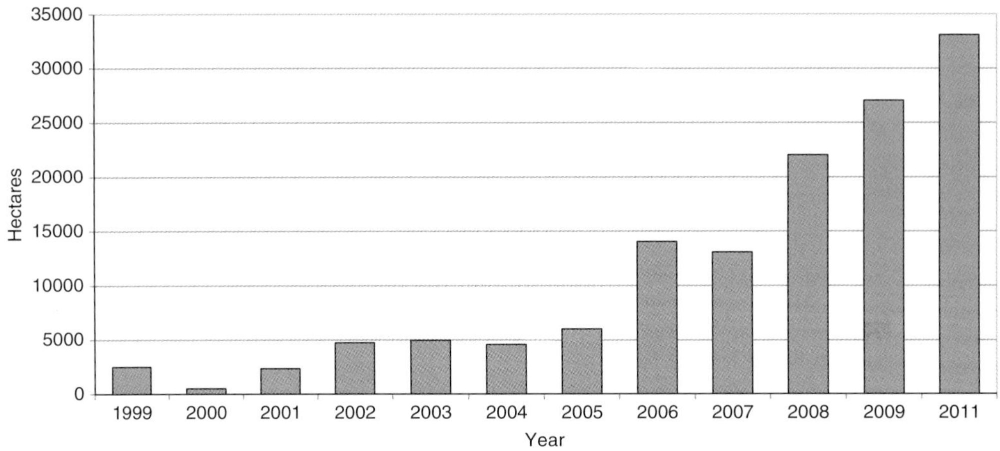

Fig. 21.1. Surface treated with *Metarhizium anisopliae* to control *Aeneolamia* sp. spittlebugs in sugarcane in Mexico (retrieved from CONADESUCA, 2011).

biocontrol. Currently > 35,000 ha are treated with *M. anisopliae*. To be able to satisfy the demand for *M. anisopliae*, several sugar mills started their own mass production (Alatorre-Rosas and Hernández-Rosas, 2015).

Fruit flies

Mexican fruit flies (*Anastrepha* spp.) are a very serious threat to many commercial crops and fruits in the country. Their direct damage is caused by oviposition of eggs in fruits and by larvae feeding on the pulp, causing early fall. Indirect damage consists of contamination of fruits by pathogenic microorganisms or decomposers. The Federal Government runs a National Campaign against Fruit Flies based on biocontrol and the sterile insect technique (SIT). The general objective of the campaign is to have historical and current information about local situations to plan field operations for control. This demands a good organizational structure and sufficient human resources and these needs are met by the financial support of the Federal Government, State and producers, the USA and donor organizations such as the UN's Food and Agriculture Organization (FAO) and the International Atomic Energy Agency. Currently, the biocontrol programme mass produces and releases 25 million individuals of the parasitoid *Diachasmimorpha longicaudata* (Ashmead) on 10,685 ha per week (F. Ramírez, Chiapas, México, 2018, personal communication). This parasitoid is capable of parasitizing fruit flies in the field even when at low fly densities. The programme has another 13 species of parasitoids under study, with *Coptera haywardi* Loia and *Fopius arisanus* Sonan showing good possibilities for mass production and releases in the short term (J. Cancino, Chiapas, México, 2018, personal communication).

Various other biological control programmes

Currently in development are control of: (i) housefly, using native species of *Spalangia* and *Muscidifurax* (Nava-Camberos and Ávila-Rodríguez, 2008); (ii) buzzer midge *Chironomus plumosus* (L.) of Palearctic origin, using the native backswimmer *Buenoa scimitra* Bare and the mosquitofish *Gambusia affinis* (Baird and Girard) (Quiroz-Martínez *et al.*, 2015); and (iii) an exotic pest, the whitefly *Aleurocybotus occiduus* Russell, through fortuitous biocontrol by native natural enemies. Vejar-Cota and Rodríguez-del-Bosque (2015) found that populations of this new invasive species are significantly reduced by natural enemies such as spiders of the families Araneidae, Lycosidae, Oxyopidae, Salticidae, Theridiidae and Thomisidae, along with predatory insects such as *Chrysoperla* sp., *Collops femoratus* Schaeffer and *C. sanguinea*, *Orius* sp., *Geocoris* sp., *Sinea* sp. and Mantidae, as well as two new parasitoids, *Encarsia longitarsis* Myartseva and *Metaphycus cereales* sp. nov. Myartseva & Ruíz.

21.3.3 Biological control in protected agriculture

The production of vegetables, fruits and ornamental plants in protected environments has strongly increased in the past 15 years (Lomelí-Flores *et al.*, 2016). Protected agriculture began in the 1960s; in 1970 about 100 ha were reported; in 2017, 25 thousand ha were registered (Bastida-Tapia, 2017). The most important crops are tomato, cucumber and pepper, occupying 70%, 16% and 10%, respectively, of the protected environment (Lomelí-Flores *et al.*, 2016). Many other crops are produced on the remaining 4% (García-Victoria *et al.*, 2011). Forty-five per cent of the protected surface consists of greenhouses, 29% use shade-houses and 16% macro-tunnels. The rest of the surface is covered with pavilions, shadow-roofs, and micro-tunnels (Lomelí-Flores *et al.*, 2016). Natural enemies used in protected cultivation are imported by commercial companies and are listed in Table 21.5. An important recent finding is that yields of pepper are on average 10% higher when using beneficial organisms compared with chemical control (García and Gastellum, 2013). During 2017, > 29,000 ha of greenhouses were estimated to have used biocontrol agents; the species most used were *Amblyseius* (*Typhlodromips*)

Table 21.5. Biological control agents used in protected agriculture in Mexico during 2017.

Pest	Crop	Biocontrol agent	Area (ha) under biocontrol
Tetranychus urticae	Tomato, bell pepper, cucumber, aubergine, strawberry, ornamental plants	*Feltiella acarisuga*	1,887
		Neoseiulus californicus	2,340
		Phytoseiulus persimilis	3,210
Aphis gossypii, *Myzus persicae*	Tomato, chilli, bell pepper, cucumber, aubergine, strawberry, raspberry, blackberry, ornamental plants	*Aphidius colemani*	3,161
		Aphidoletes aphidimyza	15
		Chrysoperla carnea s.l.	58
Aulacorthum solani, *Macrosiphum euphorbiae*, *M. rosae*	Tomato, chilli, bell pepper, cucumber, aubergine, strawberry, ornamental plants	*Aphelinus abdominalis*	158
		Aphidius ervi	378
		Aphidius matricariae	38
		Aphidoletes aphidimyza	15
		Chrysoperla carnea s.l.	38
Trialeurodes vaporariorum, *Bemisia tabaci*	Tomato, aubergine, bell pepper, chilli, cucumber, ornamental plants	*Encarsia formosa*	235
		Eretmocerus eremicus	1,918
		Typhlodromips swirskii	159
Frankliniella occidentalis, *Thrips tabaci*	Tomato, cucumber, ornamental plants	*Amblydromalus limonicus*	159
		Neoseiulus cucumeris	2,366
		Orius insidiosus	825
		Orius laevigatus	5
		Typhlodromips swirskii	9,478
Liriomyza spp.	Tomato, bell pepper, chilli, cucumber, aubergine, ornamental plants	*Diglyphus isaea*	2,907
Bradysia sp.	Tomato, bell pepper, cucumber, lettuce, ornamental plants	*Heterorhabditis bacteriophora*	199
		Steinernema feltiae	310
		Steinernema carpocapsae	9
Bactericera cockerelli	Tomato, potato, pepper	*Tamarixia triozae*	?
Mealybugs	Tomato, bell pepper, chilli, cucumber, aubergine, strawberry, raspberry, blackberry	*Cryptolaemus montrouzieri*	53

swirskii (Athias-Henriot), *P. persimilis*, *E. formosa*, *E. eremicus*, *A. aphidimyza*, *Feltiella acaraisuga* (Vallot), *N. californicus*, *N. cucumeris*, *O. insidiosus* and *A. colemani*.

García-Victoria *et al.* (2011) interviewed 6,127 greenhouse growers at various locations in Mexico, to find out whether they used chemical, biological and organic control methods as well as cultural practices. Although chemical control was still the most popular pest control method, use of biocontrol was increasing. For example, 42% of the producers in Chihuahua used beneficial organisms for pest management, 37% in Baja California Sur, followed by Sonora, Nuevo León, Sinaloa and Jalisco with, respectively, 36%, 32%, 30% and 30% (Table 21.6).

21.3.4 Mass production of biological control agents

Currently, Mexico has 65 laboratories that mass produce and commercialize 40 species of beneficial organisms: 22.5% are predators, 27.5% are parasitoids, 25% are entomopathogenic microorganisms, 15% are phytophagous or phytopathogens used to control weeds and 10% are antagonists used in disease control (Table 21.7) (Arredondo-Bernal, 2018). Forty per cent of the laboratories produce one or more *Trichogramma* species, which are the most common beneficial species used for over 50 years. It is well known that there are national and international commercial companies producing and selling biocontrol agents in Mexico, but information about their production is unclear.

The biocontrol agents provided by the 65 laboratories are produced on natural diets and are used for control of whiteflies, lepidopterans like the sugarcane stalkborer, aphids, scale, thrips, ants, mealybugs, psyllids, weevils of chilli pepper, spittlebugs, locusts and grasshoppers, phytophagous mites, stable flies and weeds. Biocontrol agents are used by, among others, producers of vegetables, papaya, *Opuntia* cacti, citrus, mango, soursop, sugarcane, jackfruit, maize, teak, coffee, grapes, strawberry, raspberry, blackberry, aubergine, forest and ornamental plants, as well as by governmental and state-owned institutions for control of various pests. Biocontrol agents listed in Table 21.8 are kept for experimentation and development of new biocontrol programmes, like the natural enemies of *D. citri* (the vector of HLB) and the neuropteran *Chrysoperla externa* (Hagen) and the coccinellid *Exochomus marginipennis* (LeConte) (Palomares-Pérez *et al.*, 2015). The 58 species of natural enemies being authorized for import and commercialization in Mexico are presented in Table 21.9 (RSPM 26, 2015).

Table 21.6. Type of pest control used in protected agriculture in Mexico (retrieved from García-Victoria *et al.*, 2011).

State	Area (ha)	Management method used (%)			
		Chemical	Biological	Organic	Cultural
Nuevo León	91	43	32	0	24
Coahuila	127	56	19	5	11
Puebla	299	87	5	2	2
Zacatecas	305	71	15	1	8
Mexico State	559	95	2	0	2
Guanajuato	574	64	23	2	6
Michoacán	637	76	23	0	1
San Luis Potosí	780	48	25	1	15
Durango	838	71	14	4	5
Sonora	1048	46	36	1	17
Chihuahua	1070	53	42	0	3
Baja California Sur	1142	56	37	0	7
Jalisco	1581	60	30	1	7
Sinaloa	2490	56	30	0	14
Baja California	2642	52	25	1	6

Table 21.7. Biological control agents mass produced in Mexican laboratories.

Biological control agent	Pest
Predators	
Chrysoperla carnea s.l. (Stephens)	Whitefly and aphids
C. comanche (Banks)	Asian citrus psyllid
C. rufilabris (Burmeister)	Citrus pests
Ceraeochrysa valida (Banks)	Asian citrus psyllid
Cryptolaemus montrouzieri (Muls.)	Pink hibiscus mealybug
Geocoris punctipes (Say)	Red spider mite, potato psyllid
Orius insidiosus Say	Whitefly, thrips
Hippodamia convergens Guérin-	Aphids
Méneville	Aphids
Cycloneda sanguinea L.	
Parasitoids	Pink hibiscus mealybug
Anagyrus kamali Moursi	Fruit flies
Diachasmimorpha longicaudata (Ash.)	Whitefly
Eretmocerus eremicus Rose and	Stored grain pests
Zolnerowich	Cane borers
Habrobracon sp.	Aphids
Lixophaga diatraeae (Towns.)	Domestic and stable flies
Lysiphlebus testaceipes (Cresson)	Asian citrus psyllid
Spalangia endius Walter	Potato psyllid
Tamarixia radiata (Waterston)	Cane borers, Lepidoptera
T. trioze (Burks)	Cane borers
Trichogramma pretiosum Riley	
T. exiguum Pinto & Plat.	Water lily
Phytophagous weed control agents	Water lily
Neochetina eichhorniae Warner	*Arundo donax* L. (giant reed)
N. bruchi Hustache	*A. donax* (giant reed)
Tetramesa romana Walker	
Rhizaspidiotus donacis (Leonardi)	Water lily
Phytopathogenic weed control agents	Water lily
Cercospora piaropi Tharp.	
Acremonium zonatum (Sawada)	Lepidoptera
Entomopathogenic bacteria	
Bacillus thuringiensis Berliner	Chilli beetle, coffee borer
Entomopathogenic fungi	Brown citrus aphid, whitefly
Beauveria bassiana (Bals.) Vuill.	Asian citrus psyllid
Isaria fumosorosea Wize	Aphids (*Aphis*, *Toxoptera*)
Isaria javanica (Bally) Samson & Hywel-	Spittlebug (*Aeneolamia*, *Prosapia*)
Jones	Grasshopper
Lecanicillium lecanii (Zimmerman)	Mites
Metarhizium anisopliae (Metchnikoff) Sorokin	Nematodes
M. acridum Driver and Milner	
Hirsutella thompsoni Fisher	Soil pests
Pochonia chlamydosporia (Goddard) Zare &	
Gams	*Phytophthora*, *Rhyzoctonia*
Entomopathogenic nematodes	*Rizoctonia*, *Alternaria* and *Pythium*
Heterorabdithis bacteriophora Poinar	*Rizoctonia*, *Alternaria* and *Pythium*
Antagonists	*Rizoctonia*, *Alternaria* and *Pythium*
Bacillus subtilis (Ehrenberg) Cohn	
Trichoderma asperellum Samuels	
T. harzianum Rifai	
T. viride (Pers.)	

Table 21.8. Natural enemies maintained in laboratories in Mexico for experimental purposes.

Biological control agent	Pest	Institution
Predators		
Exochomus marginipennis (LeCon)	*Diaphorina citri* Kuwayama	CNRCB (DGSV)
Chrysoperla externa (Hagen)	*D. citri*	CNRCB (DGSV)
Ceraeochrysa valida (Banks)	*D. citri*	CNRCB (DGSV)
Ceraeochrysa cincta (Schneider)	*Raoiella indica* Hirst	CNRCB (DGSV)
Ceraeochrysa claveri (Navás)	*R. indica*	CNRCB (DGSV)
Parasitoids		
Ganaspis *Aceratoneuromyia indica* (Silv.)	Fruit fly	Moscafrut (DGSV)
pelleranoi (Brèthes)	Fruit fly	Moscafrut (DGSV)
Coptera haywardi Loia.[a]	Fruit fly	Moscafrut (DGSV)
Diachasmimorpha tryoni (Ash.)	Fruit fly	Moscafrut (DGSV)
Dirhinus giffardii Silvestri	Fruit fly	Moscafrut (DGSV)
Doryctobracon areolatus (Szépligeti)	Fruit fly	Moscafrut (DGSV)
Doryctobracon crawfordi (Viereck)[b]	Fruit fly	Moscafrut (DGSV)
Eurytoma sivinskii Gates & Grissell[b]	Fruit fly	Moscafrut (DGSV)
Fopius arisanus (Sonan)[a]	Fruit fly	Moscafrut (DGSV)
Odontosema anastrephae Borgmeier	Fruit fly	Moscafrut (DGSV)
Opius hirsutus Tobias	Fruit fly	Moscafrut (DGSV)
Utetes anastrephae (Viereck)[b]	*Drosophila suzukii* Matsumura	CNRCB (DGSV)
Pachycrepoideus vindemmiae (Ron)	*D. suzukii*	CNRCB (DGSV)
Leptopilina boulardi Förster	*D. suzukii*	CNRCB (DGSV)
Spalangia simplex Perkins	*D. suzukii*	CNRCB (DGSV)
Trichopria drosophilae Perkins	*D. suzukii*	
Phytopathogens		
Alternaria eichhorniae Nag Raj & Ponnappa	Water hyacinth	IMTA

DGSV = General Directorate of Plant Health; IMTA = Mexican Institute of Water Technology
[a]Planned for release
[b]high-level production

21.4 New Developments in Biological Control in Mexico

In Mexico, official biocontrol programmes are implemented by SENASICA, a unit of the Federal Government, but programmes are developed also by State Governments, research institutes and producers.

21.4.1 Pest risk scenarios

Mexico has recently compiled risk scenarios for 1,271 pests potentially being a threat to Mexican agriculture, of which 1,074 are not yet present in Mexico. In 2016, the 28 species listed in Table 21.10 were priority species on inspection lists (SENASICA, 2016). Recent invasive arthropods for which biocontrol is tested include melon thrips *Thrips palmi* Karny, red palm tree mite *Raoiella indica* Hirst and potato moth *Tecia solanivora* (Povolný).

21.4.2 Biological control of new pests and diseases identified by risk scenarios

Biocontrol of diseases is an activity of increasing interest in Mexico and work with antagonists, competitors and mycopathogens against *Phytophthora capsici* Leonian has been done in chilli pepper farming. *Pochonia chlamydosporia* var. *chlamydosporia* (Goddard) Gams and Zare is used against phytopathogenic nematodes (Zavaleta-Mejía *et al.*, 2015). For *Hemileia vastatrix* Berk. & Broome, biocontrol by way of the fungal species *Lecanicillium*, *Simplicillium* and *Sporothrix* is in development (A.M. Berlanga-Padilla, Colima, Mexico, 2017, personal communication).

The red laurel ambrosia beetle *Xyleborus glabratus* Eichhoff and the polyphagous borer beetle *Ewallaceae* nr. *fornicatus* are new threats for Mexican agriculture. These insects have mycangia, which are specialized structures used for the storage, transport and transmission of fungi

Table 21.9. Species of biological control agents authorized for import into Mexico (retrieved from RSPM 26, 2015).

Parasitoids	Predatory insects	Predatory mites
Anaphes iole Girault	*Aphidoletes aphidimyza* (Rondani)	*Galendromus occidentalis* (Nesbitt)
Anagyrus kamali Moursi	*Atheta coriaria* Kraatz	*Galendromus helveolus* (Chant)
Anisopteromalus calandrae (Howard)	*Chrysoperla carnea* (Stephens)	*Iphiseius degenerans* Berlese
Aphelinus abdominalis (Dalman)	*Chrysoperla rufilabris* (Burmeister)	*Neoseiulus barkeri* Hughes
Aphidius colemani Viereck	*Cryptolaemus montrouzieri* Mulsant	*Neoseiulus californicus* (McGregor)
Aphidius ervi Haliday	*Cybocephalus nipponicus* Endrödy-Younga	*Neoseiulus cucumeris* (Oudemans)
Aphidius matricariae Haliday	*Delphastus pusillus* (LeConte)	*Phytoseiulus persimilis* Athias-Henriot
Aphytis lingnanensis Compere	*Feltiella acarisuga* (Vallot)	*Typhlodromips swirskii* (Athias-Henriot)
Aphytis melinus DeBach	*Orius insidiosus* (Say)	
Habrobracon hebetor Say	*Orius laevigatus* (Fieber)	
Coccidoxenoides perminutus Girault	*Orius tristicolor* (White)	
Cotesia flavipes Cameron	*Podisus maculiventris* (Say)	
Cotesia vestalis (= *plutellae*) (Haliday)	*Stethorus punctillum* (Weise)	
Dacnusa sibirica Telenga	*Xylocoris flavipes* (Reuter)	
Diadegma insulare (Cresson)		
Diglyphus isaea (Walker)		
Encarsia formosa Gahan		
Eretmocerus californicus Howard		
E. eremicus Rose & Zolnerowich		
E. mundus Mercet		
Leptomastix dactylopii Howard		
Muscidifurax raptor Girault & Sanders		
Muscidifurax raptorellus Kogan & Legner		
Muscidifurax zaraptor Kogan & Legner		
Nasonia vitripennis (Walker)		
Spalangia cameroni Perkins		
Spalangia endius Walker		
Spalangia nigroaenea Curtis		
Tamarixia triozae (Burks)		
Telenomus remus Nixon		
Trichogramma brassicae Bezdenko		
Trichogramma evanescens Westwood		
Trichogramma minutum Riley		
Trichogramma platneri Nagarkatti		
Trichogramma pretiosum Riley		

from host to host in order to feed (Freeman *et al.*, 2016). The species of fungi associated with these beetles are *Raffaelea lauricola* T.C. Harr., Fraedrich & Aghayeva, responsible for laurel wilt, and *Fusarium euwallaceae* S. Freeman, Z. Mendel, T. Aoki & O'Donnell, *Graphium* sp. and *Paracremonium* (*Acremonium*) *pembeum* S.C. Lynch & Eskalen (Freeman *et al.*, 2016), respectively, which have killed native and commercial trees in southern USA. *Ewallaceae* nr. *fornicatus* has been recorded in Tijuana, Baja California, on the Mexican border with California (USA) (García-Ávila *et al.*, 2016). These wood borers infest avocado trees (Peña *et al.*, 2012) and they may affect avocado production in Mexico, which is the world's largest producer with approximately 2.05 million tons harvested in 2018, providing 45.95% of the global production (SAGARPA, 2017). It can also affect

Table 21.10. Priority species on inspection list of the Mexican Government in 2016 (retrieved from SENASICA, 2016).

Common name	Scientific name
Red bay ambrosia beetle	*Xyleborus glabratus* Eichhoff
Khapra beetle	*Trogoderma granarium* Everts
Asian gypsy moth	*Lymantria dispar asiatica* L.
Cactus moth	*Cactoblastis cactorum* (Berg)
Red palm tree weevil	*Rhynchophorus ferrugineus* (Olivier)
Tomato leaf miner	*Tuta absoluta* (Meyrick)
Orange tortrix	*Argyrotaenia franciscana* (Walsingham)
Oriental fruit moth	*Grapholita molesta* (Busck)
Light brown apple moth	*Epiphyas postvittana* (Walker)
European grapevine moth	*Lobesia botrana* (Denis & Schiffermüller)
Polyphagous shot borer beetle[a]	*Euwallacea* nr. *fonicatus* (Eichhoff)
African cotton bollworm	*Helicoverpa armigera* (Hübner)
Citrus canker	*Xanthomonas citri* (Hasse)
Citrus black spot	*Guignardia citricarpa* Kiely
Citrus variegated chlorosis	*Xylella fastidiosa pauca* (Wells)
Cacao witches broom disease	*Moniliophthora perniciosa* (Stahel) Aime & Phillips-Mora
Black pod rot	*Phytophthora palmivora* (Butler)
Coffee leaf rust[a]	*Hemileia vastatrix* (Berkeley & Broome)
Pierce's disease[a]	*Xylella fastidiosa fastidiosa* Well. *et al.*
Fusariosis of the pineapple	*Fusarium guttiforme* Nirenberg & O'Donnell
Wheat stem black leaf rust	*Puccinia graminis* f. sp. race Ug99 Pers.
Moko disease of banana	*Ralstonia solanacearum* race 2 (Smith)
Bacterial banana wilt	*Xanthomonas campestris* pv. *musacearum* (Yirgou & Bradbury)
Partial bunt[a]	*Tilletia indica* Mitra
Panama disease	*Fusarium oxysporum* f. sp. *cubense* (E.F. Sm.) W.C. Snyder & H.N. Hansen raza 4 Tropical
Bacterial leaf scorch	*Xylella fastidiosa multiplex* Shaad *et al.*
Citrus leprosis[a]	*Citrus leprosis* virus C
Banana bunchy top	Banana bunchy top virus

[a]Species present only in some areas of Mexico

other host plants such as evergreens inside buildings and mesophilic mountain forests with great diversity and abundance of Lauraceae (Lorea, 2002). Biocontrol for this pest is in development, including a search for native entomopathogenic fungi of species closely related to *X. glabratus* and *E.* nr. *fornicatus*. Laboratory selection of the most virulent native isolates was followed by isolate selection based on productivity and optimization of mass production. Next, biological safety studies are planned, along with formulation tests and evaluation of entomopathogens in the field. So far, *Beauveria bassiana* Balsamo (Vuillemin) isolates have been identified. The project involves biocontrol work on *R. lauricola*, *F. euwallaceae*, *Graphium* sp. and *A. pembeum* by using endophytes.

Drosophila suzukii (Matsamura), another recent invader, attacks cranberries, raspberries, blackberries, strawberries, cherries, grapes and other crops and was detected in November 2011 in Michoacán and subsequently in Colima, Jalisco and Baja California (CABI, 2012). CNRCB used sentinel field traps with larvae and pupae of *D. suzukii* on banana to catch parasitoids and found the larval parasitoid *Leptopilina boulardi* Barbotin and the pupal parasitoids *Pachycrepoideus vindemmiae* (Rondani), *Spalangia simplex* Perkins and *Trichopria drosophilae* (Perkins) (Moreno-Carrillo *et al.*, 2015; García-Cancino *et al.*, 2015). These parasitoids are now being evaluated for control of this pest. Further, the vulnerability of adult *D. suzukii* to strains of *I. javanica* and *M. anisopliae* has been determined, showing that *I. javanica* caused the greatest death rate (85%) (Naranjo-Lázaro *et al.*, 2014). Biocontrol is currently also considered for the tomato moth *Tuta absoluta* (Meyrick), which is native to South America, for the red palm tree weevil *Rhynchophorus ferrugineus* Olivier, native to South-east Asia, for European grapevine moth *Lobesia botrana* Den. and for the corn earworm *Helicoverpa armigera* (Hübner), as preventive technology, since these pests are not yet present in Mexico.

21.4.3 Mexican legislation for biological control of agricultural pests

With the recent official publication of the regulations of the Federal Plant Health Act of 15 July 2016 (Official Journal of the Federation) (Diario Oficial de la Federación) (DOF, 2016), the Ministry of Agriculture, Livestock, Rural Development, Fisheries and Food (SAGARPA) now has the responsibility to comply with adequate and safe use of biocontrol. This regulation allows for obtaining knowledge of species with potential use as biocontrol agents of regulated pests. It also regulates the national mobilization of species collected in the field, as well as the imports of arthropods as commercial biocontrol agents, entomopathogens not formulated for research and exotic entomophagous insects. The Ministry of Environment and Natural Resources can be asked to perform a risk analysis before import of exotic beneficial organisms. Analysis of the biological characteristics of species used in federal campaigns is also regulated. Further, SAGARPA can grant certificates of origin and biological purity for biocontrol agents that Mexican producers want to export. By presidential mandate, SAGARPA will also verify the biocontrol activities in campaigns against pests implemented by SENASICA, by validating and developing technology for biocontrol of regulated pests. With this legislation, Mexico will be able to reach an adequate security level for the imports of biocontrol agents and the collection of native agents for distribution to other states. This will limit the introduction of unwanted species with negative agricultural, food production and environmental impacts.

21.4.4 The future of biological control in Mexico

Although the first cases of biocontrol in Mexico date back more than a century, the study of biocontrol agents and their practical application have increased remarkably during the past two decades (Rodríguez-del-Bosque *et al.*, 2015). The infrastructure for research and application of biocontrol is well developed, implementation of new programmes can profit from knowledge obtained from earlier successes, multidisciplinary and inter-institutional collaboration is supported and financed by funding agencies and the country has well equipped laboratories to rear natural enemies, as well as commercial suppliers. Further, the government sector (SENASICA–CNRCB) and different academic institutions are involved in the development of biocontrol programmes (Williams *et al.*, 2013) and the country has a Scientific Biocontrol Society to train and promote collaboration among researchers. It is also important that Mexico has biocontrol researchers that support the implementation of control programmes with a regional focus. Finally, international collaboration takes place with, among others, the USA and Canada. This altogether means that Mexican researchers now have access to the newest tools for proper selection of beneficial organisms, to design quality control protocols that guarantee the effectiveness of natural enemies, and molecular techniques allowing precise identification of organisms and their monitoring after field releases. Mexico is considered a megabiodiverse country with coasts along the Atlantic and Pacific Oceans. This rich biodiversity is the basis for the recent approach of the Mexican government to first explore the potential of native species as candidates for biocontrol, before introducing exotic natural enemies.

Several other factors are influencing the development of biocontrol in Mexico. The recently implemented national development plan includes an integral development policy linking environmental sustainability with costs and benefits to society. It aims at aligning and coordinating federal programmes and inducing inclusive green growth, while updating environmental legislation for efficient regulation of actions contributing to the preservation and restoration of the environment and natural resources. Further, it promotes the use and consumption of ecofriendly products, which is a stimulus for biocontrol (DOF, 2013). Classical and augmentative biocontrol programmes will be useful in realizing the goals of the national development plan, since exotic pests will continue to invade Mexico and existing native and exotic pests offer opportunities for control with native natural enemies (Rodríguez-del-Bosque *et al.*, 2015).

Development of biocontrol is also expected to be stimulated because of the requirements by export markets for products with low levels of residues and specific standards that limit the use of pesticides. In developing countries, where insecticides are expensive and pesticide resistance occurs frequently, biocontrol can play a special role that has not been sufficiently exploited (Rodríguez-del-Bosque *et al.*, 2015).

21.5 Acknowledgements

We would like to express our special thanks to M.A. Mellín Rosas, J. González, M. Palomares, V. González Padilla (National Center for Biological Control of the Plant Health Directorate) and R. Román Vásquez (Phytosanitary Protection Department of the Plant Health General Directorate), as well as F. Ramírez y Ramírez and P. Montoya (National Campaign Against Fruit Flies of the Plant Health General Directorate), who supported us with their valuable comments and their search for information making it possible to realize this chapter.

References

(References with grey shading are available as supplementary electronic material)

Alatorre-Rosas, R. and Hernández-Rosas, F. (2015) Mosca pinta, *Aeneolamia* spp. y *Prosapia* spp. (Hemiptera: Cercopidae) [Spittlebugs, *Aeneolamia* spp. and *Prosapia* spp.]. In: Arredondo-Bernal, H.C. and Rodríguez-del-Bosque, L.A. (eds.) *Casos de Control Biológico en México*. Vol. 2. Biblioteca Básica de Agricultura, Colegio de Postgraduados, Mexico, pp. 141–164.

Arredondo-Bernal, H.C. (2018) Directorio de Laboratorios reproductores y comercializadores de agentes de control biológico [Directory of Laboratories rearing and marketing biological control agents]. Available at: http://www.senasica.gob.mx/?id=3164) (accessed 21 November 2015).

Arredondo-Bernal, H.C. and Rodríguez-del-Bosque, L.A. (2008) *Casos de control biológico en México* [*Cases of biological control in Mexico*]. Ed. Mundi-Prensa, Mexico City, Mexico.

Arredondo-Bernal, H.C. and Rodríguez-del-Bosque, L.A. (2015) *Casos de control biológico en México* [*Cases of biological control in Mexico*]. Vol. 2. Biblioteca Básica de Agricultura, Colegio de Postgraduados, Mexico, pp. 89–100.

Ávila-Valdez, J. and Rodríguez-del-Bosque, L.A. (2008) Gusano tercipelo de la soya, *Anticarsia gemmatalis* (Lepidoptera: Noctuidae) [Velvetbean caterpillar, *Anticarsia gemmatalis*]. In: Arredondo-Bernal, H.C. and Rodríguez-del-Bosque, L.A. (eds) *Casos de control biológico en México*. Ed. Mundi-Prensa, Mexico City, Mexico, pp. 89–100.

Ávila-Valdez, J. and Rodríguez-del-Bosque, L.A. (2015) Impacto económico y ecológico del uso comercial del nucleopoliedrovirus de *Anticarsia gemmatalis* (AgNPV) [Lepidoptera: Noctuidae) en soya (Economic and ecological impact of the commercial use of the nucleopolyhedrovirus of *Anticarsia gemmatalis* (AgNPV) in soybean]. In: Arredondo-Bernal, H.C. and Rodríguez-del-Bosque, L.A. (eds) *Casos de control biológico en México*. Vol. 2. Biblioteca Básica de Agricultura, Colegio de Postgraduados, Mexico, pp. 93–100.

Bahena-Juárez, F. (2008) Palomilla de la papa, *Phthorimaea operculella* (Lepidoptera: Gelechiidae) [Potato tuber moth, *Phthorimaea operculella*)]. In: Arredondo-Bernal, H.C. and Rodríguez del Bosque, L.A. (eds) *Casos de control biológico en México*. Ed. Mundi-Prensa, Mexico City, Mexico, pp. 33–46.

Barrera-Gaytán, J.F., Gómez, J., Castillo, A., López, E., Herrera, J. and González, G. (2008) Broca del café, *Hypothenemus hampei* (Coleoptera: Curculionidae) [Coffee berry borer, *Hypothenemus hampei*]. In: Arredondo-Bernal, H.C. and Rodríguez-del-Bosque, L.A. (eds) *Casos de control biológico en México*. Ed. Mundi-Prensa, Mexico City, Mexico, pp. 101–120.

Barrientos-Lozano, L., Astacio, O., Álvarez, F. and Poot, O. (1992) *Manual técnico sobre la langosta voladora (Schistocerca piceifrons piceifrons (Walker 1870)) y otros acridoideos de Centro América y Sureste de México* [*Technical manual on the locust (Schistocerca piceifrons piceifrons) and other Acrididae of Central America and Southeast of Mexico*]. FAO-Organismo Internacional Regional de Sanidad Vegetal, San Salvador, El Salvador.

Barrientos-Lozano, L., Hunter, D.M., Ávila-Valdez, J., García-Salazar, P. and Hernández-Velázquez, V.M. (2004) Control biológico de la langosta voladora *Schistocerca piceifrons* (Walker) con *Metarhizium anisopliae* var. *acridum* en el noreste de México [Biological control of the locust *Schistocerca piceifrons* with *Metarhizium anisopliae* var. *acridum* in northeastern Mexico]. In: *Proceedings of the XXVII National Congress of Biological Control*. Sociedad Mexicana de Control Biológico, Los Mochis, Sinaloa, Mexico, pp. 137–142.

Bastida-Tapia, A. (2017) Evolución y Situación Actual de la Agricultura Protegida en México. [Evolution and Current Situation of Protected Agriculture in Mexico]. Sexto Congreso Internacional de Investigación en Ciencias básicas y Agronómicas. Available at: http://dicea.chapingo.mx/wp-content/uploads/2018/05/MEMORIA_MESA_3_2_CONGRESO2017.pdf (accessed 28 October 2019).

Bautista, A. and González, N. (2005) Tres dosis de *Metarhizium anisopliae* sobre la mosca pinta (*Aeneolamia* spp.) en caña de azúcar en la región de los ríos, estado de Tabasco [Three doses of *Metarhizium anisopliae* on the spittlebug (*Aeneolamia* spp.) on sugarcane in the region of the rivers, State of Tabasco]. *Universidad y Ciencia* 21(41), 37–40.

Bautista-Martínez, N., Villanueva-Jiménez, J.A. and Arredondo-Bernal, H.C. (2008) Minador de la hoja de los cítricos, *Phyllocnistis citrella* (Lepidoptera: Gracilariidae) [Citrus leafminer, *Phyllocnistis citrella*]. In: Arredondo-Bernal, H.C. and Rodríguez del Bosque, L.A. (eds) *Casos de control biológico en México*. Ed. Mundi-Prensa, Mexico City, Mexico, pp. 347–358.

CABI (Crop Protection Compendium) (2012) Data Sheets: *Drosophila suzukii*. Available at: http://www.cabi.org/cpc/default.aspx?LoadModule=datasheet&dsID=109283&CompID=1&site=161&page=868 (accessed 25 April 2015).

Cibrián-Tovar, D. (2015) Conchuela del eucalipto, *Glycaspis brimblecombei* (Hemiptera: Psyllidae) [Red gum lerp psyllid, *Glycaspis brimblecombei*]. In: Arredondo-Bernal, H.C. and Rodríguez-del-Bosque, L.A. (eds) *Casos de control biológico en México*. Vol. 2. Biblioteca Básica de Agricultura, Colegio de Postgraduados, Mexico, pp. 395–413.

Cicero-Jurado, L., Sánchez-Borja, M., Loeza-Kuk, E., Barrié, C. and Arredondo-Bernal, H.C. (2013) Porcentaje de parasitismo de *Tamarixia radiata* (Waterston) en zonas urbanas de Yucatán [Percentage of parasitism of *Tamarixia radiata* in urban areas of Yucatan]. In: Vásquez-López, A., and Pérez Pacheco, R. (eds) *Proceedings of the XXXVI National Congress of Biological Control*. Sociedad Mexicana de Control Biológico, Oaxaca de Juárez, Oaxaca, Mexico, pp. 325–328.

CNRCB (1997) *Documento interno de trabajo* [*Internal work document*]. Dirección General de Sanidad Vegetal- Secretaría de Agricultura y Desarrollo Rural (DGSV-SAGAR). México DF (Mexico City), Mexico.

CONADESUCA (2011) Reporte de avance de producción No.33 Zafra 2011–2012 [Production progress report No.33, Harvest 2011–2012]. Agreement of the Board of Directors of CONADESUCA, Reference Price for Zafra Liquidations 10/. Available at: http://www.cndsca.gob.mx/ (accessed 14 October 2012).

CONAFOR-FAO (2009) *Competitividad del sector forestal de México, tendencias y perspectivas* [*Competitiveness of the forestry sector of Mexico, trends and perspectives*]. DOI: 10.03140/RG.2.2.118771.87201. Available at: https://www.researchgate.net/publication/308951461_Competitividad_del_Sector_Forestal_de_Mexico_Tendencias_y_Perspectivas (accessed 19 April 2019).

DOF (2013) Notes. Diario Oficial de la Federación. Available at: http://dof.gob.mx/nota_detalle.php?codigo=5299465&fecha=20/05/13 (accessed 19 April 2019).

DOF (Diario Oficial de la Federación) (2016) Reglamento de la Ley Federal de Sanidad Vegetal (Regulation of the Federal Act of Plant Protection). *Secretaria de Agricultura, Ganadería, Desarrollo Rural, Pesca y Alimentación*, 15 July 2016, Cd. de México, Mexico. Available at: http://www.diputados.gob.mx/LeyesBiblio/regley/Reg_LFSV.pdf (accessed 19 April 2019).

Freeman, S., Sharon, M., Dori-Bachash, M., Maymon, M., Belausov, E., Maoz, Y. and Mendel, Z. (2016) Symbiotic association of three fungal species throughout the life cycle of the ambrosia beetle *Euwallacea* nr. *fornicatus*. *Symbiosis* 68, 115–128.

Fu-Castillo, A.A. (2008) Piojo harinoso de la vid, *Planococcus ficus* (Hemiptera: Pseudococcidae) [Vine mealybug, *Planococcus ficus*]. In: Arredondo-Bernal, H.C. and Rodríguez-del-Bosque, L.A. (eds) *Casos de control biológico en México*. Ed. Mundi-Prensa, Mexico City, Mexico, pp. 253–265.

Fu-Castillo, A.A. and Grageda-Grageda, J. (2002) Control biológico del piojo harinoso de la vid *Planococcus ficus* en Hermosillo, Sonora [Biological control of vine mealybug *Planococcus ficus* in Hermosillo, Sonora]. In: González-Hernández, H. Fu-Castillo, A.A. and Báez-S, R. (eds.) *International Symposium of Mealybugs*. XXV Congreso Nacional de Control Biológico. Sociedad Mexicana de Control Biológico, Hermosillo, Sonora, Mexico, pp. 22–27.

Fu-Castillo, A.A., Osori Acosta, G., Miranda Blanco, J.L., Márquez Cervantes, J.A. and Martínez Díaz, G. (2002) Manejo integrado de piojo harinoso de la vid [Integrated management of vine mealybug]. Instituto Nacional de Investigaciones Forestales y Agropecuarias, Campo Experimental Costa de Hermosillo, Hermosillo, Sonora, México. *Folleto Técnico* No. 24.

García, S. and Gastellum, J. (2013) Programa general de control biológico, casos de éxito [General program of biological control, successful cases]. In: Torres Ruíz, A. and Rodríguez Leyva, E. (eds) *Guía para el manejo integrado de plagas del pimiento bajo invernadero, con énfasis en el picudo del chile*. Koppert México S.A. de C.V., Quéretaro, Mexico, pp. 34–37. Available at: https://www.researchgate.net/publication/277004440

García-Ávila, C.J., Trujillo-Arriaga, F.T., López-Buenfil, J.A., González-Gómez, R., Carrillo, D., Cruz, L.F. and Acevedo-Reyes, N. (2016) First report of *Euwallacea* nr. *fornicatus* (Coleoptera: Curculionidae) in Mexico. *Florida Entomologist* 99, 555–556.

García-Cancino, M.D., González-Hernández, A., González-Cabrera, J., Moreno-Carrillo, G., Sánchez-González, J.A. and Arredondo-Bernal, H.C. (2015) Parasitoides de *Drosophila suzukii* (Matsumura) (Diptera: Drosophilidae) en Colima, México [Parasitoids of *Drosophila suzukii* (Matsumura) (Diptera: Drosophilidae) in Colima, Mexico]. *Southwestern Entomologist* 40(4), 855–858.

García-Valente, F. and Ortega-Arenas, L.D. (2008) Mosquita blanca, *Trialeurodes* vaporariorum (Hemiptera: Aleyrodidae) [Whitefly, *Trialeurodes vaporariorum*]. In: Arredondo-Bernal, H.C. and Rodríguez-del-Bosque, L.A. (eds) *Casos de control biológico en México*. Ed. Mundi-Prensa, Mexico City, Mexico, pp. 167–176.

García-Victoria, N., van der Valk, O. and Elings, A. (2011) Mexican protected horticulture: Production and market of Mexican protected horticulture described and analyzed. Report GTB-1126. Ministry of Economic Affairs, Agriculture and Innovation, Wageningen UR,

GBIF (2019) Species. Global Biodiversity Information Facility, Copenhagen. Available at: https://www.gbif.org/species/1267154 (accessed 12 April 2019).

González-Hernández, H., Lomelí-Flores, J.R. and Myartseva, S.N. (2015) Escamas (Hemiptera: Diaspididae; Coccidae) en cítricos [Scale in citrus]. In: Arredondo-Bernal, H.C. and Rodríguez-del-Bosque, L.A. (eds) *Casos de control biológico en México*. Vol. 2. Biblioteca Básica de Agricultura, Colegio de Postgraduados, Mexico, pp. 295–310.

Hernández-Velázquez, V.M. and Toriello, C. (2008) Langosta voladora, *Schistocerca piceifrons* piceifrons (Orthoptera: Acrididae) [Locust, *Schistocerca piceifrons piceifrons*]. In: Arredondo-Bernal, H.C. and Rodríguez-del-Bosque, L.A. (eds) *Casos de control biológico en México*. Ed. Mundi-Prensa, Mexico City, Mexico, pp. 47–55.

Hernández-Velázquez, V.M., Hunter, D.M., Barrientos-Lozano, L., Lezama-Gutiérrez, R. and Reyes-Villanueva, F. (2003) Susceptibility of *Schistocerca piceifrons* (Orthoptera: Acrididae) to *Metarhizium anisopliae* var. *acridum* (Deuteromycotina: Hyphomycetes): laboratory and field trials. *Journal of Orthoptera Research* 12, 89–92.

Herrera, A.L., de la Barreda, L., Bonesa, S. and Tellez, O. (1902) Virus contagioso para la destrucción de las ratas [Contagious virus for the destruction of rats]. *Boletín de la Comisión de Parasitología Agrícola* 1, 9, 345–349.

INEGI (2017) Encuesta Nacional Agropecuaria (ENA) 2017 [National Agricultural Survey (2017)]. Instituto Nacional de Estadística y Geografía. Available at: http://www.beta.inegi.org.mx/contenidos/programas/ena/2017/doc/ena2017_pres.pdf

Loera-Gallardo, J., Reyes-Rosas, M.A. and López Arroyo, J.I. (2008) Picudo del algodonero, *Anthonomus grandis* (Coleoptera: Curculionidae) [Boll weevil, *Anthonomus grandis*]. In: Arredondo-Bernal, H.C. and Rodríguez del Bosque, L.A. (eds) *Casos de control biológico en México*. Ed. Mundi-Prensa, Mexico City, Mexico, pp. 75–88.

Lomelí-Flores, J.R., Peña-Martínez, R., Villegas-Jiménez, N. and Trejo-Loyo, A.G. (2008) Pulgón del melón, *Aphis gossypii* (Hemiptera: Aphididae) [Cotton aphid, *Aphis gossypii*]. In: Arredondo-Bernal, H.C. and Rodríguez-del-Bosque, L.A. (eds) *Casos de control biológico en México*. Ed. Mundi-Prensa, Mexico City, Mexico, pp. 137–153.

Lomelí-Flores, J.R., Rodríguez-Leyva, E. and Torres-Ruíz, A. (2016) El control biológico de plagas en la agricultura protegida en México [Biocontrol of pests in protected agriculture in Mexico]. In: Lomelí-Flores, J.R. and González Hernández, H. (eds.) *XXVII National Course on Biological Control*. Sociedad Mexicana de Control Biológico, Guadalajara, Jalisco, Mexico, pp. 29–36.

López, J.L., Loera Gallardo, J., Rocha Peña, M.A., Canales, R., Hernández, I., Reyes, M.A., Berlanga, A. and Miranda, M.A. (2008) Pulgón café de los cítricos [Brown citrus aphid, *Toxoptera citricida*]. In: Arredondo-Bernal, H.C. and Rodríguez del Bosque, L.A. (eds) *Casos de control biológico en México*. Ed. Mundi-Prensa, Mexico City, Mexico, pp. 279–292.

Lorea, H.F. (2002) La familia Lauraceae en el sur de México: diversidad, distribución y estado de conservación [The Lauraceae family in southern Mexico: diversity, distribution and state of conservation]. *Boletín de la Sociedad Botánica de México* 71, 59–70.

Macías, C.P. and Ramírez, J.C. (1998) Importancia económica de la langosta y el chapulín en México [Economic importance of locusts and grasshoppers in Mexico]. In: *Proceedings of the Ith Symposium of Locust and Grasshoppers and XXI National Congress of Biological Control*. Sociedad Mexicana de Control Biológico, Río Bravo, Tamaulipas, Mexico, pp. 1–2.

Martínez-Carrillo, J.L., Nava-Camberos, U., Arredondo-Bernal, H.C. and Aguilar-Medel, S. (2015) Mosca blanca, *Bemisia tabaco* (Hemiptera: Aleyrodidae) [Whitefly, *Bemisia tabaci*]. In: Arredondo-Bernal, H.C. and Rodríguez-del-Bosque, L.A. (eds.) *Casos de control biológico en México*. Vol. 2. Biblioteca Básica de Agricultura, Colegio de Postgraduados, Mexico, pp. 269–294.

Martínez-Jiménez, M. (2008) Lirio acuático, *Eichhornia crassipes* (Water hyacinth, *Eichhornia crassipes*). In: Arredondo-Bernal, H.C. and Rodríguez-del-Bosque, L.A. (eds) *Casos de control biológico en México*. Ed. Mundi-Prensa, Mexico City, Mexico, pp. 415–423.

Michaud, J.P. and Álvarez, R. (2000) First collection of brown citrus aphid (Homoptera: Aphidiidae) in Quintana Roo. *Florida Entomologist* 83, 357–358.

Moreno-Carrillo, G., Sánchez-González, J.A. and Arredondo-Bernal, H.C. (2013) Avances en la evaluación de *Tamarixia radiata* (Waterston) (Hymenoptera: Eulophidae) en áreas urbanas de la región pacífico centro de México [Advances in the evaluation of *Tamarixia radiata* in urban areas of the central Pacific region of Mexico]. In: Vásquez-López, A. and Pérez Pacheco, R. (eds.) Memorias del XXXVI Congreso Nacional de Control Biológico. Sociedad Mexicana de Control Biológico, Oaxaca de Juárez, Oaxaca, Mexico, pp. 341–346.

Moreno-Carrillo, G., Rodríguez-Vélez, B., Sánchez-González, J.A. and Arredondo-Bernal, H.C. (2015) Trampeo y registro del parasitoide *Pachycrepoideus vindemmiae* (Rondani) (Hymenoptera: Pteromalidae) sobre *Drosophila suzukii* (Matsumura) (Diptera: Drosophilidae) en México [Trapping and recording of the parasitoid *Pachycrepoideus vindemmiae* of *Drosophila suzukii* in Mexico]. *Southwestern Entomologist* 40(1), 199–203.

Munguía-Rosales, R. (2002) Experiencias en la liberación de agentes de control biológico del pulgón café de los cítricos en la península de Yucatán, México [Experiences in the release of biological control agents of brown citrus aphid in the Yucatan Peninsula, Mexico]. In: Arredondo-Bernal, H.C., Mellín Rosas, M.A., Pérez Serrato, P. and Martínez-Soriano, J.P. (eds) *Biological Control of the Brown Citrus Aphid* Toxoptera citricida, *Vector of the Citrus Tristeza Virus*, 2ª Edición. Servicio Nacional de Sanidad, Inocuidad y Calidad Agroalimentaria (SENASICA), Dirección General de Sanidad Vegetal, México DF (Mexico City), Mexico, pp. 165–169.

Naranjo-Lázaro, J.M., Mellín-Rosas, M.A., González-Padilla, V.D., Sánchez-González, J.A., Moreno-Carrillo, G. and Arredondo-Bernal, H.C. (2014) Susceptibilidad de *Drosophila suzukii* Matsumura (Diptera: Drosophilidae) a hongos entomopatógenos [Susceptibility of *Drosophila suzukii* to entomophatogenic fungi]. *Southwestern Entomologist* 39(1), 201–203.

Nava-Camberos, U. and Ávila-Rodríguez, V. (2008) Mosca doméstica, *Musca domestica* (Diptera: Muscidae) [Housefly, *Musca domestica*]. In: Arredondo-Bernal, H.C. and Rodríguez-del-Bosque, L.A. (eds) *Casos de control biológico en México*. Ed. Mundi-Prensa, Mexico City, Mexico, pp. 395–413.

Ortíz-Garcia, S., Herrera Estrella, L., Guerrero Ortíz, S., Lopez Mungua, A., Campos Reales, N. and Castillo Villanueva, E. (2017) Food and nutritional security in Mexico. In: *Challenges and opportunities for food and nutrition security in the Americas. The view of the Academies of Sciences*. IANAS, IAP and BMBF, México DF, pp. 386–419. [Free public access of this publication in English and Spanish at www.ianas.org]

Palomares-Pérez, M., Rodríguez-Vélez, J.M., Rodríguez-Vélez, B., Marín-Jarillo, A., Sánchez-González, J.A. and Arredondo-Bernal, H.C. (2015) First record and predatory activity of *Exochomus marginipennis* (LeConte) (Coleoptera: Coccinellidae) on *Diaphorina citri* Kuwayama (Hemiptera: Liviidae). *Entomological News*, 125, 191–195.

Peña, J., Carrillo, D., Duncan, R., Capinera, J., Brar, G., Mclean, S. and Kendra, P. (2012) Susceptibility of *Persea* spp. and other Lauracea to attack by redbay ambrosia beetle, *Xyleborus glabratus* (Coleoptera: Curculionidae: Scolytinae). *Florida Entomologist* 95(3), 783–787.

Quiroz-Martínez, H., Rodríguez-Castro, V.A., Siller-Aguillón, I.A., Zepeda-Cavazos, I.G. and Martínez-Perales, J.F. (2015) Mosquito enano, *Chironomus plumosus* (Diptera: Chironomidae) [Buzzer midge, *Chironomus plumosus*]. In: Arredondo-Bernal, H.C. and Rodríguez-del-Bosque, L.A. (eds.) *Casos de control biológico en México*. Vol. 2. Biblioteca Básica de Agricultura, Colegio de Postgraduados, Mexico, pp. 385–394.

Reyes-Domínguez, E. (1996) Control biológico de plagas agrícolas en México [Biological control of agricultural pests in Mexico]. In: Zapater, M.C. (ed.) *Biological Control in Latin America*. International Organization for Biological Control, Buenos Aires, Argentina, pp. 55–62.

Rodríguez-del-Bosque, L.A. and Arredondo-Bernal, H.C. (1999) Quién es quién en control biológico en México [Who is who in biological control in Mexico]. INIFAP-CIRNE. Campo Experimental Río Bravo. *Folleto Técnico*, Núm. 23. Tamaulipas, Mexico.

Rodríguez-del-Bosque, L.A. and Vejar-Cota, G. (2008) Barrenadores del tallo (Lepidoptera: Crambidae) del maíz y caña de azúcar [Stem borers of corn and sugarcane]. In: Arredondo-Bernal, H.C. and Rodríguez-del-Bosque, L.A. (eds) *Casos de control biológico en México*. Ed. Mundi-Prensa, Mexico City, Mexico, pp. 9–22.

Rodríguez-del-Bosque, L.A, Arredondo-Bernal, H.C., Williams, T. and Barrera-Gaytán, J.F. (2015) Pasado, presente y perspectivas del control biológico en México [Past, present and perspectives of biological control in Mexico]. In: Arredondo-Bernal, H.C and Rodríguez-del-Bosque, L.A. (eds) *Casos de control biológico en México*. Vol. 2. Biblioteca Básica de Agricultura, Colegio de Postgraduados, Mexico, pp. 17–28.

Romo-Lozano, J.L., García-Jiménez, J., Cibrián-Tovar, D. and Serrano-Gálvez, E. (2007) Análisis económico del control biológico del psílido del eucalipto en la Ciudad de México [Economic analysis of the biological control of the red gum lerp psyllid in Mexico City]. *Revista Chapingo, Serie Ciencias Forestales y del Ambiente* 13(1), 47–52.

RSPM 26 (Regional Standards for Phytosanitary Measures 26) (2015) Certification of commercial arthropod biological control agents or non-*Apis* pollinators moving into NAPPO member countries. The Secretariat of the North American Plant Protection Organization. Ottawa, Ontario, Canada. Available at: http://www.nappo.org/files/5114/4908/7113/RSPM_26_rev_18-09-2015-e.pdf (accessed 25 June 2018).

Ruíz-Cancino, E., Coronado Blanco, J.M., Varela Fuentes, S.E. and Luna Salas, J.F. (2008) Negrilla de los cítricos, *Phyllocoptruta oleivora* (Acari: Eriophydae) [Citrus rust mite, *Phyllocoptruta oleivora*]. In: Arredondo-Bernal, H.C. and Rodríguez del Bosque, L.A. (eds) *Casos de control biológico en México*. Ed. Mundi-Prensa, Mexico City, Mexico, pp. 315–322

SADER (2018) Atlas Agroalimentario 2018 [Agroalimentary Atlas 2018]. Servicio de Información Agroalimentaria y Pesquera (SIAP). Available at: https://nube.siap.gob.mx/gobmx_publicaciones_siap/pag/2018/Atlas-Agroalimentario-2018. (accessed 14 January 2019).

SAGARPA (2016) Atlas Agroalimentario 2016 [2016 Agroalimentary Atlas]. Servicio de Información Agroalimentaria y Pesquera (SIAP). Available at: https://nube.siap.gob.mx/gobmx_publicaciones_siap/pag/2016/Atlas-Agroalimentario-2016 (accessed 28 October 2019).

SAGARPA (2017) Planeación Agrícola Nacional 2017-2030. Aguacate Mexicano. Available at: https://www.gob.mx/cms/uploads/attachment/file/257067/Potencial-Aguacate.pdf (accessed 28 October 2019).

Salazar-Solís, E. and Salas-Araiza M.D. (2008) Palomilla dorso de diamante, *Plutella xylostella* (Lepidoptera: Plutellidae) [Diamondback moth, *Plutella xylostella*]. In: Arredondo-Bernal, H.C. and Rodríguez-del-Bosque, L.A. (eds) *Casos de Control Biológico en México*. Ed. Mundi-Prensa, Mexico City, Mexico, pp. 155–165.

Salazar-Solís, E., Salas-Araiza, M.D. and Martínez-Jaime, O.A. (2015) Chinche de la panoja del sorgo, *Oebalus mexicana* (Hemiptera: Pentatomidae) [Brown bug, *Oebalus mexicana*]. In: Arredondo-Bernal, H.C. and Rodríguez-del-Bosque, L.A. (eds) *Casos de control biológico en México*, Vol. 2. Biblioteca Básica de Agricultura, Colegio de Postgraduados, Mexico, pp. 101–111.

Salcedo-Baca, D., Lomelí-Flores, J.R., Terrazas-González, G.H., Rodriguez-Leyva, E., Vera-Villagrán, E., Múzquiz-Fragoso, C. and Hurtado-Arellano, A. (2012) Evaluación de diseño, procesos y factibilidad económica: campaña contra la cochinilla rosada del hibisco [Evaluation of design, processes and economic feasibility: campaign against the pink hibiscus mealybug]. Instituto Interamericano de Cooperación para la Agricultura, Mexico. Available at: http://www.iica.int/es (accessed 19 April 2019).

Sánchez-González, J.A., Mellín-Rosas, M.A., Arredondo-Bernal, H.C., Vizcarra Valdez, N.I., González-Hernández, A. and Montesinos-Matías, R. (2015) Psílido asiático de los cítricos, *Diaphorina citri* (Hemiptera: Psyllidae) [Asian citrus psyllid, *Diaphorina citri*]. In: Arredondo-Bernal, H.C. and Rodríguez-del-Bosque, L.A. (eds) *Casos de control biológico en México*. Vol. 2. Biblioteca Básica de Agricultura, Colegio de Postgraduados, Mexico, pp. 339–372.

Sanidad Vegetal (2018) Octavo Informe Mensual Campaña contra Huanglongbing de los Cítricos [Eighth Monthly Report Against Citrus Huanglongbing Campaign]. Available at: https://www.gob.mx/cms/uploads/attachment/file/410960/8_Informe_nacional_HLB_2018.pdf (accessed 28 October 2019).

Santiago-Islas, T., Zamora-Cruz, A., Fuentes-Temblador, E.A., Valencia-Luna, L. and Arredondo-Bernal, H.C. (2008) Cochinilla rosada del hibiscus, *Maconellicoccus hirsutus* (Hemiptera: Pseudococcidae) [Pink hibiscus mealybug, *Maconellicoccus hirsutus*]. In: Arredondo-Bernal, H.C. and Rodríguez-del-Bosque, L.A. (eds) *Casos de control biológico en México*. Ed. Mundi-Prensa, Mexico City, Mexico, pp.177-191.

SENASICA (2009) Ficha Técnica *Schistocerca piceifrons piceifrons* Walker, Langosta centroamericana [Technical Sheet *Schistocerca piceifrons piceifrons*, Central American locust]. National Service for Agrifood Health, Safety and Quality. (Servicio Nacional de Sanidad, Inocuidad y Calidad Agroalimentaria) Available at: https://www.gob.mx/cms/uploads/attachment/file/147938/Ficha_Tecnica_Langosta.pdf (accessed 2 February 2017).

SENASICA (2012) Ficha técnica: chapulín [Technical sheet: grasshopper]. National Service for Agrifood Health, Safety and Quality. Available at: http://www.senasica.gob.mx/includes/asp/ (accessed 15 August 2018).

SENASICA (2016) Priority species. Servicio Nacional de Sanidad, Inocuidad y Calidad Agroalimentaria. Available at: http://www.gob.mx/senasica/acciones-y-programas/plagas-bajo-vigilancia (accessed 19 April 2019).

SIAP (2017) Producción agricola. Servicio de Información Agroalimentaria y Pesquero. Available at: https://nube.siap.gob.mx/cierreagricola/ (accessed 14 January 2019).

Vejar-Cota, G. and Rodríguez-del-Bosque, L.A. (2015) Mosca blanca de los cereales, *Aleurocybotus occiduus* (Hemiptera: Aleyrodidae) [Cereal whitefly, *Aleurocybotus occiduus*]. In: Arredondo-Bernal, H.C. and Rodríguez-del-Bosque, L.A. (eds) *Casos de control biológico en México*. Vol. 2. Biblioteca Básica de Agricultura, Colegio de Postgraduados, Mexico, pp. 113–122.

Williams, T., Arredondo-Bernal, H.C. and Rodríguez-del-Bosque, L.A. (2013) Biological Pest Control in Mexico. *Annual Review of Entomology* 58, 119–140.

Zavaleta-Mejía, E., Bravo-Luna, L. and Guigón-López, C. (2015) Fitopatógenos con origen en el suelo [Soil-inhabiting phytopathogens]. In: Arredondo-Bernal, H.C and Rodríguez-del-Bosque, L.A. (eds.) *Casos de control biológico en México*. Vol. 2. Biblioteca Básica de Agricultura, Colegio de Postgraduados, Mexico, pp. 65–92.

22 Biological Control in Nicaragua

Patricia Castillo*

Centro de Investigación y Reproducción en Control Biológico (CIRCB), Facultad de Ciencias y Tecnología, Universidad Nacional Autónoma de Nicaragua (UNAN)-León, Nicaragua

* E-mail: castillopatri@yahoo.es

Abstract

Biological control started in Nicaragua in 1957 with a study of parasitoids of the fall armyworm and augmentative releases of parasitoids and predators against pests in cotton. In the 1970s, cotton production was no longer profitable due to the high costs of frequent pesticide sprays. With assistance from FAO, an IPM programme for cotton pests was developed, which included biocontrol. Universities started to teach fundamentals of pest ecology and initiated prospecting for native natural enemies. Among others, a native strain of *Trichogramma pretiosum* was found and a mass rearing system was developed. The parasitoid has since been released in several crops. The predator *Chrysoperla carnea* was also mass produced and applied in the field. At the end of the 1980s, prospecting for entomopathogenic nematodes, fungi and viruses started, resulting in the production of several of these biocontrol agents, in particular five virus products for control of lepidopterans in various crops. In the mid 1990s, augmentative biocontrol of diamondback moth was developed by releasing native and exotic parasitoids. Currently there are several biofactories in Nicaragua that produce parasitoids, predators, entomopathogenic nematodes, fungi and viruses, and phytopathogenic fungi.

22.1 Introduction

Nicaragua has an estimated population of slightly more than 6 million (July 2017) (CIA, 2017). According to Huete-Perez *et al.* (2017, pp. 426–443):

> Nicaragua is the largest country in Central America with an area of 129,494 km^2 ... Historically, agriculture has been Nicaragua's main economic activity ... According to the National Agricultural Census, the total area for agriculture is 2,476,800 ha, i.e. 80% of productive families in Nicaragua earn their livelihood from family agriculture. Family agriculture is a model in Nicaraguan agriculture that contributes decisively to food sovereignty ... The role of small family farming is doubly commendable, not only because it guarantees food sovereignty, but also because it has preserved soils, water and biodiversity to guarantee its survival ... This produces items that are vital to the daily food supply of the Nicaraguan population, supplying over 60% of beans, 50% of maize, 40% of pork and 30% of domestic production of meat and milk, roots and tubers, vegetables and cacao ... Plant production in Nicaragua, considering the current context, could be grouped as crops with good profitability, such as sugarcane, groundnut, banana, coffee, tobacco, cacao (fine or aroma), oil palm, vegetables (on high lands and into greenhouse), maize (hybrids) sorghum (red-seeded hybrids) and rice (irrigated systems) ... Coffee represents around 54% of agricultural exports. On the other hand, in the second group we have those crops that are for subsistence ... such as common beans, maize (synthetic varieties), sorghum, rice (rainfall systems), vegetables (not produced into greenhouse), plantain and fruits that remain with low productivity ... despite many investments aiming to increase productivity, when compared with other countries in Central America, Nicaragua holds the lowest yields in most major crops ... Crop yields such as maize and the common bean, key components of Nicaraguans' diet, still are the least productive ones. Conversely, groundnut and sorghum are among the highest, probably caused by the high investments in inputs and technology by the private sector ... Crops such as sugarcane, groundnut, oil palm and banana, which represent around 30% of arable land, are highly mechanized, using irrigation systems and a package of inputs. However, basic grains that represent around 47% remain an old technology. Protected agriculture that incorporates modern irrigation and nutrition systems are not extensively used in Nicaragua. Tomato, potato, sweet pepper, onion, flower and cucurbit production are conducted on open fields, exposing plants to virus vectors and pathogens and producing considerable losses in production and quality ... Scattered forests and trees provide fruit, edible seeds and other wild foods, sustaining much of the food chain. The strong link between forests and food and nutrition security has sometimes been overlooked, although it has gained greater recognition in recent years, together with the importance of the protection and sustainable management of forests to ensure the food needs of a growing population ... Animal production is mainly concentrated in cattle, pig and chicken production in Nicaragua ... The livestock sector has accounted for a large share of the economy in recent years, proving to be the sector with the highest growth in exports for the 2011–2012 period, above the coffee sector ... Aquaculture and marine resources: The production of marine shrimp is conducted in ponds with capacities between 10 to 50 hectares, under intensive, semi-intensive and artisanal

using larvae from the wild. The expansion of the area of shrimp farming in the 2005 to 2014 period increased the area by 4,233 hectares over a period of nine years. On the other hand, pisciculture units are limited to the small-scale production of fish with between 0.10 and 0.2 hectares with the cultivation of introduced species (tilapia and carp) as part of economic diversification and food security.

22.2 History of Biological Control in Nicaragua

22.2.1 Period 1870–1969

In Nicaragua, ancient people recognized the value and richness of insects as natural enemies of pests and mentioned them in their native language, Nahuatl (Vaughan, 1958; Dávila, 1992). From 1879 to 1950, scientific contributions on entomology and pest management in Nicaragua were made by foreign researchers and these contributions are described in several publications (Andrews and Quezada, 1989)

In 1957–1958, M. Vaughan with the support of R. Bodán carried out the first studies on parasitoids of the fall armyworm *Spodoptera frugiperda* (J.E. Smith) and found a new species, *Rogas vaughani* Muesebeck (Vaughan, 1962). The first list of insects of importance for Nicaraguan agriculture was published in 1958 (Vaughan, 1962).

With the rise of cotton cultivation during the 1950s and 1960s, joint research was initiated with the University of California (Berkeley and Riverside, USA), and the use of the term integrated pest management (IPM) was first introduced. This collaboration led to the importation and release of the parasitic wasp *Trichogramma minutum* Riley in Nicaraguan cotton fields by E. J. Dietrich, resulting in promising biocontrol results (Vaughan, 1994). During the same period another natural enemy, the predatory ladybird *Hippodamia convergens* Guerin-Meneville, was imported from California by Vaughan to be used in cotton crops (Vaughan, 1996).

22.2.2 Period 1970–2000

During this period, Nicaragua faced an economic crisis because the cultivation of cotton was no longer profitable, due to the increasing costs of excessive applications of agrochemicals for pest control. For that reason, the government made efforts to restore the cotton industry. In 1970 the Ministry of Agriculture and Livestock (Ministerio de Agricultura y Ganaderia) (MAG) asked for technical assistance from L. Falcon and R. Daxl from the Food and Agriculture Organization of the United Nations (FAO, 1977). They started a programme to develop research, extension services and education, in relation to the study of pest population dynamics, natural pest control, cotton cropping techniques and the search for alternatives to non-chemical management of the cotton pests complex. Measures were targeted for control of the Mexican cotton boll weevil *Anthonomus grandis* Boheman (CCIPA, 1979).

From 1970 to 1985, the Autonomous National University of Nicaragua – Campus León (UNAN-León) and the Organization of American States (Organización de los Estados Americanos) (OEA) launched a Master of Science (MSc) programme on IPM, under the direction of G. León. The aim was to teach the fundamentals of pest ecology and prepare professionals trained in IPM and biocontrol, who might then solve national phytosanitary problems. The first MSc students started to evaluate native biocontrol candidates, such as the parasitoid *Trichogramma pretiosum* Riley against the cotton leafworm *Alabama argillacea* (Hübner) and the American cotton bollworm *Heliothis* (= *Helicoverpa*) *zea* (Boddie), and the parasitoid wasp of blackfly, *Encarsia opulenta* (Silvestri), to manage pests complexes in citrus crops. Also the use of the bacterium *Bacillus thuringiensis* var. Israeliensis was studied for control and eradication of the mosquito species *Anopheles albimanus* (Wiedemann).

During this same period, FAO and the United Nation Development Programme (UNDP) funded a research project and collaborated with the Nicaraguan Institute of Agricultural Technology (Instituto Nicaragüense de Tecnología Agropecuaria) (INTA) on studies about IPM in small farmers' maize crops. Many aspects of maize production were evaluated, partly by field experimentation, and elements of biocontrol were described within a maize IPM framework (van Huis, 1981).

Later, MAG provided funds to develop a mass rearing of *T. pretiosum*, supervised by E. Cano (UNAN-León). Collections of parasitized *H. zea* eggs were made in cotton crops in the field during 1983 to be used for laboratory propagation. D. Vincent (USDA Insect Identification and Beneficial Insect Introduction Institute, Maryland) helped with identification of *T. pretiosum* (Cano, 2001). Thereafter, baseline biological traits and reproductive parameters of *T. pretiosum* were obtained and life tables were constructed (Cano and Swezey, 1992). Laboratory studies with several *Trichogramma* species indicated that females could parasitize up to 14.2 eggs per day and they lived on average 13 days. Trials using the grain moth *Sitotroga cerealella* (Olivier) as factitious host for *T. pretiosum* resulted in good-quality parasitoids with a lifespan of about 35 days when reared at 26°C and a relative humidity of 76% (Cano *et al.*, 2002). These results were used to develop a novel mass-rearing system for *T. pretiosum*, based on the Flanders and Hassan (Flanders, 1929) and Morrison methodology for *S. cerealella* (Morrison *et al.*, 1978; Morrison, 1985; Cano *et al.*, 2004a, b). Biocontrol with *T. pretiosum* was assessed in crops in several agro-climatic systems in Nicaragua and 60–100% parasitism was achieved (Table 22.1). After the year 2000, mass production of *T. pretiosum* was optimized, mainly by replacing imported inputs with local ones, in order to reduce product costs. Currently, the cost of a one-square-inch (6.5 cm^2) unit with *Trichogramma* pupae is US$0.25. Parasitoid production in the biofactory is scheduled to accomplish a mean of five releases per cropping cycle to cover a maximum area of 3,500 ha.

Production of predators was initiated by the staff of UNAN-León in 1987 by mass rearing the lacewing *Chrysoperla externa* (Hagen) (Cano, 2001). This predator is reared on a diet based on *S. cerealella* eggs in semi-commercial biofactories. Field efficiency of this predator was an average reduction of 84% of aphids in water melon, 42% of whiteflies in cotton, 62% of thrips in onion and 65% of scales in banana (Cano, 2001). Later, it was found that *C. externa* could also be employed to reduce populations of the potato psyllid *Bactericera cockerelli* (Sulc) by 32% (E. Cano, Leon, 2012, personal communication). This psyllid is a vector of the tomato–potato zebra chip disease associated with the bacterium *Candidatus* Liberibacter *solanacearum* in potato.

Researchers at the National Center for Plant Protection (Centro Nacional de Protección Vegetal) (CENAPROVE) started studies in 1987 to produce the entomopathogenic organisms *B. thuringiensis*, *Steinernema feltiae* Filipjev and *Metarhizium anisopliae* (Metschnikoff), as

Table 22.1. Percentage parasitism of various pests in crops in Nicaragua by *Trichogramma pretiosum* (retrieved from CIRCB. UNAN-León; Cano *et al.*, 2004a).

Crops	Pests	Number of wasps released per ha	Number of releases per cropping cycle	Parasitism (%)
Cucurbitaceae in general	Melon worm moths, *Diaphania hyalinata* and *D. nitidalis*	70,000–350,000	4–6	65–83
Cotton	Corn earworm, *Helicoverpa zea*	70,000–350,000	6–9	95
	Cotton leafworm *Alabama argillacea*	70,000–175,000	3–6	100
Sugarcane	Sugarcane borer *Diatraea saccharalis*	70,000–210,000	4–6	95
Soybean	Velvetbean caterpillar *Anticarsia gemmatalis*	70,000–175,000	4	60
Tomato	*Heliothis* (= *Helicoverpa*) *zea*	70,000–350,000	4–6	100
Maize	Fall armyworm *Spodoptera frugiperda*	70,000–350,000	4–6	85
Sesame seeds and beans	Cabbage looper *Trichoplusia ni*	70,000–210,000	3–6	70

well as botanical insecticides based on neem extracts (Lacayo Parajón, 1987). The Center was closed in 1990 without implementing these technologies, but some of its well trained personnel founded biofactories to formulate biopesticides based on those organisms (see Section 22.3, Current Situation of Biological Control in Nicaragua).

Investigations into the use of entomopathogenic organisms were also made by UNAN-León in 1987. Collections of living material from West Nicaragua yielded 19 virus isolates infesting the lepidopteran pests *Spodoptera exigua* (Hübner), *S. sunia* (Guenée), *S. frugiperda*, *H. zea* and *Trichoplusia ni* (Hübner). These viruses were identified by P. Entwistle (Institute of Virology and Environmental Microbiology, UK), who placed them within the Family Baculoviridae, nuclear polyhedrosis viruses, and registered them as VPNSe, VPNSs, VPNSf, VPNHz and VPNTni, respectively. Rearing of the five lepidopteran host species was started for production of these pathogens. Artificial diets and rearing methods were designed, as well as virus pathogenicity and virulence, LD_{50}, types of formulations and quality-control protocols for their production. Diets were developed using 90% locally produced ingredients, such as soybeans and red beans. Host life cycles reached an average 27 ± 1.7 days in laboratory conditions (temperature 26°C; RH 75%). Viruses were produced *in vivo* by means of host-diet contamination and 5th-instar host larvae killed by the virus were used. The rate of infection ranged between 89% and 92% with a concentration of 10^8 polyhedral inclusion bodies per microlitre (Narváez and Rizo, 2000). Evaluation of pest control efficacy by these bioproducts has been made under different agroclimatic conditions and applied to various crops (Table 22.2) (Rizo and Narváez, 2001; López *et al.*, 2004).

From 1996 to 1998, researchers at UNAN studied biocontrol of the diamondback moth *Plutella xylostella* L. and released the parasitoids *Cotesia plutellae* (Kurdjumov) and *Microplitis plutellae* Muesebeck in cabbage fields in Matagalpa and Jinotega regions, resulting in 67% of parasitism. In 1998–1999, INTA introduced the parasitoid *Diadegma semiclausum* Hellen from Taiwan and reared and released it for control of *P. xylostella* in cabbage. Parasitoids reared at Universidad Nacional Agraria (UNA) were used for behavioural studies and field evaluations. This parasitoid showed an average life cycle of 31.4 days when reared at 21°C and 65% RH. Percentages of parasitism reached up to 90% in the laboratory and 70% in the field. Compared with natural parasitism of 59% by *Diadegma insulare* Cresson, the imported *D. semiclausum* increased *P. xylostella* mortality by 11% (Cerda and Miranda, 2011).

22.3 Current Situation of Biological Control in Nicaragua

In 2009, UNAN-León started colonies of native species of the genus *Orius* in West Nicaragua in collaboration with Y. Colmenárez (CABI, Trinidad & Tobago). Sample specimens were sent to the National Museum of Natural History, Smithsonian Institution (Washington, DC), which were identified by T. Henry as *Orius euryale* Herring and *O. insidiosus* Say. Population parameters of *O. euryale* were determined in the laboratory at a temperature of 27°C, 70% RH and LD

Table 22.2. Control efficacy of three nuclear polyhedrosis virus (NPV) strains (retrieved from CIRCB. UNAN-León, C. Narváez, R. Carmen and P. Castillo, unpublished data, 1994–1998).

Pest	NPV strain	Doses per ha in larval equivalents (LE)	Crops	Control efficiency
Spodoptera frugiperda	VPNSf	708 LE	Maize, sorghum, grasses	75%
Spodoptera sunia	VPNSs	212 LE	Tomato, soybean, beans, sweet pepper, sesame seeds	80%
Spodoptera exigua	VPNSe	212 LE	Onion, pepper, tomato, okra	80%

period of 11:13. Oviposition substrates were short stems of purslane *Portulaca oleraceae* L. The mean life cycle of this anthocorid species lasts 58.3 days, with *Ro* = 156.464, T = 22.6 days, and *r* = 0.22. *Orius* spp. were released to control nymphs of whiteflies and aphids, and pest populations were reduced by 48–52% (Castillo *et al.*, 2010).

UNAN-León is also working with the parasitoid *Tamarixia radiata* Waterston to control the Asian citrus psyllid *Diaphorina citri* Kuwayama, a vector of the citrus disease huanglongbing (HLB) caused by the bacterium *Candidatus* Liberibacter.

Since 2000, a boost of Nicaraguan biotechnology and availability of specialized personnel has led to the development of small biofactories that produce biological products. According to the Foundation for the Nicaraguan Agricultural and Forestry Technological Development (FUNICA, 2007), applications with microbial control agents increased from 7% in the 1990s to 12.4 % in 2007. The institutions producing biological products are listed in Table 22.3.

22.4 New Developments of Biological Control in Nicaragua

In Nicaragua, biocontrol experience and production of biocontrol agents were developed thanks to the efforts of government institutions, non-governmental organizations (NGOs), private companies and universities. Currently, 12 biocontrol products are used in both conventional and organic production. The demand for these products is still scarce, especially due to limited logistics of the biocontrol industry and difficulty in reaching farmers in rural areas. The biocontrol industry can hardly compete with the marketing networks of the synthetic pesticides industry. About ten crops have been targeted for biocontrol, including sugarcane, maize, coffee and horticultural crops. Augmentative biocontrol strategy was predominant and the biocontrol agents employed were mainly entomopathogenic organisms. After consultation with the producers about sales of biocontrol products, it can be concluded that biocontrol is used on 10,000 ha of sugarcane alone and on 484 ha of other crops (Table 22.4).

22.5 Acknowledgements

I thank the researchers from the Universidad Nacional Autónoma de Nicaragua and others who provided information for this chapter, M.G. Luna for translation of the Spanish manuscript into English, J.C. van Lenteren, Y. Colmenarez, and V.H.P. Bueno for editing.

Table 22.3. Biofactories and biological control agents produced in Nicaragua.

Institution	Region	Microbial agent
UCA Miraflor	Estelí	*Beauveria bassiana*
Universidad Nacional Agraria (UNA)	Managua	*B. bassiana*
		Metharhizum anisopliae
		Paecelomyces sp.
Universidad Nacional Autónoma de Nicaragua (UNAN)		*B. bassiana*
		Trichoderma harzianum
		Trichogramma pretiosum
		Chrysoperla externa
		Nuclear polyhedrosis virus (NPV)
		Orius insidiosus
Ingenio San Antonio	Chinandega	*M. anisopliae*
Ingenio Pantaleón	Chinandega	*Steinernema* sp.
Biotorlab	Matagalpa	*M. anisopliae*
		Trichoderma asperellum
Cooperativa de Mujeres Primero de Noviembre	Matagalpa	*B. bassiana*

Table 22.4. Crops under augmentative biological control in Nicaragua.

Crop	Pest	Biological control agent	Area (ha) under biocontrol
Maize	*Spodoptera frugiperda*	Nuclear polyhedrosis virus (NPV sf)	21
Cucurbitaceae various	*Diaphania nitidales*, *D. hyalinata*, *Aphis* sp.	*Trichogramma pretiosum* *Chrysoperla externa*	5
Water melon	Same pests	Same natural enemies	275
Okra	*Spodoptera exigua*	Nuclear polyhedrosis virus (NPVSe)	14
Onion	*S. exigua*, *Spodoptera sunia*	Nuclear polyhedrosis virus (NPVSe) Nuclear polyhedrosis virus (NPVSs)	15
Tomato	*Helicoverpa zea*	*Trichogramma pretiosum*	14
Chiltoma	*Anthonomus eugenii*	*Beauveria bassiana*	–
Coffee	*Hypothenemus hampei*	*Beauveria bassiana*	70
Vegetable seedlings	*Fusarium oxysporum*, *Pythium* sp., *Rhizoctonia solani*	*Trichoderma harzianum*	70
Sugarcane	*Aeneolamia postica*	*Metharhizium anisopliae*	10,000

References

Andrews, K.L. and Quezada, J. (1989) Manejo de Plagas Insectiles en la Agricultura: Estado Actual y Futuro [Integrated Insect Pest Management in Agriculture: Current Stage and Future]. Escuela Agrícola Panamericano El Zamorano, Honduras.

Cano, E. (2001) Cría masiva de *Trichogramma pretiosum* Riley, *Sitotroga cerealella* y *Chrysoperla externa* [Mass rearing of *Trichogramma pretiosum*, *Sitotroga cerealella* and *Chrysoperla externa*]. Avances en el Fomento de Productos Fitosanitarios No-Sintéticos, Bol. Informativo MIP/CATIE, Turrialba, Costa Rica 60, 93–96.

Cano, E. and Swezey, S. (1992) Tabla de vida en laboratorio y liberación en el campo de *Trichogramma pretiosum* Riley (Hymenoptera: Trichogrammatidae) en Nicaragua. [Laboratory life table and field releasing of *Trichogramma pretiosum* Riley (Hymenoptera: Trichogrammatidae) in Nicaragua]. *Revista Nicaragüense de Entomología* 21, 43–56.

Cano, E., Pérez, O. and Pacheco, S. (2002) Determinación del ciclo biológico de *Sitotroga cerealella* (Olivier) hospedero ficticio de *Trichogramma pretiosum* (Riley) [Determination of the biological cycle of *Sitotroga cerealella* as factitious host of *Trichogramma pretiosum*]. BSc thesis. En la cría comercial del centro de investigación reproducción de controladores biológicos, UNAN-León, Nicaragua. Available at: http://riul.unanleon.edu.ni:8080/jspui/handle/123456789/6274 (accessed 9 April 2019).

Cano, E., Carballo, M. and Salazar, D. (2004a) Control biológico de insectos mediante parasitoides [Biological control of insects by means of parasitoids]. *Control Biológico de plagas agrícolas, Serie Técnica Manual Técnico CATIE* 53, 89–122.

Cano, E., Fonseca, H. and Rostrán, A (2004b) Control de calidad y proceso de producción de *Trichogramma pretiosum* [Quality control and mass production process of *Trichogramma pretiosum*]. BSc thesis. Centro de Investigación y Reproducción de Controladores Biológicos (CIRCB), UNAN-León, Nicaragua. Available at: http://riul.unanleon.edu.ni:8080/jspui/handle/123456789/6274 (accessed 9 April 2019).

Castillo, P., Moreno, L., Gómez, Y. and Bustamante, B. (2010) Identificación, crianza y tabla de vida del depredador *Orius euryale* en condiciones de laboratorio [Identification, rearing and life table of the predator *Orius euryale* in laboratory conditions]. BSc thesis. Centro de Investigación y Reproducción de Controladores Biológicos CIRCB, UNAN-León, Nicaragua. Available at: http://riul.unanleon.edu.ni:8080/jspui/handle/123456789/6274 (accessed 9 April 2019).

Cerda, K. and Miranda, F. (2011) Introducción y evaluación del parasitoide *Diadegma semiclausum* (Hellen) para el control de *Plutella xylostella* en Nicaragua [Introduction and evaluation of the parasitoid *Diadegma semiclausum* to control *Plutella xylostella* in Nicaragua]. Universidad Nacional Agraria, Nicaragua, *La Calera* 9(12), 35–40,

CCIPA (1979) Manual de manejo integrado de plagas de algodonero de asistencia técnica. Comité de Control Integrado de Plagas del Algodonero [Committee for Integrated Pest Control in cotton]. Banco Nacional de Nicaragua, Managua, Nicaragua, pp. 369–370.

CIA (2017) The World Factbook: Nicaragua. Available at: https://www.cia.gov/library/publications/the-world-factbook/geos/nu.html (accessed 9 April 2019)

Dávila, A. (1992) *Glosario de nombres náhuatl de plantas, pájaros y algunas otras especies, con descripción de su etimología y comentarios del autor* [*Glossary of Náhuatl names for plants, birds and other species, with ethimological descriptions and commentaries from the author*]. Fondo Editorial Centro de Investigación de la Realidad de América Latina (CIRA), Managua, Nicaragua.

FAO (1977) Informe al gobierno de Nicaragua sobre control integrado de plagas del algodonero [Report to the Nicaraguan goverment on integrated pest control in cotton]. Food and Agriculture Organization, Managua, Nicaragua.

Flanders, S.E. (1929) The mass production of *Trichogramma pretiosum* Riley and observations on the natural and artificial parasitism of codling moth eggs. *Transactions of the 4th International Congress of Entomology*, August 1928, Ithaca, New York, USA, Vol. 2, pp. 110–130.

FUNICA (2007) *Avances en el Desarrollo de los Mercados Locales de Tecnologías en Nicaragua* [*Advances in the Development of Local Technology Markets in Nicaragua*]. Fundación para el Desarrollo Tecnológico Agropecuario y Forestal de Nicaragua, Managua, Nicaragua, pp. 20–36.

Huete-Perez, J.A., Hegg, M.O., Lopez, M.R., Cordoba, M., Montenegro, S., Vammen, K., Cortez, M.J. and Cornejo, A. (2017) Food and nutrition security for the sustainable development of Nicaragua. In: *Challenges and Opportunities for Food and Nutrition Security in the Americas. The View of the Academies of Sciences*. IANAS, IAP and BMBF, México DF, pp. 424-–53. [Available in English and Spanish at www.ianas.org]

Lacayo Parajón, L.I. (1987) Manejo integrado de plagas en Nicaragua: estado actual y perspectivas [Integrated pest management in Nicaragua: state of the art and perspectives]. *Memorias V Congreso Nacional y I Centroamericano, México y el Caribe de Manejo Integrado de Plagas*, pp. 312–331.

López P., Rizo, C. and Narváez, C. (2004) Control biológico de insectos mediante virus entomopatógenos [Insect biological control by means of entomopathogenic viruses]. *Control Biológico de plagas agrícolas, Serie Técnica Manual Técnico* CATIE 53, pp. 59–72.

Morrison, R., Jones, S. and López, J. (1978) A unified system for the production and preparation of *Trichogramma pretiosum* for field release. *Southwestern Entomologist* 3, 62–68.

Morrison, R. (1985) Effective mass production of eggs of the angoumois grain moth, *Sitotroga cerealella* (Olivier). *Southwestern Entomologist* Suppl. 8, 28–37.

Narváez, C. and Rizo, C. (2000) Evaluación de la patogenicidad e infectividad del virus de la polihedrosis nuclear (VPN) en *Spodoptera frugiperda* Smith (Lepidoptera: Noctuidae) [Evaluation of pathogenicity and infectiousness of nuclear polyhedrosis virus (NPV) on *Spodoptera frugiperda* Smith]. MSc thesis, Centro de Investigación y Reproducción de Controladores Biológicos CIRCB, UNAN-León, Nicaragua. Available at: http://riul.unanleon.edu.ni:8080/jspui/handle/123456789/6274 (accessed 9 April 2019).

Rizo, C. and Narváez, C. (2001) Uso y producción del Virus de la Poliedrosis Nuclear en Nicaragua [Usage and production of the Polyhedrosis Nuclear Virus in Nicaragua]. *Avances en el Fomento de Productos Fitosanitarios No-Sintéticos, Bol. Informativo MIP/CATIE*, Turrialba, Costa Rica 61, 90–96.

van Huis, A. (1981) Integrated pest management in the small farmer's maize crop in Nicaragua. Mededelingen Landbouwhogeschool Wageningen, 81–6, 1–222. Available at: http://edepot.wur.nl/290043 (accessed 9 April 2019).

Vaughan, M. (1958) Lista de insectos clasificados de Nicaragua [List of identified insects of Nicaragua]. Departamento de Entomología, Servicio Técnico Agrícola de Nicaragua. Ministerio de Agricultura y Ganadería, Managua, Nicaragua. pp. 138–142.

Vaughan, M. (1962) Especies parasíticas del gusano cogollero del maíz, *Laphigma frugiperda* (J.E. Smith) encontradas en 'La Calera' de agosto a julio de 1958 [Parasitic species of the maize worm, *Laphigma frugiperda* found in 'La Calera', from August to July 1958]. *8th Reunión Centroamericana, Proyecto Cooperativo Centroamericano de Mejoramiento del Maíz*. San José, Costa Rica.

Vaughan, M. (1994) Anales del curso y foro subregional centroamericano y del Caribe de control biológico de plagas [Annals of the course and the subregional forum of Central American and the Caribbe on biological control of pests]. León, Nicaragua, pp. 2–7.

Vaughan, M. (1996) Reseña del Manejo Integrado de Plagas en Nicaragua. [Review on IPM in Nicaragua]. Ministerio de agricultura y ganadería y Universidad Nacional Agraria de Nicaragua, Revista virtual Bio-Nica. Available at: www.bio-nica.info/biblioteca /Vaughan1996MIPNicaragua.pdf (accessed 10 August 2017).

23 Biological Control in Panama

Bruno Zachrisson[1]* and Anobel Barba[2]

[1]*Entomology and Biological Control of Insect-Pest Laboratory, Instituto de Investigación Agropecuaria de Panamá (IDIAP), Chepo Panamá;* [2]*Plant Protection Laboratory, Instituto de Investigación Agropecuaria de Panamá (IDIAP)*

* E-mail: bruno.zachrissons@idiap.gob.pa

Abstract

The first activity in biological control in Panama concerned import and release of a parasitoid of citrus blackfly in 1931. Later, the development of biocontrol programmes was targeted mainly against pests in sugarcane, vegetables, cantaloupes, watermelons, coffee and rice. In the 1970s, the releases of *Cotesia flavipes* and *Paratheresia claripalpis* for the control of *Diatraea saccharalis* guaranteed the cost effectiveness of the crop. Subsequently, *Diadegma semiclausum*, *Cotesia plutellae* and *Microplitis plutellae* were imported in the 1990s for biocontrol of *Plutella xylostella* in cruciferous crops. This was followed in the 2000s by the introduction of *Prorops nasuta* and *Phymastichus coffea* for the management of *Hyphotenemus hampei*. Prospecting programmes were executed in Panama to collect and identify natural enemies of, among others, whiteflies, leaf miners, thrips, mites and rice stink bug. Also a large number of entomopathogenic fungal strains were collected and identified. Several of the natural enemies and entomopathogens that were found to be promising as biocontrol agents are now mass produced and tested in the field. Ongoing work on artificial diets for mass rearing of pests that serve as hosts/prey for natural enemies is expected to result in cheaper production of biocontrol agents. Recently a conservation biocontrol programme was started for control of the rice stink bug by studying how weeds in and near rice fields can attract and stimulate populations of parasitoids of the stink bug.

23.1 Introduction

Panama has an estimated population of slightly more than 3,750,000 (July 2018) and its main agricultural products are bananas, rice, maize, coffee, sugarcane, vegetables, livestock and shrimp (CIA, 2019). According to Zachrisson *et al.* (2017, pp. 455 and 463):

> The country has an area of 75,845.072 km^2 ... Agricultural land covers 30.4% of the country's total area ... Panama boasts a variety of agricultural, livestock, fishery and aquaculture production systems, the most important being rainfed and irrigated rice, bovine milk and meat, swine and avian production and wild-caught fish ... Panama's forest resources are characterized by mature forest cover, intervened and secondary forests, which accounted for 61.9% of the land area in 2014.

23.2 History of Biological Control in Panama

23.2.1 Period 1880–1969

Investigations into biological control in Panama began in 1931 and were focused on managing citrus blackfly *Aleurocanthus woglumi* Ashby, with releases of the parasitoid *Eretmocerus serius* Silvestri (Rodríguez and Arredondo, 2007) in citrus groves from 1931 to 1943 (Table 23.1). This parasitoid was studied also by H.D. Smith, a US Department of Agriculture (USDA) entomologist based in Panama in 1943, who reared *E. serius* (Arredondo-Bernal and Rodríguez-del-Bosque, 2008).

23.2.2 Period 1970 –2000

During this entire period, inundative releases of *Cotesia flavipes* Cameron were made in sugarcane, resulting in high parasitism rates of *Diatraea saccharalis* Fabricius, but not of *Diatraea tabernella* (Dyar). The economic impact of *D. tabernella* exceeded that reported for *D. saccharalis* (López, 1995), but lack of information about natural enemies made biocontrol of this borer difficult. Later, integrated pest management (IPM) in sugarcane plantations made production profitable and the biocontrol component consisting of simultaneous releases of *Paratheresia claripalpis* Wulp and *C. flavipes* resulted in improved biocontrol of *Diatraea* sp.: a cumulative number of 5,721,180 of these parasitoids were released on a cumulative cultivated area of 292,930 m^2 (Rodríguez *et al.*, 2004).

In the early 1990s, *Bemisia tabaci* (Gennadius) transmitted *Begomovirus* to Cucurbitaceae and Solanaceae, in particular to melon, watermelon and tomato. Conservation biocontrol was studied as a management option for *B. tabaci*. The following natural enemies were found in the locality of Los Santos, Panama: *Encarsia* sp. *Eretmocerus* sp. *Cycloneda sanguinea* L., *Hippodamia convergens* Guérin-Méneville, *Coleomegilla maculate* De Geer, *Scymnus* sp., *Chrysopa* sp. and *Polistes panamensis* L. (Zachrisson, 1992) (Table 23.1). Other species of parasitoids were subsequently identified, among which *Encarsia pergandiella* Howard, *Signiphora* sp. and *Amitus* sp. were identified for the first time in Panama (González *et al.*, 2009). *B. tabaci* on tomato in Los Santos showed rates of parasitism between 20.4% and 14.8%.

Table 23.1. Chronology of activities related to biological pest control in Panama.

Year	Activity	Biocontrol agent	Pest	Reference
1931	Introduction from Cuba	*Eretmocerus serius*	*Alerocanthus woglumi*	Altieri *et al.*, 1989
1938/43	First and second export to Mexico	*E. serius*	*A. woglumi*	Rodríguez and Arredondo, 2007
1991	Parasitoids identified in Cerro Punta, Chiriquí	*Diglyphus* sp., *Chysocharis* sp., *Oenonogastra* sp., *Halticoptera* sp.	*Liriomyza huidobrensis*	González, 1991; Morales *et al.*,1994
1992	Parasitoids and predators identified in crops and weeds in the region of Azuero	*Encarsia* sp., *Eretmocerus* sp., *Coleomegilla maculata, Chrysopa* sp.	*Bemisia tabaci*	Zachrisson, 1992
1999	Multiplication of parasitoids	*Microplitis plutellae*, *Cotesia plutellae*	*Plutella xyllostella*	REDCAHOR, 2000
1999	Introduction from Nicaragua	*Diadegma semiclausum*	*P. xylostella*	REDCAHOR, 2000
2000	Multiplication/release/evaluation of parasitism	*D. semiclausum*	*P. xylostella*	Abrego and Polanco, 2001
2002	Identification of parasitoids in rice	*Telenomus podisi, Trissolcus basalis*	*Oebalus insularis*	Zachrisson, 2009
2003	Identification of parasitoids in melon	*Trichogramma* sp., *Conura* sp.	*Diaphania hyalinata*, *D. nitidalis*	Barba and Korytkowski, 2004
2004	Inventory of predatory mites in rice	*Proprioseiopsis* sp.	*Steneotarsonemus spinki*	Camargo *et al.*, 2009
2005	Parasitism of pest in rice	*Telenomus rowani*	*Rupela albinella*	Zachrisson, 2009
2006	Predators identified in Azuero	*Orius insidiosus*, *O.* spp.	*Thrips palmi*	Barba, 2007
2006	Import from Colombia	*Prorops nasuta*, *Phymastichus coffea*	*Hypothenemus hampei*	Pérez, 2006
2006	Use of entomopathogenic fungi	*Beauveria bassiana*	*H. hampei*	Pérez, 2006
2007	Identification of native entomopathogens	*Beauveria* sp.	*H. hampei*	Morales *et al.*, 2009
2007	Parasitoid releases on farms	*Cephalomia stephanoderis*	*H. hampei*	Pérez, 2006
2007	Parasitoids identified in tomato and pepper	*Eretmocerus* sp., *Amitus* sp., *Signiphora* sp., *Encarsia quaintancei, E. bimaculata*, *E. pergandiella*, *E. hispida*, *E. porteri*, *E. citrella*, *E. nigricephala*	*B. tabaci*, *Trialeurodes vaporariorum*	González *et al.*, 2009
2007	Identification of native entomopathogens	*M. anisopliae*, *B. bassiana*, *Paecilomyces* sp.	*Cyrtomenus bergi*	Barba *et al.*, 2009a
2009	Effect of entomopathogens on pest and natural enemy	*B. bassiana*, *M. anisopliae*	*Gynaikothrips* sp.	Hirano and Barba, 2009a
2009	Efficacy of entomopathogen	*M. anisopliae*	*D. hyalinata*	Hirano and Barba,2009b

Continued

Table 23.1. Continued.

Year	Activity	Biocontrol agent	Pest	Reference
2008	Evaluation of predation capacity	*O. insidiosus*	*T. palmi*	Hirano, 2009; Rivera, 2009
2008	Efficacy of native of entomopathogenic fungi	*B. bassiana*, *M. anisopliae*	*Anthonomus eugenii*	Barba *et al*., 2009b; Hirano and Barba, 2009b
2009	74 strains of entomopathogenic fungi identified	Entomopathogenic fungi	Insect pests	Barba, 2010
2010	Identification of predatory mites	*Neoseiulus baraki*, *N. barabesis*, *Hypoaspis* sp., *Pseudoparasitus* sp.	*Steneotarsonemus spinki*	Quirós and Rodríguez, 2010
2013	Isolation, identification and characterization of 54 native strains of entomopathogens	*Beauveria* sp., *M. anisopliae*, *Paecilomyces* sp.	Insect pests	Gutiérrez and Vega, 2013
2015	Characterization of native isolate ¨rs006¨	*Isaria javanica*	*H. hampei*	González *et al*., 2015
2015	Pathogenicity/virulence of native isolate	*Isaria* spp.	*H. hampei*	Lezcano *et al*., 2015

The diamondback moth *Plutella xylostella* L. caused serious damage on broccoli (*Brassica oleracea* L. var. *Italica*) and cauliflower (*Brassica oleracea* L. var. *Botrytis*) in Cerro Punta and Boquete, Chiriquí in Panama. Therefore, as part of an initiative by the Collaborative Vegetable Research and Development Network (Red Colaborativa de Investigación y Desarrollo de Hortalizas para América Central) (REDCAHOR), several species of parasitoids were introduced in 1998, among them *Diadegma semiclausum* Hellen, *Cotesia plutellae* Kurdjumov and *Microplitis plutellae* Muesbeck (REDCAHOR, 2000). International operations facilitated the importation of these species from Nicaragua, followed by their multiplication in the laboratory. The parasitism rate of *P. xylostella* was 78.8% under laboratory conditions and 33.1% in the field.

Severe attacks by the polyphagous leaf miner *Liriomyza huidobrensis* Blanchard on various horticultural crops were reported in Cerro Punta, Chiriqui in early 1990 (Morales *et al.*, 1994). Native parasitoids were collected in Cerro Punta and Boquete, Chiriquí, indicating the presence of species of four genera: *Diglyphus* Walker, *Chrysocharis* Foster, *Oenonogastra* Ashmead and *Halticoptera* Spinola, including *Oenonogastra microrhopalae* Ashmead, *Diglyphus isaea* Walker, *Opius dimidiatus* Ashmead and *D. websteri* Crawford, as well as other parasitoids and predators (González, 1991) (Table 23.1).

23.3 Current Biological Control Situation in Panama

Diaphania hyalinata L. and *Diaphania nitidalis* Stoll are considered key pests of the family Cucurbitaceae, including melon, watermelon and squash. Both species have similar habits and occur on cultivated and wild plants. With the support of the International Regional Agency for Agricultural Health through the Vigilancia Fitosanitaria en Cultivos de Exportación no Tradicionales (VIFINEX) project and the Central American Master's Program in Entomology of the University of Panama, the population dynamics of *D. hyalinata* and *D. nitidalis* were determined in plots planted with melon, watermelon and squash. Several parasitoid taxa were identified, including *Trichogramma* sp., *Apanteles* sp., *Conura* sp., *Copidosoma truncatellum* Dalman, *Lespesia* sp. and *Drino* sp. (Korytkowski, 2003). Barba and Korytkowski (2004) reported a 35.0% egg parasitism rate by *Trichogramma* sp. and a larval parasitism rate of 37.5% by *Stantonia* sp. and *Conura* sp. However, the impact of native natural enemies on these species is variable.

Discovery of the coffee berry borer in 2005 at the Bajo Cerrón farm in Río Sereno, Chiriquí, promoted joint actions by the Agriculture Development Ministry (Ministerio de Desarrollo Agropecuario) (MIDA) and Instituto de Investigación Agropecuaria de Panamá (IDIAP) (Pérez, 2006) to prospect for biocontrol agents in areas close to the *Hypothenemus hampei* (Ferrari) infestation foci and several strains of *Beauveria* sp. were identified (Morales *et al.*, 2009). Classical biocontrol of *H. hampei* began in 2006 with the release of 1,600,000 adult *Prorops nasuta* Waterston, imported from Colombia, on 18 farms located in Renacimiento, Chiriquí, Panama (Pérez, 2006). The efficiency of *Beauveria bassiana* (Bals.-Criv) Vuill applications and the mortality rates of releases of *P. nasuta* and *P. coffea* against the coffee berry borer were between 48% and 54%, respectively.

Thrips palmi Karni caused severe damage to Cucurbitaceae in Herrera, Panama, affecting fruit quality and yields (Vásquez and Barba, 2013). Its population dynamics were studied in crops and non-cultivated host plants (Barba, 2015; Barba and Suris, 2015), including presence of natural enemies, and a conservation biocontrol programme was implemented (Barba, 2007). Four possibly new *Orius* species (Hirano, 2009), all predators of *T. palmi*, were evaluated in the laboratory. *Orius insidiosus* Say was mass reared, using an artificial diet based on *Sitotroga cerealella* Olivier eggs in a solution of maize pollen and honey (1:1), and *Phaseolus vulgaris* L. beans were used as an oviposition substrate (Barba and Suris, 2015), which yielded promising results in the laboratory.

Oebalus insularis Stal is the main rice pest that causes direct damage and significant losses in the milky stage of rice (Zachrisson and Martinez, 2011; Zachrisson *et al.*, 2014a). In addition, even low numbers of larvae of *Rupela albinella* (Cr.) and *Spodoptera frugiperda* (J.E. Smith) may cause severe damage during the vegetative phase of rice. The egg parasitoids *Telenomus rowani* (Gaham), *Trichogramma pretiosum* Riley and *Telenomus podisi* Ashmead have been shown to be effective agents in the control of *R. albinella, S. frugiperda* and *O. insularis*, respectively (Zachrisson, 2009). In addition, new associations of *T. podisi* with *Tibraca limbativentris* (Stal) and *Euschistus* Rolston showed parasitism rates above 80% in extensive rice production areas in Coclé, Panama (Zachrisson *et al.*, 2014b). The natural parasitism by species such as *T. rowani* and *T. podisi*, which reduce *R. albinella* and *O. insularis* populations, respectively, can be promoted by conserving natural reservoirs of weed species (Zachrisson and Polanco, 2017).

The production of entomopathogenic agents, mainly *Metarhizium anisopliae* (Metchnikoff) Sorokin, *B. bassiana* and *Isaria* sp., is currently being implemented and stimulated by the development of a collection of entomopathogenic fungi located at the Central Research Center (CIAC) of the IDIAP. The use of *Isaria* sp. to manage *H. hampei*, despite the lack of personnel trained in microbial control, is also an important accomplishment. Furthermore, there are commercial companies that produce microbial control agents at the local level, strengthening these projects.

Due to increased international trade, the number of invasions by alien species, which are often polyphagous, has also increased. Given this situation, classical biocontrol programmes were strengthened, and international agreements between Empresa Brasileira de Pesquisa Agropecuária (Embrapa) and IDIAP were implemented to encourage research to identify potential invasive species from the Americas. As a result, biological studies were promoted *in situ* to understand the behaviour of these insects and their adaptation to cultivated and wild plants.

23.4 New Biological Control Developments in Panama

The development of artificial diets for the multiplication of key pests is a priority to improve and economize mass rearing of natural enemies for augmentative biocontrol programmes. Currently, artificial diets for *S. frugiperda, D. saccharalis, D. tabernella, H. hampei* and *O. insularis* are formulated and evaluated. Also, mass production of *Trichogramma* sp. and *T. podisi* in biofactories is under implementation with the goal of controlling a large number of lepidopteran (Noctuidae, Pyralidae) and heteropteran (Pentatomidae) pests that affect the main agricultural crops.

As well as developments in the area of augmentative biocontrol, Panama is working on conservation biocontrol. Host plants of the pest *O. insularis* and its parasitoids in weeds occurring

Table 23.2. Application of biological control in Panama.

Biocontrol agent	Pest / crop	Type of biocontrol[a] / since	Effect[b] / area (ha) under biocontrol[c]
Eretmocerus serius	Citrus blackfly in citrus	CBC / 1931	+ / ? terminated in 1943
Cotesia flavipes and *Paratheresia claripalpis*	Sugarcane borers in sugarcane	ABC / 1970	+ / 38,629[b]
Complex of predators and parasitoids	Whitefly in tomato, other vegetables	NC / 1990	+ / 2,863[b]
Complex of parasitoids	*Liriomyza huidobrensis* in vegetables	NC / 1991	± / 553
Complex of parasitoids	Diamondback moth	ABC / 1998	± / 0.5
Prorops nasuta	Coffee berry borer in coffee	CBC / 2006	± / 19
Complex of predators and parasitoids	Rice stink bug in rice	ConsBC / 2009	± / 177

[a]Type of biocontrol: ABC = augmentative, CBC = classical, ConsBC = conservation biocontrol; NC = natural control;
[b]Effect: + = success, ± = partial success,
[c]Area of crop harvested in 2017 according to FAO (http://www.fao.org/faostat/en/#data/qc)

in or near rice fields have been studied and the high potential of some of the weeds to be reservoirs of egg parasitoids of this rice pest has been confirmed. The weed–pest–parasitoid complex enables studies in the field of chemical ecology, particularly for determination of the role of volatile compounds emitted by the weeds to attract pentatomid egg parasitoids.

There are no complete data about crops and areas under biocontrol in Panama, as many of the projects are under development. However, the partial data available indicate that about 42,241 ha are under biocontrol (Table 23.2).

23.5 Acknowledgement

We thank the National Secretary of Science, Technology and Innovation and the National Research System (SNI-SENACYT) for supporting the biocontrol projects mentioned in this chapter.

References

(References with grey shading are available as supplementary electronic material)

Abrego, E. and Polanco, F. (2001) Introducción y evaluación de *Diadegma semiclausum* (Hellen), (Himenoptera: Ichneumonidae), a las condiciones de campo, provincia de Chiriquí, República de Panamá [Introduction and evaluation of *Diadegma semiclausum* at field conditions, Chiriqui province, Panama]. BSc thesis. Universidad de Panamá, Panamá, Panama.

Altieri, M., Trujillo, J., Campos, L., Klein, K., Gold, C. and Quezada, J. (1989) El control biológico en América Latina y su contexto histórico [Biological control in Latin America and its historical context]. *Manejo Integrado de Plagas* 12, 82–107.

Arredondo-Bernal, H.C. and Rodríguez-del-Bosque, L.A. (2008) *Casos de control biológico en México* [*Cases of biological control in Mexico*]. Ed. Mundi-Prensa, Mexico City, Mexico.

Barba, A. (2007) Determinación de las características básicas de la población del trips del melón *Thrips palmi* (Thysanoptera: Thripidae) con la finalidad de establecer una estrategia de manejo integrado [Determination of the basic characteristics of the population of *Thrips palmi* to establish an integrated management strategy]. Informe Técnico. Instituto de Investigación Agropecuaria de Panamá, Panamá, Panama.

Barba, A. (2010) Alternativas para el manejo integrado de insectos picadores chupadores que afectan el cultivo de cucurbitáceas [Alternatives for the integrated management of sucking and biting insects that affect cucurbit crops]. Informe Técnico. Instituto de Investigación Agropecuaria de Panamá, Panamá. Panama, pp. 1–2.

Barba, A. (2015) Manejo integrado de *Thrips palmi* (Thysanoptera: Thripidae) en cultivo de sandía de exportación en la región de Azuero, Panamá [IPM of *Thrips palmi* in export watermelons in the Azuero region, Panama]. PhD thesis. Universidad Agraria de la Habana, La Habana, Cuba.

Barba, A. and Korytkowski, C.A. (2004) Análisis demográfico de las poblaciones de *Diaphania hyalinata* y *D. nitidalis* (Linnaeus, 1767) (Lepidoptera: Pyralidae) asociadas a Cucurbitaceas cultivadas y silvestres en la Península de Azuero (2003–2004) [Demographic analysis of the populations of *Diaphania hyalinata* and *D. nitidalis* associated with wild and cultivated cucurbits in the Azuero Peninsula (2003–2004)]. *Revista Scientia* 18, 51–73.

Barba, A. and Suris M. (2015) Presencia de *Thrips palmi* Karny (Thysanoptera: Thripidae) en arvenses asociadas al cultivo de sandía para la región de Azuero, Panamá [Presence of *Thrips palmi* Karny in weeds associated with watermelon crop in the Azuero region, Panama]. *Revista Protección Vegetal* 30, 171–175.

Barba, A., Aguilera, V., Masachika, H. and Gordón, R. (2009a) Manejo integrado de *Anthonomus eugenii* Cano (Coleoptera: Curculionidae) en el cultivo de Ají [IPM of *Anthonomus eugenii* Cano in chili pepper]. Panamá. Boletín Técnico. Instituto de Investigación Agropecuaria de Panamá. Panamá, Panama, pp. 1–16.

Barba, A., Hernández, R. and González, A. (2009b) Identificación de agentes de control biológico del chinche de la viruela *Cyrtomenus bergi* [Identification of biological control agents of the subterranean burrower bug *Cyrtomenus bergi*]. Informe Técnico. Instituto de Investigación Agropecuaria de Panamá, Panamá, Panama, pp. 1–6.

Camargo, I., González, F., Quiróz, E., Zachrisson, B. and Von Chong, K. (2009) Manejo Integrado del complejo 'Acáro (*Steneotarsonemus spinki* Smiley) – Hongo (*Sarocladium orizae* Sawada/Gams and Hamks) Bacteria (*Burkholderia glumae*)', en el cultivo del arroz [IPM of the complex 'Mite (*Steneotarsonemus spinki*) – Fungus (*Sarocladium orizae*) – bacteria (*Burkholderia glumae*)', in rice). Boletín Técnico. Instituto de Investigación Agropecuaria de Panamá, Panamá, Panama, pp. 18–19.

CIA (2019) The World Factbook: Panama. Available at: https://www.cia.gov/library/publications/the-world-factbook/geos/pm.html (accessed 12 July 2019).

González, G. (1991) Contribución al conocimiento de los enemigos naturales de (*Liriomyza* spp.) encontrados en Cerro Punta, Boquete [Contribution to the knowledge of the natural enemies of (*Liriomyza* spp.) found in Cerro Punta, Boquete]. *Revista Ciencia Agropecuaria* 7, 59–64.

González, G., Guerra, J., Villarreal, N., Adames, K., Araúz, L. and Núñez, J. (2009) Contribución al conocimiento de los parasitoides de la mosca blanca [Contribution to the knowledge of whitefly parasitoids]. Boletín Técnico. Instituto de Investigación Agropecuaria de Panamá, Panamá, Panama, pp. 1–4.

González, G., Caballero, S., Contreras, G., Vergara, G. and Mejía, L. (2015) Caracterización morfológica y molecular del aislado endémico rs006, biocontrolador de *Hypothenemus hampei* en Panamá [Morphological and molecular characterization of the isolated endemic rs006, biocontrol agent of *Hypothenemus hampei* in Panama]. *Revista Ciencia Agropecuaria* 22, 78–85.

Gutiérrez, M. and Vega, D. (2013) Aislamiento, identificación y caracterización de cepas nativas de hongos entomopatógenos a partir de insectos plagas en cultivos agrícolas [Isolation, identification and characterization of native strains of entomopathogenic fungi from insect pests in agricultural crops]. BSc thesis. Universidad de Panamá, Panamá, Panama.

Hirano, M. (2009) Investigación e identificación de enemigos naturales de trips [Research and identification of thrips natural enemies]. Informe Técnico. Agencia Internacional de Colaboración del Japón (JICA), pp. 69–78.

Hirano, M. and Barba, A. (2009a) Efecto de los insecticidas sobre *Gynaoikothrips* sp (Thysanoptera: Phlaeothripidae) y su enemigo natural *Montandiola* sp. (Hemiptera: Anthocoridae). [Effect of insecticides on *Gynaoikothrips* sp and on its natural enemy *Montandiola* sp.]. Informe final sobre Investigaciones en Divisa. Agencia Internacional de Colaboración del Japón (JICA), pp.1–13.

Hirano, M. and Barba, A. (2009b) La eficacia de los insecticidas químicos y biológicos en condiciones de laboratorio [Efficacy of chemical and biological insecticides under laboratory conditions]. Informe final sobre Investigaciones en Divisa. Agencia Internacional de Colaboración del Japón (JICA), pp. 14–32.

Korytkowski, C.A. (2003) Monografía de Manejo Integrado de Plagas [Integrated Pest Management Monograph]. BSc thesis. Universidad de Panamá, Panamá, Panama.

Lezcano, J., Saldaña, E., Ruíz, R. and Caballero, S. (2015) Patogenicidad, virulencia del aislado de la cepa nativa de *Isaria* spp. y dos hongos entomopatógenos comerciales [Pathogenicity and virulence of isolates of native strain of *Isaria* spp. and two entomopathogenic fungi]. *Revista Ciencia Agropecuaria* 23, 20–38.

López, E. (1995) Evaluación de los niveles de daño causado por el barrenador de la caña de azúcar (*Diatraea* spp.) y su parasitismo natural (Julio–Diciembre de 1995), Ingenio Benjamín Zeleron Rivas [Evaluation of the levels of damage caused by the sugarcane borer (*Diatraea* spp.) and its natural parasitism (July–December 1995)]. BSc thesis. Universidad de Costa Rica, San José, Costa Rica.

Morales, R., Atencio, F., Lara, J. and Muñoz, J. (1994) La mosquita minadora (*Liriomyza* spp.) en Panamá [The leaf miner (*Liriomyza* spp.) in Panama]. Programa Regional Cooperativo de Papa (PRECODEPA). Boletín Técnico. Instituto de Investigación Agropecuaria de Panamá, Panamá, Panama, pp. 1–22.

Morales, R., Sánchez, E., Caballero, S. and Muñoz, J. (2009) Inventario de hongos asociados a la muerte de la broca del cafeto (*Hypothenemus hampei* Ferr.) en Renacimiento, Panamá [Inventory of fungi associated with the death of the coffee berry borer (*Hypothenemus hampei*) in Renacimiento, Panama]. Boletín Técnico. Instituto de Investigación Agropecuaria de Panamá, Panamá, Panama, pp. 1–4.

Pérez, J. (2006) Manejo de la broca del café en la República de Panamá [Management of the coffee berry borer in the Republic of Panama]. In: Barrera, J.F., García, A., Domínguez V. and Luna, C. (eds) *La Broca del café en América Tropical: Hallazgos y Enfoques* [*The coffee berry borer in Tropical America: findings and approaches*]. Sociedad Mexicana de Entomología. El Colegio de la Frontera Sur, Mexico, pp. 33–36.

Quirós, E. and Rodríguez, H. (2010) Acáros depredadores asociados a *Steneotarsonemus spinki* Smiley (Acari: Tarsonemidae) en Panamá [Predatory mites associated with *Steneotarsonemus spinki* Smiley (Acari: Tarsonemidae) in Panama)] *Revista de Protección Vegetal* 25, 1–5.

REDCAHOR (2000) Taller manejo integrado de plagas 'Combate del picudo del chili *Anthonomus eugenii*', en Cerro Punta, Panamá [Workshop on IPM 'Combat of the chili weevil *Anthonomus eugenii*', in Cerro Punta, Panama). Informe Técnico. Red Colaborativa de Investigación y Desarrollo de Hortalizas para América Central, Panamá, Panama, pp. 10–12.

Rivera, L. (2009) Capacidad reproductiva y depredadora de *Orius insidiosus* (Hemiptera: Anthocoridae) en condiciones de laboratorio [Reproductive and predation capacity of *Orius insidiosus* under laboratory conditions]. BSc thesis. Universidad de Panamá, Panamá, Panama.

Rodríguez, L.A. and Arredondo, H.C. (2007) *Teoría a Aplicación del Control Biológico* [*Theory of Application of Biological Control*]. Sociedad Mexicana de Control Biológico, Mexico.

Rodríguez, V., Chavarría, L., Gómez, I., Peñaloza, I. and Tejada, M. (2004) Desarrollo del parasitoide *Cotesia flavipes* Cameron 1891 (Hymenoptera: Braconidae) en *D. tabernella* Dyar, *D. saccharalis* Fabricius, 1794 (Pyralidae) y su efectividad en el control de *D. tabernella* [Development of the parasitoid *Cotesia flavipes* in *D. tabernella*, *D. saccharalis* and its effectiveness in the control of *D. tabernella*]. *Revista Tecnociencia* 6, 85–94.

Vásquez, J. and Barba, A. (2013) Identificación de *Thrips palmi* Karny (Thysanoptera: Thripidae) en cultivos de cucurbitáceas en Panamá [Identification of *Thrips palmi* in cucurbits in Panama]. *Mesoamerican Journal of Agronomy* 24, 45–55.

Zachrisson, B. (1992) Manejo Integrado de Mosca Blanca [Integrated white fly management]. Resumen, Jornada Agropecuaria. Centro Regional Universitario de Azuero, Chitré, Panama, pp. 20–21.

Zachrisson, B. (2009) Avances en el control biológico de plagas de arroz (*Oryza sativa*), por medio de parasitoides oofagos, en Panamá [Progress in biological control of rice pests (*Oryza sativa*), by means of eggs parasitoids in Panama]. Boletín Técnico. Instituto de Investigaciones Agropecuarias de Panamá, Panamá, Panama.

Zachrisson, B. and Martinez, O. (2011) Bioecología de *Telenomus podisi* (Ashmead), *Trissolcus basalis* (Wollaston), parasitoides del chinche del arroz (*Oebalus insularis* Kulghast), Panamá [Bioecology of *Telenomus podisi*, *Trissolcus basalis*, rice stink bug parasitoids (*Oebalus insularis*), Panama]. *Revista Tecnociencia* 13, 65–76.

Zachrisson, B. and Polanco, P. (2017) Natural parasitism of *Oebalus insularis* Stal (Heteroptera: Pentatomidae) eggs in host weeds associated with rice cultivation in Panama. *IOSR Journal of Agriculture and Veterinary Science* 10, 1–4.

Zachrisson, B., Margaría, C.B., Loiácono, M. and Martínez, O. (2014a) Parasitismo de huevos de *Tibraca limbativentris* (Hemiptera: Pentatomidae), en arroz (*Oryza sativa* L.) en Panamá [Parasitism of *Tibraca limbativentris* eggs in rice (*Oryza sativa*) in Panama]. *Revista Colombíana de Entomología* 40, 189–190.

Zachrisson, B., Polanco, P. and Martinez, O. (2014b) Desempeño biológico y reproductivo de *Oebalus insularis* Stal. (Hemiptera: Pentatomidae), en diferentes plantas hospedantes [Biological and reproductive performance of *Oebalus insulalris* on different host plants]. *Revista de Protección Vegetal* 29, 77–81.

Zachrisson, B., Camargo Buitrago, I., Him, C., Murillo, E., Cambra, R. and Arcia, D. (2017) *Food and Nutrition Security for Panama. Challenges and Opportunities for This Century*. In: *Challenges and Opportunities for Food and Nutrition Security in the Americas. The View of the Academies of Sciences*. IANAS, IAP and BMBF, México DF, pp. 454-469. [Available at: www.ianas.org]

24 Biological Control in Paraguay

Claudia Carolina Cabral Antúnez[1]*, Gloria Resquín Romero[2] and Victor Adolfo Gómez López[1]

[1]*Laboratory of Entomology, Facultad de Ciencias Agrarias, Universidad Nacional de Asunción, Paraguay;* [2]*Laboratory of Phytopathology, Facultad de Ciencias Agrarias, Universidad Nacional de Asunción, Paraguay*

* E-mail: claudia.cabral@agr.una.py

Abstract

Biological control in Paraguay started in the 1980s with the use of a baculovirus for augmentative biocontrol of the soybean caterpillar and application of parasitoids to control the sugarcane borer. Since 2000, organic production has stimulated use of biocontrol. Also in 2000, large-scale prospecting for natural enemies and microbial control agents in crops such as cotton, maize, bean, peanut, sesame, soybean and sugarcane was initiated. Many natural enemies found during these prospecting activities are currently used in conservation biocontrol programmes. Several entomopathogenic and phytopathogenic agents have been registered and are used in Paraguay.

24.1 Introduction

Paraguay has an estimated population of almost 7 million (July 2017) and its agricultural products consist of cotton, sugarcane, soybean, maize, wheat, tobacco, cassava, fruits, vegetables, beef, pork, eggs, milk and timber (CIA, 2017).

Paraguay is divided by the Río Paraguay into two geographical regions: the eastern region (Región Oriental) and the western region (Región Occidental), also known as the Chaco. The terrain consists mostly of grassy plains and wooded hills in the eastern region. To the west are mostly low marshy plains. The overall climate is tropical to subtropical. Rainfall varies strongly across the country, with substantial rainfall in the eastern portions, and semi-arid conditions in the far West. A large percentage of the population, especially in rural areas, derives its living from agricultural activity, often on a subsistence basis (Wikipedia, 2018).

Paraguay covers a total of 40.6 million hectares of land, but only one-fifth of that area is suitable for normal crop production. A 1981 agricultural census indicated that about 7% of the land was dedicated to crop production, 20% to forestry, 26% to livestock and 47% to other purposes. Agriculture provides the majority of jobs in Paraguay and is also the largest sector of export. Soybean, cotton and tobacco are the most important export crops. Cassava, maize, and beans, peanuts, sorghum, sweet potatoes, and rice are important food crops. Forests cover about one-third of Paraguay's total land area and they constitute a key economic resource. Approximately half of the forested area contains commercially valuable timber and over 45 species of wood are suitable for export, but fewer than ten species are exported in quantity. Deforestation, mainly to get firewood and to obtain more land for agricultural activities, is a serious problem. Raising and marketing livestock form a major segment of agriculture and the economy in Paraguay. Livestock output accounts for about 30% of agricultural production and 20% of the exports. Livestock consists mainly of cattle farming, followed by poultry and pig farming. Fishing is a minor industry in Paraguay. Only about 50 species of fish are eaten, dorado and pacú being the most popular (Country Studies, 2018).

24.2 History of Biological Control in Paraguay

24.2.1 Period 1970–2000

The history of biocontrol in Paraguay started in the 1980s with the use of *Baculovirus anticarsia* for control of the soybean caterpillar *Anticarsia gemmatalis* Hubner, as well as the parasitoids *Trichogramma* spp. (Resquín-Romero, 2000) and *Cotesia flavipes* Cameron (Silvie *et al.*, 2014b) for control of the sugarcane borer *Diatraea saccharalis* (F.). *Trichogramma* technology was introduced into Paraguay, as into other Latin American countries, from Colombia (García, 1996).

In the early 1980s an *Anticarsia* control programme was initiated by the Department of Itapúa and then extended to soybean producers in all regions in the mid-1980s. It is based on the use of a nuclear anti-variant polyhedrosis virus (AgMNPV). The programme was implemented with the help of Empresa Brasileira de Pesquisa Agropecuária (Embrapa), Brazil, in conjunction with Instituto Paranaense de Assistência Técnica e Extensão Rural (EMATER-PR) and some cooperatives, including the United Colonies Cooperative (Paraguay), and started for soybean farmers in the Departments Itapúa and Alto Paraná (Moscardi, 1983).

Also during this period, the biopesticide *Bacillus thuringiensis* (Bt) was used to control the cotton leafworm *Alabama argillacea* (Hueb.).

24.3 Current Situation of Biocontrol in Paraguay

24.3.1 Introduction

Interest in biocontrol has increased recently in Paraguay for several reasons. First, a greater appreciation has developed for the production of certified organic sesame by agroecological farmers and by a certain type of peasant family farming that promotes sustainable farming practices. Secondly, arthropod pests like the cotton boll weevil *Anthonomus grandis* (Boheman), *A. argillacea*, *A.gemmatalis*, *D.saccharalis* and others developed resistance to one or more pesticides, which stimulated a search for alternative management strategies. Finally, consumers increasingly demand products that are grown without application of chemical pesticides. However, despite these stimulating factors, biocontrol has showed a slow adoption by most growers.

Currently, a lot of biocontrol research for control of pests and diseases is taking place at universities, and in research centres of governmental and non-governmental organizations (NGOs). Until now, biocontrol of weeds has not been actively pursued in this country, though herbivorous *Stenopelmus rufinasus* Gyllenhal weevils were collected in Paraguay for control of the red water fern *Azolla filiculoides* Lamarck (Barreto, 2008).

24.3.2 Sampling for and identification of natural enemies of pests in different crops

During the past decades, Paraguayan researchers have been sampling for parasitoids in crops such as cotton, maize, bean, peanut, sesame, soybean and sugarcane. Important advances have been made with the identification of native and exotic natural enemies in Paraguay. The Hymenoptera listed in Table 24.1 play a fundamental role in virtually all crop ecosystems and are of substantial economic importance as biocontrol agents. In addition, the predators of various pests have been sampled and identified and they are presented in Table 24.2. Many of these parasitoids and predators are playing a role in conservation biocontrol progammes.

24.3.3 Use of microbial control agents in biological control of pests

In Paraguay, microbial control agents are applied as formulated biocontrol agents (Table 24.3), but also in a non-formulated way by farmers who collect insects infected by entomopathogens in the field and then reapply them to their crops, causing epizootics in the population of pests (Moscardi, 1983; Resquín-Romero *et al.*, 2016). Use of entomopathogenic fungi in farmers' fields has been reported by governmental and non-governmental institutions (NGOs) such as Alter Vida and the Center of Education, Training and Peasant Technology (Centro de Educación, Capacitación y Tecnología Campesina) (CECTEC) (Peralta *et al.*, 2003).

The *Anticarsia* baculovirus (the nucleopolyhedrovirus AgMNPV) and *B. thuringiensis* (Bt) are used for the biocontrol of key pests in the main agricultural crops in Paraguay, like soybean, cotton, maize and sesame. The presence of *A. gemmatalis* in most of the soybean-producing regions in this country required urgent control measures. The programme initiated during the 1980s was successfully applied in all soybean-producing areas until 1995 and its use was decreased after this period. In soybean trials on farms, application of strain 108 of a granulosis virus was as effective as that of Bt (Mayeregger de Salas, 1991).

B. thuringiensis was introduced along with the baculovirus in Paraguay to control a range of pests in cotton, soybean and maize (Table 24.3).

24.3.4 Use of antagonistic fungi and bacteria in the biological control and management of plant diseases

The reducing capacity of several disease-causing fungi is well known and *Trichoderma* spp. formulations in particular are now used on large areas in Latin America (see e.g. Chapter 6: Brazil). In Paraguay, there are examples of small-scale family farms using compost enriched with microorganisms, including antagonists, for more than 30 decades (Peralta *et al.*, 2003; Nicholls and Resquín-Romero, 2007). Since 2013, studies have been initiated on the biocontrol of two diseases, *Macrophomina phaseolina* (Tassi) and *Fusarium* spp., causing carbonaceous rot and necrotic spotting in soybean, cotton, sesame and other

Table 24.1. Parasitoids identified in major crops in Paraguay (retrieved from Silvie *et al.* (2007a), Nicholls and Resquín-Romero (2007) and updated).

Parasitoid	Host	Crop	References
Apanteles spp.	*Spodoptera frugiperda* *Pectinophora gossypiella*	Cotton Maize	Silvie *et al.*, 2014a Silvie *et al.*, 2014b
Aphelinus albipodus (Hayat and Fatima)	*Aphis gossypii*		Resquín-Romero (2000)
Aphelinus gossypii (Timberlake)	*A. gossypii*		
Aphidius spp.	*Spodoptera* sp.	Bean Cotton	Resquín-Romero 2000 Silvie *et al.*, 2014a.
Aximopsis sp.	*Eutinobothrus brasiliensis*		Silvie *et al.*, 2014b.
Brachymeria subconica Bouček	*A. argillacea*		
Brachymeria aff. compsilurae (Crawford)			
Brachymeria annulata (Fabricius)			
Brachymeria annulata (Fabricius)	*A. argillacea*		
Bracon mellitor Say	*Anthonomus grandis*	Cotton	Silvie *et al.*, 2014a
Bracon sp.	*P. gossypiella;* *Spodoptera* sp. *E. brasiliensis* *A. grandis*	Bean	Silvie *et al.*, 2014b Nicholls and Resquin-Romero, 2007 Resquín-Romero ,2000
Campoletis sp.		Cotton	Silvie *et al.*, 2014a
Catolaccus grandis Burks	*A. grandis*, *E. brasiliensis*		Silvie *et al.*, 2014b
Conura destinata (Walker)	*Lepidopteran*		
Conura immaculata (Cresson	*A. argillacea*		
Conura pulchripes (Cameron)			
Conura fulvovariegata (Cameron)	*A. argillacea*		Silvie *et al.*, 2007a Silvie *et al.*, 2014b
Copidosoma floridanum (Ashmead)	*A. argillacea*, *Pseudoplusia includens*		
Cotesia sp. *Cotesia flavipes* (Cameron)	*Diatraea saccharalis*	Sugarcane Cotton	Arias *et al.*, 2014; Silvie *et al.*, 2014a Silvie *et al.*, 2014b
Chelonus sp.			
Diadegma sp.			Silvie *et al.*, 2014a
Diplazon laetatorius (Fabricius)	*A. argillacea*; *S. frugiperda*	Cotton	Silvie *et al.*, 2014b
Eiphosoma sp., *Parania tricolor* (Szépligeti)			
Dolichogenidea sp.	*Trichoplusia* sp.		
Glyptapanteles sp.			
Epidinocarsis lopezi (De Santis)	*Phenacoccus gossypii*		
Eretmocerus sp.	*Bemisia tabaci*		
Eupelmus cushumani (Crawford)	*A. grandis*, *P. gossypiella*		
Eupelmus cushumani (Crawford)	*E. brasiliensis*		
Euplectrus comstockii (Crawford)	*A. argillacea*		
Glyptapanteles spp.; *Glyptapanteles muesebecki* (Blanchard)	*A. argillacea*	Cotton	Ortiz, 2017; Peña, 2017 Benitez, 2016; Silvie *et al.*, 2007a; Silvie *et al.*, 2014b; Resquín-Romero, 2000

Continued

Table 24.1. Continued.

Parasitoid	Host	Crop	References
Gonatocerus sp.; *Heterospilus annulicornis* Muesebeck	*E. brasiliensis*	Cotton	Silvie *et al.*, 2014a Silvie *et al.*, 2014b
Heterospilus gossypii Muesebeck			
Heterospilus hambletoni Muesebeck			
Homalotylus eytelweinii (Ratzeburg)	Coccinellids		
Hoplognathoca sp.			
Horismenus crassus Hansson	*A. argillacea*		
Horismenus sp.	*A. argillacea*		
Lysiphlebus testaceipes Cresson	*A. gossypii*	Cotton	Silvie *et al.*, 2014a Silvie *et al.*, 2014b
Microcharops sp.		Soybean	Benitez, 2016; Ortiz, 2017; Peña, 2017
Neocatolaccus longiventris (Gahan)	*E. brasiliensis*, Chrysopids *A. exotica*	Cotton	Silvie *et al.*, 2014a Silvie *et al.*, 2014b
Oomyzus sp.	*S. frugiperda*		
Pachyneuron albutium Walker	*P. gossypiella*		
Parania tricolor (Szépligeti)			
Pimpla sp.			
Oomyzus spp.		Cotton	Silvie *et al.*, 2007a
Pachyneuron spp.	*A. argillacea*		
Plagiotrypes sp; *Euderus* sp.; *Copidosoma* sp.; *Aleiodes* sp.; *Exasticolus fuscicornis* (Cameron); *Meteorus laphygmae* (Viereck); *Voria* sp., *Archytas sp.*; *Nemorilla* sp.		Soybean	Ortiz, 2017; Peña, 2017 Benitez, 2016; Montiel, 2015
Prochiloneurus sp.	*A. argillacea*	Cotton	Silvie *et al.*, 2007a
Prochiloneurus sp.	Coccinellids		Silvie *et al.*, 2014a Silvie *et al.*, 2014b
Rhaconotus sp.	*A. argillacea*		
Rhaconotus sp.	*A. argillacea*		Silvie *et al.*, 2007a
Bracon sp.			
Sarcophaga acridiorum Weyenberg *Sarcophaga caridei* Brèthes	*Schistocerca paranensis*		Silvie *et al.*, 2014a Silvie *et al.*, 2014b
Sarcophaga sp.	*Spodoptera eridania*		
Syrphophagus? nigricornis (De Santis)	*Allograpta exotica*		
Syrphophagus aphidivorus (Mayr)	*A. gossypii*		
Syrphophagus sp.	Coccinellids		
Telenomus sp. *Telenomus remus* Nixon	*S. frugiperda*; *A. argillacea,*	Soybean Maize Cotton	Silvie *et al.*, 2014a Silvie *et al.*, 2014b Nicholls and Resquin-Romero, 2007 Silvie *et al.* 2007a Resquín-Romero, 2000; Morales *et al.*, 2000
Theronia lineata (Fabricius)	*A. argillacea*, *P. gossypiella*	Cotton	Silvie *et al.*, 2014a
Triaspis sp.	*E. brasiliensis*		Silvie *et al.*, 2014b

Continued

Table 24.1. Continued.

Parasitoid	Host	Crop	References
Trichogramma pretiosum Riley	*A. argillacea*	Cotton	Silvie *et al.*, 2007a Silvie *et al.*, 2014a Silvie *et al.*, 2014b Benitez, 2000
Trissolcus leviventris (Cameron)	*Edessa* sp.		Silvie *et al.*, 2014a
Trissolcus sp.	*Nezara viridula*	Cotton	Silvie *et al.*, 2014b
Trissolcus basalis Wollaston		Soybean	Resquín-Romero, 2000
Urosigalphus eulechriopis Cushman	*E. brasiliensis*	Cotton	Montiel, 2015 Silvie *et al.*, 2014b
Winthemia sp.	*A. argillacea* *S. eridania*		
Xylocopa sp.	*A. argillacea; Spodoptera* sp.	Cotton Sesame	Resquín-Romero, 2000

Table 24.2. Predators identified in major crops in Paraguay (retrieved from G.A. Resquín-Romero (2000, unpublished results); Silvie *et al.* (2007b) and updated)

Predator	Crop	References
Acontiothespis brevipennis (Saussure)	Cotton	Silvie *et al.*, 2014a.
Acontiothespis concinna (Perty)		
Naemia (Eriopis) connexa (Germar)		
Allograpta exotica (Wiedemann)		
Apiomerus apicalis Burmeister		
Apiomerus lanipes (F.)		
Apiomerus sp.		
Apiomerus sp.	Cotton	Zárate, 2014; Nuñez, 2016 Agüero, 2016; Pereira, 2016 Godoy, 2016; Salinas, 2017
Arilus carinatus (Forster)	Cotton	Silvie *et al.*, 2014a
Aspisoma sp.	Cotton	Silvie *et al.*, 2014a
Atopozelus opsimus Elkins		
Atrachelus cinereus ssp. *crassicornis* (Burmeister)		
Brachygastra lecheguana (Latreille)		
Calosoma sp.	Cotton	Resquín-Romero, 2000
Cicindela sp.	Peanut	
Azya luteipes Mulsant	Maize	
Coccinellina sp.	Bean	
Camponotus senex Smith	Sesame	
Polibia sp.		
Ophion spp.		
Castrida alternans granulatum (Perty)	Cotton	Silvie *et al.*, 2014a; Resquín-Romero, 2000
Chrysoperla sp.		
Coccinephillus sp.	Bean Peanut Maize	Mancuello, 2015 Benítez, 2000
Condylostylus graenicheri (Van Duzee)	Cotton	Silvie *et al.*, 2014a
Condylostylus similis (Aldrich)		
Condylostylus sp.		Zárate, 2014; Nuñez, 2016 Agüero, 2016 Pereira, 2016; Godoy, 2016 Salinas, 2017

Continued

Table 24.2. Continued.

Predator	Crop	References
Condylostylus sp.	Cotton	Silvie *et al.*, 2014a
Cosmoclopius sp. aff. annulosus Stål		
Cycloneda conjugata (Mulsant)		
Cycloneda sanguinea (L.)	Sesame	Silvie *et al.*, 2014a
	Bean	Mancuello, 2015
	Corn	Nicholls and Resquín-Romero, 2007; Resquín-Romero, 2000;
Delphastus argentinicus Nunenmacher	Cotton	Silvie *et al.*, 2014a
Diomus sp. aff. tantillus		
Dissomphalus spp.	Maize	Armoa, 2017; Cárdenas, 2016
Ophion spp.		Cabral *et al.*, 2016; Montiel, 2015; González, 2015 Resquín-Romero, 2000
Doru lineare (Eschsholtz)	Cotton	Silvie *et al.*, 2014a
Doru luteipes sp.		Nuñez, 2016; Agüero, 2016 Pereira, 2016; Godoy, 2016 Salinas, 2017 Zárate, 2014
Eriopis connexa (Germar)	Cotton	Mancuello, 2015
Coleomegilla maculata De Geer		Noda *et al.*, 2002
C. quadrifasciata (Schoenheer)		Resquín-Romero, 2000
Olla v-nigrum (Mulsant)		
Franklinothrips vespiformis (Crawford)	Cotton	Silvie *et al.*, 2014a
Geocoris sp.	Bean	Agüero, 2016; Nuñez, 2016
Nabis sp.	Cotton	Zárate, 2014; Pereira, 2016
Orius sp.		Godoy, 2016; Salinas, 2017
Podisus sp.		Mancuello, 2015
Polistes sp.		Resquín-Romero, 2000
Geocoris ventralis (Fieber)	Cotton	Silvie *et al.*, 2014a
Graptocleptes bicolor (Burmeister)		
Gymnopolybia sp.		
Harmonia axyridis (Pallas)	Cotton	Mancuello, 2015
Hyperaspis sp.		Silvie *et al.*, 2014a
Olla v-nigrum (Mulsant); *Zelus* sp.;		
Hippodamia convergens Guérin-Méneville	Cotton	Silvie *et al.*, 2014a Resquín-Romero, 2000
Hiranetis sp.	Cotton	Silvie *et al.*, 2014a
Hyperaspis (Hyperaspis) festiva Mulsant		
Hyperaspis (Tenuisvalva) notata Mulsant		
Lebia sp.		
Megacephala sp.		
(Coleomegilla maculata De Geer)		
Nephus sp.	Cotton	Silvie *et al.*, 2014a
Notocyrtus dorsalis (Gray)		
Oecanthus sp.		
Olla v-nigrum (Mulsant)		
Orius insidiosus (Say)		
Orthoderella ornata Giglio-Tos		
Phelister rufinotus Marseul		
Phymata fasciata (Gray)		
Phymata sp. aff. fortificata (Herrich-Schäffer)		
Polistes canadensis (Linné)		
Polistes cavapyta Saussure		

Continued

Table 24.2. Continued.

Predator	Crop	References
Polistes versicolor (Olivier)		
Polybia ignobilis (Haliday) = *Polybia atra* Saussure		
Polybia occidentalis (Olivier)		
Polybia paulista Ihering		
Polybia scutellaris (White)		
Polybia sericea (Olivier)		
Pseudodoros clavatus (F.)		
Psyllaephagus blitens Riek		Benitez, 2016
Cryptolaemus montrouzieri Mulsant		
Pullus gilae (Casey)	Cotton	Silvie *et al*., 2014a
Pullus loewii Mulsant		
Pullus spp.		
Repipta sp.		
Arilus sp. aff. cristatus (L.)		
Scymnus sp.		Resquín-Romero, 2000
Paederus sp		Nicholls and Resquín-Romero, 2007
Forficula auricularia L.		
Pseudodorus clavatus (F.)		
Hippodamia convergens Guérin-Méneville		
Scymnus sp.	Cotton	Silvie *et al*., 2014a
Sinea sp.		
Solenopsis sp.		
Sympherobius sp.		
Synoeca cyanea (Fabricius)		
Toxomenus flaralis (F.)		
Toxomerus sp. cf. *watsoni* (Curran)		
Trox suberosus F.		
Zelus armillatus (Lepeletier and Serville)		
Zelus illotus Berg		
Zelus laticornis (Herrich-Schäffer)		
Zelus leucogrammus (Perty)		
Zelus longipes (L.)	Cotton	Silvie *et al*., 2007b
Z. laticornis, Z. ruficeps, Z. armillatus		Resquín-Romero, 2000
Zelus ruficeps Stål		
Zelus spp.		
Zeta argillaceum (L.)		

Table 24.3. Entomopathogenic organisms identified, studied and/or used as biological control agents in crops in Paraguay (retrieved from Nicholls and Resquín-Romero (2007) and updated).

Pathogen	Pest	Crop	References
Viruses			
AgMNPV *Baculovirus anticarsia*)	*Anticarsia gemmatalis*	Soybean	Kliewer and Candia, 1998
Bacteria			
Bacillus thuringiensis	*Alabama argillacea*	Bean, peanut, vorn, cotton	Ferreira Agüero *et al*., 2017; Gomez *et al*., 2017; Ruíz *et al*., 2004
Bacillus thuringiensis var. thuringiensis	Lepidopterans, dipterans	Mize, cotton	Nicholls and Resquín-Romero, 2007
Bacillus subtillis	Coleopterans, dipteran mosquitoes		

Continued

Table 24.3. Continued.

Pathogen	Pest	Crop	References
Fungi			
Entomophthora spp.	Aphids, lepidopterans, coleopterans, orthoptera	Bean, peanut, maize, cotton	Nicholls and Resquín-Romero, 2007
Neozygites fresenii	*Aphis gossypii*	Cotton	Silvie *et al.*, 2014b
Beauveria bassiana	*Lagria villosa*; *Alabama argillacea*; *Anticarsia gemmatalis*; *Spodoptera* spp.; *Anthonomus grandis; Diabrotica speciosa*; *Schistocerca piceifrons piceifrons*	Bean, peanut, corn, cotton, vegetables	Silvie *et al.*, 2014b; Peralta *et al.*, 2003; Resquín-Romero, 2000; Villamayor Morínigo, 1998
Metarhizium anisopliae, *Beauveria bassiana*, *Paecilomyces* sp.,	Lepidopterans, coleopterans, cicadelids, cercopids		Ferreira Aguero *et al.*, 2014; Alborno Jover, 2010; Grabowski Ocampos *et al.*, 2005; Peralta *et al.*, 2003; Resquín-Romero, 2000
Beauveria bassiana and *Metarhizium brunneum*	Endophytic fungi	Cantaloupe	Resquin-Romero *et al.*, 2016, 2017
Paecilomyces sp.	*Bemisia tabaci*, aphids, cicadelids	Citrus	Resquín-Romero *et al.*, 2015, 2016; Peralta *et al.*, 2003 Resquín-Romero, 2000
Nomurea releyi	*Anticarsia gemmatalis*, *Heliothis zea*, *H. virescens* *Trichoplusia ni*	Soybean	Silvie *et al.*, 2014b; Peralta *et al.*, 2003; Resquín-Romero, 2000
Verticilium lecanii	Aphids, white flies	Citrus, tomato, cassava	Resquín-Romero, 2000 Peralta *et al.*, 2003

crops. These two diseases are particularly problematic in rural family farming. The first experiments led by researchers of the Faculty of Agricultural Sciences (Facultad de Ciencias Agrarias, Universidad Nacional de Asunción) (FCA-UNA), co-funded by the National Institute of Biotechnology (Instituto Nacional de Biotecnología) (INBIO), have resulted in biocontrol of *M. phaseolina* and *Fusarium* spp., but the antagonists are not yet formulated for field use (Orrego *et al.*, 2013; Orrego and Franco, 2013). Currently, the Paraguayan Institute of Agricutural Technology (Instituto Paraguayo de Tecnología Agraria) (IPTA), the Faculty of Chemical Sciences (Facultad de Ciencias Químicas, Universidad Nacional de Asunción) (FCQ-UNA) and the FCA-UNA also deal with research on control of the two diseases. The results of the identification of potential biocontrol agents of diseases are summarized in Table 24.4.

To date, several products based on entomopathogens are available on the market. Also, baculoviruses are registered by Paraguay's National Service for Plant and Seed Quality and Health (Servicio Nacional de Calidad y Sanidad Vegetal y de Semillas) (SENAVE, 2018) (Table 24.5).

24.4 Future of Biological Control in Paraguay

The use of biocontrol in Paraguay is currently influenced by the demand for high-quality food and organic produce free of chemical residues. The increase in application of biocontrol will depend on the aid of government institutions to encourage organic production, the demand

Table 24.4. Antagonists of plant diseases identified, studied and/or used as biological control agents in crops in Paraguay (retrieved from Nicholls and Resquín-Romero (2007) and updated).

Pathogen	Disease	Crop	References
Bacteria			
Bacillus subtillis *Pseudomonas fluorescens*	*Fusarium graminearum* *Fusarium* sp. *Macrophomina phaseolina* *Alternaria solani*	Peanut Maize	Duarte Ovejero *et al.*, 2017; Rojas, 2014; Peralta *et al.*, 2003; Silvero, 2017; Resquín-Romero, 2000
Fungi			
Trichoderma spp.	*Macrophomina phaseolina* *Rhizoctonia solani* *Sclerotium rolfsi* *Sclerotinia sclerotiorum* *Fusarium* spp. *Rosellinia* sp.	Sesame Locote Peanut	Ruiz-Díaz *et al.*, 2017; Vera Centurión *et al.*, 2017; Villalba Silvero *et al.*, 2017; Cubilla Rios *et al.*, 2017; Sanabria Velázquez and Grabowski Ocampos, 2016; Galeano Alfonso, 2016; Orrego and Garcete, 2011, 2013; Garcete Gómez and Orrego Fuente, 2011; Manzur Gamarra, 2011; Peralta *et al.*, 2003; Resquín-Romero, 2000

Table 24.5. Biological control agents registered in Paraguay (retrieved from SENAVE, 2018).

Name of product	Active ingredients	Origin
BACTER PROTECT	*Bacillus subtilis*	Argentina
BIO-BAC	*Bacillus lentimorbus*, *B. thuringiensis*, *Brevundimonas vesicularis*, *Cellulomonas flavigena*, *Corynebacterium ammoniagenes*, *Pseudomonas aeruginosa*, *Rhodococcus chubuensis*	USA
BOVEMAX (EC)	*Beauveria bassiana*	Brazil
BT-2X	*Bacillus thuringiensis*	Peru
TRIFESOL 1000	*Trichoderma viride*	Colombia

of consumers for produce with minimal chemical residues, and the efficiency of pest management with biocontrol agents. The Paraguayan universities and several research institutes have trained technicians in the field of biocontrol.

As previously mentioned, biocontrol in soybean with baculoviruses has been used in Paraguay. However, the intensive use of cheap chemical synthetic pesticides replaced the use of biocontrol agents.

Successful use of biocontrol in Paraguay will only be possible with a programme that identifies crops where biocontrol agents can be used and its demand (both of the crop and the biocontrol agent) in the market. For example, biocontrol was successfully used in organic cotton, but due to the lack of demand for this crop, the programme was abandoned. Similarly, *D. saccharalis* was successfully controlled with *C. flavipes* in organic sugarcane production, but this programme was terminated as well. Currently, though, IPTA provides biocontrol agents to producers in the central region of Paraguay.

There is enough experience to apply biocontrol in Paraguay. However, it will only be possible through government institutions, by increasing the production and commercialization of biocontrol agents and the deployment to small-scale producers. Without governmental aid, the use of biocontrol in Paraguay will be very difficult. Currently, biocontrol is used on a limited scale and reliable data about areas under control are not available.

References

(References with grey shading are available as supplementary electronic material)

* All BSc and MSc theses defended at the Universidad Nacional de Asunción, San Lorenzo, Paraguay available at: http://sdi.cnc.una.py/zsgb/cliente.cgi? (accessed 5 October 2018).

Agüero, J.C. (2016) Entomofauna asociada al cultivo de soja (*Glycine max* (L.) Merril) en el Departamento de Alto Paraná [Entomofauna associated with soybean cultivation (*Glycine max*) in the Department of Alto Paraná]. BSc thesis*, Universidad Nacional de Asunción, Paraguay.

Alborno Jover, M.S. (2010) Estudio del efecto in vitro de extractos acuosos de hojas de Neem (*Azadirachta indica* A. Juss.) sobre los hongos entomopatogenos *Beauveria bassiana* (Bals.) Vuill. y *Paeciclomyces* sp. [Study of the effect *in vitro* of aqueous extracts of neem leaves on the entomophagous fungi *Beauveria bassiana* and *Paeciclomyces* sp.]. MSc thesis*, Universidad Nacional de Asunción, Paraguay.

Arias Ruiz Diaz, O.R, Gomez Lopez, V.A, and Tami Haywood S.T. (2014) Liberación de *Cotesia flavipes* (Cameron 1891) (Hymenoptera: Braconidae) para el control de *Diatraea saccharalis* en la región central del Paraguay [Release of *Cotesia flavipes* to control of *Diatraea saccharalis* in central Paraguay]. Abstract XXV, Congresso Brasiliero de Entomologia, 14–18 September 2014, Goiana, Brazil.

Armoa, NK. (2017) Parasitismo natural de larvas de *Spodoptera frugiperda* (Smith) (Lepidóptera; Noctuidae) en cultivo de maíz en el Departamento de Itapúa [Natural parasitism of larvae of *Spodoptera frugiperda* in corn culture in the Department of Itapúa]. MSC thesis*, Universidad Nacional de Asunción, Paraguay.

Barreto, R.W. (2008) Latin American weed biological control science at the crossroads. In: Julien, M.H., Sforza, R., Bon, M.C., Evans, H.C., Hatcher, P.E., Hinz., H.L. and Rector, B.G. (eds) *Proceedings of the XII International Symposium on Biological Control of Weeds*. CAB International, Wallingford, UK, pp. 109–121.

Benítez, E.A. (2000) Levantamiento, identificación, y cría de avispas parasitas (*Trichogramma* spp.) con fines de uso en control biológico de plagas [Surveying, identification, and breeding of parasitic wasps (*Trichogramma* spp.) for the purpose of use in biological pest control]. *Revista de Ciencia y Tecnología de la UNA*. Available at: http://sdi.cnc.una.py/zsgb/cliente.cgi? (accessed 5 October 2018)

Benitez, E.A. (2016) *Nuevas plagas y enemigos naturales registrados para la república del Paraguay [New pests and natural enemies registered in the republic of Paraguay]*. SENAVE/DL-DSVBM, San Lorenzo, Paraguay.

Cabral, C.C., Gómez López V.A., Cardenas S.R. and Armoa, C.C. (2016) Identificación de parasitoides en el cultivo de maíz (*Zea mays*) en Canindeyú e Itapúa [Identification of parasitoids in corn (*Zea mays*) in Canindeyú and Itapúa]. Internal report, Faculty of Agricultural Sciences. Universidad Nacional de Asunción, Paraguay.

Cárdenas, S.R. (2016) Parasitismo en larvas de *Spodoptera frugiperda* (Lepidoptera, Noctuidae) en cultivos de maíz (*Zea mays* L.) en el Departamento de Canindeyú [Parasitism of larvae of *Spodoptera frugiperda* in corn (*Zea mays*) in the Department of Canindeyú]. BSc thesis*, Universidad Nacional de Asunción, Paraguay.

CIA (2017) The World Factbook: Paraguay. Available at: https://www.cia.gov/library/publications/resources/the-world-factbook/geos/pa.html (accessed 5 October 2018).

Country Studies (2018) Paraguay: forestry and fishing. In: Hanratty, D.M. and Meditz, S.W. (eds) *Paraguay: A Country Study*. GPO for Library of Congress, Washington, DC. Available at: http://countrystudies.us/paraguay/47.htm (accessed 5 October 2018).

Cubilla Rios, A.A., Romero Rodríguez, M.C., Flores-Giubi, M.E. and Barúa, J.E. (2017) Evaluación de la producción de moléculas de alto y bajo peso molecular implicadas en la antibiosis de tres especies de *Trichoderma* contra *Macrophomina phaseolina* [Evaluation of the production of high and low molecular weight involved in antibiosis of three species of *Trichoderma* against *Macrophomina phaseolina*]. In: León, E.A.B., Leite, G.M. and González, A.L. (eds) *Libro de resúmens: Congreso Nacional de Ciencias Agrarias (4a.: 2017 Abr. 19–21; San Lorenzo, Paraguay)*. FCA/UNA, San Lorenzo, Paraguay, pp. 1066–1070. Available at: http://www.agr.una.py/fca/index.php/libros/catalog/book/303 (accessed 5 October 2018).

Duarte Ovejero, N.N., Soilán Duarte, L.C. and Grabowski Ocampos, C.J. (2017) Antagonismo de bacterias benéficas sobre hongos fitopatógenos asociados a semillas de maní (*Arachis hipogaea* L.) [Antagonism of beneficial bacteria on phytopathogenic fungi associated with peanut seeds (*Arachis hipogaea*)]. In: León, E.A.B., Leite, G.M. and González, A.L. (eds) *Libro de resúmens: Congreso Nacional de Ciencias Agrarias (4a.: 2017 Abr. 19–21; San Lorenzo, Paraguay)*. FCA/UNA, San Lorenzo, Paraguay pp. 949–952. Available at: http://www.agr.una.py/fca/index.php/libros/catalog/book/303 (accessed 5 October 2018).

Ferreira Aguero, M.A. and Oliveira Janeiro Neves, P.M. (2014) Seletividade de *Metarhizium anisopliae* a *Telenomus podisis* (Hymenoptera: Scelionidae) [Selectivity of *Metarhizium anisopliae* to *Telenomus podisi*]. *Investigación Agrararia* 16, 21–28.

Ferreira Agüero, M.A, Gusman Sosa, E., Benitez, C., Acosta Resquín, M. and Bogarin, C. (2017) Diversidad de lepidópteros plagas en soja transgénica con toxina de *Bacillus thuringiensis* en el Departamento del Amambay [Diversity of lepidopteran pests in transgenic soybean with *Bacillus thuringiensis* toxin in the Department of Amambay]. In: León, E.A.B., Leite, G.M. and González, A.L. (eds) *Libro de resúmens: Congreso Nacional de Ciencias Agrarias (4a.: 2017 Abr. 19–21; San Lorenzo, Paraguay)*. FCA/UNA, San Lorenzo, Paraguay, pp. 885–888. Available at: http://www.agr.una.py/fca/index.php/libros/catalog/book/303 (accessed 5 October 2018).

Galeano Alfonso, C. (2016) *Trichoderma* spp. como agente de biocontrolador de la fusariosis (*Fusarium* spp.) en Vinca (*Catharanthus roseus* (L.) G. Don) [*Trichoderma* spp. biocontrol agent of Fusarium wilt (*Fusarium* spp.) for rosy periwinkle (*Catharanthus roseus*]. MSc thesis*, Universidad Nacional de Asunción, Paraguay.

Garcete Gómez, J.M. and Orrego Fuente, A.L. (2011) Efecto de aislados nativos de *Trichoderma* spp. en la incidencia de *Macrophomina phaseolina* (Tassi) Goid en sésamo (*Sesamum indicum* L.) [Effect of isolated native *Trichoderma* spp. in the incidence of *Macrophomina phaseolina* in sesame (*Sesamum indicum*)]. MSc thesis*, Universidad Nacional de Asunción, Paraguay. Available at: http://www.agr.una.py/revista/index.php/ria/article/view/220 (accessed 5 October 2018).

García, R.F. (1996) El control biológico aplicado en Colombia [Applied biological control in Colombia]. In: Zapater, M. (ed.) *El Control Biológico en América Latina (Biological Control in Latin America)*. IOBC/NTRS, Buenos Aires, Argentina, pp 31–34.

Godoy, R. (2016) Entomofauna asociada al cultivo de la soja (*Glycine max* (L.) Merril) en el Departamento de Caaguazú [Entomofauna associated with the cultivation of soybean (*Glycine max*) in the Department of Caaguazú]. BSc thesis*, Universidad Nacional de Asunción, Paraguay.

Gomez Lopez, V.A., Villalba, G.E., Arias Ruiz Diaz, O.R., Ramírez, M.B. and Gaona, E.F. (2017)Toxicidad de la proteína de Bt expresada en las hojas de diferentes eventos de maíz transgénicos liberados en Paraguay sobre *Spodoptera frugiperda* [Toxicity of the protein of Bt towards *Spodoptera frugiperda* as expressed in leaves of different strains of transgenic maize]. *Revista de la Sociedad Entomológica Argentina* 76, 1–10.

González, A.B. (2015) Parasitismo de larvas de *Spodoptera frugiperda* (J. E. Smith) (Lepidoptera: Noctuidae) en cultivos de maíz en el Departamento de Caaguazú [Parasitism of *Spodoptera frugiperda* larvae in corn in the Department of Caaguazú]. BSc thesis*, Universidad Nacional de Asunción, Paraguay.

Grabowski Ocampos, C.J., Orrego Fuente, A.L. and Stauffer Bonzon, A.D. (2005) Eficiencia de sustratos sobre sobre la esporulación de hongos entomopatógenos *Beauveria bassiana*, *Metarhizium anisopliae* y *Paecilomyces* sp. [Efficacy of substrates on the sporulation of entomopathogenic fungi *Beauveria bassiana*, *Metarhizium anisopliae* and *Paecilomyces* sp.]. *Investigación Agraria* 7, 42–48.

Kliewer, I. and Candia, S. (1998) Control biológico de plagas con *Baculovirus anticarsia*: una solución económica, eficiente y ecológica en el control de la oruga verde de la soja. Ministerio de Agricultura y Ganadería, San Lorenzo (Paraguay) [Biological control of pests with *Baculovirus anticarsia*: an economical, efficient and ecological solution in the control of the green caterpillar of soybean]. Internal report of Proyecto Conservación de Suelos MAG-GTZ. Ministry of Agriculture and Livestock, San Lorenzo, Paraguay.

Mancuello, L. (2015) Uso de vegetales como forma de atrayentes de enemigos naturales en el cultivo de algodón (*Gossypium hirsutum* L.) [Use of vegetables as a way to attract natural enemies in the cultivation of cotton (*Gossypium hirsutum*)]. BSc thesis*, Universidad Nacional de Asunción, Paraguay.

Manzur Gamarra, A. (2011) Control de *Sclerotium rolfsii* con aislados nativos y comerciales de *Trichoderma* sp. *in vitro* [*Sclerotium rolfsii* control, with isolated natives and commercials of *Trichoderma* sp. *in vitro*]. MSc thesis*, Universidad Nacional de Asunción, Paraguay.

Mayeregger de Salas, S. (1991) Producción masal de baculovirus de la poliedrosis nuclear para el control de la *Spodoptera frugiperda* en el maíz [Mass production of a nuclear polyhedrosis baculovirus for the control of *Spodoptera frugiperda* in corn]. MSc thesis*, Universidad Nacional de Asunción, Paraguay.

Montiel, R.I. (2015) Parásitos en *Spodoptera frugiperda* (J. E. Smith) (Lepidoptera: Noctuidae) en el cultivo de maíz en el Departamento de Alto Paraná [Parasitism of *Spodoptera frugiperda* in corn in the Department of Alto Paraná]. BSc thesis*, Universidad Nacional de Asunción, Paraguay.

Morales, S.J., Santos Gallardo, V.J., Vásquez, C. and Ríos, Y. (2000) Patrón de emergencia, longevidad, parasitismo y proporción sexual de *Telenomus remus* (Hymenoptera: Scelionidae) con relación al cogollero del maíz [Pattern of emergence, longevity, parasitism and sex ratio of *Telenomus remus* in relation to the corn budworm]. *Bioagro* 12(2), 47–54.

Moscardi, F. (1983) Utilizacao de *Baculovirus anticarsia* no controle da lagarta da soja, *Anticarsia gemmatalis* [Use of *Baculovirus anticarsia* for control of the soybean caterpillar, *Anticarsia gemmatalis*]. Comunicado Técnico 23, Embrapa-CNPSo, Londrina. Available at: https://www.infoteca.cnptia.embrapa.br/handle/doc/445512 (accessed 5 October 2018).

Nicholls, C.I and Resquín-Romero, G. (2007) Bases Agroecológicas para la Protección de Cultivos [Agroecological basis for crop protection]. Centro de Educación, Capacitación y Tecnología Campesina (CECTEC), Pirapey km 40, Itapúa, Paraguay. Available at: https://es.calameo.com/books/0048754395cb0177db3a3 (accessed 5 October 2018).

Noda, T., Kimura, Y., Ramírez, de López M.B., and Palacio, L.C. (2002) Potential of predatory natural enemies for biological control of sap-sucking insect pests in Paraguay. *Japan Agricultural Research Quarterly* 36(1), 31–35. DOI: 10.6090/jarq.36.31

Nuñez, D.M. (2016) Entomofauna asociada al cultivo de la soja (*Glycine max* (L.) Merril) en el Departamento de San Pedro [Entomofauna associated with the cultivation of soybean (*Glycine max*) in the Department of San Pedro]. BSc thesis*, Universidad Nacional de Asunción, Paraguay.

Orrego, A.L. and Garcete, G. (2011) Efecto de aislados nativos de *Trichoderma* spp. en la incidencia de *Macrophomina phaseolina* (Tassi) Goid en sesamo (*Sesamum indicum* L.) [Effects of native isolates of *Trichoderma* spp. on the incidence of *Macrophomina phaseolina* in sesame (*Sesamum indicum*)]. *Investigación Agraria* 132, 87–93.

Orrego, A.L., Grabowski, C., Rodriguez, H.N. and Soilan, L. (2013) Aislados de *Trichoderma* spp. de diversos Departamentos de Paraguay [Isolates of various species of *Trichoderma* spp. in various Departments of Paraguay]. In: *Trichoderma* spp. Hongo biocontrolador de fitopatógenos. Editora Intelectual, v. 1, Aida Lorenza Orrego Fuente, San Lorenzo, pp. 29–37.

Orrego, A.L. and Franco, B.M. (2013) Compatibilidad *in vitro* de aislados nativos de *Trichoderma* spp. con fungicidas para el tratamiento de semillas [*In vitro* compatibility of native isolates of *Trichoderma* spp. as fungicides for seed treatment-. *Investigación Agraria* 15, 15–22.

Ortiz, N.N. (2017) Parasitismo de larvas de lepidópteros en soja (*Glycine max* (L.) Merril) en el Departamento de Alto Paraná [Parasitism of lepidopteran larvae in soybean (*Glycine max*) in the Departamento of Alto Paraná]. BSc thesis*, Universidad Nacional de Asunción, Paraguay.

Peña, S.M.G. (2017) Parasitismo natural de larvas de lepidópteros en soja *Glycine max* (L.) Merril en el Departamento de Itapúa [Natural parasitism of lepidopteran larvae in soybean *Glycine max* in the Department of Itapúa]. BSc thesis*, Universidad Nacional de Asunción, Paraguay.

Peralta, P., Lezcano, W. and Formoso, T. (2003) Huerta Ecológica Campesina [Peasant Ecological Orchard], Paraguay. Available at: https://es.calameo.com/books/004875439ac25881dc629 (accessed 5 October 2018).

Pereira, M.F. (2016) Entomofauna asociada al cultivo de la soja (*Glycine max*) en el Departamento de Caazapá [Entomofauna associated with soybean (*Glycine max*) in the Department of Caazapá]. BSc thesis*, Universidad Nacional de Asunción, Paraguay.

Resquín-Romero, G. (2000) Evaluación e identificación de los insectos dañinos y los enemigos naturales de importancia económica en la zona del Chaco Central [Evaluation of pests and natural enemies of economic importance in the Central Chaco zone]. Proyecto Estación Experimental Chaco Central (EE.CC) de Cruce de los Pioneros. MAG-GT, Ministerio de Agricultura y Ganadería, Paraguay.

Resquín-Romero, G., Garrido-Jurado, I. and Quesada-Moraga, E. (2015) Combined use of entomopathogenic fungi and their extracts improves the control of *Spodoptera littoralis* (Boisduval) (Lepidoptera: Noctuidae). *Biological Control* 92, 101–110.

Resquín-Romero, G., Garrido-Jurado, I., Delso, C., Ríos-Moreno, A. and Quesada-Moraga, E. (2016) Transient endophytic colonizations of plants improve the outcome of foliar applications of mycoinsecticides against chewing insects. *Journal of Invertebrate Pathology* 136, 23–31. DOI: 10.1016 / j.jip.2016. 03.003

Resquín-Romero, G., Garrido-Jurado, I. and Quesada-Moraga, E. (2017) Morfoanatomía de la colonización de los hongos entomopatógenos endófitos (*Beauveria bassiana* y *Metarhizium brunneum*) en plantas de melón [Morpho-anatomic colonization of entomopathogenic fungi (*Beauveria bassiana* and *Metarhizium brunneum*) in melon plants]. Congreso Nacional de Ciencias Agrarias. In: León, E.A.B., Leite, G.M. and González, A.L. (eds) *Libro de resúmens: Congreso Nacional de Ciencias Agrarias (4a.: 2017 Abr. 19–21; San Lorenzo, Paraguay)*. FCA/UNA, San Lorenzo, Paraguay, pp. 242–245. Available at: http://www.agr.una.py/fca/index.php/libros/catalog/book/303 (accessed 5 October 2018).

Rojas, J. (2014) Control biológico del tizón temprano (*Alternaria solani* Sorauer) del tomate con cepas de *Bacillus* sp. y *Pseudomonas fluorescens* [Biological control of early blight (*Alternaria solani*) of tomato with *Bacillus* sp. and *Pseudomonas fluorescens* strains]. BSc thesis*, Universidad Nacional Asunción, Paraguay.

Ruíz, E., Cabral, C. and Pino, C. (2004) Eficiencia de *Bacillus thuringiensis* Línea HD-I en el control de *Spodoptera frugiperda* (Smith), Lepidóptera: Noctuidae en condiciones de campo y de laboratorio en el cultivo de maíz dulce *Zea mays saccharata* (en línea) [Efficiency of *Bacillus thuringiensis* HD-I line in the control of *Spodoptera frugiperda* under field and laboratory conditions in the cultivation of sweet corn *Zea mays*]. *Investigación Agraria* 6, 10–14. Available at: http://www.agr.una.py/revista/index.php/ria/article/view/198/1 (accessed 5 October 2018).

Ruiz-Díaz, D., Flores-Giubi, M.E. and Barúa, J.E. (2017) Characterization of species of *Trichoderma* spp. in its efficacy for the biological control of native isolates of *Macrophomina phaseolina.* Congreso Nacional de Ciencias Agrarias. In: León, E.A.B., Leite, G.M. and González, A.L. (eds) *Libro de resúmens: Congreso Nacional de Ciencias Agrarias (4a.: 2017 Abr. 19–21; San Lorenzo, Paraguay)*. FCA/UNA, San Lorenzo, Paraguay, pp. 976–980. Available at: http://www.agr.una.py/fca/index.php/libros/catalog/book/303 (accessed 5 October 2018).

Salinas, M.A. (2017) Entomofauna asociada al cultivo de soja (*Glycine max* (L.) Merril) en el Departamento de Itapúa [Entomofauna associated with soybean (*Glycine max*) in the Department of Itapúa]. BSc thesis*, Universidad Nacional de Asunción, Paraguay.

Sanabria Velázquez, A.D. and Grabowski Ocampos, C.J. (2016) Control biológico de *Rosellinia* sp. causante de la muerte súbita en macadamia (*Macadamia integrifolia*) con aislados de *Trichoderma* spp. [Biological control of *Rosellinia* sp. causing sudden death of macadamia (*Macadamia integrifolia*) with isolates of *Trichoderma* spp.]. MSc thesis*, Universidad Nacional de Asunción, Paraguay.

SENAVE (Servicio Nacional de Calidad y Sanidad Vegetal y de Semillas) (2018) Producto agropecuario [Agricultural products]. Available at: http://secure.senave.gov.py:8443/registros/servlet/prod_agro (accessed 5 October 2018).

Silvero, S. (2017) Control bilógico de *Fusarium graminearum* causante de la pudrición en maíz con cepas de *Bacillus* sp. y *Pseudomonas fluorescens* [Biological control of corn rot caused by *Fusarium graminearum* with strains of *Bacillus* sp. and *Pseudomonas fluorescens*]. BSc thesis*, Universidad Nacional de Asunción, Paraguay.

Silvie, P.J., Delvare, G., Aberlenc, H-P., Cardozo, R. and Gomez, V. (2007a) Novos parasitóides das pragas do algodoeiro recem identificados no Paraguai [New cotton pest parasitoids recently identified in Paraguay]. In: X Simpósio de Controle Biológico, 30 June–4 July 2007, Brasilia, Brazil. DF ID – 024. Available at: https://www.researchgate.net/publication/272744046_X (accessed 5 October 2018).

Silvie, P., Aberlenc, H-P., Duverger, C., Bérenger, J.M., Cardozo, R. and Gómez, V. (2007b) *Harmonia axyridis* no Paraguai e novos predadores identificados no cultivo do algodoeiro [*Harmonia axyridis* in Paraguay and new predators identified in cotton]. In: X Simposio de controle biologico, 30 June–4 July 2007, Brasilia, Brazil.

Silvie, P., Gomez, V.A., Aberlenc, H-P., Bonfils, J., Cardozo, R. and Michel, B. (2014a) Diversité des Arthropodes rencontrés en culture cotonnière au Paraguay. 1. Insectes et acariens phytophages [Diversity of arthropods found in cotton in Paraguay. 1. Phytophagous insects and mites]. *Entomologie faunistique* [*Faunistic Entomology*] 67, 165–178.

Silvie, P.J., Delvare, G., Aberlenc, H.P., Prudent, P., Gil-Santana, H., Gomez, V.A., Cardozo, R. and Michel, B. (2014b) Diversité des Arthropodes rencontrés en culture cotonnière au Paraguay. 2. Insectes prédateurs, parasitoïdes et hyperparasitoïdes [Diversity of arthropods found in cotton in Paraguay. 2. Predators, parasitoids and hyperparasitoids]. *Entomologie Faunistique* [*Faunistic Entomology*] 67, 179–191.

Vera Centurión, A.E., Flores-Giubi, M.E. and Barua, J.E. (2017) Antagonistic capacity of native paraguayan isolates of *Trichoderma* spp. against capsicum pathogens (*Capsicum annuum* var Natalie): *Rhizoctonia solani* and *Sclerotinia sclerotiorum*. In: León, E.A.B., Leite, G.M. and González, A.L.

(eds) *Libro de resúmens: Congreso Nacional de Ciencias Agrarias (4a.: 2017 Abr. 19–21; San Lorenzo, Paraguay)*. FCA/UNA, San Lorenzo, Paraguay, pp. 985–988. Available at: http://www.agr.una.py/fca/index.php/libros/catalog/book/303 (accessed 5 October 2018).

Villalba Silvero, F.A., Flores-Giubi, M.E. and Barúa, J.E. (2017) Antagonist capacity of native Paraguayan isolates of *Trichoderma* spp. against *Macrophomina phaseolina* isolated from soybean (*Glycine max*) and sesame (*Sesamum indicum* L.). In: León, E.A.B., Leite, G.M. and González, A.L. (eds) *Libro de resúmens: Congreso Nacional de Ciencias Agrarias (4a.: 2017 Abr. 19–21; San Lorenzo, Paraguay)*. FCA/UNA, San Lorenzo, Paraguay, pp. 1018–1021. Available at: http://www.agr.una.py/fca/index.php/libros/catalog/book/303 (accessed 5 October 2018).

Villamayor Morínigo, C.J. (1998) Evaluación de la patogenicidad de *Beauveria bassiana* sobre huevos de la polilla del tomate (*Tuta absoluta*) (Meyrick, 1917) (Lepidoptera: Gelechiidae) [Assessment of the pathogenicity of *Beauveria bassiana* on eggs of the tomato borer (*Tuta absoluta*)]. MSc thesis*, Universidad Nacional de Asunción, Paraguay.

Wikipedia (2018) Paraguay. Available at: https://en.wikipedia.org/wiki/Paraguay (accessed 6 October 2018).

Zarate, A.M. (2014) Entomofauna asociada al cultivo de la soja (*Glycine max* (L.) Merril) en el Departamento de Alto Paraná [Entomofauna associated with soja (*Glycine max*) in the Department of Alto Paraná]. BSc thesis*, Universidad Nacional de Asunción, Paraguay.

25 Biological Control in Peru

Norma Mujica[1]* and Mary Whu[2]

[1]*Departamento de Entomología, Universidad Nacional Agraria La Molina, Lima, Peru;* [2]*Asociación de Control Biológico del Perú*

* E-mail: n.mujica@lamolina.edu.pe

Abstract

In Peru, the first introduction of six species of beneficials occurred in 1904 for control of cotton white scale. By 1969, more than 20 other beneficial species had been imported for control of scales, aphids and lepidopterans in important crops. Many species established and still control pests today. In 1961, the Center for Introduction and Rearing of Useful Insects was created for biocontrol of the most economically important pests. Identification of parasitoids and predators, study of their biology and development of mass-production methods were executed to optimize their use. From 1979, research concentrated on the behaviour of parasitoids and predators of mealybugs, scales and lepidopteran pests. Also, collection and investigation of entomopathogenic fungi was initiated. In the late 1990s, a National Program for Biological Control was created to intensify biocontrol in important crops through training of professionals and promotion and sale of biocontrol agents to farmers. Biocontrol agents were supplied to different agricultural valleys via a network of production laboratories, which reared 42 species of biocontrol agents to control pests in more than 45 crops. The area under augmentative biocontrol increased from 10,000 ha to 253,000 ha in a 6-year period. By 2015, 177 exotic species of biocontrol agents had been introduced into Peru and 53 of these introductions resulted in complete or substantial classical or augmentative control of the pest. In Peru, the use of biocontrol agents has grown due to the increased demand by producers for safe and healthy products for internal and external markets, by consumers for products free of pesticide residues and by the government to achieve greater sustainability in agricultural production.

25.1 Introduction

Peru has an estimated population of slightly more than 31 million (July 2017) and its main agricultural products are artichokes, asparagus, avocados, blueberries, coffee, cocoa, cotton, sugarcane, rice, potatoes, maize, plantains, grapes, oranges, pineapples, guavas, bananas, apples, lemons, pears, coca, tomatoes, mangoes, barley, medicinal plants, quinoa, palm oil, marigolds, onion, rubber, wheat, dry beans, poultry, beef, pork, dairy products, guinea pigs and fish (MINAGRI, 2015; INEI, 2018). According to Gonzales *et al.* (2017, pp. 472–490):

> Peru is one of the ten countries in the world with greatest mega-diversity ... The population uses approximately 5,000 of the country's 25,000 plant species (10% of the world total), of which at least 30% are endemic, for a variety of purposes: food (782); medicine (1,400); decoration (1,608), timber and construction (618); fodder (483) and dyes and coloring (134) ... Farmers in ancient Peru domesticated 25 species of edible roots and tubers. These crops can be of global significance such as the potato ... Pulses include the lupin ... while grains include quinoa ... and amaranth ... both with a great nutritional value ... Despite having such a vast territory and being one of the centers of origin of cultivated plants ... Peru's agricultural potential is reduced to 5.9% of the country's total area ... the area under irrigation represents 2.6 million ha (36.2%) of a total of 7.1 million ha, while 4.5 million are rainfed (63.8%) ...The agricultural sector accounts for 25% of the economically active population ... Since the 1990s, a greater impetus has been placed on the development of two types of agriculture. The first is agriculture with an export potential that still needs more state support to create technology and reach the investment levels required for the development of amaranth, canihua, tarwi, tara, heart of palm, inchi sacha, yacon, camu camu and maca. The second is non-traditional export agriculture that uses high technology and has high investment levels because of its access to credit, enabling it to develop crops such as asparagus, paprika, citrus, artichoke, mango, among others ... Main export products include asparagus, coffee, mango, olives, fresh grapes, fresh avocado, artichokes, dried peppers, beans, mandarin, ginger, beans, onion, fresh peas, quinoa, maca and blueberries ... Peru is a land of forests. A total of 57.3% of its territory is covered by this resource and by 2014 there were only 16.8 million ha of permanent production forests, 4.3 million ha of which are under management plans ... (Livestock production concerns cattle, pigs, sheep and poultry; editors) ... cattle, sheep and pigs predominate in the mountains and birds on the coast...Peru has an extensive coastline and abundant maritime resources ... Anchovy accounted for 79.9%, while the remaining species included horse mackerel, mahi mahi, mackerel and bonito ... Trout ... is raised at fish farms in the Peruvian highlands and is an important protein source for High Andean populations ... In the forest, there are fish farms that breed paiche (*Arapaima gigas* Cuvier), an emblematic fish from the Amazon measuring an average of 2.5 m and weighing 250 kg.

25.2 History of Biological Control in Peru

25.2.1 Period 1880–1969

In Peru, the first introductions of natural enemies occurred in 1904 and concerned *Aphytis diaspidis* (How), *Aphytis fuscipennis* (How), *Prospaltella berlessi* How, *Aspidiotiphagus citrinus* (Crwf), *Arrhenophagus chionaspidis* Auriv. and *Scymnus* sp. for control of the cotton white scale *Pinnaspis strachani* Ferris and Rao. These introductions were carried out by C. Towsend of the Experimental Agricultural Station of Piura in northern Peru (Wille, 1952).

During the period 1922–1969 several biocontrol successes were recorded with the introduction of various parasitoids and predators for the control of different pests: *Aphelinus mali* (Hald) for control of woolly apple aphid (*Eriosoma lanigerum* (Hausm.)); *Rodolia cardinalis* (Mulsant) for control of cottony cushion scale (*Icerya purchasi* Mskell); *Metaphycus lounsburyi* How, *Scutellista cyanea* Motsch and *Lecanobius utilis* Compere for control of olive black scale (*Saissetia oleae* Oliver); *Hippodamia convergens* Guam. for control of *Toxoptera* spp.; *Aphytis lepidosaphes* Compere for control of purple scale (*Lepidoshapes beckii* Newm.); *Metaphycus helvolus* Comp. for control of coffee hemispheric scale (*Saissetia* spp.); and *Aphytis holoxantus* DeBach for control of Florida red scale (*Chrysomphalus ficus* Ashmead) (Aguilar, 1980; Beingolea, 1990). In addition, many cases of efficient natural control exerted by species introduced involuntarily, or by native species, were achieved.

Until 1960, classical biocontrol was implemented based on a few isolated introductions. In 1961, the Center for Introduction and Rearing of Useful Insects (Centro de Introducción y Cría des Insectos Utiles) (CICIU) was created as a project within the Sub-Directorate of Plant Protection of the Directorate of Agricultural Inspection and Control of the Ministry of Agriculture, with the purpose of carrying out quarantine and introduction of exotic biocontrol agents. This formed the start of the institutionalized stage of biocontrol in Peru (Pacora, 1979; Valdivieso, 1991) with the first activities in studying, mass rearing and using beneficial insects. During the 1960s, biocontrol reached great importance in industrial crops, with releases on an average of 146,000 ha of cotton and on 144,000 ha of sugarcane. In the same decade, CICIU executed classic biocontrol of the main pests in citrus, sugarcane, cotton, alfalfa and olive (Risco, 1958, 1960, 1961; Gonzales, 1968; Herrera, 2010). Table 25.1 summarizes some of the projects executed in this period

25.2.2 Period 1970–2000

In 1975, CICIU became a supporting organ for the General Directorate of Production of the Ministry of Food, with the purpose of introducing, rearing and field releasing of predators, parasitoids and other insects useful for pest control of cultivated plants, and to develop integrated pest control systems. In 1979, by Decree Law No. 22431, CICIU became part of the National Institute of Agricultural Research (Instituto Nacional de Investigaciones Agropecuarias) (INIA), with emphasis on research on the behaviour of parasitoids and predators of mealybugs, scales and lepidopteran pests (Pacora, 1979; Valdivieso, 1991). Also, collection and investigation of entomopathogenic fungi was started, and inundative releases of *Trichogramma exiguum* Pinto & Platner for the control of *Argyrotaenia sphaleropa* (Meyrick) were implemented.

In 1995, CICIU became part of the National Service of Agrarian Health (Servicio Nacional de Sanidad y Calidad Agroalimentaria) (SENASA), and in 1998 the National Program of Biological Control (Programa Nacional de Control Biológico) (PNCB) was created with the objective of intensifying biocontrol in crops of economic importance through the training of 44 new professionals and the promotion and sale of biocontrol agents to farmers (Whu, 2016; SENASA, 2018). To meet the demand for biocontrol agents, the creation of private production laboratories in agreement with SENASA was encouraged, providing them with material, equipment and initial stock rearings. In addition, they were trained in management of mass rearings with continuous supervision and quality control of the biocontrol agents produced. During the 6 years of the programme, application of biocontrol increased from 10,000 ha per year to 253,000 ha per year. Financial support by the Development Program of Agricultural Health

Table 25.1. Biological control in Peru during the period 1880–1969.

Crop	Pest	Natural enemy	Type of biocontrol[a]	Area (ha) under biocontrol
Citrus	Cottony cushion scale *I. purchasi*	*Rodolia cardinalis*	CBC	? but still active
Citrus	Purple scale *L. beckii*	*Aphytis lepidosaphus*	CBC	? but still active
Citrus	Florida red scale *C. aonidum*	*Aphytis holoxantus*	CBC	? but still active
Citrus	Citrus aphid *Toxoptera* sp.	*Hippodamia convergens*	CBC	? but still active
Apple	Woolly apple aphid *E. lanigera*	*Aphelinus mali*	CBC	? but still active
Olive	Olive black scale, *S. oleae*	*Metaphycus lounsburyi*	CBC	? but still active
Olive	Olive black scale, *S. oleae*	*Scutellista cyanea*	CBC	? but still active
Olive	Olive black scale, *S. oleae*	*Lecanobius utilis*	CBC	? but still active
Olive	Coffee hemispheric scale, *S. coffeae*	*Metaphycus helvolus*	CBC	? but still active
Olive	Coffee hemispheric scale, *S. coffeae*	*Coccophagus rusti*	CBC	? but still active
Cotton	Tobacco budworm, *H. virescens*	*Trichogramma* sp.	ABC	146,000 1968
Cotton	Cotton aphid, *A. gossypii*	*Hippodamia convergens*	ABC	estim.
Sugarcane	Sugarcane borer, *D. saccharalis*	*Paratheresia claripalpis*[b]	ABC	44,000 1968
Sugarcane	Sugarcane borer, *D. saccharalis*	*Trichogramma minutum*	ABC	estim.

[a]Type of biocontrol: ABC = augmentive biocontrol, CBC = classical biocontrol
[b]Native parasitoid

(PRODESA) allowed actions to reduce impacts of pests on agricultural production through integrated pest management (IPM), to improve crop health and the health of Peruvian producers and consumers as a result of lower amounts of pesticides used. After this 6-year programme, demands for biocontrol agents continued and resulted in natural enemy production laboratories that are still functioning today.

Table 25.2 presents a list of biocontrol agents introduced to Peru in the period 1970–2000 (Pacora, 1979; Whu, 1987, 2016; Beingolea, 1990; SENASA, 2002). An example of a successful introduction is *Aphytis roseni* DeBach from Uganda in 1971 for control of *Selenaspidus articulatus* Morgan. This parasitoid has established in all places where it was released and reached levels of control of almost 100%. Until then, Peru had been the only country in the Americas where *A. roseni* was used in biocontrol. Like many other beneficial insects, it is very susceptible to chemicals, running the risk of extinction.

Aphidius smithi Sharma & Subba Rao was introduced from Chile in 1973 for the control of alfalfa green aphid (*Acyrthosiphon pisum* Harris), a very severe problem that causes foliage loss. It was easily multiplied in the laboratory and established in Lima, from where it was sent to all alfalfa-producing areas, with satisfactory control results. After introduction to a new area, great numbers of mummies can easily be seen adhering to the leaves, which makes harvesting of parasitoids easy and helps in their transfer to colonize new areas.

Cales noacki Howard was introduced from California between 1974 and 1975 for the control of *Aleurothrixus floccosus* Mask, the 'woolly whitefly of citrus', a very important pest in citrus. It was multiplied on citrus seedlings, introduced and successfully established in all citrus zones of the country.

The polyembryonic parasitoid *Ageniaspis citricola* Logvinovskaya was introduced from Florida in 1996–1997 for the control of the citrus leaf miner *Phyllocnistis citrella* Stainton, a pest that entered the country in 1995. A successful mass rearing and release programme was established.

Psyllaephagus pilosus Noyes was introduced from the USA in 2000 for the control of the blue

Table 25.2. List of biological control agents introduced into Peru in the period 1970–2000.

Exotic species introduced	Target pest	Type of biocontrol[a]	Country of origin	Year
Coccophagus rusti Compere*	*Saissetia* spp.	CBC	USA	1970
Anagyrus saccharicola Timb.*	*Saccaricoccus sacchari* Cock	CBC	Hawai	1970
Lindorus lophantae (Blaisd)	*Selenaspidus articulatus* Morgan	CBC	Trinidad	1971
Rhyzobius purchellus Montrozier	*Selenaspidus articulatus* Morgan	CBC	Trinidad	1971
Aphytis roseni De Bach*	*Selenaspidus articulatus* Morgan	CBC	Uganda	1971
Lindorus lophantae (Blaisd)	*Selenaspidus articulatus* Morgan	CBC	CIBC[b]	1971
Aphydius smithi Sher & Rao*	*Acyrthosiphon pisum* (Harris)	CBC	Chile	1973
Encyrtus lecaniorum (Mayr)	*Saissetia oleae* Bm	CBC	USA	1974
Cales noacki Howard*	*Aleurothrixus floccosus* (Maskell)	CBC	USA	1974–75
Trichogramma pintoi Voegele*	*Diatraea saccharalis* (Fabricius)	ABC	Trinidad	1975
Apanteles flavipes Cameron*	*Diatraea saccharalis* (Fabricius)	ABC	CIBC[b]	1975
Telenomus remus Nixon	*Spodoptera frugiperda* (J.E.Smith)	ABC	Trinidad	1975–84
Aphytis melinus De Bach	*Aspidiotus hederae* (Vallot)	CBC	USA	1977–2001
Aphytis melinus De Bach	*Chrysomphalus dictiospermi* (Morgan)	CBC	USA	1977–2001
Diachasmimorpha longicaudata (Asmead)	*Ceratitis capitata* (Wied)	ABC	Costa Rica	1978–86
Pachycrepoideus vindemiae Rondani	*Ceratitis capitata* (Wied)	ABC	Costa Rica	1978
Trichogramma brasiliensis (Ashm.)	*Alabama argillacea* Hübner	ABC	Brazil	1983
Trichogramma pretiosum Riley*	*Heliothis* spp.	ABC	USA	1984
Coccidophilus citricola Brethes	*Quadraspidiotus permiciosus* Cosmtock	CBC	Chile	1984–85
Trichogramma fuentesi Torres*	*Pectinophora gossypiella* Saunders	ABC	Mexico	1985
Spalangia endius Wlk	*Musca domestica* (Linnaeus)	ABC	Chile	1986
Elaeidobius kamerunicus Faust.*	Oil palm pollinator	ABC	Colombia	1987
Elaeidobius plagiatus (Fabricius)	Oil palm pollinator	ABC	Colombia	1987
Elaeidobius singularis (Faust)	Oil palm pollinator	ABC	Colombia	1987
Bracon kirkpatricki (Wilkonson)	*Pectinophora gossypiella* Saunders	ABC	Paraguay	1989
Muscidifurax raptorellus G & S	*Musca domestica* (Linnaeus)	ABC	USA	1991
Muscidifurax zaraptor G & S	*Musca domestica* (Linnaeus)	ABC	USA	1991
Trichogrammatoidea bactrae Nagaraja	*Pectinophora gossypiella* Saunders	ABC	USA	1991–92

Continued

Table 25.2. Continued.

Exotic species introduced	Target pest	Type of biocontrol[a]	Country of origin	Year
Trichogramma atopovirilia Oatman y Platner	*Spodoptera frugiperda* (J.E.Smith)	ABC	Colombia	1994
Ageniaspis citricola Logvinovskaya*	*Phyllocnistis citrella* Stainton	ABC	USA	1996–97
Verticillium lecanii (Zimm.)	Several	ABC	Cuba	1998
Beauveria bassiana	Several	ABC	Cuba	1998
Trichoderma harzianum (cepa nrrl-13019)*	Antagonist of phytophagous fungi	ABC	USA	1999
Trichoderma viride (cepanrrl-6418)*	Antagonist of phytophagous fungi	ABC	USA	1999
Trichoderma virens (J.H. Miller, Giddens & A.A. Foster) Arx*	*Fusarium*, *Moniliasis*	ABC	USA	1999
Trichoderma stromaticum Samuels & Pardo-Schulth.	*Moniliasis*	ABC	Brazil	1999
Encarsia formosa Gahan	*Trialeurodes vaporariorum* (Westwood)	ABC	Canada	1999
Metarhizium anisopliae var acridum (Green Muscle)	*Aleurodidae*	ABC	Brazil	2000
Heterorhabditis bacteriophora Poinar	*Spodoptera*, different Coleoptera	ABC	Venezuela	2000
Psyllaephagus pilosus Noyes	*Ctenarytaina eucalypti* (Maskell)	ABC	USA	2000

*Species that became established and did control the pest (species without * may have become established but did not successfully control the pest)
[a]Type of biocontrol: ABC = augmentative biocontrol, CBC = classical biocontrol
[b]Commowealth Institute of Biological Control (now CABI)

psyllid *Ctenarytaina eucalypti* Maskell, a serious pest in eucalyptus. This biocontrol agent originates from Australia and was introduced into the USA by Dr D. Dahlsten, who donated a small colony, with which a rearing on eucalyptus seedlings was initiated. After many difficulties (it took two years to obtain the biological material, keeping infested eucalyptus plants under laboratory conditions and the recovery of only six pairs of living individuals after shipment) it was possible to establish rearing and start releasing parasitoids. In less than six months, parasitoids had established in all areas where the blue psyllid was a problem, and control was satisfactory.

25.3 Current Situation of Biological Control in Peru

In 2005, the PNCB changed into the Subdirección de Control Biológico (SCB) (Sub-direction of Biological Control) from SENASA and was responsible for the development and application of biocontrol in the main crops and agricultural valleys of Peru. In addition, it was commissioned to promote the reduction of pesticides, toxic waste in food and environment pollution and to protect the health of the farmers (SENASA, 2005). In order to achieve these objectives, it has five units: (i) a Quarantine Unit, responsible for introducing exotic beneficial species, establish their rearing and introduction into the field; (ii) a Useful Insects Unit and (iii) a Beneficial Microorganisms Unit, responsible for the production of beneficial organisms and for the maintenance and conservation of rearing stock and strains for distribution to laboratories upon request; (iv) a Training Unit, responsible for programming, coordinating and executing training courses on biological pest control; and (v) an Integrated Pest Management Unit, which is responsible for conducting basic studies and trials for application of biocontrol agents. The biocontrol activities are carried out in most regions of Peru, corresponding to the 25 Executive Directorates of SENASA (SENASA, 2005; Whu, 2016).

By 2015, 177 exotic species of biocontrol agents had been registered in Peru and 53 of these introductions resulted in complete or substantial control of the pest. Some of the most successful introductions in the period 2001–2010 are summarized below and in Table 25.3.

Orius insidiosus Say, introduced from the USA in 2001, is an excellent predator of thrips, aphids, whiteflies, spiders and the larvae and eggs of lepidopterans. It is easily reared on eggs of *Sitrotoga cerealella* Oliver and *Ephestia kuehniella* Zeller. *Telenomus alsophilae* Viereck, an egg parasitoid for control of *Oxydia vesulia* Cramer in avocado, was introduced from Colombia in 2002. Mass rearing was developed and the parasitoid was introduced to avocado plantations by the Camposol company, resulting in sufficient control. It is a parasitoid with a long lifespan and a wide range of action in fruit trees. *Citrostichus phyllocnistoides* Narayan was introduced from Spain in 2005, through the association of citrus farmers (PROCITRUS), for control of *P. citrella* and appeared to be more effective than *A. citricola* for control of the citrus leaf miner. The predatory mites *Euseius victoriensis* Womersly, *E. stipulates* (Athias-Henriot) and *Amblyseius largoensis* (Muma) were introduced from Australia (2005), Spain (2006) and Cuba (2007), respectively, for the control of the phytophagous mites *Panonychus citri* (McGregor), *Phyllocoptruta oleivora* (Ashmead) and *Polyphagotarsonemus latus* (Banks). The SCB carried out the mass production of these predators and they were successfully transferred to citrus orchards. Currently there are 15 laboratories producing predatory mites (SENASA, 2005, 2015, 2018; Whu, 2016). *Orgilus lepidus* (Muesebeck) was introduced from Australia in 2007 for control of *Phthorimaea operculella* (Zeller); and *Anagyrus pseudococci* (Girault) from Israel (2009) for control of *Planococcus citri* Risso.

25.3.1 Augmentative biological control

Production at the Central Laboratory of the SCB

The units of 'Useful Insects' and 'Beneficial Microorganisms' of SCB are responsible for optimal mass production of biocontrol agents. The Useful Insects unit maintains stock rearings of 21 species of parasitoids, with ten species of the genus *Trichogramma* (*T. pretiosum* Riley, *T. exiguum*, *T. pintoi* Voegelé, *T. galloi* Zucchi, *T. cacoeciae* Marchal, *T. lopezandinensis* Sarmiento, *T. marandobai* Brun, Moraes e Soares, *T. dendrolimi* Matsumura, *T. brassicae* Bezdenko and *T. nerudai* (Pintureau &

Table 25.3. List of biological control agents introduced into Peru in the period 2001–2010; all species were imported for augmentative biological control projects.

Exotic species introduced	Target pest	Country of origin	Year
Orius insidiosus[a]	Thrips, aphids, whiteflies, lepidopteran eggs	USA	2001
Trichoderma stromaticum	*Crinipellis perniciosa*	Brazil	2001
Trichoderma virens	*Moniliophtora roreri*	USA	2001
Cladobotrium amazonense	*Moniliophtora roreri*	USA	2001
Aphyitis melinus	*Chrysomphalus diptyospermi*	USA	2001
Erectmocerus eremicus	*Bemisia argentifolii*	USA	2001
Phymasticus coffeae	*Hypothenemus hampei*	Colombia	2001
Trichogramma lopezandinensis	*Copitarsia turbata*	Colombia	2002
Telenomus alsphilae	*Oxydia versulia*	Colombia	2002
Citrostichus phyllocnistoides[a]	*Phyllocnictis citrella*	Spain	2005
Euseius victoriensis	*Panonychus citri*	Australia	2005
Euseius stipulatus	*Phyllocoptruta oleivora*	Spain	2006
Amblyseius largoensis	*Polyphagotarsonemus latus*	Cuba	2007
Orgilus lepidus	*Phthorimaea operculella*	Australia	2007
Anagyrus pseudococci	*Planacoccus citri*	Israel	2009

[a]Species that provided effective pest control

Gerding)), *Trichogrammatoidea bactrae* Nagaraja, *Telenomus remus* Nixon, *Coccophagus rusti* Compere, *Leptomastidea abnormis* (Girault), *Coccidoxenoides peregrinus* (Timberlake), *A. pseudococci, Muscidifurax raptorellus* Kogan & Legner, *Spalangia endius* (Walker), *Pachycrepoides vindenmiae* Rondani and *Tetrastichus howardi* (Olliff). Also, 14 species of predators are reared: *O. insidiosus, Chrysoperla externa* (Hagen), *C. asoralis* (Banks), *C. carnea* (Stephens), *Ceraeochrysa cincta* (Schneider), *Sympherobius barberi* (Banks), *Cryptolaemus montrouzieri* Mulsant, *Metacanthus tenellus* Stal, *Geocoris callosullus, Podisus nigrispinus* (Dallas), *Hyperaspis onerata* (Mulsant), *Euseius stipulatus* (Athias-Henriot), *A. largoensis* and *Amblyseius chungas* Denmark & Muma.

The Beneficial Microorganisms unit has two laboratories where the strains of entomopathogenic and antagonistic organisms are maintained, to be tested for control of various pests and diseases (SENASA 2015, 2016a, b). The entomopathogenic laboratory maintains 150 strains of fungi of the genera *Beauveria, Isaria, Hirsutella, Lecanicillium, Pochonia, Metharizium* and *Purpureocillium* (which are used for control of several pests listed in Table 25.4), six isolates of nuclear polyhedrosis viruses (NPVs) (SpeVPN *Spodoptera eridania*, SpfVPN *S. frugiperda*, SpocVPN *S. ochrea*, DijnVPN *Dione juno*, CodVPN *Copitarsia decolora* and EupeVPN *Euprosterna eleasa*) and four granulosis viruses (PoGV *Phthorimaea operculella*, EreVG *Erinnys ello*, SpeVG *Spodoptera eridania* and SpfVG *S. frugiperda*). Entomopathogenic nematodes are represented by two strains of *Heterorhabditis bacteriophora* Poinar and one strain of the genus *Steinernema*, used mainly for the control of soil pests. In the antagonist laboratory, 178 isolates of *Trichoderma* spp., four of *Clonostachys rosea* (Link) Schroers and one species of *Cladobotryum amazonensis* Bastos, Evans & Samson are maintained. Diseases controlled by the antagonistic fungi of the genus *Trichoderma* are presented in Table 25.5.

Table 25.4. List of entomopathogenic fungi maintained in the Central Biological Control Laboratory and the pests they control in various crops in Peru.

Fungus	Strains	Main target pests	Crops
Beauveria bassiana	104	*Hypothenemus hampei, Cosmopolites sordidus, Metamasius hemipterus, Orthezia olivicola, Hylesinus oleiperda, Macrosiphon euphorbiae, Anomala* spp.	Coffee, banana, crucifers, cotton, alfalfa, pasture, olive
Beauveria brongniartii	2	*Premnotrypes* spp.	Potato
Isaria farinosa	4	*Euprosterna elaeasa, Sibine* sp., *Peleopoda* spp., *Aleurodiccus juleikae, Bemisia tabaci*	Palm oil, avocado, citrus
Isaria fumosorosea	9	*Bemisia tabaci, Liriomyza huidobrensis*	Cotton, vegetables, crucifers, flowers, legumes
Hirsutella thompsonii	1	*Phyllocoptruta oleivora*	Citrus
Lecanicillium lecanii	19	*Trialeurodes vaporariorum, Hemileia vastatrix*	Tomato, flowers, coffee
Pochonia chlamydosporia	2	*Meloidogyne incognita*	Tomato, banana, kion, coffee, granadilla, citrus, flowers, onion, legumes, asparagus, garlic, vegetables
Metharizium anisoplae	6	*Cosmopolites sordidus, Metamasius hemipterus, Schistocerca interrita, Plutella xylostella, Lygirus maimon, Phyllophaga* spp., *Aeneolamia* sp.	Coffee, banana, crucifers, pasture
Purpureocillium lilacinum	3	*Meloidogyne incognita*	Tomato, banana, kion, coffee, granadilla, citrus, flowers, onion, legumes, asparagus, garlic, vegetables

Table 25.5. List of antagonistic fungi and related phytopathogens controlled in different crops in Peru.

Antagonistic fungus	Target phytopathogens	Crops
Trichoderma viride	*Rhizoctonia solani*, *Sclerotinia sclerotiorum*, *Sclerotium rolfsii*, *Phythopthora cinnamomi*, *Phythopthora*, *Armillaria mellea*, *Phytium*, *Cladosporium fulvum*, *Fusarium*	Cotton, flowers, vegetables, avocado, legumes, coffee, citrus, oil palm, tara, cocoa, ginger
Trichoderma harzianum	*Rhizoctonia solani*, *Phythopthora*, *Sclerotinia sclerotiorum*, *Sclerotium rolfsii*, *Armillaria mellea*, *Phytium*, *Botrytis cinerea*	Tomato, flowers, cacao, citrus, quinoa, avocado
Trichoderma virens	*Rhizoctonia*, *Phytium*, *Fusarium solani*, *Fusarium*, *Sclerotinia*, *Sclerotium*, *Cercospora*	Flowers, vegetables, coffee, cocoa
Trichoderma martiale	*Oidiopsis taurica*, *Phytophthora palmivora*	Cacao, flowers, cucurbits
Trichoderma asperellum	*Rhizoctonia*, *Phythopthora*, *Sclerotinia sclerotiorum*, *Sclerotium rolfsii*, *Fusarium*, *Phytium*, *Botrytis*	Avocado, citrus, tomato, olive, cucurbits

Promotion and use of biological control agents

The promotion of biocontrol in the main agricultural valleys of the country seeks to reduce the applications of agrochemicals in order to reduce pesticide residues in food and to protect the health of farmers. Biocontrol is currently widely used by agricultural companies as a basic part of an integrated pest management (IPM) strategy that allows sustainable agricultural production. The successes achieved by SENASA using biocontrol agents resulted in an increase of areas under biocontrol of crops such as sugarcane, asparagus, avocado, olive, forest, coffee, cacao, vine and vegetables. The area with biocontrol increased in 10 years from 40,844 ha in 2005 to 99,595 ha in 2015 (Table 25.5) (SENASA, 2005, 2010, 2015). Although application of biocontrol agents has increased substantially in the past 10 years, there is still a much greater proportion in use in coastal areas (75%), compared with the Sierra (20%) and forest areas (5%) (Table 25.6). The large agro-exporting and agro-industrial companies (sugarcane, asparagus, avocado, citrus and olive plantations) are located on the coast, with a high demand for biocontrol agents. This is different in the Sierra, where smaller field sizes, production destined to local markets and the lack of promotion for biocontrol agents makes their implementation difficult. The same holds for the jungle with a few exceptions, such as palm oil plantations. More than 50% of the areas under biocontrol were sugarcane fields, followed by asparagus, olive, various fruit trees and coffee (Table 25.6). In 2015 a significant increase in the use of biocontrol in avocado, cacao, legumes and chilli pepper crops took place and first-time use occurred in quinoa and grapevine. An important fact was that biocontrol generated savings between 35% and 80% compared with chemical control (Duarte, 2012). Likewise, control costs of secondary pests that appear or reappear (due to the suppression of natural enemies) when chemical control is applied were eliminated.

Production of biological control agents in the network of regional laboratories in agreement with SENASA

Biocontrol agents are supplied to different agricultural valleys in Peru through a network of production laboratories at national level. These laboratories are managed by associations of farmers, irrigation committees, universities, municipalities, private companies, etc. in different agricultural valleys. Through agreements between the Central Laboratory of Biological Control of SCB and laboratories, biocontrol agents are distributed to farmers in different regions of Peru. ‘Biological Control Promotion Agreements’ seek to initiate and stimulate research and dissemination of biocontrol and are established with institutions of higher education such as universities and technical institutes, and municipalities, and are renewed every three years. The agreements involve training of students, teachers and professionals.

Table 25.6. Area (ha) with application of biological control agents by the major regions and by crops in the years 2005, 2010 and 2015 in Peru.

	Years		
	2005	2010	2015
Major regions			
Coast and coastal desert	30,721	61,103	73,127
Sierra and Andes	8,428	14,904	20,183
Forest and Amazon	1,391	2,832	5,516
Total	40,539	78,838	98,826
Crop			
Cotton	1,994	332	714
Chilli pepper	451	449	1,206
Alfalfa	536	923	405
Aromatic herbs		50	
Rice	992	453	682
Cocoa		337	2,004
Coffee	1,383	2,449	3,551
Sugarcane	20,668	51,064	50,950
Citrus	1,798	2,307	3,730
Asparagus	2,800	8,929	12,872
Flowers		13	41
Forestry	592	1,228	2,921
Fruit trees	3,746	2,150	3,993
Vegetables	891	1,375	1,371
Maize	457	869	842
Legumes	272	156	1,022
Olive	3,224	2,970	4,219
Oil palm		21	483
Avocado		1,265	4,424
Potato	509	255	63
Pastures			69
Quinoa			832
Vine			2,856
Others	530	1,688	348
Total	40,844	79,282	99,595

Writing research theses related to biocontrol is part of these agreements. 'Biological Control Agent Production Agreements' are aimed at increasing the private production laboratories of biocontrol agents and are also renewed every three years. These can be self-supply laboratories for agricultural farms, for suppliers of other laboratories, or direct sale of biocontrol agents. Training, supervision, evaluation and technical advice on the operation of the biocontrol agents laboratories are also provided, and quality control analyses are performed every 3 months.

In 2015, 69 agreements (44 production and 25 promotional agreements) were signed and concentrated in the regions with many export crops such as Ica and La Libertad (Table 25.7) (SENASA, 2015). In order to consolidate the use of biocontrol, there are on-site and distance training courses, preparing inspectors and persons involved in Green Farm certification (see below) to learn about pest and natural enemies. Private laboratories produced 42 species of biocontrol agents to control different pests in more than 45 crops, and covering an area of 330,327 ha throughout the country (Table 25.8).

The main biocontrol agents produced were:

- *T. pretiosum*, with 1,682,904 inch^2 released to control *Diatraea saccharalis* (F.) in sugarcane, to control *Palpita persimilis* Munroe in olive, to control *Heliothis virescens* (F.) in

Table 25.7. List by Peruvian region of the main private laboratories and the agents they produced under supervision of SENASA in 2015.

Region	Main laboratories	Biological control agents
Arequipa	NOVAGRI SAC	*Trichoderma harzianum*, *Paecilomyces lilacinus*
Cusco	BIOPLAG	*Beauveria bassiana, Trichoderma viride*
Ica	AGROKASA, Alamein, BETA, BIOAGRUM, BIOGEA, CALSA, COBISA, COEXA, Copacabana, Cuatro Vientos, DM Agrícola SAC, EURO SA, Jy S del Sur, EURO SA, La Calera, OASIS OLIVES, OLIPERU, PROAGRO, Rojas Insectario	*Anagyrus* sp., *Ceraeochrysa cincta, Chrysoperla asoralis*, *Chrysoperla externa*, *Chrysoperla carnea*, *Chrysoperla* sp., *Trichogramma pretiosum*, *Paecilomyces lilacinus*, *Cryptolaemus montrouzieri*, *Euseius stipulatus*, *Euseius concordis*, *Euseius* sp., *Trichogramma exiguum*, *Trichoderma* sp., *Trichoderma viride*, *Trichogramma* sp.
La Libertad	Agentes de Biocontrol Agrícola, BIOAPLICA SAC, BIOINSA, Biológicos del Perú SAC, Biomil Solutions EIRL, Bioseguridad de Cultivos SAC, Empresa Agrícola SINTUCO SA, Insumos Biológicos Perú SAC, Laboratorio Agrícola SAC, Laboratorio de Agentes de Biocontrol Agrícola SAC, Procultivos Peru SAC, Protección de Cultivos SAC, Sociedad Agrícola Virú, SOLAGRO SAC, Valle Sol SAC	*Chrysoperla carnea*, *Chrysoperla* sp., *Billaea claripalpis*, *Trichogramma exiguum*, *Trichogramma pretiosum*, *Trichogramma galloi* , *Trichogramma pintoi*, *Paecilomyces lilacinus*, *Trichoderma* sp.
Lambayeque	AGROMIP SAC, Plantaciones el Sol, San Juan, Vista Florida INIA	*Cotesia flavipes*, *Paecilomyces lilacinus*, *Trichogramma pretiosum*, *Ceraeochrysa cincta*, *Trichogramma pintoi*
Lima-Callao	Contry Home SA, Técnicas Agrobiológicas SAC	*Chrysoperla asoralis*, *Chrysoperla externa*, *Asperillium* sp., *Beauveria bassiana*, *Trichoderma asperellum*, *Trichoderma harzianum*, *Trichoderma virens*, *Trichoderma viride*
Piura	Fundo Montelima	*Paecilomyces lilacinus*
San Martín	ACEPAT	*Beauveria bassiana*
Tacna	INPREX, Las Lagunas	*Ceraeochrysa cincta*, *Chrysoperla carnea*, *Chrysoperla* sp.

cotton and pomegranate and to control various lepidopteran pests in asparagus, lima bean, tomato, grapevine, and sweet pepper.

- *T. exiguum* with 3,270,880 inch^2 released in sugarcane, rice and maize to control *D. saccharalis*, in citrus and avocado for *A. sphaleropa*, and in blueberries for *H. virescens*.
- *Billaea claripalpis* Wulp. with 1,503,769 pairs released in sugarcane for the control of *D. saccharalis*.
- *Chrysoperla carnea* (Stephens) with 231,882 individuals released in olive for the control of *P. persimilis*, in alfalfa, tara, quinoa, sugarcane, citrus, maize and asparagus for the control of aphids, in grapevine for *Copitarsia* sp., in blueberries for *Heliothis* sp., in pomegranate for thrips, in avocado for hatched crawler scales, and in aubergine and pepper for *Spodoptera frugiperda* J.E. Smith.
- *Cotesia flavipes* Cameron on 12,415 ha in sugarcane for the control of *D. saccharalis*.

Table 25.8. Crops and area of application of the main biological control agents produced by private laboratories in Peru during 2015.

Crop	Pest	Natural enemy	Origin (Native or Exotic)	Area (ha) under biocontrol[a]
Alfalfa	Aphids	*Chrysoperla asoralis*	N	14.5
Alfalfa	Aphids	*Chrysoperla carnea*	E	328.0
Alfalfa	*Epinotia aporema*	*Trichogrammatoidea bactrae*	N	161.4
Alfalfa	*Spodoptera eridania*	*Heterorhabditis bacteriophora*	N	1.0
Alfalfa	*Spodoptera frugiperda*	*Trichogramma pintoi*	E	4,152.0
Alfalfa	*Spodoptera frugiperda*	*Trichogramma pretiosum*	N	12.0
Asparagus	Aphids	*Ceraeochrysa cincta*	N	17.0
Asparagus	Aphids	*Chrysoperla* sp.	N	43.0
Asparagus	Aphids	*Chrysoperla carnea*	E	5,482.3
Asparagus	*Bemisia tabaci*	*Isaria fumosoroseus*	N	85.8
Asparagus	*Elasmopalpus* sp.	*Beauveria bassiana*	N	11.9
Asparagus	Fungal diseases	*Trichoderma harzianum*	N	2.0
Asparagus	*Fusarium* sp.	*Trichoderma viride*	N	2.0
Asparagus	*Fusarium* sp.	*Trichoderma* sp.	N	2,105.6
Asparagus	Lepidopteran eggs	*Chrysoperla asoralis*	N	9,754.9
Asparagus	Lepidopteran eggs	*Chrysoperla externa*	N	78.0
Asparagus	Lepidopteran eggs	*Trichogramma pretiosum*	N	15,180.8
Asparagus	*Meloidogyne incognita*	*Paecilomyces lilacinus*	N	4,548.7
Asparagus	Mites	*Euseius stipulatus*	E	10.0
Asparagus	*Puccinia asparagi*	*Lecanicillium lecanii*	N	451.0
Asparagus	*Spodoptera frugiperda*	*Podisus nigrispinus*	N	247.2
Asparagus	*Spodoptera frugiperda*	*Telenomus remus*	E	10.0
Asparagus	*Spodoptera frugiperda*	*Trichogramma exiguum*	N	13.0
Asparagus	*Spodoptera* spp.	*Trichogramma pintoi*	E	5.0
Asparagus	Thrips	*Chrysoperla externa*	N	4,578.6
Asparagus	Thrips	*Orius insidiosus*	N	86.0
Asparagus	White grubs	*Beauveria bassiana*	N	2.0
Avocado	*Aleurothrixus floccosus*	*Isaria fumosoroseus*	N	21.7
Avocado	*Aleurothrixus floccosus*	*Lecanicillium lecanii*	N	8.5
Avocado	Aphids	*Chrysoperla externa*	N	476.1
Avocado	*Argirotaenia sphaleropa*	*Trichogramma exiguum*	N	278.8
Avocado	*Botrytis* sp., *Cladosporium* sp.	*Trichoderma harzianum*	N	191.3
Avocado	*Cladosporium* sp.	*Trichoderma* sp.	N	30.0
Avocado	*Fusarium*	*Trichoderma* sp.	N	1.0
Avocado	*Lasiodiplodia teobromae*	*Trichoderma harzianum*	N	275.0
Avocado	*Meloidogyne incognita*	*Paecilomyces lilacinus*	N	42.0

Continued

Table 25.8. Continued.

Crop	Pest	Natural enemy	Origin (Native or Exotic)	Area (ha) under biocontrol[a]
Avocado	Mites	*Euseius stipulatus*	E	1,539.2
Avocado	*Oligonychus punicae*	*Amblyseius chungas*	N	7.3
Avocado	*Oligonychus punicae*	*Euseius scutalis*	E	55.7
Avocado	*Phythopthora* sp.	*Trichoderma viride*	N	96.0
Avocado	*Phythopthora* sp.	*Trichoderma* sp.	N	41,475.2
Avocado	*Protopulvinaria pyriformis*	*Metaphycus helvolus*	E	7.7
Avocado	Pseudococcids	*Anagyrus pseudococci*	E	4.0
Avocado	Scales	*Chrysoperla asoralis*	N	39.9
Avocado	Scales	*Crysoperla carnea*	E	100.7
Avocado	*Spodoptera frugiperda*	*Podisus nigrispinus*	N	1.0
Avocado	Whiteflies	*Beauveria bassiana*	N	10.2
Avocado	Whiteflies	*Ceraeochrysa cincta*	N	480.3
Blueberry	*Anomala* sp.	*Heterorhabditis bacteriophora*	N	37.1
Blueberry	*Fusarium* sp	*Trichoderma viride*	N	26.0
Blueberry	*Heliothis* sp.	*Chrysoperla carnea*	E	10.0
Blueberry	*Heliothis virescens*	*Trichogramma exiguum*	N	8.2
Blueberry	*Heliothis virescens*	*Trichogramma pretiosum*	N	292.7
Blueberry	Whiteflies	*Beauveria bassiana*	N	4.0
Blueberry	White grubs	*Metarhizium anisopliae*	N	3.0
Citrus	*Alternaria* sp.	*Trichoderma viride*	N	0.7
Citrus	Aphids	*Chrysoperla asoralis*	N	125.0
Citrus	Aphids	*Chrysoperla externa*	N	690.0
Citrus	Aphids	*Chrysoperla carnea*	E	45.0
Citrus	*Argirotaenia sphaleropa*	*Trichogramma exiguum*	N	534.5
Citrus	*Botrytis cinerea*	*Trichoderma harzianum*	N	88.4
Citrus	*Icerya purchasi*	*Rodolia cardinalis*	E	208.0
Citrus	*Panonychus citri*	*Amblyseius chungas*	N	50.0
Citrus	*Panonychus citri*	*Euseius stipulatus*	E	706.7
Citrus	*Panonychus citri*	*Stethorus* sp.	N	193.8
Citrus	*Panonychus citri*	*Neoseiulus californicus*	N	30.0
Citrus	*Phytophthora* sp.	*Trichoderma* sp.	N	28.0
Citrus	*Planococcus citri*	*Sympherobius barberi*	N	0.5
Citrus	*Planococcus citri*	*Coccidoxenoides peregrinus*	N	67.5
Citrus	*Planococcus citri*	*Leptomastidea abnormis*	N	7.8
Citrus	*Planococcus citri*	*Cryptolaemus montrouzieri*	E	85.7
Citrus	*Planococcus citri*	*Anagyrus pseudococci*	E	441.5
Citrus	*Rizoctonia fusarium*	*Trichoderma harzianum*	N	389.0
Citrus	Thrips	*Orius insidiosus*	N	1.0
Citrus	Whiteflies	*Ceraeochrysa cincta*	N	27.0
Coffee	*Hypotenemus hampei*	*Beauveria bassiana*	N	222.0
Cotton	*Heliothis* eggs	*Orius insidiosus*	N	2.0
Cotton	*Heliothis virescens*	*Trichogramma pretiosum*	N	209.0

Continued

Table 25.8. Continued.

Crop	Pest	Natural enemy	Origin (Native or Exotic)	Area (ha) under biocontrol[a]
Cotton	*Pectinophora gosypiella*	*Trichogrammatoidea bactrae*	E	295.0
Maize	Aphids	*Chrysoperla carnea*	E	1,929.5
Maize	*Diatraea saccharalis*	*Trichogramma exiguum*	N	1,520.7
Maize	*Helicoverpa zea*	*Trichogramma pintoi*	E	3.5
Maize	*S. frugiperda* and *H. zea*	*Chrysoperla externa*	N	5.0
Maize	*Spodoptera frugiperda*	*Heterorhabditis bacteriophora*	N	2.4
Maize	*Spodoptera frugiperda*	*Podisus nigrispinus*	N	13.0
Maize	*Spodoptera frugiperda*	*Telenomus remus*	E	11.0
Olive	*Palpita persimilis*	*Chrysoperla carnea*	E	565.7
Olive	*Palpita persimilis*	*Eriborus* sp.	N	38.0
Olive	*Palpita persimilis*	*Trichogramma pretiosum*	N	230.1
Olive	*Palpita persimilis*, *Siphoninus phillyreae*	*Ceraeochrysa cincta*	N	11.2
Olive	*Palpita persimilis*, *Siphoninus phillyreae*	*Chrysoperla asoralis*	N	293.9
Olive	*Palpita persimilis*, *Siphoninus phillyreae*	*Chrysoperla externa*	N	643.0
Pepper	*Fusarium sp*	*Trichoderma viride*	N	312.7
Pepper	Lepidopterans	*Trichogramma pretiosum*	N	35.0
Pepper	*Liriomyza huidobrensis*	*Beauveria bassiana*	N	2.3
Pepper	*Meloidogyne incognita*	*Pochonia chla-mydosporia*	E	0.4
Pepper	*Meloidogyne incognita*	*Paecilomyces lilacinus*	N	872.6
Pepper	*Spodoptera frugiperda*	*Chrysoperla carnea*	E	0.5
Pomegranate	Aphids	*Ceraeochrysa cincta*	N	17.0
Pomegranate	Aphids	*Chrysoperla externa*	N	17.0
Pomegranate	*Fusarium spp*	*Trichoderma harzianum*	N	100.0
Pomegranate	*Heliothis virescens* eggs	*Trichogramma pretiosum*	N	150.0
Pomegranate	*Meloidogyne incognita*	*Paecilomyces lilacinus*	N	453.0
Pomegranate	Mites	*Euseius stipulatus*	E	4.0
Pomegranate	*Thrips tabaci*	*Beauveria bassiana*	N	128.0
Pomegranate	*Trips* sp., *Heliothis virescens*	*Chrysoperla asoralis*	N	121.0
Pomegranate	*Trips* sp., *Heliothis virescens*	*Chrysoperla carnea*	E	717.1
Sugarcane	Aphids	*Chrysoperla carnea*	E	677.5
Sugarcane	*Diatraea saccharalis*	*Billaea claripalpis*	N	95,770.6
Sugarcane	*Diatraea saccharalis*	*Chrysoperla externa*	N	3,221.7
Sugarcane	*Diatraea saccharalis*	*Cotesia flavipes*	N	12,415.1
Sugarcane	*Diatraea saccharalis*	*Trichogramma exiguum*	N	82,752.6
Sugarcane	*Diatraea saccharalis*	*Trichogramma pretiosum*	N	20,009.2

Continued

Table 25.8. Continued.

Crop	Pest	Natural enemy	Origin (Native or Exotic)	Area (ha) under biocontrol[a]
Sugarcane	*Fusarium* sp.	*Trichoderma viride*	N	5.6
Sugarcane	*Meloidogyne incognita*	*Paecilomyces lilacinus*	N	267.3
Sugarcane	*Phythopthora* sp.	*Trichoderma viride*	N	10.0
Tara[b]	Aphids	*Chrysoperla asoralis*	N	34.6
Tara	Aphids	*Chrysoperla externa*	N	4.0
Tara	Aphids	*Chrysoperla carnea*	E	5,048.6
Tara	*Spodoptera frugiperda*	*Trichogramma pretiosum*	N	26.0
Tomato	*Bemisia tabaci*	*Isaria fumosoroseus*	N	1.5
Tomato	*Bemisia tabaci*	*Lecanicillium lecanii*	N	11.0
Tomato	Cutworms	*Metarhizium anisopliae*	N	0.5
Tomato	*Fusarium* spp.	*Trichoderma viride*	N	85.3
Tomato	*Fusarium* spp.	*Trichoderma harzianum*	N	86.3
Tomato	Lepidopterans	*Trichogramma pretiosum*	N	509.0
Tomato	*Meloidogyne incognita*	*Pochonia chlamydosporia*	E	0.5
Tomato	*T. absoluta, E. aporema*	*Chrysoperla asoralis*	N	120.0
Vine	*Aleurodiccus*	*Ceraeochrysa cincta*	N	948.0
Vine	*Heliothis* sp. *Copitarsia* sp.	*Chrysoperla carnea*	E	455.0
Vine	Lepidopterans	*Chrysoperla asoralis*	N	135.0
Vine	Lepidopterans	*Trichogramma pretiosum*	N	15.0
Vine	*Phythopthora* sp.	*Trichoderma viride*	N	0.3
Vine	*Planococcus ficus*	*Chrysoperla externa*	N	264.0
Vine	Pseudococcids	*Anagyrus pseudococci*	E	157.0
Vine	*Bemisia tabaci*	*Beauveria bassiana*	N	62.0
Potato	*Liriomyza huidobrensis*	*Isaria fumosoroseus*	N	1.5
Potato	*Liriomyza huidobrensis*	*Lecanicillium lecanii*	N	1.5
Potato	*Pthorimaea operculella*	*Bacillus thuringiensis*	N	73.0
Quinoa	Aphids	*Chrysoperla asoralis*	N	2.5
Quinoa	Aphids	*Chrysoperla carnea*	E	121.5
Quinoa	*Fusarium* sp.	*Trichoderma viride*	N	5.0
Quinoa	*Fusarium* sp.	*Trichoderma harzianum*	N	20.0
Quinoa	Hemiptera	*Metarhizium anisopliae*	N	1.5
Quinoa	*Liorhyssus hialinus*	*Telenomus* sp.	E	32.0
Rice	*Diatraea saccharalis*	*Trichogramma exiguum*	N	211.0
Vegetables[c]	Aphids	*Beauveria bassiana*	N	0.5
Vegetables	Cutworms	*Metarhizium anisopliae*	N	0.5
Vegetables	Fungal diseases	*Trichoderma viride*	N	27.25
Vegetables	Fungal diseases	*Trichoderma harzianum*	N	13.5
Vegetables	Fungal diseases	*Trichoderma virens*	E	53.0
Vegetables	Fungal diseases	*Trichoderma* spp.	N	538.2
Vegetables	*Fusarium oxysporum*	*Trichoderma harzianum*	N	10.0
Vegetables	*Heliothis* eggs	*Orius insidiosus*	N	2.0
Vegetables	Lepidopterans	*Trichogramma pretiosum*	N	5.0

Continued

Table 25.8. Continued.

Crop	Pest	Natural enemy	Origin (Native or Exotic)	Area (ha) under biocontrol[a]
Vegetables	*Liriomyza huidobrensis*	*Isaria fumosoroseus*	N	2.0
Vegetables	*Liriomyza huidobrensis*	*Lecanicillium lecanii*	N	2.0
Vegetables	*Meloidogyne incognita*	*Pochonia chlamydosporia*	N	5.5
Vegetables	*Meloidogyne incognita*	*Paecilomyces lilacinus*	N	119.4
Vegetables	*Prodiplosis longifila*	*Metarhizium anisopliae*	N	0.5
Vegetables	Pseudococcids	*Cryptolaemus montrouzieri*	E	0.9
Vegetables	*Spodoptera frugiperda*	*Chrysoperla carnea*	E	73.3
Vegetables	*Spodoptera frugiperda*	*Telenomus remus*	E	16.0
Vegetables	Thrips	*Lecanicillium lecanii*	N	4.75
Vegetables	Thrips	*Orius insidiosus*	N	12.8
Vegetables	Thrips	*Beauveria bassiana*	N	1.65
Vegetables	White grubs	*Heterorhabditis bacteriophora*	N	0.3
Vegetables	Whiteflies	*Isaria fumosoroseus*	N	19.8
Vegetables	Whiteflies	*Beauveria bassiana*	N	27.5
Vegetables	Whiteflies	*Lecanicillium lecanii*	N	0.5
Fruits[d]	*Alternaria* sp.	*Trichoderma viride*	N	0.5
Fruits	Aphids	*Beauveria bassiana*	N	1.0
Fruits	Aphids	*Ceraeochrysa cincta*	N	29.6
Fruits	*Aphids*	*Chrysoperla asoralis*	N	1.0
Fruits	*Botrytis* sp.	*Trichoderma harzianum*	N	6.3
Fruits	*Cosmopolites sordidus*	*Beauveria bassiana*	N	0.3
Fruits	Fungal diseases	*Trichoderma viride*	N	27.4
Fruits	Fungal diseases	*Trichoderma* spp.	N	29.0
Fruits	*Heliothis virescens*	*Trichogramma pretiosum*	N	117.1
Fruits	*Meloidogyne incognita*	*Pochonia chlamydosporia*	N	2.0
Fruits	Mites	*Euseius stipulatus*	E	10.0
Fruits	*Sagalassa valida*	*Beauveria bassiana*	N	50.0
Fruits	*Trialeurodes vaporariorum*	*Isaria fumosoroseus*	N	2.0
Total				330,327.1

[a]Area of application of the main biological control agents produced by private laboratories in 2015
[b]Tara = *Caesalpinia spinosa* (Feuillée ex Molina) Kuntze
[c]Vegetables = zucchini, basil, beans, cucumber, aubergine, onion, spinach
[d]Fruits = pineapple, papaya, lucuma (*Pouteria lucuma* (Ruiz & Pav.) Kuntze), pecan, watermelon, peach, oil palm.

- *Trichoderma* sp. with 592,369 kg in avocado crops for control of *Fusarium* sp., *Cladosporium* sp. and *Phytopthora* sp., and in asparagus, capsicum, cranberry, basil, aubergine, zucchini, spinach, banana, onion, peach and mandarin for control *Fusarium* sp.

25.4 New Developments of Biological Control in Peru

In Peru, the use of biocontrol agents has increased due to the growing demand of producers for

Table 25.9. Green Farms certified by SENASA in the different regions of Peru in 2015.

Region	No.	Area (ha)	Crops
La Libertad	4	1,722	Asparagus, maize, sugarcane
Ica	4	1,512	Asparagus, citrus, pomegranate
Arequipa	4	1,218	Olive, maize, grapevine, sugarcane, rice, wheat
Lambayeque	1	650	Sugarcane
Cusco	2	54	Citrus
Moquegua	8	19	Olive
Cajamarca	1	15	Fruit trees, avocado, citrus, tara, maize, beans, alfalfa, rice, date palm, olive, vegetables
Huánuco	1	10	Sugarcane, alfalfa y coffee
Lima-Callao	1	8	Olive, aloe, grapevine, tuna (*Opuntia ficus-indica* (L.) Mill.)
Tacna	1	6	Tomato
Junín	1	5	Valencia orange
Total	28	5,218	

obtaining safe and healthy products for the internal and external market, by consumers for products free of pesticide residues, and by the government to achieve greater sustainability in agricultural production. Biocontrol within IPM contributes to the conservation of agricultural ecosystems, allows exporting companies to reduce costs, complies with international phytosanitary measures and supports the preservation of the environment. As a result of this, Peruvian agro-export companies could build a sustainable competitive advantage and seek a position as a socially responsible business (Duarte, 2012; SENASA, 2015, 2016b).

25.4.1 The role of SENASA in the promotion of biological control

SENASA plays an important role in opening markets and has achieved the entry of avocados, citrus, grapes, mangoes and asparagus in many countries around the world. Innovative actions undertaken by SENASA to strengthen and promote the use of biocontrol agents include Green Farm certification and agreements with farmers' associations with export crops.

Green Farm certification

This certification is granted by SENASA to farms whose crops are free of pesticides, where biocontrol agents are used for pest control, and consists of inspection and/or evaluation visits during one year. In 2015, Green Farm certification was granted to 28 farms producing a great diversity of crops on 5,203 ha (Table 25.9). Quinoa producers are now being trained to obtain Green Farm certification to be able to export quinoa without chemical residues (SENASA, 2015, 2016b).

Agreement with the association of citrus farmers (SENASA-PROCITRUS)

The Institutional Cooperation Agreement between the Association of Citrus Producers (PROCITRUS) and SENASA for implementation of the project 'Management of Pests in Citrus Cultivation through the use of Biological Control' was signed in 2004 and is still in force. PROCITRUS brings together approximately 90% of the country's citrus farmers with more than 7,000 ha in the main citrus-producing valleys of the regions of Piura, Lambayeque, La Libertad, Lima and Ica. It consists of activities such as: (i) preparation of evaluation sheets of pests and their beneficial fauna in citrus; (ii) field training in the recognition of pests and natural enemies of citrus (218 trained people); (iii) course of installation and rearing of beneficial insects directed at professionals and technicians of the associated farms; (iv) development of mass rearing of the predator *Stethorus* sp. for the control of red spider mite; (v) rearing of aphid parasitoids; (vi) rearing of *A. roseni* for the control of *S. articulates*; and (vii) development of ten bulletins on biocontrol agents mass rearing. Also, as part of this project, the predatory mite *E. victoriensis* was introduced from Australia for the control of the mite *P. oleivora*, and the parasitoid *Citrostichus phyllocnistoides* Narayan from Spain for the control of *Phyllocnistis citrella* Stainton. Important achievements of this project are, among others, the development of a mass-rearing method for predatory mites and the establishment of 15 laboratories dedicated to mass production of these mites in the main citrus producing areas (SENASA, 2005, 2010).

Project with the Peruvian Asparagus Institute (SENASA-IPE)

SENASA, together with the Peruvian Asparagus Institute (IPE), implemented the project 'Integrated management of asparagus pests with emphasis on *Copitarsia decolora* (Guenée) in the main agro-exporting valleys of Peru', with the aim of improving the competitiveness of fresh asparagus for international markets. Presence of *C. decolora* in the shoots of this crop species is one of the main causes of loss of quality and difficulty concerning phytosanitary conditions for export of fresh asparagus to the USA, as *C. decolora* is considered a quarantine pest/exotic pest for this country. This project became relevant because Peru has an area of more than 20,000 ha of asparagus crops and is the first exporting country of fresh asparagus to more than 40 countries worldwide. Sixty per cent of the exportation market is represented by USA and it makes up 25% of the export volume of Peruvian fresh fruits and vegetables (OCM, 2003; Duarte, 2012). The asparagus IPM proposal emphasizes biological and behavioural control, and only allows chemical control in exceptional situations. In pilot IPM areas, technical training of the field staff and quality control staff was carried out through workshops to learn to identify *C.decolora*. Further, control of thrips with the entomopathogenic fungi *Beauveria bassiana* (Bals.-Criv) Vuill. and *Lecanicillium* (*Veticillium*) *lecani* (Zimm.) Zare & Gams. and behavioural control with molasses in bottle traps to capture *C. decolora* and other cutworms have been developed (SENASA, 2005).

Plan Quinoa

This project aims to reduce economic losses by pests and improve the quality and health of the quinoa crop for export. Biocontrol actions include: (i) seed treatment with *Trichoderma viride* Pers. to prevent fungal diseases and/or use of *Heterorhabditis bacteriophora* Poinar for control of soil worms; (ii) applications of *Trichoderma harzianum* Rifai and *Bacillus subtilis* Cohn to the foliage to prevent mildew; (iii) use of *B. bassiana*, *Metarhizium anisopliae* (Metchnikoff) and *Bacillus thuringiensis* Berliner to control *Spodoptera eridania* Cram; and (iv) application of *B. thuringiensis* and releases of *C. externa* and *C. asoralis* for the control of *Eurysacca quinoa* Povolny (SENASA, 2015).

25.4.2 Biological control as the basis for large-scale sustainable agriculture

Also at the level of private initiatives there are successful examples of sustainable agricultural pest and disease management. Camposol is an agricultural company located on the northern coast of Peru where it produces asparagus, avocado, chilli pepper, mangoes and grapes, to be exported mainly to the USA and Europe. IPM strategies are implemented in each of their crops, such as: (i) rearing and release of parasitoids and predators, as well as mass production and application of microorganisms; (ii) use of different types of behavioural traps with light, water, molasses, vegetable oil and materials of different colours; (iii) application of extracts of medicinal plants that act as repellents and/or insecticides; (iv) installation of biological corridors, shrubs, hedges and windbreak edges that serve as refuges and feeding areas for natural enemies; and (v) use of selective pesticides compatible with biocontrol measures. The company has laboratories for mass production of biocontrol agents, which are supervised by SENASA. Recently the IPM methods have had a positive impact and the company has intensified its applications of plant extracts and biocontrol agents and, as a result, successfully reduced pesticide usage. The company's phytosanitary policy is committed to further implement sustainable agricultural practices allowing the development of healthy crops, minimizing environmental impact and improving the health of workers. Harvard University (USA) developed a business case of the Camposol company, presenting it as a leader in the Peruvian agribusiness sector and praising its model of sustainable agriculture on a large scale (Camposol 2010, 2011, 2017).

25.4.3 Concluding remarks

Currently, augmentative biocontrol with macrobial and microbial agents is emphasized to control the most important pests in Peru. Biocontrol agents are in growing demand, particularly from agro-export companies, to satisfy international market requirements for products free from pesticide residues and the global trend for environmental protection.

Also, conservation biocontrol is high on the agenda in Peru, though it is a more complex strategy as it involves identification of the factors limiting the effectiveness of biocontrol agents and trying to take away these limitations to improve their effectiveness. Often it involves minimizing or terminating pesticide use, use of cultural practices and provision of alternative hosts, food and appropriate microclimates to increase their development and survival. Studies of arthropods in agricultural ecosystems have shown that important pest species have many natural enemies and the importance of intricate multi-trophic relationships. This information on agroecological biodiversity promoted applications in conservation biocontrol, with the creation of biological corridors and plant shelter strips for natural enemies, and eventually in successful cases of sustainable agricultural production in Peru.

The total area under classical biocontrol in Peru is difficult to estimate. During the period 1880–1969, at least nine classical biocontrol projects were successfully implemented and the natural enemies are still present, but there are no recent estimates in which area they control pests nowadays (Table 25.1). However, if we use estimates from FAO for areas of crops harvested in 2017 (http://www.fao.org/faostat/en/#data/qc), classical biocontrol might still be at work on at least 108,000 ha of apple, citrus and olives and on large areas of alfalfa and eucalyptus. Augmentative biocontrol was applied on at least 190,000 ha during this period. From 1970 to 2000, four new classical biocontrol programmes were initiated, but no data are available about areas under control (Table 25.2). Augmentative biocontrol took place on about 250,000 ha during this period. In the past two decades, several new classical biocontrol programmes have been started, but again, areas under control are not known. A large number of augmentative biocontrol projects have recently been developed and these are applied in a number of crops on a documented area of 330,327 ha (Table 25.8). Probably, this figure is an underestimate, because there are many agricultural production and export companies that have their own mass-rearing laboratories for biocontrol agents for use on their crops and they are not required to inform SENASA about crop areas treated.

25.5 Acknowledgements

We thank H. Gómez (Subdirección de Control Biológico-Perú) for the valuable information provided.

References

Aguilar, P. (1980) Apuntes sobre el control biológico y el control integrado de las plagas agrícolas en el Perú [Notes on biological control and integrated control of agricultural pests in Peru]. *Revista Peruana de Entomología* 23(1), 83–110.

Beingolea, O. (1990) Sinopsis sobre el control biológico de plagas insectiles en el Perú, 1909–1990 [Synopsis on the biological control of insect pests in Peru, 1909–1990]. *Revista Peruana de Entomología* 33, 105–112.

Camposol (2010) Control biológico como base de una agricultura sustentable [Biological control as a basis for sustainable agriculture]. *Revista Institucional El Camposolino* 30, 14–15.

Camposol (2011) Informe de sostenibilidad 2011 [Sustainability report 2011]. Available at: https://database.globalreporting.org/reports/14366/download-report-pdf/ (accessed 28 October 2019).

Camposol (2017) La Universidad de Harvard elabora un business case de CAMPOSOL y lo presenta como parte de un curso internacional de Agronegocios [Harvard University develops a business case of CAMPOSOL and presents it as part of an international course in Agribusiness]. Available at: http://www.camposol.com.pe/prensa/noticias/ (accessed 17 July 2017).

Duarte, F. (2012) El control biológico como estrategia para apoyar las exportaciones agrícolas no tradicionales en Peru: un análisis empírico [Biological control as a strategy to support non-traditional agricultural exports in Peru: an empirical analysis]. *Contabilidad y Negocios* 7 (14), 81–100. Available at: http://www.redalyc.org/articulo.oa?id=281624914006 (accessed 22 May 2019).

Gonzales, G.F., Colarossi, A., Bernex, N., Rubin de Celis, V., Caballero-Gutierrez, L.S. and Alvarez, F. (2017) Food and nutritional security in Peru. In: *Challenges and Opportunities for Food and Nutrition Security in the Americas. The View of the Academies of Sciences.* IANAS, IAP and BMBF, Mexico City, Mexico, pp. 470–503. [Free public access of this publication in English and Spanish at www.ianas.org]

Gonzales, P. (1968) *Diatraea saccharalis* Fabr. y su control integrado en maíz, arroz y caña de azúcar en los valles de Arequipa [*Diatraea saccharalis* and its integrated control in corn, rice and sugar cane in the valleys of Arequipa]. *Revista Peruana de Entomología* 11(1), 9–17.

Herrera, J. (2010) Primera experiencia a nivel mundial del Manejo Integrado de Plagas: El caso del algodonero en el Perú [First experience in the world of Integrated Pest Management: the case of cotton in Peru]. *Revista Peruana de Entomología* 46(1), 1–8.

INEI (2018) Encuesta Nacional Agraria 2017: Características de las Pequeñas, Medianas y Grandes Unidades Agropecuarias [National Agrarian Survey 2017: characteristics of small, medium and large agricultural units]. Instituto Nacional de Estadística e Informática Available at: https://www.inei.gob.pe/media/MenuRecursivo/publicaciones_digitales/Est/Lib1593/ (accessed 5 December 2018).

MINAGRI (2015) Memoria Anual – Sector Agricultura y Riego (Annual Report – Agriculture and Irrigation Sector). Ministerio de Agricultura y Riego Available at: http://www.minagri.gob.pe/portal/download/pdf/memoria-anual-2015.pdf (accessed 20 November 2018).

OCM (2003) *Manejo Integrado de Plagas del esparrago* [*Integrated Management of asparagus pests*]. Comité de Medidas Sanitarias y Fitosanitarias, G/SPS/GEN/444. Organización Mundial de Comercio [World Trade Organization]. Available at: https://docs.wto.org/imrd/directdoc.asp?DDFDocuments/s/G/SPS/GEN/444 (accessed 25 July 2019).

Pacora, J.F. (1979) El Centro de Introducción y Cría de Insectos Útiles y los resultados de la investigación en apoyo a la producción agraria [The Center for the Introduction and Mass rearing of Useful Insects and the results of research in support of agricultural production]. *Revista Peruana de Entomología* 22, 99–102.

Risco, H. (1958) La utilización de *Parathesia claripalpis* W. para el control biológico de *Diatraea saccharalis* Fabr. con especial referencia a los resultados obtenidos en los Valles Pativilca y Huaura [The use of *Paratheresia claripalpis* for the biological control of *Diatraea saccharalis* with special reference to the results obtained in the Pativilca and Huaura Valleys]. *Revista Peruana de Entomología* 1(1), 24–29.

Risco, H. (1960) La situación actual de los barrenedores de la caña de azúcar del género *Diatraea* y otros taladradores en el Perú, Panamá y Ecuador [The current situation of the sugarcane borers of the *Diatraea* genus and other borers in Peru, Panama and Ecuador]. *Revista Peruana de Entomología* 3(1), 6–10.

Risco, H. (1961) Posibilidades de *Trichogramma minutum* Riley en el control biológico del borer de la caña de azúcar [Possibilities of *Trichogramma minutum* in the biological control of sugarcane borer]. *Revista Peruana de Entomología* 4(1), 8–11.

SENASA (2002) Memoria Anual 2002 [Annual Report 2002]. Servicio Nacional de Sanidad y Calidad Agroalimentaria. Available at: http://repositorio.senasa.gob.pe/handle/SENASA/22 (accessed 6 July 2017).

SENASA (2005) Memoria Anual 2005 [Annual Report 2005]. Servicio Nacional de Sanidad y Calidad Agroalimentaria. Available at: http://repositorio.senasa.gob.pe/handle/SENASA/37 (accessed 10 June 2017).

SENASA (2010) Memoria Anual 2010 [Annual Report 2010]. Servicio Nacional de Sanidad y Calidad Agroalimentaria. Available at: http://repositorio.senasa.gob.pe/handle/SENASA/40 (accessed 12 June 2017).

SENASA (2015) Memoria Anual 2015 [Annual Report 2015]. Servicio Nacional de Sanidad y Calidad Agroalimentaria. Available at: http://repositorio.senasa.gob.pe/handle/SENASA/43 (accessed 15 June 2017).

SENASA (2016a) Importancia del control biológico de plagas en la agricultura Peruana [Importance of biocontrol of pests in Peruvian agriculture]. Servicio Nacional de Sanidad y Calidad Agroalimentaria. Available at: https://www.senasa.gob.pe/senasacontigo/importancia-del-control-biologico-de-plagas-en-la-agricultura-peruana/ (accessed 20 June 2017).

SENASA (2016b) Acciones del SENASA en el Control Biológico y Manejo Integrado de Plagas Agrícolas en el Perú [Actions of SENASA in the biocontrol and integrated management of agricultural pests in Peru]. *Revista Comunicándonos* 8, 12–14.

SENASA (2018) Cronología del desarrollo del control biológico en el Perú [Chronology of the development of biocontrol in Peru]. Servicio Nacional de Sanidad y Calidad Agroalimentaria. Available at: http://www.senasa.gob.pe/senasacontigo/cronologia-del-desarrollo-del-control-biologico-en-el-peru/ (accessed 8 September 2018).

Valdivieso, L. (1991) Situación del control biológico en el Perú [Situation of biocontrol in Peru]. In: Gomero, L. (ed.) *Agroquímicos problema nacional-Políticas y alternativas*. IDMA, Lima, Peru, pp. 305–316.

Wille, J.E. (1952) *Entomología agrícola del Perú* [*Agricultural Entomology of Peru*]. Ministerio de Agricultura, Lima, Peru.

Whu, M. (1987) Estudios biosistemáticos de *Trichogramma* spp. [Biosystematic studies of *Trichogramma* spp.]. *Revista Peruana de Entomología* 28, 5–8.

Whu, M. (2016) El control biológico en el Perú [Biological control in Peru]. In: Huanca, J. and Alcazar, J. (eds) *Libro de resúmenes de la LVIII Convención Nacional de Entomología*. SEP, Tumbes, Peru, p. 94. Available at: https://hdl.handle.net/10568/96099 (accessed on 22 May 2019).

26 Biological Control in Puerto Rico

Mariangie Ramos[1*], Olgaly Ramos-Rodriguez[2] and Fernando Gallardo-Covas[3]

[1]*Sustainable Agriculture Program, Department of Agricultural Technology, University of Puerto Rico at Utuado, Puerto Rico;* [2]*Entomology and Pest Management, Department of Agricultural Technology, University of Puerto Rico at Utuado, Puerto Rico;* [3]*Entomology, Department of Agro-environmental Sciences, University of Puerto Rico at Mayaguez, Puerto Rico*

* E-mail: mariangie.ramos@upr.edu

Abstract

The role of natural enemies in pest control was recognized by naturalists visiting Puerto Rico around 1800 and 1870. Natural control of sugarcane borers by parasitoids and vertebrate predators of sugarcane borer was first documented in 1895. Pests in sugarcane received much attention in the early 20th century, starting with the introduction in 1912 of a ladybird for control of mealybugs. Biocontrol research in Puerto Rico peaked in the period from 1900 to 1950, when more than 75 species of parasitoids and predators of various pests were introduced. After 1970, several successful classical biocontrol projects resulted in control of water weeds, such as hornwort with grass carp, and water hyacinth, water lettuce and alligator weed with phytophagous coleopterans. Other projects concerned natural, classical, fortuitous and conservation biocontrol with predators and parasitoids of pests in citrus, coffee, cucurbits, papaya and pigeon pea. In 2014 a project was initiated for control of the Harrisia cactus mealybug. Recent interest in organic agriculture and environmentally friendly agricultural practices in Puerto Rico has resulted in biocontrol being considered more often for management of pests and diseases.

26.1 Introduction

Puerto Rico has an estimated population of slightly more than 3,350,000 and its main agricultural products are coffee, plantains, bananas, tuber crops, vegetables, pineapples and livestock (USDA, 2012). Puerto Rico is a 9,104 km^2 island in the Caribbean region. The arrival of the first human settlers is estimated to have occurred 2,500 years ago. These first inhabitants were Arawakan speakers from South America. They brought in their boats plant propagules of typical South American species such as cassava and found native plant food like products from palm.

26.2 History of Biological Control in Puerto Rico

26.2.1 Period 1800–1969

Naturalists stress the importance of predators and parasitoids

The first written reference to biocontrol in Puerto Rico comes from the chronicle of A.P. Ledrú, a botanist who came to the Caribbean with the ship *Triomphe* in 1797. He described 46 insect species, and when discussing wasps, he wrote that they are 'audacious, voracious and make war to other insects' (Ledrú, 1810). Native natural enemies were also described by a German naturalist, J. Gundlach, who lived in Cuba and visited Puerto Rico to aid in the collection and identification of animal species. In 1873, he collected material in the western part of Puerto Rico and in 1875 in the Bayamon region. He made a last trip to Puerto Rico in 1881 (Wolcott, 1948). The specimens he collected were sent to Berlin and Switzerland for identification and results were published under the title 'Fauna Puerto-Riqueña' in the *Anales de la Sociedad Española de Historia Natural, Madrid* (Annals of the Spanish Natural History Society, Madrid) (Gundlach, 1887). From May 1887 to September 1893, the Fauna Puerto-Riqueña included sections about insects. Gundlach mentioned species of chalcidid parasitic wasps belonging to genus *Smicra* (*Conura*), including *S. punctata* Fabricius and *S. emarginata* Fabricius. He noted: 'The larvae of all species of this family are raised within the body of caterpillars and larvae, or pupae and nymphs, thus being useful to Agriculture'. He also mentioned a parasitic wasp, *Evania laevigata* Olivier (*appendigaster* E. Guerin-Ménéville), attacking the egg cases (oothecae) of the cockroach.

Natural and classical biological control of pests in sugarcane

Besides these early observations made by naturalists, the history of biocontrol is closely linked to the crops that dominated the island over the past two centuries. Sugarcane was the crop receiving most technological investments in the 1880–1969 period. The first example of classical biocontrol documented in Puerto Rico was the infamous introduction of the small Indian mongoose *Herpestes auropunctatus* Hodgson to control rat populations in sugarcane plantations (for a history of the introduction see Chapter 20: Jamaica). Puerto Rican sugarcane growers acquired mongooses from a sugarcane producer in Jamaica and by 1877 these animals had established in the sugarcane regions (Long, 2003). At first, the mongooses seemed to reduce the rat populations and initial declines were observed

(Pimentel, 1955). However, the mongoose, a diurnal and omnivorous animal, proved to be a nuisance itself, damaging crops and poultry production and serving as a rabies reservoir.

Agricultural experimental stations were established in Mayaguez and Rio Piedras in 1888 by Spanish Royal Decree. F. López Tuero, a Spanish agronomist and the director of the Rio Piedras Agronomic Station, studied the productivity of several crops grown in Puerto Rico, including coffee and sugarcane. In 1894, he identified the white-grub (*Phyllophaga* spp.) as the cause of the 'sugarcane epidemic' that sugarcane producers were experiencing in the late 19th century. He communicated his findings to the general public through a series of agricultural booklets, with the aim to 'propagate the indispensable knowledge about existing crops and which farmers lack' (López Tuero, 1889). Each booklet was dedicated to a major and a minor crop. In 1895, he published his treatise on sugarcane (López Tuero, 1895), in which he mentioned *Apanteles* sp. and *Euplectrus* sp. as natural enemies of the sugarcane borer *Diatraea saccharalis* Fabricius and explained how they controlled the borer. In addition, he mentioned the importance of birds, lizards and other reptiles to reduce the incidence of insect pests.

In 1898, Puerto Rico became a territory of the USA. The US Congress authorized the establishment of the Porto Rico Agricultural Experimental Station in 1901, and this Federal Experimental Station was located initially in Rio Piedras and later transferred to Mayaguez in 1902. The first entomologist and botanist of the experimental station was O.W. Barrett, who started with a study of the mole cricket or 'changa', *Scapteriscus vicinus* Latr., because it was causing substantial damage to many crops, including sugarcane and tobacco (Barrett, 1902). He observed that there were few natural enemies of mole crickets present on the island and that most predation was done by birds and, to a lesser degree, by lizards. Later, Barrett also documented the presence of *Zagrammosoma multilineata* Ashm., a parasitoid of the coffee leafminer *Leucoptera coffeella* Guerin-Ménéville, in Puerto Rico (Barrett, 1906).

The land dedicated to sugarcane production in the island increased under the new US rule and was dominated by US companies. In 1910, the sugarcane producers established the Insular Agricultural Experiment Station in Rio Piedras, which was dedicated exclusively to studies on sugarcane production and management. The first entomologist of the station, D.L. Van Dine, introduced the Australian ladybird *Cryptolaemus montrouzieri* Mulsant into Puerto Rico in 1912 to control mealybugs in sugarcane (Van Dine, 1912). This was the earliest purposeful introduction of a natural enemy to control insect pests on the island. Van Dine also studied the maggot parasitoid *Tiphia* sp. in the US mainland and *Tiphia* sp. cocoons were later introduced in cane fields by another entomologist, G.N. Wolcott.

White grubs (*Phyllophaga* spp.) continued to be a significant problem in sugarcane plantations. Some plantations were even mined with dynamite to try to remove white grubs from the soil; after the explosion, workers were paid by the volume of grubs collected from the fields. F. Sein, the first Puerto Rican entomologist, worked for the Insular Agricultural Experiment Station. He studied a native predator of white grubs, the larvae of the 'cucubano' (the click beetle *Pyrophorus luminosus* Illiger) (Sein, 1923). Cucubano larvae were later (1936) introduced into the island of Mauritius to help in controlling another species of white grub, *Phyllophaga smithi* (Arrow), at the request of W.R. Thompson of the Imperial Institute of Entomology (now CABI) (Bartlett, 1939). Private sugar companies also hired entomologists to aid in the control of white grubs and other sugarcane pests, like H. Box, who worked at the Aguirre Sugar Company. He studied the biology of Scoliidae wasps, natural parasitoids of *Phyllophaga* in Puerto Rico (Box, 1925). In 1920, the cane toad *Rhinella marina* L. was introduced into Puerto Rico from Barbados, where it had been introduced before 1844, to manage white grubs. About 12 individuals were released in the Porto Rico Agricultural Experimental Station in Mayaguez. A second introduction of about 40 individuals from Jamaica was done at Rio Piedras in 1923–1924. The introductions were successful and after a year the toads were numerous on the island (Wolcott, 1950). After the introduction of the cane toad, cane plantations did not have significant problems with white grubs. However, it has been suggested that the decline of cucubano *P. luminosus* populations is related to the introduction of the cane toad. Later, 149 cane toads were exported from Puerto Rico to

Hawaii in 1932 (Easteal, 1981). From Hawaii, the cane toad was introduced to the Pacific Islands and Australia, where it became invasive.

Several efforts were also made to manage the sugarcane borer *D. saccharalis* using biocontrol. H. Box studied the tachinid parasitic fly *Lixophaga diatraeae* Townsend and concluded that it did not sufficiently control the sugarcane borer. Hence, he recommended the introduction of braconid parasitoids to the island (Box, 1928).The introduction of parasitoids to control the sugarcane borer continued in the 1930s (Dohanian, 1937) and also the research into its native parasitoids. G.N. Wolcott and L.F. Martorell studied the egg parasitoid *Trichogramma minutum* Riley (Wolcott and Martorell, 1943). By the 1950s, 13 species had been introduced to control the sugarcane borer, but only one, *Bassus stigmaterus* Holloway, was recovered (Gallardo-Covas, 2017). Borer infestations were low and natural parasitoids (*T. minutum*, *Tetrastichus haitiensis* Gahan and *L. diatraeae*) were effective in reducing its populations, with *L. diatraeae* exercising 20–60% of control.

Classical biological control of pests in citrus and coffee

Biocontrol was also studied in other crops and the first publication about economic entomology in Puerto Rico included biocontrol information for all pests of economic importance present on the island (Wolcott, 1924). In 1932, the coccinellid *Rodolia cardinalis* Mulsant, was introduced from Florida as a predator of the cottony cushion scale *Icerya purchasi* Mask; the scale had become a problem in citrus plantations in northern Puerto Rico (Wolcott and Sein, 1933). *R. cardinalis* and the entomopathogenic fungus *Spicaria javanica* Bally were effective in reducing *I. purchasi* populations below injury levels. In 1937, the braconid *Mirax insularis* Muesebeck was introduced into Puerto Rico from the island of Guadeloupe to control the coffee leaf miner *Leucoptera coffeella* (Guérin-Méneville & Perrottet) (Sein, 1940). The parasitoid successfully established and is currently the main parasitoid of this pest.

Biocontrol research in Puerto Rico peaked in the period from 1900 to 1950. The decade with the largest amount of published biocontrol research in Puerto Rico was the 1930s (Gallardo-Covas, 2017). This was mainly due to the arrival of several entomologists on the island to study control measures for sugarcane insect pests. A total of 75 species of parasitoids and predators of various pests were introduced into Puerto Rico from 1935 to 1950 by the Federal Agricultural Experiment Station. Biocontrol research and use were hampered by research focusing on the validation and registration of synthetic pesticides during the 1960s and 1970s.

26.2.2 Period 1970–2000

Classical biological control of weeds

Though biocontrol research for agricultural pests was limited in the 1970s, its use was promoted for the management of invasive aquatic plants in water reservoirs. The plants had been introduced as ornamental plants for ponds and became problematic in dams. Mechanical removal and herbicide use were not enough to control their growth. In 1973, the grass carp *Ctenopharyngodon idella* Valenciennes was introduced to control hornwort *Ceratophyllum demersum* L, resulting in effective control. In 1974, the Aquatic Plants Control Program was developed cooperatively by State and Federal Agencies. In 1977, the mottled water hyacinth weevil *Neochetina eichhorniae* Hustache was introduced experimentally into Puerto Rico to control the water hyacinth *Eichhornia crassipes* (Mart.) Solms. The weevil became established but did not control *E. crassipes*. Abreu and Bernier (2002) attributed the lack of control to large populations of the water hyacinth due to eutrophication of water reservoirs, compared with small populations of the weevil.

Natural and classical biological control of sugarcane rootstalk weevil

A new generation of Puerto Rican entomologists began the revival of biocontrol research for agricultural purposes in the 1980s and many efforts were dedicated to the control of the sugarcane rootstalk weevil *Diaprepes abbreviatus* Linnaeus, which is native to the Caribbean and attacks many crops. After DDT and aldrin insecticides were banned in the 1970s, *D. abbreviatus* became very problematic for sugarcane and citrus production. The organochlorine insecticide lindane was used with an emergency authorization from the US Environmental Protection Agency (EPA) to control this pest, but it was rapidly banned because pesticide traces were detected in local cow's milk (E. Abreu, Puerto Rico, 2018, personal communication). Biocontrol research became necessary to find new ways to manage this pest. Armstrong (1987) studied the native eulophid egg parasitoid *Tetrastichus haitiensis* Gahan, which parasitizes the egg masses of the weevil endophytically deposited inside the sugarcane leaves. Armstrong concluded that *T. haitiensis* had low efficiency in controlling *D. abbreviatus*, probably due to difficulty of ovipositing through sugarcane leaves and to the weevil's high reproductive potential (5,000 eggs per female). However, when *D. abbreviatus* was found attacking citrus orchards in Florida in 1964, *T. haitiensis* was imported from Puerto Rico into Florida for its control (Beavers and Selhime, 1975). Ants were also studied as potential predators for the sugarcane rootstalk weevil. Ants are commonly used in Cuba for controlling other root weevils such as the banana weevil *Cosmopolites sordidus* Germar. Seven species of ants were found preying on *D. abbreviatus* neonate larvae in Puerto Rico (Castro, 1986). Among those, *Monomorium floricola* (Jerdon) feeds on *D. abbreviatus* egg masses (Richman *et al.*, 1983). Entomopathogenic fungi were also studied as potential biocontrol agents. Out of 17 species of fungi associated with *D. abbreviatus* larvae, half of them were able to kill the larvae in pathogenicity tests (Colón, 1986). However, entomopathogenic nematodes appeared to be the most effective biocontrol agents. *Steinernema feltiae* (Filipjev), *S. glasseri* Steiner and *S. bibionis* Bovien showed activity against *D. abbreviatus* (Figueroa and Roman, 1990). Entomopathogenic nematodes were applied in inundative releases in sugarcane plantations. Due to industrialization and changes in the local economy, sugarcane production ended in Puerto Rico in 2000, with the closing of the last sugarcane mill 'Coloso'.

Natural and classical biological control of the melon worm in cucurbits

A biocontrol success was management of the melon worm *Diaphania hyalinata* Linnaeus, which feeds on the leaves of several cucurbits, including the tropical squash *Cucurbita moschata* Duch. ex Poir., in southern Puerto Rico. Tropical squash is the second most important vegetable crop in this country. In the 1990s, *D. hyalinata* populations became increasingly resistant to several insecticides and caused severe damage to squash plantations. Faunal surveys conducted in the 1980s found nine parasitoids and three predators attacking the pest (Medina *et al.*, 1989). In 1991, J. Capinera (University of Florida) sent 400 pupae of the braconid parasitoid *Cardiochiles diaphaniae* Marsh to the Department of Agroenvironmental Sciences, University of Puerto Rico at Mayaguez. Upon arrival, adults emerged and approximately 291 individuals were released at two sites in southern Puerto Rico: the Agricultural Experimental Station of Juana Diaz and a commercial farm in the municipality of Guayanilla. After this release, *C. diaphaniae* became established in squash-growing low-elevation areas of Puerto Rico. At present, *C. diaphaniae* exerts ca. 34% of parasitism and *D. hyalinata* damage to tropical squash has diminished notably (Gallardo-Covas *et al.*, 2012).

Natural, fortuitous and classical biological control of citrus blackfly and black citrus aphids in citrus

Biocontrol is often considered for the management of newly introduced pests and weeds. Several biocontrol projects for control of invasive pests were executed in the 1980s and 1990s. For example, in 1988, the citrus blackfly *Aleurocanthus woglumi* Ashby was detected in Puerto Rico (Medina *et al.*, 1991). A local survey of its natural enemies resulted in finding predatory thrips, coccinellids, anthocorids, reduviids, chrysopids, psoeids, mites and a parasitoid attacking *A. woglumi* in the island. In 1989, the parasitoids *Amytus hesperidum* Sylvestri and *Encarsia opulenta* Sylvestri were introduced to Puerto Rico from Florida. They were released a few days before Hurricane Hugo struck the island. By 1990, *A. woglumi* populations were so low that it was difficult to recover the blackfly or its parasitoids from the field (Browning, 1992). Another example was the control of the brown citrus aphid *Toxoptera citricida* Kirkaldy, a vector of citrus tristeza virus. The aphid was detected in Puerto Rico in 1992. A complex of several species of coccinellids and syrphids was found to be effectively controlling the aphids (Michaud and Browning, 1999), which can be considered for the large part as fortuitous biocontrol, as several of the coccinellid species had been introduced into Puerto Rico in the 1930s and 1940s. Michaud (1999) attributed the high abundance of coccinellids in Puerto Rico to 'higher overall plant diversity, smaller average grove size, intercropping, different climatic conditions, and greater microclimatic variation' than in Florida.

Natural and classical biological control of pink hibiscus mealybug and the papaya mealybug

The pink hibiscus mealybug *Maconellicoccus hirsutus* Green is effectively managed with biocontrol. The pest arrived in the western hemisphere in 1993. It was first observed in the Caribbean island of Grenada and continued to spread throughout other Caribbean islands. It was detected in eastern Puerto Rico in 1997, in the island municipality of Vieques (Michaud and Evans, 2000). The parasitoid *Anagyrus kamali* Moursi was imported from China and Hawaii and the parasitoid *Gyranusoidea indica* Shafee, Alam, and Agarwal from Egypt. In addition, a complex of native and introduced coccinellids was observed attacking this mealybug. Due to the effect of the imported and native natural enemies, *M. hirsutus* abundance was reduced and it did not cause as much economic loss in Puerto Rico as it did in other Caribbean islands (Michaud, 2003). Another mealybug, the papaya mealybug *Paracoccus marginatus* Williams and Granara of Willink, was also the object of biocontrol in Puerto Rico. The pest was first intercepted in 1995 (Sáez, 2000). A collaborative biocontrol programme between agriculture government agencies of the USA, Puerto Rico and the Dominican Republic was developed, and the parasitoids *Aponagyrus* spp., *Anagyrus* spp. and *Acerophagus* spp. were imported and released (Ramirez and Saez, 2002). A 97% control success was observed (Walker *et al.*, 2003).

Natural and classical biological control of various other pests

Biocontrol was also used to manage common insect pests in plantain, coffee, pigeon pea and other crops in the island during the 1980s and 1990s. For example, the banana weevil *C. sordidus* was managed effectively in experimental trials with the use of the entomopathogenic nematodes *S. feltiae*, *S. glasseri* and *S. bibionis* (Figueroa, 1990), but the method was not adopted by farmers. Faunal surveys of the parasitoids of the coffee leaf miner during 1985 and 1986 revealed that the parasitoid *M. insularis*, introduced in 1937, had become established and continues to parasitize *L. coffella* (Gallardo-Covas, 1988). A plan for augmenting *M. insularis* populations for the management of *L. coffeella* was proposed (Gallardo-Covas, 1992) and research related to this plan continues today. Faunal surveys of the natural enemies of the pulse pod borer moth *Etiella zinckenella* Treitschke during 1980 and 1981 found a complex natural enemy community that included *Polistes* wasps, *Anolis* lizards and larval parasitoids (Segarra and Barbosa, 1988). In 1987, near 2,000 individuals of *Bracon cajani* Muesebeck and *Eiphosoma dentator* Fabricius were introduced from Trinidad and Tobago to manage *E. zinckenella* on pigeon peas. *Bracon cajani* became established and decreased pod borer moth populations. Ten years after the introduction of *B. cajani*, it was observed parasitizing 85.70% of *E. zinckenella* larvae examined (Gonzalez, 2006).

Classical biological control of invasive aquatic weeds

After a 20-year hiatus, biocontrol was used again for the management of invasive aquatic plants. In 1997, the alligator weed flea beetle *Agasicles hygrophila* Selman and Vogt was introduced from Florida and effectively controlled the alligator weed *Alternanthera philoxeroides* (Mart) Griseb. (Abreu and Semidey, 1998). In March 1998, the water lettuce weevil *Neohydronomus affinis* Hustache was introduced from Florida to control the water lettuce *Pistia stratiotes* L., which was problematic in the important La Plata water reservoir that supplies water to the island's metropolitan area. The weevil became established and reduced the water lettuce population. Later, water lettuce plants were washed out of the reservoir by floods caused by Hurricane Georges in September 1998 and never grew to previous infestation levels (Abreu and Bernier, 2002). In 2002, *Neochetina bruchi* Hustache was also imported from

Florida and released in water reservoirs and creeks to control the water hyacinth *E. crassipes*. Additionally, *N. bruchi* was reared under laboratory conditions in Puerto Rico and released periodically. Approximately 58,000 adults were released around the island. A reduction in the population density and the plant size of the water hyacinth was observed (E. Abreu, Puerto Rico, 2018, personal communication).

Conservation biological control of pests in coffee plantations

During the 1970–2000 period, land use for agriculture significantly reduced, resulting in an increase in forest cover in Puerto Rico. Biologists became interested in studying the contribution of forest, biodiversity and farm tree management to natural biocontrol. In the late 1990s, researchers examining suitable habitats for the endangered Puerto Rican parrot *Amazona vittata* Boddaert began to look at the conservation potential of active and abandoned shade coffee plantations in Central Puerto Rico. The importance of coffee agroforestry systems for biodiversity conservation and insect management had been studied in other countries of Latin America before (Perfecto *et al.*, 1996). Studies in Puerto Rico revealed that species of forest-dwelling birds and *Anolis* lizards were more common in shaded than in sun-exposed plantations of Puerto Rico (Borkhataria, 2001). Furthermore, the experimental exclusion of birds and lizards in these farms resulted in higher populations of coffee-associated insects, such as planthopper *Petrusa epilepsis* Kirkaldy (Borkhataria *et al.*, 2006). In 2007, the USDA National Resource Conservation Service began promoting shade-coffee by paying incentives to coffee farmers. USDA-NRCS (2013) mentioned reduced pesticide use as one of the benefits of shade-coffee.

26.3 Current Situation of Biological Control in Puerto Rico

In the 2000s, three important insect pest species arrived to Puerto Rico: the Asian citrus psyllid, the Harrisia cactus mealybug and the coffee berry borer, for which biocontrol strategies were evaluated.

26.3.1 Fortuitous biological control of Asian citrus psyllid

The Asian citrus psyllid *Diaphorina citri* Kuwayama was first detected in Puerto Rico in 2001 (Halbert and Nuñez, 2004). The psyllid is the vector of *Candidatus* Liberibacter *asiaticus*, the sieve tubes-restricted bacterium that had been causing huanglongbing (citrus greening) in citrus trees in Puerto Rico since 2009 (Estévez de Jensen *et al.*, 2010). Its parasitoid, *Tamarixia radiata* Waterston, probably arrived with the psyllid (Pluke *et al.*, 2008). The incidence of parasitism of *T. radiata* ranged from 79% to 88%, and high rates of parasitism in the spring were followed by continuously reduced psyllid populations during the summer (Pluke *et al.*, 2008). Also, several coccinellids were observed attacking *D. citri* (Pluke *et al.*, 2005). A programme for mass production of *D. citri* parasitoids was later developed by the Puerto Rico Department of Agriculture. Research efforts to manage *D. citri* and huanglongbing continue today. Since 2014, biocontrol of *D. citri* has been focused on the evaluation, introduction and monitoring of a Pakistan strain of the parasitoid *T. radiata* at the Center for Excellence in Quarantine and Invasive Species (CEQIS, see below). This new strain is expected to adapt and show increased parasitism rates.

26.3.2 Biological control of the Harrisia cactus mealybug

The Harrisia cactus mealybug *Hypogeococcus pungens* Granara de Willink, native to South America, was first observed in Puerto Rico in 2000, feeding on *Portulaca oleracea* L. ornamentals in San Juan and later feeding on native cacti from Guanica in 2005 (Segarra-Carmona *et al.*, 2010). It attacks columnar cacti and has been used as an effective biocontrol agent of invasive cacti in Australia and South Africa. In Puerto Rico, *H. pungens* is threatening the endangered cactus *Leptocereus quadricostatus* (Bello) Britton & Rose and it is also attacking seven other cactus species. In 2010, prospecting for natural enemies in its native range in Argentina started. It became clear that *H. pungens* is a species complex in that country, different from the mealybug attacking cacti in Puerto Rico (Aguirre *et al.*,

2016). Since 2014, research has included surveys of local entomopathogenic nematodes and the rearing and release of the parasitoids already present in the island at CEQIS. A native parasitoid, *Leptomastidae* sp., was observed attacking the mealybug. Also, two species of parasitoids from Argentina have been evaluated for their effectiveness in reducing *H. pungens* and their effect on native non-target mealybugs in the Quarantine Laboratory and in greenhouses. Based on future results, permission to release the Argentine parasitoids will be requested.

26.3.3 Natural, augmentative and conservation biological control of the coffee berry borer

The coffee berry borer *Hypothenemus hampei* Ferrari was first detected in Puerto Rico in 2007. Control with insecticides was not available for farmers, since no insecticide was registered for this pest and endosulfan is prohibited. In 2008, the Puerto Rico Department of Agriculture started to apply a commercial formulation of the entomopathogenic fungus *Beauveria bassiana* (Balsamo) Vuillemin to control *H. hampei*. However, the commercial product was expensive for farmers and material produced in governmental laboratories was insufficient to meet the demands. Following the example of Colombia, research and farmers workshops educated farmers on using *B. bassiana* from infected borers in their coffee farms. Sadly, the initiative was halted by concerns of the private industry's patent rights over *B. bassiana*, a fungus that occurs naturally in the soils of Puerto Rico. The fungus was later observed to infect *H. hampei* naturally (Gallardo-Covas *et al.*, 2010). Because of federal regulations, it was also difficult to obtain permits to import *H. hampei* parasitoids from Colombia to Puerto Rico. Later, the parasitoid *Cephalonomia stephanoderis* Betrem was observed naturally occurring at low densities. Current research is focused on rearing and augmenting the populations of this parasitoid in coffee fields.

Another interesting approach for control of the coffee berry borer in Puerto Rico has been studied by I. Perfecto and J. Vandermeer (University of Michigan, USA) and their collaborators since 2010. Analysis of interactions among organisms in coffee agroecosystems showed that *Anolis* lizards are important predators of the coffee berry borer. The mean abundance of *Anolis* lizards in Puerto Rican farms is 610 per hectare, compared with 43 per hectare on Mexican farms (Monagan *et al.*, 2017). Current studies also examine the contribution of ants for *H. hampei* control. The non-native ant *Wasmania auropunctata* Roger is an effective predator of *H. hampei*, entering coffee beans and eating the borer larvae. However, this ant is a nuisance for Puerto Rican coffee pickers. In Cuba, coffee farmers approach this dilemma by moving nests of *Pheidole megacephala* (F.) next to coffee bushes right before coffee harvesting. *P. megacephala* temporarily displaces *W. auropunctata* from coffee bushes. Coffee farmers can then maintain the biocontrol exerted by *W. auropunctata* throughout the year, but do not get stung during coffee harvesting (I. Perfecto, Utuado, Puerto Rico, 2018, personal communication,). Perfecto's group is also examining natural enemies of *Hemileia vastatrix* Berk. & Broome, a disease agent of coffee rust, and found 15 mycoparasites present in coffee rust lesions (James *et al.*, 2016). Further, they are identifying two species of snails and one species of *Mycodiplosis* fly that eat *H. vastatrix* spores.

26.3.4 Establishment of Center for Excellence in Quarantine and Invasive Species

Several biocontrol initiatives are currently being developed on Puerto Rico. An important milestone for the advancement of biocontrol on Puerto Rico was the creation of CEQIS in 2014, with the aim to 'develop expertise, promote education and generate tools in quarantine and invasive species and support sound decision-making'. One of the CEQIS working areas is biocontrol, with both introduced and native natural enemies. The Center provides advice and facilities for importing biocontrol agents for research purposes, in order to meet all state and federal requirements. The Center contributes to the development of introduction and release protocols, which are verified and approved by federal and state agencies. The Center has a state-of-the-art 920 m^2 Quarantine Laboratory with four containing rooms and greenhouses. Among the

Center's projects, there is research on the three above-mentioned invasive pests.

According to information about biocontrol agents in Puerto Rico (Table 26.1) and data about crop areas harvested in 2017, it is estimated that more than 10,000 ha are under classical, fortuitous, natural and conservation biocontrol. No estimates could be obtained for the areas under biocontrol of pests occurring on ornamentals, such as the pink hibiscus mealybug, or for water weeds.

26.4 New Developments in Biological Control in Puerto Rico

In the past decade, interest in organic agriculture and environmentally friendly agricultural practices has increased in Puerto Rico. Because of this, biocontrol is being considered more often as a management alternative of pests and diseases. Organic farmers commonly include flowers and natural vegetation in their farming programmes to promote conservation biocontrol.

Table 26.1. Chronology of biological control in Puerto Rico.

Biocontrol agent	Pest / crop	Type of biocontrol[a] / since	Effect / area (ha) under biocontrol
Apanteles sp., *Euplectrus* sp.	Sugarcane borer in sugarcane	NC / 1895	+ / ?
Birds and lizards	Mole crickets, sugarcane, tobacco	NC / 1902	+ / ?
Cryptolaemus montrouzieri	Mealybugs in sugarcane	CBC / 1912	+ / ?
Pyrophorus luminosus, *Scoliid* wasps, *Rhinella marina*	White grub in sugarcane	NC+CBC / 1923	+ / ?
Lixophaga diatraeae, Trichogramma minutum, Tetrastichus haitiensis	Sugarcane borer in sugarcane	NC / 1928	+ / ?
Rodolia cardinalis	Cottony cushion scale in citrus	CBC / 1932	+ / 1,896[b]
Mirax insularis	Coffee leaf miner in coffee	CBC / 1937	+ / 6,937[b]
Ctenopharyngodon idella	Hornwort in water	CBC / 1970	+ / still active today
Neochetina eichhorniae	Water hyacinth	CBC / 1977	- /
Bracon cajani	Pulse pod borer in pigeon peas	CBC / 1987	+ / 127[b]
Steinernema spp.	Sugarcane rootstalk weevil	ABC / 1990	+ / sugarcane production terminated in 2000
Cardiochiles diaphaniae	Melon worm in cucurbits	CBC / 1991	+ / 1,266[b]
Agasicles hygrophila	Alligator weed in water	CBC / 1997	+ / ?
Neohydronomus affinis	Water lettuce	CBC / 1998	+ / ?
Complex of coccinellids	Black citrus aphid in citrus	NC+FBC / 1999	+ / 1,896[b]
Birds and lizards	Pests in coffee	ConsBC / 2001	+ / 6,937[b]
Complex of parasitoids and predators	Pink hibiscus mealybug	NC+CBC / 2002	+ / ?
Complex of parasitoids	Papaya mealybug, vars fruit	CBC / 2002	+ / 303[b]
Neochetina bruchi	Water hyacinth	CBC / 2002	+ / ?
Tamarixia radiata	Asian citrus psyllid in citrus	FBC / 2008	+ / 1,896[b]
Complex of biocontrol agents	Coffee berry borer in coffee	ABC+ConsBC / 2008	? / ?
Complex of biocontrol agents	Harrisia cactus mealybug in ornamental and wild cacti	CBC+NC / 2014	? / ?

[a]Type of biocontrol: ABC = augmentative, CBC = classical, FBC = fortuitous, ConsBC = conservation biocontrol; NC = natural control
[b]Area of crop harvested in 2017 according to FAO (http://www.fao.org/faostat/en/#data/qc)

Two private companies currently sell commercial biological agents in Puerto Rico. Farmers also purchase biocontrol commercial products through the internet, though this practice is not regulated and could have detrimental effects if new natural enemies are imported without the proper controls.

Several farmers have travelled to Cuba to learn low-input organic management practices, including the use of biocontrol. In addition, large seed companies located in the South of the island are considering biocontrol alternatives due to Bt maize resistance found in populations of *Spodoptera frugiperda* (J.E. Smith) in Puerto Rico. Biocontrol is a practice that is included in USDA-NRCS incentives for farmers, within an IPM programme. The Puerto Rico Department of Agriculture also has a Biological Control Laboratory, funded by the US Department of Agriculture (USDA), which focuses on rearing natural enemies of recently introduced pest species.

In conclusion, there are good opportunities to develop more biocontrol research in Puerto Rico. Researchers, agencies and farmers should use this opportunity to meet and develop a biological control agenda for Puerto Rico. This agenda should continue to support the use of classical biocontrol for recently introduced pests and disease agents, while also promoting conservation biocontrol. Most importantly, it should include the participation of farmers in the design of biocontrol strategies to increase its adoption and successful use.

26.5 Acknowledgements

The authors would like to thank the following persons who kindly provided information for this book: M. Davila (Univ. of Puerto Rico-Utuado), E. Abreu (retired from UPR-Isabela Agricultural Experimental Station), J.C. Verle Rodrigues (Center for Excellence in Quarantine and Invasive Species), I. Perfecto (Univ. of Michigan, USA), C. Torres (Luis Munoz Marin Foundation) and J. Collazo (North Carolina State Univ., USA), and the agency officials: J. Martinez (US Fish and Wildlife), E. Mas (USDA-NRCS), S. Cruz (Puerto Rico Dept Agric.) and N. Gabriel (USDA-APHIS).

References

Abreu, E. and Semidey, N. (1998) La liberacion de la pulga de la yerba caimán en Puerto Rico [Release of the alligator weed flea beetle in Puerto Rico]. *Agrotemas* 9, 10.

Abreu, E. and Bernier, L.S. (2002) Control biológico de plantas invasoras de Puerto Rico [Biological control of invasive plants in Puerto Rico]. Available at: http://drna.pr.gov/wp-content/uploads/2015/04/Plantas-invasoras-en-Puerto-Rico-control-biol%C3%B3gico.pdf (accessed 18 September 2018).

Aguirre, M.B., Diaz-Soltero, H., Claps, L.E. , Saracho Bottero, A., Triapitsyn, S., Hasson, E. and Logarzo, G.A. (2016) Studies on the biology of *Hypogeococcus pungens* (sensu stricto) (Hemiptera: Pseudococcidae) in Argentina to aid the identification of the mealybug pest of Cactaceae in Puerto Rico. *Journal of Insect Science* 16,1–7.

Armstrong, A.M. (1987) Parasitism of *Tetrastichus haitiensis* Gahan on egg masses of *Diaprepes abbreviatus* in Puerto Rico. *Journal of Agriculture of the University of Puerto Rico* 71, 407–409.

Barrett, O.W. (1902) The changa or mole cricket. *Puerto Rico Agricultural Experimental Station Bulletin* 2, 1–19.

Barrett, O.W. (1906) Report of the entomologist and botanist: investigations in entomology and plant pathology. *Puerto Rico Agricultural Experimental Station Report for 1905*, 22–23.

Bartlett, K.A. (1939) The results of shipments of the predatory elaterid beetle *Pyrophorus luminosus* from Puerto Rico to England. *Bulletin of Entomological Research* 30, 209–210.

Beavers, J.B. and Selhime, A.G. (1975) Further attempts to establish the weevil egg parasite, *Tetrastichus haitiensis* in Florida. *Florida Entomologist* 58, 29–31.

Borkhataria, R.R. (2001) Ecological and political implications of conversion from shade to sun coffee in Puerto Rico. MSc thesis. North Carolina State University, North Carolina.

Borkhataria, R.R., Collazo, J.A. and Groom, M.J. (2006) Additive effects of vertebrate predators on insects in a Puerto Rican coffee plantations. *Ecological Applications* 16, 696–703.

Box, H.E. (1925) Porto Rican cane-grubs and their natural enemies, with suggestions for the control of Lamellicorn larvae by means of wasp-parasites (Scoliidae). *Journal of Agriculture of the University of Puerto Rico* 9, 291–356.

Box, H.E. (1928) The introduction of braconid parasite of *Diatraea saccharalis* Fabr., into certain of the West Indian Islands. *Bulletin of Entomological Research* 18, 365–370.

Browning, H.W. (1992) Overview of biological control of Homopterous pests in the Caribbean. *Florida Entomologist* 75, 440–445.

Castro, S. (1986) Hormigas depredadoras de larvas neonatas de *Diaprepes abbreviatus* (Coleóptera: Curculionidae) [Ant predators of neonatal larvae of *Diaprepes abbreviatus*]. MSc thesis. University of Puerto Rico, Mayaguez, Puerto Rico.

Colón, I. (1986) Estudio preliminar de géneros de hongos aislados de larvas de *Diaprepes abbreviatus* en el área oeste de Puerto Rico y su evaluación como control biológico [Preliminary study of fungi genera isolated from *Diaprepes abbreviatus* in western Puerto Rico and their biological control potential]. MSc thesis. University of Puerto Rico, Mayaguez, Puerto Rico.

Dohanian, S.M. (1937) The introduction of parasites of the sugarcane borer into Puerto Rico. *Journal of Agriculture of the University of Puerto Rico* 21, 237–241.

Easteal, S. (1981) The history of introductions of *Bufo marinus* (Amphibia: Anura); a natural experiment in evolution. *Biological Journal of the Linnean Society* 16, 93–115.

Estevez de Jensen, C, Vitoreli, A. and Román, E. (2010) Citrus greening in commercial orchards in Puerto Rico. *Phytopathology* 100, S34.

Figueroa, W. (1990) Biocontrol of the banana root borer weevil, *Cosmopolites sordidus* (Germar), with steinernematid nematodes. *Journal of Agriculture of the University of Puerto Rico* 74, 15–20.

Figueroa, W. and Roman, J. (1990) Parasitism of entomophilic nematodes on the sugarcane rootstalk borer, *Diaprepes abbreviatus* L. (Coleoptera:Curculionidae), larvae. *Journal of Agriculture of the University of Puerto Rico* 74, 197–202.

Gallardo-Covas, F. (1988) Faunal survey of the coffee leaf miner, *Leucoptera coffeella*, parasitoids in Puerto Rico. *Journal of Agriculture of the University of Puerto Rico* 72, 255–264.

Gallardo-Covas, F. (1992) Augmentation of *Mirax insularis*. Alternative for population control of the coffee leaf miner, *Leucoptera coffeella*, in Puerto Rico. *Journal of Agriculture of the University of Puerto Rico* 76, 43–54.

Gallardo-Covas, F. (2017) Biological control of insect pests in Puerto Rico. *Journal of Agriculture of the University of Puerto Rico* 101, 153–163.

Gallardo-Covas, F., Hernández, E. and Pagán, J. (2010) Presencia natural del hongo *Beauveria bassiana* (Bals.) Vuill. en la broca del café *Hypothenemus hampei* (Ferrari) en Puerto Rico [Natural occurrence of the fungus *Beauveria bassiana* in the coffee berry borer in Puerto Rico]. *Journal of Agriculture of the University of Puerto Rico* 94, 195.

Gallardo-Covas, F., González, O.P. and Pérez, H. (2012) Presencia de *Cardiochiles diaphaniae* W., parasitoide de *Dyaphania hyalinata* L. en cultivos de calabaza en Puerto Rico: Veinte años después [Presence of *Cardiochiles diaphaniae,* parasitoid of *Dyaphania hyalinata* in squash fields: Twenty years later]. In: *Proceedings of the 2011 Annual Meeting of the Sociedad Puertorriqueña de Ciencias Agrícolas*. Ponce, Puerto Rico.

Gonzalez, A. (2006) Population dynamics and economic injury levels of important insects and crops in Puerto Rico. USDA-NIFA Project Report. Available at: https://reeis.usda.gov/web/crisprojectpages/0065344-population-dynamics-and-economic-injury-levels-of-important-insects-and-crops-in-puerto-rico.html (accessed 23 February 2018).

Gundlach, J. (1887) Apuntes para la fauna Puerto-Riqueña. VI. Crustáceos [Notes on the Puerto Rican fauna. VI. Crustaceans]. *Anales de la Sociedad Española de Historia Natural* 15, 115–199.

Halbert, S.E. and Nuñez, C.A. (2004) Distribution of the Asian citrus psyllid *Diaphorina citri* Kuwayama (Rhynchota: Psyllidae) in the Caribbean Basin. *Florida Entomologist* 87, 401–402.

James, T.Y., Marino, J.A., Perfecto, I. and Vandermeer, J. (2016) Identification of putative coffee rust mycoparasites via single-molecule DNA sequencing of infected pustules. *Applied Environmental Microbiology* 82 631–639.

Ledrú, A.P. (1810) Voyage aux iles de Tenerife, La Trinité, Sainte-Tomas, Sainte-Croix et Porto-Ricco. Exécuté par ordre du govermenement Francais. Tome premier. [Travel to the islands of Tenerife, Trinidad, Saint Thomas, Saint Croix and Puerto Rico. Ordered by the French Government]. Chez Arthus Bertand Library, Paris.

Long, J.L. (2003) *Introduced Mammals of the World*. CAB International, Wallingford, UK.

López-Tuero, F. (1889) *Arroz y Cacao* [*Rice and Cocoa*]. Imprenta De Acosta, San Juan, Puerto Rico.

López-Tuero, F. (1895) *La caña de azúcar en Puerto Rico, su cultivo y enfermedad* [*Sugarcane in Puerto Rico: crop and disease*]. Puerto Rico Boletín Mercantil, San Juan, Puerto Rico.

Medina, S., Abreu, E., Gallardo, F. and Franqui, R. (1989) Natural enemies of the melon worm, *Diaphania hyalinata* L. (Lepidoptera: Pyralidae), in Puerto Rico. *Journal of Agriculture of the University of Puerto Rico* 73, 313–320.

Medina, S., Segarra, A. and Franqui, R. (1991) La mosca negra de los cítricos, *Aleurocanthus woglumi* Ashby (Homoptera: Aleyrodidae) en Puerto Rico [Citrus black fly in Puerto Rico]. *Journal of Agriculture of the University of Puerto Rico* 75, 301–305.

Michaud, J.P. (1999) Sources of mortality in colonies of brown citrus aphid, *Toxoptera citricida*. *BioControl* 44, 347–367.

Michaud, J.P. (2003) Three targets of classical biological control in the Caribbean: success, contribution, and failure. Available at: http://forestpestbiocontrol.info/international_symposium/day5_pdf/michaud.pdf (accessed 18 September 2018).

Michaud, J.P. and Browning, H.W. (1999) Seasonal abundance of the brown citrus aphid, *Toxoptera citricida* (Homoptera: Aphididae) and its natural enemies in Puerto Rico. *Florida Entomologist* 82, 424–447.

Michaud, J.P. and Evans, G.A. (2000) Current status of the pink hibiscus mealybug in Puerto Rico including a key to parasitoid species. *Florida Entomologist* 83, 97–101.

Monagan, I.V., Morris, J.R., Davis Rabosky, A.R., Perfecto, I. and Vandermeer, J. (2017) Anolis lizards as biocontrol agents in mainland and island agroecosystems. *Ecology and Evolution* 7, 2193–2203.

Perfecto, I., Rice, R.A. Greenberg, R. and van der Voort, M.E. (1996) Shade coffee: a disappearing refuge for biodiversity. *BioScience* 46, 598–608.

Pimentel, D. (1955) Biology of the Indian mongoose in Puerto Rico. *Journal of Mammalogy* 38, 62–68.

Pluke, R.W.H., Escribano, A. , Michaud, J.P. and Stansly, P.A. (2005) Potential impact of lady beetles on *Diaphorina citri* (Homoptera: Psyllidae) in Puerto Rico. *Florida Entomologist* 88, 123–128.

Pluke, R.W.H., Qureshi, J.A. and Stansly, P.A. (2008) Citrus flushing patterns, *Diaphorina citri* (Hemiptera: Psyllidae) populations and parasitism by *Tamarixia radiata* (Hymenoptera: Eulophidae) in Puerto Rico. *Florida Entomologist* 91, 36–42.

Ramirez, A. and Saez, L. (2002) *Papaya mealybug (*Paracoccus marginatus*) in Puerto Rico*. Biological Control Laboratory, Dept. Agriculture of Puerto Rico, Training Workshop Papaya Mealybug Biological Control Program, Oct. 23–25, 2002. San Juan, Puerto Rico.

Richman, D.B., Buren, W.F. and Whitcom, W.H. (1983) Predatory arthropods attacking the eggs of *Diaprepes abbreviatus* (L.) (Coleoptera: Curculionidae) in Puerto Rico and Florida. *Journal of the Georgia Entomological Society* 18, 335–342.

Sáez, L. (2000) Parasitoides naturales de la chinche harinosa de la papaya, *Paracoccus marginatus* (Williams y Granara de Willink) y parasitoides naturales e importados de la chinche harinosa rosada del hibisco *(Maconellicoccus hirsutus* (Green) en dos regiones en Puerto Rico [Natural parasitoids of the papaya mealybug, and natural and imported parasitoids of pink hibiscus mealybug in two regions of Puerto Rico]. MSc thesis. University of Puerto Rico, Mayaguez, Puerto Rico.

Segarra, A.E. and Barbosa, P. (1988) Notes on the natural enemies of *Etiella zinckenella* in Puerto Rico. *Journal of Agriculture of the University of Puerto Rico* 72, 153–159.

Segarra-Carmona, A.E., Ramírez-Lluch, A., Cabrera Asencio, I. and Jiménez-López, A.N. (2010) First report of a new invasive mealybug, the Harrisia cactus mealybug *Hypogeococcus pungens* (Hemiptera: Pseudococcidae). *Journal of Agriculture of the University of Puerto Rico* 94, 183–187.

Sein, F. (1923) El cucubano, *Pyrophorus luminosus* Illiger (*Pyrophorus luminosus* click beetle). *Ciencia* 80, 1–8.

Sein, F. (1940) *Annual Report, Puerto Rico Agricultural Experiment Station 1938–39*. University of Puerto Rico, Rio Pedras, Puerto Rico. USDA, Washington, DC, pp. 50–52.

USDA (2012) Census of Agriculture. US Department of Agriculture. Available at: https://www.agcensus.usda.gov/Publications/2012/Full_Report/ Census_by_State/Puerto_Rico/ (accessed 18 September 2018).

USDA-NRCS (2013) Coffee conservation initiatives. US Department of Agriculture, National Resource Conservation Service. Available at: https://www.nrcs.usda.gov/wps/portal/nrcs/detailfull/pr/ programs/landscape/?cid=nrcseprd1291252 (accessed 18 September 2018).

Van Dine, D.L. (1912) Progress report on introduction of beneficial parasites into Porto Rico. *First Report of the Board of Commissioners of Agriculture of Porto Rico* 1, 31–47.

Walker, A., Hoy, M. and Meyerdirk, D. (2003) Papaya mealybug, *Paracoccus marginatus* Williams and Granara de Willink (Insecta: Hemiptera: Pseudococcidae). Available at: http://edis.ifas.ufl.edu/pdffiles/IN/IN57900.pdf (accessed 8 November 2018).

Wolcott, G.N. (1924) Entomología Económica Puertorriqueña [Puerto Rican Economic Entomology]. *Porto Rico Insular Experiment Station Boletin* 32, 1–176.

Wolcott, G.N. (1948) The insects of Puerto Rico. *Journal of Agriculture of the University of Puerto Rico* 20, 1–224.

Wolcott, G.W. (1950) The rise and fall of the white grub in Puerto Rico. *American Naturalist* 84, 183–193.

Wolcott, G.N. and Sein, F. (1933) A year's experience with the cottony cushion scale in Puerto Rico. *Puerto Rico Department of Agriculture Journal* 17, 199–221.

Wolcott, G.N. and Martorell, L.F. (1943) Natural parasitism by *Trichogramma minutum* of the eggs of the sugarcane north borer, *Diatraea saccharalis*, in the cane fields of Puerto Rico. *Journal of Agriculture of the University of Puerto Rico* 27, 39–83.

27 Biological Control in the Remaining Caribbean Islands

Joop C. van Lenteren[1]* and Vanda H.P. Bueno[2]

[1]*Laboratory of Entomology, Wageningen University, The Netherlands;*
[2]*Laboratory of Biological Control, Department of Entomology, Federal University of Lavras, Lavras, Minas Gerais, Brazil*

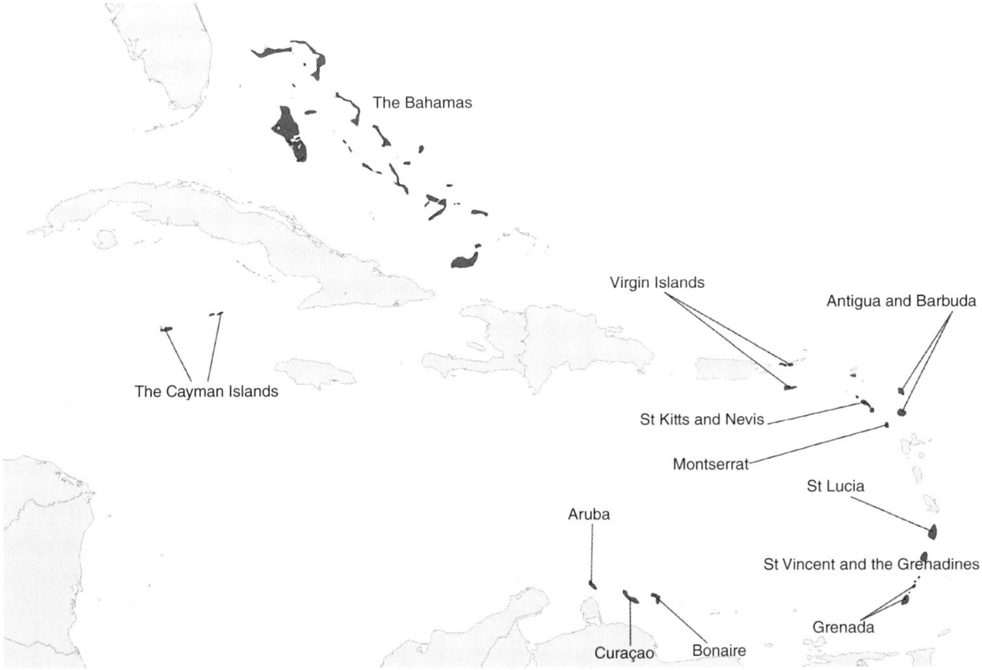

* E-mail: joop.vanlenteren@wur.nl

Abstract

Biological control activities on 18 (groups of) Caribbean islands are summarized. Many natural enemies were introduced to these islands through Trinidad and Tobago up to 1980. Also, inter-island exchange of biocontrol agents took place. The majority of projects concerned classical biocontrol, while in some cases natural, conservation and augmentation biocontrol were used. Successes were obtained with biocontrol of pests in crops such as arrowroot, citrus, coconut, cotton and sugarcane and of weeds like prickly pear and puncture vine. After 1980, the number of natural enemy introductions decreased, though the region was faced with many invasions by exotic pests, including the citrus leaf miner, citrus blackfly, papaya mealybug, giant African snail, coconut whitefly and pink hibiscus mealybug. Two large region-wide programmes resulted in successful biocontrol of the pink hibiscus mealybug and the papaya mealybug. In addition, biocontrol by a native natural enemy complex was demonstrated for the coconut whitefly and the passion vine mealybug. The Food and Agriculture Organization of the United Nations (FAO) Code of Conduct for the Import and Release of Exotic Biological Control Agents has recently been applied in the region. Farmers Field Schools, with the aim to enable farmers to use IPM and become less dependent on chemical pesticides, are being implemented.

27.1 Introduction

In this chapter we summarize information about biological control carried out in 18 (groups of) small islands in the Caribbean Sea with a total population of about 1.2 million inhabitants. Though the volume of agricultural production is small, a surprising amount of biocontrol activities have been performed on these islands in the past and currently some large Caribbean-wide projects are being implemented. The history and current situation related to status (independent or not) of the islands are rather complicated and have changed a number of times during the past decades.

- **Antigua** and **Barbuda** (autonomous country, member of British Commonwealth) have an estimated population of almost 95,000 (July 2017) and their agricultural products – mainly for local use – are cotton, fruits, vegetables, bananas, coconuts, cucumbers, mangoes and sugarcane; there is some livestock production as well (CIA, 2018a).
- **Aruba** (Netherlands) has an estimated population of slightly more than 116,500 inhabitants (July 2018) and agriculture is an unimportant economic activity with as export products small amounts of aloe, livestock and fish (CIA, 2018b).
- The **Bahamas** (autonomous country, member of British Commonwealth) have an estimated population of almost 330,000 (2017) and their main agricultural products are citrus and vegetables; also poultry and seafood are produced (CIA, 2018d).
- **Bonaire** (Caribbean Netherlands) has an estimated population of almost 20,000, with some agricultural activities such as maize cropping, and fisheries (Wikipedia, 2019a; Landenweb, 2019).
- The **Cayman Islands** (autonomous country, member of British Commonwealth) have an estimated population of almost 60,000 (2018) with very limited agricultural activities, which include vegetables and fruit; there is some livestock and turtle farming (CIA, 2018e).
- **Curaçao** (Netherlands) has an estimated population of more than 150,000 (July 2018) and agriculture is an unimportant economic activity, with as export products small amounts of aloe, sorghum, peanuts, vegetables and tropical fruit (CIA, 2018c).
- **Grenada** (autonomous country, member of British Commonwealth) has an estimated population of slightly more than 112,000 (2018) and its main agricultural activities are bananas, cocoa, nutmeg, mace, soursop, citrus, avocados, root crops, maize, vegetables; fishing also takes place (CIA, 2018f).
- **Montserrat** (UK) has an estimated population of slightly more than 5,000 (2018), with limited agricultural activities; the main products are cabbages, carrots, cucumbers, tomatoes, onions and peppers, and some livestock production takes place (CIA, 2018g).
- **Saba** (Caribbean Netherlands) has an estimated population of about 1,900 (Wikipedia, 2019b) and produces some vegetables and fruit.
- **Sint Eustatius** (Caribbean Netherlands) has an estimated population of about 4,000

(Wikipedia, 2019c) and also produces some vegetables and fruit.

- **Sint Maarten** (Netherlands) has a population of slightly more than 42,600 (Wikipedia, 2019d) and agriculture is an unimportant economic activity, with sugar as an export product.
- **St Kitts** and **Nevis** (autonomous country, member of British Commonwealth) have an estimated population of slightly more than 53,000 (2018) with very limited agricultural activities; products are rice, yams, vegetables, bananas, and there is also fishing (CIA, 2018h).
- **St Lucia** (UK) has an estimated population of slightly more than 165,000 (2018) with some agricultural activities, including production of bananas, coconuts, vegetables, citrus, root crops and cocoa (CIA, 2018i).
- **St Vincent** and the **Grenadines** (autonomous country, member of British Commonwealth) have an estimated population of almost 102,000 (2018), with some agricultural activities, including production of bananas, coconuts, sweet potatoes, spices; there is some livestock production (cattle, sheep, pigs, goats) and also fishing (CIA, 2018j).
- The **Virgin Islands** (UK and USA) have an estimated population of almost 107,000 (2018) and have few agricultural activities; main products are fruit, vegetables, sorghum and some cattle (CIA, 2018k).

27.2 History of Biological Control in the Remaining Caribbean Islands

Most of the information in this section originates from Cock (1985). As the review by Cock (1985) covers the period up to 1980, we will not split the information in two periods (1880–1969 and 1970–1999) as done in the other chapters. Section 27.3 (Current situation of biological control in the remaining Caribbean islands) will include developments since 1980.

27.2.1 Biological control of pests of citrus

On the islands discussed in this chapter, citrus production is small and mainly for local consumption. Still, several attempts to develop biocontrol took place in citrus and these are summarized below.

Citrus blackfly

BAHAMAS. Citrus blackfly *Aleurocanthus woglumi* Ashby occurred in the Bahamas prior to 1916. Releases of *Eretmocerus serius* Silv. obtained from Cuba in 1931 resulted in establishment, but control was insufficient, though in 1947 citrus blackfly seemed to be well controlled by this parasitoid. The predatory coccinellid *Catana clauseni* Chapin was also obtained from Cuba and released in 1936, but did not establish. In 1972, as a result of new blackfly outbreaks, *Encarsia opulenta* (Silv.) was obtained from Barbados and resulted in sufficient control in 1973.

CAYMAN ISLANDS. *E. serius* from Jamaica was released in the 1940s and 1950s on Grand Cayman, and more releases of *E. serius*, apparently also containing *E. opulenta*, were made in 1966. In 1970, only *E. opulenta* was found on the island.

Various whitefly species

BAHAMAS. Three aphelinids, *Encarsia meritoria* Gah., *E.* sp.nr. *variegata* How. and *Coccophagus aleurodici* Gir., and the coccinellid *Nephaspis amnicola* Wingo were obtained in 1961–1962 from Trinidad and Tobago and released in the Bahamas for whitefly control in citrus and other crops.

ST VINCENT. Miscellaneous coccinellids from Trinidad and Tobago were sent to St Vincent in 1951 for control of whitefly and coccids.

COTTONY CUSHION SCALE. Cottony cushion scale *Icerya purchasi* Mask. is present on many islands in the Caribbean region and releases have been made with the coccinellid predator *Rodolia cardinalis* (Muls.) in most of the infested islands.

ANTIGUA. Releases of *R. cardinalis* were made in 1966, 1970 and 1973 with individuals from St Kitts and Montserrat.

BAHAMAS. *R. cardinalis* releases were made from Puerto Rico in 1934, which resulted in good control.

CAYMAN ISLANDS. *R. cardinalis* was found in 1970 on Grand Cayman without records of introductions. Additional releases were made on Grand Cayman and predators were also released on Little Cayman with material from Barbados. Good control results were obtained.

MONTSERRAT. Several *R. cardinalis* releases were made on Montserrat. In 1964–1965 predators from Bermuda were released and in 1966 predators from St Kitts were reared and released. In late 1966 excellent control had been achieved.

ST KITTS. In 1966, *R. cardinalis* was introduced from Barbados into St Kitts and established. Field-collected and cage-reared individuals were released in heavily infested sites, resulting in good control. Due to flare-ups of the pest, the tachinid *Cryptochetum iceryae* (Will.) was introduced from California.

Citrus mealybug

Citrus mealybug *Planococcus citri* (Risso) is a minor pest of citrus in the Caribbean.

BAHAMAS. A first attempt to introduce *Cryptolaemus montrouzieri* Mulsant from Florida into the Bahamas in 1932 was not successful. In 1968 it was introduced again against the sugarcane mealybug *Saccharicoccus sacchari* (Ckll.) and recovered from sugarcane as well as from citrus infested with *P. citri*.

GRENADA. The coccinellid *Cryptognatha affinis* Crotch was introduced from Trinidad and Tobago into Grenada for control of *P. citri*.

MONTSERRAT. CABI UK sent *C. montrouzieri* to Montserrat in 1935 for control of *P. citri*, but it is unknown whether it established.

Citrus weevils

Citrus weevils of the genus *Diaprepes* damage roots and seedlings of citrus trees.

ST LUCIA. *Tetrastichus gala* W1k. was sent from Dominica to St Lucia in 1938 for citrus weevil control, but is it is not known if it established.

Fruit flies

Anastrepha spp. fruit flies may cause serious damage to citrus and other fruit.

ST KITTS. Shipments of *Biosteres longicaudatus* (Ashmead) and *Aceratoneuromyia indica* (Silvestri) were made in 1966 to St Kitts from Mexico via Trinidad and Tobago. Also in 1966, *Pachycrepoideus vindemmiae* (Rondani) was introduced from Trinidad and Tobago to St Kitts. In November 1966 new shipments containing *Doryctobracon cereus* (Szépligeti), *A. indica*, *B. longicaudatus* and *P. vindemmiae* were received from Mexico and Trinidad and Tobago. In 1970, *Opius* spp. and *P. vindemmiae* were sent from Trinidad and Tobago. Next, *B. longicaudatus* was introduced from Costa Rica via Trinidad and Tobago in 1973, and in 1974 from Florida. No recoveries were made.

27.2.2 Biological control of pests of coconuts

Coconuts are extensively grown on plantation scale, as well as in smallholdings and gardens in the Caribbean.

Coconut whitefly

ST VINCENT. The coconut whitefly *Aleurodicus cocois* (Curt.) occurs on several Caribbean islands. One of its natural enemies, *Encarsiella noyesi* Hayat, was first found on Trinidad and Tobago, but was also collected in other Caribbean islands, including Antigua and Grenada, and Central and South America (Boughton *et al.*, 2015). The parasitoid has been successfully used for control of the coconut whitefly on various Caribbean islands (see country-specific chapters), but introduction of *E. noyesi* to St Vincent in 1950 did not result in recovery (Cock, 1985).

Coconut mealybug

ST KITTS. Several native natural enemies of the coconut mealybug *Nipaecoccus nipae* Mask. were found on St Kitts, including the predators *Leucopis bella* Loew, an undetermined cecidomyiid, *Chrysopa* sp., *Scymnus* sp. and the parasitoid *Allotropa* sp. *Cryptolaemus montrouzieri* was introduced from India via Trinidad and Tobago between 1971 and 1973, resulting in establishment and a strong reduction of infestations. In 1971, the coccinellid predator *Hyperaspis jucunda* (Muls.) was introduced from Trinidad and Tobago, followed by shipments of the parasitoid *Pseudaphycus utilis* Timberlake in 1972, also from Trinidad and Tobago; there are no reports of establishment of the predator and the parasitoid.

Coconut scale

Many introductions of coccinellids, often from colonies started in Trinidad and Tobago, were made into Caribbean islands for the control of the coconut scale *Aspidiotus destructor* Sign. between 1937 and 1973.

ANTIGUA. After a coconut scale outbreak in 1937, *Cryptognatha nodiceps* Marshall was introduced from Trinidad and Tobago, but there are no reports of establishment.

BAHAMAS. In the Bahamas, coconut scale is attacked by the parasitoids *Aspidiotiphagus* sp. and *Encarsia* sp. and the predators *Chilocorus cacti* L. and *Diomus* sp., but control is insufficient. *Cryptognatha*, *Pseudoazya trinitatis* (Marshall) and '*Pentilia* spp.' were shipped from Trinidad and Tobago in 1961–1962, but did not establish.

CAYMAN ISLANDS. Coccinellids, including *Chnoodes* sp., *C. nodiceps*, and *P. trinitatis*, were introduced from Trinidad and Tobago in 1949, followed in 1970 by releases of *C. nodiceps* and *P. trinitatis* from Trinidad and Tobago, and *Coccidophilus cariba* Gordon (as *C. citricola* Breth.) from Nevis.

GRENADA. Many species of coccinellids, including *Chnoodes* sp., *C. nodiceps*, *P. trinitatis*, *Azya orbigera* Muls., *Chilocorus bipustulatus* L., *Exochomus bisbinotatus* Gorham, *Exoplectra dubia* Crotch, *Lotis neglecta* Muls., *Lotis nigerimma* Csy., *Pentilia* spp., *Scymnus* sp., *Telsimia nitida* Chapin, *Zenoria emarginata* Gordon and *Rhyzobius pulchellus* Montrouzier, were released between 1947 and 1970 on Grenada, mainly originating from colonies from Trinidad and Tobago. Before the first releases were made, *C. cacti* and a *Scymnus* sp. were found associated with the coconut scale in Grenada. Few, if any, of the released species were recorded during surveys.

MONTSERRAT. *Rhyzobius pulchellus* was introduced into Montserrat from Trinidad and Tobago in 1971, but establishment was not reported.

ST KITTS AND NEVIS. The coccinellids *P. trinitatis*, *C. cacti* and *Coccidophilus cariba* Gordon and the parasitoid *Aphytis* sp. were found present in 1966. In 1970, when scale damage was very serious, *C. nodiceps* was introduced from Trinidad and Tobago, resulting in a good degree of control.

ST LUCIA. After heavy coconut scale attacks, the coccinellids *C. nodiceps*, *Cryptognatha simillima* Sicardi, *P. trinitatis*, *Pentilia egena* Muls. and *Cryptognatha flaviceps* Crotch were introduced from Trinidad and Tobago in 1938–1939. In 1950, only *P. trinitatis* was found. In 1971, *R. pulchellus* was introduced from Trinidad and Tobago.

ST VINCENT. On this island only *C. cacti* and an *Azya* sp. seemed to be present on coconut before 1950. In 1951, *Chnoodes* sp., *Cladis nitidula* F., *C. similima* and *C. nodiceps*, *Lioscymnus diversipes* Champ., *L. neglecta*, *L. nigerimma*, *Pentilla* spp., *P. trinitatis*, *R. lophanthae* and *Scymnus* sp. were introduced, mainly from Trinidad and Tobago. Later, *R. pulchellus* was introduced. There are no reports of establishment.

27.2.3 Biological control of pests of other tree crops and ornamentals

Orthezia scales

GRENADA. *Orthezia* spp. scales sometimes create problems on limes and other citrus fruit on Grenada and seem to have only *Melaleucopis simmondsi* Sabrosky as a native natural enemy. Miscellaneous coccinellids from Trinidad and Tobago were released in 1952, and in 1953 *Rhinoleucophenga* sp. and *Hyperaspis donzeli* Mulsant were introduced. None of the species was recovered later in 1953.

Miscellaneous mealybugs

MONTSERRAT. *Cryptolaemus montrouzieri* was sent from Trinidad and Tobago for release against *Puto barberi* (Ckll.) on citrus. There are no reports of recovery.

Miscellaneous scale insects

BAHAMAS. The coccinellids *A. orbigera* and *Cryptolaemus affinis* Crotch were sent from Trinidad and Tobago in 1961 for control of *Pulvinaria psidii* Mask. on guava. Before they were released, the pest population had declined, so the coccinellids had little chance of establishment.

Cocoa thrips

The cocoa thrips *Selenothrips rubrocinctus* (Giard) is a pest of cocoa and cashew in the Caribbean. However, outbreaks subside when shade and windbreaks were provided as a cultural practice when producing cocoa.

GRENADA. A hyphomycete fungus *Beauveria globulifer* (Speg.) Pic., obtained from St Vincent, killed many thrips when sprayed on cocoa plants, but it provided control only when the humidity was high. The parasitoid *Goetheana parvipennis* (Gahan) was introduced from Trinidad and Tobago in 1937, but was not recovered.

Banana weevil

Bananas and plantains are grown for domestic consumption throughout the Caribbean, but Grenada, St Lucia and St Vincent are also exporting bananas. The banana weevil *Cosmopolites sordidus* (Germ.) is the most important pest and the only one against which biocontrol was attempted with the predatory beetles *Plaesius javanus* Erichs., *Dactylosternum hydrophiloides* (Macleay) and *D. abdominale* (F.) from South-east Asia, and *Hololepta (= Leionota) quadridentate* (F.), a native of Trinidad and Tobago that has adapted to feeding on the banana weevil.

GRENADA. In 1949 and 1951, *P. javanus* and *H. quadridentata* from Trinidad and Tobago were released in piles of rotting banana stems. The two species were not recovered in 1951 and later.

ST LUCIA. In the 1950s, *C. sordidus* caused serious damage to bananas and plantains. During 1950–1954, *H. quadridentata*, *P. javanus* and *Dactylosternum* sp. (presumably *D. subdepressum* Lap.) from Trinidad and Tobago were released. In 1971, a *Dactylosternum* sp. was collected which might have been native, but no histerids were recovered.

ST VINCENT. In 1951, *P. javanus* sent from Jamaica via Trinidad and Tobago were released by mistake in sugarcane fields. Although there is no record of the introduction of *H. quadridentata* in St Vincent, an adult was obtained from a banana stump and several from felled palm trunks attacked by the curculionid *Rhynchophorus palmarum* L. in 1970.

27.2.4 Biological control of pests of cotton

West Indian sea island cotton, a cultivar of *Gossypium barbadense* L, produces the highest quality lint of any commercially grown cotton. Prior to the 1940s, sea island cotton was grown extensively in many islands in the Caribbean, but its importance declined and little biocontrol of its pests was attempted. However, due to high demand for this type of cotton in the 1970s, interest in biocontrol of its pests also increased.

Cotton stainers

Cotton is attacked by cotton stainers, *Dysdercus* spp., wherever it is grown. *Dysdercus andreae* (L.) occurs in Antigua, Montserrat and Nevis, while *D. fulvoniger* Deg. ssp. *discolor* Wlk. occurs in St Vincent and Montserrat. Cotton stainers are generally minor pests in the Caribbean, but they can build up large populations where alternative host plants are present. The tachinids *Hyalomyia chilensis* Macq. and *Acaulona peruviana* Tns. were released to supplement the native *A. erythropyga* Sabrosky, but did not become established.

Green stink bug

The green stink bug *Nezara viridula* (L.) is the most serious pest on cotton in Antigua and Montserrat.

ANTIGUA. In 1949, *Trichopoda pennipes* (F.) from Florida was released, with no evidence of establishment. In 1955, *Trichopoda pilipes* (F.) reared from *N. viridula* from Montserrat was released. In 1961–1962 shipments of *Anastatus* spp. originating from Pakistan and mass produced in Trinidad and Tobago were released. In 1963, *Xenoencyrtus niger* Riek from Hawaii and reared in Trinidad and Tobago was released. No recoveries were made of any of these species.

MONTSERRAT. *N. viridula* is common in Montserrat. This may explain why the parasitoid *T. pilipes* is much more common in Montserrat than in other Caribbean islands. Among the other parasitoids in the region are the widely distributed scelionid *Trissolcus basalis* (Woll.) and the encyrtids *Ooencyrtus submetallicus* (How.) and *O. trinidadensis* J.C. Crawford. Although these parasitoids provide some level of control elsewhere, *N. viridula* is still a sufficiently serious pest of several crops to continue attempts at biocontrol. In 1961–1962 shipments of *Anastatus* spp. originating from Pakistan and mass produced in Trinidad and Tobago were released. In 1966, *Trissolcus mitsukurii* (Ashm.) from Japan and mass reared in Trinidad and Tobago was released. No recoveries of *Anastatus* and *T. mitsukurii* were made.

ST KITTS. In 1961–1962 *Anastatus* spp. and in 1966 *T. mitsukurii* were released. No recoveries of *Anastatus* and *T. mitsukurii* were made.

ST VINCENT. In 1961–1962, *Anastatus* spp. were released, but no recoveries were made.

Pink bollworm

Native Caribbean parasitoids of pink bollworm *Pectinophora gossypiella* (Saund.) include a braconid *Bracon hebetor* Say, the chalcidids *Spilochalcis torvina* (Cress.) and *Brachymeria* sp., and the bethylid *Perisierola nigrifemur* (Ashm.), but their combined parasitism is negligible.

ANTIGUA, MONTSERRAT, ST KITTS AND ST VINCENT. *Apanteles* sp. ? *angeleti*, *Bracon greeni* Ashmead and *Brachycoryphus nursei* (Cam.) were sent to Trinidad and Tobago from India, cultured and introduced into the islands in 1962–1963, but no recoveries were found in 1964.

Cotton leafworm

Cotton leafworm *Alabama argillacea* (Hb.) is attacked by the local parasitoids *Trichogramma* sp., *Apanteles* sp., *Brachymeria* sp., *Winthemia* spp. and *Sarcophaga* sp. Attempts at biocontrol have centred on the importation or manipulation of predatory vespid wasps, *Polistes* spp.

ST VINCENT. The Jack Spaniard wasp of St Vincent, *Polistes cinctus cinctus* Lepeletier, is an effective predator of the cotton leafworm. Initially it was assumed that the St Vincent race of this wasp was a more effective predator strain than strains on other islands and it was introduced into several islands. Later it was argued that the better control obtained on St Vincent results from the constant supply of food for *P. c. cinctus* in the form of the arrowroot leaf roller *Calpodes ethlius* (Stoll). If this view is correct, then the introductions discussed below are unlikely to lead to improved control of *A. argillacea*. *P. c. cinctus* has been encouraged in St Vincent by the erection of shelters near the cotton fields under which it can nest, which is an example of conservation biocontrol.

ANTIGUA. In 1910 and subsequently, attempts were made to establish *Polistes cinctus barbadensis* Richards from Barbados and *P. c. cinctus* from St Vincent, but they failed to establish.

GRENADA. *P. c. cinctus* was plentiful at one time but became scarce and fresh stock was introduced from St Vincent. In 1919, it was reported that the wasp was almost extinct and later it was no longer found.

MONTSERRAT. *P. c. cinctus* was introduced from St Vincent in 1910, although this subspecies was already present. In 1918, it was reported to be controlling the pest in the area where it was first established, but later control appeared insufficient. The species was found well established during surveys in 1961, 1966 and 1973.

ST KITTS. An unsuccessful attempt was made to introduce *P. c. cinctus* into St Kitts from St Vincent in 1919. However, in 1978 this subspecies was recorded on the island.

ST LUCIA. *P. c. cinctus* was introduced in 1916 from St Vincent, reported as established in some areas in 1934 and recovered in 1970.

27.2.5 Biological control of pests of cruciferous crops

Cabbage is an important crop in the region, while cauliflower and Chinese cabbage are also grown. Main pests of these crops in the Caribbean are all Lepidoptera, with cabbage budworm *Hellula phidilealis* (Wlk.) and the diamondback moth *Plutella xylostella* (L.) as the two most important. On some islands the cabbage butterfly *Ascia monuste* (L.) and the cabbage loopers *Trichoplusia ni* (Hb.) (= *Plusia brassicae* Ril.) and *Chrysodeixis includens* (Wlk.) are also important.

Diamondback moth

Diamondback moth *P. xylostella* is attacked by a variety of local parasitoids, including braconids *Apanteles* sp. and *Apanteles (Rhygoplitis) aciculatus* (Ashm.); trichogrammatid egg parasitoids *Trichogramma* sp. and *Trichogramma brasiliensis* (Ashm.); a facultative hyperparasitic eulophid *Tetrastichus sokolowskii* Kurd.; and a preferentially hyperparasitic chalcidid *Spilochalcis hirtifemora* (Ashm.). The local parasitoids failed to control diamondback moth, particularly when pesticides were regularly applied. *Apanteles (Cotesia) plutellae* Kurd was obtained from India in 1970, *Diadegma eucerophaga* Horstmann and *A. plutellae* from the Netherlands in 1971, and *D. varuna* Gupta and *Diadromus collaris* (Grav.) from India in 1972. Both *Diadegma* spp. failed to develop more than one generation in the laboratory in Trinidad and Tobago, but the two strains of *A. plutellae* and *D. collaris* were mass produced and distributed throughout the area, including Antigua, Grenada, Montserrat, St Kitts and Nevis, St Lucia and St Vincent. It seems, however, that the parasitoids hardly suppressed diamondback moth.

ANTIGUA. The Indian strain of *A. plutellae* was introduced in 1970–1972, the Dutch strain in 1972 and *D. collaris* in 1972. *A. plutellae* was recovered in 1971, indicating at least temporary establishment.

GRENADA. The Indian strain of *A. plutellae* was introduced in 1970–1972, the Dutch strain in 1972 and *D. collaris* in 1972. No recovery data are available.

MONTSERRAT. *A. plutellae* was released in 1971 and recovered in 1971 and 1973, suggesting that it had established.

ST KITTS AND NEVIS. *A. plutellae* was recovered from both St Kitts and Nevis in 1973 following releases of the Indian and Dutch strain in 1970–1972, but *D. collaris* released in 1972–1973 was not recovered.

ST LUCIA. *A. plutellae* was recovered in 1973 at its release site of the previous years, but *D. collaris* released in 1972 was not recovered.

ST VINCENT. The Indian strain of *A. plutellae* was introduced in 1970–1972, the Dutch strain in 1972 and *D. collaris* in 1972. Recovery surveys made in 1972 did not result in finding any of the two parasitoids.

Cabbage butterfly

Cabbage butterfly *A. monuste* is causing damage to crucifers in the Caribbean (for more details see Chapter 3: Barbados).

ST LUCIA. A single consignment of *Pteromalus puparum* L. was sent from Barbados and released in 1950, but no recoveries have been made.

27.2.6 Biological control of pests of sugarcane

In most of the Caribbean, sugarcane was the dominant agricultural crop during the 19th century and first half of the 20th century. Although production decreased and was no longer continued on some of the islands, it is still the most widely grown crop in the region as a whole.

West Indian cane fly

West Indian cane fly *Saccharosydne saccharivora* (Westw.) has a well developed natural enemy complex in the Caribbean, but irregular outbreaks occur leading to attempts to establish new natural enemies.

BAHAMAS. West Indian cane fly was widespread but of minor importance. Small releases were made of *Anagrus flaveolus* Waterhouse from Jamaica and Trinidad and Tobago in 1968 and from Barbados in 1972, but the parasitoid was not recovered.

Sugarcane froghopper

GRENADA. Sugarcane froghopper *Aeneolamia* spp. occurs as pests of sugarcane on Grenada, where *Metarhizium anisopliae* (Metschnikoff) Sorokin was found to infect *Aeneolamia varia* var. *saccharina* (Dist.), resulting in its control.

Yellow sugarcane aphid

The yellow sugarcane aphid *Sipha flava* (Forbes) occurs throughout the Caribbean and is an occasional pest of various Gramineae.

BAHAMAS. *Coccinella septempunctata* L. obtained from Pakistan was released in 1968 to supplement local natural enemies, but no recoveries were made.

Sugarcane mealybugs

The sugarcane mealybugs *Saccharicoccus sacchari* (Ckll.). and *Dysmicoccus boninsis* (Kuway.) occur wherever sugarcane is grown.

BAHAMAS. Since the natural enemies already present did not provide sufficient control, coccinellids *Nephus* sp., *Hyperaspis* sp. and *C. montrouzieri* were imported from India in 1968, and additional *Hyperaspis trilineata* Muls. obtained from Barbados in 1968–1969 were liberated. *Hyperaspis trilineata* was recovered several times during 1968 and 1969, but did not become abundant. *C. montrouzieri* became established and spread rapidly, and to prevents its elimination when cane fields were burnt prior to harvesting, coccinellids were collected and transferred to other fields.

ST KITTS. *Hyperaspis trilineata* adults were imported from Barbados in 1958. In 1966 recoveries were made and they were widespread in 1967, contributing to *S. sacchari* control. The encyrtid *Anagyrus saccharicola* Timb. obtained from East Africa was introduced via Trinidad and Tobago to St Kitts in 1971 and established, according to recoveries made in 1975 and 1978.

White grub larvae of beetles

ANTIGUA. The brown hard-back beetles *Phyllophaga antiguae* (Arr.) and *Phyllophaga* sp. caused severe damage to sugarcane prior to 1915. Planting of *Cordia curassavica* (Jacq.) R. & S. (= *interrupta*), which provide nectar to adults of the native tiphiid, *Tiphia parallela* F. Sm., a well known parasitoid of *Phyllophaga* spp., resulted in a reduction of beetle populations. This can be considered as an example of conservation biocontrol.

Sugarcane borers

Sugarcane borers, *Diatraea* spp., are the most widespread and in many places the most important pests of sugarcane in the Caribbean. *Diatraea saccharalis* (F.) is the most prominent of these and occurs throughout the region except the Bahamas, where it is replaced by *D. lineolata* (Wlk.). Two other species, *Diatraea centrella* (Moschler) (= *canella* Hamps.) and *D. impersonatella* (Wlk.), occur on some of the islands discussed in this chapter. Biocontrol of *Diatraea* spp. proceeded in three phases. During the first half of the 20th century, native tachinid *Diatraea* parasitoids with limited ranges within the Neotropical Region were redistributed, often leading to a degree of control. From 1929 to 1959, extensive trials were made with inundative releases of *Trichogramma* spp., principally in Barbados and Guyana. From 1950, when CABI set up its laboratory in Trinidad and Tobago, a number of exotic stem borer parasitoids were imported from Africa and Asia, and one of these, *Cotesia flavipes* (Cam.), proved particularly successful. Tables 11, 12 and 13 in Cock (1985) summarized the many introductions of natural enemies of sugarcane borers in the Caribbean.

ANTIGUA. *Ipobracon grenadensis* Ashm. and *Agathis stigmatera* Cress. from Guyana were introduced in 1926 but did not establish. In 1932 the tachinid parasitoid *Lixophaga diatraeae* (Townsend) was imported from Cuba, bred in the laboratory, released in several places, established, and sugarcane damage due to *D. saccharalis* decreased. However, in 1936, *D. saccharalis* damage increased again and additional releases were made with *L. diatraeae* from St Kitts. The parasitoid established but provided insufficient control, so inoculative seasonal releases were made annually from 1937 to the 1960s, when commercial sugarcane production ceased. *Paratheresia claripalpis* Wulp. from Trinidad and Tobago and

Lydella minense (Townsend) from Guyana were released in 1937–1938, and small releases of *C. flavipes* were made in 1970.

BAHAMAS. When commercial large-scale plantings of sugarcane started in 1965, damage during this period was caused by *D. lineolate*. In 1967 it was confirmed that *D. centrella* was present and caused serious damage. Chemical control failed and 20 species of hymenopteran parasitoids of *Diatraea* spp. and of allied stem borers were introduced in an extensive biocontrol programme during 1968–1970 (Table 14 in Cock, 1985). Also the tachinids *Jaynesleskia jaynesi* Townsend and *Palpozenillia diatraeae* Townsend were released in this period. None of the hymenopteran or tachinid species established. The sugarcane industry was closed down in 1970.

GRENADA. During 1950–1954, *L. minense*, *P. claripalpis* and *L. diatraeae* were shipped from Trinidad and Tobago and released. *Telenomus alecto* Crawford, obtained from Barbados and bred in Trinidad and Tobago, was also shipped. Other releases have been made since 1961, but none of the introduced parasitoids established.

MONTSERRAT. *Lixophaga diatraeae* was introduced from St Kitts in March 1935 for control of *D. saccharalis*, but there are no reports of recoveries. Later, other parasitoid species were introduced (Table 14 in Cock, 1985). During a recovery survey in 1965, only *L. diatraeae* was found, but no recoveries of this species were made during a survey in 1966.

ST KITTS AND NEVIS. *D. saccharalis* is the only borer present and was, prior to parasitoid introductions, attacked by *Trichogramma exiguum* Pinto & Platner, *Agathis stigmatera* (Cresson) and the fungus *Cordyceps barberi* Giard, but control was insufficient. From 1932 to 1934 *L. diatraeae* was introduced and spread over the island. Sugarcane borer injury decreased to values below the economic threshold and remained at that level. From 1970 to 1973 *L. minense*, *C. flavipes*, *Pediobius furvus* (Gah.) and *Apanteles sesamiae* Cam. were introduced, but only *C. flavipes* established. Borer damage appeared to have increased in 1978, supposedly due to use of new sugarcane varieties, pre-harvest burning and drought. In 1982 the Caribbean Agricultural Research and Development Institute (CARDI) started the release of *Allorhogas* sp..

ST LUCIA. In the 1930s infestation by *D. saccharalis* and *D. centrella* was high. In 1933, *L. diatraeae* were obtained from Antigua. Parasitoids were recovered from *D. saccharalis*, but not from *D. centrella*. The parasitoid did not sufficiently control *D. saccharalis*. *Lydella minense* was obtained from Guyana in 1934, and establishment and spread occurred rapidly. Borer damage caused by *D. saccharalis* decreased to below the economic threshold, but for *D. centrella* the reduction was less spectacular.

ST VINCENT. *D. saccharalis* predominates in plant cane and *D. centrella* usually in ratoons. Native parasitoids of borers found on the island were *T. exiguum*, *T. alecto* and *A. stigmatera*. Between 1940 and 1961 several parasitoids were introduced (Table 14 in Cock, 1985), but no recoveries were made. *Cotesia flavipes* obtained from Barbados was released in 1978 and 1982, and recoveries were made in 1979 and 1980.

27.2.7 Biological control of pests of other vegetable and field crops

Phytophagous snails

BAHAMAS. Phytophagous snails, especially *Zachrisia auricoma* (Ferussac) and *Bulimulus sepulcralalis* (Poey), occur on these islands. In 1961–1962 small shipments of the predacious snail *Euglandina rosea* (Ferussac), originating from East Africa, were obtained from Bermuda and liberated for control of *Z. auricoma* and *B. sepulcralis*; it was recovered in 1970. *Gonaxis quadrilateralis* (Preston) (origin East Africa) individuals sent to Trinidad and Tobago from Hawaii in 1968 were forwarded to the Bahamas for liberation. There are no records of recoveries.

Pigeon pea pod borers

Of the various species of pigeon pea pod borers occurring in the Caribbean, *Ancylostomia siercorea* (Zell.) is the most serious pest.

BAHAMAS. Thousands of various species of parasitoids of *A. stercorea* were sent from Trinidad and Tobago (Table 19 in Cock, 1985) and liberated into heavily attacked pigeon pea sites from 1952 to 1970. Although conditions for establishment seemed ideal, not a single recovery was made.

Arrowroot leaf roller

The arrowroot leaf roller *Calpodes ethlius* (Stoll) is a pest of arrowroot *Maranta arundinacea* L., which is grown for its fine-grained starch. St Vincent is the principal world producer, although it has also been grown in Barbados, Bermuda, Dominica, Jamaica and St Lucia, and small amounts are still grown in Antigua and Montserrat.

ST VINCENT. Here, the principal egg parasitoid of the leaf roller is a dark 'race' of *Trichogramma 'minutum* Riley', which might be the same race found parasitizing *C. ethlius* over a wide area in the Americas. Also, an eulophid, *Ardalus scutellatus* (How.), was recorded from St Vincent. Other parasitoid records from St Vincent include a chalcidid, two tachinids and two sarcophagids. In 1951–1952, *Ooencyrtus* sp., *Apanteles talidicica* Wlkn. and *A. scutellatus* were introduced from Trinidad and Tobago. *Ooencyrtus* established well. The chief predators of *C. ethlius* in St Vincent are 'Jack Spaniard' wasps, *Anolis* lizards and birds. The wasp *P. cinctus cinctus* is very active wherever there is a heavy leaf roller attack. Some farmers erected roofed shelters on wooden stands scattered through the fields to afford nesting sites for the wasps, thus encouraging the growth of the predator population, an example of conservation biocontrol.

Armyworms

In the Caribbean region, armyworms of five species of *Spodoptera* and two species of *Heliothis* are of economic importance: *S. frugiperda* (J.E. Smith), *S. latifascia* (Wlk.), *S. dolichos* (F.), *S. eridania* (Cram.), *S. sunia* (Gn.), *H. zea* (Boddie) and *H. virescens* (F.). In addition *S. exigua* (Hb.) is extending its range and will probably spread through the Caribbean. The larvae of all these species feed on a variety of crops and are general defoliators. *S. frugiperda* and *H. zea* are particularly associated with maize, *S. eridania* with cabbage and tomato, *H. virescens* with pigeon pea, and *S. frugiperda*, *S. latifascia*, *S. eridania* and *S. sunia* with cotton. Other crops that are attacked include asparagus, beans, beet, carrot, cauliflower, aubergine, okra, onion, peppers, potato, pumpkin, sugarcane, sweet potato and tobacco. *Spodoptera ornithogalli* (Gn.) does not occur in the region, except in Antigua. Many species of parasitoids of armyworms were received by CABI in Trinidad and Tobago from Asia, Europe and North and South America from 1970 to 1980 and several of these species were introduced into other Caribbean islands. After 1980, biocontrol attempts were made by CARDI.

ANTIGUA, BAHAMAS, GRENADA, ST KITTS, ST LUCIA, ST VINCENT. Several species of parasitoids have been introduced to the islands (Tables 24 and 25 in Cock, 1985), but only *Telenomus remus* Nixon established on Antigua, Dominica and St Vincent and no recoveries of other species have been made.

MONTSERRAT. *T. remus* was released on Montserrat in 1973 by CABI and later by CARDI, and established.

27.2.8 Biological control of forestry pests

Mahogany shoot borer

The mahogany shoot borer *Hypsipyla grandella* (ZeM.) is a pest of mahogany (*Swietenia macrophylla* G. King and *S. mahagoni* Jacq.), cedar (*Cedrela odorata* L.) and crappo (*Carapa guianensis* Aubl.), which are forest trees of commercial value in the Caribbean.

GRENADA, ST KITTS, ST LUCIA AND ST VINCENT. Surveys undertaken by CABI revealed several natural enemies on these islands (five braconids, two ichneumonids, two trichogrammatids, two tachinids and a mermithid), but these did not sufficiently control *H. grandella*. Several species of parasitoids were imported from Asia from 1968 to 1982 (Cock, 1985) and released. During surveys in 1970–1972, none of the released parasitoids was recovered.

27.2.9 Biological control of pests of humans and domestic animals

House and stable flies

House flies (*Musca* spp.) and stable flies (*Stomoxys* spp.) breed in large numbers in the accumulated excrement found in stables and poultry pens.

GRENADA. In 1951, a shipment of *Pachylister chinensis* Quensel, a histerid native to Malaysia and Indonesia, was sent from Trinidad and Tobago for release.

ST KITTS. *Muscidifurax uniraptor* Kogan & Legner (1966) and *Pachycrepoideus vindemmiae* Rond. (1966 and 1970) originating from California were released against house flies and fruit flies (*Anastrepha* spp.), but no recoveries were made.

ST VINCENT. *Pachylister chinensis* from Trinidad and Tobago was released in 1951. No records about establishment are available.

Mosquitoes vectoring human diseases

SINT MAARTEN. Gerberg and Visser (1978) reported: 'A preliminary field trial on the Caribbean island of St Maarten demonstrated the feasibility of using a predator mosquito larva, *Toxorhynchites brevipalpis* Theob. as a biocontrol agent for *Aedes aegypti* L. Sixteen days after the introduction of *T. brevipalpis* eggs into *A. aegypti* breeding containers, all of the 21 houses sampled no longer had *A. aegypti* breeding.' The authors concluded that the predator could be mass reared in sufficient quantities, predator eggs could be transported by air and would hatch into larvae in the climate of St Maarten. They anticipated that flooding St Maarten with predator eggs at 4-week intervals might well suppress and possibly eradicate *A. aegypti*.

27.2.10 Introduction of vertebrate natural enemies into the Caribbean

Several vertebrates have been introduced into the Caribbean, but this is no longer recommended, due to their wide food range, resulting in negative non-target effects such as preying on beneficial and economically important organisms, as well as on other valued species. Species that were not introduced for pest or weed control but that have resulted in non-target effects were discussed in Cock (1985).

Giant toad

The first introduction of the giant toad *Bufo marinus* (L.) was into Barbados from Guyana in about 1830 to control white grubs in sugarcane (see Chapter 3: Barbados). The toad was then moved from Barbados to most of the Caribbean islands. According to Cock (1985), the toads do consume large numbers of insects in cane, and although this doubtless includes many beneficial insects, on balance they are probably beneficial.

Small Indian mongoose

The small Indian mongoose *Herpestes auropunctatus* (Hodgson) was introduced into Jamaica in 1872 from India to control rats. It had spread to many of the other Caribbean islands by the end of the century. It established on Antigua, Grenada, St Kitts and Nevis, St Lucia, St Vincent and some of the Virgin Islands, among others. Early reports suggested a substantial reduction in rats and their damage, but mongoose quickly showed negative non-target effects. Lizards of *Ameiva* spp. occur widely in the Caribbean and are considered beneficial insect predators. Wherever the mongoose became established, these lizards became rare or extinct. The effects on ground-nesting birds have been similar. There is little evidence with regard to the effect on snakes. Mongooses are also widely recognized as pests of poultry and cause considerable concern as vectors of rabies. On the whole, the introduction of the mongoose has been harmful.

27.2.11 Biological control of weeds

Prickly pear

Prickly pear (*Opuntia* spp.) has been controlled in several world regions after introduction of cochineal insects of *Dactylopius* spp. originating from North America and a pyralid *Cactoblastis cactorum* (Berg) originating from Argentina. In the Caribbean, *C. cactorum* was introduced first to Nevis and then released in Antigua, the Cayman islands and Montserrat, and spread naturally to other islands in the Caribbean, including the Bahamas.

ANTIGUA. *C. cactorum* was brought from Nevis in 1960 and successfully established on *Opuntia triacantha* (Willd.) Sweet, which it controlled effectively.

CAYMAN ISLANDS. *C. cactorum* was introduced into the Cayman Islands from Nevis and Antigua in 1970 and became well established on *Opuntia dillenii* (Ker-Gawler) Haworth .

MONTSERRAT. *O. triacantha* was successfully controlled after the introduction of *C. cactorum* in 1960 from Nevis.

ST KITTS AND NEVIS. Three species, *O. dillenii*, *Opuntia lindheimeri* Engelmann and *O. triacantha*, caused most problems in Nevis. Of these, *O. triacantha* was the most serious, as it displaced pasture grass and the spines caused injury to livestock. *C. cactorum (ex* Argentina), *Dactylopius opuntiae* (Ckll.) *(ex* USA) and *D. austrinis* De Lotto (as *D.* sp. nr. *confusus* (Ckll.) *(ex* USA) were introduced into Nevis in 1957. *C. cactorum* became established and abundant, proving particularly effective against *O. triacantha.* By 1964 *O. triacantha* was scarce in most areas, the other two species were gone from pastures, but persisted along roadsides, and the programme was considered an outstanding success. Neither species of *Dactylopius* became established. In 1964, *C. cactorum* was observed in St Kitts having spread naturally across the 4-mile gap between St Kitts and Nevis

Love vine

Love vine (*Cuscuta* spp.) is a considerable nuisance to ornamental garden plants and occasionally becomes a pest of economic plants such as citrus and mango.

BAHAMAS. The dipteran leaf miner *Melanagromyza cuscutae* Hering and a seed-feeding weevil *Smicronyx roridus* Mshl. were imported from Pakistan in 1966 and 1968, but no recoveries were made.

Puncture vine

Puncture vine *Tribulus cistoides* L. is a palearctic weed that became established in the Caribbean. Biocontrol is based on the use of the weevils *Microlarinus lareynii* (Duv.) and *M. lypriformis* (Woll.). These two weevils were successfully introduced from Italy into California and Hawaii.

ST KITTS AND NEVIS. Puncture vine became established in St Kitts in the 1950s and for several years was considered an attractive ornamental. However, in 1964 when patches of the weed were reported in pasture lands and along the roads, questions about its potential as a weed and methods of control were raised. Attempts of biocontrol in St Kitts were started in 1966; the stem weevil *M. lypriformis* obtained from Hawaii in November 1966 rapidly became established and within 4 months every stem sampled near the release site was infested. Within a year solid stands of the weed had disappeared from pasture lands and it has been almost completely replaced by grass and other weeds. In 1971, *T. cistoides* comprised less than 5% of the ground cover at sites where in 1966 it covered over 80%. In an attempt to reduce further the reproductive potential of *T. cistoides*, the seed weevil *M. lareynii* was obtained from California in 1968. Although present in 1969, it was not recovered in October 1969 or March 1971. *T. cistoides* was first noticed at the airport of Nevis in 1968 and liberations of *M. lypriformis* were made by placing infested plants from St Kitts near the infestation. Early in 1969, *M. lypriformis* was well-established and sufficiently controlled puncture vine.

27.2.12 Remaining Caribbean Islands as source of biological control agents

Cock (1985) mentioned only a few cases of the Remaining Caribbean Islands being a source for export of natural enemies to countries outside the Caribbean. Many indigenous Caribbean natural enemies from other islands have been sent from Trinidad and Tobago by CABI to countries around the world.

Natural enemies of the green stink bug

The green stink bug *N. viridula* is a widespread pest in the Caribbean and export of indigenous parasitoids resulted in some biocontrol success outside the Caribbean area.

ANTIGUA AND MONTSERRAT. The tachinids *T. pilipes* collected in Montserrat and *T. basalis* collected in Antigua and Montserrat were shipped via CABI Trinidad and Tobago to Australia, Hawaii and the USA several times in the period 1952–1979. In Hawaii, *T. pilipes* established and became common. *Trissolcus basalis* also became very common, parasitizing over 90% of the green stink bug eggs. Control below the economic threshold has been achieved in several crops.

Weed biological control agents

Many aquatic and terrestrial weeds presently found throughout the tropics originated in the Neotropical region and also a number of the most successful examples of biological weed control involve species that originated in this region. In the Caribbean, it is Trinidad and Tobago that have been an important source of weed biocontrol agents (see Chapter 29: Trinidad and Tobago; and Cock, 1985). The Cock (1985) review does not mention any of the Remaining Caribbean Islands as sources of weed biocontrol agents.

27.2.13 Conclusions about biological control in the Remaining Caribbean Islands up to 1980

Substantial or complete biocontrol successes in the Remaining Caribbean Islands are summarized in Table 27.1. According to Cock (1985, p. 179):

> The projects which achieved these successes can be divided into three types: first, those using known biological control agents which have been consistently successful when used on the same pest problem elsewhere, second, the trial of agents as available from similar hosts, habitats or crops in ... the 'hit or miss approach', and third, projects based on extensive research and foreign exploration to find suitable control agents.

From the successes listed in Table 27.1, the control of the insect pests *A. woglumi*, *A. destructor* and *I. purchasi* and of the weeds *Opuntia* spp. and *Tribulus terrestris* L. are examples of the use of biocontrol agents of known effectiveness, while the programmes against *D. saccharalis* were based on extensive research. Other successes obtained in the Caribbean region, such as the control of *A. cocois*, *P. xylostella* and *S. frugiperda*, were the result of introductions based upon minimal research.

Related to releases made in the Caribbean, Cock (1985) discussed the important issue of predictability of success of biocontrol introductions, a topic also addressed in the section on 'Finding, evaluation and utilization of biological control agents' in Chapter 1 of this volume. Cock (1985) mentions that:

> It has often been suggested that if a biological control agent has been proven effective in one situation, then it is likely to be effective against the same pest elsewhere. The results of the programme against *Diatraea* spp. show that this is not always the case, and different parasitoids have been effective on different islands in a totally unpredictable manner. This result provides some justification for the 'hit or miss approach'. Until reliable predictions can be made as to the effectiveness of different types of agents, it is difficult to say that the hit or miss approach is not justified where funds are limited. However, substantial research will be needed to clarify why some agents are effective while others are not, and until such data are available, it is difficult for biological control to develop into a predictive science.

Still, when looking at the successes, one may conclude that they were often only obtained after long-term research. For example, in the project of sugarcane borer management in the Caribbean, attempts to develop biocontrol have been going on since the 1930s and more than 50 species of parasitoids have been studied as possible candidates for release (Baker *et al.*, 1992).

27.3 Current Situation of Biological Control in the Remaining Caribbean Islands

Many biocontrol programmes were implemented before the 1980s in the Caribbean, but only a few were conducted from 1980 to 1990. From 1990, the region was faced with a number of invasions by exotic pests, including the citrus leaf miner, citrus blackfly, papaya mealybug, giant African snail, coconut whitefly and pink hibiscus mealybug (Kairo *et al.*, 2003a). Invasive species have long been a challenge to Caribbean agriculture, but the problem has been amplified in recent times as the movement of goods and people has

Table 27.1. Biological control successes achieved with arthropod natural enemies during the period 1880–1980 in the Remaining Caribbean islands (retrieved from Cock, 1985).

Biological control agent	Pest/weed	Remaining Caribbean Island
Cotesia flavipes (Cam.)	*Diatraea saccharalis* (F.)	Antigua
Cactoblastis cactorum (Berg)	*Opuntia* spp.	Many islands
Cryptognatha nodiceps (Mshl.)	*Aspidiotus destructor* Sign.	St Kitts and Nevis
Cryptolaemus montrouzieri Mulsant	*Nipaecoccus nipae* Mask.	St Kitts
Encarsia opulenta (Silv.), *Eretmocerus serius* Silv.	*Aleurocanthus woglumi* Asby	Bahamas, Cayman Islands
Lixophaga diatraea Tns.	*Diatraea saccharalis* (F.)	Antigua, St Kitts
Lydella minense Tns.	*Diatraea saccharalis* (F.)	St Lucia
Trichogramma spp.	*Diatraea lineolate* (Wlk.)	Bahamas
Metarhizium anisopliae	*Aeneolamia* spp.	Grenada
Microlarinus lypriformis (Woll.)	*Tribulus cistoides* L.	St Kitts and Nevis
Native natural enemy complex	*Saccharosydne saccharivora* (Westw.)	Several islands
Polistes cinctus cinctus Lepeletier	*Alabama argillacea* (Hb.)	St Vincent
Polistes cinctus cinctus Lepeletier	*Calpodes ethlius* (Stoll)	St Vincent
Rodolia cardinalis (Muls.)	*Icerya purchasi* Mask.	Bahamas, Cayman Islands, Montserrat, St Kitts and Nevis
Tiphia parallela F. Sm.	*Phyllophaga antiguae* (Arr) and *Phyllophaga* sp.	Antigua

increased strongly. In a report prepared by CABI (Kairo *et al.*, 2003b), a shortlist of 23 major invasive species threats was presented occurring in five or more (up to 16 for some species) Caribbean countries. Kairo *et al.* (2003b) concluded that an approach which minimizes the entry of alien species or allows for early detection before establishment and spread would considerably reduce the overall cost of elimination or management. However, such an approach demanded cooperation among the many countries in the Caribbean region concerning trading of agricultural commodities as well as tourist activities, an issue not easily resolved. Still, at least two early detection programmes involving region-wide collaboration appear to have resulted in successful biocontrol: that of the pink hibiscus mealybug *Maconellicoccus hirsutus* (Green) and the papaya mealybug *Paracoccus marginatus* Williams & Granara de Willink. In both programmes, the introduction of exotic parasitoid species resulted in mealybug population density reductions ranging from 82% to 97%. Early programme development allowed for quick technology transfer to newly infested Caribbean islands and to the USA mainland within 30 days of being found infested (Meyerdirck and DeChi, 2003).

Examples of other invasive species that have entered and for which biocontrol attempts have been made are the whitefly *Bemisia tabaci* (Gennadius), the citrus leaf miner *Phyllocnistis citrella* Stainton and the citrus blackfly *A. woglumi*, the imported red fire ant *Solenopsis invicta* Buren, the coconut whitefly *Aleurodicus pulvinatus* (Maskell), the red palm mite *Raoiella indica* Hirst and the melon thrips *Thrips palmi* Karny. The most important successes are listed in Table 27.2.

27.3.1 Biological control of pests of citrus

Citrus leaf miner

BAHAMAS. *Ageniaspis citricola* Logviniskaya was introduced into the Bahamas in 1996 for control of *P. citrella* (Kairo *et al.*, 2003a).

Citrus blackfly

ST KITTS AND NEVIS. Colmenarez *et al.* (2018) mentioned that in the 1990s Dominica, Guyana,

Table 27.2. Biological control successes achieved after 1980 in the Remaining Caribbean Islands.

Biological control agent	Pest	Remaining Caribbean Island
Amitus hesperidum, Encarsia perplexa	*Aleurocanthus woglumi* Ashby	St Kitts and Nevis
Anagyrus kamali Moursi	*Maconellicoccus hirsutus* (Green)	Several islands
Apoanagyrus spp., *Anagyrus* spp., *Acerophagus* spp.	*Paracoccus marginatus* Williams & Granara de Willink	Several islands
Native natural enemy complex	*Aleurodicus pulvinatus* (Maskell)	Several islands
Native natural enemy complex	*Planococcus minor* Maskell	Several islands

French Guiana and St Kitts and Nevis experienced the resurgence of citrus blackfly *A. woglumi*, resulting in serious problems in citrus. A classical biocontrol programme was initiated to control the pest, with the introduction of *Amitus hesperidum* Silvestri and *Encarsia perplexa* Huang and Polaszek, resulting in control of the pest (White *et al.*, 2005).

27.3.2 Biological control of pests of coconut

Coconut whitefly

NEVIS. The coconut whitefly *A. pulvinatus* is a serious pest of coconut and many ornamental species and caused serious problems on Nevis, though on several other Caribbean islands it is kept under natural biocontrol. Natural enemies were also found on Nevis, but they did not sufficiently control the whitefly. In an FAO-supported project, a survey for natural enemies was carried out in Trinidad and Tobago in 1995. The natural enemy complex attacking *Aleurodicus* spp. includes species of Aphelinidae in two genera (*Encarsiella* and *Encarsia*), one Encyrtidae (*Metaphycus* sp.) and several coccinellid species of the genus *Nephaspis* (Kairo *et al.*, 2001); *Encarsiella* sp. D was introduced into Nevis in 1998 and has established.

Red palm mite

The red palm mite *R. indica* was first reported in the Caribbean in 2004. It is now widely distributed in the region and may cause severe damage to Araceae (in particular, coconut *Cocos nucifera* L.), but also to Musaceae and other plants. *Amblyseius largoensis* (Muma) is very common on coconut palms and may play a role in reducing palm mite populations. Also, entomopathogenic fungi, possibly *Hirsutella* spp., have been found to infect the palm mite. However, natural biocontrol by these beneficial organisms is insufficient. Colmenarez *et al.* (2014) made an inventory of entomopathogenic fungi associated with red palm mite in the Caribbean, including Antigua, St Kitts and Nevis. Of the 27 fungal isolates identified, three belonging to *Simplicillium* sp., representing a possible undescribed taxon, and *Penicillium* sp. might be interesting for future evaluation. *Simplicillium* isolates (formerly identified as *Verticillium*) have been used as commercial biocontrol agents to control pests such as whiteflies, thrips and aphids. According to Colmenarez *et al.* (2014), it seems that the level of control by these naturally occurring entomopathogenic fungi is low.

Coconut scale

Many introductions of coccinellids were made into Caribbean islands for the control of the coconut scale *Aspidiotus destructor* Sign. between 1937 and 1973, but with little success, except in St Kitts and Nevis (see Section 27.2). The predatory beetle *Cybocephalus nipponicus* Endrödy-Younga of Asia was introduced into North America from 1989 and released in Florida in 1999 (Smith and Cave, 2006). It has been reported from the Cayman Islands and St Kitts and Nevis, among others, attacking the coconut scale (Smith and Cave, 2007).

27.3.3 Biological control of pests of other tree crops and ornamentals

Papaya mealybug

The papaya mealybug *P. marginatus* causes serious damage to tropical fruit, especially papaya and hibiscus, and its host range includes at least

55 plant species. In the Remaining Caribbean Islands it occurs on Antigua, Montserrat, Nevis, St Kitts, Sint Maarten and the Virgin Islands. This mealybug is supposed to have been introduced to the Caribbean around 1993, since when it has spread over most of the Caribbean archipelago (Miller *et al.*, 1999; Meyerdirck and DeChi, 2003). The potential for successful biocontrol was rated high, as the mealybug seemed under natural control in Mexico and five parasitoids had already been collected there: *Anagyrus californicus* (Compere), *Anagyrus loecki* Noyes and Menezes, *Acerophagus papayae* Noyes and Schauff, *Pseudleptomastix mexicana* Noyes and Schauff. and *Pseudaphycus angelicus* (Howard). The parasitoids were screened by the US Department of Agriculture's Agriculture Research Service (USDA-ARS) and an environmental assessment was made. The parasitoids were then shipped to Puerto Rico, where they were reared and field released, resulting in complete biocontrol (see Chapter 26: Puerto Rico). A similar success was obtained in the Dominican Republic (see Chapter 12: Dominican Republic) (Meyerdirck and DeChi, 2003). *Cryptolaemus montrouzieri* was introduced into Antigua in 1998 for control of the papaya mealybug (Kairo *et al.*, 2003a).

Pink hibiscus mealybug

Pink hibiscus mealybug *M. hirsutus* attacks the new flush growth, young shoots, flowers and fruits of a wide range of plants, particularly those in the family Malvaceae, but also in crops like cacao, okra, mango, plums, sorrel and soursop *Annona muricata* L. and trees such as samaan, teak and blue mahoe (Kairo *et al.*, 2000; Clarke-Harris and Lauckner, 2005). The pest was first reported in the Caribbean in 1994 in Grenada and by the beginning of 2001 it had spread to over 25 territories, from Guyana and Venezuela in the south to Bahamas (see also country-specific chapters), with 28 Caribbean territories having the pest in 2003 (Clarke-Harris and Lauckner, 2005). A programme was developed for the introduction, multiplication and release of three ladybird beetles, *C. montrouzieri*, *Scymnus coccivora* Aiyyar and *Nephus regularis* Sicard. These ladybirds were imported into the region in 1996 and local natural enemies were studied as well (Clarke-Harris and Lauckner, 2005). Biocontrol for the pink hibiscus mealybug was first developed in St Kitts and Nevis between 1995 and 1997 and was eventually based on releases of *Anagyrus kamali* Moursi. This parasitoid has since been transferred to several of the Remaining Caribbean Islands (Meyerdirck and DeChi, 2003). In all islands where biocontrol was implemented, this resulted in successful control, which is summarized in Kairo *et al.* (2000).

ANTIGUA. *A. kamali* was introduced into Antigua in 2001 for control of hibiscus mealybug (Kairo *et al.*, 2003a).

BAHAMAS. The Bahamas were found to be infested in November 2000. After *A. kamali* parasitoid introduction, the mealybug population density was reduced within a year by 82% (Meyerdirck and DeChi, 2003).

CURAÇAO. *A. kamali* was introduced into Curaçao in 1999 for control of hibiscus mealybug (Kairo *et al.*, 2003a).

GRENADA. The pink hibiscus mealybug appeared in the Caribbean for the first time in Grenada in 1994 and the first biocontrol project was an FAO Technical Cooperation Project started in 1995, followed by a 15-country FAO funded project (Kairo *et al.*, 2000). Interestingly, it was one of the first projects for which ISPM No. 3 (see below) was applied. While there were risks associated with some of the agents, particularly the generalist mealybug predator *C. montrouzieri*, these were considered of less importance than those posed by the pest and thus introductions were made in 1997 (Cock, 2002). In 1998, the predator *S. coccivora* was introduced. Also the parasitoids *A. kamali* (1996; origin China and Hawaii) and *Gyranusoidea indica* Shafee, Alam & Argarwal (1998, origin Egypt) were introduced, sourced through the USDA. Complete biocontrol of the pink hibiscus mealybug was obtained after these natural enemy releases.

MONTSERRAT. *A. kamali* was introduced into Montserrat in 1997 (Kairo *et al.*, 2003a).

ST KITTS. *A. kamali* and *S. coccivora* (both in 1996) and *C. montrouzieri* (1997) were introduced into St Kitts (Kairo *et al.*, 2003a). By January of 1998, the parasitoid *A. kamali* had reduced the mealybug population by 92% (Meyerdirck and DeChi, 2003).

ST LUCIA. *A. kamali* and *C. montrouzieri* were introduced into St Lucia in 1997 (Kairo *et al.*, 2003a).

ST VINCENT. *A. kamali* was introduced into St Vincent in 1997 (Kairo *et al.*, 2003a).

VIRGIN ISLANDS. Pink hibiscus mealybug was first found in 1997 and after releases of parasitoid *A. kamali*, the mealybug populations decreased by more than 90% (Meyerdirck and DeChi, 2003).

Passion vine mealybug

The passion vine mealybug *Planococcus minor* Maskell, native to Asia, is a polyphagous pest with a host range exceeding 200 plant species and has been recorded from several countries in the Caribbean (Kairo *et al.*, 2008), including Antigua, Grenada, St Lucia and the Virgin Islands. In the Caribbean, it seems that this mealybug is restricted to only a few plants, with cocoa as important host. Currently, mealybug populations are low at all locations and the pest is attacked by several native predators and parasitoids (Table 2 in Kairo *et al.*, 2008; Roda *et al.*, 2013).

27.3.4 Biological control of pests of cotton

Whiteflies

In the Caribbean, three species of whitefly, *Trialeurodes abutilonea* (Haldeman), *T. vaporariorum* (Westwood) and *B. tabaci*, are known to be able to transmit geminiviruses in addition to the direct damage they may cause. *Bemisia tabaci* is the most damaging of these three, feeding on more than 500 species of plants, including at least 17 crops in the Caribbean. In 2002, CARDI initiated a project on integrated pest management (IPM) of whitefly, including biocontrol components. Information provided in Clarke-Harris and Lauckner (2005) showed that research on whitefly biocontrol was done in Grenada and St Kitts and Nevis.

27.3.5 Biological control of pests of cruciferous crops

Diamondback moth

ST KITTS. Based on a 2-year field study on St Kitts, Yencho *et al.* (1987) reported that presence of high densities of the parasitoid *A. plutella* delayed the time to the first pesticide spray and that parasitism levels exceeding 25–35% may control diamondback moth.

27.3.6 Biological control of pests of sweet potato

Sweet potato weevils

Sweet potato *Ipomoea batatas* (L.) Lam. is one of the most important food crops in developing countries, including much of the Caribbean basin. Insect pests rank as one of the top three production problems for sweet potato, and the sweet potato weevil *Cylas formicarius* (Summers) is the most important pest in the region. In some Caribbean islands (e.g. St Vincent), the West Indian sweet potato weevil *Euscepes postfasciatus* (Fairmaire) is the predominant weevil species (Jackson *et al.*, 2002). Jackson *et al.* (2002) provided a list of other pests occurring on sweet potato in the Caribbean region and gave an overview of IPM methods to control pests on this crop, including biocontrol. They remarked that natural biocontrol agents should be conserved through judicious use of pesticides. Predatory ants, nematodes and entomopathogens (especially *B. bassiana* [Bals.] Vuill. and *M. anisopliae*) were mentioned as potentially effective agents against weevils in sweet potato (Jackson *et al.*, 2002).

27.3.7 Biological control of pests of other vegetable and field crops

Phytophagous mites

Frank *et al.* (1992) published distribution records for *Oligota minuta* Cameron, a polyphagous predator of arthropods, including tetranychid mites such as cassava green mite *Mononychellus tanajoa* (Bondar). They mentioned that this predator was found in Antigua, the Bahamas, Montserrat, Nevis and the Virgin Islands, among others.

27.3.8 Biological control of pests of humans and domestic animals

Mosquitoes vectoring human diseases

ANTIGUA. Fifteen Caribbean strains of copepods, including a strain from Antigua, were assessed for their predation ability against mosquito larvae by Rawlins *et al.* (1997) in order to find a biocontrol tool for the dengue vector *Ae. aegypti*. *Mesocyclops* sp. from Antigua showed a high percentage of predation of *A. aegypti*, but not of *Culex quinquefasciatus* Say. According to Rawlins *et al.* (1997), the availability of mosquito-larvivorous copepods in the Caribbean region offered a good promise for control of *Ae. aegypti* now that appropriate strains of *Macrocyclops* and *Mesocyclops* had been identified.

Fire ants

The fire ants *Solenopsis richteri* Forel and *Solenopsis invicta* Buren occur in a number of Caribbean islands, including Antigua, the Bahamas and the Virgin Islands, and it is foreseen that more islands will be infested in the future (Williams and deShazo, 2004). Fire ants are aggressive when their nests are disturbed and cause painful stings to humans and other animals. In South America, 40 species of parasitoids, pathogens, predators and competitors are believed to be the major controls of fire ant density and a review of natural enemies of fire ants was published by Williams *et al.* (2003). In the USA, biocontrol of fire ants has recently been evaluated; and decapitating parasitic flies native to South America (*Pseudacteon tricuspis* Borgmeier, *Pseudacteon curvatus* Borgmeier and *Pseudacteon litoralis* Borgmeier) as well as a protozoan pathogen (*Thelohania solenopsae* Knell, Allen & Hazard) might be useful for fire ant control in the Caribbean (Williams and deShazo, 2004).

27.4 New Developments of Biological Control in the Remaining Caribbean Islands

Two recent developments related to biocontrol in the Remaining Caribbean Islands – regulations for import of exotic biocontrol agents and the implementation of Farmers Field Schools – are summarized below.

27.4.1 The effect of regulations on implementation of biological control in the Caribbean

During the past 40 years, guidelines and regulations have been developed concerning import and release of exotic natural enemies, which are discussed in Chapters 1 and 32. Kairo *et al.* (2003a) evaluated the effect of one of these guidelines, the Code of Conduct for the Import and Release of Exotic Biological Control Agents, which was endorsed by members of the FAO in 1995 and became the International Standards for Phytosanitary Measures (ISPM) No. 3 under the International Plant Protection Convention in 1996 (IPPC, 1996; FAO, 1997). The Code of Conduct was developed as a result of growing awareness that introduction of exotic natural enemies without proper evaluation of the risks might result in negative impacts to the environment and beneficial and other valued organisms. The Code was considered particularly important for countries with limited expertise in biocontrol; and the preparation of a dossier prior to each biocontrol agent introduction according to the Code was considered essential. The latest update of ISPM 3, now 'Guidelines for the export, shipment, import and release of biological control agents and other beneficial organisms', was published in 2017 (IPPC, 2017).

Kairo *et al.* (2003a, p. 15N) used several biocontrol projects from the Caribbean region to review the use of the Code of Conduct during the first years after its implementation, and they concluded that:

> Either ISPM No. 3 or similar national procedures were applied in most cases to support decisions regarding import and release of exotic biological control agents since 1996. It has provided a mechanism for formalizing current good practice and provided internationally accepted

standards to countries with little experience in implementing biological control. It provides a good basis for facilitation of regional projects.

27.4.2 Implementation of Farmers Field Schools in the Caribbean region

Farmers Field Schools (FFSs) use participatory and ecological approaches for field testing and local adaptation of innovative practices and knowledge in different technical areas. They have played and still play an important role in enabling farmers to apply IPM and biocontrol, thus becoming less dependent on chemical pesticides in several parts of the world. The FFS philosophy, its way of work and evaluation of several large projects are described in Van den Berg and Jiggins (2007) and Lopez and Ramroop (2014). FFSs were introduced into the Caribbean during 2002–2003 to address the indiscriminate use of toxic pesticides and the consequent negative impact on the environment and human health. Since then, they have spread within the Caribbean (Paul, 2016; Paul *et al.*, 2016). The FAO Sub-regional Office for the Caribbean in Barbados recently analysed progress made by FFSs in the Caribbean. Paul *et al.* (2016, p. 1) summarized the main results as follows:

> In 2002, Trinidad and Tobago was the first country to be introduced to the FFS methodology, with the implementation of a Training of Master Trainers for participants from six countries (Dominica, Dominican Republic, Haiti, Jamaica, Suriname and Trinidad and Tobago) under an EU-funded Regional Pilot Project. In 2003, the six countries embarked on a Training of Trainers under the same project. Over the next 3-4 years, field schools were organized in some of the project countries (e.g. Dominica, Suriname and Trinidad and Tobago). Guyana successfully mobilized funding from the Guyana Rice Development Board (GRDB) to launch a commodity (rice) FFS in June 2003. St Lucia launched a FFS-TOT as part of an EU-funded project implemented by FAO. Antigua did likewise in 2013 with the launch of the Zero Hunger Challenge Initiative and St Kitts and Nevis became engaged in May 2015, through the FAO project TCP/STK/3501.

Since the start of FFSs in the Caribbean, thousands of farmers have been trained. Lopez and Ramroop (2014) provided information on seven case studies, including the effect of FFSs on IPM, and concluded (p. 21) that:

> ... field schools are still a relatively new concept in some Caribbean countries. The process of developing an understanding and trust amongst the different stakeholders is gradual, but the approach is gaining support from government and communities. The promotion of farmer participatory approaches is in keeping with the commitment of most countries for an improved approach to pest management and to the delivery of agricultural extension services to farmers in the Caribbean.

27.4.3 Final remarks

Chapter 1 presented an overview of organizations that study, implement or coordinate biocontrol activities in the region. In the Caribbean basin, the following organizations are involved: the Trinidad and Tobago Station of the Centre for Agriculture and Biosciences International (CABI), the Caribbean Agricultural Research and Development Institute (CARDI), the Inter-American Institute for Cooperation on Agriculture (IICA), the Tropical Agriculture Research and Higher Education Center (CATIE), the International Regional Organization for Plant Protection and Animal Health (OIRSA), and the United Nations Food and Agriculture Organization Regional Office for Latin America and the Caribbean (FAO).This list of organizations is impressive, but taking the large number of invasive organisms that have established in the Caribbean into account, and the exotic species expected to establish in the near future (Kairo *et al.*, 2003b; Clarke-Harris and Lauckner, 2005), available funding for development of IPM and biocontrol programmes seems insufficient for finding sustainable pest control solutions. Other factors hampering implementation of biocontrol are the poorly funded and limited size of extension services, which results in farmers obtaining pest control information only from chemical companies. Also, pesticide regulations are often not enforced. The result is that pesticide misuse and overuse are common. An important recent factor stimulating implementation of IPM and biocontrol in the Caribbean region concerns the strict export demands for the North American and European market, which prohibit the use of large groups of pesticides.

References

(References with grey shading are available as supplementary electronic material)

Baker, P.S., Khan, A., Mohyuddin, A.I. and Waage, J.K. (1992). Overview of biological control of Lepidoptera in the Caribbean. *Florida Entomologist* 74(4), 477–483.

Boughton, A.J., Mendez, M.A., Francis, A.W., Smith, T.R., Osborne, L.S. and Mannion, C.M. (2015) Host stage suitability and impact of *Encarsia noyesi* (Hymenoptera: Aphelinidae) on the invasive rugose spiraling whitefly, *Aleurodicus rugioperculatus* (Hemiptera: Aleyrodidae), in Florida. *Biological Control* 88, 61–67.

CIA (2018a) The World Factbook: Antigua and Barbuda. Available at: https://www.cia.gov/library/publications/the-world-factbook/geos/ac.html (accessed 22 July 2019).

CIA (2018b) The World Factbook: Aruba. Available at: https://www.cia.gov/library/publications/the-world-factbook/geos/aa.html (accessed 22 July 2019).

CIA (2018c) The World Factbook: Curacao. Available at: https://www.cia.gov/library/publications/the-world-factbook/geos/uc.html (accessed 22 July 2019).

CIA (2018d) The World Factbook: Bahamas. Available at: https://www.cia.gov/library/publications/the-world-factbook/geos/bf.html (accessed 22 July 2019).

CIA (2018e) The World Factbook: Cayman Islands. Available at: https://www.cia.gov/library/publications/the-world-factbook/geos/cj.html (accessed 22 July 2019).

CIA (2018f) The World Factbook: Grenada. Available at: https://www.cia.gov/library/publications/the-world-factbook/geos/gj.html (accessed 22 July 2019).

CIA (2018g) The World Factbook: Montserrat. Available at: https://www.cia.gov/library/publications/the-world-factbook/geos/mh.html (accessed 22 July 2019).

CIA (2018h) The World Factbook: St Kitts and Nevis. Available at: https://www.cia.gov/library/publications/the-world-factbook/geos/sc.html (accessed 22 July 2019).

CIA (2018i) The World Factbook: St Lucia. Available at: https://www.cia.gov/library/publications/the-world-factbook/geos/st.html (accessed 22 July 2019).

CIA (2018j) The World Factbook: St Vincent. Available at: https://www.cia.gov/library/publications/the-world-factbook/geos/vc.html (accessed 22 July 2019).

CIA (2018k) The World Factbook: Virgin Islands. Available at: https://www.cia.gov/library/publications/the-world-factbook/geos/vq.html (accessed 22 July 2019).

Clarke-Harris, D.O. and Lauckner, F.B. (2005) Research efforts to deal with invasive species in the Caricom region. *Proceedings of the Caribbean Food Crops Society* 41(1), 125–150.

Cock, M.J.W. (ed.) (1985) *A Review of Biological Control of Pests in the Commonwealth Caribbean and Bermuda up to 1982*. Technical Communication No. 9, Commonwealth Institute of Biological Control. Commonwealth Agricultural Bureaux, Farnham Royal, UK.

Cock, M.J.W. (2002) Risks of non-target impact versus stakeholder benefits in classical biological control of arthropods: selected case studies from developing countries. In: Van Driesche, R. (ed.) *Proceedings of the First International Symposium on Arthropod Biological Control*, Honolulu, Hawaii, USA, 13–18 January 2002. USDA Forest Service, Morgantown, West Virginia, pp. 25–33.

Colmenarez, Y., Moore, D., Polar, P. and Vasquez, C. (2014) Population trends of the red palm mite, *Raoiella indica* Hirst (Acari: Tenuipalpidae) and associated entomopathogenic fungi in Trinidad, Antigua, St Kitts and Nevis and Dominica. *Acarologia*, 54(4), 433–442. DOI: 10.1051/acarologia/20142141

Colmenarez, Y.C., Corniani, N., Mundstock, J.S., Sampaio, M.V. and Vásquez, C. (2018) Use of Parasitoids as a Biocontrol Agent in the Neotropical Region: Challenges and Potential. DOI: 10.5772/intechopen.80720. Available at: https://www.intechopen.com/online-first/use-of-parasitoids-as-a-biocontrol-agent-in-the-neotropical-region-challenges-and-potential (accessed 25 July 2019).

FAO (1997) Code of Conduct for the Import and Release of Exotic Biological Control Agents. *Biocontrol News and Information* 18, 119N–124N

Frank, J.H., Bennett, F.D. and Cromroy, H.L. (1992) Distribution and prey records for *Oligota minuta* (Coleoptera: Staphelinidae), a predator of mites. *Florida Entomologist* 75(3), 376–380.

Gerberg, E.J. and Visser, W.M. (1978) Preliminary field trial for the biological control of *Aedes aegypti* by means of *Toxorhynchites brevipalpis*, a predatory mosquito larva. *Mosquito News* 38(2), 197–200.

IPPC (1996) *Code of Conduct for the Import and Release of Exotic Biological Control Agents*. International Standards for Phytosanitary Measures No. 3. International Plant Protection Convention. Food and Agriculture Organization of the United Nations (FAO), Rome.

IPPC (2017) *Guidelines for the export, shipment, import and release of biological control agents and other beneficial organisms*. International Standards for Phytosanitary Measures No. 3. International Plant Protection Convention. Food and Agriculture Organization of the United Nations (FAO), Rome. Available at: https://www.ippc.int/static/media/files/publication/en/2017/05/ ISPM_03_2005_En_2017-05-23_PostCPM12_InkAm.pdf (accessed 17 May 2019).

Jackson, D.M., Bohac, J.R., Dalip, K.M., Lawrence, J., Clarke-Harris, D., McComie, L., Gore, J., McGlashan, D., Chung, P., Edwards, S., Tolin, S. and Edwards, C. (2002) Integrated Pest Management of sweetpotato in the Caribbean. Proceedings of the 1st International Symposium on Sweetpoptato. *Acta Horticultura* 583, 143–154.

Kairo, M.T.K., Pollard, G.V., Peterkin, D.D. and Lopez, V.F. (2000) Biological control of the hibiscus mealybug, *Maconellicoccus hirsutus* Green (Hemiptera: Pseudococcidae) in the Caribbean. *Integrated Pest Management Reviews* 5, 241–254.

Kairo, M.T.K., Lopez, V.F., Pollard, G.V. and Hector, R. (2001) Biological control of the coconut whitefly, *Aleurodicus pulvinatus*, in Nevis. *Biocontrol News and Information* 22, 45N–50N.

Kairo, M.T.K., Cock, M.J.W. and Quinlan, M.M. (2003a) An assessment of the use of the Code of Conduct for the Import and Release of Exotic Biological Control Agents (ISPM No. 3) since its endorsement as an international standard. *Biocontrol News and Information* 24, 15N–27N.

Kairo, M.T.K., Cheeseman, O.D., Ali, B.S., Haysom, K. and Murphy, S.T. (2003b) Dangerous invasive species threatening or with a foothold in the Caribbean. *Proceedings of the Caribbean Food Crops Society* 39(1), 12–22.

Kairo, M.T.K., Francis, A. and Roda, A. (2008) Developing strategic research for biological control of new pest threats: the passion vine mealybug, *Planococcus minor* as a case study. *Proceedings of the Caribbean Food Crops Society* 44(1), 118–123.

Landenweb (2019) Bonaire. Available at: https://www.landenweb.nl/bonaire/economie/ (accessed 21 July 2019).

Lopez, V. and Ramroop, D. (2014) Escuelas de campo de agricultores en el Caribe [Farmer Field Schools in the Caribbean]. In: Sacedo, S. and Guzmán, L. *Agricultura familiar en América Latina y el Caribe*. Organización de las Naciones Unidas para la Alimentación y la Agricultura (FAO) Santiago, Chile, pp. 233–252. Available at: http://www.fao.org/docrep/019/i3788s/i3788s.pdf (accessed 17 May 2019).

Meyerdirck, D.E. and DeChi, L.W. (2003) Models for minimizing risks of dangerous pests: the pink hibiscus mealybug and papaya mealybug. *Proceedings of the Caribbean Food Crops Society* 39(1), 12–22.

Miller, D.R., Williams, D.J. and Hamon, A.B. (1999) Notes on a new mealybug (Hemiptera: Coccoidea: Pseudococcidae) pest in Florida and the Caribbean: the papaya mealybug, *Paracoccus marginatus* Williams and Granara de Willink. *Insectamundi* 13(3–4), 179–181.

Paul, R. (2016) *Stocktaking of Farmer Field Schools in the Caribbean*. Food and Agriculture Organization Sub-regional Office for the Caribbean, Barbados.

Paul, R., Lopez, C. and Allara, M. (2016) *Stocktaking of farmer field schools in the Caribbean*. Paper presented at the 52nd CFCS Annual Meeting, Guadeloupe, July 10–16, 2016.

Rawlins, S.C., Martinez, R., Wiltshire, S., Clarke, D., Prabhakar, P. and Spinks, M. (1997) Evaluation of Caribbean strains of *Macrocyclops* and *Mesocyclops* (Cyclopoidae: Cyclopidae) as biological control tools for the dengue vector *Aedes aegypti*. *Journal of the American Mosquito Control Association* 13, 18–23.

Roda, A., Francis, A., Kairo, M.T.K and Culik, M. (2013) *Planococcus minor* (Hemiptera: Pseudococcidae): bioecology, survey and mitigation strategies. In Peña, J.E. (ed.) *Potential Invasive Pests of Agricultural Crops*. CAB International, Wallingford, UK, pp. 288–300.

Smith, T.R. and Cave, R.D. (2006) The life history of *Cybocephalus nipponicus* a predator of the cycad aulacaspis scale, *Aulacaspis yasumatsui* (Homoptera: Diaspididae). *Proceedings of the Entomological Society of Washington* 108, 905–916.

Smith, T.R. and Cave, R.D. (2007) The Cybocephalidae (Coleoptera) of the West Indies and Trinidad. *Annals of the Entomological Society of America* 100, 164–172.

van den Berg, H. and Jiggins, J. (2007) Investing in farmers – the impacts of farmer field schools in relation to Integrated Pest Management. *World Development* 35(4), 663–686.

White, G.L., Kairo, M.T.K. and Lopez, V. (2005) Classical biological control of the citrus blackfly *Aleurocanthus woglumi* by *Amitus hesperidum* in Trinidad. *BioControl* 50, 751–759.

Wikipedia (2019a) Bonaire. Available at: https://en.wikipedia.org/wiki/Bonaire (accessed 29 July 2019).

Wikipedia (2019b) Saba. Available at: https://en.wikipedia.org/wiki/Saba (accessed 29 July 2019).

Wikipedia (2019c) Sint Eustatius. Available at: https://en.wikipedia.org/wiki/Sint_Eustatius (accessed 29 July 2019).

Wikipedia (2019d) Sint Maarten. Available at: https://en.wikipedia.org/wiki/Sint_Maarten (accessed 29 July 2019).

Williams, D.F., Oi, D.H., Porter, S.D., Pereira, R.M. and Briano, J.A. (2003) Biological control of imported fire ants. *American Entomologist* 49, 144–155.

Williams, D.F. and deShazo, R.D. (2004) Biological control of fire ants: an update on new techniques. *Annals of Allergy, Asthma & Immunology* 93, 15–22.

Yencho, G.C., Blanchette, T. and Dolphin, E. (1987) Diamondback moth, *Plutella xylostella* (L.), control studies on cabbage in St Kitts, 1987. *23rd Annual Meeting of the Caribbean Food Crops Society*, Antigua, pp. 190–196.

28 Biological Control in Suriname

Alies van Sauers-Muller* and Maitrie Jagroep

Agricultural Experiment Station, Ministry of Agriculture, Animal Husbandry and Fisheries, Paramaribo, Suriname

* E-mail: aliesmuller@yahoo.com

Abstract
Research on biological control in Suriname started in the 1910s with reports of native entomopathogenic fungi attacking pests in coffee, followed by native predators attacking pests in cotton in the 1920s. Many other native natural enemies of crop pests were identified in the period up to the 1970s and prospecting for natural enemies has been a continuous activity in Suriname. An example of successful biocontrol is that of the coconut caterpillar, which is attacked by various native parasitoids, a fungus and several predators. A number of other pests hardly pose a problem for the crops they attack, due to the presence of native natural enemies. In cassava, for example, the green cassava mite seldom causes damage to the crop. Also in citrus, several scales are kept under natural control by parasitoids, a ladybird and a mite, resulting in little use of pesticides. A recent example of classical biocontrol is that of the pink hibiscus mealybug, which was rapidly brought under control after its introduction by the release of a predatory beetle and a parasitoid. Recently, much work has been done on IPM of the invasive carambola fruit fly, which may include a biocontrol element involving the release of exotic parasitoids.

28.1 Introduction

Suriname has an estimated population of almost 600,000 (July 2017) (CIA, 2017). Its main agricultural, fishery and forestry products are: rice, bananas, coconuts, plantains, peanuts, vegetables, beef, chickens, shrimp, tuna and lumber.

28.2 History of Biological Control in Suriname

28.2.1 Period 1880 –1969

Suriname was a Dutch colony from since 1667 until independence in 1975. Agricultural research started in December 1904 with the foundation of the Agricultural Research Station, where many researchers worked on agricultural pest problems. We summarize this work below and in Table 28.1, as it may be of use in future biocontrol programmes in Latin America and the Caribbean.

Van Dinther (1960) listed a number of pathogens, predators and parasitoids that had been recorded by numerous researchers over the years in Suriname, and also mentioned natural enemies that have a significant impact on pests (Table 28.1). Some projects mentioned in Table 28.1 are summarized below.

Natural control of hawk moth in tomato and pepper

The eggs of the hawk moth *Manduca (Protoparce) sexta paphus* (Cr.) may become parasitized up to 99% by *Telenomus connectans* Ashm. Only one parasitoid emerges for each hawk moth egg.

Natural control of palm caterpillar in coconut palm

The palm or coconut caterpillar *Brassolis sophorae sophorae* (L.) is predated upon by the birds greater ani *Crotophaga major* Gmelin and smooth-billed ani *Crotophaga ani* L. (kaw-foetoeboi, or kaw-futuboi) and also by the monkey *Saimiri sciureus* (L.). The egg parasitoid *Telenomus nigrocoxalis* Ashm. is responsible for regularly attacking the egg clusters of this moth. Two to seven parasitoids developed in one egg, from which they emerge through a common exit hole. The larval parasitoids *Winthemia pinguis* (Fabricius) and the pupal parasitoid *Brachymeria incerta* (Cress) have also been noted as natural enemies of the palm caterpillar and parasitized up to 15% of the pupae during a serious outbreak in 1953.

Prospecting for natural enemies in sugarcane

The tachinid *Lydella* (= *Metagonistylum*) *minense* Tns. was discovered as a parasitoid from the sugarcane borer *Diatraea saccharalis* (F.) in the lower Amazon region in 1932. This so-called 'Amazon fly' was then successfully introduced into Guyana in 1933 for biocontrol of the sugarcane borer (see Chapter 17: Guyana). The Amazon fly was subsequently distributed to important sugarcane-growing areas in Barbados, Cuba, Puerto Rico, Trinidad, Guadeloupe and St Lucia and in the USA in Louisiana and Florida. Although it became an important parasitoid for control of the sugarcane borer in some countries, in Suriname it was found only incidentally. In the period 1967–1968 an inventory of parasitoids of *D. saccharalis*. and *Eodiatraea*

Table 28.1. Biological control studies in different production systems in Suriname.

Crop / Plant	Pest	Natural enemy	Control effect	Reference
Amaranthus	*Prodenia (=Spodoptera) eridania* (Cr.)	*Archytas divisa* (Walk.)	Not known	Van Dinther (1960)
Avocado, *Terminalia catappa*	*Megalopyge lanata* (Stoll.), vars Megalopygidae	*Archytas vernalis* Curran, *Pammaecerus leptotrichopus* (B.-B.)	Not known Not known Good control	Van Dinther (1960) Van Dinther (1960) Van Dinther (1960)
Bamboo	*Dinoderus minutus* (F.)	*Doryctes parvus* Mues.	Not known	Van Dinther (1960)
Cabbage	*Lipaphis erysimi* (Kltb.)	*Cycloneda sanguinea* (L.), *Mesograpta* sp.	Not known	Van Dinther (1960)
Cassava	*Erinnyis alope* (Drury), *Erinnyis ello*	*Telenomus* sp. probably *dilophonotae* Cam, *Derostenus* sp.	Good control	Van Dinther (1960)
Cassava	*Iatrophobia brasiliensis* (Rübs)	*Tetrastichus* sp.	Not known	Van Dinther (1960)
Cassava	*Mononychellus tanajoa* (Bondar)	*Oligota minuta* Cameron	Good control	De Voogd (1978)
Cassava	*Phlyctaenodes bifilalis* Hamp.	*Apanteles* sp.	Not known	Van Dinther (1960)
Citrus	*Chrysomphalus ficus* Ashm.	*Aphytis chrysomphali* Mercet	Up to 50%	Van Brussel and Bhola (1970)
Citrus	*Phyllocoptruta oleivora* (Ashm.)	*Hirsutella thompsonii* Fisher	Partly good control	Van Brussel (1975)
Citrus	*Toxoptera citricidus* (Kirk.)	*Cycloneda sanguinea* (L.) , *Baccha* sp. with hyperparasitoid, *Protolaccus bacchadis* Burks	Not known	Van Dinther (1960)
Cocoa	*Rutela lineola* (L.)	*Pitangus sulphuratus sulphuratus* ('grietjebie')	Not known	Geijskes (report 1951: 65; in Van Dinther 1960)
Cocoa	*Selenothrips rubrocinctus* (Giard)	*Chrysopa* spp. *Termatophylidea opaca* Carvalho	Not known	Reyne (1921) in Van Doesburg (1964)
Cocoa	*Zetesima baliandra* (Meyr.)	*Spilochalcis* sp.	1% parasitism	Van Dinther (1960)
Coconut	*Aspidiotus destructor* Sign.	*Azya trinitatis* Marshall, *Cryptognatha auriculata* Muls, *Pentilia castanea* Muls, *Scymnus* sp.	Good control	Van Dinther (1960)
Coconut	*Brassolis sophorae sophorae* (L.)	*Crotophaga major* Gmelin, *Crotophaga ani* (kaw-foetoeboi), *Saimiri sciureus* *Telenomus nigrocoxalis* Ashm. *Winthemia pinguis* (F.) *Brachymeria incerta* (Cress) *Anastatus* sp. *Eupelmela* sp	Excellent control by total complex 75% control Both eupelmids 10% control	Van Dinther (1960) Van Dinther (1960) Van Sauers-Muller (1988) /Van Dinther (1960) Van Dinther (1960) Van Sauers-Muller (1988) Van Sauers-Muller (1988)
Coconut	*Castnia daedalis*	*Ooencyrtus* sp. *Neoaplectana carpocapsae* Weiser, *Heterorhabditis bacteriophora* Poinar, *Heterorhabditis* sp.	Not known Good in lab, bad in field	Van Dinther (1960) Segeren-van den Oever *et al.* (1984)

Coconut	*Nasutitermus ephratae* (Holmgr.)	*Micranchenus lineola* (F)	Not known	Van Dinther (1960)
Coconut	*Sibine fusca* (Stoll)	*Yahuartachina* sp.n.	Not known	Van Dinther (1960)
		Chistolia sp.	Not known	Van Sauers-Muller (1988)
		Theronia sp.	Not known	Van Dinther (1960)
		Fornicia sp.	Minor importance	Van Dinther (1960)
Cotton	*Aphis gossypii* Glov.	*Cycloneda sanguinea* (L.)	Not known	Fernandes (report 1924:24) VD[a]
		Psyllobora divisa (F.)		Van Dinther (1960)
		Coleomegilla maculata (De G.)		Fernandes (report 1924:24) VD[a]
Coffee	*Coccus viridis* (Green)	*Cephalosporium lecanii* Zimm.	Not known	Kuyper (1913a:43) VD[a] Reyne (1929:1028) VD[a]
Coffee	*Ischnapsis longirostris* (Sign)	fungus, *Microcera* sp.	Not known	Stahel, (report 1914: 11) VD[a]
Cotton	*Dysdercus* spp	*Pitangus sulphuratus sulphuratus* (L), grietjebie	Not known	Fernandes (1926c) VD[a]
		Crotophaga ani L., kawfoetoeboi.		Fernandes (1926c) VD[a]
Grassland	*Mocis latipes* (Guen.)	*Polistes canadensis infuscatus* Lep., *Polistes versicolor vulgaris* Beq., *Polybia chrysothorax* (Web.), *Polybia liliacea* (F.), *Polybia rejecta* (F.), *Polybia sericea* (Oliv.), *Polybia striata* (F.)	Some control	Van Dinther (1960)
Guava	*Anastrepha striata* Schiner	*Opius cereus* Gahan	Insufficient control	Van Brussel and van Vreden (1966)
Hibiscus, fruit trees	*Maconellicoccus hirsutus* (Green)	*Cryptolaemus montrouzieri* Muls., *Anagyrus kamali* Moursi	Good control	FAO (2006)
Maize	*Dysmicoccus brevipes* (Ckll.)	*Hyperaspis festiva* Muls	Not known	Van Dinther (1960)
Papaya	*Pseudaulacaspis pentagona* (Targ)	*Azya trinitatis* Marshall	Not known	Van Dinther (1960)
		Coccidophilus sp.		Van Dinther (1957)
		Oeneis nigrans Muls , *Pentilia castanea* Muls, *Nodita* sp. , *Chrysopa* sp. near Silvana Navas		Van Dinther (1960)
Peanut, soybean	*Anticarsia gemmatalis* Hbn	*Zelomorpha* sp.	Not known	Van Dinther (1960)
Peanut	*Stegasta basqueella* (Chamb.)	*Apanteles* sp.	Not known	Van Dinther (1960)
Rice	*Hydrellia* sp.	*Opius* sp., *Hexacola* sp., *Eupteromalis* sp.	Not known	Van Dinther (1960)
Rice	*Oebalus poecilus* (Dall.)	*Telenomus* sp.	Some control	Van Dinther (1960)
Rice	*Pomacea dolioides* Reeve	*Rostrhamus sociabilis* (Vieillot)	Good control	Haverschmidt and Mees (1994)

Continued

Table 28.1. Continued.

Crop / Plant	Pest	Natural enemy	Control effect	Reference
Rice	*Rupela albinella* (Cr)	*Telenomus* sp.	Good control	Van Dinther (1960)
		Heterospilus sp.	Not known	Van Dinther (1960)
		Venturia ovivenans nov. spec.	Not known	Zwart (1973)
		Strabotes rupelae nov. spec.	Not known	Zwart (1973)
Rice	*Vehilius celeus* Mab.	*Dusona* sp., *Elachertus* sp.	Some control	Van Dinther (1960)
Rice, vegetables	*Laphygma* (= *Spodoptera*) *frugiperda* (J.E.Smith)	*Polistes canadensis infuscatus* Lep., *Polistes versicolor vulgaris* Beq., *Polybia chrysothorax* (Web.), *Polybia liliacea* (F.), *Polybia rejecta* (F.), *Polybia sericea* (Oliv.), *Polybia striata* (F.), *Pitangus sulphuratus sulphuratus*, *Crotophaga ani*,	Some control	Van Dinther (1960)
		Archytas marmoratus (Tns.)		(*A. piliventris* Van Der Wulp) VD[a] Van Dinther (1960)
		Apanteles marginiventris (Cress.), *Meteorus laphygmae* Vier.		Geijskes (1950) VD[a]
Soursop	*Cerconota anonella* (Sepp)	*Apanteles* sp.	Not known	Van Dinther (1960)
Sugarcane	*Aeneolamia flavilatera* (Urich)	*Salpingogaster nigra* Schiner	Not known	Geijskes (report 1951: 50) VD[a]
Sugarcane	*Diatraea saccharalis* (Fabricius)	Eggs: unknown parasitoid		
		Agathis stigmaterus (Cresson)	Low parasitism	Van Brussel (1968)
		Lydella (Metagonistylum) minense Tns.		
		Iphiaulax grenadensis (Ashm.)	Low parasitism	Van Dinther (1960)
		Leskiopalpus diadema Wied.		Van Brussel (1968)
		Megaselia sp.		Van Brussel (1968)
Sugarcane	*Eodiatraea centrella* Mosch	*Agathis stigmatera* Cress., *Leskiopalpus diadema* Wied.	Some control	Van Brussel (1968)
		Megaselia sp.		
Sugarcane	*Longiunguis sacchari* (Zhntn)	*Cycloneda sanguinea* (L)	Good control	Fernandes (report 1924:25) VD[a]
		Coleomegilla maculata (De G.)		Fernandes (report 1924:25) VD[a]
Tomato, pepper	*Protoparce sexta paphus* (Cr.)	*Telenomus connectans* Ashm.	Good control	Van Dinther (1960)
Vegetables	*Bemisia tabaci* (Gennadius)	*Encarsia* sp., Coccinellid species related to *Delphastus*	Not known	Wijngaarde (2002)

[a]VD = Van Dinther (1960) quotes this author from a local report.

centrella Mosch was taken (Van Brussel, 1968). For both pests, the parasitoids *Leskiopalpus diadema* Wied., *Agathis stigmatera* Cress and *Megaselia* sp. were found.

Native predators of cocoa thrips in cocoa

During investigations in 1960 on the cocoa thrips *Selenothrips rubrocinctus* (Giard), a small mirid was found living on the underside of mature cocoa leaves, mostly in the thrips colonies or in their immediate vicinity. The spot is often covered with a fine web, making the mirids difficult to detect. Only thrips nymphs were predated upon. The study did not investigate the importance of this predator for thrips control; however, a single mirid could eliminate a small thrips colony. Reyne (1921) also found larvae of several *Chrysopa* spp. as predators on the cocoa thrips.

Native parasitoid of the paddy bug in rice

Eggs of the paddy bug *Oebalus poecilus* (Dallas) in rice were parasitized just after they had been laid by a *Telenomus* sp. Parasitized eggs turn black, while the green non-infested eggs gradually change to red.

Native parasitoid of the guava fruit fly

The braconid wasp *Opius cereus* Gahan was found as a parasitoid of the guava fruit fly *Anastrepha striata* Schiner (Van Brussel and van Vreden, 1966). Less than 1% of the larvae were parasitized, which indicates that the parasitoid is of minor economic importance for the control of this fruit fly.

28.2.2 Period 1970–2000

Natural control of citrus pests

Rust mite *Phyllocoptruta oleivora* (Ashm.) is known as a common pest in citrus species. Research by Van Brussel (1975) showed that population levels increased at the beginning of the dry season. Maximum counts were reached in 4–5 weeks, but dropped again to a low level in a similar period. Although dry seasons coincide with lowest relative humidity, the rust mite populations were not much affected by relative humidity, but by the increase in levels of the fungus *Hirsutella thompsonii* Fisher. Low mite counts during the rainy season were not entirely attributed to this entomopathogenous fungus either, despite the favourable moist conditions for fungal growth. They were the result of larval mortality, which increased when larvae were wetted and a water film was present on the food plant. A moist substrate interferes with molting and oviposition.

The Florida red scale *Chrysomphalus ficus* Ashm. was found to be parasitized by *Aphytis chrysomphali* Mercet (Van Brussel and Bhola, 1970). In the dry season of 1967 up to 50% of the scales were parasitized. Furthermore, dead scales were frequently covered with an orange-coloured fungus. Adults of the ladybird beetle *Cycloneda sanguinea* (L.) and a not-yet-identified predatory mite have also been observed to feed upon the crawlers.

Augmentative biological control attempts of the coconut stem borer with nematodes

Imported strains of the nematodes *Neoaplectana carpocapsae* Weiser and *Heterorhabditis bacteriophora* Poinar, and a *Heterorhabditis* sp. found in compost in Suriname, were tested by Segeren-van den Oever *et al.* (1984) for their efficacy as biocontrol agents of the oil palm and coconut stem borer *Castnia dedalus*. In laboratory experiments 25–69% of treated *Castnia* larvae were killed by *Heterorhabditis* sp. in petri dishes containing one *Castnia* larva and 70% were killed by *N. carpocapsae* (Mexican strain). When nematodes were applied to jars with a *Castnia* larva in coconut wood pulp, larval mortality was 5–62% for *Heterorhabditis* sp. and only up to 11% for *H. bacteriophora*. Spraying with a suspension of 2.4 million and 5 million nematodes of *N. carpocapsae* per oil palm tree did not control *Castnia*. Applications of 1 million *Heterorhabditis* sp. per coconut tree killed none of the *Castnia* larvae found. Though the protected environment of the leaf axils seemed quite favourable in terms of temperature and humidity, the nematodes did not survive in the tree sufficiently long to kill the stem borer.

Natural control of the coconut caterpillar in coconut

Van Sauers-Muller (1988) determined the importance of parasitoids in the control of *B. sophorae sophorae*, the coconut caterpillar. A complex of parasitoids attack this insect. The most important of these is *T. nigrocoxalis*, which accounted for 75% of parasitized eggs in one plantation. As there was a great difference between the amount of control exercised by parasitoids on several plantations, further research is needed to increase control of *Brassolis* in coconut and oil palm by parasitic wasps and fungi. Apart from *Telenomus*, the other egg parasitoids encountered were *Anastatus* sp. and *Eupelmela* sp. All egg parasitoids together parasitized up to 85% of *Brassolis* eggs. Pupal parasitoids found at Jarikaba plantation (Saramacca) were *Spilocalcis* sp. and *Brachymeria* sp. (reported by Van Dinther (1960) as *Brachymeria incerta* Cress.). A fungus attacking larvae and pupae was not identified.

Natural control of the green cassava mite in cassava

The green cassava mite *Mononychellus tanajoa* (Bondar) is a common mite in Suriname on cassava (de Voogd, 1978). However, the amount of damage is usually negligible, probably due to the presence of an important predator, the rove beetle *Oligota minuta* Cameron. Predatory mites were also found, but are still unidentified.

Natural control of Pomacea *snails in rice*

The snail kite *Rostrhamus sociabilis* (Vieillot), which occurs in Suriname, is an important predator of the freshwater snail *Pomacea dolioides* Reeve, a pest in rice in the seedling stage (Haverschmidt and Mees, 1994).

Native parasitoids of the rice stem borer in rice

Venturia ovivenans nov. spec. and *Strabotes rupelae* nov. spec., reared from the rice stem borer *Rupela albinella* (Cr.), were described by Zwart (1973). *Strabotes* was found to be an ectoparasitoid of the last-instar larva and it was also found sometimes feeding upon the pupa.

28.3 Current Situation of Biological Control in Suriname

28.3.1 Classical biological control of the pink hibiscus mealybug

The pink hibiscus mealybug *Maconellicoccus hirsutus* (Green) was first identified in Suriname in 2001, although it already had arrived in Grenada, the Caribbean, in 1994. Initially, 300 predatory beetles (*Cryptolaemus montrouzieri* Muls.) and 300 parasitic wasps (*Anagyrus kamali* Moursi) were released, followed at a later stage with the release of another 6,000 wasps, which proved more effective at controlling *M. hirsutus* at low densities (FAO, 2006). The hibiscus mealybug never reached high population numbers, probably due to the fact that at the time of introduction some natural enemies arrived with the pest, and the early release of natural enemies in Suriname have contributed to keeping numbers down. Some local predators have also been observed feeding on the mealybug. At present, only occasional infestations appear in the dry season, rapidly followed by a decrease when natural enemies take their toll.

28.3.2 Native natural enemies of *Bemisia* whiteflies

During a study to evaluate the effect of certain biological and chemical pesticides on whitefly (Wijngaarde, 2002), two natural enemies for *Bemisia tabaci* Gennadius were found: a parasitoid, *Encarsia* sp., and a predatory beetle from the family Coccinellidae, related to *Delphastus*.

28.3.3 IPM of the carambola fruit fly

Much work was recently done on biocontrol and the male annihilation technique (MAT) of the carambola fruit fly (*Bactrocera carambolae* Drew and Hancock) in Suriname, Guyana, French Guiana and Brazil. The carambola fruit fly (CFF) was thought to have been introduced into Suriname in the late 1960s or 1970s. It was first collected in Paramaribo in 1975, but went unnoticed in the insect collection of the Agricultural Experiment Station in Suriname. At the time of detection in Suriname in 1986, it was realized that this species was an introduction in the Americas, supposedly from South-east Asia, and an extensive fruit collection programme was set up to determine the hosts of what was at that time still thought to be the Oriental fruit fly. This was followed by trapping programmes and eradication trials. An eradication programme based on the MAT was developed and funded by IFAD (International Fund for Agricultural Development), the Netherlands, France and the USA and began officially in 1998, covering Guyana, Suriname, French Guiana and Brazil (Midgarden *et al.*, 2016). As a result of MAT the distribution of *B. carambolae* was reduced to limited areas of Suriname and French Guiana by 2001, and Guyana was declared free of this pest.

The main tool in the eradication project was the MAT, applied by placing small blocks impregnated with the lure Methyl Eugenol and Malathion in the field. The number of blocks was initially 4 per hectare, based on the MAT for Oriental fruit fly *Bactrocera dorsalis* (Hendel). Although CFF is part of the Oriental fruit fly complex, it is a different species (Schutze *et al.*, 2015) and this became clear in the treatment: the number of blocks per hectare had to be increased fourfold to be effective. Apart from MAT, fruits were destroyed, trees were pruned, and in 'hot spot' areas with remaining fly populations, bait sprays were applied to kill the female flies. In 2002, funding for the eradication programme was reduced and eventually terminated. This resulted in expansion of the distribution of CFF, with detections as far South-east as Curralinho, in the Para State of Brazil, and as far North as Orlando, Florida, in the USA, and re-infestation of over 50% of the regions in Guyana. Termination of the programme before completion of eradication of CFF has resulted in increased cost to South American agriculture and increased risk to Central and North America and the Caribbean. A coordinated programme amongst infested countries could still mitigate the risk of spread of *B. carambolae* in the region (Midgarden *et al.*, 2016).

Extensive host surveys of CFF have been implemented in Suriname (van Sauers-Muller, 2005), French Guiana (Vayssières *et al.*, 2013) and Brazil (Lemos *et al.*, 2014), and these were compared with host surveys in the area of origin, South-east Asia (Indonesia, Thailand and Malaysia; Allwood *et al.*, 1999). The main hosts in Suriname were cultivated hosts originating from Asia, especially *Averrhoa carambola* L. (carambola) and *Syzygium samarangense* (Blume) Merr. & Perry (curacao apple, java apple). In Brazil, *Psidium guajava* L. (guava) was found to be an important host (Lemos *et al.*, 2014), partly due to the fact that the other main hosts were much rarer than in Suriname. Currently Brazil and Suriname emphasize the importance of collecting wild fruit species, to determine whether CFF is adapting to native fruit. Although Vayssières *et al.* (2013) did not find wild fruit species infested by CFF, it is important to continue monitoring, because both Lemos *et al.* (2014) and van Sauers-Muller often found wild fruits infested by CFF near infested cultivated hosts.

In its area of origin, CFF is parasitized by *Fopius arisanus* Sonan (Vijaysegaran, 1984) and this parasitoid has been introduced for control of *B. dorsalis* (Hendel) in Hawaii (Clausen *et al.*, 1965). Even though at least nine species of parasitoids have been found on *Anastrepha* fruit flies in Suriname, with *Doryctobracon areolatus* Szepligeti being the most abundant, none of these species has so far been found to develop in CFF (A. van Sauers-Muller, Paramaribo, 2016, personal communication). During the eradication project, biocontrol was not considered, as the aim was to eradicate the pest. Now that it seems that CFF is about to remain in South America after termination of the eradication programme, options for biocontrol need to be considered. *Diachasmimorpha longicaudata* Ashmead was introduced into the Oyapoque area in Brazil, on the border with French Guiana, to reduce the spread of CFF. During a survey reported in Vayssières *et al.* (2013), this parasitoid was found in the St Georges area in French Guiana, along the border with Brazil, as well in CFF pupae and in pupae of *Anastrepha* species, with up to 6% parasitism recorded in 2003. Although the native parasitoid *Aganaspis pelleranoi* (Brethes) was found associated with CFF, it apparently does not (yet) have an important reducing effect on CFF populations. However, the braconid egg parasitoid *F. arisanus* is able to reach levels of over 75% parasitism in CFF in the area of origin, Malaysia (Vijaysegaran, 1984) and levels of parasitism between 60% and 79% in an area where the parasitoid was introduced

Table 28.2. Overview of biological control in Suriname.

Crop	Pest	Natural enemy	Type of biocontrol[a] / success	Area (ha) under biocontrol
Avocado, *Terminalia catappa*	Megalopygidae	*Pammaecerus leptotrichopus*	NC / +[b]	?
Cassava	*Erinnyis alope*, *Erinnyis ello*	*Telenomus* sp. probably *dilophonotae*, *Derostenus* sp.	NC / +	?
Cassava	*Mononychellus tanajoa*	*Oligota minuta*	NC/ +	?
Citrus	*Chrysomphalus ficus*	*Aphytis chrysomphali*	CBC / ±	?
Citrus	*Phyllocoptruta oleivora*	*Hirsutella thompsonii*	ABC / ±	?
Cucumber, vegetable gourd, yard-long bean and hyacinth bean in rotation	Various	Various	ABC / +	0.5[c]
Coconut	*Aspidiotus destructor*	*Azya trinitatis*, *Cryptognatha auriculata*, *Pentilia castanea*, *Scymnus* sp.	NC / +	?
Coconut	*Brassolis sophorae sophorae*	*Telenomus nigrocoxalis*, *Anastatus* sp., *Eupelmela* sp, *Spilocalcis* sp., *Brachymeria* sp. probably *incerta*, *Crotophaga major*, *Crotophaga ani*, *Saimiri sciureus*	NC / +	700[c]
Grassland	*Mocis latipes*	*Polistes canadensis infuscatus*, *Polistes versicolor vulgaris*, *Polybia chrysothorax*, *Polybia liliacea*, *Polybia rejecta*, *Polybia sericea*, *Polybia striata*	NC / –	?
Hibiscus, fruit trees	*Maconellicoccus hirsutus*	*Cryptolaemus montrouzieri*, *Anagyrus kamali*	NC / +	?
Pineapple	Various	Various	NC / +	52[c]
Rice	*Oebalus poecilus*	*Telenomus* sp	NC / –	?
Rice	*Pomacea dolioides*	*Rostrhamus sociabilis* (Vieillot)	NC / +	?
Rice	*Rupela albinella*	*Telenomus* sp.	NC / +	?
Rice	*Vehilius celeus*	*Dusona* sp., *Elachertus* sp	NC / –	?
Rice, vegetables	*Laphygma* (= *Spodoptera*) *frugiperda*	*Polistes canadensis infuscatus*, *Polistes versicolor vulgaris*, *Polybia chrysothorax*, *Polybia liliacea*, *Polybia rejecta*, *Polybia sericea*, *Polybia striata*, *Pitangus sulphuratus sulphuratus*, *Crotophaga ani*, *Archytas marmoratus*	NC / –	?
Sugarcane	*Eodiatraea centrella*	*Agathis stigmatera*, *Leskiopalpus diadema*, *Megaselia* sp.	NC / –	?
Sugarcane	*Longiunguis sacchari*	*Cycloneda sanguinea*, *Coleamegilla maculata*	NC / +	?
Tomato, pepper	*Protoparce sexta paphus*	*Telenomus connectans*	NC / +	?

[a]Type of biocontrol: ABC = augmentative biocontrol, CBC = classical biocontrol; NC = natural control
[b]Success: + = good control, ± = substantial contribution to control, – = some contribution to control
[c]A. van Sauers-Muller, Paramaribo, 2018, personal communication

(Hawaii) on Oriental fruit fly (Clausen *et al.*, 1965). *F. arisanus* was introduced in Costa Rica in the period 1955–1958 for control of the medfly, *Ceratitis capitata* (Wiedemann). However, in a recovery attempt from 1958 to 1962, only one specimen of *F. arisanus* was recovered (Wharton *et al.*, 1981). A study in the same area in 1979 showed establishment of this parasitoid, but it only parasitized *Anastrepha* species up to 0.10% (Wharton *et al.*, 1981). Many other parasitoids were found during the study of Wharton *et al.*, including *D. areolatus* and *Opius bellus* Gahan, two parasitoids also found in Suriname. It would be important to introduce the non-endemic fruit fly parasitoid *F. arisanus* into Suriname for control of CFF, whereby the FAO code of conduct for the importation and release of exotic biocontrol agents (FAO, 1996) should be followed carefully.

28.3.4 Natural control of pineapple pests

In the district of Para, close to various indigenous villages, a total of over 50 ha pineapples is grown without the use of pesticides. Since the project is only a few years old, there is still much to learn about natural control of pests in this crop.

In conclusion, both historically and currently, quite a lot of research has been done on biocontrol in Suriname. The areas with classical biocontrol by natural enemies introduced in the period 1900–1969 are not well documented. Current information (Table 28.2) indicates that at least 750 ha are under biocontrol.

28.4 New Developments of Biological Control in Suriname

28.4.1 Biological control of carambola fruit fly

Plans exist to introduce *F. arisanus* for biocontrol of the carambola fruit fly *B. carambolae*, as eradication efforts are expensive and not enforceable if not all affected countries surrounding Suriname participate. An introduction of the egg parasitoid *F. arisanus* could reduce infestation levels to as much as 50% of its present level, thereby reducing infestation levels in the least susceptible hosts, which could then be produced for the market.

28.4.2 Prospecting for native natural enemies of aphids

Recent research in controlling aphids (*Aphis gossypii* Glover) with biopesticides produced from herbs showed the existence of several fungi, parasitoids and predators that were preventing the build-up of the experimental aphid population. The fungi, parasitoids and predators are now being identified for further tests as future biocontrol agents (A. van Sauers-Muller, Paramaribo, April 2019, personal communication).

28.5 Acknowledgements

S. Khodabaks is acknowledged for providing information about the coconut culture in Suriname.

28.6 References

Allwood, A.J., Chinajariyawong, A., Kritsaneepaiboon, S., Drew, R.A.I., Hamacek, E.L. *et al.* (1999) Host plant records for fruit flies (Diptera: Tephritidae) in Southeast Asia. *The Raffles Bulletin of Zoology*, Supplement 7, 1–92.

CIA (2017) The World Factbook: Suriname. Available at: https://www.cia.gov/library/publications/the-world-factbook/geos/ns.html (accessed 23 October 2019).

Clausen, C.P., Clancy, D.W. and Chock, Q.C. (1965) Biological control of the Oriental fruit fly (*Dacus dorsalis* Hendel) and other fruit flies in Hawaii. US Department of Agriculture *Technical Bulletin* 1322, 1–102.

De Voogd, W.B. (1978) Mijten en hun natuurlijke vijanden op Cassave in Suriname. [Mites and their natural enemies on Cassava in Suriname]. MSc thesis. Agricultural Experiment Station, Paramaribo, Suriname.

FAO (1996) *Code of conduct for the import and release of exotic biological control agents*. International Standard of Phytosanitary Measures 3. Available at: http://www.fao.org/3/a-j5365e.pdf (accessed 18 October 2019).

FAO (2006) *Assistance for the management of the pink hibiscus mealybug. Suriname*. Technical Cooperation Programme, AG: TCP/SUR/2903. Food and Agriculture Organization of the United Nations, Rome, Italy, pp. 1–13.

Haverschmidt, F. and Mees, G.F. (1994) *Birds of Suriname*. Rev. Ed. Vaco, Paramaribo, Suriname.

Lemos, L. do N., Adaime, R., Jesus-Barros, C.R. de and De Deus, E. da (2014) New hosts of *Bactrocera carambolae* (Diptera: Tephritidae) in Brazil. *Florida Entomologist* 97(2), 841–843.

Midgarden, D., van Sauers-Muller, A., Signoretti Godoy, M.J. and Vayssières, J-F. (2016) Overview of the program to eradicate *Bactrocera carambolae* in South America. In: Ekesi, S., Mohamed, S.A. and De Meyer, M. (eds.) *Fruit Fly Research and Development in Africa – toward Sustainable Management Strategy to Improve Horticulture*. Springer, Dordrecht, Netherlands, p. 705–736.

Reyne, A. (1921) De cacaothrips (*Heliothrips rubrocinctus* Giard) [The cacao thrips (*Heliothrips rubrocinctus*)]. *Bulletin Departement Landbouw Suriname* 44, 1–214.

Schutze, M.K., Aketarawong, N., Amornsak, W., Armstrong, K.F., Augustinos, A.A. *et al*. (2015) Synonymization of key pest species within the *Bactrocera dorsalis* species complex (Diptera: Tephritidae): taxonomic changes based on a review of 20 years of integrative morphological, molecular, cytogenetic, behavioural and chemoecological data. *Systematic Entomology* 40, 456–471. DOI: 10.111/syen.12133.

Segeren-van den Oever, H.A., Sanchit-Bekker, M.L. and van Sauers-Muller, A. (1984) Biological control of *Castnia dedalus* with insect parasitic nematodes. *Surinam Agriculture* 32(2), 45–50.

Van Brussel, E.W. (1968) Entomologisch onderzoek [Entomological research]. *Surinam Agriculture* 17(1), 1–82.

Van Brussel, E.W. (1975) Interrelations between citrus rust mite, *Hirsutella thompsonii* and greasy spot on citrus in Surinam. Agricultural Experiment Station Surinam, *Bulletin* 98, 1–66.

Van Brussel, E.W. and Bhola, B. (1970) Biology and control of the Florida red scale, *Chrysomphalus ficus* Ashm. *Surinam Agriculture* 18(2), 64–76.

Van Brussel, E.W. and van Vreden, G. (1966) Studies on the biology, damage and control of the guava fruit fly, *Anastrepha striata* Schiner, in Suriname. *Surinam Agriculture* 16(3), 110–122.

Van Dinther, J.B.M. (1957) *Pseudaulacaspis pentagona* Targ. as a papaya pest. *Entomologische Berichten* 17, 165–168.

Van Dinther, J.B.M. (1960) Insect pests of cultivated plants in Surinam. *Bulletin Agricultural Experiment Station Suriname* 76, 1–159.

Van Doesburg, F.H. Jr (1964) *Termatophylidea opaca* Carvalho, a predator of thrips. (Hem. – Het). *Entomologische Berichten* 24, 1.XII.1964.

Van Sauers-Muller, A. (1988) The occurrence of parasites of *Brassolis sophorae* in Suriname, with special reference to the egg-parasite *Telenomus nigrocoxalis*. *Surinam Agriculture* 36(4), 11–15.

Van Sauers-Muller, A. (2005) Host plants of the Carambola fruit fly, *Bactrocera carambolae* Drew & Hancock (Diptera: Tephritidae), in Suriname, South America. *Neotropical Entomology* 34(2), 203–214.

Vayssières, J-F., Cayol, J-P., Caplong, P., Séguret, J., Midgarden, D. *et al*. (2013) Diversity of fruit fly (Diptera: Tephritidae) species in French Guiana: their main host plants and associated parasitoids during the period 1994–2003 and prospects for management. *Fruits* 68, 219–243.

Vijaysegaran, S. (1984) The occurrence of Oriental fruit fly on starfruit in Serdang and the status of its parasitoids. *Journal of Plant Protection in the Tropics* 1(2), 93–98.

Wijngaarde, J. (2002) Biologische bestrijding van de witte vlieg, *Bemisia tabaci* (Gennadius) in tomaat, *Lycopersicon esculentum* (Mill.) [Biological control of *Bemisia tabaci* on tomato, *Lycopersicon esculentum*]. BSc thesis. ADEK University of Suriname.

Wharton, R.A., Gilstrap, F.E., Rhode, R.H. Fischel-M, M. and Hart, W.G. (1981) Hymenopterous egg-pupa, and larval-pupal parasitoids of *Ceratitis capitata* and *Anastrepha* spp. (Diptera: Tephritidae) in Costa Rica. *Entomophaga* 26 (3), 285–290.

Zwart, K.W.R. (1973) New Ichneumonidae, parasitic upon the rice borer *Rupela albinellla* (Cr.) (Lep., Pyralidae) in Surinam, with a key to the species of *Strabotes* (Hym., Ichneumonidae). *Entomologische Berichten* 33, 1.XII.1973.

29 Biological Control in Trinidad and Tobago

Ayub Khan[1]* and Wendy-Ann P. Isaac[2]

[1]Department of Life Sciences, University of the West Indies, St Augustine Trinidad; [2]Department of Food Production, University of the West Indies, St Augustine Trinidad

* E-mail: ayub.khan@sta.uwi.edu

Abstract
Since 1918, Trinidad and Tobago has been a rich source of biological control agents and has shipped numerous natural enemies both regionally and internationally. Successful classical biocontrol programmes using predominantly predators and parasitoids began in earnest in the 1970s, initially for *Aeneolamia varia saccharina* and then for *Diatraea saccharalis* control in sugarcane. Several other pests, including citrus blackfly *Aleurocanthus woglumi*, diamondback moth *Plutella xylostella* and pink hibiscus mealybug *Maconellicoccus hirsutus*, were successfully managed using biocontrol. The use of microbial agents, particularly entomopathogenic fungi, has also been assessed. *Metarhizium anisopliae* has caused high nymphal and adult mortality in *A. varia saccharina*, while *Paecilomyces tenuipes* has caused as much as 67% larval mortality in *P. xylostella*. Over the past 10 years there has been an increase in the number of exotic insect species in Trinidad and Tobago, and classical biocontrol has played and is expected to keep playing an important role in managing these invasive pests.

29.1 Introduction

Trinidad and Tobago have an estimated population of slightly more than 1,200,000 (July 2017) and its main agricultural products are cocoa, dasheen, pumpkin, cassava, tomatoes, cucumbers, aubergine, hot pepper, pommecythere, coconut water and poultry (CIA, 2017).

29.2 History of Biological Control in Trinidad and Tobago

29.2.1 Period 1870–1969

Perhaps the first record of classical biocontrol in Trinidad was the introduction of the mongoose *Herpestes auropunctatus* Hodgson to control rats and, to a lesser extent, snakes in sugarcane in 1870, but this first attempt failed and nine individuals from Calcutta, India, were introduced into Jamaica in 1872 (Espeut, 1882). This release was successful and further introductions to other Caribbean islands were largely made from this Jamaican population. The Trinidad population was introduced from St Lucia (Anonymous, 1920). The mongoose became a pest shortly after its introduction to the island and lizards that had previously kept the population of a serious pest, the sugarcane froghopper *Aeneolamia varia saccharina* (Dist.), under control were being consumed in large numbers by the mongoose and consequently led to a decrease in sugarcane yields (Anonymous, 1920). Meyers (1935) observed that the ground lizard *Ameiva atrigularis* L. was still common in mixed rainforest and in the vicinity of houses but rare in sugarcane fields 'owing to the inroads of the mongoose'. Urich (1931) noted that a bonus was paid to those who caught a total of 30,895 mongooses between 1902 and 1908. By 1918 the Government of Trinidad and Tobago had introduced an act (Mongoose Act Chapter 67:55) banning the further importation of mongoose and permitting destruction by authorized personnel of mongooses present on the islands. A fine of TT$1000.00 was instituted against those who contravened any part this act. This may in part have led to the increased numbers of the animal killed (142,324 between 1927 and 1930) (Urich, 1931). It is unfortunate that this example is frequently cited as one of the major blunders of classical biocontrol to the detriment of the numerous successes.

Bennett (1990, 1991) stated that, apart from classical biocontrol, several other biocontrol tactics were used in Trinidad against insect pests in the early 1900s. Conservation biocontrol was used in various ways: construction of bird roosts and planting of bamboo clumps to encourage predacious birds close to sugarcane fields to control the giant moth borer *Telchin licus* (Drury) (= *Castnia licoides* (Boisduval)), use of plants as nectar sources for attracting arthropod natural enemies and even the erection of wasp shelters in proximity to crops plagued by insect pests. Augmentation biocontrol was also used and an example is the breeding and release of the predator *Salpingogaster nigra* Schiner to control the sugarcane froghopper *A. varia saccharina*. Another example of augmentation biocontrol is the attempt to use microbial control against the froghopper by culturing and application of the green muscardine fungus *Metarhizium anisopliae* (Metchnikoff) Sorokin using techniques developed in Trinidad (Bennett, 1990, 1991).

29.2.2 Period 1970–2000

Biological control programmes attempted in Trinidad and Tobago for some exotic arthropod species during the period 1970–2000 are listed in Table 29.1 and summarized in this section.

Sugarcane moth borer

Three species of *Diatraea* are pests of sugarcane in Trinidad: *Diatraea centrella* (Möschl.), *D. impersonatella* (Walker) and *D. saccharalis* Fabricius, with the last causing the greatest economic losses. Little attention was paid to biocontrol of *D. saccharalis* prior to the early 1970s, primarily due to the low levels of damage attributed to this pest, probably as a result of suppression by its local natural enemies: *Paratheresia claripalpis* (Wulp), *Miobiopsis diadema* (Wiedemann), *Agathis stigmatera* (Cresson) and *Ipobracon grenadensis* Ashmead (Cock, 1985). The introduction of the apparently susceptible but higher-yielding sugarcane variety B41227 in 1970, in combination with the concomitant excessive use of synthetic insecticides for control of the sugarcane froghopper, resulted in a 19.3–23% upsurge of bored internodes by *D. saccharalis* (Cock, 1985; Bennett, 1990). Biocontrol of *D. saccharalis* in Trinidad was thus restarted in earnest in 1974 when *Cotesia* (= *Apanteles*) *flavipes* Cameron, a gregarious larval endoparasitoid of graminaceous stalk borers in Asia, and *Lixophaga diatraeae* (Townsend), a larviparous parasitoid, were released in sugarcane fields. Following the release of 175,332 adult *C. flavipes* and of 12,717 *L. diatraeae* in sugarcane fields in Trinidad during the period 1974–1979, and the subsequent recovery of both species, DesVignes (1980) concluded that both species were established, with field parasitism rates as high as 90% (mean 71.3%) for *C. flavipes* during the period 1980–1981 (DesVignes, 1985). Bennett (1978) attributed the success of *C. flavipes* to its short life cycle and absence of a pre-oviposition period, especially when frequent aerial insecticide applications were made for the control of *A. varia saccharina*.

Diamondback moth

Plutella xylostella L. is one of the most serious pests of brassicas worldwide. Despite being first reported in Trinidad in 1945 (Lamont and Callan, 1950), it did not achieve pest status until 1970 (Yaseen, 1974). Based on life-table studies conducted by James and Khan (2008), they concluded that on reaching stable age distribution the egg, larval, pupal and adult stages would contribute 83.76%, 15.79%, 0.32% and 0.13%, respectively, to the total population. After reaching

Table 29.1. Selected exotic arthropod pests and their natural enemies in Trinidad and Tobago.

Pest	First recorded	Natural enemies	CBC/ABC[a]	References
Bemisia tabaci	1988	*Paecilomyces fumosoroseus*	ABC	Jones, 1994
Thrips tabaci	1988	*Orius insidiosus, Nabis sordidus*	ABC	Jones, 1994
		Coleomegilla maculata	ABC	Jones, 1994
Aleurocanthus woglumi	1996	*Encarsia perplexa*	CBC	White *et al.*, 2005
		Amitus hesperidum	CBC	White *et al.*, 2005
Toxoptera citricida	1985	*Lysiphlebus testaceipes*	ABC	Yokomi *et al.*,1994
Maconellicoccus hirsutus	1995	*Cryptolaemus montrouzieri*	CBC	Pollard, 1995
		Anagyrus kamali	CBC	Persad and Khan, 2002
Plutella xylostella	1945	*Cotesia plutellae*	CBC	Lamont and Callan, 1950; Yaseen, 1974
Diatraea saccharalis	1974	*Cotesia flavipes*	CBC	DesVignes, 1985
	1974	*Lixophaga diatraeae*	CBC	DesVignes, 1985

[a]Type of biocontrol: ABC = augmentative biocontrol, CBC = classical biocontrol

a stable age distribution, approximately 99.7% of the population exists in the immature stages and this emphasizes the importance of management of the immature stages of this pest. Over the period 1970–1977, releases were made of two parasitoids: (i) *Cotesia* (= *Apanteles*) *plutellae* (Kurdjumov), 2,133 individuals from India and 215 from the Netherlands (1971–1972); and (ii) 8,431 individuals of *Tetrastichus* (= *Oomyzus*) *sokolowski* Kurd. from the Leeward Islands were introduced into Trinidad to control *P. xylostella* (Cock, 1985). Both species were established in Trinidad but were not providing sufficient control, probably due to high pesticide use in brassicas (Cock, 1985).

Sugarcane froghopper

Despite being recorded on sugarcane (as *Tomaspis saccharina*) in Trinidad as early as 1890 (Williams, 1919), the sugarcane froghopper *A. varia saccharina* did not achieve pest status until an outbreak in the village of Chaguanas, Trinidad, in 1906 (Williams, 1919). Eggs of this pest are oviposited around the sugarcane roots in soil. One generation of *A. varia saccharina* is completed in approximately 60 days (Gopee, 1979). Since eggs are located in the soil and nymphs are normally covered in spittle, chemical control efforts are rendered ineffective against these stages. For this reason, biocontrol of *A. varia saccharina* was attempted for the first time in 1911 with the introduction of *Castolus plagiaticollis* Stål. from Mexico (Bennett, 1985). Since then several local and exotic natural enemies have been assessed for their potential to control *A. varia saccharina* in Trinidad (Table 29.2). Apart from *Metarhizium anisopliae* Metchnikoff Sorokin, none of these gave satisfactory levels of control and a method for its mass production was developed by Rorer in 1910, who also demonstrated that field application of spores resulted in high nymphal and adult mortality (Rorer, 1913).

More recently Allard *et al.* (1990) demonstrated the effectiveness of a local strain of *M. anisopliae* against *A. varia saccharina*. The company Caroni Limited Trinidad has since developed a commercial formulation of this bio-insecticide for control of *A. varia saccharina*. Mass produced *M. anisopliae* spores were applied to fields four times per froghopper brood in 1995 and although frequent applications of spores were necessary to obtain economic control of this pest (Ali, 1996), the results have since been very encouraging. Apart from causing high nymphal and adult mortality, *M. anisopliae*-infected females produced significantly fewer eggs (32.66 eggs) compared with healthy adults (66.28) (Ali, 2001).

29.3 Current Situation of Biological Control in Trinidad and Tobago

29.3.1 Current biological control projects in Trinidad and Tobago

Citrus blackfly

Citrus blackfly *Aleurocanthus woglumi* Ashby was first recorded in the Caribbean in Jamaica in 1913 by Ashby (1915). It eventually spread to the Bahama Islands, Cuba, Costa Rica and Panama (Dietz and Zetek, 1920) and then to Barbados, Belize, Bermuda, Cayman Islands, Colombia, Dominica, Dominican Republic, Ecuador, El Salvador, Florida, Guatemala, Guyana,

Table 29.2. Natural enemies assessed for control of *Aeneolamia varia saccharina* in Trinidad (retrieved from Cock, 1985).

Natural enemy	Stage attacked	References
Anagrus urichi	Egg	Urich, 1915; Williams, 1921; Pickles, 1933
Asarkina ericetorum	Nymphs	Cock, 1985
Carabunia waterstoni	Nymphs	Cock, 1985
Castolus plagiaticollis	Nymphs/Adults	Cock, 1985
Crotophaga ani	Nymphs/Adults	Cock, 1985
Metarhizium anisopliae	Nymphs/Adults	Urich, 1915; Williams, 1921; Pickles, 1933
Oligosita giraulti	Egg	Kershaw, 1913
Salpingogaster nigra	Nymphs	Urich, 1915; Williams, 1921; Pickles, 1933

Haiti, Mexico, Netherland Antilles, Peru, Puerto Rico, Suriname and Venezuela (IIE, 1995). It was recorded for the first time in Trinidad in 1998 (Parkinson *et al.*, 2002). Two parasitoids, *Amitus hesperidum* Silvestri and later *Encarsia perplexa* Huang and Polaszek, were released in Trinidad in 2000. Experiments by White *et al.* (2005) in Trinidad indicated that *A. hesperidum* reduced the populations of *A. woglumi* by more than 98% within a few months.

Diamondback moth

Earlier we concluded that use of parasitoids to control *P. xylostella* was not successful. Therefore, another option was tested: the management of *P. xylostella* by using entomopathogenic fungi. Towards the end of 2002, a fungus identified by CAB International, United Kingdom as *Paecilomyces tenuipes* (Peck) Samson (IMI 189/02) was found infecting *P. xylostella* larvae inside cabbage heads for the first time in Trinidad (Lopez *et al.*, 2003). Subsequent studies have indicated that this fungus caused highest mortality to third-instar *P. xylostella* larvae at 25°C in approximately 3 days. Mycosis for larvae and pupae under field conditions was 66.7% and 57.0%, respectively, 7 days post fungal application (Baksh and Khan, 2012). The fungus is still present in a small farming area where it was first isolated.

Pink hibiscus mealybug

The accidental introduction of the pink hibiscus mealybug *Maconellicoccus hirsutus* (Green) into Grenada in 1994, Trinidad in 1995 and subsequently to other locations in the Caribbean and South, Central and North America (Pollard, 1995, 1998) led to the implementation of one of the most outstanding classical biocontrol programmes over the past few decades. The biology of *M. hirsutus* has been studied in various countries, including India (Mani, 1989) and Trinidad (Persad and Khan, 2002). Based on its cryptic nature, the presence of its wax covering and diverse host range, classical biocontrol was considered to be the most sustainable strategy for management of this pest. Therefore, the request by the Ministry of Agriculture, Land and Marine Resources (MALMR), Trinidad, to CAB International led to the initial release of 600 adults of the parasitoid *Anagyrus kamali* Moursi from China in 1996 (Ram *et al.*, 1997). Almost simultaneously with the release of *A. kamali*, the predators *Cryptolaemus montrouzieri* Mulsant and *Scymnus coccivora* Aiyyar from India by MALMR through the efforts of Caribbean Agricultural Research Development Institute (CARDI) were released in 1996 (Gautam *et al.*, 1996; McComie *et al.*, 1996). The concomitant release of both a predator and a parasitoid for management of *M. hirsutus* in Trinidad raised the concern of *C. montrouzieri* consuming *M. hirsutus* mummies parasitized by *A. kamali*. This concern was not realized, as studies indicated that there was no significant difference between the consumption rate of parasitized and unparasitized *M. hirsutus* by *C. montrouzieri* (Persad and Khan, 2007). The concerted efforts of the MALMR's extensive public information campaign and research coupled with the research initiatives of CAB International, CARDI and the University of the West Indies all led to the successful completion of this classical biocontrol programme.

29.3.2 Trinidad as a source of natural enemies for biological control

Trinidad's proximity to and past geological association with the South American mainland explains in large measure the diverse nature of its flora and fauna. It is therefore not surprising that numerous natural enemies of pests have been obtained from the island. Since its inception in Trinidad in 1946, the Commonwealth Institute of Biological Control (CIBC and later CAB International) has shipped numerous species of natural enemies from Trinidad to various parts of the world to manage a diverse array of pest insects (Table 29.3).

29.4 New Developments of Biological Control in Trinidad and Tobago

Much has already been written on the influence of climate change and global warming and their concomitant impacts on increased spread of invasive species, especially insects (e.g. Masters and Norgrove, 2010; Cohen *et al.*, 2015). Over the past 10 years there has been an increase in

Table 29.3. List of selected natural enemies shipped from Trinidad (retrieved from CIBC, 1983–1998; Cock, 1985; Gungah *et al.*, 2005; Julien and Griffiths, 1998).

Natural enemy	Pest controlled	Country	Year
Apanteles etiellae	*Ancylostomia stercorea*	Bahamas	1968–1970
Apanteles etiellae	*Fundella pellucens*	Barbados	1968–1970
Azya orbigera	*Aspidiotus destructor*	Grenada	1949–1951
Azya orbigera	*Pulvinaria psidii*	Bermuda	1956–1957
Chnoodes sp.	*Pulvinaria psidii*	Cayman Islands	1949
Copidosoma truncatellum	*Trichoplusia ni*	Cape Verde	1987
Cybocephalus sp.	*Pseudaulacaspis pentagona*	Bermuda	1955
Cryptognatha affinis	*Planococcus citri*	Grenada	1961
Cryptognatha nodiceps	*Planococcus citri*	Bahamas 1961, Cayman Islands 1949, Grenada 1947, Jamaica 1958, St Kitts 1970, St Vincent 1951	1947–1970
Eiphosoma dentator	*Hellula phidilealis*	Barbados	1982
Eiphosoma dentator	*Maruca testulalis*	Mauritius	1971
Bracon cajani	*Maruca testulalis*	Fiji	1971
Apanteles etiellae	*Maruca testulalis*	Sri Lanka	1974, 1977
Apanteles etiellae	*Maruca testulalis*	Cape Verde	1985
Encarsiella noyesi	*Aleurodicus cocois*	Barbados	1950–1951
Encarsiella noyesi	*Aleurodicus cocois*	Nevis	1998
Encarsiella noyesi	*Aleurodicus pulvinatus*	Nevis	1998
Exochomus sp.	*Pulvinaria psidii*	Bermuda	1951
Hyperaspis jucunda	*Pseudococcus longispinus*	Bermuda	1955
Liothrips urichi	*Clidemia hirta*	Fiji	1930
Liothrips mikaniae	*Mikania micrantha*	Malaysia	1990
Liothrips mikaniae	*Mikania micrantha*	Solomon Islands	1988
Mescinia parvula	*Chromolaena odorata*	Guam	1986
Melanagromyza eupatoriella	*Chromolaena odorata*	Guam	1986
Metarhizium anisopliae	*Aeneolamia varia saccharina*	Guyana	1918
Metrogaleruca obscura	*Cordia curassavica*	Mauritius	1948–1949
Metrogaleruca obscura	*Cordia curassavica*	Sri Lanka	1980
Eurytoma attiva	*Cordia curassavica*	Sri Lanka	1980
Neochetina eichhorniae	*Eichhornia crassipes*	Zambia	1971
Nephaspis bicolor	*Aleurodicus dispersus*	Hawaii	1979
Nephaspis bicolor	*Aleurodicus dispersus*	Mauritius	2003
Ooencyrtus submetallicus	*Nezara viridula*	Australia	1952
Ooencyrtus submetallicus	*Nezara viridula*	Hawaii	1962
Tamarixia leucaenae	*Heteropsylla cubana*	China, Malaysia, Nepal	1992–1993
Psyllaephagus yaseeni	*Heteropsylla cubana*	Tanzania, Kenya	1995
Telenomus remus	*Spodoptera* spp.	Honduras	1985
Trissolcus basalis	*Nezara viridula*	Australia	1952
Trissolcus basalis	*Nezara viridula*	Hawaii	1962
Pachycrepoideus vindemiae	*Anastrepha obliqua*	Belize	1969
Paratheresia claripalpis	*Diatraea saccharalis*	Antigua	1937
Paratheresia claripalpis	*Diatraea saccharalis*	Dominica	1952
Paratheresia claripalpis	*Diatraea saccharalis*	Grenada	1950–1954

Continued

Table 29.3. Continued.

Natural enemy	Pest controlled	Country	Year
Pareuchaetes pseudoinsulata	*Chromolaena odorata*	Sri Lanka	1972
Pareuchaetes pseudoinsulata	*Chromolaena odorata*	Malaysia	1970–1974
Pareuchaetes pseudoinsulata	*Chromolaena odorata*	Guam	1985
Pseudogonatopus saccharivorae	*Saccharosydne saccharivora*	Jamaica	1954
Anagrus flaveolus	*Saccharosydne saccharivora*	Jamaica	1954
Synopeas sp.	*Contarinia lycopersici*	Barbados	1974–1975

Table 29.4. Crops and estimated areas under biological control in Trinidad and Tobago.

Crop	Pest	Natural enemy	Origin	Type of biocontrol[a]	Area (ha) under biocontrol
Brassicas	*Plutella xylostella*	*Cotesia plutellae*	India	CBC	50
Citrus	*Aleurocanthus woglumi*	*Amitus hesperidum*	USA	CBC	2,000
		Encarsia perplexa	USA	CBC	2,000
	Toxoptera citricida	*Lysiphlebus testaceipes*	Native	ABC	2,000
Sugarcane	*Diatraea saccharalis*	*Cotesia flavipes*	Pakistan	CBC	15,000
		Lixophaga diatraeae	India	CBC	15,000
	Aeneolamia varia saccharina	*Metarhizium anisopliae*	Native	ABC	15,000
Tomato	*Bemisia tabaci*	*Paecilomyces fumosoroseus*	Native	ABC	200
Eggplant	*Thrips palmi*	*Orius insidiosus*	Native	ABC	150
Various crops	*Maconellicoccus hirsutus*[b]	*Anagyrus kamali*	China	CBC	340
Plants	*Maconellicoccus hirsutus*	*Cryptolaemus montrouzieri*	India	CBC	340
Forest spp. / horticultural plants	*Maconellicoccus hirsutus*[b]	*Anagyrus kamali*	China	CBC	15,000
		Cryptolaemus montrouzieri	India	CBC	15,000

[a]Type of biocontrol: ABC = augmentative, CBC = classical
[b]Pink hibiscus mealybug, *Maconellicoccus hirsutus*, occurs on the entire island and was brought under complete CBC

the number of exotic insect species in Trinidad and Tobago and the Caribbean. Several of these pest species have been successfully managed by classical biocontrol and, as has been indicated previously in this chapter, classical biocontrol has played an important role in Trinidad and Tobago. In Table 29.4, estimates are provided about crops receiving biocontrol, indicating that a total area of about 33,000 ha is under biocontrol.

Agriculture is consumer driven, and as consumers demand higher-quality foods with reduced synthetic pesticides, farmers are becoming more receptive to alternatives (e.g. biocontrol) to solely chemical control. Conservation and augmentative releases of natural enemies (including entomopathogens) are important priorities for farmers. Biocontrol thus appears to have a bright future both locally and in the wider Caribbean.

References

(References with grey shading are available as supplementary electronic material)

Ali, B.S. (1996) Biocontrol of froghopper, *Aeneolamia varia saccharina* Dist. (Cercopidae). *Caroni Research Station Technical Report* 33, 82–83.

Ali, B.S. (2001) Effect of *Metarhizium anisopliae* var. *anisopliae* Metschnikoff (Sorokin) (Deuteromycotina:-Hyphomycetes) on fecundity of the sugarcane froghopper, *Aeneolamia varia saccharina* Dist. (Cercopidae) in Trinidad. *Proceedings of the 27th West Indies Sugar Technologists Conference*. Port of Spain, Trinidad and Tobago.

Allard, G.B., Chase, C.A., Heale, J.B., Isaac, J.E. and Prior, C. (1990) Field evaluation of *Metarhizium anisopliae* (Deuteromycotina: Hyphomycetes) as a mycoinsecticide for control of sugarcane froghopper, *Aeneolamia varia saccharina* (Hemiptera:Cercopidae). *Journal of Invertebrate Pathology* 55, 41–46.

Anonymous (1920) Mongoose in Trinidad – a query whether the skins of the animals have any value. *The New York Times*, 12 September 1920.

Ashby, S.F. (1915) Notes on diseases of cultivated crops observed in 1913–1914. *Bulletin of the Department of Agriculture, Jamaica* 2, 299–327.

Baksh, A. and Khan, A. (2012) Pathogenicity of *Paecilomyces tenuipes* to diamondback moth *Plutella xylostella* at three temperatures in Trinidad. *International Journal of Agriculture and Biology* 14, 261–265.

Bennett, F.D. (1978) A comparison of the reproductive strategies and certain other biological characteristics of *Apanteles* spp. and the tachinid parasites of *Diatraea saccharalis* (Fabr.). *Proceedings of the International Society of Sugarcane Technologists* 16, 523–527.

Bennett, F.D. (1985) Biological control of sugarcane froghoppers – a review and recommendations for future investigations. In: *22nd Sugar Technologists Conference*, Sugarcane Association of the Caribbean, Trinidad and Tobago, pp. 44–57.

Bennett, F.D. (1990) An overview of classical biological control in the Caribbean and some examples of the utilization of entomophagous insects. In: Pavis, O. and Kermarrec, A. (eds) *Caribbean Meetings on Biological Control*, Guadeloupe, French Antilles, pp. 25–35.

Bennett, F.D. (1991) Biological control strategies in Trinidad and Tobago. *Journal of the Agricultural Society of Trinidad and Tobago* 88, 67–71.

CIA (2017) The World Factbook: Trinidad and Tobago. Available at: https://www.cia.gov/library/publications/the-world-factbook/geos/td.html (accessed 19 January 2018)

CIBC (1983–1998) *CAB International Annual Reports*. CAB International, Wallingford, UK.

Cock, M.J.W. (ed.) (1985) *A Review of Biological Control of Pests in the Commonwealth Caribbean and Bermuda up to 1982*. Technical Communication No. 9, Commonwealth Institute of Biological Control. Commonwealth Agricultural Bureaux, Farnham Royal, UK.

Cohen, J.E., Clark-Harris, D.O., Khan, A. and Isaac, W.A.P. (2015) Sustainable management of invasive species for small island developing states under changing climates. In: Ganpat, W.G. and Isaac, W.A.P. (eds) *Impacts of Climate Change on Food Security in Small Island Developing States*. IGI Global, Hershey, Pennsylvania, pp. 312–360.

DesVignes, W.G. (1980) *Strategy for the control of small moth borer by parasite introduction and periodic release of* Apanteles flavipes *(Cam.) on sugarcane in Trinidad*. Report of Caroni Research Station, Trinidad.

DesVignes, W.G. (1985) Aspects of the biological control of *Diatraea* spp. on sugarcane in Trinidad. PhD thesis. University of the West Indies, St Augustine, Trinidad.

Dietz, H.F. and Zetek, J. (1920) The blackfly of citrus and other subtropical plants. *United States Department of Agriculture Bulletin* 885, 55.

Espeut, W.B. (1882) On the acclimatization of the Indian mongoose in Jamaica. *Proceedings of the Zoological Society of London* 50, 712–714.

Gautam, R.D., deChi, W. and Maraj, C. (1996) Preliminary studies on the inoculative releases of exotic ladybirds, *Cryptolaemus montrouzieri* Mulsant and *Scymnus coccivora* Aiyyar against the hibiscus mealybug *Maconellicoccus hirsutus* (Green) in County St. George. *Proceedings of the 1st seminar on the Hibiscus Mealybug*, Centeno, Trinidad and Tobago, pp. 20–24.

Gopee, T.J. (1979) Research on the chemical control of froghopper *Aeneomalia varia saccharina* (Dist.). *Proceedings of seminar on Insect Pest Control: Research programmes and technical problems in operations*. Caroni Research Station, Trinidad and Tobago, pp. 10–19.

Gungah, B., Seewooruthun, S.I., Nundloll. P. and Rambhunjun, M. (2005) Biological control of the spiraling whitefly, *Aleurodicus dispersus*. In: *MAS 2005*. Ministry of Agro-Industry and Fisheries, Food and Agricultural Research Council, Mauritius, pp. 307–311.

IIE (1995) *Distribution Maps of Pests*, Series A No. 91 (3rd Revision). CAB International, Wallingford, UK.

James, M. and Khan, A. (2008) Fecundity life tables for *Plutella xylostella* L. (Lepidoptera: Plutellidae) at two temperatures in Trinidad and its implications in Integrated Pest Management. *Annals of Plant Protection Sciences* 16, 269–273.

Jones, M.T. (1994) *Thrips palmi* in the Eastern Caribbean with special reference to Trinidad. In: Walmsley, D. (ed.) *Integrated Pest Management: New Strategies for the Caribbean farmer*. Caribbean Agricultural Research and Development Institute (CARDI)/CTA, Trinidad, pp. 37–45.

Julien, M.H. and Griffiths, M.W. (eds) (1998) *Biological Control of Weeds: a World Catalogue of Agents and their Targets*. CAB International, Wallingford, UK.

Kershaw, J.C. (1913) *Froghoppers*. Special Circular 5, Department of Agriculture, Trinidad and Tobago.

Lamont, N. and Callan, E. McC. (1950) Moths new to Trinidad, BWI. *Zoologica* 35, 197–207.

Lopez, V., Ganpat, W., Dowlat, P. and Kairo, M. (2003) *Cabbage Pests in Trinidad and Tobago*. CABI Caribbean and Latin American Regional Centre, Trinidad and Tobago.

Mani, M. (1989) A review of the pink mealybug – *Maconellicoccus hirsutus*. *Insect Science and its Application* 10, 157–167.

Masters, G. and Norgrove, L. (2010) *Climate change and invasive alien species*. CABI Working Paper 1. Available at: https://www.cabi.org/Uploads/CABI/expertise/invasive-alien-species-working-paper.pdf (accessed 19 January 2018).

McComie, L.D., Gosine, S. and Siew, P. (1996) The effect of *Cryptolaemus montrouzieri* (Mulsant) on the hibiscus mealybug *Maconellicoccus hirsutus* (Green) on hibiscus plants in Trinidad. *Tropical Fruits Newsletter* 23, 7–10.

Meyers, J.G. (1935) Second report of an investigation into the biological control of West Indian insect pests. *Bulletin of Entomological Research* 26, 181–252.

Parkinson, K., Seales, J. and White, G. (2002) The distribution of citrus blackfly *Aleurocanthus woglumi* Ashby (Homoptera:Aleyrodidae) in Trinidad. *Journal of the Agricultural Society of Trinidad and Tobago* 91, 23–29.

Persad, A. and Khan, A. (2002) Comparison of life table parameters for *Maconellicoccus hirsutus* Green and its natural enemies *Anagyrus kamali* Moursi, *Cryptolaemus montrouzieri* Mulsant and *Scymnus coccivora* Ayyar. *BioControl* 47, 137–149.

Persad, A. and Khan, A. (2007) Effects of four host plants on biological parameters of *Maconellicoccus hirsutus* Green (Homoptera:Pseudococcidae) and efficacy of *Anagyrus kamali* Moursi (Hymenoptera: Encyrtidae). *Journal of Plant Protection Research* 47, 35–42.

Pickles, A. (1933) Entomological contributions to the study of the sugarcane froghopper. I. The study of biotic factors of control. *Tropical Agriculture* 10, 222–233.

Pollard, G.V. (1995) Pink hibiscus mealybug in the Caribbean. *Caraphin News*, 12, 1–2.

Pollard, G.V. (1998) *Update on new pest introductions*. Circular Letter No. 3/98. United Nations Food and Agriculture Organization (FAO) Regional Office for Latin America.

Ram, P., Betrand, C., Lopez, V., Morais, M. and Peterkin, D. (1997) A perspective on the release of *Anagyrus kamali* (Hymenoptera: Encyrtidae) by the Ministry of Agriculture, Land and Marine Resources. *Proceedings of the 1st seminar on the Hibiscus Mealybug*, Centeno, Trinidad and Tobago, pp. 50–55.

Rorer, J.B. (1913) The green muscardine fungus and its use in cane fields. *Circular of the Department of Agriculture, Trinidad and Tobago* 8, 1–14.

Urich, F.W. (1915) Insects affecting sugarcane in Trinidad. *Bulletin of the Department of Agriculture, Trinidad and Tobago* 14, 156.

Urich, F.W. (1931) The mongoose in Trinidad. *Tropical Agriculture* 8, 95–97.

White, G.L., Kairo, M.T.K. and Lopez, V. (2005) Classical biological control of the citrus blackfly *Aleurocanthus woglumi* by *Amitus hesperidum* in Trinidad. *BioControl* 50, 751–759.

Williams, C.B. (1919) The relation of root fungus to froghopper blight of sugarcane in Trinidad. *Bulletin of the Department of Agriculture, Trinidad and Tobago* 18, 52–56.

Williams, C.B. (1921) Report of the froghopper blight of sugarcane in Trinidad. *Memoirs of the Department of Agriculture, Trinidad and Tobago* 1, 1–179.

Yaseen, M. (1974) Biology, seasonal incidence and parasites of *Plutella xylostella* (L.) in Trinidad and the introduction of exotic parasites into the Lesser Antilles. In: Brathwaite, C.W.D., Phelps, R.H. and Bennett, F.D.

(eds.) *Crop Protection in the Caribbean*. St Augustine Publications, University of the West Indies, pp. 237–244.

Yokomi, R.K., Lastra, R., Stoetzel, M.B., Lee, R.F., Garnsey, S.M. *et al*. (1994) Establishment of the brown citrus aphid (Homoptera:Aphididae) in Central America and the Caribbean basin and transmission of citrus tristeza virus. *Journal of Economic Entomology* 87, 1078–1085.

30 Biological Control in Uruguay

César Basso[1]*, Adela Ribeiro[2], Ximena Cibils[3], Willy Chiaravalle[4] and Karina Punschke[5]

[1]Unidad de Entomología, Facultad de Agronomía, Universidad de la República, Montevideo, Uruguay; [2]Unidad de Entomología, Estación Experimental 'Dr. M.A. Cassinoni', Facultad de Agronomía, Universidad de la República, Paysandú, Uruguay; [3]Entomología, Protección Vegetal, Instituto Nacional de Investigación Agropecuaria, Colonia, Uruguay; [4]Entoagro. Roberto Koch, Montevideo, Uruguay; [5]Registro de Agentes de Control Biológico, División Control de Insumos, Dirección General de Servicios Agrícolas, Ministerio de Ganadería, Agricultura y Pesca, Montevideo, Uruguay

* E-mail: cbasso.bis@gmail.com

Abstract

The first reported case of biological control in Uruguay was an attempt to import the parasitoid *Encarsia berlesei* from Italy to manage the white peach scale in 1912, which failed due to high mortality during the long boat trip. Later introduction of the same parasitoid (in 1913) and the predator *Lindorus lophanthae* (in 1915) resulted in permanent control of peach scale. In the early 20th century, Uruguay was a pioneer in South America in the successful introduction of natural enemies of pests recently arrived in the country and was also a provider of biocontrol agents to other countries in the region by re-exporting these exotic species. Throughout this century the introduction and colonization of biocontrol agents continued. In the 1980s and 1990s, national production of entomopathogens and parasitoids was initiated. During this period (and until today), large-scale prospecting projects were executed to find and identify native natural enemies and microbial control agents in several important crops. The largest areas under classical biocontrol are currently in pine and eucalyptus plantations. Recently, government regulations for the registration and control of biocontrol products have been established in Uruguay. The first commercial biocontrol products on the market are used to manage pests in horticultural crops in greenhouses and for field crops such as soybeans, intended for local consumption and for export. Uruguay aspires to be recognized for the production of high-quality food. Biocontrol helps to realize this aspiration, because it contributes to food safety and adds to environmental protection.

30.1 Introduction

The population of Uruguay is slightly more than 3.36 million (July 2017 estimate) and the main agricultural products are cellulose, beef, soybeans, rice, wheat, dairy products, fish, lumber, tobacco and wine (CIA, 2017). According to Cabrera *et al.* (2017, pp. 533–535):

> Uruguay is a country with a total area of 176,216 km^2 with a variety of natural resources that allow the development of agricultural and forestry activities in over 90% of its territory ... The country has abundant natural resources, arable soils, and available water and natural meadows, which has contributed to its thriving agricultural and livestock industries. Uruguay is a net exporter of agricultural products. Over 80% of the territory is devoted to livestock production, while extensive crops and commercial forestry occupy the majority of the remaining area. A key aspect of agriculture in Uruguay is the significance of mixed agricultural-livestock production systems in parts of the country where soil is suitable for crop growing. The coexistence of livestock (mainly beef cattle) and crops in crop-pasture rotations is an important aspect of agriculture in Uruguay. Whereas the area under soybean cultivation was insignificant before 2000, it has become the largest crop in recent years, occupying more than one million hectares. In terms of value, soybean is currently the main export crop, ... with beef for first place in the country's exports ...

30.2 History of Biological Control in Uruguay

30.2.1 Period 1880–1969

During the early 20th century several natural enemies were introduced in Uruguay, resulting in successful biocontrol of a number of insect pests (DeBach, 1975), making Uruguay a pioneer country in South America. In this period, the main difficulty in realizing introductions was to provide adequate conditions for the survival of natural enemies during the long boat transfers, which led to repeated failures.

According to Trujillo Peluffo (1963), the first reported case of an introduction into Uruguay in 1912 concerned the parasitoid *Encarsia berlesei* (Howard), a natural enemy of the white peach scale *Pseudaulacaspis pentagona* (Targioni-Tozzetti). Uruguayan members of the Agricultural Defense Committee (a governmental unit) responded rapidly to the appearance and dispersal of this pest and implemented a short-term chemical control strategy and a long-term biocontrol approach based on the introduction of *E. berlesei*, due to its recent control success in Italy. The first shipments of parasitoids from Florence, Italy, in 1912 failed due to high insect mortality during shipment. In 1913, a Uruguayan scientist travelled to Florence, took parasitoids back home and established an *E. berlesei* colony in August 1913. By December 1913, adult parasitoids had been released and recaptured in Uruguayan fields, after presumed rapid establishment.

In 1915, another natural enemy of the peach scale, the ladybird beetle *Lindorus lophanthae* (Blaisdell), was introduced. Since then, *P. pentagona* populations have been successfully maintained under the economic threshold.

The cottony cushion scale *Icerya purchasi* Maskell was first reported in Uruguay in 1917, reaching pest status in 1918. Chemical and mechanical control (cutting and burning of infested material) were tried, but did not lead to eradication of the pest. In 1919, after previous unsuccessful introduction from Portugal in 1917, *Rodolia cardinalis* (Mulsant) originating from France was released, enabling successful control of the cushion scale and the white peach scale. Uruguay became the first South American country to introduce *R. cardinalis*, and later, colonies of *R. cardinalis* were sent to Brazil (1920), Argentina (1922) and Spain (1922), where they successfully established. Also *Cryptochetum iceryae* (Williston) was introduced for control of *I. purchasi*.

Eriosoma lanigerum (Hausmann) (previously *Schizoneura lanigera* Hausmann) is an aphid pest that devastated apple production in Uruguay during the 1920s; the year of its establishment in Uruguay is unknown. In early 1921 the parasitoid *Aphelinus mali* Haldeman was imported from the USA. It took a few years before the parasitoid established, but it succeeded and has been present in Uruguay ever since.

In April 1921, the San José scale *Comstockaspis perniciosus* (Comstock) was reported from fruit trees in Minas Gerais, Brazil, preceding its presence in California, Argentina and Uruguay. Extensive checks in Uruguay showed that this pest was not yet present, thus import of potentially infested fruit was prohibited. However, *C. perniciosus* was found in 1923 on plum and apple trees in a farm in East Uruguay. Supposedly, infested apple trees were imported by a farmer from New Zealand, where the scale had already been reported. First, eradication with pesticides was tried. Subsequently, in early 1924, several coccinellid species, including *Chilocorus stigma* (Say) and *Cycloneda oculata* (Thumberg), were imported from the USA. Of the introduced species, only *Chilocorus bivulnerus* (Mulsant) survived the journey, and in 1926, after successful establishment and control of the scale in Uruguay, live specimens of this coccinellid were shipped to Argentina.

Several other cases of biocontrol were reported in the first half of the 20th century. In 1941, *Anaphes nitens* (Girault) specimens were shipped to Uruguay from South Africa to control *Gonipterus gibberus* Boisduval, the eucalyptus weevil, which was first reported in 1937 in eucalyptus plantations in Uruguay. Presumably the weevil entered Uruguay through Argentina. The parasitoid established and is still present in eucalyptus plantations. To improve control, *A. nitiens* is currently collected, multiplied under laboratory conditions and periodically mass released in different eucalyptus plantations in Uruguay. In 1948, the parasitoid *Tetrastichus giffardianus* Silvestri was introduced from Brazil into Uruguay for control of the Mediterranean fruitfly *Ceratitis capitata* (Wiedemann), already known to have been present in Uruguayan grapes since 1927. The parasitoid established in Uruguay but control was often insufficient and management of *C. capitata* cannot rely only on *T. giffardianus*.

Uruguay also pioneered work on entomopathogens as biocontrol agents. In 1911, *Coccobacillus acridiorum* D'Herelle, the causal organism of an epizootic disease of the locust *Schistocerca cancellata* (Serville), was imported from Mexico. In both the 1911 and later the 1937 introductions, failures were reported during years with large locust infestations, as the pathogen was not effective under Uruguayan field conditions. In 1937, *Bacillus thuringiensis* (Berliner) strains were imported for control of *Colias lesbia* (Fabricius) in forage crops such as alfalfa. Although it was promising in experimental assays, no further information of its use has been reported to date.

30.2.2 Period 1970–2000

Characterization of natural enemy complexes

Initially Uruguay mainly focused on classical biocontrol, with later cases of augmentative biocontrol. Also, many researchers devoted their life work to characterization of the beneficial entomofauna in Uruguay's major crops and pastures. For example, Ruffinelli and Carbonell (1944) published the 'First systematic list of insects related to national agriculture'. In 1953 they published the 'Second list of insects and other arthropods of economic importance in Uruguay'.

Other authors, such as Parker *et al.* (1951) and Silveira Guido and Ruffinelli (1956), made catalogues of the natural enemies found in the country. From 1988 to 1992, the natural enemy complexes of *Pseudaletia adultera* (Schaus) and *Faronta albilinea* (Hübner), lepidopteran pests of winter crops and grass, were surveyed during a collaborative project between the National Institute of Agricultural Research (INIA) and the Republic University (UdelaR) of Uruguay. The survey revealed the natural presence of *Rogas* spp., *Campoletis* spp. and Tachinidae in *F. albilinea* populations, while *Apanteles* spp., *Rogas* spp., *Campoletis* spp., *Ophion* spp. and Tachinidae naturally occurred in *P. adultera* populations (Ribeiro and Zerbino, 1994).

Following this successful collaboration, INIA and UdelaR focused on surveying and identifying the natural enemy complex occurring in the *Crocidosema aporema* (Walsingham) forage legume–soybean system from 1992 to 2009. *C. aporema* is a native devastating pest of legumes (Alzugaray *et al.*, 1999; Ribeiro *et al.*, 2015) and its natural enemy complex was successfully identified. In addition, the viruses attacking *Rachiplusia nu* (Guenée), also an important pest of legume crops, were characterized in 1996 (Aznárez, 1996) and parasitoids belonging to Tachinidae (*Voria* spp.), Braconidae (*Rogas* spp.), Ichneumonidae, Chalcidoide and Encytridae were found naturally occurring in fields with *R. nu* (Aznárez, 1996).

Augmentative biological control

An inundative control programme was executed between 1988 and 1993 with releases of *Trichogramma galloi* Zucchi against *Diatraea saccharalis* (Fabricius) in sugarcane, treating up to 500 ha until the programme was discontinued in 1993 (Basso and Morey, 1991; Basso and Franco, 1995). Further, the parasitoid *Cotesia flavipes* Cameron was introduced from Brazil in 1987, mass produced and periodically released in an inundative way for the control of *D. saccharalis* in sugarcane until 1993 (Chiaravalle, 1996). Later, *Trichogramma pretiosum* Riley and *Trichogramma exiguum* Pinto & Platner were evaluated for control of *Bonagota cranaodes* (Meyrick) and *Argyrotaenia sphaleropa* (Meyrick) in vines (Basso and Pintureau, 1998; Basso *et al.*, 1999) and in apple (1998–1999). *Trichogramma pretiosum* was also evaluated against *Alabama argillacea* (Hübner) in cotton (1998) and *C. aporema* in lotus seedlings (2001) (Basso *et al.*, 2006), but these programmes were also discontinued.

The nematode *Beddingia siricidicola* Bedding was imported from Chile in 1987, being a specialist natural enemy of the exotic pine wood wasp *Sirex noctilio* Fabricius, which invaded Uruguay in 1980 (Bianchi, 1992). Uruguay periodically imports and releases this nematode in pine plantations across the country today.

From 1989 to 1997, the Department of Biological Control was part of the General Directorate of Agricultural Services (DGSA) of the Ministry of Livestock, Agriculture and Fisheries (MGAP). Its main activity was to develop a microbial control agent based on a baculovirus for control of *R. nu*, the sunflower caterpillar. Funding came from the MGAP and from the 'Organic Producers of Uruguay' group, and scientific support was provided by Dr Flavio Moscardi of the National Center for Soya Research, Embrapa, Brazil. With initial inoculum from Brazil, the Department of Biological Control worked for 6 years to develop a host mass rearing, using *R. nu* on an artificial diet and mass production of the virus, and tested its effectiveness in the field (Chiaravalle *et al.*, 1987; Aznárez, 1996). The formulated biopesticide was first applied to 600 ha of sunflower using an aircraft in 1994 (Chiaravalle, 1996). Although results were promising, the programme was not continued.

Multiplication of the polyhedrosis virus of *Anticarsia gemmatalis* Hübner based on initial inocula from Brazil started in 1986. During the 1986–1987 field season, 50 ha of soybean were sprayed with polyhedrosis virus-infested *A. gemmatalis* macerated caterpillars suspended in water. Experimental applications at field level were later done with the granulose virus of *Cydia pomonella* (L.) with inocula from the USA (Chiaravalle *et al.*, 1992). In 1992, the dung beetles *Onthophagus taurus* (Schreber) and *Euoniticellus fulvus* Goeze were successfully introduced by INIA from Australia and went through a process of study and quarantine. After quarantine rearing, large numbers of beetles were released, but establishment failed.

30.3 Current Situation of Biological Control in Uruguay

30.3.1 Characterization of natural enemy complexes

The year 2001 was a milestone in the study of biocontrol in Uruguay, because the manual *Natural Enemies: Illustrated Manual for Agriculture and Forestry* was published, featuring natural enemy pictures and identification notes, biology and habits, etc. (Bentancourt and Scatoni, 2001). With this work, survey studies of natural enemies became increasingly important in Uruguay. From 2004 to 2007, UdelaR entomologists made an inventory of biocontrol agents of *Piezodorus guildinii* (Westwood), the most serious stink bug pest of legumes. *Telenomus podisi* (Ashmead), *Trissolcus brochymenae* (Ashmead), *Trissolcus basalis* (Wollaston), *Trissolcus urichi* (Crawford), *Trissolcus teretis* (Johnson), and several species of Encyrtidae were among the species reported. Among them, only *T. podisi* has potential to significantly reduce *P. guildinii* eggs. Egg parasitism reached 66.5% and 99.65% of this mortality was caused by *T. podisi* (Ribeiro and Castiglioni, 2008). Additionally, the entomopathogenic fungi *Beauveria bassiana* (Balsamo) Vuillemin and *Metarhizium anisopliae* (Metschnikoff) were isolated from nymphs and adults of *P. guildinii* (Ribeiro and Castiglioni, 2008; Castiglioni *et al.*, 2010). *T. podisi* was proposed for use in inundative release in soybean and currently mass rearing and release experiments are carried out at the Dr. M. Cassinoni Experimental Station of the UdelaR. Additionally, studies on behaviour and biology of this parasitoid and its relationship with the host have been performed (Borghi and Cano, 2014; Armand Pilón, 2017).

From 2005 to 2007, the natural enemy complex of the soybean caterpillar *A. gemmatalis* and the legume aphid species complex were surveyed by INIA and UdelaR (Alzugaray *et al.*, 2010; Silva, 2016; Kucharski, 2017). In 2008, the book *Soybean Insects in Uruguay: illustrated pest and natural enemy recognition manual* was published (Ribeiro *et al.*, 2008).

Finally, since 2015 the UdelaR has established a close collaboration with the University of California at Berkeley (USA) to work on the influence of landscape structures (simple and complex) on communities of beneficial insects, focusing mainly on sorghum and soybean systems. These studies arose from the idea that Uruguayan agricultural–pastoral systems have great potential for the conservation of natural enemies, which is supported by previously mentioned results from characterization of natural enemy complexes (Ribeiro 2004, 2010).

30.3.2 Classical biological control

In 2000, *Ageniaspis citricola* Logvinovskaya was introduced from Argentina, mass produced and released several times until 2006 for the control of *Phyllocnistis citrella* Stainton in citrus (Buenahora *et al.*, 2001). Additionally, *Citrostichus phyllocnistoides* (Narayanan) was introduced in 2004 from Argentina and in 2005 from Spain, mass produced and released until 2006 to control *P. citrella*. *A. citricola* populations can be found in Uruguay annually, mostly during the fall months (April–June), but *C. phyllocnistoides* is rarely found in the field. The control efficiency of these two species has not been evaluated.

For the control of *Thaumastocoris peregrinus* Carpintero & Dellapé, detected in eucalyptus in Uruguay in 2008 (Martínez and Bianchi, 2010), *Cleruchoides noackae* Lin & Huber was introduced from Brazil in 2013 (INIA, 2013). The parasitoid established, but it is still periodically released to improve effectiveness.

The most recent introduction was the parasitoid *Tamarixia radiata* (Waterston) from Mexico in the fall of 2016 for control of *Diaphorina citri* Kuwayama on citrus. After successful mass rearing, the first pilot release was carried out during the summer (January–March) of 2017.

30.3.3 Augmentative biological control

Inoculative augmentative control, which mainly focuses on introducing a small number of natural enemies during critical periods, aiming for their establishment in the crop resulting in pest suppressions throughout the season, was experimentally evaluated by releasing

Encarsia formosa Gahan to control the whitefly *Trialeurodes vaporariorum* Westwood in tomatoes during the first decade of the 21st century. Augmentative releases of *E. formosa* were never performed commercially (Grille and Basso, 2001).

After sampling of native *Trichogramma* species (Basso and Pintureau, 2004; Grille *et al.*, 2009) and experimental small-scale releases against different pests carried out by the Entomology Unit of the Faculty of Agronomy of the UdelaR, inundative releases of large numbers of *T. pretiosum* as a commercial product (registered as Tricholine Maxi) have been made since 2010. The product uses small capsules for control of *A. gemmatalis*, *R. nu* and *C. aporema*. This programme covered 300 ha of soybean in 2017 and 1,300 ha in January–February 2019. This project was a partnership between UdelaR, the French company Bioline and local soybean producers. Positive results were obtained, which will stimulate an increase in use of *Trichogramma*. This will be of great environmental and commercial impact, since soybean crops occupy at present more than 1.2 million hectares in Uruguay.

Since 2011, commercial consignments of *Amblyseius swirskii* Athias-Henriot from Belgium and *Orius insidiosus* (Say) (initially from Belgium and currently from Argentina) for the control of *Bemisia tabaci* (Gennadius) and *Frankliniella occidentalis* (Pergande) in pepper have been tested in a partnership between UdelaR, INIA and Biobest (Belgium) with Brometan (Argentina) (Buenahora and Basso, 2015). At the moment, these natural enemies are used on approximately 30–35 ha of greenhouse-grown sweet pepper per year.

As of 2016, there has been evaluation of experimental applications of *Isaria fumosorosea* Wize to control *B. tabaci* in pepper and *T. vaporariorum* in tomato, and of *B. bassiana* to control *T. vaporariorum* in tomato, produced at vegetable farms. Currently 0.5 ha of pepper and 1 ha of tomato are being treated with entomopathogens. Although the programme is in its initial phase, it is expected to increase because of a growing demand for these microbials.

In conclusion, currently more than 1,100,000 ha of forest and food crops are under classical and augmentative biocontrol in Uruguay (Table 30.1).

30.4 New Developments of Biological Control in Uruguay

In Uruguay, three major recent advances in biocontrol should be highlighted: (i) the creation of the Bioinsumos platform; (ii) the creation of the Center for Forest Bioservices (CEBIOF); and (iii) the mandatory regulation and registration of products formulated with microorganisms (entomopathogens) and entomophagous arthropods for the control of agricultural pests and diseases by MGAP.

The 'Bioinsumos' platform is based at INIA. Bioinsumos collects and receives microbial strains and maintains a large collection of potential microbes to be used as bio-fertilizers and biocontrol agents of insect pests and diseases. These strains represent capital for the development of biological inputs such as: (i) biological fixation of nitrogen in crops and forage species; (ii) development of bio-fertilizers based on microorganisms that produce phosphate; (iii) development of inoculants based on microorganisms for disease suppression during crop establishment; (iv) development of fungal biocontrol agents for soil insects; (v) formulation of entomopathogenic fungi for foliar use to control aphids, whitefly and stink bugs; and (vi) identification of bioactive compounds to control post-harvest diseases. Thus, Bioinsumos serves as a bank that adequately preserves the national microbial diversity of Uruguay.

CEBIOF was created in 2012 as part of a project funded by the National Agency for Research and Innovation (ANII) focused on strengthening of scientific and technological services in Uruguay. CEBIOF is an association between three partners: INIA, UdelaR and the forestry producers association (SPF). CEBIOF's main goal is to operate and provide services to the forestry sector, mainly as a platform for research activities. Services offered by CEBIOF are: (i) genetic traceability of plant germplasm, so that certain eucalyptus clones are planted at correct locations; (ii) artificial inoculation of aggressive disease-causing pathogens to determine clone susceptibility or tolerance/resistance to key diseases in eucalyptus and pine; (iii) multiplication and provision of the parasitoid *A. nitiens* for control of the eucalyptus weevil, *Gonipterus* spp., in commercial plantations; and (iv) seed cleaning and seed viability testing.

Table 30.1. Areas under biological control in Uruguay (retrieved from C. Basso (Montevideo, Uruguay, April 2019, personal communication) and Buenahora and Basso, 2015).

Crop	Pest	Natural enemy	Area (ha) under biocontrol
Classical biological control			
Stone fruits	*Pseudaulacaspis pentagona*	*Encarsia berlesei*	1,870[a]
Stone fruits	*Pseudaulacaspis pentagona*	*Rhizobius lophantae*	1,870[a]
Pome fruits, stone fruits	*Cydia molesta*	*Ascogaster quadridentata* *Macrocentrus ancylivorus*	5,481[a]
Pome fruits, stone fruits	*Comstockaspis perniciosa*	*Chilocorus stigma* *Rhizobius lophantae*	5,481[a]
Apple	*Eriosoma lanigerum*	*Aphelinus mali*	2,703[a]
Citrus	*Icerya purchasi*	*Rodolia cardinalis*	15,101[a]
Citrus	*Phyllocnistis citrella*	*Ageniaspis citricola*	15,101[a]
Sugarcane	*Diatraea saccharalis*	*Cotesia flavipes*	7,100[a]
Pine	*Sirex noctilio*	*Beddingia siricidicola* *Ibalia leucospoides*	257,687[a]
Eucalyptus spp	*Phoracantha semipuntata*	*Avetianella longoi*	726,323[a]
Eucalyptus spp	*Gonipterus scutellatus*	*Anaphes nitens*	726,323[a]
Eucalyptus grandis	*Ctenarytaina spatulata*	*Psyllaeaphagus pilosus*	250,569[a]
Eucalyptus globulus	*Ctenarytaina eucalypti*	*Psyllaeaphagus pilosus*	309,088[a]
Eucalyptus tereticornis *Eucalyptus camaldulensis*	*Glycaspis brimblecombei*	*Psyllaeaphagus bliteus*	722[a]
Augmentative biological control			
Soybean	*Crocidosema aporema, Rachiplusia nu, Anticarsia gemmatalis*	*Trichogramma pretiosum*	1,300
Pepper	*Bemisia tabaci*	*Amblyseius swirskii*	35
Pepper	*Frankliniella occidentalis*	*Orius insidiosus*	35
Pepper	*Bemisia tabaci*	*Isaria fumosorosea*	0.5
Tomato	*Bemisia tabaci*	*Isaria fumosorosea*	1.0
Eucalyptus	*Gonipterus scutellatus*	*Anaphes nitens*	20

[a]For these crops, the total area in Uruguay is given, because the natural enemies introduced for control of pests in these crops are commonly found throughout the crop areas.

CEBIOF thus offers innovative services for forestry, but also acts as an example on how an association of different institutes stimulates proper execution of biocontrol programmes. Currently, CEBIOF is no longer operating (Pilar Gasparri, Montevideo, 2018, personal communication).

Mandatory regulation and registration of biocontrol products was taken up in 2007 by DGSA of MGAP. In 2013–2014, Uruguay's MGAP approved standards for the registration and control of biocontrol products. Since 2007, DGSA has authorized the manufacturing, formulation, release, commercialization and use of biocontrol agents. Authorization is provided pending evaluation of scientific data demonstrating that the biocontrol agent is effective for the intended purpose and does not pose risks to the environment or to human, animal or plant health. The current normative framework includes registration procedures of commercial insecticides including biocontrol agents, both microorganisms (entomopathogens) and parasitoids or predators (entomophagous arthropods), for agricultural use. The above-mentioned process for the registration for microorganisms (entomopathogens) or biocontrol agents (entomophagous arthropods) took 3 years (2007–2010) to come into practice.

Today, there are nine commercially available products based on microorganisms or arthropods in Uruguay. Three of these, Trichosoil and Rizoderma (*Trichoderma harzianum* Rifai) and Baktillis (*Bacillus subtilis* Cohn), are formulated as fungicides for control of various diseases in tomato, onion, garlic, lettuce, pepper, wheat, barley and eucalyptus. Five of these are registered as insecticides: Madex (Granulovirus) for control of *C. pomonella* in apple; Madex Twin (another strain of granulosis virus) to control *C. pomonella* and *Grapholita molesta* (Busck) in apple, peach, pear and plum; Swirskii System (*A. swirskii*) to control *B. tabaci* in pepper; Orius System (*O. insidiosus*) for control of *F. occidentalis* in pepper; and Crebio 1 (*B. bassiana*) to control *Atta* spp. and *Acromyrmex* spp. in many crops. Finally, Nemix C (*B. subtilis* and *Bacillus licheniformis* (Weigmann) Chester) is a nematicide to control *Meloidogyne* spp. in tomato. Complete registration information was provided and evaluated for the above-mentioned nine products.

Uruguay, a country with a relatively small geographical area, aspires to be distinguished in the world by the production of high-quality agricultural food. Biocontrol of pests and diseases can contribute to this goal, since it not only avoids toxicity to humans and environmental pollution caused by chemical insecticides, but also can contribute to increasing the value of products, though products with transparent traceability are required. In a number of Uruguayan situations, reduction of pesticide use combined with an increase in use of biocontrol seems realistic, such as horticultural crops grown in greenhouses, soybean destined for human consumption and for export, and in forestry. There are, however, some major constraints, which include: (i) the low price of chemical pesticides; (ii) the misconception of farmers that a crop needs to be completely free of pests, resulting in overuse of pesticides, and (iii) unavailability of a diversity of biocontrol agents on the Uruguayan market. Still, during the past few years, there has been much progress in thinking about more usage of biocontrol. There is greater concern in governmental circles and among the public about pollution of food and the environment; consumers are beginning to demand healthier food and global markets are willing to pay higher prices for products without pesticide residues. Thus we, the authors of this chapter, are optimistic about the future of biocontrol in Uruguay.

References

(References with grey shading are available as supplementary electronic material)

Alzugaray, R., Zerbino, M.S., Stewart, S., Ribeiro, A. and Eilenberg, J. (1999) Epizootiologia de hongos Entomophthorales. Uso de *Zoophthora radicans* (Brefeld) Batko (Zygomicotina: Entomophthorales) para el control de *Epinotia aporema* (Wals.) (Lepidoptera: Tortricidae) en Uruguay. [Use of *Zoophthora radicans* to control *Epinotia aporema* in Uruguay]. *Revista de la Sociedad Entomológica Argentina* 58, 307–311.

Alzugaray, R., Ribeiro, A., Silva, H., Stewart, S., Castiglioni, E., Bartaburu, S. and Martínez, J.J. (2010) Prospección de agentes de mortalidad natural de áfidos en leguminosas forrajeras en Uruguay [Determination of aphid mortality caused by biological agents in forage legumes in Uruguay]. *Agrociencia (Uruguay)* 14, 27–35.

Armand Pilón, A. (2017) Biología y potencial reproductivo del parasitoide de huevos *Telenomus podisi* (Ashmead) (Hymenoptera: Scelionidae) de *Piezodorus guildinii* (Westwood) (Hymenoptera: Pentatomidae) [Biology and reproductive potential of the *Piezodorus guildinii* egg parasitoid *Telenomus podisi*]. BSc thesis. Universidad de la República, Facultad de Agronomía, Montevideo, Uruguay.

Aznárez, G.L. (1996) Eficiencia y persistencia de un virus de la polihedrosis nuclear en el control de la «lagarta del girasol» *Rachiplusia nu* (Lepidoptera: Noctuidae) en condiciones de campo [Efficiency and persistence of a nuclear polyhedrosis virus in the control of 'sunflower lizard' *Rachiplusia nu* under field conditions]. BSc thesis. Universidad de la República, Facultad de Agronomía, Montevideo, Uruguay.

Basso, C. and Franco J. (1995) Determinación del momento de control de *Diatraea saccharalis* (F.) en caña de azúcar por medio de *Trichogramma* en el Uruguay [Determining the optimal control moment of *Diatraea saccharalis* in sugar cane using *Trichogramma* in Uruguay]. *Boletín de investigación* 39.

Basso, C. and Morey, C. (1991) Biological control of the sugarcane borer *Diatraea saccharalis* (Fabricius, 1798) (Lepidoptera: Pyralidae) with *Trichogramma* spp. in Uruguay. *Les Colloques de l'INRA* 56, 165–169.

Basso, C. and Pintureau, B. (1998) Biological control using *Trichogramma* in Uruguay, first results and prospects. *Mitteilungen aus der Biologischen Bundesanstalt für Land-und Forstwirtschaft Berlin-Dahlem* 356, 93–96.

Basso, C. and Pintureau, B. (2004) Las especies de *Trichogramma* del Uruguay (Hymenoptera: Trichogrammatidae) [*Trichogramma* spp. of Uruguay]. *Revista de la Sociedad Entomológica Argentina* 63, 71–80.

Basso, C., Grille, G. and Pintureau, B. (1999) Eficacia de *Trichogramma exiguum* Pinto & Platner y de *T. pretiosum* Riley en el control de *Argyrotaenia sphaleropa* (Meyrick) y de *Bonagota cranaodes* (Meyrick) en la vid en el Uruguay [*Trichogramma exiguum* and *T. pretiosum* efficacy of control against *Argyrotaenia sphaleropa* and *Bonagota cranaodes* in Uruguay]. *Agrociencia (Uruguay)* 3, 20–26.

Basso, C., Grille, G., Alzugaray, R. and Pintureau, B. (2006) Comparative study of the effects of *Trichogramma pretiosum* (Hym.,Trichogrammatidae) releases and Triflumuron applications on *Epinotia aporema* (Lep., Tortricidae) in birdsfoot trefoil seedbeds. *Boletín de Sanidad Vegetal, Plagas* 32, 563–571.

Bentancourt, C.M. and Scatoni, I.B. (2001). *Enemigos naturales: Manual ilustrado para la agricultura y la forestación* [*Biological Control Agents: Illustrated Manual for Agriculture and Forestry*]. Universidad de la República, Facultad de Agronomia, Montevideo, Uruguay.

Bianchi, M. (1992) Situación del *Sirex noctilio* F. y otros insectos plaga forestales en Uruguay [*Sirex noctilio* and other forest pest insects in Uruguay: present status]. In: *Conferencia Regional da Vespa da Madeira Sirex noctilio na America do Sul (Florianópolis SC, 1992)*. Abstract, Florianópolis SC, Brasil, pp. 65–71.

Borghi, C. and Cano, F. (2014) Efecto de la edad y el número de hembras de *Telenomus podisi* Ashmead en la parasitación de huevos de *Piezodorus guildinii* (Westwood) de diferentes edades [Effect of age and number of females of *Telenomus podisi* in the parasitization of *Piezodorus guildinii* eggs of different ages]. BSc thesis. Universidad de la República, Facultad de Agronomía, Montevideo, Uruguay.

Buenahora, J. and Basso, C. (2015) Utilización de *Amblyseius swirskii*, un enemigo natural clave para el manejo integrado de plagas en el cultivo de pimiento en invernadero [Use of *Amblyseius swirskii*, a key natural enemy for integrated pest management in greenhouse pepper cultivation]. In: *4tas. Jornadas de enfermedades y plagas en cultivos bajo cubierta*. Universidad Nacional de La Plata, Facultad de Ciencias Agrarias y Forestales, La Plata, Argentina.

Buenahora, J., Bentancourt, C., Scatoni, I., Asplanato, G., Paullier, J., Pazos, J., Pintos, J. and González, A. (2001) Enemigos naturales del minador de las hojas de los cítricos *Phyllocnistis citrella* Stainton (Lep. Gracillaridae) [Natural enemies of citrus leafminer *Phyllocnistis citrella*]. In: *8° Congreso de Horticultura*. Salto, Uruguay.

Cabrera, M.C., Astigarraga, L., Borsani, O., Camussi, G., Caputi, P. *et al.* (2017) Uruguay, a world food producer: toward sustainable production from a food and nutrition security perspective. In: *Challenges and Opportunities for Food and Nutrition Security in the Americas. The View of the Academies of Sciences*. IANAS, IAP and BMBF, México DF, pp. 532–565. [Free public access of this publication in English and Spanish at www.ianas.org]

Castiglioni, E., Ribeiro, A., Alzugaray, R., Silva, H., Ávila, I. and Loiácono, M. (2010) Prospección de parasitoides de *Piezodorus guildinii* (Westwood) (Hemiptera: Pentatomidae) en el litoral Oeste del Uruguay [Prospecting of parasitoids of *Piezodorus guildinii* in the West coast of Uruguay]. *Agrociencia (Uruguay)* 14, 22–25.

Chiaravalle, W. (1996) Informe sobre el avance del control biológico en Uruguay (1989–1991) [Biological control progress in Uruguay (1989–1991)]. In: Zapater, M. (ed.) *El control biológico en América Latina*. IOBC/SRNT, Buenos Aires, Argentina, pp. 93–98.

Chiaravalle, W., Parra, J.R. and Moscardi F. (1987) Biología comparada de *Pseudoplusia includens* (Walker, 1857) (Lep.-Noctuidae) em dietas naturais e artificiais [Comparative biology of *Pseudoplusia includens* (Walker, 1857) (Lep.-Noctuidae) in natural and artificial diets]. In: *XI Congreso Brasileiro de Entomología*. Campinas, San Pablo, Brazil. Abstract, p. 43.

Chiaravalle, W., Ferreiro, A., Aznarez, G. and Casella, E. (1992) Control Biológico de *Cydia pomonella* con su virus de granulosis [Biological control of *Cydia pomonella* with its granulosis virus]. In: *I Congreso Iberoamericano de Horticultura*. Montevideo, Uruguay. Abstract.

CIA (2017) The World Factbook: Uruguay. Available at: https://www.cia.gov/library/publications/the-world-factbook/geos/uy.html (accessed 29 July 2019).

DeBach, P. (1975) Control biológico de las plagas de insectos y malas hierbas [Biological control of insects and weeds]. CECSA, México DC, Mexico.

Grille, G. and Basso, C. (2001) Relevamiento de especies de 'moscas blancas' y sus parasitoides en cultivos de interés hortícola en Uruguay [Survey of whitefly species and their parasitoids in crops of horticultural interest in Uruguay]. In: *8º Congreso de Horticultura*. Salto, Uruguay. Abstract, p. 53.

Grille, G., Basso, C. and Pintureau, B. (2009) Discovery of *Trichogramma colombiensis* Velásquez de Ríos and Terán, 1995, in Uruguay. *Agrociencia (Uruguay)* 13, 36–37.

INIA (2013) Lanzamiento de importante herramienta para combatir plagas forestales: avispa parasitoide [Launching of an important tool to fight forest pests: parasitoid wasp]. *Revista INIA*, 33, 59.

Kucharski, A. (2017) Fluctuaciones de poblaciones de áfidos y sus enemigos naturales en alfalfa [Fluctuations of populations of aphids and their natural enemies in alfalfa]. BSc thesis. Universidad de la República, Facultad de Agronomía, Montevideo, Uruguay.

Martínez, G. and Bianchi, M. (2010) Primer registro para Uruguay de la chinche del eucalipto, *Thaumastocoris peregrinus* Carpintero y Dellappé, 2006 (Heteroptera: Thaumastocoridae) [First record for Uruguay of the eucalyptus bug, *Thaumastocoris peregrinus*]. *Agrociencia (Uruguay)* 14, 15–18.

Parker, H.L., Berry, P.A. and Silveira Guido. A. (1951) Host-parasite and parasite-host list of insects reared in the South American parasite laboratory during the period 1940–1946. *Revista de la Asociación de Ingenieros Agrónomos* 92, 15–112.

Ribeiro, A. (2004) Características de las poblaciones de insectos en los sistemas agrícola-pastoriles [Characteristics of insect populations in agricultural–pastoral systems]. *Cangüé* 26, 11–14.

Ribeiro, A. (2010) Prospección de agentes para el control natural de plagas en sistemas agrícola-pastoriles [Prospecting of natural control agents of pests in agricultural–pastoral systems]. In: Altier, N., Rebuffo, M. and Cabrera, K. (eds) *Enfermedades y plagas en pasturas*. *INIA Serie Técnica*, 183, 05–110.

Ribeiro, A. and Castiglioni, E. (2008) Caracterización de las poblaciones de enemigos naturales de *Piezodorus guildinii* (Westwood) (Hemiptera: Pentatomidae) [Characterization of the populations of natural enemies of *Piezodorus guildinii*]. *Agrociencia (Uruguay)* 12, 48–56.

Ribeiro, A. and Zerbino, M.S. (1994) Factores naturales de mortalidad de larvas de *Pseudaletia adultera* y *Faronta albilinea* (Lepidoptera: Noctuidae) [Natural mortality factors of larvae of *Pseudaletia adultera* and *Faronta albilinea*]. In: *4° SICONBIOL. Simpósio de Controle Biológico*. EMBRAPA/CPACT, Pelotas, Brazil. Abstract.

Ribeiro, A., Castiglioni, E. and Silva, H. (2008) *Insectos de la soja en Uruguay: manual ilustrado de reconocimiento de plagas y enemigos naturales* [*Soybean Insects in Uruguay: Illustrated Manual of Recognition of Pests and Natural Enemies*]. Universidad de la República, Facultad de Agronomía, Montevideo, Uruguay.

Ribeiro, A., Silva, H., Castiglioni, E., Bartaburu, S. and Martínez, J. (2015) Control natural de *Crocidosema* (*Epinotia*) *aporema* (Walsingham) (Lepidoptera: Tortricidae) por parasitoides y hongos entomopatógenos en *Lotus corniculatus* y *Gycine max* [Natural control of *Crocidosema (Epinotia) aporema* by parasitoids and entomopathogenic fungi in *Lotus corniculatus* and *Gycine max*]. *Agrociencia (Uruguay)* 19, 36–41.

Ruffinelli, A. and Carbonell, C.S. (1944) Primera lista sistemática de insectos relacionados con la agricultura nacional [First systematic list of insects related to national agriculture]. *Revista de la Asociación de Ingenieros Agrónomos* 1, 13–32.

Ruffinelli A. and Carbonell C.S. (1953) Segunda lista de insectos y otros artrópodos de importancia económica en el Uruguay [Second list of insects and other arthropods of economic importance in Uruguay]. *Revista de la Asociación de Ingenieros Agrónomos* 94, 33–82.

Silva, H. (2016) Descripción cuantitativa de una red trófica de tres niveles: leguminosas-áfidos-parasitoides y entomopaógenos [Quantitative description of a trophic network of three levels: legumes-aphids-parasitoids and entomoparogens]. MSc thesis. Universidad de la República, Facultad de Agronomía, Montevideo, Uruguay.

Silveira Guido, A. and Ruffinelli, A. (1956) Primer catálogo de los parásitos y predatores encontrados en el Uruguay [First catalog of parasites and predators found in Uruguay]. *Boletín* 32. Universidad de la República, Facultad de Agronomía, Montevideo, Uruguay.

Trujillo Peluffo, A. (1963) *Breve historia entomológica uruguaya* [*Brief Entomological History of Uruguay*]. Universidad de la República, Facultad de Agronomía, Montevideo, Uruguay. [A copy of this publication can be obtained at informacion@fagro.edu.uy]

31 Biological Control in Venezuela

Carlos Vásquez[1]*, Francisco Ferrer[2], Yelitza C. Colmenarez[3] and José Morales Sanchez[4]

[1]*Technical University of Ambato, Faculty of Agricultural Sciences, Campus Querochaca, Province of Tungurahua, Ecuador;* [2]*Independent Entomologist Advisor, Barquisimeto, Lara State, Venezuela;* [3]*CABI Brazil, UNESP – Fazenda Experimental Lageado – Foundation for Agricultural and Forestry Studies and Research, Botucatu, São Paulo, Brazil;* [4]*Universidad Centroccidental Lisandro Alvarado. Department of Biological Sciences, Barquisimeto, Lara State, Venezuela*

* E-mail: ca.vasquez@uta.edu.ec

Abstract

Venezuela has a long history in application of pest biological control. The first attempts were recorded at the end of the 19th century with the use of *Scelio famelicus* for control of the migratory locust *Schistocerca paranensis* and continued in the middle of the 20th with the introduction of *Rodolia cardinalis* to control the citrus scale *Icerya purchasi*, *Aphelinus mali* for control of *Eriosoma lanigera* and the Amazonian fly *Lydella minense* to control the sugarcane borer *Diatraea saccharallis*. Although chemical control has been widely used by Venezuelan farmers, biocontrol has been implemented in some crops such as sugarcane and maize partially due to the development of artificial diets that allowed the mass production of some parasitoids such as *Trichogramma* spp., *Telenomus remus*, *Cotesia flavipes* and the predators *Chrysoperla carnea* and *Orius tristicolor*. This review shows data on emblematic cases of the utilization of biocontrol in Venezuela and also the prospect of using other invertebrate organisms (arthropod, nematodes) as well as microorganisms in different crops. Finally, since Venezuela has a very rich biodiversity, it is presumed that a large number of entomopathogens, predators and parasitoids remain to be discovered and evaluated for potential use in biocontrol programmes, increasing the possibilities of the use of biocontrol agents in new crops, and a large range of studies to be done.

31.1 Introduction

Venezuela has an estimated population of about 31.3 million (July 2017) and its main agricultural products are maize, sorghum, sugarcane, rice, bananas, vegetables, coffee; beef, pork, milk, eggs; fish (CIA, 2017). According to Tapia *et al.* (2017, pp. 568, 572, 586 and 587):

> Venezuela has an area of 916.445 km^2 divided into 23 States, a Capital District, 235 islands and 71 islets and keys in the Caribbean Sea ... Venezuela is one of the world's ten most biologically diverse countries, ranking sixth in America. Preliminary results indicate 15,353 plant species, 261 families and 2,482 genera. Approximately 2,964 of these species are endemic. Some important genera for food and agriculture have their center of origin and biodiversity in the South-American region and include cultivated, wild and semi-domesticated species ... As for fauna, at least 203 species of terrestrial and aquatic vertebrates (5% of total Venezuela) are used as a food source by the country's rural and indigenous communities ... Four major direct threats to biodiversity have been identified: 1. Destruction, fragmentation and degradation of ecosystems; 2. Introduction, establishment and invasion of exotic species; 3. Unsustainable use of biological diversity; and 4. Introduction of genetically modified organisms ... The agricultural area (2011 figures) accounts for 24% of the country; while 84.71% of the farmland is pasture, 12.2% arable land and 3% used for permanent crop...Venezuelan agricultural systems are diverse, ranging from family and semi-commercial agriculture to various types of plantations. In 2015 ... the harvested area was 1.7 million hectares, its main items comprising production of 15.3 million Mg (Mega-grams). Highest volumes correspond to sugarcane, fruits and cereals, which occupy the largest surface, maize being the main crop. Regarding animal production, there were 497,219,975 heads of cattle, and 2,807 million liters of milk, and 2,906 million eggs were produced. Aquaculture production reported a total of 254.901 Mg. ... By 2013, there were 400 (forested, eds.) areas with 67.9 million ha, including national parks, forest reserves, natural monuments, protected forest areas and wildlife refuges. Areas assigned for the sustainable management of Forest Heritage occupy 16.3 million ha, with forest reserves covering 12.8 million ha (79%) and Forested Areas Under Protection, 3.4 million ha (21%) ... The state is the sole supplier of certain staple foods, since it nationalized input and seed distribution. Falling oil prices and dependence on imports weakened agricultural production, which, coupled with the lack of imports, produced shortages and scarcity (>50%) at critical levels of certain items and regulated foods that are the main contributors of energy and nutrients. This is compounded by the lack of investment in infrastructure, restrictions on access to foreign exchange for inputs, seeds, machinery, equipment and spare parts; pricing below production costs, and legal uncertainty over property and personal insecurity. Agribusiness, agricultural research and talent training have all been affected.

Venezuela exhibits wide ecological diversity that allows the production of a large number of agricultural crops (Laurentín, 2015) and also serves as a niche for a great diversity of insect species that, in some cases, can be serious pests and in other cases function as effective natural enemies of native or exotic pests. Economic losses caused by pests can reach up to 40%, resulting in the use of synthetic pesticides (Martínez, 2010). However, insecticides are applied in quantities greater than required and often with wrong application techniques, causing pesticide residue accumulation in food, soil and water at levels higher than allowable (0.2 mg l^{-1} in Venezuela), thus becoming a public health problem (Martínez, 2010; Chirinos and Geraud-Pouey, 2011). These facts highlight the need for more sustainable control methods in order to reduce effects of insecticides on the health of agricultural workers and consumers. The implementation of sustainable strategies should begin with the training of farmers in rational application of pesticides, but also encourage, among other integrated pest management (IPM) methods, the use of biocontrol agents and exploitation of natural resistance of plants.

31.2 History of Biological Control in Venezuela

31.2.1 Period 1880–1969

The first experiments on biocontrol in Venezuela date from 1884 with the proposal of A. Ernst, a German naturalist, botanist and zoologist, to use the parasitoid *Scelio famelicus* Riley for biocontrol of the migratory locust *Schistocerca paranensis* (Burmeister) (Guagliumi, 1962). In 1941, C. Ballou introduced the predatory coccinellid *Rodolia cardinalis* (Mulsant) to control the citrus scale *Icerya purchasi* Maskell, and *Aphelinus mali* (Haldeman) against *Eriosoma lanigera* (Hausmann) (Ferrer, 2001; Sampaio *et al.*, 2009). Between 1949 and 1950, H. Box introduced the Amazonian fly *Lydella minense* (Townsend) to control the sugarcane borer *Diatraea saccharallis* (F.) (Ferrer, 2001; Fundación Polar, 2003). This led to the construction of biological laboratories in sugar mill companies for the mass rearing of this dipteran parasitoid. In 1964, at the Central Rio Turbio, the Institute for the Promotion of Sugar Production (IFPA) was founded, which mass reared the Amazonian fly and released it in sugarcane fields. During this period, Venezuela also provided several natural enemies to Latin American and Caribbean countries, among others for control of *Diatraea* spp. (Cock, 1985).

31.2.2 Period 1970–2000

In 1972, the University of the Central-Western Region (Lara State) began the study of artificial diets for rearing larvae of *D. saccharalis* as a host of *L. minense*. Also, in 1972, the Araure Experimental Station and the National Cotton Association (ANCA) founded *Trichogramma* spp. rearing laboratories for control of lepidopteran pests. The Venezuelan Sugarcane Producers Association (UPAVE) organized a national seminar on sugarcane borer species in 1973, resulting in, among others, the development of a mass-rearing method for Tachinidae and Trichogrammatidae on artificial diets. An agreement between the private sector (the Portuguesa Sugar Central), the public sector (FONAIAP, National Agricultural Research Fund, which now is INIA: National Institute of Agricultural Research) and IFPA was established in 1973 and resulted in the production of *Trichogramma* wasps and first tests to produce and use *Metarhizium anisopliae* (Metchnikoff) Sorokin to control froghopper *Aeneolamia varia* (F.).

Also in 1973, a private company, the Biological Control Service Company (SERVBIO), was founded in Lara State by F. Ferrer, which offered IPM programmes for maize, sugarcane, cotton, vegetables, oil palm, tobacco, etc. In SERVBIO, artificial diets and mass-rearing technology for biocontrol agents have been developed since 1977, e.g. the production of *L. minense*, *Cotesia flavipes* (Cameron), *Telenomus remus* Nixon (introduced from Trinidad in 1979), *Trichogramma* spp., *Chrysoperla carnea* (Stephens) (introduced from Colombia between 1990 and 2000), *Orius tristicolor* White (introduced from Peru), *Spalangia endius* Walker and *Muscidifurax raptor* Girault & Sanders, and the nematode *Heterorhabditis bacteriophora* Poinar (introduced from Cuba in 2000). From 1977 to 2008, SERVBIO employed around 150 professionals, after which it became part of the network of laboratories created under the auspices of the INSAI (see below).

In 1975 two important events occurred concerning biocontrol in Venezuela. The first was the introduction of the parasitoid *Prospaltella opulenta* Silvestri (= *Encarsia opulenta*) from Mexico by the Service for the Farmer Foundation (FUSAGRI), the Central University of Venezuela (UCV) and the National Center for Agricultural Research (CENIAP), for control of citrus blackfly *Aleurocanthus woglumi* Ashby. Citrus blackfly was first found in 1965 on orange leaves (*Citrus sinensis* (L.) Osbeck) from Trujillo State and its presence was officially confirmed in 1967. By 1974, the citrus blackfly was widely distributed across the country on several citrus species (Ángeles *et al.*, 1971, 1972). Release of *P. opulenta* in the Central-Western region of Venezuela and Lara State resulted in complete control of citrus blackfly within 18 months after its release (Chávez, 1980). Parasitoids were released also in Aragua and Yaracuy States, but initially failed to establish because the approximately 120,000 *P. opulenta* adults introduced into the country were kept by customs officials for 2 days, and after the quarantine period only 20,000 adults survived, which were released during the rainy season, when attacks by entomopathogenic fungi had already decreased blackfly populations (Geraud F., Maracaibo, Venezuela, August 2018, personal communication,). But in 1976, about 40,000 *P. opulenta* parasitoids were introduced again and released in Aragua, Carabobo and Yaracuy States, resulting in permanent establishment and successful control (Boscán *et al.*, 1979). In 2011 there was a total of 25,000 ha planted with orange in Venezuela (Aular and Casares, 2011), of which 70% are estimated to have effective control of blackfly by *P. opulenta.*

The second event concerned the introduction of the parasitoid *C. flavipes* for control of *Diatraea* spp. group in 1975 by F. Ferrer. He obtained the parasitoid from CABI (previously known as Commonwealth Biological Control Institute) in Trinidad and Tobago. Artificial diets for rearing *D. saccharalis* were developed and a laboratory for biocontrol was established to provide natural enemies for the sugarcane factory in Portuguesa State. The first trials for management, rearing and release of *C. flavipes* were made from 1975 to 1981. The first signs of establishment of *C. flavipes* were reported 6 years after the initial releases in Ureña (Táchira State) and Cariaco (Sucre State). Specimens collected in these States were later used for mass production in the SERVBIO laboratory and successfully distributed in all sugarcane fields in Venezuela (Ferrer and Guédez, 1990; Linares and Ferrer, 1990; Linares and Yépez, 1992). Since then, it has effectively controlled all of the *Diatraea* species, resulting in increased economic benefits. The Sugar Foundation for Development, Productivity and Research (Fundacaña) has since developed more biocontrol projects for pests in sugarcane and has production laboratories for beneficial organisms in Chivacoa (Yaracuy State) and Papelon (Portuguesa State). Fundacaña mass produces *C. flavipes* for control of the sugarcane borer complex, and the entomopathogenic nematode *H. bacteriophora* for control of the sugarcane froghopper (*Aeneolamia* sp.), resulting in a considerable increase in sugarcane production (Fundacaña, 2018). In 2000, sugarcane was cultivated on 115,000 ha (Marin, 2007), and although the size of the area treated every year with *C. flavipes* and *H. bacteriophora* varied across locality and years, it is estimated that around 40% of the total area planted in the country was treated with *C. flavipes*. Biocontrol was not applied in some rural areas, due to limited availability of the parasitoid. Around 30% of the sugarcane was treated with *H. bacteriophora* in 2005–2007.

In 1984, an international seminar, sponsored by the Venezuelan Distributor of Sugar and the Venezuelan Sugar Producers Association, was held about sugarcane froghoppers and sugarcane borers in Barquisimeto (Lara State). A group coordinated by J. Salazar aimed to establish guidelines for control of both pest species and continued its work until 1994. The seminar was, among others, attended by F. Bennett of CAB International, who recommended the use of entomopathogenic nematodes of the genus *Heterorhabditis* for control of the froghopper (Ferrer *et al.*, 2004).

31.3 Current Situation of Biological Control in Venezuela

31.3.1 Perception of biological control by farmers

The successful production of *L. minense* in IFPA laboratories and use in sugarcane fields for *D. saccharalis* control, the mass rearing and use of parasitoids such as *Trichogramma* spp., *T. remus*, *C. flavipes* and the predators *C. carnea* and *O. tristicolor*, as well as the production of nematodes and microorganisms for effective control of insect pests in different crops, confirm that the development of biocontrol in Venezuela is well established. Still, it is thought that government agencies such as the Biological Laboratories Network, belonging to INSAI, and the University Extension Services should make a greater effort in promoting biocontrol, in order to encourage farmers to make more use of this control method (Altieri *et al.*, 1989). According to Guillén *et al.* (2008), biocontrol is not always popular among farmers because, on the one hand, certain levels of pest infestation must be tolerated to maintain the natural enemy populations; on the other hand, farmers perceive that the control is erratic since it depends on the environmental conditions. These ideas show the lack of knowledge about this pest management strategy at the farmers' level. Guillén *et al.* (2008) determined that tomato growers in Lara State did not know, or distrusted or were disinterested, about the advantages of biocontrol. Similarly, in Guárico State, 10% of maize growers revealed a low preference for biologically based pest management strategies, due to insufficient availability and access to biological products: 85% reported lack of knowledge about biocontrol agents to be used for each pest species, while 5% showed dissatisfaction with biocontrol effectiveness (Méndez and Páez, 2014). This highlights the need to intensify training programmes resulting in meaningful learning by small farmers.

31.3.2 Development of laboratories for mass production of biological control agents

Between 2004 and 2007, the Department of Entomology and Zoology of the West-Central University (Universidad Centroccidental 'Lisandro Alvarado', UCLA) carried out studies on taxonomy and distribution of *Trichogramma atopovirilia* Oatman and Platner, *T. exiguum* Pinto and Platner and *T. pretiosum* Riley. In 2005, a government initiative started to organize a network of biocontrol laboratories with expert advice from the Plant Health Research Institute (INISAV) from Cuba, together with officials from the Autonomous Service of Agricultural Health (SASA, currently INSAI) of the Ministry of Agriculture and Lands of Venezuela. These laboratories started production of biocontrol agents in 2007, aiming to promote agroecological approaches to sustainable crop production (Fig. 31.1). In 2008, SERVBIO became part of this and nowadays is known as 'Jacinto Lara' Laboratory. The species reared in these laboratories include five parasitoids, three predators, four species of entomopathogenic fungi, one bacterium and one nematode. Among the parasitoids, *Trichogramma* species are mostly reared, including *Trichogramma pintoi* Voegelé, *T. galloi* Zucchi and *Trichogramma* sp., which are produced in two, one and eight laboratories, respectively. Laboratories further produce the parasitoids *T. remus* (three laboratories) and *Anagyrus kamali* Moursi (one laboratory) (Fig. 31.1). Also the following predators are mass reared: *C. externa* (three laboratories), *Cryptolaemus montrouzieri* Mulsant (one laboratory) and *Orius insidiosus* (Say) (one laboratory) (Fig. 31.1). In addition, *Bacillus thuringiensis* (Berliner) is produced in nine laboratories, entomopathogenic fungi in several laboratories, *Beauveria bassiana* (Ball.-Criv) Vuill in 12 laboratories, *Trichoderma harzianum* Rifai in six laboratories, *M. anisopliae* in four laboratories and *Lecanicillium lecanii* (Zimm.) Zare & Gams in seven laboratories, and a nematode of the genus *Heterorhabditis* in three laboratories. As a result of this network of laboratories, pest management in crops such as potato, bell pepper, tomato, maize, sugarcane and others is now being done mainly by the release of biocontrol agents, resulting in production of crops of better quality.

An overview of natural enemies of major pests in Venezuela is provided in Table 31.1. A summary of use of biocontrol is given in Table 31.2, but it should be stressed that due to the current developments in Venezuela, it was difficult to obtain reliable data of the area under biocontrol. Below, some of the larger biocontrol programmes in Venezuela are summarized.

31.3.3 Research and application of *Trichogramma* species

First attempts to rear *Trichogramma* were made during the 1970s, in FONAIAP (currently INIA), mainly for control of cotton pests. A laboratory for mass rearing was established at the INIA-Anzoátegui facilities (Giraldo, 1988). This laboratory provided services to farmers in the States of Anzoátegui, Bolívar and Monagas

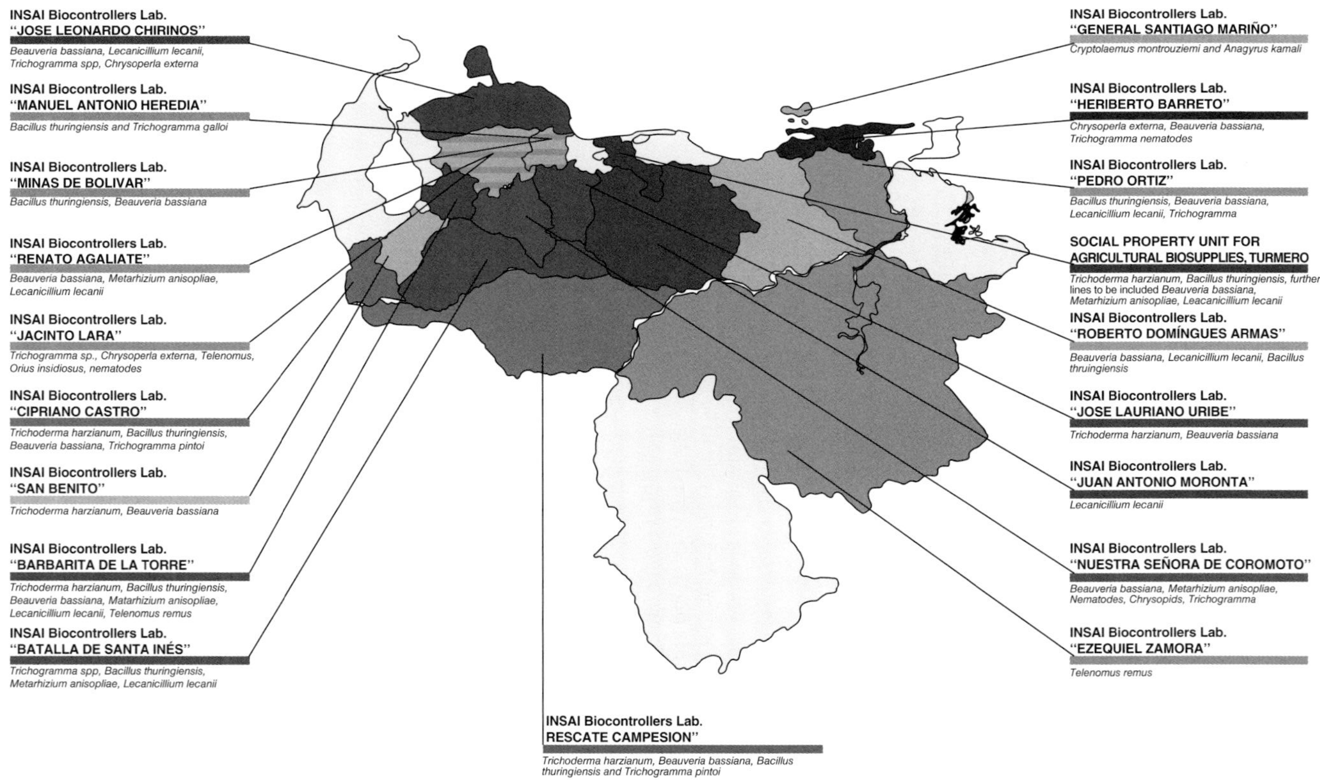

Fig. 31.1. National Institute of Integral Agricultural Health (INSAI) laboratories for mass production of biological control agents.

Table 31.1. Natural enemies of major pests reported in Venezuela (retrieved from Fundación Polar, 2003).

Order	Species	Prey/ host
PREDATORS		
Orthoptera		
	Phlugis teres DeGeer	*Saccharosydne saccharivora* and *Aeneolamia* spp.
Dermaptera		
	Forficulidae	Aphididae, Delphacidae and lepidopterans
Dictyoptera		
	Acanthops falcata Stål	Many pest species
	Acontiothespis multicolor (Saussure)	
Coleoptera		
	Calosoma alternans Fabricius	*Spodoptera frugiperda, Mocis latipes* etc.
	Azya orbigera Mulsant, *Azya* sp.	*Pseudaulacaspis pentagona, Coccus viridis*
	Chilocorus cacti (L)	Scale species
	Coleomegilla maculata DeGeer	Aphids, scales, eggs and lepidopterans
	Cryptognatha auriculata Mulsant	*Pseudaulacaspis pentagona*
	Cycloneda sanguinea (L.)	Lepidopterans, aphids, scales,
	Hippodamia convergens Guérin	Aphids, lepidopterans
	Hyperaspis trilineata Mulsant	*Saccharosydne saccharivora* Westwood
	Rodolia cardinalis Mulsant	*Icerya purchasi*
.	*Scymnus* spp.	Aphids, scales and mites
	Paederus spp.	Lepidopterans
Diptera		
	Ocyptamus clava (Fabricius)	Aphids
	Mesograpta basilaris Wiedemann	*Sipha flava* and other aphid species
	Salpingogaster nigra Schiner	*Aeneolamia* spp.
Hymenoptera		
	Larra americana Saussure	Gryllotalpidae
	Polistes versicolor Olivier	Lepidopterans, *Heliothis* complex, *Spodoptera frugiperda*
	Polybia nigra Saussure	*Heliothis* complex and other Lepidoptera
PARASITOIDS		
Diptera		
	Sarcodexia sternodontis Townsend	*Alabama argillacea*, *Eodiatraea* sp., *Diatraea* spp., *Spodoptera frugiperda, Mocis latipes* etc.
	Sarcophaga caridei Brethes	*Schistocerca* spp.
	Archytas marmoratus (Townsend)	*Helicoverpa zea* larvae, *Heliothis* complex, *Spodoptera frugiperda*
	Eucelatoria armigera (Coquillet)	*Heliothis* complex
	Euphorocera floridensis Townsend	*Alabama argillacea*, *Anticarsia gemmatalis* and other Lepidoptera
	Lydella minense (Townsend)	*Diatraea* spp.
	Paratheresia claripalpis (Wulp)	*Diatraea* spp.
	Trichopoda sp.	*Sphictyrtus* sp.
	Winthemia pinguioides (Townsend)	*Caligo memnon* C. & R. Felder and other Lepidoptera
Hymenoptera		
	Prospaltella opulenta Silvestri	*Aleurocanthus woglumi*
	Cotesia spp.	Lepidopterans
	Aphidius colemani Viereck	Aphididae species

Continued

Table 31.1. Continued.

Order	Species	Prey/ host
	Cardiochiles nigriceps Viereck	*Heliothinae* complex
	Chelonus insularis Cresson	*Spodoptera frugiperda*
	Diaeretiella rapae (McIntosh)	Aphids on cruciferous plant
	Lysiphlebus testaceipes (Cresson)	*Aphis* and *Toxoptera* species
	Meteorus laphygmae Viereck	*Spodoptera frugiperda*
	Rogas gossypii Muesebeck	*Alabama argillacea*
	Brachymeria incerta (Cresson)	Lepidopterans
	Brachymeria ovata (Say)	
	Spilochalcis dux (Walker)	*Spodoptera frugiperda*, *Mocis latipes*,
	Spilochalcis fulvomaculata (Cameron)	*Diatraea* spp. and other Lepidoptera
	Copidosoma truncatellum (Dalman)	Plusiinae, *Trichoplusia ni* etc.
	Gahaniella saissetia Timberlake	*Coccus hesperidum*
	Ooencyrtus trinidadensis Crawford	*Erinnyis ello* (L.)
	Achrysochari sp.	*Erinnyis ello* (L.)
	Euplectrus plathypenae Howard	*Spodoptera frugiperda*, *Spodoptera* spp. and *Heliothis* complex
	Coccophagus sp.	*Coccus hesperidium*
	Isosomodes sp.	*Bucrates capitatus* De Geer
	Agrothereuthes diatraeae Myers	*Diatraea* spp.
	Eiphosoma spp.	Lepidopterans
	Anagrus flaveolus Waterhouse	*Saccharosydne saccharivora* and *Delphax maidis*
	Telenomus alector (Crawford)	*Diatraea rosa* and *D. saccharalis*
	Telenomus remus Nixon	*Spodoptera frugiperda*
	Campsomeris servillei (Guérin-Menéville)	*Podischnus agenor*
	Trichogramma minutum Riley, *T.* spp.	Lepidoptera including: *Diatraea* spp., *Heliothis* complex, *Alabama argillacea*

Table 31.2. Biological control agents used in Venezuela and areas under biocontrol.

Natural enemy / exotic (ex), native (na)	Pest and crop	CBC/ABC[a]	Area (ha) under biocontrol
Rodolia cardinalis / ex	Cottony cushion scale, citrus	CBC 1941	?, but still present
Prospaltella opulenta / ex	Citrus blackfly, citrus	CBC 1975	17,500
Aphelinus mali / ex	Woolly apple aphid, fruit	CBC 1941	?, but still present
Lydella minense / ex	Sugarcane borer, sugarcane	ABC 1949	?, but still present
Cotesia flavipes /ex	Sugarcane borer, sugarcane	ABC 1975	46,000
Heterorhabditis bacteriophora / ex	Froghopper, sugarcane	ABC 1985	34,500
Trichogramma spp / na	Lepidopteran pests, cotton	ABC 1970	?, but still in use
Trichogramma sp. / na	Sugarcane borer, sorghum	ABC 2001	?, but still in use
Trichogramma sp. / na	Tobacco hornworm, tobacco	ABC 2001	?, but still in use
Trichoderma spp. / na	Phytopathogenic fungi	ABC 1978	?, but still in use

[a]CBC = classical biological control, ABC = augmentative biological control

until its closure due to the failure of agriculture in the region (Bertorelli and Rengifo, 2008).

Trichogramma spp. were released in sorghum and tobacco to control *D. saccharalis* and *Manduca sexta* (L.) (Ferrer, 2001). In April 2004, *T. pretiosum* parasitoids were recovered from *D. saccharalis* eggs in sorghum in El Tigre, Anzoátegui State. Identification of *Trichogramma* species was initially based on the external anatomy of male genitalia, limiting its application in morphologically similar species (Velásquez and Terán, 2003). Recently, other morphological characters such as morphology of wings and antennae, as well as molecular techniques, can be used for identification, making it possible to identify a number of species currently present in Venezuela (Table 31.3). Studies are still needed on other species present and their potential role in pest control.

Morales *et al.* (2007) reported *T. pretiosum*, *T. exiguum* and *T. atopovirilia* from Lara State. Velásquez and Terán (2003) extensively studied the distribution, hosts and plants associated with *Trichogramma* species for the North-western region of Guárico State. Additionally, several biological aspects and the effectiveness of *T. atopovirilia* using *Spodoptera frugiperda* (J.E. Smith) as host were studied by Morales *et al.* (2004), the host preference of *T. atopovirilia* and *T. pretiosum* on *Helicoverpa zea* (Boddie) by Navarro and Marcano (1999) and the influence of the temperature on the development, reproduction and longevity of *T. pretiosum* using *Sitotroga cerealella* (Olivier) as host by Berti and Marcano (1997).

31.3.4 Research and application of *Cotesia flavipes* and *Lydella minense*

The larval endoparasitoid *C. flavipes* has been successfully used again since 1988 in biocontrol

Table 31.3. *Trichogramma* species reported in Venezuela.

Species	Host	Host plant	Location	Reference
T. atopovirilia	*Spodoptera frugiperda* *Sitotroga cerealella*	*Malachra* spp. *Zea mays* *Capsicum annuum*	Las Lajas (Guárico) Humocaro Alto, Sabana Grande (Lara) Tintinal (Lara)	Velásquez and Terán, 2003 Morales *et al.*, 2007 Morales *et al.*, 2010
T. bennetti	Non-identified Lepidoptera species	*Spiracantha cornifolia*	Finca Las Guacamayas, (Guárico)	Velásquez and Terán, 2003
T. bruni	Noctuidae	*Spiracantha cornifolia*	Las Lajas, (Guárico)	Velásquez and Terán, 2003
T. diazi	Noctuidae	*Malachra* spp.	Las Lajas (Guárico)	Velásquez and Terán, 2003
T. exiguum	*Spodoptera frugiperda*	*Zea mays*	Humocaro Alto (Lara)	Morales *et al.*, 2007
T. fuentesi	*Helicoverpa zea*	*Zea mays*	San José de Tiznados (Guárico)	Velásquez and Terán, 2003
T. lasallei	*Anomis* sp.	*Malachra* spp.	Las Lajas (Guárico)	Velásquez and Terán, 2003
T. obscurum	*Dione juno*	*Passiflora edulis*	San Juan de los Morros (Guárico)	Velásquez and Terán, 2003
T. pretiosum	*Helicoverpa zea* *Phthorimaea operculella* *Sitotroga cerealella* *Spodoptera frugiperda* *Agraulis vanillae*	*Lycopersicum esculentum* Papa almacenada *Passiflora edulis*	El Sombrero (Guárico) Sanare (Lara) Humocaro Bajo (Lara) Anzoátegui, Humocaro Bajo y Humocaro Alto (Lara), Sabana Grande (Lara)	Velásquez and Terán, 2003 Morales *et al.*, 2007 Morales *et al.*, 2010
T. terani	Noctuidae	*Sida* spp.	Las Lajas (Guárico)	Velásquez and Terán, 2003
Trichogramma sp1	*Utetheisa ornatrix*	Leguminosa	El Pampero (Lara)	Morales *et al.*, 2010
Trichogramma sp2	*Erinnyis* sp.	*Ricinus communis*	El Pampero (Lara)	Morales *et al.*, 2010

programmes of *Diatraea* sugarcane borers, is easy to rear in the laboratory and adapts rapidly to climatic conditions of different regions (Linares and Ferrer 1990; Linares *et al.*, 1998). In Venezuela, a complex of sugarcane borers (*D. busckella* Dyar & Heinrich, *D. centrella* Molschl., *D. rosa* Heinrich and *D. impersonatella* (Walker)) constitute the most important pests in sugarcane, after the sugarcane froghopper *A. varia* (Ramón *et al.*, 2008). The ability of *C. flavipes* to parasitize a wide range of hosts puts it ahead of the Amazonian fly *L. minense*, which had been traditionally used but has been replaced due to its specificity for *D. saccharalis* (Ramón *et al.*, 2008). However, when only *D. saccharalis* infests sugarcane fields, *L. minense* seems to be the more efficient biocontrol agent (Weir and Sagarzazu, 1998). Massive releases of both parasitoids have reduced sugarcane borer populations, particularly by *L. minense*, which has controlled this pest for 50 years (Weir *et al.*, 2007). Levels of parasitism of 43–50% of *Diatraea* larvae by *C. flavipes* were found in sugarcane fields in Sucre and Táchira States (Linares and Ferrer, 1990), while in the laboratory 6.3–62.5% parasitism was found, due to age-related variation in both the parasitoid and the host (Hernández, 2010).

31.3.5 Research on coccinellids, syrphids and chrysopids

Coccinellids

Solano *et al.* (2016) studied the biology of *Cycloneda sanguinea* (L.), a predator of, among others, *Aphis craccivora* Koch, and showed that L4 and adult females are the most voracious stages. Solano *et al.* (2010) had earlier demonstrated that *A. craccivora* allowed the rapid development of *Olla v-nigrum* (Mulsant), which indicates that this aphid might be a good prey for mass production, although this coccinellid has not been observed preying on this aphid in the field, so this should be further investigated. Angulo *et al.* (2011) concluded that *Menochilus sexmaculatus* (F.) might be an important biological agent for control of *A. craccivora*. Torres and Marcano (2007) studied the development of *C. montrouzieri* feeding on *Maconellicoccus hirsutus* (Green) at a range of temperatures. They reported, inter alia, that adults showed malformation when reared at temperatures of 30°C and 35°C, indicating that high temperatures negatively affect development. The invasive species *Harmonia axyridis* (Pallas) was found feeding on the aphid *Rhopalosiphum maidis* (Fitch) in Venezuela in 2014 (Solano and Arcaya, 2014) on *Euphorbia pulcherrima* Willd and maize *Zea mays* L., in Aragua and Lara States. According to the authors, studies are needed on its establishment and impact on the ecological balance of Coccinellidae species present in the country.

Syrphids

Recently the biocontrol potential of some species of Syrphidae has been evaluated. Arcaya *et al.* (2013) recorded 2,424 specimens of Syrphinae grouped in 12 genera and 40 species in the collection of the Museum of Entomology José M. Osorio (MEJMO) of the Universidad Centroccidental Lisandro Alvarado. *Toxomerus* was the genus with the largest number of species (17 spp.) and the best-represented species were *Pseudodoros clavatus* (Fabricius), followed by *Allograpta exotica* Wiedemann, *Ocyptamus gastrostactus* (Wiedemann), *Ocyptamus dimidiatus* (Fabricius), *Toxomerus floralis* (Fabricius) and *Ocyptamus stenogaster* (Williston). This is an important research database for teaching and training on the use of these biological agents in agriculture. Later, Morales *et al.* (2014) reported that *Ornidia obesa* F. (occurring in Lara, Falcón, Trujillo and Yaracuy States) and *Quichuana picadoi* Knab (occurring in Lara, Portuguesa, Trujillo and Yaracuy States), *Ornidia major* Curran and *Copestylum isabellina* (Williston) were recorded in Lara and Falcón States, while *Rhingia nigra* Mcquart was reported in Lara and Trujillo States and *Copestylum rurale* Curran in Lara and Yaracuy States. *Copestylum sica* (Curran), *Copestylum musicanum* (Curran), *Copestylum punctiferum* (Bigot), *Lejops mexicanus* (Macquart), *Nausigaster meridionalis* Townsend, *Palpada mexicana* (Macquart), *Palpada pusila* (Macquart), *Palpada ruficeps* (Macquart), *Palpada solennis* (Walter) and *Sphiximorpha barbipes* (Loew) were only collected in Lara State. Arcaya *et al.* (2004) reported nine aphid species that could serve as food for the syrphid *P. clavatus* larvae, including *A. craccivora*.

Chrysopids

Finally, Giffoni *et al.* (2007) evaluated the duration of the biological cycle of *Chrysoperla externa* (Hagen) fed with *S. cerealella*, *A. craccivora*, *Aphis nerii* Boyer de Fonscolombe, *Thrips tabaci* Lyndeman and *Tetranychus cinnabarinus* (Boisduval) and observed that the predator only completed its development when fed with *S. cerealella* and *A. craccivora*, suggesting that these species could potentially be controlled by the predator.

31.3.6 Research on entomopathogenic nematodes

The use of entomopathogenic nematodes for control of insect pests has not received much attention in Venezuela. Poinar and Linares (1985) reported natural parasitism rates of up to 50% in *A. varia* nymphs and adults by the nematode *Hexamermis dactylocercus* Poinar Jr. and Linares in Guanare (Portuguesa State). Besides, species from *Heterorhabditis* and *Steinernema* have been recommended for control of the banana weevil borer *Cosmopolites sordidus* (Germar) (Rosales and Suárez, 1998) and pineapple weevil *Metamasius dimidiatipennis* Champeon (García *et al.*, 2013). More recently, natural populations of *H. amazonensis* Andaló *et al.* have been found in plantain–maize plantations and pastures in Zulia State and in natural pastures in Barinas State (Morales *et al.*, 2016), opening new fields of research to evaluate the efficiency of native nematode strains and species for pest control.

31.3.7 Research and application of microbial control agents

Studies of fungal pathogens are mainly focused on the use of *Trichoderma* species.

Several species of *Trichoderma* that were used for control of phytopathogenic fungi were identified (C. Zambrano, Barquisimeto, Venezuela, August 2018, personal communication). García *et al.* (2006) developed a biofungicide based on *Trichoderma harzianum* Rifai (obtained by fermentation and formulated as wettable powder at 2×10^{12} cfu) which showed an antagonistic activity of more than 25% against *Rhizoctonia solani* Kühn, *Sclerotium cepivorum* Berk., *Sclerotium rolfsii* Sacc., *Fusarium* sp., *Plasmodiophora brassicae* Wor. and *Phytophthora* sp. on potato and other solanaceous crops, and, among others, garlic, cruciferous crops, legumes, banana, coffee and tobacco. Salazar *et al.* (2012) conducted a study to identify *Trichoderma* species isolated from soil samples of bean, corn, tomato and sorghum in the Central region of the country (Aragua, Carabobo and Guárico States) to evaluate its antagonism against *Macrophomina phaseolina* (Tassi) Goid. Twelve *Trichoderma* isolates were obtained, of which nine were identified as *T. harzianum* and the other three were *T. crassum*, *T. koningiopsis* Samuels, C. Suarez & H.C. Evans and *T. longibrachiatum* Rifai. They found high *in vitro* variability in effectiveness of *Trichoderma* isolates to control *M. phaseolina*. *Trichoderma crassum* Bisset was the most effective in controlling *M. phaseolina*, since it showed the highest inhibition of sporulation and growth (91.02% and 91.50%, respectively) compared with other isolates. García *et al.* (2012) observed that potato plants var. Andinita treated with *Trichoderma* showed a significant increase in yield, as well as lower incidence and severity of damage caused by *R. solani* on tubers.

A few studies have been done with other antagonists. Pineda and Díaz-Polanco (1981) found that *Penicillium notatum* Thom showed antagonistic activity against *S. rolfsii* under natural field conditions. Treatment with *P. notatum* reduced damage caused by *S. rolfsii* to 9.2% in kidney-bean (*Phaseolus vulgaris* L.), compared with 49.2% damage of untreated control plants. Labrador (2011) showed *in vitro* and *in vivo* antagonistic effects of *T. crassum* and *Bacillus subtilis* Cohn on *Phytophthora palmivora* Butler, supporting the feasibility of applying combinations of these two biocontrol agents against *P. palmivora*.

Y. Hernandez and colleagues (2011) evaluated *in vitro* antibacterial activity of *Pseudomonas fluorescens* Flugge, *Pseudomonas putida* Trevisan, *Bacillus licheniformis* (Weigmann) Chester, *B. subtilis* and *Bacillus megaterium* de Bari against different phytopathogenic species such as *Burkholderia glumae* (Kurita and Tabei) Urakami *et al.*, *Pantoea agglomerans* (Ewing and Fife) Gavini *et al.*, *Xanthomonas phaseoli* (Smith) Gabriel *et al.*, *Xanthomonas campestris* (Pammel) Dowson and *Ralstonia solanacearum* (Smith). The results showed that *B. licheniformis* and *B. megaterium* isolates were much more effective than the other species, especially in rice, bean, onion, tomato or potato, inferring a possible synergistic effect (Y. Hernández, Caracas, Venezuela, August 2018, personal communication). Future studies are needed to explain the mechanisms of action of these bacteria.

31.4 New Developments of Biological Control in Venezuela

All of the initiatives for biocontrol in Venezuela are being encouraged by current laws promoting sustainable agriculture and production of healthy food (Articles 117, 127 and 305, 306, 307 in the Constitution) and agroecology (Article 48, Integral Agricultural Health law). However, the endorsement of norms for registration of biocontrol agents is still needed. They are now being evaluated by the National Institute of Integral Agricultural Health. As a result, most of the products registered in the country have been certified abroad, i.e. *B. thuringiensis* and more recently the resistance inducers like *Trichoderma*.

Biocontrol has been practised in Venezuela since the end of the 19th century. The success caused by the introduction of the tachinid Amazonian fly for control of the sugarcane borer around 1950 strongly stimulated new research into and application of biocontrol and resulted, for example, in the creation of laboratories in sugarcane-producing areas. Then, after the foundation of the SERVBIO laboratory in the 1970s, biocontrol of pests in various other crops began. In addition to SERVBIO, public universities have played an important role by their contributions to studies related to biocontrol. Finally, the Biological Laboratories Network was founded in 2000 as a response to the high demand for IPM services in many crops, consolidating sustainable agriculture in Venezuela.

Research on biocontrol in Venezuela has had an academic approach and only a few studies have addressed practical applications. The results obtained should be transferred to farmers through agricultural technology transfer programmes in order to solve problems and generate income. Such programmes have been developed by some of the following organizations. National government initiatives, such as the Venezuelan Agrarian Corporation (CVA), the National Institute for Agricultural Research (INIA), the National Educational Training Institute (INCE) and the Development Fund for Agriculture, Fishing, Forestry and Related Activities (FONDAFA) have promoted production of organic crops and in doing so they have increased the demand for IPM services, including biocontrol. The aim is to create awareness among farmers about the benefits of using biocontrol. For this purpose it is vital to establish efficient communication between the different institutions, farmers and providers of biocontrol agents, in order to offer a wide range of self-sustaining alternatives for pest management with the best economic and the lowest environmental impact.

Venezuela has a very rich biodiversity, with large numbers of species of parasitoids, predators and entomopathogens still available for evaluation and use in biocontrol programmes for new invasive pests as well as for those already present. Due to the success of important earlier programmes implemented in the country and increased understanding about how to use biocontrol, farmers are at present more interested in the use of biocontrol agents. However, it is still necessary to increase efforts to incorporate biocontrol as part of current IPM strategies, in order to ensure significant acceptance of biocontrol by farmers. Considering the current economic situation in Venezuela, implementation of biocontrol is becoming increasingly important due to the shortage of agrochemicals as a result of lack of foreign currency. In this respect, the production of biocontrol agents represents a positive opportunity for a source of work in agribusiness for Venezuelan agronomists.

Acknowledgement

CABI is an international intergovernmental organization, and Y.C. Colmenarez gratefully acknowledges the core financial support from CABI's member countries (see https://www.cabi.org/what-we-do/how-we-work/cabi-donors-and-partners/ for full details).

References

Altieri, M.A., Trujillo, J., Campos, L., Kelin-Koch, C., Gold, C. and Quezada J. (1989) El control biológico clásico en América Latina en su contexto histórico [Classical biological control in Latin America in its historical context]. *Manejo Integrado de Plagas* 12, 82–107.

Ángeles, N.J., Dedordy, J.R., Paredes, P.P. and Requena, J.R. (1971). Mosca prieta (*Aleurocanthus woglumi* Ashby) de los cítricos en Venezuela [Citrus black fly (*Aleurocanthus woglumi*) in Venezuela]. *Agronomía Tropical* 21, 71–75.

Ángeles, N.J., Dedordy, J.R., de Martínez, N.B., Paredes, P.P. and Requena, J.R. (1972) Aportes en el estudio de hospederas de la 'mosca prieta de los cítricos', *Aleurocanthus woglumi* Ashby, en Venezuela [Contribution to the study of host plants of the 'citrus black fly', *Aleurocanthus woglumi*, in Venezuela]. *Agronomía Tropical* 22, 549–553.

Angulo, J., Arcaya, E. and González, R. (2011) Aspectos biológicos de *Menochilus sexmaculatus* (F.) (Coleoptera: Coccinellidae) alimentado con *Aphis craccivora* Koch (Hemiptera: Aphididae) [Biological aspects of *Menochilus sexmaculatus* feeding on *Aphis craccivora*]. *Boletín del Centro de Investigaciones Biológicas* 45, 423–431.

Arcaya, E., Díaz, F. and Marcos-García, M.A. (2004) Nuevos áfidos presa de *Pseudodoros clavatus* (Fabricius, 1794) (Diptera, Syrphidae) potencial agente de control biológico [New aphid prey of *Pseudodoros clavatus,* potential biocontrol agent]. *Boletín de la Asociación Española de Entomología* 28, 245–249.

Arcaya, E., Mengual, X., Pérez-Bañón, C. and Rojo S. (2013) Registros y distribución de sírfidos depredadores (Diptera: Syrphidae: Syrphinae) en el Estado Lara, Venezuela [Reports and distribution of predatory syrphids in Lara State, Venezuela]. *Bioagro* 25, 143–148.

Aular, J. and Casares, M. (2011) Consideraciones sobre la producción de frutas en Venezuela [Consideration about Venezuelan fruit production]. *Revista Brasileira de Fruticultura*, Special volume, 187–198.

Berti, J. and Marcano, R. (1997) Influencia de la temperatura sobre el desarrollo, reproducción y longevidad de una raza uniparental de *Trichogramma pretiosum* Riley (Hymenoptera: Trichogrammatidae) [Influence of temperatura on development, reproduction and longevity of a uniparental strain of *Trichogramma pretiosum*]. *Boletín de Entomología Venezolana* 12, 7–15.

Bertorelli, M. and Rengifo, R. (2008) Producción masiva de *Trichogramma* spp., en Anzoátegui, Venezuela y su importancia como alternativa ecológica en el control de plagas [Mass production of *Trichogramma* spp., in Anzoátegui, Venezuela and its importance as an ecological alternative in pest control]. *Agronomía Tropical* 58, 21–26.

Boscán, M.N. de, Terán, J. and Geraud, F. (1979) Enemigos naturales de la mosca prieta de los cítricos *Aleurocanthus woglumi* Ashby (Homoptera: Aleyrodidae) en Venezuela [Natural enemies of the cirtrus black fly *Aleurocanthus woglumi* in Venezuela]. *Agronomía Tropical* 29, 453–458.

Chávez, H.A. (1980) Biological control of the citrus black fly (*Aleurocanthus woglumi* Ashby) by *Prospaltella opulenta* Silv. in Central-Western Venezuela with a review of the pest's invasion of the Western Hemisphere and suppression by introduced parasites. PhD thesis, University of Florida, USA.

Chirinos, D. and Geraud-Pouey, F. (2011) El manejo de plagas agrícolas en Venezuela: análisis y reflexiones sobre algunos casos [Agricultural pest management in Venezuela: analysis and reflections about some cases] *Interciencia* 36, 192–199.

CIA (2017) The World Factbook: Venezuela. Available at: https://www.cia.gov/library/publications/the-world-factbook/geos/ve.html (accessed 26 August 2018)

Cock, M.J.W. (1985) *A Review of Biological Control of Pests in the Commonwealth Caribbean and Bermuda up to 1982*. Technical Communication No. 9, Commonwealth Institute of Biological Control. Commonwealth Agricultural Bureaux, Farnham Royal, UK.

Ferrer, F. (2001) Biological control of agricultural insect pests in Venezuela: advances, achievements, and future perspectives. *BioControl* 22, 67–74

Ferrer, F.R. and Guédez, E. (1990) Evaluaciones de *Diatraea* spp (Lepidoptera: Pyralidae): el efecto de *Cotesia flavipes* Cameron (Hymenoptera: Braconidae) como nuevo parasito en el área de influencia de la Azucarera Río Turbio [Evaluation of *Diatraea* spp.: the effect of *Cotesia flavipes* as a new parasitoid in the area of influence of the Sugarcane Factory Turbio River]. *Caña de Azúcar* 8, 37–57

Ferrer, F., Arias, M., Trelles, A., Palencia, G., Navarro, J.M., and Colmenárez, R. (2004) Posibilidades del uso de nematodos entomopatógenos para el control de la candelilla de la caña de azúcar, *Aeneolamia varia* (Homoptera: Cercopidae) [Possibilities of the use of entomopathogenic nematodes to control the sugarcane froghopper, *Aeneolamia varia*]. *Revista Manejo Integrado de Plagas y Agroecología* 72, 39–43.

Fundacaña (2018) Manejo integrado de plagas [Integrated pest management]. Available at: https://fundacana.org.ve/sitio/entomologia/ (accessed 26 August 2018).

Fundación Polar (2003) *Entomofauna Agrícola Venezolana*. Maracay, Venezuela.

García, R., Riera, R., Zambrano, C. and Gutiérrez L. (2006) Desarrollo de un fungicida biológico a base de una cepa del hongo *Trichoderma harzianum* proveniente de la Región Andina Venezolana [Development of a biological fungicide based on a strain of the fungus *Trichoderma harzianum* from the Venezuelan Andean region]. *Fitosanidad* 10, 115–121.

García, M., Torres, A. and Ochoa A. (2013) Evaluación de nematodos entomopatógenos para el control del picudo de la piña en el Estado Táchira Venezuela [Evaluation of entomopathogenic nematodes for the control of the pineapple weevil in the State of Táchira, Venezuela]. *Agronomía Tropical* 63, 5–14.

García, R.G., Arcia, M.A., Pérez, M.R. and Riera, R.F. (2012) Efecto de *Trichoderma* sobre el desarrollo de papa y el biocontrol de *Rhizoctonia* bajo tres tiempos de inicio de aplicación [Effect of *Trichoderma* on potato development and biocontrol of *Rhizoctonia* at three moments of application]. *Agronomía Tropical* 62, 77–95.

Giffoni, J., Valera, N., Díaz, F. and Vásquez, C. (2007) Ciclo biológico de *Chrysoperla externa* (Hagen) (Neuroptera: Chrysopidae) alimentada con diferentes presas [Biological cycle of *Chrysoperla externa* feeding on several preys]. *Bioagro* 19, 109–113.

Giraldo, H. (1988) *Manejo Integrado de Plagas. Experiencias de control biológico en Venezuela* [*IPM. Experiences with biocontrol in Venezuela*]. Programa Agropecuario CORPOVEN-FONAIAP, Caracas, Venezuela.

Guagliumi, P. (1962) *Las Plagas de la Caña de Azúcar en Venezuela* [*Pests in Sugarcane in Venezuela*]. Tomo II. Centro de Investigaciones Agronómicas- Ministerio de Agricultura y Cría. Maracay, Venezuela.

Guillén, L., Alcalá, D., Fernández, S., Pire, A. and Alvarez, C. (2008) Percepción de los agricultores sobre el manejo integrado de plagas en el cultivo de tomate (*Lycopersicum esculentum* Mill.) [Perception of farmers about IPM in tomato crops (*Lycopersicum esculentum*)]. *Revista de la Facultad de Agronomía* 25, 222–242.

Hernández, D. (2010) Estudio de algunos aspectos biológicos de *Cotesia flavipes* Cameron (Hymenoptera: Braconidae) parasitoide de *Diatraea saccharalis* Fabricius (Lepidoptera: Crambidae) [Study of some biological aspects of *Cotesia flavipes* parasitoid of *Diatraea saccharalis*]. *Entomotropica* 25, 69–81.

Labrador, Y. (2011) Evaluación de la capacidad biocontroladora *in vitro* e *in vivo* de *Trichoderma crassum* Bissett y *Bacillus subtilis* (Ehrenberg) Cohn sobre *Phytophthora palmivora* (Butler) Butler causante de la pudrición parda de la mazorca de cacao (*Theobroma cacao* L.) [*In vitro* and *in vivo* evaluation of the biocontrol agents *Trichoderma crassum* and *Bacillus subtilis* for control of *Phytophthora palmivora* causing brown rot of cocoa pod (*Theobroma cacao*)]. MSc thesis. Universidad Central de Venezuela, Aragua, Venezuela.

Laurentín, H. (2015) Desempeño de la agricultura venezolana en el contexto de la soberanía alimentaria nacional [Performance of the Venezuelan agriculture in the context of national food sovereignty]. *Agroalimentaria* 21, 97–114.

Linares, B. and Ferrer, F. (1990) Introducción de *Cotesia flavipes* Cameron (Hym.: Braconidae) para el control de *Diatraea* spp. (Lep.: Pyralidae) en Venezuela [Introduction of *Cotesia flavipes* for control of *Diatraea* spp. in Venezuela]. *Revista de Caña de Azúcar* 8, 5–11.

Linares, B. and Yépez, G. (1992) Presencia de *Cotesia flavipes* (Cameron) (Hymenoptera: Braconidae) parásito de los taladradores de la caña de azúcar, *Diatraea* spp. (Lepidoptera: Pyralidae) en Venezuela [Presence of *Cotesia flavipes*, parasitoid of the sugarcane borer *Diatraea* spp. in Venezuela]. *Boletín de Entomología Venezolana* 7, 81–83.

Linares, B., Salazar, J.A. and Márquez, O. (1998) Técnica de muestreo masivo, nueva alternativa para evaluar la distribución de parasitoides de *Diatraea* y los porcentajes de parasitismo en campos cañeros [Mass sampling technique, a new alternative to evaluate the distribution of *Diatraea* parasitoids and the percentages of parasitism in sugarcane fields]. *Revista de Caña de Azúcar*, 16, 20–26.

Marin, C.A. (2007) Indicadores básicos de producción del cultivo de caña de azúcar, Estado Lara, periodo 2000-2005 [Basic indicators of production of sugarcane, Lara State, period 2000–2005]. BSc thesis. Universidad Centroccidental Lisandro Alvarado, Estado Lara, Venezuela.

Martínez, N. (2010) Manejo integrado de plagas: una solución a la contaminación ambiental [IPM: a solution to environmental pollution]. *Comunidad y Salud* 8, 73–82.

Méndez, A. and Páez, D. (2014) Percepción de los productores de maíz sobre los insectos plaga y su manejo en áreas de la parroquia Valle de la Pascua, Estado Guárico, Venezuela [Perception of corn producers on insect pests and their management in areas of the Valle de la Pascua parish, Guárico State, Venezuela]. *Fitosanidad* 18, 109–113.

Morales, J., Vásquez, C., Gallardo, J., Gutiérrez, F., Ríos, Y. and Pérez, N. (2004) Potencial biológico de *Trichogramma atopovirilia* (Hymenoptera: Trichogrammatidae) como parasitoide de la polilla de los granos [Biological potential of *Trichogramma atopovirilia* as parasitoid of the angoumois grain moth]. *Bioagro* 16, 197–204.

Morales, J., Vásquez, C., Pérez, N., Valera, N., Ríos, Y., Arrieche, N. and Querino, R. (2007) Especies de *Trichogramma* (Hymenoptera: Trichogrammatidae) parasitoides de huevos de Lepidópteros en el Estado Lara, Venezuela [*Trichogramma* species, parasitoids of lepidopteran eggs in Lara State, Venezuela]. *Neotropical Entomology* 36, 542–546.

Morales, J., Vásquez, C., Valera, N., Arrieche, N., Arcaya, E., and Querino, R. (2010) Nuevos registros y distribución de especies de *Trichogramma* (Hymenoptera: Trichogrammatidae) en el Estado Lara, Venezuela [New reports and distribution of *Trichogramma* species in Lara State, Venezuela]. *Bioagro* 22, 159–162.

Morales, J., González, R. and Arcaya, E. (2014) Especies de Eristalinae (Diptera: Syrphidae) presentes en Estados del Centro-Occidente de Venezuela [Species of Eristalinae found in the Central-western States of Venezuela]. *Bioagro* 26, 63–68.

Morales, N., Morales-Montero, P., Puza, V. and San-Blas E. (2016) First report of *Heterorhabditis amazonensis* from Venezuela and characterization of three populations. *Journal of Nematology* 48, 139–147.

Navarro, R. and Marcano, R. (1999) Preferencia de *Trichogramma pretiosum* Riley y *T. atopovirilia* Oatman y Platner por huevos de *Helicoverpa zea* (Boddie) de diferentes edades [Preference of *Trichogramma pretiosum* and *T. atopovirilia* for eggs of *Helicoverpa zea* from different ages]. *Boletín de Entomología Venezolana* 14, 87–93.

Pineda, J.B. and Díaz-Polanco, C. (1981) Control biológico de *Sclerotium rolfsii* Sacc. en *Phaseolus vulgaris* mediante la utilización de *Penicillium notatum* Westl [Biological control of *Sclerotium rolfsii* in *Phaseolus vulgaris* by using *Penicillium notatum*]. *Agronomía Tropical* 31, 265–281.

Poinar, G.O. and Linares, B. (1985) *Hexamermis dactylocerus* sp. n. (Mermithidae: Nematoda) a parasite of *Aeneolamia varia* (Cercopidae: Homoptera) in Venezuela. *Revue de Nematologie* 8, 109–111.

Ramón, M., Mauriello, F., Graterol, Y., Giraldo-Vanegas, H., Mendoza, C., Pérez, M. and Izarraga, R. (2008) Asociación entre las características varietales y el daño ocasionado por el taladrador de la caña de azúcar, en el Estado Portuguesa, Venezuela [Association between the varietal characteristics and the damage caused by the sugarcane borer, in Portuguesa State, Venezuela]. *Agronomía Tropical* 58, 111–116.

Rosales, L.C. and Suárez, Z. (1998). Nematodos entomopatógenos como posibles agentes de control del gorgojo negro del plátano *Cosmopolites sordidus* (Germar, 1824) (Coleoptera: Curculionidae) [Entomopathogenic nematodes as possible control agents of the banana weevil *Cosmopolites sordidus*]. *Boletín Entomología Venezolana* 13, 123– 140.

Salazar, L.A., Aponte, G.Y., Alcano, M.J., Sanabria, N.H. and Guzmán, J.J. (2012) Importancia de las especies de *Trichoderma* para el control de *Macrophomina phaseolina* en las áreas agrícolas del Estado Aragua, Venezuela [Importance of *Trichoderma* species for the control of *Macrophomina phaseolina* in the agricultural areas of Aragua State, Venezuela]. *Agronomía Tropical* 62, 7–15.

Sampaio, M.V., Bueno, V.H.P., Silveira, L.C.P. and Auad, A.M. (2009) Biological control of insect pests in the tropics. In: Del Claro, K., Oliveira, P.S. and Rico-Gray, V. (eds) *Tropical Biology and Conservation Management*, Volume 3. Encyclopedia of Life Support Systems (EOLSS) Publishers/UNESCO, Oxford, UK, pp. 28–70.

Solano, Y. and Arcaya, E. (2014) Primer registro de *Harmonia axyridis* (Pallas, 1773) (Coleoptera: Coccinellidae) en Venezuela [First record of *Harmonia axyridis* in Venezuela]. *Entomotropica* 29, 57–61.

Solano, Y., Valera, N. and Vásquez, C. (2010) Aspectos biológicos de *Olla v-nigrum* (Mulsant) (Coleoptera: Coccinellidae) alimentándose sobre *Aphis craccivora* (Koch) (Hemiptera: Aphididae) [Some biological aspects of *Olla v-nigrum* preying on *Aphis craccivora*]. *Boletín del Centro de Investigaciones Biológicas* 44, 251–260.

Solano, Y., Delgado, N., Morales, J. and Vásquez, C. (2016) Functional response of *Cycloneda sanguinea* (L.) (Coleoptera: Coccinellidae) to the black pea aphid, *Aphis craccivora* Koch (Hemiptera: Aphididae). *Entomotropica* 31, 311–318.

Tapia, M.S., Puche, M., Pieters, A., Marrero, J.F., Clavijo, S., Socorro, A.A.G., Machado-Allison, C., Raffalli, S., Herrera, M., Jiménez, , M., Oletta, J.F., Comerma, J., Silva, O., Barrios, M., Ortiz, A., Córcega, E., Soto, E., Pinto, L., Vargas, D., Garcia, V., Rey, J.C., Aciego, J.C., Mendoza, N, Fernández, G.

and Bisbal, F. (2017) Food and nutrition security in Venezuela. The agrifood abduction of a country: vision and commitment. In: *Challenges and Opportunities for Food and Nutrition Security in the Americas. The View of the Academies of Sciences*. IANAS, IAP and BMBF, México DF, pp. 566–607. [Free public access to this publication in English and Spanish at www.ianas.org]

Torres, F. and Marcano, R. (2007) Efecto de la temperatura en el desarrollo de *Cryptolaemus montrouzieri* Mulsant (Coleoptera: Coccinellidae) utilizando como presa *Maconellicoccus hirsutus* Green (Hemiptera: Pseudococcidae) [Effect of temperature on the development of *Cryptolaemus montrouzieri* using *Maconellicoccus hirsutu* as prey]. *Entomotropica* 22, 17–25.

Velásquez, M. and Terán J. (2003) Los *Trichogramma* (Hymenoptera: Trichogrammatidae) de la región Noroccidental del Estado Guárico, Venezuela [The *Trichogramma* species of North-eastern Guárico State, Venezuela]. *Entomotropica* 18, 127–145.

Weir, E.H. and Sagarzazu, L. (1998) Interspecifíc competition between *Metagonystilum minense* (Diptera: Tachinidae) and *Cotesia flavipes* (Hymenoptera: Braconidae) parasitoids of sugarcane borers (*Diatraea* spp., Lepidoptera: Pyralidae). *Revista de Biología Tropical* 46, 1135–1139.

Weir, E.H., Contreras, W. and Weir, K. (2007) Biological control of *Diatraea* spp. (Lepidoptera: Pyralidae) in Central Venezuela. *International Journal of Tropical Biology* 55, 655–658.

32 The Uptake of Biological Control in Latin America and the Caribbean

Joop C. van Lenteren[1]* and Matthew J.W. Cock[2]

[1]Laboratory of Entomology, Wageningen University, The Netherlands; [2]CABI, Egham, UK

Abstract

Biological control started to be used in the 1880s in Latin America and the Caribbean and has since developed into a widely applied pest management method. Currently almost 32 million hectares are under classical, more than 31 million hectares under augmentative and hundreds of thousands of hectares under conservation biocontrol. Achievements in this region have been impressive and are documented in this chapter. Several factors frustrate the implementation of biocontrol on an even larger area. The most important are the dominance of the pesticide industry, the negative effect of pesticides on biological and natural pest control, governmental 'subsidies' to keep chemical control cheap, the lack of funding for research and implementation of biocontrol, and an expensive, time-consuming regulatory framework. However, inherent positive characteristics of biocontrol contribute to sustainable pest management, a healthier and biodiverse environment, pesticide-free food and improved yields. These characteristics, together with the large-scale natural enemy prospecting programmes, the documentation of the many cases of natural control and the successful regional collaboration on area-wide control of new invasive pests, point at a bright future for biocontrol in Latin America and the Caribbean.

32.1 Introduction

Since its first use in 1884, biological control with arthropod natural enemies and microbial control agents has seen a strong increase in the number of species of biocontrol agents used and areas treated in Latin America and the Caribbean. Until 1970, mainly classical biocontrol was used, though in some countries augmentative biocontrol was employed as well; later, conservation biocontrol was applied and natural control was documented. Most first uses of biocontrol involved the release of invertebrate natural enemies, usually insect predators or parasitoids, but microbial agents and vertebrates were also used. Another characteristic of this early period in the Latin American region is that biocontrol was mainly aimed at insect control, with control of weeds and diseases of plants developing later.

Van Lenteren and Bueno (2003) concluded from data in the few published reviews concerning biocontrol in South and Central America and the Caribbean that in the period from 1880 to 1970, 16 countries used classical biocontrol. Well known examples of early classical biocontrol for South America are the introduction of *Rodolia cardinalis* (Mulsant) for control of cottony cushion scale *Icerya purchasi* Maskell, the release of a species of *Encarsia* for control of the

* E-mail: joop.vanlenteren@wur.nl

white peach scale *Pseudaulacaspis pentagona* (Targ-Tozz) and the introduction of *Aphelinus mali* (Hald.) for control of woolly apple aphid *Eriosoma lanigerum* (Hausmann), which have usually led to substantial or complete control in a number of countries. Examples of successful classical biocontrol for Central America and the Caribbean during this period are the complete biocontrol of the citrus blackfly *Aleurocanthus woglumi* Ashby as a result of inoculative releases with the parasitoid *Eretmocerus serius* Silv. and/or *Amitus hesperidium* Silv. in Cuba, Costa Rica, Mexico and Panama; and the use of tachinid and hymenopteran parasitoids to control sugarcane borers on different Caribbean islands. Only four countries applied augmentative biocontrol before the 1970s and areas under biocontrol were seldom provided. Augmentative biocontrol concerned inundative releases with egg parasitoids of the genera *Trichogramma* and *Telenomus* to control pests in sugarcane.

Based on data in papers published since 1970, and after obtaining information by contacting researchers in the region, van Lenteren and Bueno (2003) stated that by the year 2000, 16 countries still used classical biocontrol. Due to lack of quantitative data, a reliable estimate of the total area under classical biocontrol could not be made. The number of countries using augmentative biocontrol in this period had increased to 17, with a total estimated area under augmentative biocontrol about 4,350,000 ha (see Table 1.1 in Chapter 1).

Information provided in the country-specific chapters of this book shows that nowadays all forms of biocontrol and all types of biocontrol agents are applied in Latin America and the Caribbean (Table 32.1). The estimates for the number of countries with classical, augmentative and conservation biocontrol will be reliable. However, the number of countries with natural control (19) is underestimated, as many countries have not yet tried to document which potential pests are kept below densities causing damage. The area under augmentative biocontrol has strongly increased since 2000 and now exceeds 31 million hectares. The first estimate for classical biocontrol for Latin America and the Caribbean shows an area of more than 30 million hectares being protected.

In this chapter the achievements in biocontrol in Latin America and the Caribbean will be summarized. Next, the use of classical biocontrol in this region will be illustrated with data obtained from the BIOCAT databank (Cock *et al.*, 2016). Then, factors limiting and stimulating biocontrol in the region will be presented and, finally, the future of biocontrol in the region is sketched based on remarks made in the country chapters.

32.2 Achievements

The major biocontrol activities during different periods for each of the Latin American and Caribbean countries are summarized in Table 32.2.

Table 32.2 provides information about similarities and differences in biocontrol approaches in various countries. The similarities concern the import and release of many of the same natural enemies in classical biocontrol projects during the early period of biocontrol in the region. Often, these natural enemies had been used successfully in Asia, Europe and North America and were introduced without additional research in Latin America and the Caribbean. Currently, area-wide biocontrol programmes aim at releasing the same natural enemies of several recently introduced invasive pests. These programmes are based on more

Table 32.1. Use of different forms of biological control and area under biocontrol in Latin America and the Caribbean.

	No. of countries using biocontrol / area (ha) under biocontrol			
Period	Classical	Augmentative	Conservation	Natural
1895–1969	16 / ?	4 / ?		
1970–1999	16 / ?	17 / 4,350,000		
2000–2018	29 / 30,747,889	27 / 31,381,131	13 / 447,114	19 / 2,001,846

Table 32.2. Summary of achievements in biocontrol during three periods, with future plans and factors limiting and stimulating biocontrol for each Latin American and Caribbean country.

Period 1800–1969	Period 1970–1999	2000–now	Future	Factors limiting and *stimulating* biocontrol
Argentina: Many CBC introductions; few ABC projects; provider of natural enemies, particularly for weed BC, with as major success of control of prickly pear in Australia	Many CBC introductions; few ABC projects, few successes; start of weed BC research; important provider of weed BC agents	Continuation of CBC of arthropods in agriculture, forestry, and of weeds; increase in ABC activities; initiation of work on ConsBC; important provider of weed BC agents	Anticipated increase of all forms of BC in all areas of agriculture, forestry and of weed BC ***Specific for Argentina: use of weed CBC and provider of weed BC agents***	Dominance of CC Limited collaboration BC research and application; limited development of practical BC; limited funding for BC *Good BC expertise available*
Barbados: Many CBC introductions, successes in sugarcane and citrus; few ABC projects, success in cotton; provider of many BC agents to other islands in the region	Several new CBC projects, e.g. of pests in vegetables; few ABC projects	CBC in sugarcane and vegetables; new CBC successes in palm plantations, citrus orchards, and of pink hibiscus mealybug on various plants; ABC cotton; many NC successes	Anticipated increase in BC activities due to new invasive pests ***Specific for Barbados: documentation of NC, large scale use of CBC***	Many new invasive species *Need for biocontrol of new invasive pests*
Belize: Several CBC attempts without success; documentation NC in sugarcane	Several CBC attempts without success	Successes with CBC of Asian citrus psyllid and pink hibiscus mealybug	Continuation CBC pink hibiscus mealybug and Asian citrus psyllid; initiation of ABC with microbials of pests in sugarcane ***Specific for Belize: use of CBC***	*Presence of international institute OIRSA*
Bolivia: Several unsuccessful CBC attempts; important CBC success with *Rodolia cardinalis;* ABC of sugarcane borers	Several attempts with CBC and ABC; successes with ABC in sugarcane, citrus, potato storage, coffee; documentation of NC in several crops	Successes with several large ABC and CBC projects in citrus, coffee, potato, sugarcane, quinoa, and soybean; documentation of NC in several crops	Anticipated increase in production and use of local microbial agents; documentation and use of NC ***Specific for Bolivia: ABC with local microbials, documentation of NC***	Dominance of CC Lack of funding for BC, lack of collaboration among BC experts, lack of education in BC and lack in transfer of knowledge *Recent positive results in BC projects*

Table 32.2. Continued.

Period 1800–1969	Period 1970–1999	2000–now	Future	Factors limiting and *stimulating* biocontrol
Brazil: Unsuccessful attempts of CBC of white peach scale, coffee berry borer, woolly apple aphid, Mediterranean fruit fly and oriental fruit moth; successful CBC of Rhodes grass scale; ABC of sugarcane borer with native natural enemies; early use of microbials for pest and disease control	Successful CBC of wheat aphids, cassava mealybug, citrus leaf miner, and Sirex woodwasp; ABC with natural enemies of citrus mealybugs, flies in poultry pens, fruit flies in vars fruit crops, lepidopterans in eucalyptus, and Sirex woodwasp in pine; ABC with microbials of spittle-bugs, soybean caterpillar, mate tree borer, and soil-borne diseases; large scale local production of arthropod and microbial ABCs	CBC and ABC with natural enemies of pests in eucalyptus and pine. ABC with natural enemies of lepidopteran pests in sugarcane, soybean, cotton, corn, beans, millet and tomatoes, of stink bugs in soybean, of spider mites, fungus gnats and thrips in cotton, soybean, fruit orchards, ornamentals and vegetables, cotton bollworm, Asian citrus psyllid, aphids, thrips, lepidopterans and spider mites in greenhouse crops, various scales in various crops. ABC with microbials of pests in coffee, eucalyptus, pine, of mate tree borer, Asian citrus psyllid, soil nematodes in soybean, corn and coffee, rubber lace bug, cotton bollworm, fall army worm in corn, soil-borne diseases in many crops; prospecting for weed BC agents	Anticipated development of new ABC projects with natural enemies and microbials; strong growth of ABC industry; development of more efficient BC agent production, quality control, shipment, release and monitoring methods; progress in technology transfer from research to application, improved collaboration and networking; development of ConsBC ***Specific for Brazil: early and current large scale use of ABC with natural enemies and microbials for pest and diseases; large scale local production of BC agents, many BC research projects***	Dominance of CC industry Poor transfer of BC technology to farmer Poor quality of non-registered BC products and with insufficient farmer guidance *Rich biodiversity as source for BC agents* *Good expertise in BC, many researchers* *Large production units for BC agents, strong private industry* *International market demands for food with low pesticide residue levels* *Improved BC agent registration procedures*
Chile: Many successful early CBC projects, e.g. control of olive black scale, scales in citrus, avocado and other fruit, woolly apple aphid; use of ABC with microbials to control coleopterans and lepidopterans in various crops; weed BC; provider of BC agents	Successful CBC of wheat aphids, of the eucalyptus psyllid, and of weeds in various crops; CBC and ABC of pine shoot moth; ABC of diseases in fruit; prospecting for and commercialization of microbial agents and nematodes for ABC of coleopterans in various crops, and of pine shoot moth	CBC of walnut aphids, codling moth and woolly apple aphid in apple, wood wasps and eucalyptus weevil in forestry; attempts to use microbials for ABC of lepidopterans and coleopterans in various crops and forestry; ABC of *T. absoluta* with predator, of diseases in various crops and forestry; development of technology transfer models; prospecting for and evaluation of microbial agents	Continue prospecting for microbials for pest and disease control; improved formulations of BC agents of diseases; improved monitoring and release of natural enemies; improved BC training of agricultural technicians ***Specific for Chile: many early CBC projects, ABC with local microbials of pests and diseases, large scale prospecting for and production of microbials***	Insufficient quality control of BC products Access and Benefit Sharing regulations *Improved formulations for microbials* *Presence of BioControl Technological Center* *National and international pesticide regulations* *Expertise in research and application of BC* *Organic production for international market*

Colombia: Successful CBC of woolly apple aphid and cottony cushion scale; use of microbials to control locusts	ABC of pests in cassava, cotton, maize, sorghum, soybean, sugarcane and tomato, in forestry, and in greenhouse vegetables and ornamentals; CBC and ABC of coffee berry borer in coffee; use of microbials for ABC of pests and diseases; prospecting, evaluation and production of microbials	Documentation of NC in cassava; FBC and CBC of Columbian fluted scale; ConsBC in sugarcane, chilli pepper, oil palm, coffee and ornamentals; ABC of pests in cassava, citrus (e.g. Asian citrus psyllid), coffee (coffee berry borer), cotton, maize, oil palm, greenhouse vegetables and ornamentals, potato, sorghum, sugarcane, rice, and forestry, of flies in livestock production and vectors of human diseases; prospecting, evaluation and production of microbials	Anticipated production and use of entomopathogenic nematodes for root borer control, improved formulation technology for *Trichoderma*, development of quality control of BC agents. ***Specific for Colombia: ABC projects, many BC activities in greenhouse ornamentals; prospecting for and production of microbials; use of ConsBC***	Dominance of CC industry Complicated national legislation for import of exotic BCs Access and Benefit Sharing regulations Poor quality of unregistered BC products *Expertise in research and application of BC* *Good infrastructure and several centres for BC research* *Local production of BC agents* *Compulsory use of some forms of BC*
Costa Rica: Successful CBC in citrus, coffee and sugarcane	ABC in avocado, pineapple and cotton; ABC and CBC in sugarcane; CBC, ABC and NC in citrus; ConsBC in palm plantations and in ornamental crops; NC of pests in banana, coffee, cashew, and timber; prospecting for and development of mass production of natural enemies and microbial agents	Start of CBC of Asian citrus psyllid, ConsBC and ABC in palm plantations; CBC with ABC, NC and ConsBC in citrus, coffee and sugarcane; NC and ABC in banana; prospecting for natural enemies and microbial agents	Anticipated improved and increased local production and use of microbials; improved quality control methods for BC agents ***Specific for Costa Rica: ABC with local microbials, large scale prospecting, documentation of NC, use of ConsBC***	Aggressiveness of CC Poor quality of local BC products *Use of organic agriculture* *Export demands for agricultural products* *Establishment of national programme for BC*

Continued

Table 32.2. Continued.

Period 1800–1969	Period 1970–1999	2000–now	Future	Factors limiting and *stimulating* biocontrol
Cuba: Successful CBC of citrus blackfly with parasitoid; ABC of sugarcane borer with native parasitoid	ABC of many pests and several diseases in many crops with predators, parasitoids and microbial agents; ConsBC of sweet potato weevil with predatory ants; creation of hundreds of centres for mass production of entomophagous species; prospecting for microbial control agents	ABC with *Trichogramma* spp. of various lepidopterans in several crops; study and large scale use of ConsBC; local production of parasitoids, predators, entomopathogenic nematodes and fungi, nematopathogenic fungi and bacteria, and phytopathogenic fungi; strong governmental support for BC	Continued support for development of new control agents, for better formulations of microbial control agents, and for more efficient production to replace imported products. Increase use of ConsBC ***Specific for Cuba: ConsBC, many local centres for mass production of BC agents.***	*Awareness among farmers and community of contribution of BC to economy, ecology, environment and society* *Strong governmental support for BC* *Many local centers for production of BC agents* *Appreciation / broad use of BC by farmers*
Dominica: Several unsuccessful CBC projects in citrus, banana, coffee, vegetables; successful CBC of sugarcane borers	Successful CBC of citrus blackfly; unsuccessful ABC of diamondback moths and armyworms in various crops	Continuation of successful CBC programmes in citrus and sugarcane		
Dominican Republic: CBC with coccinellids, e.g. *Cryptolaemus* and *Rodolia*, for control of cottony cushion scale; introduction of mongoose for rat control and toads for pests in sugarcane; documentation of fungal pathogens of weeds	ABC of sugarcane pests, of whitefly in vegetables, of coffee berry borer and of rice stalk stink bug; CBC, ABC and NC of citrus pests, CBC of vector transmitting snails; testing and use of microbials; testing of weed BC; prospecting for and mass rearing of natural enemies	CBC of papaya mealybug, pink hibiscus mealybug, *Anastrepha* fruit flies, and pigeon pea pod fly; NC of diamondback moth, red palm mite and other mites, many pests of oriental vegetables, and of exotic pests in *Ficus* and *Cycas;* NC and ConsBC of pests in organic fruit and coffee; ABC with microbials	Anticipated increased funding for BC by the Ministry of Agriculture; increased field testing of BC within IPM projects ***Specific for the Dominican Republic: prospecting, demonstration of NC***	Lack of resources for BC research and application, lack of expertise in BC Dominance of CC *Export demands for products without pesticides* *Organic production, GAP labels*

Ecuador: CBC of woolly apple aphid in apple, *Icerya* sp., purple scale, and citrus blackfly in citrus, and of sugarcane borer in sugarcane	CBC of cottony cushion scale and citrus leaf miner in citrus, of white rice borer in rice, of coffee berry borer in coffee, of white mango scale in mango; ABC of leafhoppers in sugarcane, of soil-borne pests in vegetables, of lepidopteran pests in maize, soybean, sugarcane, banana and cotton; prospecting for natural enemies of whiteflies, and of *Tuta absoluta*	CBC of citrus leaf miner, fruit flies in tropical fruit and scales in mango. CBC of cottony cushion scale on the Galapagos islands; ABC of pests in banana, of soil-borne pests and diseases in broccoli, pod rot disease in cacao, diseases and pests in oil palm, rice, pineapple, in ornamentals and vegetables in greenhouses, of diseases in papaya, and of pests in sugarcane; prospecting for BC agents, including entomopathogenic fungi and nematodes; improvement of formulations of microbial control agents	Continued governmental support for local training and production of BC agents Regulation for and registration of BC agents Increase of CBC of invasive weeds and insects, and of ABC of agricultural pests on the Galapagos islands ***Specific for Ecuador: many CBC and ABC projects, disease BC in many crops; BC of invasive species in natural ecosystems; prospecting for BC agents***	*Governmental infrastructure BC research, training and extension, prospecting for BC agents, and collection of BC agents* *Governmental, private national and international producers of BC agents* *Demand for pesticide-residue free food by farmers, consumers and the international market*
El Salvador	NC in fruit trees and coconut palm; CBC and FBC in citrus; NC and CBC in cotton, corn and bean; ABC of lepidopterans and mosquitos; prospecting for nematophagous fungi	Microbial control of soil diseases, and of lepidopterans, coleopterans and nematodes in several crops; control of mosquitos with tilapia fish	***Specific for El Salvador: demonstration of NC in several crops***	
French Guiana, Guadeloupe and Martinique: Successful CBC of sugarcane borers	Successful CBC of sugarcane borers, pink hibiscus mealybug, Asian citrus psyllid, and citrus blackfly; first attempts at CBC of fruit flies	Successful CBC of sugarcane borers, pink hibiscus mealybug, Asian citrus psyllid, and citrus blackfly; ABC in vegetable crops; prospecting for and use of natural enemies in ConsBC in various crops	Anticipated studies on new exotic species for CBC of mango mealybug, and for improved ABC of the Asian citrus psyllid ***Specific for French territories: prospecting and use of ConsBC***	

Continued

Table 32.2. Continued.

Period 1800–1969	Period 1970–1999	2000–now	Future	Factors limiting and *stimulating* biocontrol
Guatemala	ABC of lepidopterans in cotton, corn and vegetables, of coffee berry borer and nematodes in coffee; CBC of citrus blackfly	NC and ABC in coffee; CBC of fruit flies; ABC in cotton; ABC of vector of malaria with microbials; testing of microbials against spittle bugs in pastures	Anticipated start with CBC of Asian citrus psyllid ***Specific for Guatemala: IPM including NC and ABC of coffee pests***	Few BC experts, limited funding, expensive registration of BC agents *Export demands for agricultural products* *Health risks of pesticides for workers and consumers* *Positive experiences with BC*
Guyana: Successful CBC of sugarcane borer, and ConsBC of lepidopteran in rice; prospecting for natural enemies of various pests	Successful ABC of lepidopteran in palm, and CBC of hibiscus mealybug	CBC in sugarcane; CBC attempts for control of fruit fly; ABC in palm; ABC attempts for control of red palm mite; ConsBC of rice pests	Continued studies for ABC of red palm mite ***Specific for Guyana: prospecting and ConsBC***	
Haiti: Partial successful CBC of sugarcane borer, successful CBC citrus blackfly	Attempts for CBC of coffee berry borer	Successful CBC of pink hibiscus mealybug; prospecting for natural enemies	Continued studies of CBC of fluted scale in peanuts and other crops	Lack of finances for CC
Honduras: Prospecting for BC agents	CBC of weed; CBC attempts for control of lepidopterans in various crops; ABC of lepidopterans, whiteflies; microbial control of lepidopterans; ConsBC in various crops; creation of teaching and research Center for Biological Control in Central America; large scale prospecting	ABC with microbials in various crops; mass production of natural enemies (including nematodes) for ABC in vegetables, sweet potato, coffee berry borer, plantain; training of many BSc students; development and production of microbial control agents for disease and pest control	Continued positive attitude towards BC, and training and production facilities ***Specific for Honduras: prospecting, teaching and research in BC, development and production of microbial agents for disease and pest control***	*Positive attitude towards BC* *Good training facilities* *Good local production facilities for BC agents*

Jamaica: CBC of rats, citrus black fly, banana weevil, cocoa thrips, various other pests and weeds; NC and CBC of sugarcane and coconut pests; provider of BC agents	CBC attempts for control of fruit flies; NC and CBC of sugarcane borer, of lepidopterans in cruciferous crops and of pine mites; ABC of sweet potato weevil; NC of whiteflies and coffee leaf miner, NC and ABC of citrus root weevils and of coffee berry borer; prospecting	NC and CBC of brown citrus aphid, NC of false Colorado beetle in gully bean, ensign scale in various crops, of lime swallow tail, of red palm mite; CBC of pink hibiscus mealybug; NC and ABC of citrus root weevil; NC and FBC of papaya mealybug; ABC and FBC of Asian citrus psyllid; ABC of coffee berry borer, sweet potato weevil, and beet armyworm	Anticipated reduction in use of high risk pesticides necessitates construction of new BC facility; development of ConsBC; new invasive pest are targets for BC ***Specific for Jamaica: early and recent CBC successes, documentation many NC cases, prospecting***	Agriculture still highly dependent on pesticides Increased registration of lower risk class pesticides, including biopesticides. *Available expertise, infrastructure and funding for BC research and application* *New invasive pest offer possibilities for BC*
Mexico: Many successful early CBC projects in alfalfa, apple, banana, bean, citrus, cotton, mango, sugarcane; ABC in pastures and sugarcane; construction of mass production centres for natural enemies in the 1960s	CBC in citrus, coffee, corn, cotton, fruit, forest, potatoes, and of water hyacinth; ABC of pests in coffee, cotton, cruciferous crops, fruit, forest, sorghum, sugarcane, vegetables and ornamentals; construction of 20 regional centres for mass rearing of natural enemies and entomopathogenic fungi; 65 private insectaries	BC of many hemipteran pests. CBC in citrus, corn, cotton, eucalyptus, mango, strawberry and of weeds in water; ABC in citrus, corn, cotton, cruciferous crops, fruit, grape, ornamentals vegetables (field and greenhouses), sorghum, soybean, and sugarcane, weeds in wetlands, and grasshoppers; mass production of 40 species of BC agent in 65 laboratories	Development of risk scenarios for more than 1200, pests; development of BC for new pests and diseases, e.g. laurel ambrosia beetle, *Drosophila suzukii;* increase in use of BC; promotion of use of eco-friendly products ***Specific for Mexico: many ABC and CBC successes, good infrastructure for research and application of BC, risk scenarios for new pests, pro-active development of BC***	*Good governmental infrastructure and support for research, production and application of BC* *Risk scenarios for new pests and diseases, development of BC for new pests and diseases* *Specific pesticide residue requirements for export markets*
Nicaragua: ABC of pests in cotton; prospecting	IPM in cotton; BC studies of pests in cotton and citrus; studies of microbials for control of mosquitos; mass production of *Chrysoperla* and *Trichogramma*, and microbial agents; ABC attempts to control diamondback moth	ABC of whiteflies, aphids and pests in sugarcane; CBC of Asian citrus psyllid; construction of biofactories for mass production of *Trichogramma* and native *Orius* spp.; prospecting	Anticipated production of entomopathogens ***Specific for Nicaragua: prospecting, local mass production of BC agents***	Dominance of CC Poor selling and logistic mechanism for BC agents *Local production of BC agents*

Continued

Table 32.2. Continued.

Period 1800–1969	Period 1970–1999	2000–now	Future	Factors limiting and *stimulating* biocontrol
Panama: CBC of citrus blackfly	ABC of sugarcane borer and diamond back moth; prospecting	CBC of coffee berry borer, ConsBC of thrips in cucurbits and lepidopterans in rice; prospecting, production and application of microbial control agents and natural enemies	Anticipated development of artificial media to economize mass rearing of natural enemies ***Specific for Panama: prospecting***	
Paraguay	ABC of soybean caterpillar with baculovirus, and sugarcane borer with parasitoids	ConsBC in several crops; ABC with locally produced microbials in various crops for pest and disease control; large scale prospecting for predators, parasitoids and microbial control agents	***Specific for Paraguay: large scale prospecting, local mass production of microbial control agents***	Dominance of cheap CC Limited governmental support for BC *Production of organic food* *Pesticide resistance against important pests* *Appreciation of pesticide free food*
Peru: CBC woolly apple aphid in apple, cottony white scale in cotton, hemispheric scale and olive blackfly in olive, cottony cushion scale, purple scale, Florida red scale and aphids in citrus; ABC of cotton aphid and tobacco budworm in cotton, and sugarcane borer in sugarcane; NC of pests in several crops; creation of centre for introduction and rearing of useful insects	CBC of West Indian red scale, citrus woolly whitefly, and citrus leaf miner in citrus, alfalfa green aphid in alfalfa, blue psyllid in eucalyptus; ABC of sugarcane borers in sugarcane, fruit flies in fruit, pink bollworm in cotton, and house flies; development of BC of diseases; increased role of governmental centre for research and application of BC; creation of private BC agent production laboratories; governmental financial support to reduce negative impacts of CC	Many ABC projects for control of pests in cotton, sugarcane, asparagus, avocado, olive, pomegranate, forest, coffee, cacao, vine, vegetables, and quinoa; continued strong role of governmental centre for research, mass production and application of BC; production of many predators, parasitoids, entomopathogenic and antagonistic agents in network of regional laboratories; large collection of microbial agents; agro-exporting companies with high demand for BC; demonstration that BC is considerably cheaper than CC and prevents secondary pests	Increase in certification of pesticide free food; governmental agreements with association of citrus farmers to use BC, with association of asparagus farmers and large private exporting companies to use IPM, including BC; development of ConsBC ***Specific for Peru: strong governmental support and infrastructure for BC; many CBC and ABC projects; certification of food produced under BC***	*Early and continuous governmental infrastructure for research and mass rearing of BC agents* *Private production labs supported by governmental center for BC* *Governmental financial support to reduce impacts of CC* *Large agro-exporting companies with high demand for BC* *Demonstration that BC is cheaper than CC and prevents secondary pests* *Certification of pesticide free products*

Puerto Rico: NC and CBC of several pests in sugarcane; CBC in citrus and coffee	CBC of water weeds; NC, FBC and CBC of several pests in citrus, sugarcane, of melon worms in cucurbits, of pink hibiscus mealybug and papaya mealybug; ConsBC of pests in coffee	FBC of Asian citrus psyllid; development of CBC of Harrisia cactus mealybug; NC, ABC and ConsBC of coffee berry borer	Anticipated increase in BC use due to growth of organic and environmentally-friendly agriculture; development of more ConsBC ***Specific for Puerto Rico: CBC successes, documentation of NC, ConsBC***	*Good infrastructure for BC research* *Organic and environmentally-friendly agriculture stimulate research in BC*
Remaining Caribbean Islands: Many unsuccessful CBC releases; CBC successes of pests in arrowroot, citrus, coconut, cotton, sugarcane, and prickly pear and puncture vine weeds; ConsBC of cotton leafworm in cotton, of white grub larvae in sugarcane, and arrowroot leaf roller in arrowroot; ABC of sugarcane borers in sugarcane; demonstration of NC of West Indian cane fly in sugarcane	CBC of cottony cushion scale and citrus blackfly in citrus, and of coconut mealybug and coconut scale in coconuts; reduction in CBC attempts since 1980	Successful region wide CBC programmes for control of the pink hibiscus mealybug and the papaya mealybug; demonstration of NC of the coconut whitefly and the passion vine mealybug; use of FAO Code of Conduct for the Import and Release of Exotic Biological Control Agents; implementation of Farmers Field Schools to enable farmers to use IPM and become less dependent on CC	Continuation of Farmer Field Schools; increased role of implementation of region-wide BC programmes ***Specific for the Remaining Caribbean islands: many early CBC projects, recent region-wide collaboration resulting in CBC successes***	Many small islands with small and diverse crop areas Many new invasive pests Limited extension service Domination of CC industry *Presence of several region-wide organizations assisting in development of BC* *Implementation of Farmer Field Schools*
Suriname: Prospecting and identification of natural enemies responsible for NC of several pests	Attempts of ABC with nematodes; demonstration of NC of coconut pests, green cassava mite and *Pomacea* snails; prospecting	CBC of pink hibiscus mealybug; research on BC of Carambola fruit fly; prospecting	Continued prospecting and efforts to control Carambola fruit fly. ***Specific for Suriname: prospecting, documentation of NC***	

Continued

Table 32.2. Continued.

Period 1800–1969	Period 1970–1999	2000–now	Future	Factors limiting and *stimulating* biocontrol
Trinidad and Tobago: CBC of various pests, ConsBC with birds and insects for control of sugarcane pests, ABC with predator and microbial to control sugarcane froghopper; important provider of BC agents in the region	Successful ABC and CBC of sugarcane pests; unsuccessful CBC of cabbage pests; important provider of BC agents for the region	Successful CBC of citrus blackfly and pink hibiscus mealybug	Increase of accidental import of exotic pests creates need for BC ***Specific for Trinidad and Tobago: good BC research infrastructure, important provider of BC agents in the region***	*Consumer appreciation for food with low pesticide residues* *Good BC research infrastructure* *BC appreciated by farmers*
Uruguay: Successful CBC of white peach scale, cottony cushion scale, woolly apple aphid, San Jose scale, eucalyptus weevil; unsuccessful ABC with entomopathogens; prospecting for native natural enemies; provider of natural enemies for the region	ABC of lepidopterans in sugarcane, vine and cotton with *Trichogramma*, of pine wood wasp with nematodes, of the sunflower caterpillar and the soybean caterpillar with viruses; prospecting for native BC agents including microbials	CBC in eucalyptus and pine; CBC attempts in citrus; ABC in vegetables and soybean. Start of centre for forest research, including BC; research of ABC with parasitoids and microbials of stinkbugs; development of ConsBC in soybean and sorghum	Start of mandatory registration for BC agents; creation of centre for collection and storage of microbials; increase in ABC of diseases ***Specific for Uruguay: many early and current CBC successes in fruit orchards and forestry, prospecting***	Farmers adhere to low price CC Insufficient BC agents available *Well established research and application network for BC* *Consumer concern about CC* *Demands for residue-free food for export*

Venezuela: Several CBC attempts in various crops; successful ABC of sugarcane borer; mass production of *Lydella*	Successful CBC of citrus blackfly; ABC of sugarcane borers with parasitoids and nematodes; improved mass production of *Lydella*, *Trichogramma* mass rearing for ABC of lepidopteran pests, and *Metarhizium* for froghopper control; development of IPM for many crops, including mass rearing of natural enemies for ABC by private company	CBC in citrus; continuation of ABC research of sugarcane pests; establishment of network of 19 regional laboratories producing BC agents and promoting agro-ecological production methods resulting in application of ABC in many crops; prospecting and study of BC potential of BC agents including antagonists for disease control	New laws promoting sustainable agriculture ***Specific for Venezuela: many ABC projects; research* on Trichogramma; *prospecting; testing and use of disease antagonists; network of BC laboratories***	Lack of transfer of knowledge from research to application Shortage of chemical pesticides *Laws promoting sustainable agriculture* *Network of BC laboratories*

Abbreviations: ABC = augmentative biocontrol, BC = biocontrol, CBC = classical biocontrol, CC= chemical control, ConsBC = conservation biocontrol, FBC = fortuitous biocontrol, NC = natural control extensive local research and involve pre-release risk evaluations, as well post-release assessment of the effect of introductions.

32.2.1 Examples of early use of the same natural enemies in many countries in the region

From 1900, a number of natural enemies were introduced into Latin America and the Caribbean and resulted in permanent, classical biocontrol of important pests. Notable examples are as follows.

- Importation from various sources of the parasitoids *Aphytis diaspidis* (How), *Aphytis fuscipennis* (How), *Encarsia* (= *Prospaltella*) *berlesei* How, *Aspidiotiphagus citrinus* (Crwf), *Arrhenophagus chionaspidis* Auriv. and the predator *Scymnus* sp. into Peru in 1904 for control of the cotton white scale *Pinnaspis strachani* Ferris and Rao. Later, these natural enemies were used in several other countries for control of similar pests in various crops.
- The predator *Cryptolaemus montrouzieri* Mulsant was imported into Puerto Rico from the USA in 1912 to control mealybugs in sugarcane. This predator was later introduced from the USA or from a regional country into many other Caribbean and Latin American countries for control of various pests in various crops.
- Introduction of *E. berlesei* from Italy into Uruguay in 1913 for control of white peach scale *P. pentagona*. The parasitoid was then sent from Uruguay to other Latin American countries.
- The predator *R. cardinalis*, native from Australia, imported into Uruguay in 1919 from France for control of cottony cushion scale *I. purchasi*. The predator was then introduced into other Latin American countries and the Caribbean, but later obtained from other sources as well.
- Importation of the parasitoid *A. mali* from the USA into Uruguay in 1921 for control of woolly apple aphid *E. lanigerum* and later introduced from Uruguay into other countries in Latin America.
- Introduction of the predator *Chilocorus bivulnerus* (Mulsant) from the USA into Uruguay in 1924 for control of San José scale *Comstockaspis perniciosus* (Comstock). The predator was later introduced from Uruguay into other countries in Latin America, but was on other occasions introduced from the USA as well.
- The parasitoids *E. serius*, *Encarsia opulenta* (Silv.), *A. hesperidum* and *Encarsia perplexa* Huang and Polaszek were imported from India into Cuba in 1930 for control of citrus blackfly *A. woglumi*. One or more of these species were then exported in 1930 to Costa Rica, Haiti and Jamaica and in 1931 to the Bahamas and Panama. Later, one or more of these species of parasitoids were introduced into other countries in the region.
- The predator *Cryptolaemus montrouzieri* and the parasitoid *Leptomastidea abnormis* (Girault) were introduced in 1931 into Chile from the USA for control of *Planococcus citri* (Risso).
- Importation of the parasitoid *Metaphycus helvolus* (Compere) in 1931 into Chile from the USA to control olive black scale *Saissetia oleae* (Olivier). The parasitoid was introduced into several other Latin American countries for control of olive black scale or other scale species.
- Introduction of the pyralid herbivore *Cactoblastis cactorum* (Berg) originating from Argentina into Nevis in 1957 for control of *Opuntia* spp. cacti, then released in Antigua, the Cayman Islands and Montserrat. It later spread naturally to other islands in the Caribbean, including the Bahamas.
- A special case is that of the biocontrol of sugarcane borers, *Diatraea* spp. Initially several tachinid parasitoids native to Latin America and the Caribbean were found (*Lixophaga diatraeae* (Tns.) and *Lydella minense* Tns.). Natural control of sugarcane borers by *L. diatraeae* parasitoids was documented as early as 1930 in Jamaica and later in Cuba. Also in Puerto Rico the important role of natural control of borers by native natural enemies was demonstrated, in this case by three native parasitoids *Trichogramma minutum* Riley, *Tetrastichus haitiensis* Gahan and *L. diatraeae*. *L. minense* was originally found in the Amazon area of Brazil and introduced into Guyana in 1932, where it reduced sugarcane borer

populations. The tachinid parasitoid species were redistributed over the region, often resulting in establishment and reduction in borer numbers. Later, in the 1950s, mass rearing and augmentative releases of native tachinids were initiated in Cuba, a practice followed by other countries in the region. In this period, extensive trials were made with inundative releases of native *Trichogramma* spp., e.g. in Barbados and Guyana. Since 1950, a number of exotic stem borer parasitoids have been imported from Africa and Asia by CABI into Trinidad and Tobago, and one of these, *Cotesia flavipes* (Cam.) originating from Asia, has been particularly successful in borer control. This parasitoid has been distributed to most countries in the region that face sugarcane borer problems and is currently released augmentatively on millions of hectares of sugarcane (see e.g. Chapter 6: Brazil).

32.2.2 Recent examples of use of the same natural enemies in the region

In the Caribbean and Central America, several area-wide biocontrol projects have recently been realized or are in the implementation phase. The natural enemies used in these projects are also used in a number of Central and South American countries. Examples of these projects are as follows.

- Introduction of the parasitoid *Anagyrus kamali* Moursi from China and the predators *C. montrouzieri* and *Scymnus coccivora* Aiyyar from India for control of the pink hibiscus mealybug *Maconellicoccus hirsutus* (Green). This mealybug, native to Asia, was accidentally introduced into Grenada in 1994, then into Trinidad and Tobago in 1995, and next into other locations in the Caribbean and South, Central and North America. Parasitoids and predators were shipped from Trinidad and Tobago to many countries in the region. Classical biocontrol of the pink hibiscus mealybug is considered one of the highlights of recent biocontrol.
- Importation and release of the parasitoids *Anagyrus loecki* Noyes and Menezes, *Acerophagous papaya* Noyes and Schauff and *Pseudleptomastrix mexicana* Noyes and Schauff for control of papaya mealybug *Paracoccus marginatus* Williams and Granara de Willink. This pest originates from Mexico and was first detected in the Caribbean in 1993. Natural enemies are reared at the USDA APHIS parasitoid-rearing facility in Puerto Rico (see Chapter 26: Puerto Rico), among other locations, and have been introduced with success into many countries in the region.
- Introduction of *Tamarixia radiata* Waterston, native to Asia, for control of the Asian citrus psyllid, *Diaphorina citri* Kuwayama. This pest is native to southern Asia and is a vector of the currently most serious citrus disease worldwide, referred to as citrus greening or huanglongbing. The parasitoid has been imported into and is mass reared in many countries in the region and has successfully reduced citrus psyllid populations.
- Natural control of the red palm mite *Raoiella indica* Hirst. This pest is native to Asia; it was accidentally introduced into the Caribbean in 2004 and now also occurs in South America. Barbados, the Dominican Republic and Jamaica, among others, have documented the role of native organisms (a predatory mite, coccinellid and neuropteran predators and acaropathogenic fungi) in reduction of this pest (see country-specific chapters).

32.2.3 Differences in use of biocontrol in the region

Although there are many similarities in biocontrol programmes applied throughout the region, there are also a number of interesting differences, which are summarized below and become obvious when looking at Table 32.2.

Classical biological control

In the early period of biocontrol up to 1970, many countries in the region imported natural enemies that had been shown to be successful in other areas of the world. Some countries have been relatively inactive in classical biocontrol, e.g. Cuba, Dominica, French territories and Suriname. From 1970 to 1999, import of natural enemies for classical biocontrol strongly decreased in many countries, with the exception of Argentina, Mexico, Peru, the Caribbean and Venezuela (see Section 32.3 for details). Currently, classical biocontrol is documented as being used in all countries in the region, with the exception of Paraguay.

Augmentative biological control

Only a few countries were involved in augmentative biocontrol in the early period. Brazil, Chile, Colombia, Cuba, Mexico and Peru played a major role during this time and were using arthropod natural enemies as well as microbial control agents. Many countries started with a few augmentative biocontrol projects during the period 1970–1999. The countries that applied a lot of augmentative biocontrol in the early period continued to do so, with Ecuador as a new country with many new projects. The spectrum of biocontrol agents used showed an impressive growth in diversity. Also, the area on which augmentative biocontrol of pests, and in particular control of diseases, was applied strongly increased. Many countries started to mass produce biocontrol agents. Currently, augmentative biocontrol is used in most countries in the region, but is limited in others (e.g. El Salvador, Paraguay and Suriname) and not used in Belize and a number of Caribbean islands. Local mass production of biocontrol agents for pest and disease control occurs in many countries.

Conservation biological control

Large differences exist among countries in the use of conservation biocontrol. Early application of conservation biocontrol by promoting the presence of beneficial birds, lizards and arthropods took place in Guyana and on a number of Caribbean islands (see country-specific chapters). During the period 1970–1999, particularly Cuba developed important conservation biocontrol projects, and also Costa Rica, Honduras and Puerto Rico started to use this form of biocontrol in several crops. Currently, Colombia, the Dominican Republic, French islands, Panama, Paraguay and Uruguay also use conservation biocontrol, in addition to the countries that started with this type of biocontrol in previous periods. Still, quite a large number of countries in the region do not study or apply conservation biocontrol.

Natural control

Early documentation of natural control of pests in sugarcane was reported by Belize, Jamaica, Puerto Rico and the Remaining Caribbean islands, and for several other pests in various crops in Peru and Suriname (see country-specific chapters). In the period 1970–1999, the following additional countries documented cases of natural control for a number of pests in different crops: Bolivia, Costa Rica, El Salvador. Recently, several other Caribbean islands and Colombia documented natural control.

Biological control of weeds

Many aquatic and terrestrial weeds that are at present found throughout the world originated in the Neotropical region. Also, a number of the most successful examples of biological weed control involve species that originated in this region, but only a few countries have been playing a role in weed biocontrol research. In Latin America, Argentina has been an important provider of weed biocontrol agents since 1899, when a phytophagous coleopteran was sent to the USA for control of snake weed, followed by many other agents for control of weeds all over the world (see Chapter 2: Argentina). In the Caribbean, Trinidad and Tobago have been an important source of weed biocontrol agents (see Chapter 29: Trinidad and Tobago; and Cock, 1985). Today, biocontrol of weeds in the region is applied in only Argentina, Chile, Honduras, Mexico and Puerto Rico, but it is studied in Brazil, Ecuador (Galapagos) and Suriname, among others.

Biological control of pests in forests

Relatively few countries use biocontrol in forestry. Uruguay is a pioneer country in forest pest biocontrol and currently also Argentina, Brazil, Chile, Colombia, Mexico and Peru apply biocontrol agents to control pests in forests.

Biological control in natural areas

Only the chapters for Chile and Ecuador mention projects about biocontrol in natural areas.

In Chile (Chapter 7), two weed species have been brought under classical biocontrol in nature: phytophagous *Chrysolina hyperici* (Foster) beetles have been released for control of St John's wort *Hypericum perforatum* L.; and the phytopathogen *Phragmidium violaceum* (Schulz) Winter was applied for control of the weedy shrub *Rubus ulmifolius* (Schott.). On the Galapagos islands in Ecuador (Chapter 13), the invasive cottony cushion scale *I. purchasi*, which was seriously affecting threatened endemic plant species, was successfully controlled by the predator *R. cardinalis*. The success of this programme resulted in ideas for biocontrol of other invasive plant and insect species in the Galapagos islands.

32.2.4 Developments of particular interest in Latin America and the Caribbean

Early and continued large-scale prospecting for natural enemies, pathogens and antagonists for pest, disease and weed control

Prospecting for biocontrol agents started before 1900 and the first biocontrol agent was exported from the region in 1899 (see Chapter 2: Argentina). After identification of many arthropod natural enemies, a large number of microbial agents were isolated. The many prospecting projects resulted in: (i) documentation of natural control; (ii) development of conservation biocontrol; (iii) identification of biocontrol agents that could be used in augmentative and classical biocontrol within and outside the region; and (iv) local large-scale mass production of arthropod natural enemies, pathogens and antagonists.

Early and continued documentation of natural control and use of conservation biocontrol

In Section 32.2.3, the use of natural and conservation biocontrol was summarized. Compared with other world regions, the documentation and use of these two activities started early and is used in a growing number of countries in this region. Some interesting early examples, with more information in the country-specific chapters, are as follows.

- Use of the native parasitoid *Scelio famelicus* Riley for control of the migratory locust *Schistocerca paranensis* (Burmeister) in Venezuela in 1884.
- Demonstration of natural control of the sugarcane borer *Diatraea saccharalis* Fabricius by *Apanteles* sp. and *Euplectrus* sp. in Puerto Rico in 1895; realization of the importance of birds, lizards and other reptiles to reduce pests.
- Demonstration of the role of predacious birds for control of the giant moth borer *Telchin licus* (Drury) (= *Castnia licoides* (Boisduval)) in sugarcane in Trinidad in the early 1900s. Construction of bird roosts and planting of bamboo clumps to encourage birds close to sugarcane fields. Use of plants as nectar sources for attracting arthropod natural enemies and the construction of predatory wasp shelters close to crops.
- Use of the native Jack Spaniard wasp *Polistes cinctus cinctus* Lepeletier, predator of the cotton leaf worm *Alabama argillacea* (Hb.) and other pests in St Vincent since 1910, and later on other Caribbean islands. Populations of the wasp have been encouraged in the Caribbean by the construction of shelters near the cotton fields under which they can nest.
- Demonstration of effectiveness of insectivorous birds in reduction of *Spodoptera frugiperda* Smith populations in rice nurseries in Guyana in the 1910s, in particular when erecting perches in the fields for birds to sit on.
- Demonstration of epizootics in populations of the grasshopper *Schistocerca americana* (Drury) caused by the native entomopathogenic bacterium *Coccobacillus acridiorum* D'Herelle in Mexico in 1911. Shortly after this observation, the bacterium was used in Argentina and Colombia to cause epizootics in grasshopper and locust populations.

Provider of biological control agents

Like some other countries worldwide, many countries in this region have provided biocontrol agents to other countries in the region, or to other world regions. Trinidad and Tobago imported many species for arthropod control and distributed these imported species and native species across the Caribbean, as well as providing them to several South and Central American countries (see Chapter 29: Trinidad and Tobago). For a long time Argentina has been playing a very important role as provider of weed biocontrol agents to other world regions (see Chapter 2: Argentina), while Trinidad and Tobago have also provided several weed biocontrol agents to other countries.

Governmental support and guidance for development of IPM and biocontrol

In most countries in the region, governmental support for the development of IPM and biocontrol is limited. There are some countries in the region that are showing an active approach in sustainable pest management by strategies to reduce chemical control and replace it with IPM and biocontrol programmes. Examples are Bolivia, Cuba, Jamaica and Peru.

Proactive approach with regard to control of potential invading organisms

Often, pest control activities are planned only after a new pest has entered a country. This then usually results in attempts to eradicate the pest, which generally involves frequent chemical pesticide applications that rarely have the intended effect of eradication. Mexico has taken the initiative to compile risk scenarios for more than 1,200 potential pests, of which more than 1,000 are not yet present in the country. For some of these species, biocontrol programmes are being designed. Jamaica is following a similar approach.

Impressive areas under classical biological control

Worldwide, information about areas of pests, diseases and weeds under classical biocontrol is scattered and incomplete. The same holds for several countries treated in this book, though this information could be obtained for a number of countries and the most impressive cases, i.e. those with areas of more than 100,000 ha under biocontrol, are:

- millions of hectares of weeds in pastures, crops and nature (Argentina, Chile);
- millions of hectares with papaya mealybug and pink hibiscus mealy bug in ornamentals and various other crops and plantings (Jamaica, Mexico);
- millions of hectares of pests in cassava (Brazil);
- hundreds of thousands to millions of hectares of insect pests in pine and eucalyptus forests (Argentina, Brazil, Chile, Uruguay);
- hundreds of thousands of hectares of insect pests of citrus (Argentina, Brazil, Mexico, Peru);
- hundreds of thousands of hectares of spittlebugs in pastures (Mexico);
- hundreds of thousands to a million of hectares of wheat aphids in wheat (Brazil, Chile);
- hundreds of thousands of hectares of Rhodes grass scale in pastures (Brazil);
- hundreds of thousands of hectares of white rice borer in rice (Ecuador);
- hundreds of thousands of hectares of sugarcane borers in sugarcane (Ecuador); and
- hundreds of thousands of hectares of mealybugs in various crops (Chile).

Impressive areas under augmentative biological control

Areas under augmentative biocontrol are usually better documented than those for classical biocontrol, both for the world and for the Latin American and Caribbean region. When data for this region are compared with information presented about worldwide use of augmentative control (van Lenteren *et al.*, 2019), the conclusion is that application is by far the largest in Latin America. The major augmentative projects, i.e. those with more than 100,000 ha under biocontrol, in the region are:

- millions of hectares of Asian citrus psyllid in citrus (Brazil);
- millions of hectares of coffee berry borer in coffee (Brazil);
- millions of hectares of lepidopterans in maize (Brazil);
- millions of hectares of soil-borne nematodes in corn (Brazil);
- millions of hectares of cotton boll worm in cotton (Brazil);
- millions of hectares of hemipterans and lepidopterans in soybean (Bolivia, Brazil, Cuba);

- millions of hectares of soil-borne diseases in soybean (Brazil);
- millions of hectares of sugarcane borers and spittlebugs in sugarcane (Brazil, Colombia);
- hundreds of thousands of hectares of mate tree borer in mate (Brazil);
- hundreds of thousands of hectares of whitefly in soybean (Brazil);
- hundreds of thousands of hectares of soil-borne nematodes in soybean (Brazil);
- hundreds of thousands of hectares of lepidopterans and coleopterans in various crops (Brazil);
- hundreds of thousands of hectares of spittlebugs in sugarcane (Bolivia, Dominican Republic);
- hundreds of thousands of hectares of potato weevils in potato (Bolivia);
- hundreds of thousands of hectares of lepidopterans in quinoa (Bolivia);
- hundreds of thousands of hectares of hemipterans and lepidopterans in various crops (Cuba);
- hundreds of thousands of hectares of rice stem stink bug in rice (Dominican Republic);
- hundreds of thousands of hectares of sugarcane aphid in sorghum (Mexico);
- hundreds of thousands of hectares of sugarcane borer in sugarcane (Mexico);
- hundreds of thousands of hectares of aphids and budworms in cotton (Peru); and
- hundreds of thousands of hectares of pests and diseases in sugarcane (Peru).

32.2.5 Achievements in areas under biocontrol in Latin America and the Caribbean

An important indicator for achievements is the area under biocontrol in a certain country. Unfortunately, even when exhaustive attempts were made to obtain those figures for each of the countries, for some countries data were simply not available and for several others data were incomplete. It was particularly difficult to obtain data for conservation and natural biocontrol. Although papers quite often mentioned that classical biocontrol programmes successfully developed in the period 1880–1969 were still functioning well, specification of the areas on which these programmes still worked was lacking. In such cases attempts were made to estimate the areas under biocontrol by using information about areas harvested in 2016 or 2017 published by FAO (http://www.fao.org/faostat/en/#data/qc). Data for augmentative programmes were easier to obtain, but for conservation and natural biocontrol authors often only mentioned that these types of control occurred (eight countries for conservation biocontrol and 12 for natural control) and did not specify on how many hectares. Areas under biocontrol are summarized in Table 32.3. The data indicate that classical biocontrol is applied on 30,747,889 ha, augmentative biocontrol on 31,381,131 ha, conservation biocontrol on 447,114 ha and natural control on 2,001,846 ha. As already said above, the figure for natural control will be vastly underestimated, as this form of control was not considered or studied by many countries. However, it is interesting to see that prospecting for natural enemies and determining their role in natural control is now on the research agenda of several countries (see Table 32.2).

When calculating total areas under biocontrol, attempts were made to minimize overestimates by checking whether pests in a certain crop were under more than one type of biocontrol. In these cases, the largest area under biocontrol for a certain type of biocontrol was taken as the estimate. For example, under the heading 'classical biocontrol' in Table 32.3, data are included for fortuitous biocontrol but these are not included in the estimate of 30,747,889 ha for classical biocontrol. In cases where more species of natural enemies are used for classical biocontrol of different pest species, only the estimate for the biocontrol agent applied on the largest area is used in the estimate for that crop. However, for augmentative biocontrol, the estimate of the total area treated given above (31,381,131 ha) is corrected for cases where more than one type of augmentative biocontrol is used to control more than one pest, but not for overlap with classical biocontrol. When correcting this figure for overlap with classical biocontrol, i.e. cases where classical biocontrol is active in controlling a certain pest in the total area of a certain crop and where augmentative biocontrol is applied for control of other pests in that crop, the area only under

Table 32.3. Types of biocontrol and surface areas (ha) currently treated in Latin America and the Caribbean.

Country	Classical biocontrol (CBC) + (Fortuitous biocontrol (FBC))	Augmentative biocontrol (ABC)	Conservation biocontrol (ConsBC)	Natural control (NC)	Country surface (ha)[a]	Inhabitants[b]	Biocontrol (ha): per ha land / per inhabitant
Argentina	4,298,000	26,778			278,040,000	44,000,000	0.015 / 0.098
Barbados	3,069	300			43,000	290,000	0.078 / 0.012
Belize	16,000				2,297,000	360,000	0.007 / 0.044
Bolivia	54,000	412,000 (+ 1,112,800 overlap with NC)		1,558,000	109,858,000	11,000,000	0.018 / 0.184
Brazil	3,012,000	21,762,000 (+ 3,711,000 overlap within ABC)			851,577,000	207,360,000	0.029 / 0.119
Chile	7,726,465	62,197			75,670,000	17,800,000	0.103 / 0.438
Colombia	4,000 (+ 4,000 overlap with FBC)	378,896	+ (?)	+ (?)	114,174,900	47,700,000	0.003 / 0.008
Costa Rica	+ (?)	15,650	+ (?)	+ (?)	5,110,000	4,930,000	0.003 / 0.003
Cuba	723,000	2,221,306	140,000	140,000	10,988,000	11,150,000	0.293 / 0.224
Dominica	2,750			+ (?)	75,000	73,900	0.037 / 0.037
Dominican Rep.	169,691 (+ 67,704 overlap with FBC)	290,451 (+ 66,000 overlap with CBC)		+ (?)	4,867,000	10,735,000	0.095 / 0.043
Ecuador	558,853	66,293	150,000	+(?)	25,637,000	16,300,000	0.030 / 0.048
El Salvador	1,500 (+ 1,500 overlap with FBC)	+ (?)		755	2,104,000	6,000,000	0.001 / 0.000
French Guiana	20,903	+ (?)	+ (?)		9,280,700	1,125,160	0.002 / 0.019
Guatemala	1,500	19,976		275,000	10,889,000	15,500,000	0.027 / 0.019
Guyana	44,000	12,000	150,000		21,497,000	740,000	0.010 / 0.278
Haiti	56,967			+(?)	2,775,000	10,900,000	0.021 / 0.005
Honduras	1	25,400	+ (?)	+(?)	11,249,000	9,000,000	0.002 / 0.003
Jamaica	1,059,676 (+ 1,023,923 overlap with FBC)	7,846		23,923	1,099,000	2,700,000	0.971 / 0.404
Mexico	11,810,404	763,000		+(?)	196,437,500	124,100,000	0.064 / 0.101
Nicaragua	+ (?)	10,484			13,037,000	6,000,000	0.001 / 0.002
Panama	19	38,630	177	3,416	7,542,000	3,750,000	0.006 / 0.011
Paraguay		+ (?)	+ (?)		40,675,200	7,000,000	? / ?
Peru	108,507	330,327	+ (?)	+ (?)	128,522,000	31,000,000	0.003 / 0.014

Continued

Table 32.3. Continued.

Country	Classical biocontrol (CBC) + (Fortuitous biocontrol (FBC))	Augmentative biocontrol (ABC)	Conservation biocontrol (ConsBC)	Natural control (NC)	Country surface (ha)[a]	Inhabitants[b]	Biocontrol (ha): per ha land / per inhabitant
Puerto Rico	10,529 (+ 1,896 overlap with FBC)	+ (?)	6,937	+ (?)	887,000	3,350,000	0.020 / 0.005
Remaining Caribs.	+ (?)	+ (?)	+ (?)	+ (?)	1,744,200	1,200,000	? / ?
Suriname	+ (?)	0.5		752	16,382,000	600,000	0.000 / 0.001
Trinidad and Tobago	32,390	350	+ (?)	+ (?)	513,000	1,200,000	0.064 / 0.010
Uruguay	1,016,165	1,356			17,622,000	3,360,000	0.058 / 0.303
Venezuela	17,500	46,000			91,205,000	31,300,000	0.001 / 0.002
Total	**30,747,889** (31,846,912 with FBC)	**31,381,131** (26,491,331 without overlap with CBC)	**447,114**	**2,001,846**			

[a]Country surface areas for 2017 or 2018 based on World Bank data (https://data.worldbank.org/indicator/ag.srf.totl.k2)
[b]Inhabitants mainly based on data from Central Intelligence Agency (CIA) (https://www.cia.gov/library/publications/the-world-factbook/geos/)

augmentative biocontrol is 26,491,331 ha. Areas treated with products based on *B. thuringiensis* are not included in the estimates, because these products do not contain a living organism and thus are not considered to be biocontrol agents.

More data for each country (crop, pest, type of biocontrol used and areas under biocontrol for each type of biocontrol) are provided as supplementary material. Also, a list of all organisms (biocontrol agents, pests, crops, weeds, etc.) with author name, order, family and common name of the organism, and the country chapters in which these organisms are mentioned is provided as supplementary material. A preliminary check of the list with all organisms mentioned in the book indicates that 715 species of parasitoids, 436 species of predatory arthropods and 204 pathogens are mentioned in relation to biocontrol projects in Latin America and the Caribbean. This list with names of all organisms can be used, for example, to check where certain natural enemies and microbial agents have been found and whether they have been successfully applied in biocontrol programmes. Providing current locations of biocontrol agents as done in this list may help in simplifying issues related to Access and Benefit Sharing regulations (see Section 32.4.1 on factors limiting development and implementation of biocontrol, below).

Six countries (Argentina, Brazil, Chile, Jamaica, Mexico and Uruguay) apply classical biocontrol on more than 1 million hectares, while four countries (Cuba, Dominican Republic, Ecuador, Peru) use classical biocontrol on more than 100,000 ha (Table 32.3). Two countries (Brazil and Cuba) apply augmentative biocontrol on more than 1 million hectares, while five countries (Bolivia, Colombia, Dominican Republic, Mexico, Peru) use augmentative biocontrol on more than 100,000 ha. Three countries (Cuba, Ecuador and Guyana) use conservation biocontrol on more than 100,000 ha, while natural control is documented for more than 1 million hectares in Bolivia and more than 100,000 ha in two countries: Cuba and Guatemala.

Although it was said earlier that a country's area under biocontrol is an important indicator for the country's achievements, this statement needs some qualification. Surface areas differ considerably among the countries in Latin America and the Caribbean. For example, Barbados consists of only 43,000 ha, while Brazil covers 851,577,000 ha (Table 32.3). When using total surface areas for each country, use of biocontrol per hectare of the total surface ranks highest and above 0.1 ha per hectare for Chile, Cuba, Jamaica and Barbados, whereas the Dominican Republic, Mexico, Trinidad and Tobago and Uruguay have more than 0.05 ha per hectare under biocontrol. However, total country surfaces do not provide the best figures for a comparison of achievements, as some countries have vast areas where agriculture is not practised or even possible. Thus, many corrections should be made to these total country surface values for better comparisons. Another way to rank country achievements in biocontrol is to calculate the area under biocontrol per inhabitant. This results in nine countries (Argentina, Bolivia, Brazil, Chile, Cuba, Guyana, Jamaica, Mexico, Trinidad and Tobago) with at least 0.1 ha of biocontrol per inhabitant (Table 32.3). Each way of ranking has its advantages and disadvantages, but the most important conclusion that can be drawn from the data in Table 32.3 is that Latin America and the Caribbean are currently world leaders in biocontrol.

32.3 BIOCAT Data on Classical Biological Control in Latin America and the Caribbean

Summary tables and lists for classical biocontrol introductions for the region were compiled for the use of insects as classical biocontrol agents against insect pests. The numbers were generated from the BIOCAT database (Cock *et al.*, 2016), based on the corrected version used by Cock (2019), which can be referred to as BIOCAT 2010.3. Each combination of biocontrol agent, primary target, release country and date or period of release was treated as one introduction. Introductions that were reported to have achieved at least partial control were classed as 'successful'. It should be noted that BIOCAT 2010.3 is not up to date and the numbers should not be relied upon for the past decade or two. Nevertheless, the numbers capture the great majority of insect classical biocontrol introductions and provide valid comparisons between countries, regions, numbers of targets and biocontrol agents, establishments and successes.

Table 32.4 summarizes the records for control of insect pests by insect natural enemies in the region in classical biocontrol. The country chapters in this book mention classical biocontrol projects, but in a qualitative way. The table illustrates that many introductions have been made and that 30% of the introductions resulted in establishment, which is an impressively high rate. Also the high percentage of target pest species controlled (34%) and the percentage of successful agent species (15%) support the conclusion that classical biocontrol has been a profitable pest management method in this region and that many countries made use of it.

In Table 32.5, the introduced species of natural enemies that contributed to classical biocontrol in the region are listed. Most of these species are, not surprisingly, also mentioned in the country-specific chapters. Exceptions are three species that, according to the BIOCAT database (based on Zúñiga, 1985), were introduced

Table 32.4. Summary of records included in BIOCAT2010.3 concerning classical biological control of insects by insect biocontrol agents in Latin America and the Caribbean.

Category	Number
No. of introductions (total records)[a]	964
No. of establishments (excluding temporary)	287
No. of primary pest targets	118
No. of agent species	387
No. of countries	36
No. of successful biological control agent introductions	128
No. of successful biological control agent species	57
No. of successful programmes	103
No. of different pest species controlled	40
No. of countries reporting at least one success	29

[a]Each agent/target country/year is a separate introduction/establishment, e.g. an introduction of the same biocontrol agent from six countries at the same time counts as one introduction, but two introductions of the same biocontrol agent 10 years apart count as two introductions.

into Chile more than 50 years ago and contributed to success in forestry pest biocontrol, but are not mentioned in the Chilean chapter: *Habrolepis dalmanni* (Westwood) for complete control of *Asterodiaspis quercicola* (Bouché) in 1928; *Leucopis obscura* Haliday for partial control of *Pineus boerneri* Annand in 1945; and *Rhaphitelus maculatus* Walker for partial control of *Scolytus rugulosus* Müller in 1915–1916. The number of insect natural enemy species used in classical biocontrol (57) is much lower than the total number of arthropod natural enemy species mentioned in the country chapters (1150). This large difference is to be expected and can be explained as follows. The list for the country chapters: (i) includes all arthropod natural enemies mentioned, both the effective and ineffective agents, while Table 32.5 only includes successful species; (ii) includes many predatory mites, whereas Table 32.5 does not; and (iii) includes species used in all types of biocontrol, where Table 32.5 is limited to classical biocontrol. Particularly this last reason explains a large part of the difference, because in augmentative biocontrol many different species of natural enemies have been tried and are used in different countries (see, for example, Chapter 6: Brazil and Chapter 25: Peru). In classical biocontrol it is common to use the same natural enemies in different countries against the same pest (see the supplementary table with numbers of introductions of insect biocontrol agents per 10-year period for each country in Latin America and the Caribbean). Many more parasitoids have been successfully used than predators.

In Table 32.6 the names of the pest species are presented that were controlled as a result of the introduction of natural enemy species mentioned in Table 32.5.

A list of countries that have reported at least one insect pest successfully control by classical biocontrol and a table with numbers of introductions of insect biocontrol agents each decade for each country in Latin America and the Caribbean are presented as supplementary material.

Ten species of entomopathogens have been released in Latin America and the Caribbean for use in classical biocontrol (Table 32.7 and Hajek *et al.*, 2016). Seven of the entomopathogens became established in at least one country and ten

Table 32.5. List of species of insect natural enemies that have contributed to a classical biological control success somewhere in Latin America and the Caribbean (retrieved from BIOCAT 2010.3).

Biological control agent	Guild	Family	Primary target	No. of countries with success
Aceratoneuromyia indica	Parasitoid	Eulophidae	*Anastrepha ludens*	1
Adalia bipunctata	Predator	Coccinellidae	Aphids	1
Aganaspis pelleranoi	Parasitoid	Figitidae	*Anastrepha striata*	1
Amitus hesperidum	Parasitoid	Platygastridae	*Aleurocanthus woglumi*	2
Amitus spiniferus	Parasitoid	Platygastridae	*Aleurothrixus floccosus*	1
Anagyrus kamali	Parasitoid	Encyrtidae	*Maconellicoccus hirsutus*	11
Anagyrus saccharicola	Parasitoid	Encyrtidae	*Saccharicoccus sacchari*	1
Aphelinus mali	Parasitoid	Aphelinidae	*Eriosoma lanigerum*	8
Aphidius matricariae	Parasitoid	Braconidae	Aphids	1
Aphidius smithi	Parasitoid	Braconidae	*Acyrthosiphon pisum*	2
Aphytis holoxanthus	Parasitoid	Aphelinidae	*Chrysomphalus aonidum*	4
Aphytis lepidosaphes	Parasitoid	Aphelinidae	*Lepidosaphes beckii*	6
Aphytis melinus	Parasitoid	Aphelinidae	*Aonidiella aurantii*	1
Aphytis roseni	Parasitoid	Aphelinidae	*Selenaspidus articulatus*	1
Aphytis sp.	Parasitoid	Aphelinidae	*Lepidosaphes gloverii* and *Aonidiella aurantii*	2
Billaea claripalpis	Parasitoid	Tachinidae	*Diatraea saccharalis*	2
Cales noacki	Parasitoid	Aphelinidae	*Aleurothrixus floccosus*	1
Chilocorus cacti	Predator	Coccinellidae	*Bambusaspis bambusae*	1
Cladis nitidula	Predator	Coccinellidae	*Bambusaspis bambusae*	1
Coccophagus gurneyi	Parasitoid	Aphelinidae	*Pseudococcus calceolariae*	1
Comperiella bifasciata	Parasitoid	Encyrtidae	*Aonidiella aurantii*	1
Copidosoma floridanum	Parasitoid	Encyrtidae	*Trichoplusia ni*	1
Cotesia flavipes	Parasitoid	Braconidae	*Diatraea saccharalis*	3
Cotesia glomerata	Parasitoid	Braconidae	*Pieris brassicae*	1
Cotesia sesamiae	Parasitoid	Braconidae	*Diatraea saccharalis*	1
Cotesia vestalis	Parasitoid	Braconidae	*Plutella xylostella*	1
Cryptognatha nodiceps	Predator	Coccinellidae	*Aspidiotus destructor*	1
Cryptolaemus montrouzieri	Predator	Coccinellidae	*Maconellicoccus hirsutus*	11
Encarsia berlesei	Parasitoid	Aphelinidae	*Pseudaulacaspis pentagona*	4
Encarsia clypealis	Parasitoid	Aphelinidae	*Aleurocanthus woglumi*	1
Encarsia noyesi	Parasitoid	Aphelinidae	*Aleurodicus cocois*	1
Encarsia opulenta	Parasitoid	Aphelinidae	*Aleurocanthus woglumi*	5
Encarsia smithi	Parasitoid	Aphelinidae	*Aleurocanthus woglumi*	1
Encarsia sp.	Parasitoid	Aphelinidae	*Lepidosaphes beckii*	1

Continued

Table 32.5. Continued.

Biological control agent	Guild	Family	Primary target	No. of countries with success
Eretmocerus serius	Parasitoid	Aphelinidae	*Aleurocanthus woglumi*	6
Gyranusoidea indica	Parasitoid	Encyrtidae	*Maconellicoccus hirsutus*	1
Habrolepis dalmanni[a]	Parasitoid	Encyrtidae	*Asterodiaspis quercicola*	1
Hippodamia convergens	Predator	Coccinellidae	*Schizaphis graminum*	1
Leucopis obscura[a]	Predator	Chamaemyiidae	*Pineus boerneri*	1
Lixophaga diatraeae	Parasitoid	Tachinidae	*Diatraea saccharalis*	4
Lysiphlebus testaceipes	Parasitoid	Braconidae	*Schizaphis graminum*	1
Lydella minense	Parasitoid	Tachinidae	*Diatraea saccharalis*	4
Metaphycus anneckei	Parasitoid	Encyrtidae	*Saissetia oleae*	1
Metaphycus helvolus	Parasitoid	Encyrtidae	*Saissetia* spp.	2
Neodusmetia sangwani	Parasitoid	Encyrtidae	*Antonina graminis*	1
Psyllaephagus pilosus	Parasitoid	Encyrtidae	*Ctenarytaina eucalypti*	1
Pteroptrix smithi	Parasitoid	Aphelinidae	*Chrysomphalus aonidum*	1
Rhaphitelus maculatus[a]	Parasitoid	Pteromalidae	*Scolytus rugulosus*	1
Rodolia cardinalis	Predator	Coccinellidae	*Icerya purchasi*	11
Telenomus alsophilae	Parasitoid	Platygastridae	*Oxydia trychiata*	1
Telenomus remus	Parasitoid	Platygastridae	*Spodoptera frugiperda*	2
Telenomus rowani	Parasitoid	Platygastridae	*Rupella albinella*	1
Trichogramma minutum	Parasitoid	Trichogrammatidae	*Diatraea saccharalis*	1
Trichogramma pretiosum	Parasitoid	Trichogrammatidae	*Diatraea saccharalis*	1
Trichogramma semifumatum	Parasitoid	Trichogrammatidae	*Diatraea saccharalis*	1
Trichogramma sp.	Parasitoid	Trichogrammatidae	*Diatraea saccharalis*	1
Tytthus mundulus	Predator	Miridae	*Perkinsiella saccharicida*	1

[a]Species not mentioned in Chilean chapter.

target pest species were brought under classical biocontrol. Correspondence between species of entomopathogens mentioned in the country-specific chapters and Table 32.7 is poor and six of the ten entomopathogens are not mentioned in the country-specific chapters: *Hirsutella thompsonii* (Argentina), *Neozygites parvispora* (Barbados), *Metarhizium anisopliae* (Guyana), *Aschersonia aleyrodis* (Virgin Islands), *Romanomermis culicivorax* (Colombia and Puerto Rico), *Romanomermis iyengari* (Cuba) and *Trichoplusia ni* nucleopolyhedrovirus (TnNPV) (Colombia). The difference for the first four species may be explained by the fact that they did not control the pest and are therefore not mentioned in the respective country-specific chapters. The two species of the genus *Romanomermis* might not have been mentioned in the country-specific chapters for Cuba and Puerto Rico because they concern agents controlling mosquitoes vectoring human diseases, a topic that may have escaped the attention of the authors.

The number of releases of insect natural enemies per decade is given in Fig. 32.1 and shows an increase during the initial four decades, which is interrupted as a result of the Second World War, but the increase continues in the 1950s, peaks during the 1970s and then dramatically drops. The peak in the 1960s and 1970s reflects a period when classical biocontrol agents were easily transported by air and introduced rather indiscriminately without careful pre-release studies, combined with the import and release of known biocontrol agents

Table 32.6. Pest target species reported to have been successfully controlled by classical biological control somewhere in Latin America and the Caribbean (retrieved from BIOCAT 2010.3)

Target pest	Family	Crop	No. of biocontrol agents involved in success	No. of territories reporting success
Acyrthosiphon pisum	Aphididae	Peas, lucerne	1	2
Aleurocanthus woglumi	Aleyrodidae	Citrus	4	9
Aleurodicus cocois	Aleyrodidae	Coconut, palms	1	1
Aleurothrixus floccosus	Aleyrodidae	Citrus	2	2
Anastrepha ludens	Tephritidae	Fruit	1	1
Anastrepha striata	Tephritidae	Citrus	1	1
Antonina graminis	Pseudococcidae	Pasture grasses	1	1
Aonidiella aurantii	Diaspididae	Citrus	3	2
Aspidiotus destructor	Diaspididae	Coconut	1	1
Asterodiaspis quercicola	Asterolecaniidae	Oak	1	1
Bambusaspis bambusae	Asterolecaniidae	Bamboo	2	1
Chrysomphalus aonidum	Diaspididae	Citrus	2	4
Ctenarytaina eucalypti	Psyllidae	Eucalyptus	1	1
Diatraea rosa	Crambidae	Sugarcane	1	1
Diatraea saccharalis	Crambidae	Sugarcane, maize	8	10
Eriosoma lanigerum	Aphididae	Apple	1	8
Icerya montserratensis	Monophlebidae	Citrus	1	1
Icerya purchasi	Monophlebidae	Citrus, etc.	1	11
Lepidosaphes beckii	Diaspididae	Citrus	2	7
Lepidosaphes gloverii	Diaspididae	Citrus	1	1
Maconellicoccus hirsutus	Pseudococcidae	Various trees and ornamentals	3	11
Nipaecoccus nipae	Pseudococcidae	Various trees	1	1
Oxydia trychiata	Geometridae	Cypress	1	1
Perkinsiella saccharicida	Cicadellidae	Sugarcane	1	1
Pieris brassicae	Pieridae	Brassicas	1	1
Pineus boerneri	Cicadellidae	*Pinus radiata*	1	1
Planococcus citri	Pseudococcidae	Fruit trees, ornamentals	1	1
Plutella xylostella	Plutellidae	Brassicas	1	1
Pseudaulacaspis pentagona	Diaspididae	Mulberry, peach	1	4
Pseudococcus calceolariae	Pseudococcidae	?	1	1
Rupella albinella	Crambidae	Rice	1	1

Continued

Table 32.6. Continued.

Target pest	Family	Crop	No. of biocontrol agents involved in success	No. of territories reporting success
Saccharicoccus sacchari	Pseudococcidae	Sugarcane	1	1
Saissetia coffeae	Coccidae	Citrus	1	1
Saissetia oleae	Coccidae	Citrus	2	2
Schizaphis graminum	Aphididae	Wheat, cereals	2	2
Scolytus rugulosus	Curculionidae	Peach	1	1
Selenaspidus articulatus	Diaspidae	Citrus	1	1
Spodoptera frugiperda	Noctuidae	Various arable	1	1
Spodoptera spp.	Noctuidae	Various arable	1	1
Trichoplusia ni	Noctuidae	Various arable	1	1
'Aphids'	Aphididae	?	2	1

Table 32.7. Summary of records concerning use of entomopathogens in classical biological control in Latin America and the Caribbean (retrieved from Hajek *et al.*, 2016).

Country	Releases	Biocontrol agent	Established	Target species	Targets controlled
Bacteria	0	0	0	0	0
Fungi					
Argentina	3	*Fusarium coccophilum*	1	*Aonidiella aurantia*	1
		Hirsutella thompsonii		*Eriophyes sheldoni* and *Phyllocoptruta oleivora*	
Barbados	1	*Neozygites parvispora*	0	*Thrips tabaci*	0
Guyana	1	*Metarhizium anisopliae*	1	*Aeneolamia flavilatera*	0
US Virgin Islands	1	*Aschersonia aleyrodis*	0	*Aleurodicus cocois* & *Aleurothrixus floccosus*	0
Microsporidia					
Argentina	1	*Paranosema locustae*	1	*Dichroplus maculipennis, D. elongatus, D. pratensis* and *Scotussa lemniscata*	1
Nematodes					
Argentina	2	*Deladenus siricidicola*	1	*Sirex noctilio*	1
Brazil	2	*Deladenus siricidicola*	1	*Sirex noctilio*	1
Chile	1	*Deladenus siricidicola*	1	*Sirex noctilio*	1
Colombia	1	*Romanomermis culicivorax*	1	*Anopheles nyssorhynchus albimanus*	1
Cuba	1	*Romanomermis iyengari*	1	*Anopheline spp.* and *Culicine spp.*	4
El Salvador	1	*Romanomermis culicivorax*	0	*Anopheles nyssorhynchus albimanus* and *A. punctipennis*	0
Puerto Rico	1	*Romanomermis culicivorax*	1	*Scapteriscus didactylus* and *S. abbreviatus*	1
Uruguay	1	*Deladenus siricidicola*	1	*Sirex noctilio*	1
Viruses					
Colombia	1	*Trichoplusia ni* nucleopolyhedrovirus (TnNPV)	1	*Trichoplusia ni*	1

used successfully in other parts of the world. Interest dropped off in the 1980s when all the obvious biocontrol agents had been introduced and few major new targets for classical biocontrol were spreading in the region. Figure 32.2 presents the number of classical biocontrol successes with insect natural enemies per decade and shows a similar trend as Fig. 32.1, but with the difference of an extra peak during the 1990s, reflecting the 11 successful programmes against *Maconellicoccus hibiscus* around the Caribbean.

32.4 Factors Limiting and Stimulating Biological Control in Latin America and the Caribbean

32.4.1 Factors limiting development and implementation of biological control

Authors of the country chapters mention many causes that may frustrate development and implementation of biocontrol. Factors limiting biocontrol are partly summarized in Table 32.2.

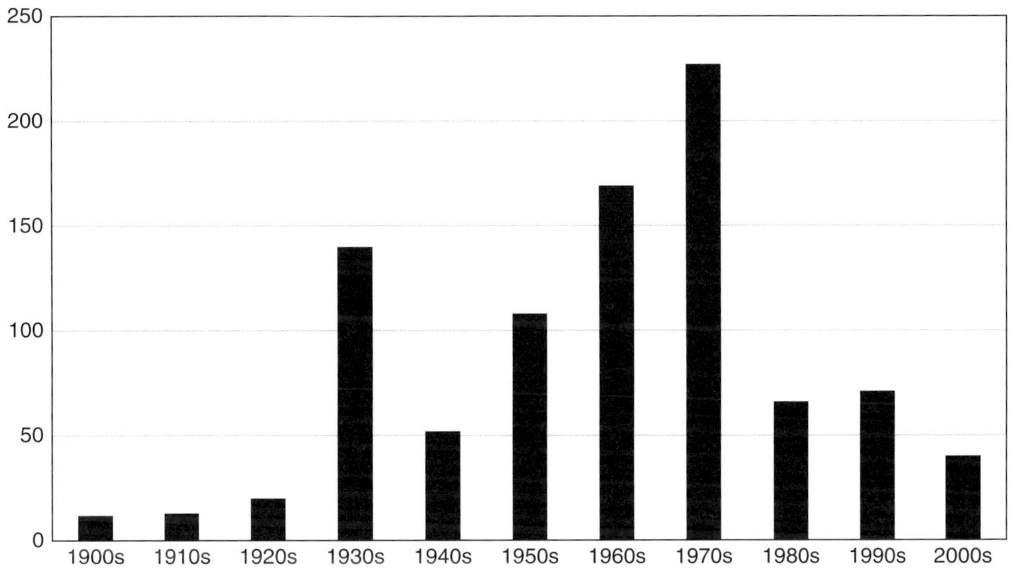

Fig. 32.1. Number of releases per decade of insect biocontrol agents for classical biological control of insect pests in Latin America and the Caribbean (retrieved from BIOCAT 2010.3).

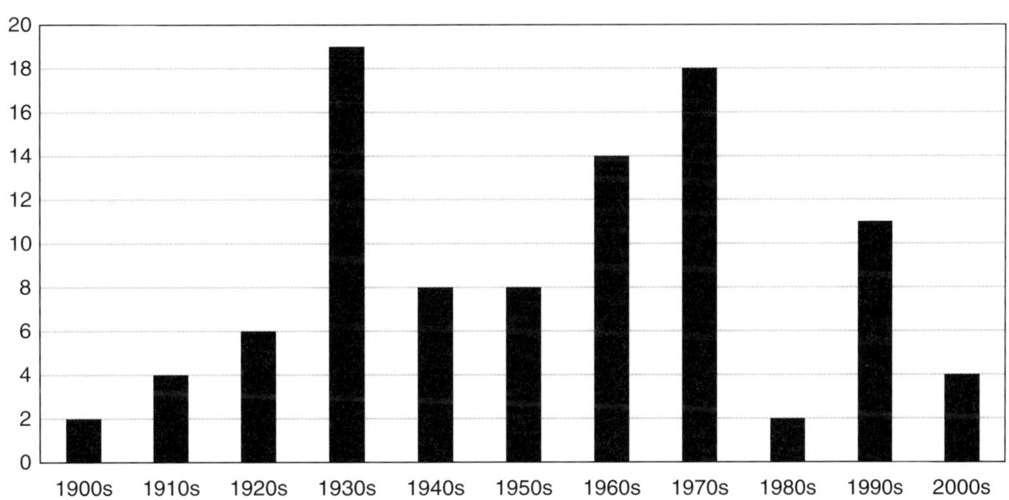

Fig. 32.2. Number of classical biological control successes of insect pests by insect biocontrol agents per decade in Latin America and the Caribbean (retrieved from BIOCAT 2010.3) Each target successfully controlled counts once for each country and is allocated to the decade when the first biocontrol agent contributing to that success was released.

The factor most often mentioned is the dominating role that the pesticide industry and their sales persons play. Pesticide companies strongly lobby at governmental levels for faster pesticide registration procedures and continuously stress the need to apply pesticides in order to be able to feed a growing human population. Crop protection advisors prefer to advise (over-) use of pesticides, as they are sold per unit of volume and their income is based on volumes sold. Another reason why they prefer to advise the use of pesticides is that the profit margins of synthetic pesticides are in the order of 25% or more, while they are only about 5% for biocontrol agents. Further, sales people and extension service personnel providing information about pest management often discourage the application of biocontrol by referring to it as being too complicated to use. Next, most people now involved in crop protection matters – from the highest level in ministries of agriculture down to farmers and their workers – have been raised under the mantra that 'the only good insect is a dead insect, and chemical control is able to kill insects fast and cheaply'. The result is that many crop protection advisors and farmers are pesticide addicted after having heard this mantra for 70 years, and this addiction appears to be hard to heal by the relatively small numbers of experts in the field of biocontrol.

Another factor often mentioned in the chapters in this book is also related to chemical pesticide use and concerns obstructing the ecosystem function of natural pest control as a result of pesticide applications. The effect of this is that other primary pests need to be sprayed as well, and that secondary pests, which are kept at low densities by naturally occurring beneficial organisms, start to create serious problems. The dramatic negative effect of pesticides in the reduction of biodiversity, including beneficial species like biocontrol agents, has been documented during the past decades and culminated in the 2019 assessment by the Intergovernmental Science-Policy Platform on Biodiversity and Ecosystem Services (IPBES, 2019). Many policy makers and pesticide-industry representatives repeat over and again that agriculture has to feed some 10 billion people by the year 2050 and that, for this reason, a substantial increase in food production is needed, which can only be achieved using synthetic pesticides. Van Lenteren *et al.* (2018) discussed, documented and explained why this reasoning is simplistic, erroneous and misleading. Related to the topic of this book – biocontrol – the above reasoning is simplistic, because by stating that synthetic pesticides are needed for increased food production, the presence of a multitude of other approaches to pest, disease and weed control illustrated in this book and in many other publications is ignored.

A third factor related to pesticides hampering the implementation of biocontrol and other non-chemical pest management methods is the unrealistically low cost of chemical products. Application of the 'true cost' principle for pesticides would strongly increase the market for biocontrol. Pesticides are currently 'subsidized' by governments because the industry is not held responsible for human illnesses and deaths as a result of chronic exposure to pesticides. Also, the pesticide industry does not have to provide funding for repairing damage done to the environment, such as reduction of biodiversity and inhibition of the functioning of the ecosystem services of pest control, pollination and cleaning of water. The pesticide costs related to all these harmful effects are externalized and paid for by society, while the pesticide industry only picks up the economic benefits without being responsible for the true costs. Realistic pricing involving these true costs would result in chemical pesticides being two to four times more expensive (van Lenteren *et al.*, 2018), with the positive effect of fairer competition with non-chemical alternative control methods such as biocontrol.

Looking ahead, pest management in Latin America and the Caribbean (as elsewhere) will have to become more sophisticated, adaptable, flexible and sustainable than simply applying synthetic pesticides. This capacity will need to be developed in a changing world, where consumers will want and expect healthier produce, while climate change has the potential to substantially disrupt pest management (and society). Climate-smart pest management (Heeb *et al.*, 2019) incorporating all the potential of biocontrol will be required. Researchers will have to develop new effective IPM strategies, extension staff will need to understand the IPM strategies in order to promote them and farmers will want access to advice and information on IPM, including biocontrol, delivered by the most suitable methods using appropriate information communication technology. Inevitably, it will take time to make this transition. Universities and technical training facilities should have the wherewithal (resources, suitable staff, information and material) to produce the next generation of researchers and extension staff who understand the role of biocontrol and IPM. They will then be able to generate and apply appropriate strategies to optimize agricultural and forestry production and protect the environment and biodiversity. Demonstrating how well beneficial agents can control pests, diseases and weeds, as is done in this book (published both in Spanish and in

English), will be a valuable resource for this and should help to promote a wider appreciation for and application of biocontrol as part of IPM.

Many countries in the region mention lack of funding for research in biocontrol as a main reason for the few projects realized and this particularly plays a role in Central America and the Caribbean. Sometimes limited funding is temporarily provided when a new invasive pest is observed, resulting in ad hoc trials to apply biocontrol, but if quick success is not obtained these projects are prematurely terminated and, often incorrectly, the conclusion is drawn that biocontrol does not work. Other countries mention that research funding is available, but that collaboration among universities, institutes, extension services and farmers needs to be improved (e.g. Argentina, Bolivia, Brazil and Venezuela).

Time-consuming and expensive registration procedures for natural enemies and microbial control agents are also often mentioned as a limiting or even prohibiting factor, because many of the relatively small biocontrol agent producers are not able to bear those high registration costs. Chapter 1 of this book mentions that procedures for registration vary widely in Latin America and the Caribbean, and this is also the case for regulations for import and release, and for demands for environmental risk assessments of biocontrol agents. Colmenarez *et al.* (in preparation) propose the formation of a regional platform for harmonization of procedures related to biocontrol.

A recent regulation concerning prospecting for exotic natural enemies resulted in an almost complete stop in searching for exotic control agents. Under the Convention on Biological Diversity (CBD, 1993), countries have sovereign rights over their genetic resources and agreements governing the access to these resources and the sharing of benefits arising from their use need to be established between involved parties, i.e. Access and Benefit Sharing (ABS) (Cock *et al.*, 2010). The Nagoya Protocol, which came into force in October 2014, is a supplementary agreement to the CBD and provides a framework for the implementation of ABS procedures. The consequence of this protocol is that currently a permission to survey and sample potential biocontrol agents can only be granted by the country that has the rights over the genetic resources, and that collection of new natural enemies has become increasingly difficult or impossible in countries that have first accidentally exported the pest, a situation which seems quite unreasonable.

Linked to the issues raised by the Nagoya Protocol is the need for taxonomic expertise to support agriculture and pest management (Lyal *et al.*, 2008). Apart from issues relating to the identity of pests and their infraspecific variation, there is the challenge of the taxonomy of the biocontrol agents, both those used for classical introductions and inundative releases, and the indigenous natural control agents found in farmers' fields. Names are needed to understand what is happening, recognize the different species and provide clarity regarding their roles and functions. Taxonomists able to support this work are few in number and often scattered across museums and universities, both within Latin America and the Caribbean and across the world. It is important that these taxonomists are supported, their numbers increased and key linkages with pest management built and sustained. Molecular tools to extract genetic information are a valuable tool for taxonomists today, enabling them to achieve far more in less time, but this depends on the ability to move specimens and genetic information freely between taxonomists and researchers and across national borders. Regulations put in place to implement the Nagoya Protocol are intended to ensure equitable sharing of the benefits arising from access to genetic resources, but already they are impeding important taxonomic and applied pest management research based on biological specimens (i.e. genetic resources). It is important that these regulations facilitate rather than impede these public-good activities, as was specified in the Nagoya Protocol itself.

Another reason for initial reservations in using biocontrol relates to the opinion that biocontrol often fails. This opinion seems to be supported by the various unsuccessful cases mentioned in many of this book's chapters. It is a fact that a lot of early attempts to introduce and establish natural enemies did not result in establishment of the natural enemy, and establishment often did not lead to control of the target pest. Early failures to obtain establishment might have been caused by a multitude of reasons, such as: (i) poor and long transport conditions resulting in high mortality and low quality of the imported and released natural enemies; (ii) bureaucratic and time-consuming import procedures at custom offices leading to death of the biocontrol agents; (iii) too low numbers of released individuals; (iv) inadequate synchronization between release of the natural enemy and presence of the pest; (v) lack of preceding research on agroecosystem characteristics where biocontrol agents should be functioning, including knowledge about already present natural enemies that might prevent establishment; and (vi) insufficient climate matching. Early failures to establish or have impact were also the result of the 'hit-and-miss' or 'shotgun' approach often followed until the 1990s: 'release every biocontrol agent that you can get hold of, and something should work'. However, the FAO Guidelines for the export, shipment, import and release of biocontrol agents and other beneficial organisms (IPPC, 2017), which were first issued in 1996 (IPPC, 1996) now help to guide many Latin American and Caribbean countries (Kairo *et al.*, 2003). Risk assessment procedures, as well as new approaches for agent selection that are explained in Chapter 1 of this book, allow for early exclusion of ineffective or problematic candidate species.

The increased investment in evaluating the risks and potential of new biocontrol agents, particularly for insect biocontrol agents to control insects, has reduced the number of biocontrol agent species introduced, but increased the establishment and success rate (Cock *et al.*, 2016). In contrast, from very early on, weed biocontrol involved the careful selection and detailed evaluation of potential biocontrol agents, leading to higher establishment and success rates than for insect biocontrol (Schwarzländer *et al.*, 2018). Although establishment and success rates in biocontrol have increased over time, it is, surprisingly, not uncommon to read in papers written by biocontrol experts that this success rate is still low. Actually, the success rate is surprisingly high: a comparison of the success rate and costs involved in the identification of a marketable synthetic pesticide (1:140,000 and US$256 million) and the success rate and costs involved in finding an effective natural enemy for augmentative biocontrol (1:10 and US$2 million) demonstrates the good success rate and low costs at which a success is obtained with biocontrol (data for 2010). Details and references illustrating this issue are given in van Lenteren (2012).

Factors stimulating development and implementation of biological control

Compared with chemical pesticides, biocontrol agents show a number of inherent characteristics that make them preferred pest management tools now and in the future.

- They are less detrimental to the health of farm workers and those living in farming communities.
- Use of biocontrol does not imply a 'crop re-entry period' as is the case with most pesticides, and lack of this re-entry period makes continuous harvesting possible, which is particularly important with fresh products such as vegetables, fruit and ornamentals where market prices may vary strongly even during a day.
- They are more sustainable, as there has been no development of resistance against arthropod biocontrol agents.
- They do not cause phytotoxic damage to plants and, as a result, farmers report better yields and healthier crops after switching to biocontrol.
- Use of biocontrol results in residue-free or residue-poor products and, increasingly, produce will only be bought by exporters and local retailers when residue levels are well below the legal maximum residue levels (MRLs).
- Finally, reduction in or absence of chemical pesticide applications as a result of using biocontrol leads to increased biodiversity in and around crops.

Many of these inherent characteristics are not well known or explained to farmers and consumers, and several chapters in this book refer to this lack of knowledge. Documentation of the many successes obtained with biocontrol, as done in this book, as well as providing data showing that biocontrol can be cheaper than chemical control (see Chapter 6: Brazil and Chapter 25: Peru), may encourage farmers, consumers and policy makers to accept this sustainable pest management method.

Interesting specific factors for the Latin American and Caribbean region that stimulate implementation of biocontrol are the many prospecting projects and the large number of documented cases of natural pest control. Prospecting gives an insight into the many potential biocontrol candidates (e.g. Chapter 19: Honduras, Chapter 20: Jamaica and Chapter 24: Paraguay) and regional prospecting has resulted in many natural enemies and microbial agents that are now mass produced and released for augmentative biocontrol. Documentation of natural control not only illustrates the positive role of the ecosystem service of pest control, but also helps in understanding why excessive pesticide sprays interfere with this free pest control service and may result in causation of secondary pests.

Many countries mention that new invasive pests initiate new biocontrol research. As mentioned earlier, an interesting approach to try to solve problems caused by new invaders is used in Mexico, by first determining pest risk scenarios and then reviewing possibilities for biocontrol before these pests cause problems. Recent pest invasions in the region, e.g. pink hibiscus mealybug and papaya mealybug, have shown that regional cooperation can result in area-wide solutions using biocontrol. Regional cooperation seems a very positive option to stimulate biocontrol.

At the governmental level, adoption and funding of IPM and biocontrol approaches have a strong positive influence on application of biocontrol, as is explained for example in Chapter 5: Bolivia, Chapter 10: Cuba and Chapter 25: Peru. When such approaches are combined with concurrent measures to reduce chemical pesticide use as employed in Peru, the positive effect on application of biocontrol is even larger. Another important promoter of increased use of biocontrol is the availability of fast-track and priority registration of low-risk pest control agents such as biocontrol agents and a special registration procedure for microbial agents.

The chapters clearly show that organic food production and production of food under some form of certification for the export market is strongly growing in the Caribbean and in Central and South America. These forms of production stimulate the use of biocontrol. Also, local concerns about food safety and environmental impact issues in relation to synthetic pesticide use are mentioned in many chapters, resulting in a preference for low- or no-residue produce and a demand for biocontrol.

More efficient biocontrol agent selection methods (Chapter 1) and regional development of improved and more stable formulations for microbial biocontrol agents, cheaper and better quality production of natural enemies, and better application methods for biocontrol agents (e.g. equipment to release biocontrol agents in crops by use of drones or unmanned airplanes) also contribute to growth in uptake of biocontrol.

In conclusion: it is clear that the dominance of the chemical industry will not diminish soon, but governments, researchers, farmers and consumers are becoming increasingly aware of the negative effects of excessive pesticide use on the biosphere and its inhabitants, and this, together with the many positive effects of biocontrol, will result in a growth of this pest management method.

32.5 Future of Biological Control in Latin America and the Caribbean

In the chapters of this book, many largely varying issues are mentioned about the future of biocontrol in the region.

- Many countries anticipate a growth in application of all forms of biocontrol, because of the result of prospecting projects, increase in documentation and use of natural control, realization of more conservation biocontrol programmes as a result of better understanding of the role of insects in agroecosystem functioning, a large expansion of augmentative biocontrol after demonstration of large successes obtained in, for example, Brazil, and the long-term successes realized with classical biocontrol. Export demands for pesticide-poor or pesticide-free products are expected to boost all types of biocontrol.
- A lot of countries mention the future need of biocontrol to manage the many new invasive pests.
- A number of countries expect to increase governmental or industrial mass production of native natural enemies and, particularly, microbial agents to be able to meet

growing demands. Production of pathogens and antagonists for biocontrol of diseases is explicitly mentioned in several chapters (e.g. Brazil, Chile, Colombia and Uruguay). Also, future implementation of quality-control methods for biocontrol agents is thought by many countries to lead to better products, resulting in increasing sales.

- Anticipated continuation of current and future governmental support and guidance in development of biocontrol is mentioned by many countries.
- Several countries recognize an urgent need for harmonization of regulations for import and release of exotic biocontrol agents, as well as registration procedures. Current procedures are very diverse, resulting in the need to provide different documents for each country, leading to high costs that are difficult to cover for most of the small biocontrol companies. Harmonization is not expected to occur soon, but would certainly stimulate future application of biocontrol. A related problem that needs urgent action is to find a solution for the impasse in foreign exploration due to the Access and Benefit Sharing (ABS) issue that developed after acceptance of the Nagoya protocol (see Chapter 1 and above). The IOBC Global Commission on ABS (IOBC, 2019a) made an appeal to develop ABS regulations that support the biocontrol sector by facilitating the exchange of biocontrol agents, including development of clear guidelines. The Commission prepared best-practice guidelines for exchange of biocontrol genetic resources to assist the biocontrol community to demonstrate due diligence in complying with ABS requirements (Mason *et al.*, 2018), which will hopefully facilitate the resumption of foreign prospecting for biocontrol agents.
- The Caribbean region and Central American countries refer to the importance of future cooperation in development and application of biocontrol, especially for classical control of new area-wide invasive pests. Strengthening of the role of international institutions mentioned in Chapter 1 in coordination of biocontrol projects involving similar pests, diseases and weeds occurring in a number of countries in the region, and improved cooperation between countries, for example within the Neotropical Regional Section of IOBC (IOBC, 2019b), might help in realizing more biocontrol projects with the limited amount of national funding. A priority activity for regional cooperation in collaboration with FAO might be to compile a list of future invading organisms into the region and to identify potential biocontrol agents for management of these organisms.
- Several countries mention the need for training of technicians and farmers in the use of biocontrol. One option might be to do this via the farmer-participatory Farmers Field Schools system, which is currently implemented on several Caribbean islands with the aim of increasing the capability of farmers to use IPM and biocontrol and become less dependent on chemical control. See Chapter 27 (Remaining Caribbean Islands) for examples.
- Currently, few countries in Latin America and the Caribbean study or implement biocontrol of weeds and vectors of human and animal diseases, while much knowledge is available about biocontrol of these organisms elsewhere in the world. These are clearly areas for future attention in the region. The same holds, though for fewer countries, for biocontrol of plant diseases.
- Although a number of countries face serious economic problems that have repercussions on research, agriculture and forestry remain priority activities in these countries and in the region as a whole. Modernization of agriculture and forestry is proceeding rapidly in Latin America and the Caribbean, and the many examples of cutting-edge biocontrol projects provided in this book assist in this modernization process. Biocontrol further contributes to sustainable food production and protection of biodiversity of the region's many now-threatened natural areas.

32.6 Acknowledgements

Vanda H.P. Bueno is thanked for compiling the multipage list provided as supplementary material with names of all organisms (biocontrol agents, pests, crops, weeds, etc.) with author name, order, family and common name of the organism, and the country chapters in which these organisms are mentioned. CABI is an international intergovernmental organisation, and M.J.W. Cock gratefully acknowledges the core financial support from CABI's member countries (see https://www.cabi.org/what-we-do/how-we-work/cabi-donors-and-partners/ for full details).

References

Cock, M.J.W. (ed.) (1985) *A Review of Biological Control of Pests in the Commonwealth Caribbean and Bermuda up to 1982*. Technical Communication No. 9, Commonwealth Institute of Biological Control. Commonwealth Agricultural Bureaux, Farnham Royal, UK.

Cock, M.J.W. (2019) Unravelling the status of partially identified insect biological control agents introduced to control insects: an analysis of BIOCAT2010. *BioControl* 64, 1–7. DOI: 10.1007/s10526-018-09921-1

Cock, M.J.W., van Lenteren, J.C., Brodeur, J., Barratt, B.I.P., Bigler, F. *et al.* (2010) Do new access and benefit sharing procedures under the convention on biological diversity threaten the future of biological control? *BioControl* 55, 199–218.

Cock, M.J.W., Murphy, S.T., Kairo, M.T.K., Thompson, E., Murphy, R.J. and Francis, A.W. (2016) Trends in the classical biological control of insect pests by insects: an update of the BIOCAT database. *BioControl* 61, 349–363. DOI: 10.1007/s10526-016-9726-3

CBD (Convention on Biological Diversity) (1993) *Convention on Biological Diversity (with annexes)*. Concluded at Rio de Janeiro on 5 June 1992. United Nations – Treaty Series 1760(30619), pp. 142–382.

Hajek, A.E., Gardescu, S. and Delalibera Jr, I. (2016) *Classical Biological Control of Insects and Mites: A Worldwide Catalogue of Pathogen and Nematode Introductions*. FHTET-2016-06. Forest Health Technology Enterprise Team, Morgantown, West Virginia. Available at: https://www.fs.fed.us/foresthealth/technology/pub_titles.shtml (accessed 17 May 2019).

Heeb, L., Jenner, E. and Cock, M.J.W. (2019) Climate-smart pest management: building resilience of farms and landscapes to changing pest threats. *Journal of Pest Science* 92(3), 951–969. DOI: 10.1007/s10340-019-01083-y

IOBC (2019a) IOBC Global Commission on ABS. Available at: http://www.iobc-global.org/global_comm_bc_access_benefit_sharing.html (accessed 5 August 2019).

IOBC (2019b) International Organisation for Biological Control: Neotropical Regional Section. Available at: www.iobcntrs.org (accessed 5 August 2019)

IPBES (Intergovernmental Science-Policy Platform on Biodiversity and Ecosystem Services) (2019) Assessment Reports on Biodiversity and Ecosystem Services. Available at: https://www.ipbes.net/assessment-reports (accessed at 31 May 2019).

IPPC (1996) *Code of Conduct for the Import and Release of Exotic Biological Control Agents*. International Standards for Phytosanitary Measures No. 3. Food and Agriculture Organization of the United Nations (FAO), Rome.

IPPC (2017) *Guidelines for the export, shipment, import and release of biological control agents and other beneficial organisms*. International Standards for Phytosanitary Measures No. 3. Food and Agriculture Organization of the United Nations (FAO), Rome. Available at: https://www.ippc.int/static/media/files/publication/en/2017/05/ ISPM_03_2005_En_2017-05-23_PostCPM12_InkAm.pdf (accessed at 17 May 2019).

Kairo, M.T.K., Cock, M.J.W. and Quinlan, M.M. (2003) An assessment of the use of the Code of Conduct for the Import and Release of Exotic Biological Control Agents (ISPM No. 3) since its endorsement as an international standard. *Biocontrol News and Information* 24, 15N–27N.

Lyal, C., Kirk, P., Smith, D. and Smith, R. (2008) The value of taxonomy to biodiversity and agriculture. *Biodiversity* 9, 8–13.

Mason, P.G., Cock, M.J.W., Barratt, B.I.P, Klapwijk, J., van Lenteren, J.C., Brodeur, J., Hoelmer, K.A. and Heimpel, G.E. (2018) Best practices for the use and exchange of invertebrate biological control genetic resources relevant for food and agriculture. *BioControl* 63, 149–154. DOI: 10.1007/s10526-017-9810-3.

Schwarzländer, M., Hinz, H.L., Winston, R.L. and Day, M.D. (2018) Biological control of weeds: an analysis of introductions, rates of establishment and estimates of success, worldwide. *BioControl* 63, 319–331. DOI: 10.1007/s10526-018-9890-8.

van Lenteren, J.C. (2012) The state of commercial augmentative biological control: plenty of natural enemies, but a frustrating lack of uptake. *BioControl* 57, 71–84 + supplemental material. DOI: 10.1007/s10526-011-9395-1

van Lenteren, J.C. and Bueno, V.H.P. (2003) Augmentative biological control of arthropods in Latin America. *BioControl* 48, 123–139.

van Lenteren, J.C., Bolckmans, K., Köhl, J., Ravensberg, W.J. and Urbaneja, A. (2018) Biological control using invertebrates and microorganisms: plenty of new opportunities. *BioControl* 63, 39–59 + supplemental material. DOI: 10.1007/s10526-017-9801-4

van Lenteren, J.C., Bueno, V.H.P and Klapwijk, J. (2019) Augmentation biological control. In: Mason, P.G. (ed.) *Biological Control: A Global Endeavour*. CSIRO Publishing, Clayton, Australia.

Zúñiga, E. (1985) Ochenta años de control biologico en Chile. Revision historica y evaluacion de los proyectos desarrolados (1903–1983) [Eighty years of biological control in Chile. Historical review and evaluation of project undertaken (1903–1983)]. *Agricultura Técnica* 45(3), 175–183.

Index

A document with all scientific names of organisms mentioned in the book, together with authors, orders and families, common names and countries in which they occur is available as supplementary electronic material. Names of all pests and plants are mentioned in this index, as well as the names of the most important biocontrol agents.

ANTARCTICA IN INTERNATIONAL LAW

Antarctica, one of the world's last great wildernesses, presents special challenges for international law. Fears that Antarctica would become a front in the Cold War catalysed agreement on the 1959 Antarctic Treaty which neither legitimized nor challenged the existing sovereign claims to the continent. The unique Antarctic Treaty System has provided the foundation for peaceful, harmonious and effective governance. There are, however, new anxieties about the frozen continent and the Southern Ocean. Antarctica already feels the effects of climate change and ocean acidification. Claimant states assert rights to the Antarctic continental shelf and interest in Antarctic resources grows. Tourism brings new environmental and safety risks. China and other powers are increasing their activities, with some questioning the consensus of the western-dominated 'Antarctic club'. Security concerns are increasingly discussed, despite Antarctica's dedication to peaceful purposes. This book brings together the main primary international materials concerning the regulation and governance of Antarctica, including multilateral and bilateral treaties, United Nations materials, 'soft laws', and judicial decisions. It covers the spectrum of Antarctic issues from environmental protection to scientific cooperation to tourism. As it shows, Antarctic law has constantly adapted to meet new challenges and is a sophisticated, inclusive, dynamic and responsive regime.

DOCUMENTS IN INTERNATIONAL LAW

GENERAL EDITOR: PROFESSOR STEFAN TALMON
Director at the Institute of Public International Law, University of Bonn, and Supernumerary Fellow of St. Anne's College, Oxford

ALSO IN THIS SERIES

THE LEGAL ORDER OF THE OCEANS
Vaughan Lowe and Stefan Talmon (2009)

THE IRAN NUCLEAR ISSUE
Yaël Ronen (2010)

TERRORISM
Ben Saul (2012)

BASIC DOCUMENTS ON INTERNATIONAL INVESTMENT PROTECTION
Martins Paparinskis (2012)

THE SETTLEMENT OF INTERNATIONAL DISPUTES: BASIC DOCUMENTS
Christian Tams and Antonios Tzanakopoulos (2012)